Lecture Notes in Computer
Edited by G. Goos, J. Hartmanis, and

Springer
*Berlin
Heidelberg
New York
Hong Kong
London
Milan
Paris
Tokyo*

Carlos M. Fonseca   Peter J. Fleming
Eckart Zitzler   Kalyanmoy Deb
Lothar Thiele (Eds.)

Evolutionary Multi-Criterion Optimization

Second International Conference, EMO 2003
Faro, Portugal, April 8-11, 2003
Proceedings

 Springer

Series Editors

Gerhard Goos, Karlsruhe University, Germany
Juris Hartmanis, Cornell University, NY, USA
Jan van Leeuwen, Utrecht University, The Netherlands

Volume Editors

Carlos M. Fonseca
University of the Algarve, Faculty of Sciences and Technology
Centre for Intelligent Systems
Campus de Gambelas, 8005-139 Faro, Portugal
E-mail: cmfonsec@ualg.pt

Peter J. Fleming
University of Sheffield, Department of Automatic Control and Systems Engineering
Mappin Street, Sheffield S1 3JD, UK
E-mail: P.Fleming@sheffield.ac.uk

Eckart Zitzler
Lothar Thiele
Swiss Federal Institute of Technology, Department of Electrical Engineering
Computer Engineering and Networks Laboratory
Gloriastraße 35, 8092 Zürich, Switzerland
E-mail: {zitzler,thiele}@tik.ee.ethz.ch

Kalyanmoy Deb
Indian Institute of Technology, Department of Mechanical Engineering
Kanpur Genetic Algorithms Laboratory
Kanpur, UP 208 016, India
E-mail: deb@iitk.ac.in

Cataloging-in-Publication Data applied for
A catalog record for this book is available from the Library of Congress.

Bibliographic information published by Die Deutsche Bibliothek.
Die Deutsche Bibliothek lists this publication in the Deutsche Nationalbibliografie;
detailed bibliographic data is available in the Internet at <http://dnb.ddb.de>.

CR Subject Classification (1998): F.2, G.1.6, G.1.2, I.2.8

ISSN 0302-9743
ISBN 3-540-01869-7 Springer-Verlag Berlin Heidelberg New York

This work is subject to copyright. All rights are reserved, whether the whole or part of the material is concerned, specifically the rights of translation, reprinting, re-use of illustrations, recitation, broadcasting, reproduction on microfilms or in any other way, and storage in data banks. Duplication of this publication or parts thereof is permitted only under the provisions of the German Copyright Law of September 9, 1965, in its current version, and permission for use must always be obtained from Springer-Verlag. Violations are liable for prosecution under the German Copyright Law.

Springer-Verlag Berlin Heidelberg New York
a member of BertelsmannSpringer Science+Business Media GmbH

http://www.springer.de

© Springer-Verlag Berlin Heidelberg 2003
Printed in Germany

Typesetting: Camera-ready by author, data conversion by PTP-Berlin GmbH
Printed on acid-free paper   SPIN: 10925867   06/3142   5 4 3 2 1 0

Preface

The 2nd International Conference on Evolutionary Multi-Criterion Optimization (EMO 2003) was held on April 8–11, 2003, at the University of the Algarve in Faro, Portugal. This was the second international conference dedicated entirely to this important topic, following the very successful EMO 2001 conference, which was held in Zürich, Switzerland, in March 2001. EMO 2003 was co-located with the IFAC International Conference on Intelligent Control Systems and Signal Processing (ICONS 2003), exposing EMO to a wider audience of scientists and engineers.

The EMO 2003 scientific program included two keynote addresses, one given by David Schaffer on optimization and machine learning in industry, and another delivered by Pekka Korhonen on multiple-criteria decision making. In addition, three tutorials were presented, one on multicriterion decision analysis by Valerie Belton, another on multiobjective evolutionary algorithms by Kalyanmoy Deb, and a third on multiple objective metaheuristics by Andrzej Jaszkiewicz. The President of the Portuguese Innovation Agency, Prof. João Silveira Lobo, and the President of the Portuguese Foundation for Science and Technology, Prof. Fernando Ramoa Ribeiro, attended the opening ceremony and the closing ceremony, respectively.

In response to the call for papers, 100 full-length papers were submitted from 27 countries. Fifty-six papers were accepted for presentation at the conference after thorough reviewing by members of the program committee and they are contained in this volume.

April 2003   Carlos M. Fonseca, Peter J. Fleming, Eckart Zitzler,
                              Kalyanmoy Deb, and Lothar Thiele

Organization

EMO 2003 was organized by CSI (Centre for Intelligent Systems, University of the Algarve, Portugal) with the support of EvoNet (European Network of Excellence in Evolutionary Computing).

General Chairs

Carlos M. Fonseca	Universidade do Algarve, Portugal
Peter J. Fleming	University of Sheffield, UK
Eckart Zitzler	ETH Zürich, Switzerland

Program Committee

Enrique Baeyens	University of Valladolid, Spain
Peter Bentley	University College London, UK
Jürgen Branke	University of Karlsruhe, Germany
Nirupam Chakraborti	IIT, Kharagpur, India
Carlos Coello Coello	CINVESTAV-IPN, Mexico
David Corne	University of Reading, UK
William Crossley	Purdue University, US
Dragan Cvetkovic	Soliton Associates, Canada
Kalyanmoy Deb	IIT, Kanpur, India
Rolf Drechsler	University of Bremen, Germany
Kary Främling	Helsinki University of Technology, Finland
António Gaspar-Cunha	Universidade do Minho, Portugal
Tushar Goel	John F. Welch Technology Centre, India
Thomas Hanne	Fraunhofer Institute for Industrial Mathematics, Germany
Alberto Herreros	University of Valladolid, Spain
Jeffrey Horn	Northern Michigan University, US
Evan J. Hughes	Cranfield University, UK
Hisao Ishibuchi	Osaka Prefecture University, Japan
Yaochu Jin	Honda R&D Europe, Germany
Joshua Knowles	IRIDIA, Belgium
Petros Koumoutsakos	ETH Zürich, Switzerland
Rajeev Kumar	IIT, Kharagpur, India
Gary B. Lamont	Air Force Institute of Technology, US
Marco Laumanns	ETH Zürich, Switzerland
Daniel Loughlin	MCNC Environmental Modeling Center, US
Carlos Mariano Romero	IMTA, Mexico

Martin Middendorf — University of Leipzig, Germany
Tadahiko Murata — Kansai University, Japan
Shigeru Obayashi — Tohoku University, Japan
Pedro Oliveira — Universidade do Minho, Portugal
Geoffrey T. Parks — Cambridge University, UK
Ian C. Parmee — University of the West of England, UK
Amrit Pratap — California Institute of Technology, US
Ranji Ranjithan — North Carolina State University, US
Katya Rodriguez-Vazquez — IIMAS-UNAM, Mexico
Günter Rudolph — Parsytec AG, Germany
J. David Schaffer — Philips Research, US
Marc Schoenauer — INRIA, France
Hans-Paul Schwefel — Universität Dortmund, Germany
El-ghazali Talbi — LIFL, France
Kay Chen Tan — National University of Singapore, Singapore
Lothar Thiele — ETH Zürich, Switzerland
Dirk Thierens — Utrecht University, The Netherlands
Thanh Binh To — Universität-Otto-von-Guericke-Magdeburg, Germany
David Van Veldhuizen — Wright-Patterson AFB, US
James F. Whidborne — Kings College London, UK

Local Organizing Committee

Carlos M. Fonseca, CSI
António E. Ruano, CSI
Maria G. Ruano, CSI
Fernando Lobo, CSI

Pedro M. Ferreira, CSI
Daniel Castro, CSI
Susy Rodrigues, FEUA

EMO Steering Committee

David Corne — University of Reading, UK
Kalyanmoy Deb — IIT Kanpur, India
Peter J. Fleming — University of Sheffield, UK
Carlos M. Fonseca — Universidade do Algarve, Portugal
J. David Schaffer — Philips Research, US
Lothar Thiele — ETH Zürich, Switzerland
Eckart Zitzler — ETH Zürich, Switzerland

Acknowledgements

Invited Speakers

We thank the keynote and tutorial speakers for their talks given at the conference.

Keynote Speakers

Pekka Korhonen	Helsinki School of Economics and Business Administration, Finland
J. David Schaffer	Philips Research, US

Tutorial Speakers

Valerie Belton	University of Strathclyde, UK
Kalyanmoy Deb	IIT, Kanpur, India
Andrzej Jaszkiewicz	Poznan University of Technology, Poland

Local Sponsors

Support by the following organizations and companies is gratefully acknowledged.

Universidade do Algarve

- Reitoria
- Faculdade de Ciências e Tecnologia
- Faculdade de Economia
- Departamento de Engenharia Electrónica e Informática

Fundação para a Ciência e a Tecnologia
Fundação Calouste Gulbenkian
Fundação Oriente
Fundação Luso-Americana para o Desenvolvimento
Câmara Municipal de Faro
Câmara Municipal de Tavira
Governo Civil de Faro
PROAlgarve
Turismo do Algarve
Águas de Monchique
Banco Português do Atlântico
Epaminondas
Illy Cafés
Parmalat S.A.
Refrige
SINFIC
Staples Office Center

Table of Contents

Objective Handling and Problem Decomposition

The Maximin Fitness Function; Multi-objective City and
Regional Planning 1
   Richard Balling

Conflict, Harmony, and Independence: Relationships in
Evolutionary Multi-criterion Optimisation........................... 16
   Robin C. Purshouse, Peter J. Fleming

Is Fitness Inheritance Useful for Real-World Applications? 31
   Els Ducheyne, Bernard De Baets, Robert De Wulf

Use of a Genetic Heritage for Solving the Assignment Problem
with Two Objectives 43
   Xavier Gandibleux, Hiroyuki Morita, Naoki Katoh

Fuzzy Optimality and Evolutionary Multiobjective Optimization 58
   M. Farina, P. Amato

IS-PAES: A Constraint-Handling Technique Based on Multiobjective
Optimization Concepts 73
   Arturo Hernández Aguirre, Salvador Botello Rionda,
   Giovanni Lizárraga Lizárraga, Carlos A. Coello Coello

A Population and Interval Constraint Propagation Algorithm 88
   Vincent Barichard, Jin-Kao Hao

Multi-objective Binary Search Optimisation 102
   Evan J. Hughes

Covering Pareto Sets by Multilevel Evolutionary Subdivision Techniques . 118
   Oliver Schütze, Sanaz Mostaghim, Michael Dellnitz, Jürgen Teich

An Adaptive Divide-and-Conquer Methodology for Evolutionary
Multi-criterion Optimisation.. 133
   Robin C. Purshouse, Peter J. Fleming

Multi-level Multi-objective Genetic Algorithm Using Entropy to
Preserve Diversity.. 148
   S. Gunawan, A. Farhang-Mehr, S. Azarm

Solving Hierarchical Optimization Problems Using MOEAs.............. 162
   Christian Haubelt, Sanaz Mostaghim, Jürgen Teich, Ambrish Tyagi

Multiobjective Meta Level Optimization of a Load Balancing
Evolutionary Algorithm.. 177
  David J. Caswell, Gary B. Lamont

Algorithm Improvements

Schemata-Driven Multi-objective Optimization 192
  Skander Kort

A Real-Coded Predator-Prey Genetic Algorithm for
Multiobjective Optimization.. 207
  Xiaodong Li

Towards a Quick Computation of Well-Spread Pareto-Optimal Solutions . 222
  Kalyanmoy Deb, Manikanth Mohan, Shikhar Mishra

Trade-Off between Performance and Robustness: An Evolutionary
Multiobjective Approach 237
  Yaochu Jin, Bernhard Sendhoff

Online Adaptation

The Micro Genetic Algorithm 2: Towards Online Adaptation in
Evolutionary Multiobjective Optimization........................... 252
  Gregorio Toscano Pulido, Carlos A. Coello Coello

Self-Adaptation for Multi-objective Evolutionary Algorithms 267
  Dirk Büche, Sibylle Müller, Petros Koumoutsakos

MOPED: A Multi-objective Parzen-Based Estimation of Distribution
Algorithm for Continuous Problems 282
  Mario Costa, Edmondo Minisci

Test Problem Construction

Instance Generators and Test Suites for the Multiobjective
Quadratic Assignment Problem 295
  Joshua D. Knowles, David W. Corne

Dynamic Multiobjective Optimization Problems: Test Cases,
Approximation, and Applications 311
  M. Farina, Kalyanmoy Deb, P. Amato

No Free Lunch and Free Leftovers Theorems for Multiobjective
Optimisation Problems 327
  David W. Corne, Joshua D. Knowles

Performance Analysis and Comparison

A New MOEA for Multi-objective TSP and Its Convergence
Property Analysis 342
    Zhenyu Yan, Linghai Zhang, Lishan Kang, Guangming Lin

Convergence Time Analysis for the Multi-objective Counting
Ones Problem 355
    Dirk Thierens

Niche Distributions on the Pareto Optimal Front..................... 365
    Jeffrey Horn

Performance Scaling of Multi-objective Evolutionary Algorithms 376
    V. Khare, X. Yao, Kalyanmoy Deb

Searching under Multi-evolutionary Pressures 391
    Hussein A. Abbass, Kalyanmoy Deb

Minimal Sets of Quality Metrics 405
    A. Farhang-Mehr, S. Azarm

A Comparative Study of Selective Breeding Strategies in
a Multiobjective Genetic Algorithm 418
    Andrew Wildman, Geoff Parks

An Empirical Study on the Effect of Mating Restriction on the
Search Ability of EMO Algorithms 433
    Hisao Ishibuchi, Youhei Shibata

Alternative Methods

Using Simulated Annealing and Spatial Goal Programming for
Solving a Multi Site Land Use Allocation Problem 448
    Jeroen C.J.H. Aerts, Marjan van Herwijnen, Theodor J. Stewart

Solving Multi-criteria Optimization Problems with
Population-Based ACO 464
    Michael Guntsch, Martin Middendorf

A Two-Phase Local Search for the Biobjective Traveling
Salesman Problem 479
    Luis Paquete, Thomas Stützle

Implementation

PISA — A Platform and Programming Language Independent
Interface for Search Algorithms 494
    Stefan Bleuler, Marco Laumanns, Lothar Thiele, Eckart Zitzler

A New Data Structure for the Nondominance Problem in
Multi-objective Optimization 509
  Oliver Schütze

The Measure of Pareto Optima 519
  M. Fleischer

Distributed Computing of Pareto-Optimal Solutions with
Evolutionary Algorithms 534
  Kalyanmoy Deb, Pawan Zope, Abhishek Jain

Applications

Multiobjective Capacitated Arc Routing Problem 550
  P. Lacomme, C. Prins, M. Sevaux

Multi-objective Rectangular Packing Problem and Its Applications 565
  Shinya Watanabe, Tomoyuki Hiroyasu, Mitsunori Miki

Experimental Genetic Operators Analysis for the Multi-objective
Permutation Flowshop.. 578
  Carlos A. Brizuela, Rodrigo Aceves

Modification of Local Search Directions for Non-dominated
Solutions in Cellular Multiobjective Genetic Algorithms for Pattern
Classification Problems 593
  Tadahiko Murata, Hiroyuki Nozawa, Hisao Ishibuchi, Mitsuo Gen

Effects of Three-Objective Genetic Rule Selection on the
Generalization Ability of Fuzzy Rule-Based Systems................... 608
  Hisao Ishibuchi, Takashi Yamamoto

Identification of Multiple Gene Subsets Using Multi-objective
Evolutionary Algorithms 623
  A. Raji Reddy, Kalyanmoy Deb

Non-invasive Atrial Disease Diagnosis Using Decision Rules:
A Multi-objective Optimization Approach............................. 638
  Francisco de Toro, Eduardo Ros, Sonia Mota, Julio Ortega

Intensity Modulated Beam Radiation Therapy Dose Optimization with
Multiobjective Evolutionary Algorithms 648
  Michael Lahanas, Eduard Schreibmann, Natasa Milickovic,
  Dimos Baltas

Multiobjective Evolutionary Algorithms Applied to the
Rehabilitation of a Water Distribution System: A Comparative Study ... 662
  Peter B. Cheung, Luisa F.R. Reis, Klebber T.M. Formiga,
  Fazal H. Chaudhry, Waldo G.C. Ticona

Optimal Design of Water Distribution System by Multiobjective
Evolutionary Methods 677
  Klebber T.M. Formiga, Fazal H. Chaudhry, Peter B. Cheung,
  Luisa F.R. Reis

Evolutionary Multiobjective Optimization in Watershed Water
Quality Management 692
  Jason L. Dorn, S. Ranji Ranjithan

Different Multi-objective Evolutionary Programming Approaches for
Detecting Computer Network Attacks 707
  Kevin P. Anchor, Jesse B. Zydallis, Gregg H. Gunsch,
  Gary B. Lamont

Safety Systems Optimum Design by Multicriteria
Evolutionary Algorithms.. 722
  David Greiner, Blas Galván, Gabriel Winter

Applications of a Multi-objective Genetic Algorithm to
Engineering Design Problems 737
  Johan Andersson

A Real-World Test Problem for EMO Algorithms 752
  A. Gaspar-Cunha, J.A. Covas

Genetic Methods in Multi-objective Optimization of Structures
with an Equality Constraint on Volume 767
  J.F. Aguilar Madeira, H. Rodrigues, Heitor Pina

Multi-criteria Airfoil Design with Evolution Strategies 782
  Lars Willmes, Thomas Bäck

Visualization and Data Mining of Pareto Solutions Using
Self-Organizing Map 796
  Shigeru Obayashi, Daisuke Sasaki

Author Index 811

The Maximin Fitness Function; Multi-objective City and Regional Planning

Richard Balling

Department of Civil and Environmental Engineering, Brigham Young University, Provo, UT, 84602, USA

Abstract. The maximin fitness function can be used in multi-objective genetic algorithms to obtain a diverse set of non-dominated designs. The maximin fitness function is derived from the definition of dominance, and its properties are explored. The modified maximin fitness function is proposed. Both fitness functions are briefly compared to a state-of-the-art fitness function from the literature. Results from a real-world multi-objective problem are presented. This problem addresses land-use and transportation planning for high-growth cities and metropolitan regions.

1 Introduction

This paper has two objectives. The first objective is to present the maximin fitness function [1] to the EMO community. This function is a simple, elegant fitness function that can be used in multi-objective evolutionary optimization. The second objective of this paper is to present results from a real-world application of multi-objective evolutionary optimization. This application is in the area of land use and transportation planning for high-growth cities and metropolitan region.

2 Derivation of the Maximin Fitness Function

Consider a problem with m minimized objectives. Let us assume that each objective is scaled by dividing its unscaled value by an appropriate positive constant. Consider a genetic algorithm where the generation size is n. Let f_k^i be the scaled value of the kth objective for the ith design in a particular generation. The jth design <u>weakly dominates</u> the ith design if:

$$f_k^i \geq f_k^j \text{ for all k from 1 to m} \qquad (1)$$

The jth design <u>dominates</u> the ith design if, in addition to Equation (1), we have:

$$f_k^i > f_k^j \text{ for at least one k from 1 to m} \qquad (2)$$

The only way to satisfy Equation (1) and not satisfy Equation (2) is to have:

$$f_k^i = f_k^j \text{ for all k from 1 to m} \qquad (3)$$

In this paper, two designs satisfying Equation (3) will be referred to as <u>duplicate</u> designs. Even though duplicate designs are equal in objective space, they may be different in variable space. Two designs that do not satisfy Equation (3) will be referred to as <u>distinct</u> designs. If all designs in a generation are distinct, then weak domination is equivalent to domination.

Now it is possible to define the maximin fitness function. Equation (1) is equivalent to:

$$\min_{k}\left(f_k^i - f_k^j\right) \geq 0 \qquad (4)$$

The ith design in a particular generation will be weakly dominated by another design in the generation if:

$$\max_{j \neq i}\left(\min_{k}\left(f_k^i - f_k^j\right)\right) \geq 0 \qquad (5)$$

The maximin fitness of design i is defined to be:

$$\text{fitness}^i = \max_{j \neq i}\left(\min_{k}\left(f_k^i - f_k^j\right)\right) \qquad (6)$$

In Equation (6), the min is taken over all the objectives from 1 to m, and the max is taken over all designs in the generation from 1 to n except design i. The maximin fitness function given by Equation (6) is attractive because of its simplicity. It is easily implemented as three nested loops. The outer loop over i ranges from 1 to n. The middle loop over j ranges from 1 to n. The inner loop over k ranges from 1 to m. Thus, the number of comparisons is mn^2.

3 Properties of the Maximin Fitness Function

Any design whose maximin fitness is greater than zero is a dominated design. Any design whose maximin fitness is less than zero is a non-dominated design. A design whose maximin fitness is equal to zero is a weakly-dominated design, and is either a dominated design or a duplicate non-dominated design.

Consider the designs A, B, and C of generation I in Figure 1. The maximin fitness of each design has been calculated and written beside each design in parentheses. Since all three designs are non-dominated, there maximin fitnesses are negative. The same is true for the three designs in generation II. However, the equally-spaced designs in generation I have the same maximin fitness while in generation II, the closely-spaced designs B and C have higher fitness than design A. Since maximin fitness is minimized in a genetic algorithm, the maximin fitness rewards diversity and penalizes clustering of non-dominated designs. Thus, niching techniques are not needed with the maximin fitness function.

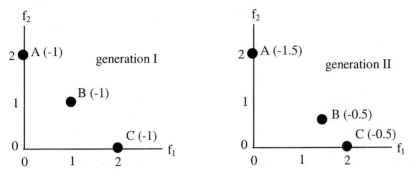

Fig. 1. Two Generations of Three Designs

One might jump to the conclusion that maximin fitnesses of equally-spaced non-dominated designs are equal. However, this conclusion is false as shown by the sample generations in Figure 2 where the maximin fitness is written beside each design.

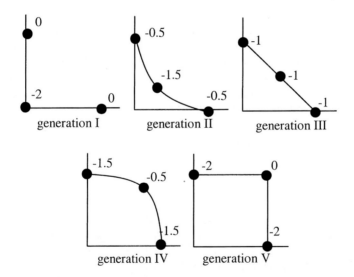

Fig. 2. Some Sample Generations

In generations II, III, and IV in Figure 2, the designs are equally spaced on the non-dominated front. However, the maximin fitnesses are equal only for generation III where the non-dominated front is linear. In generation II, the non-dominated front is convex, and the maximin fitness favors the design in the middle of the front. Increasing the convexity to the limit produces generation I where the lower left design is non-dominated, and the other two designs are dominated. In generation IV, the non-dominated front is concave, and the maximin fitness favors the designs on the ends of the front. Increasing the concavity to the limit produces generation V where the upper right design is dominated, and the other two designs are non-dominated. The property that the maximin fitness of non-dominated distinct designs is less than zero, and the maximin fitness of dominated designs is greater than or equal to zero is manifest in all the generations of Figure 2.

The generations in Figure 2 illustrate that the maximin fitness function is a continuous function of objective values. Other fitness functions based on ranking, sorting, or scoring give equal fitness to all non-dominated designs, and are therefore discontinuous functions of objective values. Specifically, these other fitness functions would give a higher fitness to the non-dominated design at the origin in generation I of Figure 2 than to the other two dominated designs. But if the design at the origin were moved infinitesimally upward and rightward, they would give equal fitness to all three designs since they are non-dominated. Thus, small changes in objective values produce jumps in fitness. This is not the case with the maximin fitness function.

Consider the generation I in Figure 3 where the maximin fitness is written beside each design in parentheses. Designs A, B, and C are non-dominated, and their maximin fitnesses are negative. Design D is dominated, and its maximin fitness is positive. In generation II, we add design E. Its maximin fitness is less than that of design D because it is closer to the non-dominated front. Thus, for dominated designs, the maximin fitness is a metric of distance to the non-dominated front. Note also, that the presence of design E did not change the maximin fitness of design D. This is because for a dominated design, the max function in Equation (6) is always controlled by a non-dominated design. This means that the maximin fitness function does not penalize the fitness of dominated designs due to clustering. Finally, note that the presence of design E changed the maximin fitness of design B. This suggests that for a non-dominated design, the max function in Equation (6) may

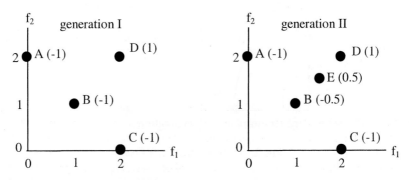

Fig. 3. Two More Generations

be controlled by a dominated design. This means that the maximin fitness function penalizes the fitness of non-dominated designs due to clustering from both dominated or non-dominated designs.

Let's summarize the properties of the maximin fitness function:

1) The maximin fitness of dominated designs is greater than or equal to zero. The maximin fitness of distinct non-dominated designs is less than zero.

2) The maximin fitness function penalizes clustering of non-dominated designs. In the limit, the maximin fitness of duplicate non-dominated designs is zero.

3) The maximin fitness function rewards designs at the middle of convex non-dominated fronts, and designs at the extremes of concave non-dominated fronts. The maximin fitness function is a continuous function of objective values.

4) The maximin fitness of dominated designs is a metric of distance to the non-dominated front.

5) The max function in the maximin fitness of a dominated design is always controlled by a non-dominated design and is indifferent to clustering. The max function in the maximin fitness of a non-dominated design may be controlled by a dominated or a non-dominated design.

4 The Modified Maximin Fitness Function

The idea behind the modified maximin fitness function is that the fitness of non-dominated designs should not be penalized by clustering from dominated designs. The reason for this is that the user is interested in a diverse set of non-dominated designs, and dominated designs will eventually be discarded and should not affect the fitness of non-dominated designs. The formula for the modified maximin fitness function is:

$$\text{fitness}^i = \max_{j \neq i, j \in P} \left(\min_k \left(f_k^i - f_k^j \right) \right) \tag{7}$$

The only difference between Equation (7) and Equation (6) is that in Equation (7), the max function ranges over designs in the set P rather than over the entire generation. The set P is the set of non-dominated designs in the generation. If there is only one design in set P, then the fitness of that design is assigned an arbitrary negative number. Even though the max function is taken over fewer designs, the modified maximin fitness function requires more computational effort than the maximin fitness function because it requires that the set P be identified before fitness evaluation. In our implementation of the modified maximin fitness function, if there are duplicate non-dominated designs, only one of them will be included in set P. This means that the fitness of the included duplicate will be negative, and the fitnesses of the other duplicates will be zero.

5 Comparison with Other Fitness Functions

The performances of the maximin fitness function and the modified maximin fitness function were compared to the performances of two other fitness functions: the ranking fitness function and the NSGAII fitness function.

The ranking fitness function is based on Goldberg's idea of non-dominated sorting [2]. For a given generation, the non-dominated set is identified and assigned a rank of one. If there are duplicate non-dominated designs, they are all given a rank of one. All rank one designs are then temporarily deleted from the generation, and the non-dominated designs in the remaining generation are identified and assigned a rank of two. Rank two designs are then temporarily deleted from the generation, and the non-dominated designs in the remaining generation are identified and assigned a rank of three. This process continues until all designs in the generation have been assigned a rank. The fitness is equal to the rank. Thus, all designs with the same rank have the same fitness. This fitness function is indifferent to clustering, and cannot be regarded as a state-of-the-art method unless niching methods are used. Nevertheless, the ranking fitness function used in this comparison did not include niching methods.

The NSGAII fitness function is a state-of-the-art method [3]. This fitness function first finds the ranking fitness of every design in the generation. Then the designs of each rank are further sorted according to the crowding distance. Thus, all designs with rank 3 are assigned real-valued fitnesses ranging from 3 to 3.9, where 3 corresponds to the largest crowding distance and 3.9 corresponds to a zero crowding distance. The crowding distance is a measure of clustering. Suppose that for a particular rank, there are r designs. Crowding distances for each of these r designs are calculated by initializing them to zero, looping through the minimized objectives, and accumulating contributions from each objective. To determine the contribution from a particular objective, the r designs are ordered by their objective values from lowest to highest. Ties are broken arbitrarily. Let i be the order number (from 1 to r) of a particular design. If i is greater than 1 and less than r, the crowding distance d^i for design i receives the following contribution for objective k:

$$d^i = d^i + f_k^{i+1} - f_k^{i-1} \tag{8}$$

If i is equal to 1 or r, the crowding distance d^i for design i receives the following contribution for objective k:

$$d^i = d^i + \text{(large value)} \tag{9}$$

In the comparison, the same genetic algorithm and algorithm parameters were used for all four fitness functions. The generation size was n = 100. Tournament selection was used with a tournament size of 3. Single-point crossover was used with a crossover probability of 0.7. The mutation rate was taken as 0.01. The algorithm was an elitist algorithm that generated a child generation of size n from a parent generation of size n via repeated selection, crossover, and mutation. The child generation and the parent generation were then combined into a generation of size 2n.

The fitness function was then evaluated for each design in this combined generation. Then the n designs with lowest fitness were retained as the next parent generation while the other n designs were discarded. The fitnesses of the retained n designs were not recomputed.

The first test problem was Schaffer's unconstrained, two-objective, single-variable problem, the most-studied test problem in the literature [4]:

Minimize: $f_1 = x^2$

Minimize: $f_2 = (x - 2)^2$ (10)

The single design variable x ranged from -100 to 100 and was represented by 50 binary genes. Both objective functions were scaled by dividing by 10000. The non-dominated solution to this problem is a convex curve in unscaled objective space

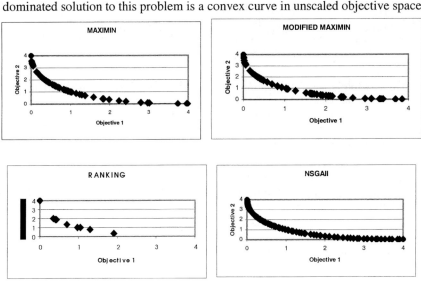

Fig. 4. First Test Problem Iteration 100

running from $(f_1,f_2) = (0,4)$ to $(f_1,f_2) = (4,0)$. Generation 100 from each run is plotted in unscaled objective space in Figure 4. The NSGAII fitness function produced the best results, and the maximin and modified maximin fitness functions produced reasonably good results. As expected, the ranking fitness function resulted in some clustering on the non-dominated front.

6 City Planning

The next test problem is from a real-world application. Future land-use and transportation plans were sought for the high-growth twin cities of Provo and Orem in the state of Utah, USA. Planners divided the cities into 195 zones. The future land use for each zone had to be selected from among 12 different land uses:

FARM	agricultural
VLDR	very low density residential
LDR	low density residential
MDR	medium density residential
HDR	high density residential
CBD	central business district
SC	shopping center
GC	general commercial
LI	light industrial
HI	heavy industrial
MIX	mixed residential and commercial
UNIV	university

Planners for both cities specified which of the above land uses were allowed for each zone. Some of the zones were not allowed to change at all from their status quo land use. For those zones that were allowed to change, a single gene was assigned to the zone, and the gene was represented by an integer whose base ranged from 2 to 12 depending on the number of allowed land uses for the zone.

Forty-five major streets were identified throughout both cities. The future classification of each street had to be selected from among 11 different street classes:

C2	two-lane collector
C3	three-lane collector
C4	four-lane collector
C5	five-lane collector
A2	two-lane arterial
A3	three-lane arterial
A4	four-lane arterial
A5	five-lane arterial
A6	six-lane arterial
A7	seven-lane arterial
F1	freeway

Streets that are currently collectors were allowed to change to collector or arterial classes with the same or higher number of lanes. Streets that are currently arterials were allowed to change to arterial classes with the same or higher number of lanes. Streets that are currently freeways were not allowed to change. Again, for those streets that were allowed to change, a single gene was assigned to the street, and the gene was represented by an integer whose base ranged from 2 to 10 depending on the number of allowed street classes for the street.

Two objectives and one constraint were identified for this problem. The constraint required that a feasible future plan must have enough residentially-zoned land to house a projected future population of 327000. This constraint is easily evaluated by summing over the residential zones the product of area and density. A plan satisfying this constraint is called a feasible plan. During tournament selection, the winner was the feasible plan with lowest fitness. If there were no feasible plans in the tournament, the plan with the lowest constraint violation was the winner.

The first objective was to minimize traffic congestion. To evaluate this objective, a sophisticated traffic model was implemented that analyzed both the peak commute period and the off-peak period. For each period, the model produced home-to-work, home-to-nonwork, and non-home-based trips depending on the land uses of the zones. These trips were attracted to other zones according to a gravity model, and assigned to streets according to a multi-path assignment model. As the number of trips assigned to a street during the peak commute period approached the street capacity, the speed on the street was degraded forcing trips to be rerouted to other streets. The first objective was the minimization of the sum of the travel times of all trips in a 24-hour day.

The second objective was the minimization of change from the status quo. Change was calculated by summing over the zones the number of people currently living or employed in the zone multiplied by a change factor, and summing over the streets the number of people living or employed on the street multiplied by a change factor. The change factor for a zone / street comes from a change factor matrix where rows correspond to the status quo land use / street class, and columns correspond to the planned future land use / street class. Change factors in the matrix are greatest when the change is the greatest. For example, changing very low density residential (VLDR) to heavy industrial (HI) would yield a large change factor while changing a 4-lane collector to a 5-lane collector would yield a small change factor.

The execution of the traffic model to evaluate the travel time objective for a single future plan required about 10 seconds on a Dell Latitude laptop computer with a Pentium III processor and 128M RAM. The calculation time for the change objective and housing constraint was negligible. Therefore, the calculation of the objectives for a single generation of 100 plans required 1000 seconds = 16 minutes. Execution of the genetic algorithm for 100 generations required 1600 minutes = 26 hours. The same elitist genetic algorithm was used as before with the same parameter settings (generation size = 100, tournament size = 3, crossover probability = 0.7, number of crossover points = 1, mutation rate = 0.01). The random starting generation, generation 100 using the modified maximin (MMM) fitness function, and generation 100 using the NSGAII fitness function are shown in Figure 5. Using an informal visual inspection of the results in Figure 5, the modified maximin fitness function produced slightly better results than the NSGAII algorithm on this problem.

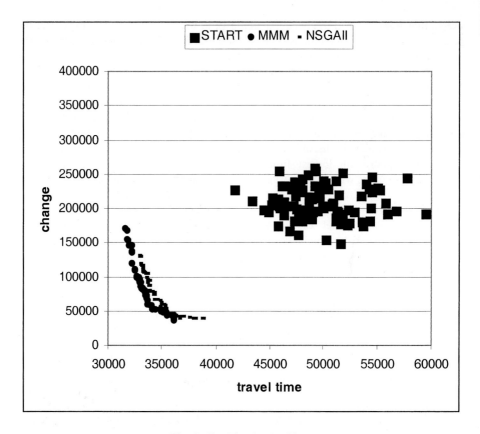

Fig. 5. City Planning Problem

The genetic algorithm was re-executed from a random starting generation seeded with two designs. These two designs were obtained from separate runs of the genetic algorithm with just one objective at a time. Thus, the first seeded design was obtained by executing 100 generations of the genetic algorithm with travel time as the single objective to obtain a minimum travel time design. Since the change function can be evaluated quite rapidly, the second seeded design was obtained by executing 10000 generations of the genetic algorithm with change as the single objective to obtain the minimum change design. These two designs were inserted into the starting generation, and the genetic algorithm was executed for 100 generations with both objectives. The results are plotted in Figure 6. Note that the non-dominated front is much broader than in Figure 5. Note also that the travel time was reduced a little further from the minimum travel time seed. The NSGAII results are more uniformly spread out, but the modified maximin results visually appear to be slightly more optimal.

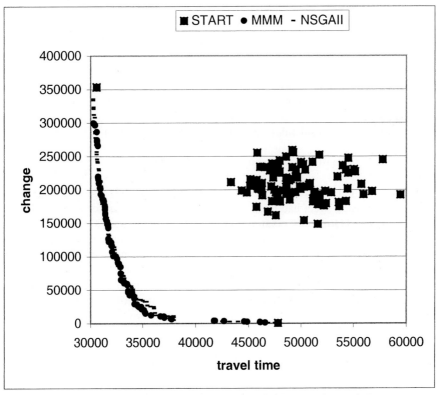

Fig. 6. City Planning Problem with Seeded Starting Generation

7 Regional Planning

Future land-use and transportation plans were sought for the four-county Salt Lake metropolian region in the state of Utah, USA. The developable land in the region was divided into 343 districts. These regional districts were much larger in area than city zones, and it was inappropriate to specify a single land use for an entire district. Therefore, the future land use percentages for each district had to be selected from among 16 different scenarios. These scenarios were typical of scenarios currently existing in the region:

```
predominantly open space scenarios
    #1     5% residential    27% commercial    68% open
    #2    12% residential     6% commercial    82% open
    #3    34% residential     0% commercial    66% open
    #4    43% residential    10% commercial    45% open
predominantly residential scenarios
    #5    62% residential     0% commercial    38% open
    #6    89% residential     7% commercial     4% open
    #7    72% residential    15% commercial    13% open
    #8    88% residential     3% commercial     9% open
```

predominantly commercial scenarios
#9   4% residential   80% commercial   16% open
#10 17% residential 65% commercial   17% open
#11 12% residential 74% commercial   14% open
#12 26% residential 41% commercial   31% open
mixed residential / commercial scenarios
#13 57% residential 26% commercial   17% open
#14 60% residential 33% commercial   7% open
#15 49% residential 31% commercial   20% open
other
#16 airport and university

A total of 260 inter-district streets were identified for the region. The future classification of each street had to be selected from among the same 11 street classes used in the Provo / Orem city planning problem. Three constraints were imposed. A future plan must have enough residential land to house a projected future population of 2,401,000. A future plan must have enough commercial land to employ a projected future job force of 1,210,000. A future plan must have at least 165,000 acres of open space, representing 20% of the developable land. As in the Provo / Orem city planning problem, the first objective was the minimization of the sum of the travel times of all trips in a 24-hour day, and the second objective was the minimization of change from the status quo.

The genetic algorithm was executed with the modified maximin fitness function. The algorithm parameters and procedures were the same as in the Provo / Orem city planning problem. A convex set of non-dominated plans were obtained in the final generation. Table 1 gives data for four plans: the status quo, the minimum change plan, the minimum travel time plan, and an arbitrarily selected compromise plan in the middle of the convex set of non-dominated plans. The status quo plan does not satisfy the minimum housing and employment constraints. The minimum change plan satisfies these constraints with the least amount of change from the status quo. Change is measured in terms of number of people affected. The minimum travel time plan affects 18 times as many people as the minimum change plan, but it cuts the travel time, and thus the air pollution, in half. The compromise plan affects six times as many people as the minimum change plan and has a travel time 30% greater than the minimum travel time plan. The status quo plan has the most open space, but the open space acreage in the other three plans is well above the minimum requirement.

Table 1.

	status quo	minimum change	minimum travel time	compromise
change	0	59,934	1,119,385	359,597
travel time	1,349,617	2,025,681	984,436	1,278,768
housing	1,742,914	2,401,937	2,401,360	2,404,375
employment	995,293	1,210,048	1,466,150	1,433,446
open space	349,583	248,541	247,840	235,941

The streets in the minimum change plan were unchanged from the status quo. The streets in the minimum travel time plan were changed to seven-lane arterials since this street type has maximum capacity and speed.

The land use in the status quo, minimum change, and minimum travel time plans is shown in Figure 7. There are four figures for each plan. The first shows those districts that were assigned predominantly open space scenarios, the second shows those districts that were assigned predominantly residential scenarios, the third shows those districts that were assigned predominantly commercial scenarios, and the fourth shows those districts that were assigned mixed residential / commercial scenarios. The main difference between the status quo and the minimum change plan is that a significant amount of open space land was converted to residential land and to commercial land in order to meet the minimum housing and employment requirements. This is what has actually occurred in the metropolitan region over the past few decades. In other words, the region has been accommodating growth in a manner that affects the fewest number of people. Note that the minimum travel time plan has less open space land, less predominantly residential land, and less predominantly commercial land than the status quo plan, but it has significantly more mixed residential / commercial land. This shows that if people really want to cut their travel times in half, significant land use changes are needed. Specifically, residential and commercial land use must be integrated throughout the region rather than concentrated in a few "downtown" areas.

8 Conclusions

The maximin fitness function may be derived directly from the definition of dominance. Its form is elegant, its implementation is simple, and its evaluation is efficient. It rewards both dominance and diversity without the use of niching techniques. Unlike most other fitness functions used in multi-objective optimization, the maximin fitness function is a continuous function of the objectives. The modified maximin fitness function requires fewer comparisons than the maximin fitness function, but it requires that the non-dominated designs in the generation be identified first.

A brief performance comparison was made between the maximin and modified maximin fitness functions and a state-of-the-art fitness function (NSGAII). They were compared on a standard two-objective problem from the literature, and a real-world application problem. The comparative results in this paper are not extensive enough to be conclusive. They do, however, show that in some situations, the maximin and modified maximin fitness functions perform as well or better than other state-of-the-art multi-objective fitness functions, and are worthy of further study.

The value of seeding the starting generation with optimal designs considering each objective separately was investigated on the real-world application problem. This practice led to a wider non-dominated front, and thus proved beneficial.

14   R. Balling

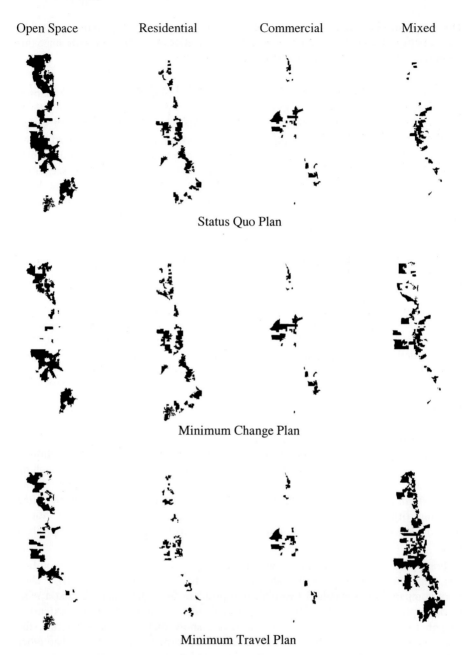

Fig. 7. Optimum Land Use Patterns

The paper demonstrates a new approach to city and regional planning. Genetic algorithms can be used to search over thousands of plans to find a non-dominated set of feasible plans representing a tradeoff between competing objectives.

Results from the regional planning problem show that change is minimized in high-growth areas by converting open space land to residential and commercial land. Travel time is minimized by converting open space, residential, and commercial land to mixed residential / commercial land.

Acknowledgments. This work was funded by the USA National Science Foundation under Grant CMS-9817690, for which the author is grateful.

References

1. Balling, R. J., Pareto sets in decision-based design. *Journal of Engineering Valuation and Cost Analysis*, 3:189–198 (2000)
2. Goldberg, D.E., *Genetic Algorithms for Search, Optimization, and Machine Learning*, Addison-Wesley, Reading, MA, USA (1989)
3. Deb, K., Agrawal, S., Pratap, A., and Meyarivan, T., A fast elitist non-dominated sorting genetic algorithm for multi-objective optimization: NSGA-II. *Proceedings of Parallel Problem Solving from Nature VI Conference (PPSN-VI)*:849–858 (2000)
4. Schaffer, J.D., *Some Experiments in Machine Learning Using Vector Evaluated Genetic Algorithms*, Ph.D. Dissertation, Vanderbilt University, Nashville, TN, USA (1984)

Conflict, Harmony, and Independence: Relationships in Evolutionary Multi-criterion Optimisation

Robin C. Purshouse and Peter J. Fleming

Department of Automatic Control and Systems Engineering
University of Sheffield, Mappin Street, Sheffield, S1 3JD, UK.
{r.purshouse, p.fleming}@sheffield.ac.uk

Abstract. This paper contributes a platform for the treatment of large numbers of criteria in evolutionary multi-criterion optimisation theory through consideration of the relationships between pairs of criteria. In a conflicting relationship, as performance in one criterion is improved, performance in the other is seen to deteriorate. If the relationship is harmonious, improvement in one criterion is rewarded with simultaneous improvement in the other. The criteria may be independent of each other, where adjustment to one never affects adjustment to the other. Increasing numbers of conflicting criteria pose a great challenge to obtaining a good representation of the global trade-off hypersurface, which can be countered using decision-maker preferences. Increasing numbers of harmonious criteria have no effect on convergence to the surface but difficulties may arise in achieving a good distribution. The identification of independence presents the opportunity for a divide-and-conquer strategy that can improve the quality of trade-off surface representations.

1 Introduction

Theoretical evolutionary multi-criterion optimisation (EMO) studies generally consider a small number of objectives or criteria. The bi-criterion case is by far the most heavily studied. EMO applications, by contrast, are frequently more ambitious, with the number of treated criteria reaching double figures in some cases [1, pp207-290]. Hence, there is a very clear need to develop an understanding of the effects of increasing numbers of criteria on EMO. The recently proposed set of benchmark problems, which are scalable to any number of conflicting criteria, represent an important early step towards this aim [2].

This paper establishes a complementary platform for research into increasing numbers of criteria via consideration of the different types of pair-wise relationships between the criteria. A classification of possible relationships is offered in Sect. 2, and the notation used in the paper is introduced. Conflict between criteria is discussed in Sect. 3, whilst Sect. 4 considers harmonious objectives. The aim of a multi-objective evolutionary algorithm (MOEA) is generally regarded as to generate a sample-based representation of the Pareto optimal front, where the samples lie close to the true front and are well distributed across the front. The effects of increasing numbers of each type of criteria on both aspects of the quality of the trade-off surfaces produced are described, together with a review of methods for dealing with the difficulties that arise. The case where criteria can be optimised independently of each other is introduced in Sect. 5. Qualitative studies of pair-wise relationships between criteria are not uncommon in the EMO community, especially in the case of real-world applications. These are discussed in Sect. 6, alongside similar quantitative methodologies from the multi-criterion decision-making (MCDM) discipline. Conclusions are drawn in Sect. 7.

Some of the concepts described in this paper are illustrated using an example result from a recently proposed multi-objective genetic algorithm (MOGA) [3] solving the 3-criterion *DTLZ2* benchmark problem [2]. The equations for this test function are provided in Definition 1. Note that all criteria are to be minimised.

Definition 1. 3-criterion DTLZ2 test function [2].

Min. $z_1(\mathbf{x}) = \left[1 + g(x_3,...,x_{12})\right]\cos(x_1 \pi/2)\cos(x_2 \pi/2)$,

Min. $z_2(\mathbf{x}) = \left[1 + g(x_3,...,x_{12})\right]\cos(x_1 \pi/2)\sin(x_2 \pi/2)$,

Min. $z_3(\mathbf{x}) = \left[1 + g(x_3,...,x_{12})\right]\sin(x_1 \pi/2)$,

where $0 \leq x_i \leq 1$, for $i = 1, 2, ..., 12$,

and $g(x_3,...,x_{12}) = \sum_{i=3}^{12}(x_i - 0.5)^2$.

2 Relationships between Criteria

2.1 Classification

In theoretical EMO studies, the criteria are generally considered to be in some form of conflict with each other. Thus, in the bi-criterion case, the optimal solution is a one-dimensional (parametrically speaking) trade-off surface upon which conflict is always observed between the two criteria. However, other relationships can exist between criteria and these may vary within the search environment. A basic classification of possible relationships is offered in Fig. 1. These relationships are explained in the remainder of the paper.

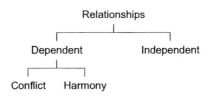

Fig. 1. Classification of relationships between criteria

The dependency classifications are not necessarily mutually exclusive. For example, in the case of three conflicting criteria, there may be regions where two criteria can be improved simultaneously at the expense of the third. This is illustrated in Fig. 2 for the final on-line archive of a MOGA solving the 3-criterion DTLZ2 problem (see Definition 1). For example, ideal performance in z_2 and z_3 (evidence of harmony) can be achieved at the expense of nadir performance in z_1 (evidence of conflict), as indicated by the left-most criterion vector in the figure. However, on the far right of the figure, z_1 and z_3 are now in harmony and are both in conflict with z_2. Thus, the nature of the relationships change across the Pareto front.

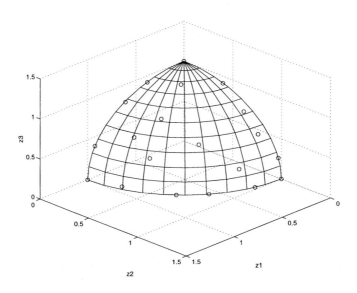

Fig. 2. Final on-line archive of MOGA (depicted as circles) solving DTLZ2, superimposed on the global trade-off surface

2.2 Notation

The following notation is used in the remainder of the paper: M is the number of criteria to be considered in the optimisation procedure, Z is the set of all realisable criterion vectors $\underline{z} \in \square^M$, and Z_R is a particular region of interest in criterion-space, $Z_R \subseteq Z$. If $Z_R = Z$ then the relationship is said to be *global*, otherwise it is described as *local*. The case $Z_R = Z^*$, where Z^* is the Pareto optimal set, may be of particular interest since these are typically the relationships that will be presented to the decision-maker (DM). The DM is the entity that expresses preferences within the optimisation and selects an acceptable solution from the set returned by the optimisation process. The DM is usually one or several humans.

Let i and j be indices to particular criteria: $i, j \in [1,...,M]$. Let a and b be indices to individual criterion vector instances: $a, b \in \left[1 \ldots |Z_R|\right] : \underline{z}^a, \underline{z}^b \in Z_R$. Also let (a,b) denote a pair of instances for which $a \neq b$. Minimisation is assumed throughout the paper without loss of generality.

The dependency relationships that can be identified via pair-wise analysis are summarised in Fig. 3. They are based on the position of criterion vector $\underline{z}^b$ relative to the position of $\underline{z}^a$. These relationships are explored in more detail in Sect. 3 and Sect. 4 to follow.

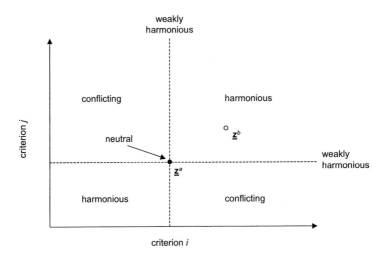

Fig. 3. Dependency relationship regions between a pair of criteria, i and j, identified using the location of sample vector $\underline{z}^b$ relative to that of $\underline{z}^a$

3 Conflicting Criteria

3.1 Definitions of Conflict

A relationship in which performance in one criterion is seen to deteriorate as performance in another is improved is described as *conflicting*. This is summarised by Definition 2 below and can be related to the $\underline{z}^b$-relative-to-$\underline{z}^a$ regions marked as such in Fig. 3.

Definition 2. Criteria i and j exhibit evidence of conflict according to the condition $\left(z_i^a < z_i^b\right) \wedge \left(z_j^a > z_j^b\right)$. If $\not\exists(a,b)$ for which the condition holds then there is *no conflict*, if $\exists(a,b)$ then there is *conflict*, whilst if the condition holds $\forall(a,b)$ then there is *total conflict*.

Note that no attempt has been made to define intermediate levels of conflict (or harmony, as discussed in Sect. 4) since this requires further DM preference information.

3.2 Effect on EMO

For M conflicting criteria, an $(M-1)$-dimensional trade-off hypersurface exists in criterion-space. The number of samples required to achieve an adequate representation of the surface is exponential in M. Given a finite population, an evolutionary optimiser will encounter intractable difficulties in representing the surface when large numbers of conflicting criteria are considered. Even if such a representation were possible, the value to the DM of such a large number of candidate solutions is questionable.

Deb [4, pp400-405] has shown that the proportion of locally non-dominated criterion vectors in a finite randomly-generated sample becomes very large as the number of criteria increases (interestingly, it would appear to be possible to optimise all of the criteria independently in this particular study). Similar results were reported in [5b] for the final non-dominated set of a real-world MOEA application. Since dominance is used to drive the search towards the true Pareto front, there may be insufficient selective pressure to make such progress. The use of a large population can help reduce the proportion, but this is impractical for many real-world problems in which evaluation of a single candidate solution is very time-consuming. Also, the benefit would appear to become progressively weaker as the number of criteria increases.

Many MOEAs use some method of population density estimation, either to modify the selection probability of an individual or as part of the on-line archive acceptance procedure, to achieve a good distribution of solutions. Density estimation also becomes increasingly difficult as the dimensionality of the problem increases (the number of criteria, for density estimation in criterion-space). Due to the 'curse of dimensionality' (the sparseness of data in high dimensions), the ability to fully explore surfaces in greater than five dimensions is regarded as highly limited [6]. Statisticians generally use dimensionality reduction techniques prior to application of the estimator. This assumes that the 'true' structure of the surface is of lower dimension, but the potential for reduction may be limited for a trade-off surface in which all criteria are in conflict with each other.

3.3 Remedial Measures

Preferences. The exploitation of DM preferences is arguably the current best technique for handling large numbers of conflicting criteria. In this case, the aim of EMO is to achieve a good representation of trade-off regions of interest to the DM (essentially limiting the ambition of the optimiser by requiring it to represent only a subspace of the trade-off hypersurface). This sub-section provides a brief overview of preference articulation within MCDM, before examining two popular techniques that use DM preferences to handle large numbers of conflicting criteria.

Preference articulation overview. Preference information can be classified into four distinct types:

- *definitions* of the criteria and problem domain,
- *requirements* of a solution,
- *abstract judgements* on the relative importance of criteria,
- *specific judgements* on a set of candidate solutions.

Definitions of the problem domain and the criteria to be optimised are essential to the optimisation process. However, this information is not necessarily readily available a priori and may contain a degree of uncertainty. Problem domain definitions include limits on the range of decision variables. Development of a good set of criteria is critical to the performance of the optimiser.

Requirements of a solution tend to be specified at criterion level, resulting in the formulation of an ideal or acceptable global solution. Definitions of unacceptability may also be given. In this approach, the DM could specify goals for particular objectives. These goals may be a requirement that must be met (a hard constraint) or may represent softer aspirations.

The DM may wish to provide abstract information about the relative importance of criteria. This may include precise, or perhaps rather vague, information about the level of trade-off that the DM is willing to accept between two criteria (for example, an acceptance of 'an improvement of $\bullet_1$ in criterion i in exchange for a detriment of $\bullet_2$ in criterion j'). Priority information may also be provided, in which the DM expresses a preference for some criteria being more important than others. This information may be imprecise. It may also be qualitative ('much more important') or quantitative ('twice as important'). Priority information can be used to build a partial ordering of criteria, perhaps to be optimised lexicographically, or to determine weights in an aggregation of criteria.

Given a set of candidate solutions, the DM may be able to express preferences for some candidate solutions over others (perhaps allowing a partial ordering of potential solutions to be generated). Again, this information may be qualitative or quantitative, and is likely to be somewhat imprecise.

Preference articulation schemes are generally classified according to when the preference data is elicited from the DM:

- *a priori*, in which preference data is incorporated prior to execution of the optimiser,
- *progressive*, in which the information is requested and exploited during the optimisation process,
- *a posteriori*, where a solution is chosen from a group of results returned by the completed optimisation process.

Refer to [1, pp321-344] for a review of preference articulation schemes in the EMO literature.

Aggregation. One method for reducing the number of conflicting performance criteria is to combine several of them into a single optimisation criterion. Aggregation may be achieved by means of a weighted-sum, or a more complicated function. In this approach, the DM pre-specifies the trade-offs between the combined subset of criteria. This eliminates the requirement for the optimiser to represent this portion of the global trade-off surface. The inherent disadvantage of the approach is that the DM must be able to specify the required trade-off a priori. Also, any adjustments to the preferences will require a complete re-run of the optimisation. Nevertheless, this may be an appropriate technique, especially when faced with very large numbers of criteria.

Goals and Priorities. Greater flexibility can be achieved in terms of criterion reduction by exploiting goal values and priorities for various criteria, if these can be elicited from the DM. The *preferability* relation developed in [5a] unifies various classical operations research (OR) schemes based on goals and priorities and applies them within the context of EMO. In essence, the method adaptively switches on or off different criteria, from the perspective of the dominance relation, for each pair of vectors considered. The iterative nature of the EA paradigm can be exploited to update the preferences as information becomes progressively available to the DM.

Dimension Reduction. Existing dimensionality reduction techniques could be used to transform criterion-space into a lower dimension. This could be done prior to the optimisation, based on some preliminary analysis, or could be updated on-line as the MOEA evolves. The key benefit of the latter approach is that, as the MOEA progressively identifies the trade-off surface, the reduction is performed on a space more relevant to both the EA and the DM. If the reduction is to be performed iteratively then the balance between capability and complexity of the applied technique must be considered. For example, curvilinear component analysis [7] has good general applicability but a significant computational overhead, whilst principal components analysis [8] has the opposite features.

Dimension reduction methods can be applied directly to the density estimation process to preserve trade-off diversity in information-rich spaces. However, since the methods do not respect the dominance relation, they cannot be used directly in the Pareto ranking process without modification.

Visualisation. Note that the ability to visualise the developing trade-off surface becomes increasingly difficult as the number of criteria increases. The method of parallel coordinates is a popular countermeasure for large numbers of criteria. Scatter-plots with brushing and glyph approaches, such as Chernoff faces [9], are amongst the possible alternatives [6][10]. Parallel coordinates and scatter-plots are both closely linked to the concepts of conflict and harmony described in this paper, and are discussed further in Sect. 6.

4 Harmonious Criteria

4.1 Definitions of Harmony

A relationship in which enhancement of performance in a criterion is witnessed as another criterion is improved can be described as *harmonious*. If performance in the criterion is unaffected, the relationship is described as *weakly harmonious*. Complete definitions are provided below and can be related to the relevant $\underline{z}^b$-relative-to-$\underline{z}^a$ regions and lines in Fig. 3.

Definition 3. Levels of harmony are determined by the condition $\left(\underline{z}_i^a < \underline{z}_i^b\right) \wedge \left(\underline{z}_j^a < \underline{z}_j^b\right)$. If $\not\exists (a,b)$ for which the condition holds then there is *no harmony*, if $\exists (a,b)$ then there is *harmony*, and if the condition holds $\forall (a,b)$ then there is *total harmony*.

Definition 4. Levels of weak harmony are determined by the condition $\left[\left(\underline{z}_i^a < \underline{z}_i^b\right) \wedge \left(\underline{z}_j^a = \underline{z}_j^b\right)\right] \vee \left[\left(\underline{z}_i^a = \underline{z}_i^b\right) \wedge \left(\underline{z}_j^a < \underline{z}_j^b\right)\right]$. If $\not\exists (a,b)$ for which the condition holds then there is *no weak harmony*, if $\exists (a,b)$ then there is *weak harmony*, and if the condition holds $\forall (a,b)$ then there is *total weak harmony*.

Definition 5. Neutrality is determined by the condition $\left(\underline{z}_i^a = \underline{z}_i^b\right) \wedge \left(\underline{z}_j^a = \underline{z}_j^b\right)$. If $\not\exists (a,b)$ for which the condition holds then there is *no neutrality*, if $\exists (a,b)$ then there is *neutrality*, and if the condition holds $\forall (a,b)$ then there is *total neutrality*.

Harmonious relationships have been observed in several EMO application papers, where they are indicated by non-crossing lines between pairs of criteria on a parallel coordinates plot (see Sect. 6), including the following:

- passenger cabin acceleration versus control voltage in electromagnetic suspension controller design for a maglev vehicle [11],
- gain margin versus phase margin, and 70% rise time versus 10% settling time, in the design of a Pegasus low-pressure spool speed governor [5b].

4.2 Effect on EMO

In either form of total harmony, one of the criteria can be removed without affecting the partial ordering imposed by the Pareto dominance relation on the set Z_R of candidate solutions. This type of relationship has received some consideration in the classical OR community, usually for $Z_R = Z^*$, where one member of the criterion pair is known variously as *redundant*, *supportive*, or *nonessential* [12][13][14]. It remains an open question whether or not to include redundant criteria in the optimisation process. Reasons to keep such criteria include:

- knowledge of the relationship may be of interest to the DM, especially if the rate of harmonious behaviour changes over the course of the search space,
- the relationship may not be apparent from a random finite sample of the search space,
- inclusion does not, necessarily, harm the search,
- the DM may be more comfortable with the inclusion of the criterion.

Reasons to remove redundant criteria include:

- to eliminate the extra burden on the DM, who must inspect and make decisions on matters that do not affect the search and may be misleading,
- to reduce the computational load, in terms of both performance evaluations and comparisons.

The inclusion of a redundant criterion does not affect the partial ordering of candidate solutions imposed by the Pareto dominance operator. Thus, progress towards the global Pareto front is unaffected. It is, however, possible that such an inclusion could affect the diversity in the representation of the trade-off hypersurface. This depends on the definition of distance between criterion vectors used by the density estimator. For example, any procedure using Euclidean distances or the *NSGA-II* crowding algorithm [15] could suffer from potential bias. Consider the case of three criteria: where z_1 and z_2 totally conflict, z_1 and z_3 totally conflict, and z_2 and z_3 are in total harmony. The resulting trade-off surface is one-dimensional, and can be represented by the conflict between z_1 and z_2. A uniform distribution in the Euclidean sense may not be arrived at across the normalised trade-off surface, even if such a distribution is achievable, because the Euclidean distance calculation is biased in favour of z_2 since $\{z_2, z_3\}$ has greater influence on the Euclidean distance measure than z_1. Thus, a diversity preservation technique would bias in favour of diversity in z_2 on the trade-off surface. The overall effect of this depends on the trade-off surface in question: sometimes, good diversity in z_2 will naturally lead to good diversity in z_1 but this is not guaranteed to be the case.

4.3 Remedial Measures

Redundant criteria may be identified by using the sample set contained within the EA population for each criterion and looking for large positive correlations between the data sets for each pair of criteria. Redundant criteria may be removed if this is felt appropriate for the problem in-hand. Alternatively, the criteria may be selectively ignored in the density estimation process (and also the ranking process in order to reduce the number of unnecessary comparisons) and yet still be presented to the DM.

5 Independent Criteria

5.1 Independence in the Context of EMO

In this paper, *independence* refers to the ability to decompose the global optimisation problem into a group of sub-problems that can be solved separately from each other. Thus, different criteria and decision variables will be allocated to different sub-problems.

In the context of the relationship between a pair of criteria, independence means that the criteria can, in theory, be optimised completely separately from each other. As with a harmonious relationship, it is possible to make improvements to both criteria simultaneously (from the perspective of the complete solution). The difference between independence and harmony is that appropriate adjustments must be made to two distinct parts of the complete solution in the former case, whilst in the latter case a single good decision modification for one of the criteria will *naturally* produce improvement in the second criterion.

If two criteria are independent then they do not form part of the same trade-off surface. Thus multiple, distinct, trade-off surfaces exist, each of which should be represented separately for inspection by the DM.

5.2 Effects of Independence on EMO

Consider a global problem, **p**, comprised of n independent sub-problems $[p_1,\ldots,p_n]$ with associated independent sets of criteria $[z_1,\ldots,z_n]$ and independent sets of decision variables $[\mathbf{x}_1,\ldots\mathbf{x}_n]$. If advance knowledge of these sets is available then the global problem can be decomposed into the groups of sub-problems prior to optimisation. Then a proportion of the total available resources (candidate solution evaluations) could be exclusively allocated to the optimisation of each sub-problem. Both a global approach and the aforementioned *divide-and-conquer* method should yield the same solution of n independent trade-off surfaces. From an EMO perspective, it then becomes a matter of interest as to which technique produces superior results in terms of trade-off surface quality. Is the effort expended identifying and exploiting the correct decompositions rewarded with improved results?

In the first study of its kind, an attempt has been made to answer this question in [16]. The study demonstrated that, for a simple test problem, a divide-and-conquer strategy could substantially improve MOEA performance. A priori decompositions were evaluated in criterion-space, decision-space, and both spaces simultaneously. Parallel EA models were applied to each sub-problem. All three methods led to significantly higher-quality trade-off surfaces than the global approach, with both-space decomposition proving the most attractive. Given that it may not be possible to accurately identify the sub-problems in advance of the optimisation, an on-line adaptive divide-and-conquer strategy for MOEAs was also proposed and evaluated in [16]. Bivariate statistical tests for independence were applied to the population sample data in order to identify the independence relationship.

6 Existing Methods for Identifying Pair-Wise Relationships

This paper considers the relationships that exist between pairs of criteria, by comparing pairs of criterion vectors. In this approach, composite relationships must be inferred from these simpler relations. However, the pair-wise methodology is very popular in multivariate studies and forms a good foundation for analysis, with many qualitative and quantitative techniques based on this approach. Methods that are closely linked to the definitions of conflict and harmony described earlier are discussed in the remainder of this section.

6.1 Qualitative Methods

The method of parallel coordinates, first described in [17] and subsequently applied to EMO in [18], reduces an arbitrary high-dimensional space to two-dimensions. The parallel coordinates representation of the on-line archive of Fig. 2 is shown in Fig. 4. Criterion labels are located at discrete intervals along the horizontal axis (and these should be interchangeable). Normalised performance in each criterion is indicated on the vertical axis. A particular criterion vector is displayed by joining the performance levels in all adjacent criteria by straight lines. Then, considering two criterion vector instances for a pair of criteria, the lines representing the two instances will cross if conflict is exhibited according to Definition 2 or will fail to cross if harmony is observed according to Definitions 3 or 4 (in the case of Definition 5, the lines will be superimposed). Thus, the magnitude of conflict is heuristically visualised as 'many' crossing lines.

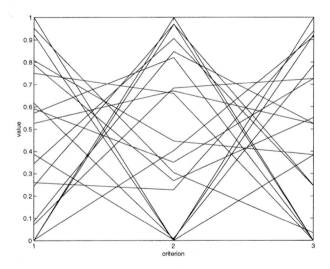

Fig. 4. Parallel coordinates representation of the data shown in Fig. 2

Wegman [19] presents some valuable insights and extensions towards using the parallel coordinates representation as a high-dimensional data analysis tool. Statistical interpretations of the plots are possible, with features such as marginal densities, correlations, clusters, and modes proving readily identifiable. Parallel coordinates plots can suffer from over-plotting for large data sets and thus a density plot variant is also presented in the paper to overcome this.

Another popular method of pair-wise visualisation, which in its full form presents more simultaneous comparisons than the standard parallel coordinates plot, is the scatterplot matrix [10]. Such a plot for the MOGA on-line archive of Fig. 2 is shown in Fig. 5. Each element of the matrix of plots shows a particular bi-criterion section of the trade-off surface. For example, the upper central plot shows z_2 on the horizontal axis and z_1 on the vertical axis. It can sometimes be difficult to extract information from these plots, especially as the number of criteria increases. Highlighting of a particular criterion vector instance or group of instances – a technique known as *brushing* – can often aid higher-order understanding. The filled circle in Fig. 5 indicates one particular vector.

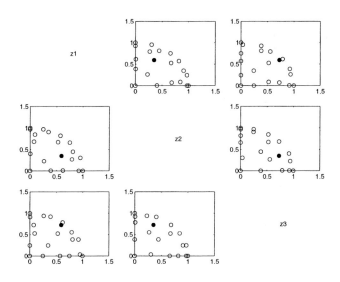

Fig. 5. Scatterplot matrix representation of the data shown in Fig. 2

6.2 Quantitative Methods

Several pair-wise methods exist for quantifying conflict between criteria that use similar concepts to the parallel coordinates notion of crossing lines. The *Kendall sample correlation statistic* measures the difference between the number of *concor-*

dant samples (as one variable increases/decreases, the other follows suit) and the number of *discordant* samples (as one variable increases/decreases, the other does the opposite) [20]. Thus, discordance produces crossing lines whilst concordance does not. Fuzzy measures of conflict also use this type of approach: see, for example, [21].

Schroder [22] developed a technique based directly on the method of parallel coordinates. In this approach, each criterion range is partitioned into a number of equally sized regions. The level of conflict is then defined as a weighted-sum of the crossings between pairs of regions (rather than actual solutions), where the weights are based on the separation between the regions. Crossings between distant regions are argued to be indicative of strong conflict. The method also normalises conflict levels with respect to population density. This may be appropriate if one particular region is thought to be over-sampled by the search, but may not in general be correct because if vectors are similar in two criteria this does not necessarily mean they are from the same region of the global trade-off surface. The method also requires additional preference information, unlike the previous techniques, since it is not based purely on ordinal data. However, more information can, potentially, be extracted using this method.

7 Conclusions

EMO applications have long considered the simultaneous optimisation of large numbers of criteria. However, EMO algorithm developers have tended to concentrate almost entirely on the bi-criterion domain. Thus, there is a present lack of understanding of how MOEAs cope with larger numbers of criteria. This paper has sought to lay foundations for future work in this direction by considering how increasing numbers of criteria affect MOEA search performance.

Three relationships – conflict, harmony, and independence – have been identified. It has been demonstrated how the relationship between two criteria can contain elements of both conflict and harmony, resulting from interaction with other criteria.

It has been argued that increasing numbers of conflicting criteria will severely hamper the ability of an MOEA to represent the global trade-off surface. Thus, in general, the oft-stated EMO aims of closeness to and diversity across the entire Pareto front could be little more than a pipedream. It is difficult to see how, when confronted with large numbers of criteria, increased requirements for preference information can be avoided. Even if an MOEA was capable of adequately representing the entire, complicated, high-dimensional trade-off surface, this amount of information is surely of little benefit to the human DM, who is faced with the task of selecting a single solution. The dimension of the problem must be kept reasonably low for human DM analysis to remain tractable: twelve criteria has been suggested as an upper limit [23]. Aggregation of criteria, whilst anathematic to many EMO researchers, may thus be necessary.

Even if the number of criteria is limited by DM considerations, it may still be prudent to consider further dimensionality reduction in the context of the MOEA search. This can be achieved by removing some criteria from certain comparisons (and thus preserving the dominance relation) or by applying some form of transformation to a new set of coordinates (the standard dimension reduction approach). The utility of methods based on the latter approach is limited because they do not respect the dominance relation. They may, however, be used to help achieve good diversity in information-rich spaces (where 'information' is defined according to the chosen method).

Interactive preference articulation schemes, such as [5a], are particularly valuable in problems with large numbers of conflicting criteria and fit very nicely within the iterative EA framework. In such schemes, the attention of the optimiser is focused on various sub-regions of the trade-off surface as the search progresses. This is beneficial to the DM who may only be interested in learning about certain trade-offs within the global problem. The progressive nature of the scheme suits the often changing aspirations of the DM as more knowledge is uncovered. The main drawback of this approach is that it can be rather DM-intensive.

To summarise, increasing numbers of conflicting criteria in a problem transforms the aim of EMO from identification of a globally optimal solution set towards assisting the DM in learning about the trade-offs between criteria and finding an *acceptable* solution.

Harmonious criteria, and the special case of redundancy, do not have the same severe impact on EMO as does conflict. Convergence to the Pareto front is unaffected by increasing numbers of totally harmonious criteria. Issues surrounding distribution of solutions across the surface do require some care however. The decision on whether to eliminate any identified redundant criteria from the search is perhaps best left to the discretion of the analyst and the DM.

The existence of independence within the global problem leads to multiple, separate, trade-off surfaces. If independence can be identified then the deployment of a divide-and-conquer strategy could potentially improve EMO performance [16].

The recently proposed suite of scalable test problems provides the opportunity to explore in full the behaviour of the MOEAs in a controlled and tractable manner [2]. Advancements in techniques for performance analysis and visualisation are required to aid understanding of the results from such problems. Assessment of the utility of contemporary dimension reduction methods for EMO is a rich area for future research. Alternative methods to the standard sample-based approach of describing the trade-offs may prove useful both as a decision aid and within the context of the search itself.

In conclusion, the simultaneous consideration of many criteria is arguably the greatest challenge facing the EMO community at the present time.

Acknowledgments. The authors would like to thank the anonymous reviewers for their valuable comments and advice given on the original submissions of this paper and [16]. Robin Purshouse also wishes to acknowledge the UK Engineering and Physical Sciences Research Council (EPSRC) for research studentship support.

References

1. Coello, C.A.C., Van Veldhuizen, D.A., Lamont, G.B.: Evolutionary Algorithms for Solving Multi-Objective Problems. Kluwer Academic Publishers, New York Boston Dordrecht London Moscow (2002)
2. Deb, K., Thiele, L., Laumanns, M., Zitzler, E.: Scalable Multi-Objective Optimization Test Problems. Proceedings of the 2002 IEEE Congress on Evolutionary Computation (CEC 2002), Vol. 1. (2002) 825–830

3. Purshouse, R.C., Fleming, P.J.: Why use Elitism and Sharing in a Multi-Objective Genetic Algorithm? Proceedings of the Genetic and Evolutionary Computation Conference (GECCO 2002). (2002) 520–527
4. Deb, K.: Multi-Objective Optimization using Evolutionary Algorithms. Wiley, Chichester New York Weinheim Brisbane Singapore Toronto (2001)
5. Fonseca, C.M., Fleming, P.J.: Multiobjective optimization and multiple constraint handling with evolutionary algorithms (a) Part I: A unified formulation (b) Part II: Application example. IEEE Transactions on Systems, Man, and Cybernetics, Part A: Systems and Humans **28** (1998) 26–37 and 38–47
6. Scott, D.W.: Multivariate Density Estimation: Theory, Practice, and Visualization. Wiley, New York Chichester Brisbane Toronto Singapore (1992)
7. Demartines, P., Hérault, J.: Curvilinear Component Analysis: a Self-Organising Neural Network for Non-linear Mapping of Data Sets. IEEE Transactions on Neural Networks **8** (1997) 148–154
8. Jolliffe, I.T.: Principal Component Analysis. Springer, New York (1986)
9. Chernoff, H.: The Use of Faces to Represent Points in k-Dimensional Space Graphically. Journal of the American Statistical Association **68** (1973) 361–368
10. Cleveland, W.S.: Visualizing Data. Hobart Press, Summit (1993)
11. Dakev, N.V., Whidborne, J.F., Chipperfield, A.J., Fleming P.J.: Evolutionary H-infinity design of an electromagnetic suspension control system for a maglev vehicle. Proceedings of the Institution of Mechanical Engineers Part I **211** (1997) 345–355
12. Agrell, P.J.: On redundancy in multi criteria decision-making. European Journal of Operational Research **98** (1997) 571–586
13. Carlsson, C., Fullér, R.: Multiple Criteria Decision Making: The Case for Interdependence. Computers & Operations Research **22** (1995) 251–260
14. Gal, T., Hanne, T.: Consequences of dropping nonessential objectives for the application of MCDM methods. European Journal of Operational Research **119** (1999) 373–378
15. Deb, K., Pratap, A., Agarwal, S., Meyarivan, T.: A Fast and Elitist Multiobjective Genetic Algorithm: NSGA-II. IEEE Transactions on Evolutionary Computation **6** (2002) 182–197
16. Purshouse, R.C., Fleming, P.J.: An Adaptive Divide-and-Conquer Methodology for Evolutionary Multi-Criterion Optimisation. This volume. (2003)
17. Inselberg, A.: The Plane with Parallel Coordinates. The Visual Computer **1** (1985) 69–91
18. Fonseca, C.M., Fleming, P.J.: Genetic algorithms for multiobjective optimization: Formulation, discussion and generalization. Proceedings of the Fifth International Conference on Genetic Algorithms. (1993) 416–423
19. Wegman, E.J.: Hyperdimensional Data Analysis Using Parallel Coordinates. Journal of the American Statistical Association **85** (1990) 664–675
20. Kendall, M.G.: A new measure of rank correlation. Biometrika **30** (1938) 81–93
21. Lee, J., Kuo, J.-Y.: Fuzzy decision making through trade-off analysis between criteria. Journal of Information Sciences **107** (1998) 107–126
22. Schroder, P.: Multivariable control of active magnetic bearings. Doctoral dissertation, University of Sheffield. (1998)
23. Bouyssou, D.: Building criteria: a prerequisite for MCDA. In: Bana e Costa, C.A. (ed.): Readings in Multiple Criteria Decision Aid. Springer-Verlag, Berlin Heidelberg New York (1990) 58–80

Is Fitness Inheritance Useful for Real-World Applications?

Els Ducheyne[1], Bernard De Baets[2], and Robert De Wulf[1]

[1] Lab. of Forest Management and Spatial Information Techniques
[2] Department of Applied Mathematics, Biometrics and Process Control
Ghent University, Coupure links 653, 9000 Gent, Belgium
Els.Ducheyne@rug.ac.be,
http://fltbwww.rug.ac.be/forman

Abstract. Fitness evaluation in real-world applications often causes a lot of computational overhead. Fitness inheritance has been proposed for tackling this problem. Instead of evaluating each individual, a certain percentage of the individuals is evaluated indirectly by interpolating the fitness of their parents. However, the problems on which fitness inheritance has been tested are very simple and the question arises whether fitness inheritance is really useful for real-world applications. The objective of this paper is to test the performance of average and weighted average fitness inheritance on a well-known test suite of multiple objective optimization problems. These problems have been generated as to constitute a collection of test cases for genetic algorithms. Results show that fitness inheritance can only be applied to convex and continuous problems.

1 Introduction

In many real-world applications of evolutionary algorithms, the fitness of an individual has to be derived using complex models or computations. Especially in the case of multiple objective optimization problems (MOOP), the time needed to evaluate these individuals increases exponentially (Chen et al. 2002) due to 'the curse of dimensionality' (Bellman, 1961). This will lead to a slower convergence of the evolutionary algorithms.

Fitness inheritance is an efficiency enhancement technique that has been originally proposed by Smith (1995) to improve the performance of genetic algorithms. Sastry et al. (2001) and Chen et al. (2002) have developed analytical models for fitness inheritance. In fitness inheritance an offspring receives a fitness value that is inferred from the fitness values of its parents instead of through complete fitness evaluation.

The objective of this paper is to evaluate whether the existing fitness inheritance techniques are capable of solving multiple objective optimization problems. To this end, the different techniques are applied to a test bed proposed by Zitzler (1998,1999,2000). A second objective is to test whether there exists an optimal proportion of inheritance as was reported by Sastry et al. (2001) and Chen et al. (2002).

In section 2, the existing work on single and multiple objective fitness inheritance is reviewed. The test problems and their characteristics are also presented in this section. The experiments and results are discussed in Section 3. Finally, the conclusions are drawn in Section 4. Indications for future work are also given in this section.

2 Background

2.1 Single Objective Fitness Inheritance

Smith et al. (1995) were the first to introduce the technique of fitness inheritance. They used two types of inheritance: the first type is by taking the average value of the fitness values of the two parents, the second by taking the weighted average according to the similarity between the offspring and their parents. They found that the weighted average resulted in a better performance of the genetic algorithm. They gave some theoretical foundation for the success of fitness inheritance in comparison to the classical genetic algorithm for the One Max problem and for an aircraft routing problem using the building block hypothesis. Their results indicated that fitness inheritance could enhance GA performance.

Sastry et al. (2001) investigated the time to convergence, population sizing and the optimal proportion of inheritance for the One Max problem. They found that the time until convergence is:

$$t_{conv} = \frac{\pi}{2I}\sqrt{\frac{l}{1-p_i}} \tag{1}$$

with I the selection intensity, l the length of the chromosome and p_i the fraction of the individuals that inherit their value. The population size can be written as:

$$n = -\frac{2^{k-1}\log(\psi)\sqrt{\pi}}{(1-p_i^3)}\sqrt{\sigma_f^2} \tag{2}$$

where k is the size of the building block, ψ is the failure rate and σ_f^2 is the variance of the noisy fitness functions. They finally determined that the optimal proportion of inheritance p_i^* lies between $54\% - 55.8\%$. By building these analytical models, Sastry et al. were the first to give a strong theoretical foundation for fitness inheritance.

2.2 Multiple Objective Fitness Inheritance

Chen et al. (2002) extended this analytical model for multiple objective problems. They included an extra parameter M to account for the number of niches in the multiple objective problem. Their model for time to convergence is:

$$t_{conv} = \frac{\pi}{2I}\sqrt{\frac{l}{1-p_i}}\sqrt{1+\frac{M-1}{l}} \qquad (3)$$

The population size can be determined as follows:

$$n = -\frac{2^{k-1}\log(\psi)M\sqrt{\pi}}{(1-p_i^3)}\sqrt{\sigma_f^2 + \sigma_N^2} \qquad (4)$$

where σ_N^2 is the noise variance from the other niches.

2.3 Zitzler's Test Suite

Zitzler (1999) presented a test suite of problems that pose certain difficulties to multiple objective genetic algorithms. This test bed has been used extensively for the comparison of new algorithms. As this set is regarded as a benchmark in many papers, it is used in this paper to examine the behavior of genetic algorithms without and with fitness inheritance. In this paper, we categorize the different functions in 3 categories: convex functions (functions 1, 4 and 5), non-convex functions (functions 2 and 6) and discontinuous functions (function 3).

- Test function 1 has a convex Pareto-optimal front:

$$f_1(x_1) = x_1$$
$$g_1(x_2,\ldots,x_n) = 1 + \frac{9 \cdot (\sum_{i=2}^{n} x_i)}{n-1} \qquad (5)$$
$$h_1(f_1, g_1) = 1 - \left(\frac{f_1}{g_1}\right)^2$$

  where $n = 30$ and $x_i \in [0,1]$. The Pareto-optimal front is formed when g_1 equals 1.
- Test function 2 is the non-convex counterpart of test function 1:

$$h_2(f_1, g_1) = 1 - \sqrt{\frac{f_1}{g_1}} \qquad (6)$$

  where $n = 30$ and $x_i \in [0,1]$. The Pareto-optimal front is formed when g_1 equals 1.
- Test function 3 tests whether a genetic algorithm is able to cope with discreteness in the Pareto-optimal front:

$$h_3(f_1, g_1) = 1 - \sqrt{\frac{f_1}{g_1}} - \left(\frac{f_1}{g_1}\right)\sin(10\pi f_1) \qquad (7)$$

  where $n = 30$ and $x_i \in [0,1]$. The Pareto-optimal front is formed when g_1 equals 1. The sine function introduces discontinuity in the Pareto-optimal front but not in the objective space.

- Test function 4 checks the capability of a genetic algorithm to deal with multimodality:

$$g_4(x_2, \ldots, x_n) = 1 + 10(n-1) + \sum_{i=2}^{n} \left(x_i^2 - 10\cos(4\pi x_i) \right) \quad (8)$$

$$h_4(f_1, g_4) = 1 - \sqrt{\frac{f_1}{g_4}}$$

where $n = 10$, $x_1 \in [0, 1]$ and $(x_2, \ldots, x_n) \in [-5, \ldots, +5]$. The Pareto-optimal front is formed with g_4 equals 1.
- Test function 5 investigates the behavior of a genetic algorithm if it encounters a misleading problem. This class of problems is referred to as deceptive problems.

$$f_5(x_1) = 1 + u(x_1)$$

$$g_5(x_2, \ldots, x_n) = \sum_{i=2}^{n} v(u(x_i)) \quad (9)$$

$$h_5(f_5, g_5) = \frac{1}{f_5}$$

where $u(x_i)$ gives the number of ones in the bit vector x_i (unitation)

$$v(u(x_i)) = \begin{cases} 2 + u(x_i) & \text{if } u(x_i) < 5 \\ 1 & \text{if } u(x_i) = 5 \end{cases} \quad (10)$$

and $n = 10$, $x_1 \in \{0,1\}^{30}$ and $(x_2, \ldots, x_n) \in \{0,1\}^5$. The true Pareto-optimal front is formed when g_5 equals 10 while the best deceptive Pareto-optimal set includes all solutions **x** for which $g(x_2, \ldots, x_n) = 11$.
- Test function 6 includes two difficulties caused by the non-uniformity of the objective space. Firstly, the Pareto-optimal solutions are non-uniformly distributed along the Pareto-front; secondly, the density of the solutions is least near the Pareto-optimal front and highest away from the front:

$$f_6(x_1) = 1 - \exp(-4x_1)\sin^6(6\pi x_1)$$

$$g_6(x_2, \ldots, x_n) = 1 + \left(\frac{9 \cdot (\sum_{i=2}^{n} x_i)}{n-1} \right)^{0.25} \quad (11)$$

$$h_6(f_6, g_6) = 1 - \left(\frac{f_6}{g_6} \right)^2$$

where $n = 130$ and $x_i \in [0, 1]$. The Pareto-optimal front is formed when g_6 is equals to 1 and is non-convex.

2.4 Comparing Multiple Objective Optimizers

The comparison of different techniques of multiple objective evolutionary algorithms is very difficult. Zitzler et al. (2002) give a mathematical background as to why the quality assessment of multiple objective optimizers is difficult. They conclude that by using unary operators such as the hypervolume metric (Zitzler, 1999) or the generational distance (Van Veldhuizen, 1999),it is impossible to formulate strong statements. A statement of the type "one algorithm is not worse than another" is the strongest statement that can be made using the current unary operators. However, it is still possible to use unary operators to focus on certain qualities of one technique over the other. In this paper we will apply the hypervolume metric C (Zitzler, 1999) because this metric gives an idea about the spread and distance to the Pareto-front. When performing multiple runs, it is possible to give an indication of the variance between the different Pareto-sets using the average distance to the Pareto-optimal set X_p, also referred to as the generational distance (Deb, 2001). By using these metrics it is not the aim of the authors to state that one technique is better than another, the only statement we want to make is that for the given unary operator one algorithm scores better than the other.

3 Experiments and Results

The experiments were performed using a selectorecombinative GA with binary tournament selection, one point crossover with crossover probability of 0.8 and a uniform mutation rate of 0.01. The population size was set to 100 and the number of generations was set to 200. These settings are the same as in Zitzler (1999) with as only difference the number of generations. The encoding of the decision vector was the same as in Zitzler: an individual is represented as a bit vector where each parameter x_i is represented by 30 bits. The parameters $x_2, \ldots, x_n$ for test function t_5 only comprise 5 bits. The crowding distance assignment procedure was used for sharing, so no parameters were set for the sharing. The fitness assignment applied was proposed by Deb et al. (2000) and is equal to the number of solutions that dominate the current solution. All experiments were repeated 10 times. Since we chose the same settings as Zitzler is it possible to check the implementation of our own genetic algorithm as well as to test the behavior of the fitness inheritance strategies.

In our experiments, we used the two fitness inheritance strategies originally proposed by Smith et al. (1995), namely the average and the weighted average fitness values. The similarity between offspring and parents was easily derived. As one point crossover was used, the length until the cutting point determined the similarity between child and parent.

3.1 How Does the Genetic Algorithm with Fitness Inheritance Cope with the Test Suite?

In Figs. 1 to 6 the non-dominated sets achieved by the evolutionary algorithm without fitness inheritance, as well as the non-dominated set of solutions for the two evolutionary algorithms with average or weighted average fitness inheritance, are presented for one representative function per category. The optimal Pareto-front is drawn for comparison purposes. For each of the test functions, the first objective $F_1 = f$ is depicted on the X-axis. The Y-axis represents a function of the two other objectives:

$$F_2 = \frac{h \times g}{g^*} \qquad (12)$$

where g^* is the value where the Pareto-optimal front is reached for the other objectives.

As we do not only want to look at the distance of the Pareto-front to the true Pareto-front but also want to investigate the robustness that results when doing multiple runs, we present all non-dominated solutions achieved by the different algorithms for all repetitions for these representative functions. With robustness, the authors mean the variance that results from the repetitions within an algorithm.

In what follows, the unary metrics are statistically analyzed with the significance level is always set to 95%.

3.2 Convex Functions

In Figs. 1 and 2, we find that the standard genetic algorithm is capable of finding a Pareto-front very close to the optimal Pareto-front for function 1. We also can conclude that the two inheritance strategies are able to give a good approximation of the true Pareto-front. On the other hand, we see in Fig. 2 that the variance is higher when applying the fitness inheritance. When calculating the generational distance for each of the runs, however, we find that there is no significant difference between the different algorithms for this measure. There is also no difference between the hypervolume metrics calculated for the first function. The standard GA as well as the inheritance techniques encounter problems when trying to solve the multimodal as well as the deceptive function.

As the fourth function of the test set is convex, the overall Pareto-front as well as the scatter plot exhibit similar behavior as in the first function. The variance between the non-inherited approach and the weighted average is not significantly different, there is also no difference between the inheritance strategies but the classical GA is significantly different from the average inheritance. The overall Pareto-surface dominates the two other surfaces. The average weighted approach also dominates the non-weighted approach.

The fifth function also exhibits the same behavior as the other two convex functions.

3.3 Non-convex Functions

In Fig. 3 we see that the algorithms based on fitness inheritance have a hard time solving non-convex optimization problems. The generational distance between the standard GA and the other strategies is significantly different (Fig. 4), but

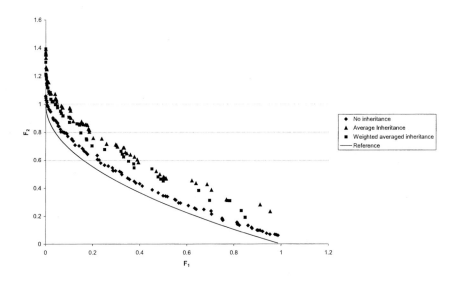

Fig. 1. The overall Pareto-front for test function 1. On the X-axis the $F_1 = f_1$, on the Y-axis $F_2 = h_1 * g_1$

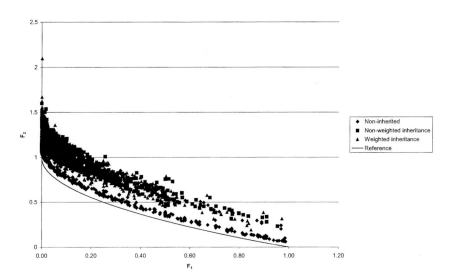

Fig. 2. Scatter plot of all non-dominated solutions found in 10 repetitions for the first test function

there is no difference between the two inheritance strategies according to the results of a non-parametric Kruskal-Wallis test. We also observe that the solutions are not uniformly distributed for the inheritance techniques whereas the

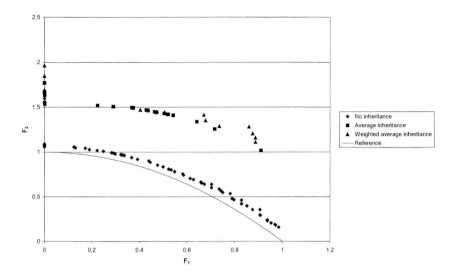

Fig. 3. The overall Pareto-front for test function 2. On the X-axis the $F_1 = f_1$, on the Y-axis $F_2 = h_2 * g_1$

classical GA has a more uniform distribution. This is confirmed when calculating Zitzler's hypervolume metric. The same conclusions can be drawn for function 6. The non-parametric analysis of the hypervolume metrics indicates that the standard GA is significantly better than the two other techniques. Again we see that the variance resulting from multiple runs for the inheritance GAs is not larger than when using the standard GA.

3.4 Discontinuous Functions

The inheritance techniques are unable to solve the third discontinuous problem (Figs. 5 and 6). The generational distance for the fitness inheritance techniques are significantly worse than for the standard genetic algorithm. Both inheritance techniques give the same result but their Pareto-front approaches the true Pareto-front in a linear way. The solutions are also unevenly distributed along the fronts. Apparently, the noise resulting from the linear interpolation of the fitness values of the offspring results in a high disturbance of the genetic algorithm.

3.5 Variable Inheritance Probability

Both Sastry et al. (2001) and Chen et al. (2002) reported that there exists an optimal proportion of inheritance for speed up. In this section we present the results of different proportions of inheritance and try to discern whether the same is true for the test suite. When we look at Fig. 7 we see that for the

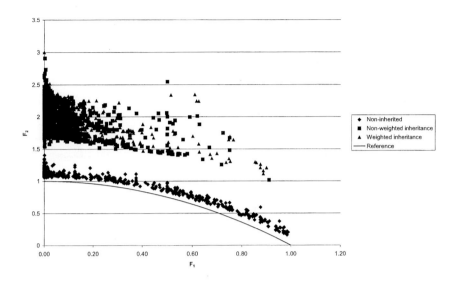

Fig. 4. Scatter plot of all non-dominated solutions found in 10 repetitions for the second test function

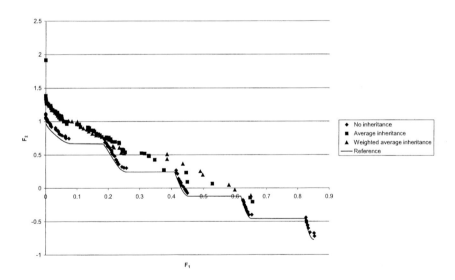

Fig. 5. The overall Pareto-front for test function 3. On the X-axis $F_1 = f_1$, on the Y-axis $F_2 = h_3 * g_1$

first (and most easy) convex function the Pareto-fronts are interdispersed. The hypervolume metric proves this as there is no significant difference at the 95% significance level. This means in fact that it is possible to decode only 10% of the

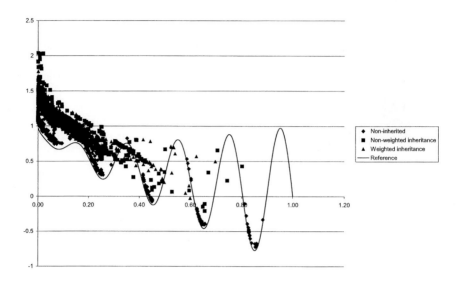

Fig. 6. Scatter plot of all non-dominated solutions found in 10 repetitions for the third test function

individuals and still get the same result as when decoding the full population size, for convex functions that are similar to function 1. This is indeed a higher inheritance than was discovered by Sastry et al. and by Chen et al.. In some respect, this corresponds to the findings of Smith et al, who reported that even the full evaluation of just 1 individual for a single-objective problem results in an near-optimal solution, that is not significantly different from the optimum reached when all individuals are evaluated.

4 Conclusions

It can be concluded that the fitness inheritance efficiency enhancement techniques can be used in order to reduce the number of fitness evaluation provided that the Pareto-front is convex and continuous. If the surface is not convex, the fitness inheritance strategies fail to reach the true Pareto-optimal front. If a decision about which inheritance strategy must be taken, it is safer to choose the weighted average approach. For most of the functions there is no significant difference between the two, but for some such as the multimodal function, the weighted average performs better in terms of variance. As for the inheritance probability it can be stated that the rate of inheritance can be set high, even at a rate of 90% and still yield no significant difference for the hypervolume metric for a convex function. This means that for real-world applications that have a similar Pareto-front as the first test function the speed-up achieved by using fitness inheritance is very high.

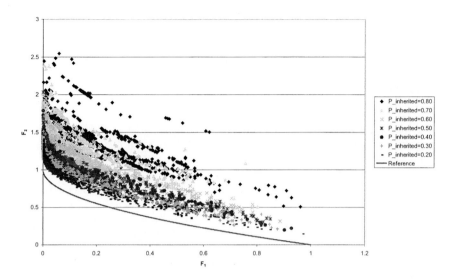

Fig. 7. A scatter plot of all non-dominated solutions found in 10 repetitions for a variable inheritance probability for test function 1. The inheritance probability ranges between 10 − 90%

If real-world practioners want to use fitness inheritance, it is advisable to check beforehand what the nature (convex, non-convex,...) of the Pareto-front is. A possible way of doing this, is by solving the problem with a classical multiple objective GA for a low number of generations and then switch to fitness inheritance techniques if there is sufficient indication that the Pareto-front is convex and continuous.

In future work, it would be interesting to build a more robust foundation as to why the fitness inheritance fails when solving non-convex functions. From our observations, it is also clear tat other techniques than fitness inheritance should be explored for reducing the computational burden, such as local fitness recalculation, caching of previous solutions, etc.

The authors will furthermore increase the number of fitness evaluations for these type of functions to see at what level the fitness inheritance strategies do work.

Acknowledgements. Funding for this research was provided by the Research Fund of Ghent University under grant no. 011D5899. The authors would like to thank the anonymous reviewers for their useful comments.

References

Bellman R.: Adaptive control processes: a guided tour. Princeton University Press, New York (1961)

Chen J. H., Goldberg D, E., Ho S.Y., Sastry K.: Fitness Inheritance in Multi-objective Optimization. In: Langdon W. B., Cantú-Paz E., Mathias K., Roy R., Davis D., Poli R., Balakrishnan K., Honavar V., Rudolph G., Wegener J., Bull L., Potter M. A., Schultz A. C., Miller J. F., Burke E., Jonoska N. (eds.): GECCO 2002: Proceedings of the Genetic and Evolutionary Computation Conference, Morgan Kaufmann, New York (2002) 319–326

Deb K.: Multi-Objective Optimization using Evolutionary Algorithms. Wiley and Sons, England (2001)

Deb K., Agrawal S., Pratap A., Meyarivan T.: A fast elitist non-dominated sorting algorithm: NSGA-II. In:Schoenauer M., Deb K., Rudolph G., Yao X., Lutton E., Merelo J. J., Schwefel H.: Proceedings of the Parallel Problem Solving from Nature VI (PPSN-VI), 849–868

Sastry K., Goldberg D. E., Pelikan M.: Don't Evaluate, Inherit. In: Spector L., Goodman E. D., Wu A., Langdon W. B., Voigt H.-M., Gen M., Sen S., Dorigo M., Pezesh S., Garzon M. H., Burke E. (eds.): GECCO 2001: Proceedings of the Genetic and Evolutionary Computation Conference, Morgan Kaufmann, San Francisco (2001) 551–558

Smith R. E., Dike B. A., Stegmann S. A.: Fitness Inheritance in Genetic Algorithms. In: Proceedings of the 1995 ACM Symposium on Applied Computing, February 26-28, ACM, Nashville (1995)

Van Veldhuizen, D.: Multiobjective Evolutionary Algorithms: Clasisfications, Analyses and New Innovations, Unpublished doctoral dissertation, Ar Force Institute Technology, Dayton (1999)

Zitzler E.: Evolutionary Algorithms for Multiobjective Optimization: Methods and Applications, Unpublished doctoral dissertation, Institut für Technische Informatik und Kommunikatinsnetze, Switzerland (1999)

Zitzler E., Deb K., Thiele L.: Comparison of Multiobjective Evolutionary Algorithms: Empirical Results. Evolutionary Computation 8 (2000) 173–195

Zitzler E., Laumanns K., Thiele L., Fonseca C. M., Grunert da Fonseca V.: Why quality assessment of Multiobjective Optimizers is Difficult. In: Langdon W. B., Cantú-Paz E., Mathias K., Roy R., Davis D., Poli R., Balakrishnan K., Honavar V., Rudolph G., Wegener J., Bull L., Potter M. A., Schultz A. C., Miller J. F., Burke E., Jonoska N. (eds.): GECCO 2002: Proceedings of the Genetic and Evolutionary Computation Conference, Morgan Kaufmann, New York (2002) 666–673

Zitzler E., Thiele L.: Multiobjective Optimization Using Evolutionary Algorithms—A Comparative Study. In: Eiben A. E. (eds.): Parallel Problem Solving from Nature V, Springer-Verlag, Amsterdam (1998) 292–301

Use of a Genetic Heritage for Solving the Assignment Problem with Two Objectives

Xavier Gandibleux[1], Hiroyuki Morita[2], and Naoki Katoh[3]

[1] LAMIH/ROI – UMR CNRS 8530
Université de Valenciennes, Campus "Le Mont Houy"
F-59313 Valenciennes cedex 9, France
Xavier.Gandibleux@univ-valenciennes.fr
[2] Osaka Prefecture University
1-1 Gakuen-cho, Sakai
Osaka 599-8231, Japan
Morita@eco.osakafu-u.ac.jp
[3] Kyoto University
Yoshidahonmachi,Sakyo ward
Kyoto 606-8501, Japan
Naoki@archi.kyoto-u.ac.jp

Abstract. The paper concerns a multiobjective heuristic to compute approximate efficient solutions for the assignment problem with two objectives. The aim here is to show that the genetic information extracted from supported solutions constitutes a useful genetic heritage to be used by crossover operators to approximate non-supported solutions. Bound sets describe one acceptable limit for applying a local search over an offspring. Results of extensive numerical experiments are reported. All exact efficient solutions are obtained using Cplex in a basic enumerative procedure. A comparison with published results shows the efficiency of this approach.

1 Introduction

Assignment problem (AP) is a well-known fundamental combinatorial optimization problem. Several practical situations can be formulated as an AP. In addition, AP is a sub-problem of more complicated ones, like transportation problems, distribution problems or traveling salesman problems. The AP mathematical structure is simple, and efficient algorithms exist to solve it in the single objective case.

In a multiple objective framework, the assignment problem with two objectives ($biAP$) can be formulated as follows, where c_{il}^q are non-negative integers and $X = (x_{11}, \ldots, x_{nn})$:

$$\left[\begin{array}{ll} "\min" z^q(X) = \sum_{i=1}^{n}\sum_{l=1}^{n} c_{il}^q x_{il} & q = 1, 2 \\ \sum_{i=1}^{n} x_{il} = 1 & l = 1, \ldots, n \\ \sum_{l=1}^{n} x_{il} = 1 & i = 1, \ldots, n \\ x_{il} \in \{0, 1\} & \end{array}\right] \quad (biAP)$$

A solution $X^\star$ of problem $(biAP)$ is *efficient* if no other feasible solution X exists, such that $z^q(X) \leq z^q(X^\star)$, $q = 1, 2$ with at least one strict inequality. Let E denote the set of efficient solutions of problem $(biAP)$. E is partitioned into two subsets : the set SE of *supported* efficient solutions which are optimal for a parameterized single objective problem, and the set $NE = E \setminus SE$ of *non-supported efficient solutions*. These non-supported efficient solutions are necessarily located in the triangles generated in the objective space by two successive supported efficient solutions. The main purpose of this paper is to develop a multiobjective heuristic to compute approximate solutions for biAP.

According to [1], the first papers concerning multiple objective AP only deal with SE, using convex combinations of objectives, or goal programming. Exact algorithms to determine the whole set E have been proposed [4,5]. They make use of single objective methods and the duality properties of the assignment problem. Only one multiobjective meta-heuristic (MOMH) method has been applied for the $(biAP)$ problem : the MultiObjective Simulated Annealing (MOSA) method, an extension of simulated annealing to deal with multiple objectives [5]. MOSA does not use any of the particularities of the $(biAP)$. The results reported in [6] show the limit of such a general purpose MOMH; quickly MOSA encounters difficulties producing good approximations for $(biAP)$. It is able to detect some exact efficient solutions only when the instance size is less than or equal to 30×30 variables. We underlined the advantage of using good genetic information in a previous paper [2] using the biobjective knapsack problem as an experimental support. Such "good genetic information" can be deduced from the set of supported efficient solutions or from some approximations (the current approximation of the efficient set, or basic approximations obtained with a simple greedy algorithm).

It is well-known that some SE solutions for $(biAP)$ can be computed easily by relaxing the integrality constraints and solving the problem with a simplex method, or using a specific solving method like the Hungarian method. In this paper, we present an EMO heuristic which makes use of the genetic information available inside the supported solutions for approximating the non-supported solutions. This heuristic is able to quickly compute good approximations of NE solutions for the assignment problem with two objectives. Our method is influenced by several keys related to MOMH and multiobjective combinatorial optimisation (MOCO). These influential ideas are mentioned below.

Ideas from MOMH:

- *A population based method.* The population is composed of potential efficient solutions $\hat{E} = SE \cup \widehat{NE}$ where $\widehat{NE}$ is an approximation of NE.
- *Pheromones from ant colony systems.* The frequency of assignments (i, j) in SE solutions is summarized in a roulette wheel.
- *Evolutionary operators.* Crossover (XO) and mutation (M) produce one offspring from parent individuals.
- *Memetic algorithm.* A Local search (LS) is performed on each potential good individual.
- *Genetic refreshment.* Periodically the genetic information is refreshed. This is performed when $\hat{E}$ has been significantly renewed, and corresponds to an important evolution of the population.

Ideas from MOCO:

- *Weighted Sum Method.* Partial or complete SE set can be computed easily with an exact method. The scalarized problem

$$\min\left\{\sum_{q=1}^{2} \lambda_q z^q(x) : x \in X\right\} \qquad (biAP_\lambda)$$

has to be solved for all $\lambda \in \mathbb{R}^2$ with $0 \leq \lambda_q \leq 1$ and $\sum_{q=1}^{2} \lambda_q = 1$. The weighted sum method finds supported efficient solutions, but of course no unsupported ones.

- *Bound sets.* Bound set results are used to limit the computing effort produced. Any individual outside the limit defined is killed. Else a local search is performed.

Due to the simplicity of our method, one generation consumes few CPUt, especially when the individual is not good. In this case, no LS is performed. Consequently, our method can be very aggressive (i.e. performing many generations). In contrast with other MultiObjective evolutionary algorithms (MOEA), we do not perform any direction search to drive the approximation process, or require any ranking method (there is no fitness measure). This is a remarkable fact, given that direction search and ranking are often criticized, the former for its aggregation of objectives, and the latter for the required computing effort. Finally, a solution too far from the exact efficient frontier will never be introduced in the approximation; all solutions produced are located in the triangles defined by two supported solutions.

Section 2 briefly develops the enumerative procedure to compute the efficient solutions. The main principles of our heuristic and its algorithmic description are given in section 3. The full numerical results are reported in section 4, and compared with previously published results. A discussion and a conclusion are provided in the last section.

2 Generation of Supported Solutions

A basic enumerative procedure is implemented to compute the efficient solutions. A dichotomic scheme using an MIP solver such as Cplex or LPsolve has been designed. The solver is invoked for solving a $(biAP_\lambda)$ with two additional constraints $c^1 x < z^1(x^{(B)})$ and $c^2 x < z^2(x^{(A)})$, where $x^{(A)}$ and $x^{(B)}$ are two optimal solutions of $(biAP_\lambda)$, for reducing the feasible domain in objective space (see figure 1). With this principle, if more than one optimal solution for $(biAP_\lambda)$ exists, only one solution will be computed.

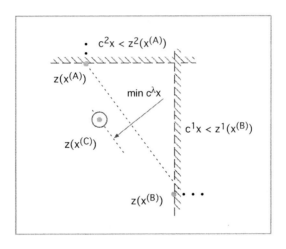

Fig. 1. Separation principle for computing all efficient solutions

---

Algorithm 1 Compute all efficient solutions and return the supported ones

EnumerativeMethod : **procedure** $(SE \uparrow)$ **is**
--| Compute lexicographically $x^{(1)}$ and $x^{(2)}$, the extreme optimal solutions for
--| permutations (z^1, z^2) and (z^2, z^1) of the objectives.
  $x^{(1)} \leftarrow$ SolveLexicographic$(z^{1|2} \downarrow)$
  $x^{(2)} \leftarrow$ SolveLexicographic$(z^{2|1} \downarrow)$
  $E \leftarrow \{x^{(1)}, x^{(2)}\}$
--| Compute all intermediate solutions between $x^{(1)}$ and $x^{(2)}$.
--| Update E with all solutions generated.
  solveRecursion$(x^{(1)} \downarrow, x^{(2)} \downarrow, E \updownarrow)$
--| Extract all SE from the set E
  $SE \leftarrow$ extractSEsolutions(E)
end EnumerativeMethod

Algorithm 2 Compute intermediate efficient solutions

```
solveRecursion : procedure ( x^(A) ↓ , x^(B) ↓ , E ↕) is
--| Compute the optimal solutions x^(C) of (P_λ) : min{λ_1 z^1(x) + λ_2 z^2(x) | x ∈ X}
--| where λ_1 = z^2(x^(A)) − z^2(x^(B)), and λ_2 = z^1(x^(B)) − z^1(x^(A)).
    x^(C) ← SolveP_λWithAddConst(λ ↓, z^1(x^(B)) ↓, z^2(x^(A)) ↓ )
    if exist(x^(C)) then
        E ← E ∪ {x^(C)}
        solveRecursion(x^(A) ↓ , x^(C) ↓ , E ↕)
        solveRecursion(x^(C) ↓ , x^(B) ↓ , E ↕)
    end if
end solveRecursion
```

Obviously, any specific algorithm for the assignment problem, such as the Hungarian method, is a useful alternative to the MIP solver to compute only the supported solutions. However, without a sensitivity analysis, only extreme SE solutions will be computed. All other SE solutions resulting from a linear combination of two SE solutions are left out.

3 Description of Our EMO|AP Heuristic

The following small problem (5 × 5 variables denoted 2AP5-1A20.dat in section 4) is used as didactical example in the continuation :

$$c^1 = \begin{bmatrix} 13 & 14 & 7 & 2 & 11 \\ 5 & 10 & 11 & 7 & 10 \\ 7 & 19 & 9 & 16 & 19 \\ 3 & 19 & 10 & 0 & 6 \\ 12 & 9 & 2 & 4 & 15 \end{bmatrix} \quad c^2 = \begin{bmatrix} 1 & 13 & 15 & 18 & 3 \\ 0 & 3 & 17 & 8 & 6 \\ 9 & 5 & 8 & 0 & 4 \\ 18 & 19 & 3 & 19 & 7 \\ 2 & 3 & 19 & 12 & 15 \end{bmatrix}$$

An assignment is coded as a permutation. For example, the permutation (3,1,0,4,2) means $i = 3$ is assigned to $j = 0$, $i = 1$ is assigned to $j = 1$, and etc., which gives the following bipartite graph :

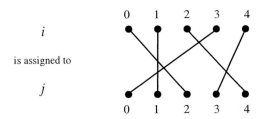

Fig. 2. Illustration using a bi-partite graph

Three supported efficient solutions have been computed for this instance :

Table 1. SE solutions for 2AP5-1A20.dat

SE	x_0	x_1	x_2	x_3	x_4	z^1	z^2
1	3	1	0	4	2	27	56
2	4	0	3	2	1	51	9
3	3	0	2	4	1	31	36

The principle of the method is simple. A genetic map extracted from the genetic heritage given by supported solutions is elaborated and used intensively to discover the non-supported solutions or approximations of them. Using the genetic map, a crossover operator is defined on two randomly selected individuals. The genes common to these two individuals are copied in the child; other genes are determined using the genetic map available in the roulette wheels. A mutation operator is proposed to introduce diversity. Offsprings which are too far from the efficient frontier, and thus not good potential candidates for generating new potential efficient solutions, are killed. An upper bound set is used to decide whether or not the candidate is good. In the positive case, a two-level local search is performed. The approximation $\widehat{E}$ also contributes to the evolution mechanism. The genetic map is refreshed periodically using the current set of potential efficient solutions.

3.1 The Genetic Map

The genetic heritage is derived from all SE solutions (or a subset of them). As in the principle of pheromones in ant colonies, the occurrence frequency of assignment "i in position j" in vector X for SE solutions is computed and stored in a roulette wheel. A roulette wheel is elaborated for each position, and the genetic map is composed of roulette wheels. For our didactical example, the five roulette wheels shown in figure 3 are built. The map represents the genetic heritage of our population. This genetic map provides useful information for fixing certain bits when crossover operators are performed on parents.

3.2 Border between Good Potential Individuals and Others

Lower bound and upper bound sets (respectively squares and bullets in Figure 4) are determined in a straightforward manner for the assignment problem [3]. The lower bound set (LBS) is provided by the supported solutions obtained by linear relaxation. The upper bound set (UBS) is defined by the set of "local nadir points", where one nadir point is derived from two adjacent SE solutions. Consequently, one admissible area is identified (marked in grey on Figure 4), where all new individuals are acceptable solutions for applying an aggressive local search procedure. Formally, the performance of an individual cannot be dominated by one nadir point. This helps to save some computing effort by avoiding a local search on an individual with little chance of generating a potential efficient solution because of its poor performance.

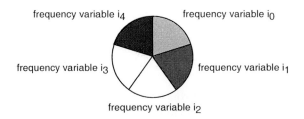
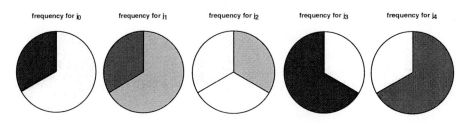

Fig. 3. Occurrence frequency of each variable in each position

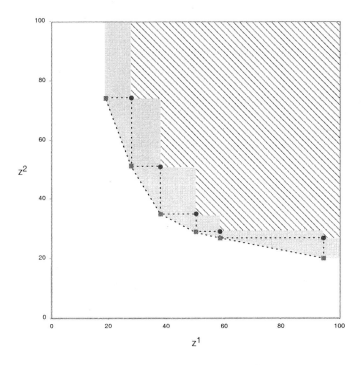

Fig. 4. Lower bound set (*squares*), upper bound set (*bullets*) and the acceptable area where a local search procedure is applied to new individuals. Triangles with dashed lines denote areas where NE solutions can exist

3.3 Algorithms

Notations
- $\widehat{E}$: set of potential efficient solutions (also called elite solutions)
- $\widehat{e}$: a set of potential efficient solutions obtained after application of a local search.
- $I1, I2, I3$ and $I4$: four individuals.
- n : number of genes for a individual
- UBS : upper bound set. LBS : lower bound set.

Parameters
- Parameter 1: a given number of generations
- Parameter 2 : a refreshment value

Algorithm 3 The generation loop

$\widehat{E} \leftarrow SE$ --| initial elite set $\widehat{E} = SE \cup \widehat{NE}$ where $\widehat{NE} = \emptyset$
generationProcess :
for a given number of generations **loop**

--| **Selection of individuals (parents)**
Select $I1$ and $I2 \in \widehat{E}$ at random ($I1 \neq I2$)

--| **Offsprings of individuals**
--| $I3$ (child) is built using the genetic information available in $I1$, $I2$ and $\widehat{E}$
for j in $1..n$ **loop**
--| genes common to parents are copied in child
if $I1_j = I2_j$ **then** $I3_j \leftarrow I1_j$ **endIf**
--| others genes : value is selected using the wheel
if $I1_j \neq I2_j$ **then** $I3_j \leftarrow$ useGeneticInformation(j) **endIf**
endFor
if $I3$ is not dominated by one nadir point $\in UBS$ **then**
  TwoLevelsLS $(I3, \widehat{E})$
endIf

--| **Mutation of individual**
$I4 \leftarrow$ Mutate $I1$ as follow :
- M, a list of b ($3 \leq b \leq n$) assignments selected randomly from $I4$ to de-assign
- fix randomly these b assignments from M, giving $I4$

if $I4$ is not dominated by one nadir point $\in UBS$ **then**
  TwoLevelsLS $(I4, \widehat{E})$
endIf

--| **Evolution of population : Refresh the genetic map**
Update the genetic information after a determined number of generations. Roulette wheels are rebuilt using $\widehat{E}$, the current set of elite solutions.

endLoop generationProcess

Algorithm 4 useGeneticInformation

useGeneticInformation : **function**($j \downarrow$)

    $i \leftarrow$ selectAssignmentUsingRouletteWheel(j)
    if assignment (i,j) is not feasible **then**
        LFA $\leftarrow$ list of feasible assignment in j
        $i \leftarrow$ selectFeasibleAssignmentRandomly(LFA)
    endIf
    return i

endFunc useGeneticInformation

Example :

j	=	0	1	2	3	4
$I1$	=	3	1	0	4	2
$I2$	=	3	0	2	4	1
$I3$	=	3	*	*	4	*

step 1 ::=   position j_1 :
            wheel$_{j_1}$ = $\{0, 1\}$
            $\Rightarrow$ 1 is selected (for example)

step 2 ::=   position j_2 :
            wheel$_{j_2}$ = $\{0, 2, 3\}$
            but 3 is not feasible $\rightarrow$ wheel$_{j_2}$ = $\{0, 2\}$
            $\Rightarrow$ 2 is selected (for example)

step 3 ::=   position j_4 :
            wheel$_{j_4}$ = $\{1, 2\}$
            but 1, 2 are not feasible $\rightarrow$ wheel$_{j_4}$ = $\emptyset$ $\rightarrow$ feasible list = $\{0\}$
            $\Rightarrow$ 0 selected

Algorithm 5 Looking for local elite solutions

TwoLevelsLS : **procedure** ($I \downarrow$, $\widehat{E} \updownarrow$)

    if I is an elite solution **then**
        $\widehat{E} \leftarrow$ Add/update the elite set $\widehat{E}$ with I
    endIf

    $\widehat{e} \leftarrow \emptyset$ --| Aggressive local search
    SeekInNeighborhood (I , $\widehat{e}$, $\widehat{E}$)
    for $I \in \widehat{e}$ **loop**
        SeekInNeighborhood (I , $\widehat{e}$, $\widehat{E}$)
    endFor

    $\widehat{E} \leftarrow$ Add/update the elite set $\widehat{E}$ with $\widehat{e}$

endProc TwoLevelsLS

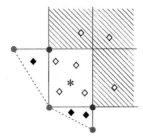

Fig. 5. Neighbors generated by the TwoLevelsLS procedure. First LS starting from the solution "$\star$" produces square solutions. A LS will be performed again on all square solutions identified as elite (*filled squares*) (algorithm 6)

Algorithm 6 Seek all neighbor solutions for I

SeekInNeighborhood : **procedure** ($I \downarrow$, $\widehat{e} \updownarrow$, $\widehat{E} \downarrow$)

    for j_1 **in** $1..n - 1$ **loop**
        for j_2 **in** $j_1 + 1..n$ **loop**
            $I' \leftarrow \text{swap}(I_{j_1}, I_{j_2})$! giving a new neighbor !
            if I' is an elite solution **then**
                $\widehat{e} \leftarrow$ Add/update the local elite set $\widehat{e}$ with I'
            endIf
        endFor
    endFor

endProc SeekInNeighborhood

---

4 Numerical Experiments

The characteristics of the computer used are as follows; CPU : Intel Pentium III 600Mhz / RAM : 192 Mb / OS : Red Hat 7.3 / Language : C / Compiler : gcc-2.95.3 / Optimizer option : -O3

4.1 Numerical Instances

A library of numerical instances for MOCO problems is available on the internet at www.terry.uga.edu/mcdm/. One series of fifteen ($biAP$) instances is given. The name of an instance provides the following characteristics: the number of objectives, the problem, the size, the series, the objective type, and the range of coefficients in $[1, value]$. For example, "2AP5-1A20" is a biobjective assignment problem with 5×5 variables from the series 1; the coefficients of the objective function are generated randomly (type A) in the range $[1, 20]$.

In all of the following, the instances used are biobjective, with objective coefficients generated randomly (type A) in the range $[1, 20]$: "2AP*-1A20". Instances 2AP5-1A20 through 2AP50-1A20 have been used in [6].

4.2 Quality Measure of $\widehat{E}$

To measure the quality of the approximation, we reuse some of the detection indicators and the distances between $\widehat{E}$ and E defined by Ulungu [5] : M_1, M_2, D_1, D_2. In our case however, M_2 makes no sense because the elite solution set is composed of solutions located in triangles, and hence $M_2 = 100\%$ at any time. For this reason, we introduce an alternative measure M_3.

- M_1, the proportion of exact efficient solutions contained in the potential efficient solution set $\widehat{E}$: $M_1 = |\widehat{E} \cap E|/|E|$
- M_2, the proportion of solutions in $\widehat{E}$ which are at least located in the triangles generated in the objective space by two successive supported efficient solutions : $M_2 = |T(\widehat{E})|/|\widehat{E}|$ where $T(\widehat{E})$ is the subset of $\widehat{E}$ whose solutions are included in the existence zones for efficient solutions.
- M_3, the proportion of exact non-supported efficient solutions contained in of $\widehat{NE}$: $M_3 = |\widehat{NE} \cap NE|/|NE|$

If $d(X,Y) = \sum_{q=1}^{2} w^q |z^q(X) - z^q(Y)|$ where w^q is a weight which takes the variation range of criterion z^q into account, and $d(\widehat{E}, Y) = \min_{X \in \widehat{E}} d(X,Y)$ is the distance between $Y \in E$ and the closest solution in $\widehat{E}$. We can consider

- D_1, an average distance between $\widehat{E}$ and E: $D_1(\widehat{E}, E) = 1/|E| \sum_{Y \in E} d(\widehat{E}, Y)$
- D_2, a worst case distance between $\widehat{E}$ and E: $D_2(\widehat{E}, E) = \max_{Y \in E} d(\widehat{E}, Y)$

4.3 Previous Results

Results obtained with an exact method (the 2-phase method) and the improved MOSA method [6] are reported in table 2 (computer used : DEC3000 alpha).

Table 2. Results published by [6]

instance	Two phases method		MOSA			
	$CPUt$	E	$CPUt$		$M1$	$M2$
$2AP5 - 1A20$	5	8	5		87.5	100
$2AP10 - 1A20$	10	16	75		56.2	91.6
$2AP15 - 1A20$	14	39	97		25.6	67.8
$2AP20 - 1A20$	61	54	63		3.7	36.8
$2AP25 - 1A20$	102	71	106		0.0	12.5
$2AP30 - 1A20$	183	88	226		3.4	42.8
$2AP35 - 1A20$	384	82	186		0.0	17.0
$2AP40 - 1A20$	1203	126	139		0.0	0.0
$2AP45 - 1A20$	3120	113	306		0.0	4.9
$2AP50 - 1A20$	3622	156	246		0.0	0.0

Comment: CPU time used by the 2-phase method increases exponentially with the size of the problem. It is a strong limit for solving larger instances. MOSA needs much less CPU time but approximation quality quickly becomes very poor. Interesting approximations can be expected only for small instances.

4.4 Results Obtained with Cplex

(computer used : Compaq AlphaServer DS20 biprocessor EV6 / 500MHz)
Comment: E was computed for each numerical instances, and all NE solutions were removed, leaving only the supported ones.

The number of efficient solutions obtained with Cplex is larger than the sets reported by [6]. The feasibility of all our efficient assignments, which were verified several times, would seem to put the validity of the results published in [6] into question. In terms of CPU time, Cplex appears efficient for solving all these instances.

Table 3. Results obtained with Cplex

instance	CPUt	E	SE	NE
2AP5 − 1A20	0.08	8	3	5
2AP10 − 1A20	0.41	16	6	10
2AP15 − 1A20	2.50	39	12	27
2AP20 − 1A20	6.24	55	13	42
2AP25 − 1A20	10.15	74	25	49
2AP30 − 1A20	20.33	88	27	61
2AP35 − 1A20	21.91	81	27	54
2AP40 − 1A20	40.39	127	54	73
2AP45 − 1A20	43.52	114	43	71
2AP50 − 1A20	69.62	163	67	96
2AP60 − 1A20	94.98	128	44	84
2AP70 − 1A20	189.24	174	60	114
2AP80 − 1A20	386.17	195	69	126
2AP90 − 1A20	360.52	191	83	108
2AP100 − 1A20	587.85	223	101	122

4.5 Convergence Analysis

The convergence of our method was analyzed using the instance 2AP50-1A20. For this problem, the cardinality of solutions is : $E = 163; SE = 67; NE = 96$. The value of the parameters is "number of generations : 1 000 000 iterations" and "periodicity of refreshment : each 200 000 iterations".

Table 4. Convergence analysis using 2AP50-1A20

generations	CPUt	$\widehat{E}$	NDe	NDne	WND	M1	M3	D1	D2
500	0.34	140	103	36	35	63.19	37.50	0.162	3.454
1000	0.45	144	108	41	34	66.26	42.71	0.159	3.454
5000	1.13	147	112	45	33	68.71	46.88	0.153	3.454
10000	2.18	150	117	50	31	71.78	52.08	0.145	3.454
50000	9.43	157	128	61	27	78.53	63.54	0.131	3.454
100000	19.14	161	133	66	26	81.60	68.75	0.126	3.454
500000	94.48	165	140	73	23	85.89	76.04	0.124	3.454
1000000	190.48	165	140	73	23	85.89	76.04	0.122	3.454

Comment: Very quickly, our method generated a very good approximation of the non-supported solution set (NDe: non-dominated efficient solutions; $NDne$: non-dominated NE solutions; WND: weakly non-dominated). After 100 000 iterations performed in less than 20 seconds, more than 80% of the efficient solutions were identified. There is no need to go further in iterations, because the detection ratio does not improve significantly, even after several hundreds of thousands generations.

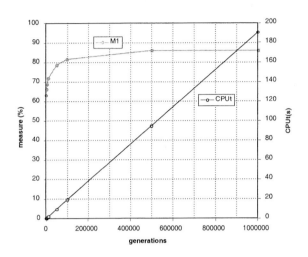

Fig. 6. Convergence analysis for 2AP50-1A20.dat

4.6 Impact of Component Analysis

Four configurations of the heuristic were tested using 2AP50-1A20 by partially disabling certain components of the heuristic (XO: crossover/M: mutation/LS: local search). This was done in order to study their impact on the generation process. The value of the parameters is "number of generations : 1 000 000 iterations" and "periodicity of refreshment : each 200 000 iterations".

Table 5. Impact of component analysis using 2AP50-1A20

Configuration	$CPUt$	$M1$	$NDne$	$M3$	$D2$
$XO + M + LS$	191.08	85.89	73	76.04	3.454
$XO + LS$	167.01	83.44	69	71.88	3.454
$M + LS$	25.79	64.42	38	39.58	3.454
LS	2.66	63.19	36	37.50	3.454

Comment: There was not any significant difference between configurations $XO + M + LS$ and $XO + LS$, indicating that the impact of mutation operator on generation process is low. In the comparison of XO/LS and M/LS, the crossover operator appears to be the most useful operator. Clearly, Local search LS alone is not effective.

4.7 Performance for a Fixed Number of Iterations

The approximation for all numerical instances are reported. Resolution was repeated 9 times using different seed values. The $min/mean/max$ values observed are shown in table 6. The value of the parameters is "number of generations: 250 000 iterations" and "periodicity of refreshment: each 100 000 iterations".

Table 6. Performance for a fixed number of iterations

		5	10	15	20	25	30	35	40
$CPUt$	min	0.95	1.19	16.13	30.55	27.58	28.88	32.27	32.49
	$mean$	1.01	1.30	16.41	30.80	27.88	29.78	33.83	32.93
	max	1.06	1.39	16.68	31	28.46	30.28	34.71	33.11
$M1$	min	100	100	94.87	87.27	86.49	85.23	82.72	81.89
	$mean$	100.00	100.00	95.44	91.31	87.24	88.76	86.83	85.56
	max	100	100	97.44	96.36	89.19	92.05	91.36	89.76
$NE(PE)$	min	5	10	25	35	39	48	40	50
	$mean$	5.00	10.00	25.22	37.22	39.56	51.11	43.33	54.67
	max	5	10	26	40	41	54	47	60
$M3$	min	100	100	92.59	83.33	79.59	78.69	74.07	68.49
	$mean$	100.00	100.00	93.41	88.62	80.72	83.79	80.25	74.88
	max	100	100	96.3	95.24	83.67	88.52	87.04	82.19
$D2$	min	0.00	0.00	0.00	0.00	0.71	0.51	1.14	1.02
	$mean$	0.00	0.00	0.91	0.84	1.63	0.51	1.14	1.04
	max	0.00	0.00	1.02	1.12	1.82	0.51	1.14	1.11

		45	50	60	70	80	90	100
$CPUt$	min	35.23	47.18	58.58	83.17	104.78	130.47	173.45
	$mean$	35.74	47.61	59.66	83.68	105.68	131.12	174.42
	max	36.08	48.42	60.27	84.68	106.8	132.47	175.41
$M1$	min	72.81	80.37	56.25	56.9	56.92	63.87	56.95
	$mean$	75.83	82.07	58.07	59.77	59.89	65.04	58.45
	max	80.7	84.05	60.16	62.07	63.08	66.49	60.09
$NE(PE)$	min	40	64	28	39	42	39	26
	$mean$	43.44	66.78	30.33	44.00	47.78	41.22	29.33
	max	49	70	33	48	54	44	33
$M3$	min	56.34	66.67	33.33	34.21	33.33	36.11	21.31
	$mean$	61.19	69.56	36.11	38.60	37.92	38.14	24.04
	max	69.01	72.92	39.29	42.11	42.86	40.74	27.05
$D2$	min	1.52	3.45	2.16	2.51	1.14	3.14	3.23
	$mean$	2.30	3.45	2.37	2.51	1.50	3.14	3.23
	max	3.00	3.45	3.20	2.51	1.55	3.14	3.23

Comment: Our results indicate performances outranking those produced by the MOSA method. With 35 × 35 variables, MOSA is unable to find any exact efficient solution, while our method detects 86% (mean) of the exact solutions. Because different computers were used, CPUt can not be comparable. Seeds have no real impact on the performance values observed, showing the good stability of our method. For instances of 60 × 60 and larger, the quality of approximation decreased. We suppose that the genetic information is not homogeneous enough to draw only one genetic map. Working on subregions, with a principle of clusters could provide better results for instances of this size.

5 Conclusion and On-going Developments

With the hypothesis "the genetic information contained in efficient solutions is similar", we presented an evolutionary multiobjective heuristic using a genetic map. The ideas are simple and easy to implement. The method does not need any kind of ranking method, or a scalarisation of objectives to approximate the efficient solution set. Numerical experiments on the biobjective assignment problem underline the effectiveness of the method, especially as compared to the MOSA method.

On-going developments revolve around the use of the principle of clusters to deal more efficiently with larger problems. Also two static parameters currently drive the method. In the future, we plan to design an advanced stopping condition in order to dynamically manage the total number of generations and the refresh mechanism. More intensive numerical experiments will complete these first results. In particular, we plan to experiment with problems containing several objective range values and objective types (random, patterns, etc.). Obviously, Cplex will be removed and a dedicated algorithm for computing some (or all) supported solutions will be used. Finally, there is no specific difficulty in applying our method to problems with three objectives. We also plan to evaluate the method performance on this class of problems.

References

1. M. Ehrgott and X. Gandibleux, Multiobjective Combinatorial Optimization. In *Multiple Criteria Optimization: State of the Art Annotated Bibliographic Survey* (M. Ehrgott and X. Gandibleux Eds.), pp.369–444, Kluwer's International Series in Operations Research and Management Science : Volume 52, Kluwer Academic Publishers, Boston, 2002.
2. X. Gandibleux, H. Morita, N. Katoh, The Supported Solutions used as a Genetic Information in a Population Heuristic. In *Evolutionary Multi-Criterion Optimization* (E. Zitzler, K. Deb, L. Thiele, C. Coello, D. Corne Eds.). pp.429–442, Lecture Notes in Computer Sciences 1993, Springer, 2001.
3. M. Ehrgott and X. Gandibleux, Bounds and Bound Sets for Biobjective Combinatorial Optimization Problems. In *Multiple Criteria Decision Making in the New Millennium* (M. Koksalan and St. Ziont Eds.), pp.241–253, Lecture Notes in Economics and Mathematical Systems 507, Springer, 2001.
4. R. Malhotra, H.L. Bhatia and M.C. Puri, Bi-criteria assignment problem, *Operations Research*, 19(2): 84–96,1982.
5. Ulungu – Ekunda Lukata, *Optimisation Combinatoire multicritère: Détermination de l'ensemble des solutions efficaces et méthodes interactives*, Université de Mons-Hainaut, Faculté des Sciences, 313 pages, 1993.
6. D. Tuyttens, J. Teghem, Ph. Fortemps and K. Van Nieuwenhuyse, Performance of the MOSA method for the bicriteria assignment problem, *Journal of Heuristics*, 6 pp. 295–310, (2000).

Fuzzy Optimality and Evolutionary Multiobjective Optimization

M. Farina and P. Amato

STMicroelectronics Srl,
Via C. Olivetti, 2, 20041, Agrate (MI), IT.
{marco.farina,paolo.amato}@st.com

Abstract. Pareto optimality is someway ineffective for optimization problems with several (more than three) objectives. In fact the Pareto optimal set tends to become a wide portion of the whole design domain search space with the increasing of the numbers of objectives. Consequently, little or no help is given to the human decision maker. Here we use fuzzy logic to give two new definitions of optimality that extend the notion of Pareto optimality. Our aim is to identify, inside the set of Pareto optimal solutions, different "degrees of optimality" such that only a few solutions have the highest degree of optimality; even in problems with a big number of objectives. Then we demonstrate (on simple analytical test cases) the coherence of these definitions and their reduction to Pareto optimality in some special subcases. At last we introduce a first extension of (1+1)ES mutation operator able to approximate the set of solutions with a given degree of optimality, and test it on analytical test cases.

1 Introduction

The application of Pareto optimum definition to optimization problems with a high number of objectives is somewhat unsatisfactory. This happens for two reasons. First, when there are more than three objectives the visualization of Pareto front must be carefully considered. Second, the set of solution classified as Pareto optimal can be a relevant fraction of the whole objective search space. Consequently, there we have little help in our effort to find the solution which is most suitable for the given problem. Many real life optimization problems have several (more than three) objectives; examples may be:

- the optimal design of electromagnetic devices where electrical efficiency, weight, cost and electric or magnetic field properties have to be considered [1,3],
- combustion process and engine optimization where efficiency, NO_x emissions, soot emissions, noise are considered [10,17]
- the aerodynamic shape optimization of supersonic wings where transonic and supersonic drag coefficients together with bending and twisting momenta [7]
- the paper machine optimization where up to five or six objective can be considered [11]

Due to the number of examples that may be considered and due to the unsolved difficulties that are encountered when more than three objectives are considered, the treatment of several objective problems is probably one of the most actual open issues in practical multiobjective evolutionary optimization.

In [8] and [13] the authors proposed a hierarchy of optimality definitions extending the Pareto one. These definitions are based on fuzzy logic [12] a key tool for the treatment of uncertainty and partial truth, and for the processing of large class of data.

The main idea behind the given definitions is to introduce different degree of optimality (each degree defining its own front). The lowest degree (let's say 0) corresponds to Pareto optimality. The highest degree (let's say 1) corresponds to a strong definition of optimality. The set of points classified as optimal by the latter are, thereafter, a small subset (eventually a single point) of the Pareto front. This classification is obtained by considering in the dominance relation (i) how many objectives a solution improves with respect to the other solution, and (ii) the size of each improvement. In fact, in this way, it is possible not only to discriminate between dominated and non-dominated (optimal) solution, but also between "less" and "more" optimal solutions.

In this work we show how these definitions can be exploited in continuous optimization problems. In particular we introduce an evolution strategy algorithm for approximating the optimal front associated to a certain degree of optimality.

2 Limits and Drawbacks of Pareto Optima Definition

The following multi-objective optimization problem is considered [14,9,15]:

Definition 1 (Multi-criteria optimization problem). *Let* $\mathbf{V} \subseteq K_1 \times K_2 \times \ldots \times K_N$ *and* $\mathbf{W} \subseteq O_1 \times O_2 \times \ldots \times O_M$ *be vector spaces, where the* K_i, O_j *(with* $i = 1, \ldots, N$ *and* $j = 1, \ldots, M$*) are (continuous or finite) fields and* $N, M \in \mathbb{N}$*, and let* $\mathbf{g} : \mathbf{V} \mapsto \mathbb{R}^p, \mathbf{h} : \mathbf{V} \mapsto \mathbb{R}^q$ *and* $\mathbf{f} : \mathbf{V} \mapsto \mathbf{W}$ *be three mappings, where* $p, q \in \mathbb{N}$*. A Non-linear constrained multi-criteria (minimum) optimization problem with* M *objectives is defined as:*

$$\min_{\mathbf{v} \in \mathbf{V}} \mathbf{f} \triangleq \{f_1(\mathbf{v}), \ldots, f_M(\mathbf{v})\} \quad \text{subject to} \begin{cases} \mathbf{g}(\mathbf{v}) \leq 0 \\ \mathbf{h}(\mathbf{v}) = 0. \end{cases}$$

Definition 2 (Design and objective search space). *We call* Design domain search space Ω *and* objective domain search space Ω_O *the following two set:*

$$\Omega = \{\mathbf{v} \in \mathbf{V} \mid \mathbf{g}(\mathbf{v}) \leq 0 \wedge \mathbf{h}(\mathbf{v}) = 0\}, \quad \Omega_O = \{\mathbf{f}(\mathbf{v}) \in \mathbf{W} \mid \mathbf{v} \in \Omega\}.$$

We consider multi-objective non-linear constrained optimization in a continuous search space are considered; Ω and Ω_O are thus continuous spaces.

The Pareto definition of optimality in a multi-criteria decision making problem can be unsatisfactory due to essentially two reasons:

P1 the number of improved or equal objective values is not taken into account,
P2 the (normalized) size of improvements is not taken into account.

This issues are essential decision elements when looking for the best solution, and they are implicitly included in the common-sense notion of optimality.

The limit of Pareto definition when the first issue (P1) is considered can be viewed in the schema shown in figure 1. Since the Pareto dominance gives a partial order of solutions in criteria space[1], when a vector (a candidate for optimal solution) in the criteria space is considered, all other possible solution can belong to one of the following three different set: better solutions, worse solutions and equivalent[2] solutions.

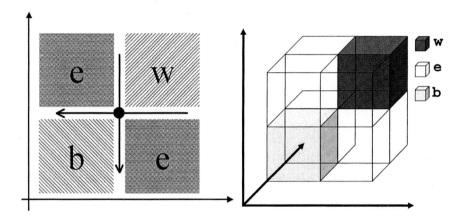

Fig. 1. Schematic view of Pareto-dominance based partial order in 2D and 3D problems when a candidate solution is considered (•): equal (e), better (b) and worse (w) solution regions are shown

Figure 1 shows such sets for 2 and 3 criteria problems. Let e the portion of the M-dimensional criteria domain space containing all the points that the P-dominance concept classify as equivalent to a given one. The portion e increases with the increasing of the number M of objectives as follows:

$$e = \frac{2^M - 2}{2^M} \qquad (1)$$

Thus when M tends to infinity, e tends to 1 (i.e., it is the whole space). This fact is general and problem independent and when a single problem is considered and Ω and Ω_O are introduced, always e tends to 1 when M tends to infinity,

---

[1] Indeed the partial ordering induced by Pareto dominance is a weak ordering. In fact, in general, an algebraic structure equipped with this partial order is not a lattice.
[2] Since Pareto dominance does not induce an equivalent relation, it should be better to call them "indifferent" solutions.

eventually with a behavior different from equation 1. From this it derives that Pareto definition is ineffective for a large number of objectives, even without considering the second aforementioned issue.

In the following sections we will give two more general definition of optimum for a multi-criteria decision making problem, taking into account one issue at a time. As we shall see, Pareto optimum definition is a special case of both definitions.

3 Two Generalizations of Pareto Definitions

In this section we give a brief description of two optimality definitions that soundly extend the Pareto one. For a deeper discussion about these definitions see [8,13].

3.1 Taking into Account the Number of Improved Objectives: k-Optimality

In Pareto optimality definition two candidate solutions $\mathbf{v}_1$ and $\mathbf{v}_2$ are equivalent if at least in one objective the first solution is better than the second one, and at least in one objective the second one is better than the first one (or if they are equal in all the objectives). Indeed a more general definition, able to cope with a wider variety of problems, should take into account in how many objectives the first candidate solution is better than the second one and viceversa. To do so, we introduce the following functions which associate to every couple of points in Ω a natural number.

$$n_b(\mathbf{v}_1, \mathbf{v}_2) \triangleq |\{i \in \mathbb{N} | i \leq M \wedge f_i(\mathbf{v}_1) < f_i(\mathbf{v}_2)\}|$$
$$n_e(\mathbf{v}_1, \mathbf{v}_2) \triangleq |\{i \in \mathbb{N} | i \leq M \wedge f_i(\mathbf{v}_1) = f_i(\mathbf{v}_2)\}|$$
$$n_w(\mathbf{v}_1, \mathbf{v}_2) \triangleq |\{i \in \mathbb{N} | i \leq M \wedge f_i(\mathbf{v}_1) > f_i(\mathbf{v}_2)\}|$$

For every couple of points $\mathbf{v}_1, \mathbf{v}_2 \in \Omega$, the function n_b computes the number of objectives in which $\mathbf{v}_1$ is better than $\mathbf{v}_2$, n_e computes the number of objectives in which they are equal, and n_w the number of objectives in which $\mathbf{v}_1$ is worse than $\mathbf{v}_2$. To lighten the mathematical notation, from now on we will consider a generic couple of points and we will write simply n_b, n_e and n_w instead of $n_b(\mathbf{v}_1, \mathbf{v}_2), n_e(\mathbf{v}_1, \mathbf{v}_2)$ and $n_w(\mathbf{v}_1, \mathbf{v}_2)$.

A moment's reflection tell us that the following inequalities holds:

$$n_b + n_w + n_e = M \quad 0 < n_b, n_w, n_e < M$$

We are now able to give a first new definition of dominance and optimality namely k-dominance and k-optimality:

Definition 3 (k-dominance). $\mathbf{v}_1$ *is said to k-dominate* $\mathbf{v}_2$ *if and only if:*

$$\begin{cases} n_e < M \\ n_b \geq \dfrac{M - n_e}{k+1}, \end{cases} \quad (2)$$

where $0 \leq k \leq 1$.

As can be easily seen, definition 3 with $k = 0$ corresponds to Pareto-dominance. Ideally k can assume any value in $[0, 1]$, but because n_b has to be a natural number only a limited number of optimality degree need to be considered. In fact in equation (2) the second inequality is equivalent to $n_b \geq \left\lceil \frac{M-n_e}{k+1} \right\rceil$. With this new dominance definition the following new optimality can be defined:

Definition 4 (k-optimality). $\mathbf{v}^*$ *is k-optimum if and only if there is no $\mathbf{v} \in \Omega$ such that $\mathbf{v}$ k-dominates $\mathbf{v}^*$*

The terms "k-dominance" and "k-optimality" derives respectively from the fact that the former is a loose version of Pareto dominance (1-dominance), while the latter is a strong version of Pareto optimality (0-optimality). We can now easily extend concepts of $\mathcal{S}_P$ and $\mathcal{F}_P$ in the following way:

Definition 5 (k-optimal set and front). *We call k-optimal set ($\mathcal{S}_k$) and k-optimal front ($\mathcal{F}_k$) the set of k-optimal solutions in design domain and objective domain respectively.*

Several $\mathcal{S}_k$ sets and $\mathcal{F}_k$ fronts are thus introduced, one for each value of k. Let us refer to Pareto optimal front as $\mathcal{F}_P$, and to Pareto optimal set as $\mathcal{S}_P$. Then it is evident that $\mathcal{S}_0 = \mathcal{S}_P$ and $\mathcal{F}_0 = \mathcal{F}_P$.

3.2 Taking into Account the Size of Improvements: Fuzzy Optimality

A natural way of extending the notion of k-dominance and k-optimality is to introduce fuzzy relations instead of crisp ones. As first step, to take into account to which degree in each objective function a point $\mathbf{v}_1$ is different from (or equal to) a point $\mathbf{v}_2$, we will consider fuzzy numbers and fuzzy arithmetic. As second step, we will consider the dominance relation itself as a fuzzy relation.

Fuzzy Numbers. A standard way to introduce fuzzy arithmetic on a given universe (here the objective domain search space Ω_O), is to associate to each of its point a triple of fuzzy sets — one for equality (fuzzy number), one for "greater than" and one for "less than". Figures 2 and 3 shows two possible definitions of the fuzzy sets for "equal to 0", "greater than 0" and "less than 0". For coherence with the terminology used so far, we refer to their respective membership function as μ_e, μ_w (where w means "worst", remember that we are talking about minimization problems) and μ_b.

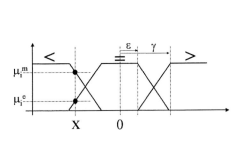

Fig. 2. Linear membership for i-th objective to be used in equation 3 for a fuzzy definition of $=, <$ and $>$: ε and γ are parameters to be chosen by the decision maker

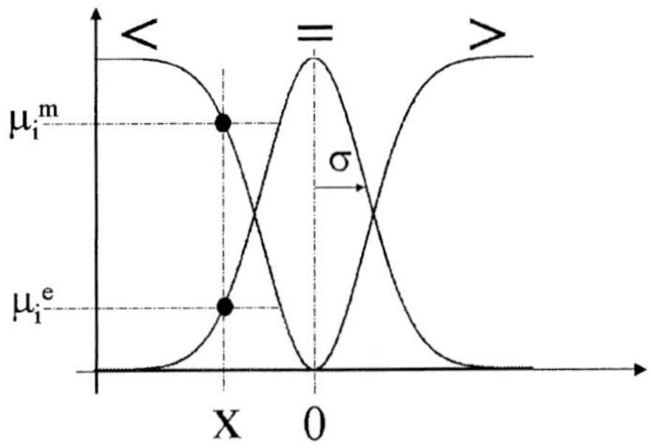

Fig. 3. Gaussian membership for i-th objective to be used in equation 3 for a fuzzy definition of $=, <$ and $>$: σ is s parameter to be chosen by the decision maker

The fuzzy definition of n_b, n_w and n_e (now with superscript F) are the following:

$$n_b^F(\mathbf{v_1}, \mathbf{v_2}) \triangleq \sum_{i=1}^{M} \mu_b^{(i)}(f_i(\mathbf{v_1}) - f_i(\mathbf{v_2}))$$

$$n_w^F(\mathbf{v_1}, \mathbf{v_2}) \triangleq \sum_{i=1}^{M} \mu_w^{(i)}(f_i(\mathbf{v_1}) - f_i(\mathbf{v_2}))$$

$$n_e^F(\mathbf{v_1}, \mathbf{v_2}) \triangleq \sum_{i=1}^{M} \mu_e^{(i)}(f_i(\mathbf{v_1}) - f_i(\mathbf{v_2}))$$

In order n_b^F, n_w^F and n_e^F to be a sound extension of n_b, n_w and n_e, the membership functions $\mu_b^{(i)}, \mu_w^{(i)}$ and $\mu_e^{(i)}$ must satisfy Ruspini condition (i.e, in each point they must sum up to 1) [2]. In fact, under this hypothesis the following holds:

$$n_b^F + n_e^F + n_w^F = \sum_{i=1}^{M}(\mu_i^b + \mu_i^w + \mu_i^e) = M$$

In the figures 2 and 3, two possible different membership shapes are considered: linear and gaussian. Both of them are characterized by parameters defining the shape, ε_i and γ_i for the linear one and σ_i for the gaussian one. Although the definition of such parameters is to be carefully considered, their intended meaning is clear. They thus can be derived from the human decision-maker knowledge on the problem.

- ε_i defines in a fuzzy way the practical meaning of equality and it can thus be considered the tolerance on the i-th objective, that is the interval within which an improvement on objective i is meaningless.

- γ_i can be defined as a relevant but not big size improvement for objective i.
- σ_i evaluation requires a combination of the two aforementioned concepts of maximum imperceptible improvement on objective i (ε_i) and γ_i; the following formula for the membership μ_e can be used:

$$\mu_e^{(i)} = \exp\left(-\frac{\ln \chi}{\varepsilon_i^2}(f_1^i - f_2^i)^2\right) \quad (3)$$

where χ is an arbitrary parameter ($0.8 < \chi < 0.99$) and where μ_b and μ_w can be computed univocally in order to satisfy Ruspini's condition.

With such a fuzzy definition of n_b^F, n_e^F and n_w^F, both k-dominance and k-optimality definition can be reconsidered.

k-optimality with Fuzzy Numbers of Improved Objectives (Fuzzy Optimality). A first extension of the definitions of k-dominance and k-optimality can be given if n_b, n_e and n_w are replaced by n_b^F, n_e^F and n_w^F in definitions 3 and 4 as follows:

Definition 6 (k_F-dominance). $\mathbf{v}_1$ *is said to k_F-dominate $\mathbf{v}_2$ if and only if:*

$$\begin{cases} n_e^F < M \\ n_b^F \geq \dfrac{M - n_e^F}{k_F + 1}, \quad 0 \leq k_F \leq 1 \end{cases} \quad (4)$$

Definition 7 (k_F-optimality). $\mathbf{v}^*$ *is k_F-optimum if and only if there is no $\mathbf{v} \in \Omega$ such that $\mathbf{v}$ k_F-dominates $\mathbf{v}^*$*

The parameter k (now called k_F) has the same meaning as in the previous case ($0 \leq k_F \leq 1$) but now a continuous degree of optimality and dominance are introduced (k_F-dominance and k_F-optimality). An extension of $\mathcal{S}_k$ and $\mathcal{F}_k$ can be defined as follows:

Definition 8 (k_F-optimal set and front). *We call $\mathcal{S}_{k_F}$ and $\mathcal{F}_{k_F}$ the set of k_F-optimal solutions in design domain and objective domain respectively.*

Each $\mathcal{S}_{k_F}$ can be viewed as the k_F-cut[3] of a fuzzy set O for the notion of "optimality in design domain search space". Then the whole membership function for this fuzzy set is implicitly defined by its k_F-cuts. In fact let $\mathbf{V}$ be the search space, then for all $\mathbf{v} \in \mathbf{V}$ its degree of optimality is given by $O(\mathbf{v}) = \sup_\alpha \{\alpha \in [0,1] | \mathbf{v} \in \mathcal{S}_\alpha\}$. An example of fuzzy set for optimality is given in figure 5. The same reasoning can be applied to define the fuzzy sets for optimality in objective space and the fuzzy sets for the dominance relation.

---
[3] Remember that an α-cut of a fuzzy set A on an Universe U is the crisp set $\{u \in U | A(u) \geq \alpha\}$.

Remark. It is it possible to give a further extension of the notion of optimality. In fact A more general procedure can be introduced by fuzzifying not only the quantities n_b^F, n_e^F and n_w^F, but also the dominance relation itself. The resulting notion of optimality has been introduced and discussed by the authors in [13]. It is this last definition that is properly referred to as *fuzzy optimality*. However in the rest of the paper we will sometime use this term to indicate k_F-optimality.

4 Test Cases

This last section is devoted to two examples showing the validity of the introduced definitions. They both are continuous constrained multiobjective optimization problems ($\mathbb{R}^N \to \mathbb{R}^M$).

4.1 A First Test Case: 6 Objectives

In the first test case (figure 4 and 5) a 6D problem with only two design variables is considered with parabolic function located in an asymmetric way on a rhombus borders.

$$\begin{cases} \min_{\mathbf{v} \in \mathbb{R}^2} & \mathbf{f} = \{f_1(\mathbf{v}), ..., f_j(\mathbf{v}), ..., f_6(\mathbf{v})\} \\ & f_j(\mathbf{v}) = (v_1 - c_{1,j})^2 + (v_2 - c_{2,j})^2 \\ s.t. & -1 \leq v_1, v_2 \leq 1 \end{cases} \quad (5)$$

The coordinates for points $(c_{1,j}, c_{2,j})$ are marked with black bullets and the search space is shown with bright gray dots. We apply k-optimality definition; as can be seen the region of Pareto optimal solution ($\mathcal{S}_P = \mathcal{S}_0$) is quite big with

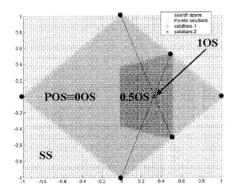

Fig. 4. k-optimality classification on problem (5). Six parabolic functions are centered on • and three different degree of k-optimal solutions are shown together with the whole search space

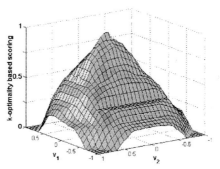

Fig. 5. *Fuzzy* optimality classification on problem (5) with linear membership

respect to the search space. If k is increased (k=.5) a smaller region is selected up to a single optimum for k=1. We thus have four regions, one included in the other: $SSP \subset S_P \equiv S_0 \subset S_{\frac{1}{2}} \subset S_1$ where SSP is the search space. If we move from a solution f^1 to a solutions f^2 both belonging to S_P we can expect one of the two to be better than the other in at least one objective; if on the other hand we move from a solution f^1 to a solutions f^2 both belonging to $S_{\frac{1}{2}}$ we can expect one of the two to be better than the other in at least two objective. The same may in general hold for S_1, but in this example S_1 is a single solution. When fuzzy-optimality is considered on the same test problem a continuous classification of solution can be obtained and is shown in figure 5; the sets S_{k_F} are infinite and corresponds to any real value of $k_F \in [0:1]$ (any k_F-cut of μ_O surface).

4.2 Second Test Case: M Objectives

As a second test case (figure 6 and 7) the classification ability of *fuzzy*-dominance and *fuzzy*-optimality is shown on the following more general $\mathbb{R}^N \to \mathbb{R}^M$ multi-objective optimization problem on a box-like constrained search space:

$$\begin{cases} \min_{\mathbf{v} \in \mathbb{R}^N} & \mathbf{f} = \{f_1(\mathbf{v}), ..., f_i(\mathbf{v}), ..., f_M(\mathbf{v})\} \\ & f_i(\mathbf{v}) = \sum_{k=1}^{np} \exp(-c_{i,k} \sum_{j=1}^{N} (v_j + p_{k,j})^2) \\ & i = 1:M \\ s.t. & L_j \leq v_j \leq U_j \quad j = 1:N \end{cases} \qquad (6)$$

In this example, the number of objectives is increased with respect to the previous one and the shape is a little bit more complex. A case with M=12 and N=2 is first considered. A sampling of 400 points on the search space (candidate solutions) is considered and classified. It is easy to see that almost all points in the search space (a box $\in \mathbb{R}^2$) are Pareto-optimal (see figure 6 where the S_P is shown on the ground level). No (or at most very poor) decision could thus be taken with Pareto optimality.

Classically multiobjective search problems are tackled *via* an equivalent scalar function such as weighted sum of objectives [14],[5]. The limitations of such an approach are the following: only one special Pareto-optimal solution can be computed, preference and tolerance on objectives are difficult to be expressed in a clear way. As an example, figure 7 shows a classification of solutions in the search space for problem (6) with a normalized sum of objectives. As can be seen the peak value identifies a special Pareto-optimal solution and there is no way to identify a bigger subset of the Pareto-optimal set.

On the other hand the proposed optimality definition can give a continuous classification of solution, and consequently any number of Pareto optimal solutions with a clearly defined degree of stronger (with respect to P-optimality) optimality can be easily computed. This possibility is clearly shown in figure 6

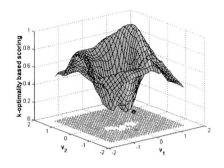
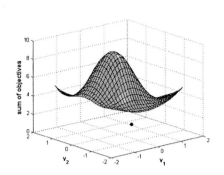

Fig. 6. *Fuzzy*-optimality based classification with linear membership and corresponding maximum point (•); linear membership with $\varepsilon_i = 0.01$ and $\gamma_i = 0.2$, $1 < i < 11$ have been used

Fig. 7. Classical normalized sum of objectives on problem (6) and corresponding minimum point (•)

where a continuous classification similar to the one in figure 5 is obtained. Moreover, the point corresponding to the minimum of the sum function is coincident with the maximum of the *fuzzy*-optimality-based classification; both points are shown under 3D surfaces. The normalized number of solution $s(k_F)$ satisfying *fuzzy*-optimality at degree k_F is plotted against k_F in figure 8 for problem (6). As

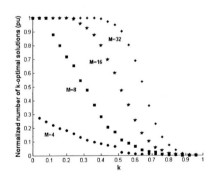
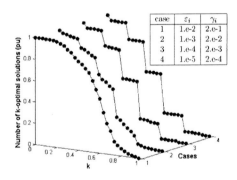

Fig. 8. Normalized number of *fuzzy*-optimal solutions versus the optimality degree k_F for different number of objectives

Fig. 9. From $fuzzy_\alpha$-optimality to k-optimality for a 4-objectives problem (6); membership parameters for the four cases are listed in the table

can be seen the $s(k_F)$ behavior depends on the problem complexity. In case of a problem with 32 objectives an high value of k_F is required in order to find some solutions belonging to $\mathcal{S}_{k_F}$. We now point out with an example (problem (6),

M=6) that the crisp k-optimality can be seen as the limit of *fuzzy*-optimality when $\varepsilon_i, \gamma_i, \sigma_i \to 0$, that is when crisp memberships are considered. This can be easily seen form figure 9 where four different $s(k)$ functions are plotted for the values of membership parameters shown in the table. From a continuous *fuzzy*-optimality based classification (case 1), a crisp k-optimality based classification (case 4) can be obtained with membership parameters close to zero; some intermediate levels are also shown.

An example of fuzzy-dominance membership function with respect to one fixed solution is shown in figure 10; as can be seen the region of the most fuzzy-dominated solution with respect to the fixed one (that is fuzzy-dominated with the highest value of k_F) correspond to the maximum region of the surface. An example of fuzzy-optimality membership building through k_F-cuts can be shown in figure 11; as can be seen the obtained maximum optimality degree is 0.5 and not 1. This is a numerical effect due to the poor sampling of the search space (a coarse sampling has been considered for better figure rendering purposes).

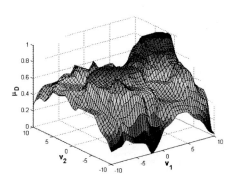
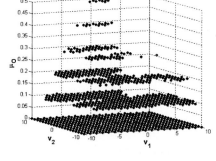

Fig. 10. Membership $\mu_D(v_1, v_2)|_{\tilde{v}}$ dominance degree with respect to a fixed solution $\tilde{v}$

Fig. 11. *Fuzzy*-optimality membership building through k_F-cuts, a sampling of 400 solutions in the search space is considered

A comparison of two k_F-cuts (corresponding to two different degrees of optimality) and the crisp Pareto optimal front is also shown in figure 12. As can be seen the fuzzy optimality definition is able to select properly, among Pareto optimal solutions, some "more optimal" solutions corresponding to the degree k_F of optimality. Moreover for low values of k_F, the crisp Pareto Optimal front (which is obtained via a different procedure for comparison and checking purposes) can be properly reconstructed.

5 Is *Fuzzy*-Optimality Useful for 2D and 3D Cases?

Though being specifically developed for treatment of many (> 3) objectives problems, *fuzzy*-optimality is meaningful even in case of 1 or 2 objectives only.

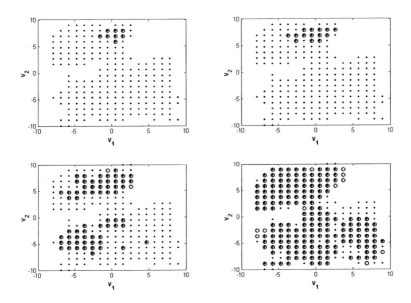

Fig. 12. Comparison of k_F-cuts (o) for a fuzzy-optimality membership at different k_F values and crisp Pareto Optimal front (·), a sampling of 400 solutions in the search space is considered

In such cases, no reduction of Pareto Optimal front is possible because only one degree of optimality can be considered ($\lceil M/2 \rceil < 2$) and k-optimality coincides with Pareto optimality. Nevertheless *fuzzy*-optimality can be used for the evaluation of a larger front taking into account the tolerances on objectives; an example is shown in figures 13 and 14.

As can be seen, when including tolerances (coming from the decision maker's perception) in the objectives comparison between candidate solution, the crisp Pareto optimal set and front are a smaller part of the fuzzy sets both in objective and design space. The size and the shape of the $\mathcal{S}_{k_F}$ set and $\mathcal{F}_{k_F}$ front (where $k_F = 0$) depends on ε and γ values. We point out with this example that $\mathcal{S}_P = \mathcal{S}_0$ if k-optimality is used but $\mathcal{S}_P \subseteq \mathcal{S}_0$ if fuzzy optimality is considered.

We stress that here we consider the objectives function as crisp, (i.e. that there is no uncertainty (fuzziness) in each objectives function). Thus the fuzzy front in objective space in figure 13 descends only from the fuzzification of the comparison between objective values. Of course it is possible to obtain different fuzzy front by considering fuzzy objective functions.

6 Toward *Fuzzy*-Optimal Sets Approximation via Evolutionary Algorithms

We have so far shown that, given a MCDM or a MO problem we can classify candidate solutions with fuzzy-optimality and build N-dimensional memberships

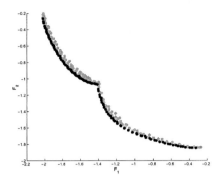 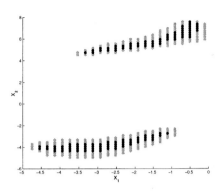

Fig. 13. *Fuzzy* front $\mathcal{F}_{k_F}$ in objective space ($k_F = 0$) (green •) and Pareto optimal front (black •) for a 2 objective 2 variable problem (6)

Fig. 14. *Fuzzy* set $\mathcal{S}_{k_F}$ in design space ($k_F = 0$) (green •) and Pareto optimal set (black •) for a 2 objective 2 variable problem (6)

for optimality and dominance (once one solution is fixed). From a practical point of view we now need a tool that can give us a satisfactory approximation of an *fuzzy*-optimal set once a desired value for k_F is fixed, without classifying all solutions in the search space. A huge variety of methods is available in literature for the approximation via evolutionary computation of the Pareto optimal front. From a general point of view any method based on Pareto-dominance [16] can be extend including fuzzy-dominance. As an example we consider a (1+1)ES algorithm for Pareto optimal front approximation [4,6] and we modify as follows the mutation operator in order the algorithm to converge towards the *fuzzy*-optimal set $\mathcal{S}_{k_F}$ and *fuzzy*-optimal front $\mathcal{F}_{k_F}$ at a given value of k_F. Given the objective values $\mathbf{f}_P$ for the parent P and for the sun S ($\mathbf{f}_S$):

previous mutation operator

- if $n_b(f_P, f_S) = M$ accept the sun as a new parent else discard, it
- go to next generation.

new mutation operator

- if $n_b(f_P, f_S) >= M - k_F M$ accept the sun as a new parent else discard it,
- go to next generation.

Several algorithms (n) with this modification are then run in parallel in order to have the population evolution. The introduced larger acceptance criterion for mutation leads to the convergence of the algorithm towards the desired k_F-optimal set (smaller than the Pareto-optimal one). An example is given in figure 15 where a comparison of k_F-optimal set from sampling (•), k_F-optimal set from evolutionary algorithm (◦) and Pareto Optimal set (·) is given in the design space. As can be seen the k_F-optimal set is correctly sampled though being a disconnected set.

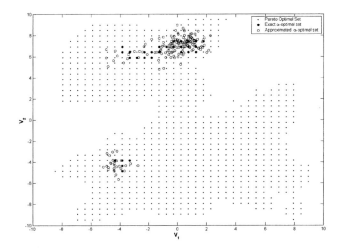

Fig. 15. Comparison of *fuzzy*-optimal set from sampling (•), *fuzzy*$_\alpha$-optimal set from evolutionary algorithm (o) and Pareto Optimal set (·)

7 Conclusion

Via simple and general examples, in this paper we have shown that some possible weakness of Pareto optimality (weakness becoming evident in the application to optimization problems with more than three objectives) can be avoided by considering other more general definitions.

In particular it seems that fuzzy logic can be profitably used in the generalization of the notion of Pareto optimality. In fact it easily and naturally allows considering the number of improved objectives (not only if at least one objectives is improved without worsening the others) and the size of improvement in each objective.

Though being developed for more than three criteria the proposed fuzzy optimality definition is meaningful for 2D and 3D cases for different reasons that are discussed in some details. In order to apply the definition to real-life decision making problems, the use of evolutionary algorithms for the computation of *fuzzy*-optimality based subset has also been shown.

References

1. P. Alotto, A. V. Kuntsevitch, Ch. Magele, C. Paul G. Molinari, K. Preis, M. Repetto, and K. R. Richter. Multiobjective optimization in magnetostatics: A proposal for benchmark problems. *Technical report, Institut für Grundlagen und Theorie Electrotechnik, Technische Universität Graz, Graz, Austria,* http://www-igte.tu-graz.ac.at/team/berl01.htm, 1996.
2. P. Amato and C. Manara. Relating the theory of partitions in mv-logic to the design of interpretable fuzzy systems. In J. Casillas, O. Cordón, F. Herrera, and L. Magdalena, editors, *Trade-off between Accuracy and Interpretability in Fuzzy Rule-Based Modeling.* Springer-Verlag, Berlin, to be published.

3. P. Di Barba and M. Farina. Multiobjective shape optimisation of air cored solenoids. *COMPEL International Journal for computation and mathematics in Electrical and Electronic Engineering*, 21(1):45–57, 2002. in press.
4. P. Di Barba, M. Farina, and A. Savini. Multiobjective Design Optimization of Real-Life Devices in Electrical Engineering: A Cost-Effective Evolutionary Approach. In Eckart Zitzler, Kalyanmoy Deb, Lothar Thiele, Carlos A. Coello Coello, and David Corne, editors, *First International Conference on Evolutionary Multi-Criterion Optimization*, pages 560–573. Springer-Verlag. Lecture Notes in Computer Science No. 1993, 2001.
5. C. Carlsson and R. Fuller. Multiobjective optimization with linguistic variables. *Proceedings of the Sixth European Congress on Intelligent Techniques and Soft Computing (EUFIT98, Aachen, 1998, Verlag Mainz, Aachen,*, 2:1038–1042, 1998.
6. Lino Costa and Pedro Oliveira. An Evolution Strategy for Multiobjective Optimization. In *Congress on Evolutionary Computation (CEC'2002)*, volume 1, pages 97–102, Piscataway, New Jersey, May 2002. IEEE Service Center.
7. Masashi Morikawa Daisuke Sasaki, Shigeru Obayashi, and Kazuhiro Nakahashi. Constrained Test Problems for Multi-objective Evolutionary Optimization. In Eckart Zitzler, Kalyanmoy Deb, Lothar Thiele, Carlos A. Coello Coello, and David Corne, editors, *First International Conference on Evolutionary Multi-Criterion Optimization*, pages 639–652. Springer-Verlag. Lecture Notes in Computer Science No. 1993, 2001.
8. M. Farina and P. Amato. A fuzzy definition of "optimality" for many-criteria decision-making and optimization problems. *submitted to IEEE Trans. on Sys. Man and Cybern.*, 2002.
9. T. Hanne. *Intelligent strategies for meta multiple criteria decision making.* Kluwer Academic Publishers, Dordrecht, THE NETHERLANDS, 2001.
10. Ryoji Homma. Combustion process optimization by genetic algorithms - reduction of co emission via optimal post-flame process. *Energy and Environmental Technology Laboratory, Tokyo Gas Co.,Ltd.*, 9, 1999.
11. P. Tarvainen J. Hamalainee, R.A.E Makinen. Optimal design of paper machine headboxes. *International Journal of Numerical Methods in Fluids*, 34:685–700, 2000.
12. G. J. Klir and Bo Yuan editors. *Fuzzy Sets, Fuzzy Logic and Fussy Systems, selected papers by L.A. Zadeh*, volume 6 of *Advances in Fuzzy Systems - Applications and Theory*. World Scientific, Singapore, 1996.
13. M.Farina and P. Amato. On the optimal solution definition for many-criteria optimization problems. In *Proceedings of the NAFIPS-FLINT International Conference2002, New Orleans, June, 2002*, pages 233–238. IEEE Service Center, June 2002.
14. K. Miettinen. *Nonlinear Multiobjective Optimization.* Kluwer Academic Publishers, Dordrecht, THE NETHERLANDS, 1999.
15. P.P. Chakrabarti P. Dasgupta and S. C. DeSarkar. *Multiobjective Heuristic Search.* Vieweg, 1999.
16. N. Srinivas and Kalyanmoy Deb. Multiobjective Optimization Using Nondominated Sorting in Genetic Algorithms. *Evolutionary Computation*, 2(3):221–248, Fall 1994.
17. http://energy.bmstu.ru/e02/diesel/d11eng.htm.

IS-PAES: A Constraint-Handling Technique Based on Multiobjective Optimization Concepts

Arturo Hernández Aguirre[1], Salvador Botello Rionda[1], Giovanni Lizárraga Lizárraga[1], and Carlos A. Coello Coello[2]

[1] Center for Research in Mathematics (CIMAT)
Department of Computer Science
Guanajuato, Gto. 36240, MEXICO
{artha,botello,giovanni}@cimat.mx
[2] CINVESTAV-IPN
Evolutionary Computation Group
Depto. de Ingeniería Eléctrica
Sección de Computación
Av. Instituto Politécnico Nacional No. 2508
Col. San Pedro Zacatenco
México, D. F. 07300
ccoello@cs.cinvestav.mx

Abstract. This paper introduces a new constraint-handling method called Inverted-Shrinkable PAES (IS-PAES), which focuses the search effort of an evolutionary algorithm on specific areas of the feasible region by shrinking the constrained space of single-objective optimization problems. IS-PAES uses an adaptive grid as the original PAES (Pareto Archived Evolution Strategy). However, the adaptive grid of IS-PAES does not have the serious scalability problems of the original PAES. The proposed constraint-handling approach is validated with several examples taken from the standard literature on evolutionary optimization.

1 Introduction

Evolutionary Algorithms (EAs) in general (i.e., genetic algorithms, evolution strategies and evolutionary programming) lack a mechanism able to bias efficiently the search towards the feasible region in constrained search spaces. Such a mechanism is highly desirable since most real-world problems have constraints which could be of any type (equality, inequality, linear and nonlinear). The success of EAs in global optimization has triggered a considerable amount of research regarding the development of mechanisms able to incorporate information about the constraints of a problem into the fitness function of the EA used to optimize it [4,17]. So far, the most common approach adopted in the evolutionary optimization literature to deal with constrained search spaces is the use of penalty functions. When using a penalty function, the amount of constraint violation is used to punish or "penalize" an infeasible solution so that feasible solutions are favored by the selection process. Despite the popularity of penalty functions, they have several drawbacks from which the main one is that they require a careful fine tuning of the penalty factors that indicates the degree of penalization to be applied [4]. Recently, some researchers have suggested the use of multiobjective optimization concepts to handle constraints in EAs (see for example [4]). This paper introduces a new approach that is based on an evolution strategy that was originally proposed for multiobjective optimization: the Pareto Archived Evolution Strategy (PAES) [14]. Our approach can be used to handle constraints both of single- and multiobjective optimization problems and does not present the scalability problems of the original PAES. The remainder of this paper is organized as follows. Section 2 gives a formal description of the general problem that we want to solve. Section 3 describes the previous work related to our own. In Section 4, we describe the main algorithm of IS-PAES. Section 5 provides a comparison of results and Section 6 draws our conclusions and provides some paths of future research.

2 Problem Statement

We are interested in the general non-linear programming problem in which we want to:

$$\text{Find } x \text{ which optimizes } f(x) \tag{1}$$

subject to:

$$g_i(x) \leq 0, \quad i = 1, \ldots, n \tag{2}$$

$$h_j(x) = 0, \quad j = 1, \ldots, p \tag{3}$$

where x is the vector of solutions $x = [x_1, x_2, \ldots, x_r]^T$, n is the number of inequality constraints and p is the number of equality constraints (in both cases, constraints could be linear or non-linear).

If we denote with $\mathcal{F}$ to the feasible region and with $\mathcal{S}$ to the whole search space, then it should be clear that $\mathcal{F} \subseteq \mathcal{S}$.

For an inequality constraint that satisfies $g_i(x) = 0$, then we will say that is active at x. All equality constraints h_j (regardless of the value of x used) are considered active at all points of $\mathcal{F}$.

3 Related Work

Since our approach belongs to the group of techniques in which multiobjective optimization concepts are adopted to handle constraints, we will briefly discuss some of the most relevant work done in this area. The main idea of adopting multiobjective optimization concepts to handle constraints is to redefine the single-objective optimization of $f(x)$ as a multiobjective optimization problem in which we will have $m + 1$ objectives, where m is the total number of constraints. Then, we can apply any multiobjective optimization technique [9] to the new vector $\bar{v} = (f(x), f_1(x), \ldots, f_m(x))$, where $f_1(x), \ldots, f_m(x)$ are the original constraints of the problem. An ideal solution x

would thus have $f_i(x)=0$ for $1 \leq i \leq m$ and $f(x) \leq f(y)$ for all feasible y (assuming minimization).

Three are the mechanisms taken from evolutionary multiobjective optimization that are more frequently incorporated into constraint-handling techniques:

1. Use of Pareto dominance as a selection criterion.
2. Use of Pareto ranking [12] to assign fitness in such a way that nondominated individuals (i.e., feasible individuals in this case) are assigned a higher fitness value.
3. Split the population in subpopulations that are evaluated either with respect to the objective function or with respect to a single constraint of the problem. This is the selection mechanism adopted in the Vector Evaluated Genetic Algorithm (VEGA) [23].

We will now provide a brief discussion of the different approaches that have been proposed in the literature adopting the three main ideas previously indicated.

3.1 COMOGA

Surry & Radcliffe [24] used a combination of the Vector Evaluated Genetic Algorithm (VEGA) [23] and Pareto Ranking to handle constraints in an approach called COMOGA (Constrained Optimization by Multi-Objective Genetic Algorithms).

In this technique, individuals are ranked depending of their sum of constraint violation (number of individuals dominated by a solution). However, the selection process is based not only on ranks, but also on the fitness of each solution. COMOGA uses a non-generational GA and extra parameters defined by the user (e.g., a parameter called ϵ is used to define the change rate of P_{cost}). One of these parameters is P_{cost}, that sets the rate of selection based on fitness. The remaining $1 - P_{cost}$ individuals are selected based on ranking values. P_{cost} is defined by the user at the begining of the process and it is adapted during evolution using as a basis the percentage of feasible individuals that one wishes to have in the population.

COMOGA was applied on a gas network design problem and it was compared against a penalty function approach. Although COMOGA showed a slight improvement in the results with respect to a penalty function, its main advantage is that it does not requiere a fine tuning of penalty factors or any other additional parameter. The main drawback of COMOGA is that it requires several extra parameters, although its authors argue that the technique is not particularly sensitive to their values [24].

3.2 VEGA

Parmee & Purchase [18] proposed to use VEGA [23] to guide the search of an evolutionary algorithm to the feasible region of an optimal gas turbine design problem with a heavily constrained search space. After having a feasible point, they generated an optimal hypercube around it in order to avoid leaving the feasible region after applying the genetic operators. Note that this approach does not really use Pareto dominance or any other multiobjective optimization concepts to exploit the search space. Instead, it uses VEGA just to reach the feasible region. The use of special operators that preserve feasibility makes this approach highly specific to one application domain rather than providing a general methodology to handle constraints.

Coello [7] used a population-based approach similar to VEGA [23] to handle constraints in single-objective optimization problems. At each generation, the population was split into $m + 1$ subpopulations of equal fixed size, where m is the number of constraints of the problem. The additional subpopulation handles the objective function of the problem and the individuals contained within it are selected based on the unconstrained objective function value. The m remaining subpopulations take one constraint of the problem each as their fitness function. The aim is that each of the subpopulations tries to reach the feasible region corresponding to one individual constraint. By combining these different subpopulations, the approach will reach the feasible region of the problem considering all of its constraints simultaneously.

This approach was tested with some engineering problems [7] in which it produced competitive results. It has also been successfully used to solve combinational circuit design problems [8]. The main drawback of this approach is that the number of subpopulations required increases linearly with the number of constraints of the problem. This has some obvious scalability problems. Furthermore, it is not clear how to determine appropriate sizes for each of the subpopulations used.

3.3 MOGA

Coello [6] proposed the use of Pareto dominance selection to handle constraints in EAs. This is an application of Fonseca and Fleming's Pareto ranking process [11] (called Multi-Objective Genetic Algorithm, or MOGA) to constraint-handling. In this approach, feasible individuals are always ranked higher than infeasible ones. Based on this rank, a fitness value is assigned to each individual. This technique also includes a self-adaptation mechanism that avoids the usual empirical fine-tuning of the main genetic operators. Coello's approach uses a real-coded GA with universal stochastic sampling selection (to reduce the selection pressure caused by the Pareto ranking process).

This approach has been used to solve some engineering design problems [6] in which it produced very good results. Furthermore, the approach showed great robustness and required a relatively low number of fitness function evaluations with respect to traditional penalty functions. Additionally, it does not require any extra parameters. Its main drawback is the computational cost ($O(M^2)$, where M is the population size) derived from the Pareto ranking process.

3.4 NPGA

Coello and Mezura [5] implemented a version of the Niched-Pareto Genetic Algorithm (NPGA) [13] to handle constraints in single-objective optimization problems. The NPGA is a multiobjective optimization approach in which individuals are selected through a tournament based on Pareto dominance. However, unlike the NPGA, Coello and Mezura's approach does not require niches (or fitness sharing [10]) to maintain diversity in the population. The NPGA is a more efficient technique than traditional multiobjective optimization algorithms, since it does not compare every individual in the population with respect to each other (as in traditional Pareto ranking), but uses only a sample of the population to estimate Pareto dominance. This is the main advantage of this approach with respect to Coello's proposal [6]. Note however that Coello and Mezura's approach requires an additional parameter called S_r that controls the diversity of the population. S_r indicates the proportion of parents selected by four comparison criteria described below. The remaining $1 - S_r$ parents will be selected by a pure probabilistic approach. Thus, this mechanism is responsible for keeping infeasible individuals in the population (i.e., the source of diversity that keeps the algorithm from converging to a local optimum too early in the evolutionary process).

This approach has been tested with several benchmark problems and was compared against several types of penalty functions. Results indicated that the approach was robust, efficient and effective. However, it was also found that the approach had scalability problems (its performance degrades as the number of decision variables increases).

3.5 Pareto Set and Line Search

Camponogara & Talukdar [3] proposed an approach in which a global optimization problem was transformed into a bi-objective problem where the first objective is to optimize the original objective function and the second is to minimize:

$$\Phi(\mathbf{x}) = \sum_{i=1}^{n} \max(0, g_i(\mathbf{x})) \qquad (4)$$

Equation (4) tries to minimize the total amount of constraint violation of a solution (i.e., it tries to make it feasible). At each generation of the process, several Pareto sets are generated. An operator that substitutes crossover takes two Pareto sets S_i and S_j, where $i < j$, and two solutions $x_i \in S_i$ and $x_j \in S_j$, where x_i dominates x_j. With these two points a search direction is defined using:

$$d = \frac{(x_i - x_j)}{|x_i - x_j|} \qquad (5)$$

Line search begins by projecting d over one variable axis on decision variable space in order to find a new solution x which dominates both x_i and x_j. At pre-defined intervals, the worst half of the population is replaced with new random solutions to avoid premature convergence. This indicates some of the problems of the approach to maintain diversity. Additionally, the use of line search within a GA adds some extra computational cost.

The authors of this approach validated it using a benchmark consisting of five test functions. The results obtained were either optimal or very close to it. The main drawback of this approach is its additional computational cost. Also, it is not clear what is the impact of the segment chosen to search on the overall performance of the algorithm.

3.6 Pareto Ranking and Domain Knowledge

Ray et al. [19] proposed the use of a Pareto ranking approach that operates on three spaces: objective space, constraint space and the combination of the two previous spaces.

This approach also uses mating restrictions to ensure better constraint satisfaction in the offspring generated and a selection process that eliminates weaknesses in any of these spaces. To maintain diversity, a niche mechanism based on Euclidean distances is used. This approach can solve both constrained or unconstrained optimization problems with one or several objective functions.

The main advantage of this approach is that it requires a very low number of fitness function evaluations (between 2% and 10% of the number of evaluations required by the homomorphous maps of Koziel and Michalewicz [15], which is one of the best constraint-handling techniques known to date). The technique has some problems to reach the global optima, but it produces very good approximations considering its low computation cost. The main drawback of the approach is that its implementation is considerably more complex than any of the other techniques previously discussed.

3.7 Pareto Ranking and Robust Optimization

Ray [20] explored an extension of his previous work on constraint-handling [19] in which the emphasis was robustness. A robust optimized solution is not sensitive to parametric variations due to incomplete information of the problem or to changes on it. This approach is capable of handling constraints and finds feasible solutions that are robust to parametric variations produced over time. This is achieved using the individual's self-feasibility and its neighborhood feasibility. The results reported in two well-known design problems [20] showed that the proposed approach did not reach solutions as good as the other techniques with which it was compared, but it turned out to be less sensitive to parametric variations, which was the main goal of the approach. In constrast, the other techniques analyzed showed significant changes when the parameters were perturbed. The main drawback of this approach is, again, its relative complexity (i.e., its difficulty to implement it), and it would also be desirable that the approach is further refined so that it can get closer to the global optimum than the current available version.

4 IS-PAES Algorithm

All of the approaches discussed in the previous section have drawbacks that keep them from producing competitive results with respect to the constraint-handling techniques that represent the state-of-the-art in evolutionary optimization. In a recent technical report [16], four of the previous techniques (i.e., COMOGA [24], VEGA [7], MOGA [6] and NPGA [5]) have been compared using Michalewicz's benchmark [17] and some additional engineering optimization problems. Although inconclusive, the results indicate that the use of Pareto dominance as a selection criterion gives better results than Pareto ranking or the use of a population-based approach. However, in all cases, the approaches analyzed are unable to reach the global optimum of problems with either high dimensionality, large feasible regions or many nonlinear equality constraints [16].

In contrast, the approach proposed in this paper uses Pareto dominance as the criterion selection, but unlike the previous work in the area, a secondary population is used in this case. The approach, which is a relatively simple extension of PAES [14] provides, however, very good results, which are highly competitive with those generated with an approach that represents the state-of-the-art in constrained evolutionary optimization.

IS-PAES has been implemented as an extension of the Pareto Archived Evolution Strategy (PAES) proposed by Knowles and Corne [14] for multiobjective optimization. PAES's main feature is the use of an adaptive grid on which objective function space is located using a coordinate system. Such a grid is the diversity maintenance mechanism of PAES and it's the main feature of this algorithm. The grid is created by bisecting k times the function space of dimension $d = g+1$. The control of 2^{kd} grid cells means the allocation of a large amount of physical memory for even small problems. For instance, 10 functions and 5 bisections of the space produce 2^{50} cells. Thus, the first feature introduced in IS-PAES is the "inverted" part of the algorithm that deals with this space usage problem. IS-PAES's fitness function is mainly driven by a feasibility criterion. Global information carried by the individuals surrounding the feasible region is used to concentrate the search effort on smaller areas as the evolutionary process takes place. In consequence, the search space being explored is "shrunk" over time. Eventually, upon termination, the size of the search space being inspected will be very small and will contain the solution desired (in the case of single-objective problems. For multi-objective problems, it will contain the feasible region). The main algorithm of IS-PAES is shown in Figure 1.

```
maxsize: max size of file
c: current parent ∈ X (decision variable space)
h: child of c ∈ X, a_h: individual in file that dominates h
a_d: individual in file dominated by h
current: current number of individuals in file
cnew: number of individuals generated thus far
current = 1; cnew=0; c = newindividual() ; add(c)
While cnew≤MaxNew do
    h = mutate(c); cnew+ =1;
    if (c≼h) then exit loop
    else if (h≼c) then { remove(c); add(g); c=h; }
    else if (∃ a_h ∈ file | a_h ≼ h) then exit loop
    else if (∃ a_d ∈ file | h ≼ a_d) then
        add( h ); ∀ a_d { remove(a_d); current− =1 }
    else test(h,c,file)
    if (cnew % g==0) then c = individual in less densely populated region
    if (cnew % r==0) then shrinkspace(file)
End While
```

Fig. 1. Main algorithm of IS-PAES

The function **test(h,c,file)** determines if an individual can be added to the external memory or not. Here we introduce the following notation: $x_1 \square x_2$ means x_1 is located in a less populated region of the grid than x_2. The pseudo-code of this function is depicted in Figure 2.

```
if (current < maxsize) then add(h)
    if (h □ c) then c = h
else if (∃a_p∈file | h □ a_p) then { remove(a_p); add(h) }
    if (h □ c) then c = h;
```

Fig. 2. Pseudo-code of **test(h,c,file)**

4.1 Inverted "Ownership"

IS-PAES handles the population *as part of* a grid location relationship, whereas PAES handles a grid location *contains* population relationship. In other words, PAES keeps a list of individuals on either grid location, but in IS-PAES either individual knows its position on the grid. Therefore, building a sorted list of the most dense populated areas of the grid only requires to sort the k elements of the external memory. In PAES, this procedure needs to inspect every location of the grid in order to produce an unsorted list, there after the list is sorted. The advantage of the inverted relationship is clear when the optimization problem has many functions (more than 10), and/or the granularity of the grid is fine, for in this case only IS-PAES is able to deal with any number of functions and granularity level.

4.2 Shrinking the Objective Space

Shrinkspace(file) is the most important function of IS-PAES since its task is the reduction of the search space. The pseudo-code of **Shrinkspace(file)** is shown in Figure 3.

The function **select(file)** returns a list whose elements are the best individuals found in *file*. The size of the list is 15% of *maxsize*. Since individuals could be feasible, infeasible or only partially feasible, the list is generated by discarding from the file the worst elements based on constraint violation. Notice that **select(file)** does not use a greedy

```
x_pob: vector containing the smallest value of either x_i ∈ X
x̄_pob: vector containing the largest value of either x_i ∈ X
select(file); getMinMax( file, x_pob, x̄_pob )
trim(x_pob, x̄_pob )
adjustparameters(file);
```

Fig. 3. Pseudo-code of **Shrinkspace(file)**

approach (e.g., searching for the best feasible individuals at once). Instead, individuals with the highest amount of constraint violation are removed from the file. Thus, the resulting list contains: 1) only the best feasible individuals, 2) a combination of feasible and partially feasible individuals, or 3) the "best" infeasible individuals. Function **trim**($\underline{x}_{pob}$, $\overline{x}_{pob}$) shrinks the feasible space around the potential solutions enclosed in the hypervolume defined by the vectors $\underline{x}_{pob}$ and $\overline{x}_{pob}$. Thus, the function **trim()** (see Figure 4) determines the new boundaries for the decision variables.

---

n: size of decision vector;
$\overline{x}_i$: actual upper bound of the i_{th} decision variable
$\underline{x}_i$: actual lower bound of the i_{th} decision variable
$\overline{x}_{pob,i}$: upper bound of i_{th} decision variable in population
$\underline{x}_{pob,i}$: lower bound of i_{th} decision variable in population
$\forall i : i \in \{1, \ldots, n\}$
$\quad slack_i = 0.05 \times (\overline{x}_{pob,i} - \underline{x}_{pob,i})$
$\quad width\_pob_i = \overline{x}_{pob,i} - \underline{x}_{pob,i}; width_i^t = \overline{x}_i^t - \underline{x}_i^t$
$\quad deltaMin_i = \frac{\beta * width_i^t - width\_pob_i}{2}$
$\quad delta_i = \max(slack_i, deltaMin_i);$
$\quad \overline{x}_i^{t+1} = \overline{x}_{pob,i} + delta_i;\ \underline{x}_i^{t+1} = \underline{x}_{pob,i} - delta_i;$
$\quad$ if ($\overline{x}_i^{t+1} > \overline{x}_{original,i}$) then
$\quad\quad \underline{x}_i^{t+1} -= \overline{x}_i^{t+1} - \overline{x}_{original,i};\ \overline{x}_i^{t+1} = \overline{x}_{original,i};$
$\quad$ if ($\underline{x}_i^{t+1} < \underline{x}_{original,i}$) then $\overline{x}_i^{t+1} += \underline{x}_{original,i} - \underline{x}_i^{t+1};$
$\quad\quad \underline{x}_i^{t+1} = \underline{x}_{original,i};$
$\quad$ if ($\overline{x}^{t+1} > \overline{x}_{original,i}$) then $\overline{x}_i^{t+1} = \overline{x}_{original,i};$

Fig. 4. Pseudo-code of **trim**

The value of β is the percentage by which the boundary values of either $x_i \in X$ must be reduced such that the resulting hypervolume is a fraction α of its initial value. In our experiments, $\alpha = 0.90$ worked well in all cases. Clearly, α controls the shrinking speed, hence the algorithm is sensitive to this parameter and it can prevent it from finding the optimum solution if small values are chosen. In our experiments, values in the range [85%,95%] were tested with no visible effect in the performance. Of course, α values near to 100% slow down the convergence speed. The last step of **shrinkspace()** is a call to **adjustparameters(file)**. The goal is to re-start the control variable σ using: $\sigma_i = (\overline{x}_i - \underline{x}_i)/\sqrt{n}\ i \in (1, \ldots, n)$ This expression is also used during the generation of the initial population. In that case, the upper and lower bounds take the initial values of the search space indicated by the problem. The variation of the mutation probability follows the exponential behavior suggested by Bäck [1].

5 Comparison of Results

We have validated our approach with several problems used as a benchmark for evolutionary algorithms (see [17]) and with several engineering optimization problems taken from the standard literature. In the first case, our results are compared against a technique called "stochastic ranking" [22], which is representative of the state-of-the-art in constrained evolutionary optimization. This approach has been found to be equally good or even better in some cases than the homomorphous maps of Koziel and Michalewicz [15].

5.1 Examples

The following parameters were adopted for IS-PAES in all the experiments reported next: $maxsize = 200$, $bestindividuals = 15\%$, $slack = 0.05$, $r = 400$. The maximum number of fitness function evaluations was set to 350,000, which is the number of evaluations used in [22]. We used ten (out of 13) of the test functions described in [22], due to time limitations to perform the experiments. The test functions chosen, however, contain characteristics that are representative of what can be considered "difficult" global optimization problems for an evolutionary algorithm.

Table 1. Values of ρ for the ten test problems chosen

TF	n	Type of function	ρ	LI	NI	NE
g01	13	quadratic	0.0003%	9	0	0
g02	20	non linear	99.9973%	2	0	0
g03	10	non linear	0.0026%	0	0	1
g04	5	quadratic	27.0079%	4	2	0
g06	2	non linear	0.0057%	0	2	0
g07	10	quadratic	0.0000%	3	5	0
g08	2	non linear	0.8581%	0	2	0
g09	7	non linear	0.5199%	0	4	0
g10	8	linear	0.0020%	6	0	0
g11	2	quadratic	0.0973%	0	0	1

To get a better idea of the difficulty of solving each of these problems, a ρ metric (as suggested by Koziel and Michalewicz [15]) was computed using the following expression:

$$\rho = |F|/|S| \qquad (6)$$

where $|F|$ is the number of feasible solutions and $|S|$ is the total number of solutions randomly generated. In this work, we generated $S = 1,000,000$ random solutions. The different values of ρ for each of the test functions chosen are shown in Table 1, where n is the number of decision variables, LI is the number of linear inequalities, NI the number of nonlinear inequalities and NE is the number of nonlinear equalities.

Table 2. Results produced by our IS-PAES algorithm

TF	optimal	Best	Mean	Median	Worst	Std Dev
g01	-15.0	-14.9997	-14.494	-14.997	-12.446	9.3×10^{-1}
g02	-0.803619	-0.803376	-0.793281	-0.793342	-0.768291	9.0×10^{-3}
g03	-1.0	-1.000	-1.000	-1.000	-1.000	9.7×10^{-5}
g04	-30665.539	-30665.539	-30665.539	-30665.539	-30665.539	0.0
g06	-6961.814	-6961.814	-6961.813	-6961.814	-6961.810	8.5×10^{-5}
g07	24.306	24.338	24.527	24.467	24.995	1.7×10^{-1}
g08	-0.095825	-0.095825	-0.095825	-0.095825	-0.095825	0.0
g09	680.630	680.630	680.631	680.631	680.634	8.1×10^{-4}
g10	7049.331	7062.019	7342.944	7448.014	7588.054	1.4×10^{2}
g11	0.750	0.750	0.750	0.750	0.751	2.6×10^{-4}

From Tables 2 and 3 we can see that the proposed approach is highly competitive. The discussion of results for each test function is provided next:

For **g01** the best solution found by IS-PAES was: x = {1, 0.999999939809, 0.999997901977, 1, 0.999981406123, 1, 1, 0.999999242667, 0.999981194574, 2.99987534752, 2.99995011286, 2.99993014684, 0.999982112914} with $F(\mathbf{x}) = -14.99968877$. In this case, IS-PAES was less consistent than stochastic ranking in finding the global optimum, mainly because the approach was trapped in a local optimum in which $F(\mathbf{x}) = -13$ during 20% of the runs.

Table 3. Results produced by the stochastic ranking algorithm [22]

TF	optimal	Best	Mean	Median	Worst	Std Dev
g01	-15.0	-15.0	-15.0	-15.0	-15.0	0.0
g02	-0.803619	-0.803515	-0.781975	-0.785800	-0.726288	2×10^{-2}
g03	-1.0	-1.000	-1.000	-1.000	-1.000	1.9×10^{-4}
g04	-30665.539	-30665.539	-30665.539	-30665.539	-30665.539	2.0×10^{-5}
g06	-6961.814	-6961.814	-6875.940	-6961.814	-6350.262	1.6×10^{2}
g07	24.306	24.307	24.374	24.357	24.642	6.6×10^{-2}
g08	-0.095825	-0.095825	-0.095825	-0.095825	-0.095825	2.6×10^{-17}
g09	680.630	680.630	680.656	680.641	680.763	3.4×10^{-2}
g10	7049.331	7054.316	7559.192	7372.613	8835.655	5.3×10^{2}
g11	0.750	0.750	0.750	0.750	0.750	8.0×10^{-5}

For **g02** the best solution found by IS-PAES was: x = {3.14860401788, 3.10915903011, 3.08909341555, 3.05835689132, 3.04000196011, 3.00100530894, 2.94955289769, 2.94207158769, 0.49907406319, 0.486231653274, 0.49055938302, 0.492879188045, 0.481722447567, 0.471623533316, 0.452037376504, 0.442565813637, 0.451211591495, 0.437863945589, 0.444359423833, 0.437834075871} with $F(\mathbf{x}) = -0.803375563$. As we can see, the best result found by stochastic ranking was better than the best result found by IS-PAES. However, the statistical performance measures of

IS-PAES were better (particularly the standard deviation which is significantly lower), which seems to indicate that our approach had more robustness in this problem.

The best solution found by IS-PAES for **g03** was: x = {0.316965968, 0.315664596, 0.314608242, 0.315958975, 0.315915392, 0.317873891, 0.316867036, 0.314518512, 0.314381436, 0.319636209} with $F(\mathbf{x}) = -1.000421429$. In can be clearly seen in this case that both IS-PAES and stochastic ranking had an excellent performance.

The best solution found by IS-PAES for **g04** was: x = {78, 33.00000002, 29.99525605, 45, 36.77581285 } with $F(\mathbf{x}) = -30665.53867$. The behavior of IS-PAES in this test function was practically the same as stochastic ranking.

For **g06**, the best solution found by IS-PAES was: x = {14.0950000092, 0.842960808844} with $F(\mathbf{x})$ = -6961.813854. Note that both approaches reach the global optimum in this case, but IS-PAES is more consistent, with very small variations in the results and a much lower standard deviation than stochastic ranking.

Stochastic ranking was clearly better in all aspects than IS-PAES for **g07**. The best solution found by IS-PAES was: x = {2.16996489702, 2.36701436984, 8.76882720318, 5.07418756668, 0.943992761955, 1.32027308617, 1.31870032997, 9.82673763033, 8.26988778617, 8.36187863755} with $F(\mathbf{x}) = 24.33817628$.

For **g08**, the best solution found by IS-PAES was: x = {1.227971353, 4.245373368} with $F(\mathbf{x}) = -0.095825041$. Both algorithms had the same performance in this test function.

Both algorithms reached the global optimum for **g09**, but IS-PAES had better statistical measures. The best solution found by IS-PAEs was: x = {2.326603718, 1.957743917, -0.468352679, 4.349668424, -0.621354832, 1.047882344, 1.590801921} with $F(\mathbf{x}) = 680.6304707$.

Except for the best solution found (which is better for stochastic ranking), the statistical measures of IS-PAES are better than those of stochastic ranking for **g10**. The best solution found by IS-PAEs was: x = {105.6345328, 1179.227593, 6070.09281, 122.497943, 257.1979828, 277.4889774, 265.2967614, 357.197398} with $F(\mathbf{x}) = 7062.019117$.

Finally, for **g11** both algorithms had a very good performance. The best solution found by IS-PAES was: x = {2.326603718, 1.957743917, -0.468352679, 4.349668424, -0.621354832, 1.047882344, 1.590801921} with $F(\mathbf{x}) = 0.749913273$.

5.2 Optimization of a 49-Bar Plane Truss

The engineering optimization problem chosen is the optimization of the 49-bar plane truss shown in Figure 5. The goal is to find the cross-sectional area of each member of the truss, such that the overall weight is minimized, subject to stress and displacement constraints. The weight of the truss is given by $F(\mathbf{x}) = \sum_{j=1}^{49} \gamma A_j L_j$, where A_j is the cross-sectional area of the j_{th} member, L_j is the corresponding length of the bar, and γ is the volumetric density of the material. We used a catalog of *Altos Hornos de México, S.A.*, with 65 entries for the cross-sectional areas available for the design. Other relevant data are the following: Young modulus = $2.1 \cdot 10^6$ kg/cm^3, maximum allowable stress = 3500.00 kg/cm^2, $\gamma = 7.4250 \cdot 10^{-3}$, and a horizontal load of 4994.00 kg applied to the nodes: 3, 5, 7, 9, 12, 14, 16, 19, 21, 23, 25 y 27. We solved this problem for two cases:

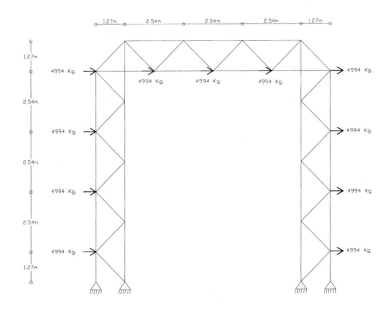

Fig. 5. 49-bar plane truss used as an engineering optimization example

1. **Case 1. Stress and displacement constraints:** Maximum allowable stress = 3500.00 kg/cm^2, maximum displacement per node = $10cm$ A total of 72 constraints, thus 73 objective functions.
2. **Case 2. Real-world problem:** The design problem considers traction and compression stress on the bars, as well as their proper weight. Maximum allowable stress = 3500.00 kg/cm^2, maximum displacement per node =10 cm. A total of 72 constraints, thus 73 objective functions.

The average result of 30 runs for each case are shown in Table 4. We compare IS-PAES with previous results reported by Botello [2] using other heuristics with a penalty function [21] (SA: Simulated Annealing, GA50: Genetic Algorithm with a population of 50, and GSSA: General Stochastic Search Algorithm with populations of 50 and 5).

We can see in this case that IS-PAES produced the lowest average weight for CASE 1, and the second best for CASE 2.

Table 4. Comparison of different algorithms on the 49-bar struss, cases 1 and 2.

Algorithm	CASE 1: Avg. Weight (Kg)	CASE 2: Avg. Weight (Kg)
IS-PAES	725	2603
SA	737	2724
GA50	817	2784
GSSA50	748	2570
GSSA5	769	2716

6 Conclusions and Future Work

We have introduced a constraint-handling approach that combines multiobjective optimization concepts with an efficient reduction mechanism of the search space and a secondary population. We have shown how our approach overcomes the scalability problem of the original PAES from which it was derived, and we also showed that the approach is highly competitive with respect to the state-of-the-art technique in the area.

As part of our future work, we want to refine the mechanism adopted for reducing the search space being explored, since in our current version of the algorithm, convergence to local optima may occur in some cases due to the high selection pressure introduced by such mechanism.

The elimination of the parameters required by our approach is another goal of our current research. Finally, we also intend to couple the mechanisms proposed in this paper to other evolutionary multiobjective optimization approaches.

Acknowledgments. The first author acknowledges support from CONACyT project I-39324-A. The second author acknowledges support from CONACyT project No. 34575-A. The last author acknowledges support from NSF-CONACyT project No. 32999-A.

References

1. Thomas Bäck. *Evolutionary Algorithms in Theory and Practice*. Oxford University Press, New York, 1996.
2. Salvador Botello, José Luis Marroquín, Eugenio Oñate, and Johan Van Horebeek. Solving Structural Optimization problems with Genetic Algorithms and Simulated Annealing. *International Journal for Numerical Methods in Engineering*, 45(8):1069–1084, July 1999.
3. Eduardo Camponogara and Sarosh N. Talukdar. A Genetic Algorithm for Constrained and Multiobjective Optimization. In Jarmo T. Alander, editor, *3rd Nordic Workshop on Genetic Algorithms and Their Applications (3NWGA)*, pages 49–62, Vaasa, Finland, August 1997. University of Vaasa.
4. Carlos A. Coello Coello. Theoretical and Numerical Constraint Handling Techniques used with Evolutionary Algorithms: A Survey of the State of the Art. *Computer Methods in Applied Mechanics and Engineering*, 191(11-12):1245–1287, January 2002.
5. Carlos A. Coello Coello and Efrén Mezura-Montes. Handling Constraints in Genetic Algorithms Using Dominance-Based Tournaments. In I.C. Parmee, editor, *Proceedings of the Fifth International Conference on Adaptive Computing Design and Manufacture (ACDM 2002)*, volume 5, pages 273–284, University of Exeter, Devon, UK, April 2002. Springer-Verlag.
6. Carlos A. Coello Coello. Constraint-handling using an evolutionary multiobjective optimization technique. *Civil Engineering and Environmental Systems*, 17:319–346, 2000.
7. Carlos A. Coello Coello. Treating Constraints as Objectives for Single-Objective Evolutionary Optimization. *Engineering Optimization*, 32(3):275–308, 2000.
8. Carlos A. Coello Coello and Arturo Hernández Aguirre. Design of Combinational Logic Circuits through an Evolutionary Multiobjective Optimization Approach. *Artificial Intelligence for Engineering, Design, Analysis and Manufacture*, 16(1):39–53, 2002.
9. Carlos A. Coello Coello, David A. Van Veldhuizen, and Gary B. Lamont. *Evolutionary Algorithms for Solving Multi-Objective Problems*. Kluwer Academic Publishers, New York, May 2002. ISBN 0-3064-6762-3.

10. Kalyanmoy Deb and David E. Goldberg. An Investigation of Niche and Species Formation in Genetic Function Optimization. In J. David Schaffer, editor, *Proceedings of the Third International Conference on Genetic Algorithms*, pages 42–50, San Mateo, California, June 1989. George Mason University, Morgan Kaufmann Publishers.
11. Carlos M. Fonseca and Peter J. Fleming. Genetic Algorithms for Multiobjective Optimization: Formulation, Discussion and Generalization. In Stephanie Forrest, editor, *Proceedings of the Fifth International Conference on Genetic Algorithms*, pages 416–423, San Mateo, California, 1993. University of Illinois at Urbana-Champaign, Morgan Kauffman Publishers.
12. David E. Goldberg. *Genetic Algorithms in Search, Optimization and Machine Learning*. Addison-Wesley Publishing Company, Reading, Massachusetts, 1989.
13. Jeffrey Horn, Nicholas Nafpliotis, and David E. Goldberg. A Niched Pareto Genetic Algorithm for Multiobjective Optimization. In *Proceedings of the First IEEE Conference on Evolutionary Computation, IEEE World Congress on Computational Intelligence*, volume 1, pages 82–87, Piscataway, New Jersey, June 1994. IEEE Service Center.
14. Joshua D. Knowles and David W. Corne. Approximating the Nondominated Front Using the Pareto Archived Evolution Strategy. *Evolutionary Computation*, 8(2):149–172, 2000.
15. Slawomir Koziel and Zbigniew Michalewicz. Evolutionary Algorithms, Homomorphous Mappings, and Constrained Parameter Optimization. *Evolutionary Computation*, 7(1):19–44, 1999.
16. Efrén Mezura-Montes and Carlos A. Coello Coello. A Numerical Comparison of some Multiobjective-based Techniques to Handle Constraints in Genetic Algorithms. Technical Report EVOCINV-03-2002, Evolutionary Computation Group at CINVESTAV-IPN, México, D.F. 07300, September 2002. available at: http://www.cs.cinvestav.mx/~EVOCINV/.
17. Zbigniew Michalewicz and Marc Schoenauer. Evolutionary Algorithms for Constrained Parameter Optimization Problems. *Evolutionary Computation*, 4(1):1–32, 1996.
18. I. C. Parmee and G. Purchase. The development of a directed genetic search technique for heavily constrained design spaces. In I. C. Parmee, editor, *Adaptive Computing in Engineering Design and Control-'94*, pages 97–102, Plymouth, UK, 1994. University of Plymouth, University of Plymouth.
19. Tapabrata Ray, Tai Kang, and Seow Kian Chye. An Evolutionary Algorithm for Constrained Optimization. In Darrell Whitley et al., editor, *Proceedings of the Genetic and Evolutionary Computation Conference (GECCO'2000)*, pages 771–777, San Francisco, California, 2000. Morgan Kaufmann.
20. Tapabrata Ray and K.M. Liew. A Swarm Metaphor for Multiobjective Design Optimization. *Engineering Optimization*, 34(2):141–153, March 2002.
21. Jon T. Richardson, Mark R. Palmer, Gunar Liepins, and Mike Hilliard. Some Guidelines for Genetic Algorithms with Penalty Functions. In J. David Schaffer, editor, *Proceedings of the Third International Conference on Genetic Algorithms (ICGA-89)*, pages 191–197, San Mateo, California, June 1989. George Mason University, Morgan Kaufmann Publishers.
22. T.P. Runarsson and X. Yao. Stochastic Ranking for Constrained Evolutionary Optimization. *IEEE Transactions on Evolutionary Computation*, 4(3):284–294, September 2000.
23. J. David Schaffer. Multiple Objective Optimization with Vector Evaluated Genetic Algorithms. In *Genetic Algorithms and their Applications: Proceedings of the First International Conference on Genetic Algorithms*, pages 93–100. Lawrence Erlbaum, 1985.
24. Patrick D. Surry and Nicholas J. Radcliffe. The COMOGA Method: Constrained Optimisation by Multiobjective Genetic Algorithms. *Control and Cybernetics*, 26(3):391–412, 1997.

A Population and Interval Constraint Propagation Algorithm

Vincent Barichard and Jin-Kao Hao

LERIA - Faculty of Sciences - University of Angers
2, Boulevard Lavoisier, 49045 Angers Cedex 01, France
Vincent.Barichard@info.univ-angers.fr
Jin-Kao.Hao@univ-angers.fr

Abstract. We present *PICPA*, a new algorithm for tackling constrained continuous multi-objective problems. The algorithm combines constraint propagation techniques and evolutionary concepts. Unlike other evolutionary algorithm which gives only heuristic solutions, *PICPA* is able to bound effectively the Pareto optimal front as well as to produce accurate approximate solutions.

1 Introduction

The multi-objective combinatorial optimization problems aim to model real world problems that involve many criteria and constraints. In this context, the optimum solution searched is not a single value but a set of good compromises or "trade-offs" that all satisfy the constraints.

A constrained continuous multi-objective problem (CCMO) can be defined as follows:

$$\text{CCMO} \begin{cases} \min & f_i(\vec{x}) \quad i = 1, ..., o \\ \text{s.t.} & C_l(\vec{x}) \geq 0 \quad l = 1, ..., m \\ & \vec{x} \in \mathbb{R}^n \end{cases}$$

where n is the number of variables, $\vec{x}$ is a decision vector, o the number of objectives and m the number of constraints of the problem.

Over the past few years, many researchers developed some evolutionary algorithms which tackle multi-objective optimization problems. They demonstrated the advantage in using population-based search methods [11,9,17,10,20,19,3].

Unfortunately, when the constraints become difficult to satisfy, or when the feasible objective space is not connected, multi-objective evolutionary algorithms hardly converge to the whole Pareto optimal front. Furthermore, these algorithms don't give any bounds of the Pareto optimal front.

In this paper, we present *PICPA*, the "Population and Interval Constraint Propagation Algorithm" which is able to produce high quality approximate solutions while giving guaranteed bounds for the Pareto optimal front. These bounds allow us to know if the solutions found are close to or far away from the optimal front. *PICPA* combines "Interval Constraint Propagation" (ICP) techniques

[2,4] with evolutionary concepts (population and Pareto selection process). Experimental evaluations of *PICPA* on some well known test problems show its effectiveness.

The paper is organized as follows: in the next section we briefly introduce the general principles of the ICP techniques. Then we present our *PICPA* algorithm in Section 3. Experimental results are the subject of the Section 4. Finally, conclusions and perspectives are given in the last section.

2 Interval Constraint Propagation

In this section, we explain briefly the basic idea of Interval Constraint Propagation (ICP). ICP combines interval computation [15] and constraint propagation [13] in order to solve non linear systems of equations and inequations. ICP algorithms have been first presented by Cleary [2] and Davis [4].

2.1 Interval Representation of Variables and Constraints

Each variable of the problem is represented by an interval, and is linked to other variables by constraints. Let us consider two variables x and y, and assume that they belong to some prior feasible value domains:

$$x \in [x^-, x^+] = [-2, 2]$$
$$y \in [y^-, y^+] = [-8, 9]$$

Let us consider now the constraint: $y = x^3$. Because of its definition, we can consider the cubic constraint as a subset of $\mathbb{R}^2$:

$$cubic = \{(x, y) \in \mathbb{R}^2 \mid y = x^3\}$$

The cubic constraint is a binary constraint since it takes into account two variables x and y. In our example, the cubic constraint can be used to *remove some inconsistent values* in the domain of y. Indeed we see that: $\forall x \in [-2, 2], \quad x^3 \leq 8$. Therefore, all the values of the domain of y that are greater than 8 can be safely removed (cf. the hatched area of figure 1).

In a more formal way, the cubic constraint allows us to contract the domain of x and y thanks to the following projection operator:

$$\begin{cases} [x^-, x^+] \longleftarrow [x^-, x^+] \cap (\sqrt[3]{[y^-, y^+]}) \\ [y^-, y^+] \longleftarrow [y^-, y^+] \cap ([x^-, x^+]^3) \end{cases}$$

$$\text{where } \sqrt[3]{[y^-, y^+]} = [\sqrt[3]{y^-}, \sqrt[3]{y^+}]$$
$$\text{and } \quad [x^-, x^+]^3 = [(x^-)^3, (x^+)^3]$$

More generally, for each *primitive constraint*, there is a projecting procedure allowing to contract each variable domain. For more information about these projection procedure, the reader is invited to consult a textbook, for example [12].

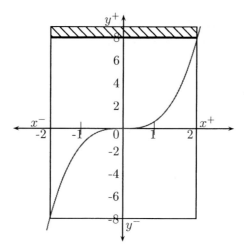

Fig. 1. Sample of the projection of the cubic constraint

2.2 To Reach a Fix Point

For a given set of constraints, an iterative application of the corresponding projection procedures over the constraints will lead to a state where no variable domain can be further reduced. That is, a fix point is reached. Notice that such a fix point doesn't constitute a solution because the variables are not instanced yet. To get a solution, other techniques have to be jointly applied.

In constraint programming, many algorithms for reaching fix points have been developed. We can mention various local (arc) consistency algorithms for discrete problems [13,14], and ICP algorithms for continuous problems [1].

By considering a projection procedure as a local consistency procedure, it can be incorporated in a fix point algorithm. Thus efficient contracting algorithms were developed for the continuous constrained problems.

2.3 Discussion

An ICP algorithm is a polynomial time procedure which reduces the domains of the variables of the problem. This kind of algorithm reduces the search space by removing inconsistent values of the variables. It does not delete any solution of the problem. However, such an application of ICP leads only to an approximation of the solution. In order to increase the precision of the approximation, one may bisect any variable domain and apply ICP to the two different alternatives. If we iterate this process, we increase the precision but we get an exponential number of bisections, and so a huge resolution time. In practice, this approach is not usable for problems with a high number of variables.

In constrained multi-objective problems, we have to satisfy constraints but also to minimize (maximize) some objective functions. So, even with a small number of variables and objectives, the problem cannot be processed simply with a classical bisection algorithm.

3 Population and Interval Constraint Propagation Algorithm (*PICPA*)

The concept of population is very suitable in a multi-objectives context. Indeed, as the Pareto optimal front is most of the time a set of solutions, each individual of the population can hopefully become a particular solution of the Pareto optimal front. As a result, the whole population will be an approximation of the targeted Pareto optimal front.

Many population-based multi-objective optimization algorithms have thus been developed. We can mention, among others, *NSGA* [17], *SPEA* [20], *MOSA* [19], *MOTS* [10], *M-PAES* [3].

In this section, we present the *PICPA* algorithm which combines interval constraint propagation with a population.

3.1 A New Dominance Relation Definition

PICPA uses a new dominance relation: the Point with Set dominance (PS-dominance). Let us recall first the classical dominance relation between two points:

Definition 1 (Dominance). *Let us consider a minimization problem. For any two vectors $\vec{x}$ and $\vec{y}$:*

- $\vec{x}$ *equals* $\vec{y}$, *iff* $\forall i \in \{1, 2, ..., n\} : x_i = y_i$
- $\vec{x}$ *dominates* $\vec{y}$, *iff* $\forall i \in \{1, 2, ..., n\} : x_i \leq y_i$ *and* $\exists j \in \{1, 2, ..., k\} : x_j < y_j$
- $\vec{x}$ *is dominated by* $\vec{y}$, *iff* $\forall i \in \{1, 2, ..., n\} : x_i \geq y_i$ *and* $\exists j \in \{1, 2, ..., n\} : x_j > y_j$
- $\vec{x}$ *is non dominated by* $\vec{y}$, *and* $\vec{y}$ *is non dominated by* $\vec{x}$ *otherwise*

We can now introduce the Point with Set Dominance definition:

Definition 2 (PS-Dominance). *For any vector $\vec{x}$ and any vector set $\{\vec{y}\}$, $|\{\vec{y}\}| > 0$:*

- $\vec{x}$ *PS-equals* $\{\vec{y}\}$, *iff* $\forall \vec{y} \in \{\vec{y}\} : \vec{x}$ *equals* $\vec{y}$
- $\vec{x}$ *PS-dominates* $\{\vec{y}\}$, *iff* $\forall \vec{y} \in \{\vec{y}\} : \vec{x}$ *dominates* $\vec{y}$
- $\vec{x}$ *is PS-dominated by* $\{\vec{y}\}$, *iff* $\forall \vec{y} \in \{\vec{y}\} : \vec{x}$ *is dominated by* $\vec{y}$
- $\vec{x}$ *is PS-non-dominated by* $\{\vec{y}\}$, *otherwise*

The time complexity of the PS-Dominance is in $\mathcal{O}(n \times |\{\vec{y}\}|)$ as we have to test $\vec{x}$ with each element of $\{\vec{y}\}$.

Afterwards, we use $[y]$ to denote any interval of $\mathbb{R}$ and $\overrightarrow{[y]}$ to denote any *interval* vector of $\mathbb{R}^n$. Furthermore, an *interval* vector of $\mathbb{R}^n$ may be also called a box of $\mathbb{R}^n$. So, we can use the PS-Dominance relation between any *decision* vector $\vec{x}$ of $\mathbb{R}^n$ and any box $\overrightarrow{[y]}$ of $\mathbb{R}^n$ (see figure 2). In this case, the time complexity of the PS-Dominance is in $\mathcal{O}(n)$. Indeed, we only need to test the dominance between $\vec{x}$ and the point located at the bottom left corner of $\overrightarrow{[y]}$.

Let us consider a box $\overrightarrow{[y]}$ of $\mathbb{R}^2$ and $\vec{x}$ a point of $\mathbb{R}^2$. In a minimization problem, we may have the examples given at Figure 2.

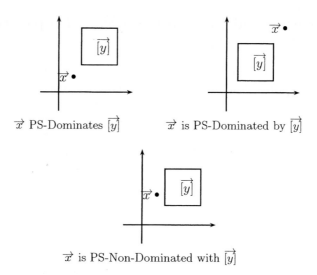

Fig. 2. Samples of some PS-Dominance cases

3.2 Representation of the Search and Objective Space

In most of the population-based algorithms, an individual or configuration of the population is a decision vector, each variable of which being given a single value. Under this representation, each individual corresponds to a particular point in the objective space.

In our approach, each individual is also a vector, but each variable is now represented by an *interval* instead of a single value. Consequently, each individual of the population corresponds now to a *box* $\overrightarrow{[x]}$ of $\mathbb{R}^o$ (o is the number of objectives of the problem) in the objective space.

Consider a problem with two variables x_1, x_2 and two objectives f_1, f_2. A population of three individuals might be represented as in Figure 3.

In this representation, we consider that the whole population gives a sub-paving (i.e. union of non-overlapping boxes) of $\mathbb{R}^o$. As a consequence, the matching boxes of $\mathbb{R}^n$ may overlap (see figure 3).

3.3 Pareto Selection of Boxes

Consider a set of individuals represented as introduced in Section 3.2. We ensure that all the feasible configurations are contained in the sub-paving described by the population. As a consequence, the Pareto optimal front is also enclosed by the population.

In order to remove the individuals which do not contain any solution of the Pareto optimal set, we apply the following Pareto selection procedure:

1. Try to instantiate all the individuals of the population with bounded local search effort.

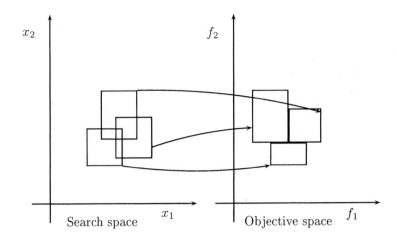

Fig. 3. Sample of the representation with intervals

2. Apply the PS-Dominance relation 3.1 to remove the individuals which are dominated by another instantiated individual.

At this stage, we get a reduced population of individuals, and we ensure that the union of these boxes (or individuals) contains the whole Pareto optimal set.

We give now some elements of the proof: let $\{\vec{f}\}$ be the feasible objective space, $\vec{y}$ a point of the Pareto optimal front and $[\vec{y}]$ a box that contains $\vec{y}$:

$$\text{therefore, } \nexists \vec{x} \in \{\vec{f}\} \text{ and } \vec{x} \text{ dominates } \vec{y}$$
$$\text{so, } \nexists \vec{x} \in \{\vec{f}\} \text{ and } \vec{x} \text{ PS-dominates } [\vec{y}]$$

As a result, $[\vec{y}]$ cannot be PS-Dominated and won't be removed from the population.

Figure 4 shows an example of a Pareto selection process with the PS-Dominance relation. We see that the hatched boxes can be safely removed because they are PS-Dominated by some feasible points of the objective space.

3.4 The *PICPA* Algorithm

PICPA combines an ICP process (see Section 2) with a Pareto interval selection (see Section 3.3) into a single algorithm.

PICPA uses a population of variable size whose maximum is a parameter to be fixed. *PICPA* starts with a single individual $[\vec{x}]$ where each variable x_i is initialized with its interval (value domain). We suppose that each variable is locally consistent. If this is not the case, a first ICP process may be applied to reach a fix point. Clearly this individual corresponds to a box of $\mathbb{R}^o$.

Let $[\vec{f}]$ be this box. Take an objective f_i and bisect its value interval $[f_i]$. Such a bisection triggers two applications of the ICP process to contract variable

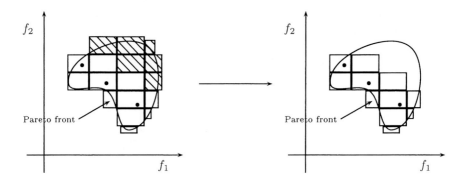

Fig. 4. A Pareto selection sample

intervals on the individual $\overrightarrow{[x]}$, leading to two new individuals $\overrightarrow{[x']}$ and $\overrightarrow{[x'']}$. These individuals replace the parent individual. These two individuals are composed of reduced intervals and correspond thus to two reduced boxes $\overrightarrow{[f']}$ and $\overrightarrow{[f'']}$ of $\mathbb{R}^o$. This "bisection-contraction" process continues until the number of individuals in the population reaches its allowed maximum. Notice that if any variable domain is reduced to the empty set by the ICP process, the underlying individual will be not inserted into the population.

Once the population reached its full size, an instantiation process will try to create a complete feasible instantiation for each $\overrightarrow{[x]}$. That is, a particular value of the interval will be searched for each individual which satisfies the problem constraints. Since this stage may require a huge processing time, PICPA uses a *search effort* parameter as a stop criterion to stop the instantiation process. Thus, after this stage, some individuals will be completely instantiated, leading to a real vector $\overrightarrow{x}$ while others remain uninstantiated. Notice that the result of this instantiation is recorded in a separate data structure and the individuals in the current population will not be altered.

At this point, we execute a Pareto selection mechanism (cf. section 3.3) to eliminate dominated individuals from the population. Since the population is reduced, we start again the above "bisection-contraction" process to extend the population to its maximal size.

The PICPA algorithm stops if one of the following cases occurs:

1. an empty population is encountered, in this case, the problem is proved to be infeasible, consequently has no solution;
2. the Pareto selection process cannot remove any individuals from the population. In this case, the individuals of the population constitute the bounds of the problem. The individuals which are successfully instantiated (recorded in a separate structure) give an approximate solution of the Pareto optimal set.

The skeleton of the PICPA is shown in algorithm 1.

- Initialize the population with a single locally consistent individual
- While $0 < |Population| < MaxPopulationSize$ do
  - While $0 < |Population| < MaxPopulationSize$ do
    1. Select an individual (parent) and bisect it according to one objective, leading to two distinct individuals (children)
    2. Contract the children
    3. Update the population:
       a) Remove the father
       b) Add the locally consistent children

  EndWhile
  - Potential instantiation of each individual
  - PS-Dominance selection process

EndWhile

Algorithm 1: Skeleton of the PICPA algorithm

As we see on algorithm 1, only two parameters (size of population and search effort) are required by PICPA.

PICPA has several advantages compared to other population-based algorithms. Firstly, it requires a small number of parameters. Secondly, it can sometimes answer "No" when the problem is infeasible. Thirdly, it gives in a single run bounds and an approximation of the Pareto optimal front.

3.5 Discussion

PICPA ensures that the Pareto optimal front will be enclosed in the returned population. As PICPA mainly bisects in the objective space, it is not very sensitive to the increase of the number of variables. The originalities of PICPA are:

1. to bound the number of bisections thanks to the population size
2. to bisect mainly in the objectives space
3. to apply a Pareto selection in order to converge to the Pareto optimal front

The computational effort of PICPA can be tuned thanks to the population size parameter. Indeed, larger population sizes lead to higher solution precisions. It is true that increasing the population size will increase the computing time. But this gives us a guaranteed way to get better approximation of the Pareto optimal front.

4 Experimental Results

This section gives experimental results of PICPA on some famous test problems. Given the deterministic nature of PICPA, the quality of solutions of PICPA can be directly assessed with respect to the final bounds found. To show its practical performance however, we contrast the results of PICPA with those of

NSGA-IIc [7][1]. Notice that the version of the NSGA-IIc algorithm used here gives better results than those given in [7]. For these test experiments, the following parameter settings are used:

- for NSGA-IIc, we used the settings given in [8], i.e. simulated binary crossover [6] with $n_c = 20$ and the polynomial mutation operator with $n_m = 20$. A crossover probability of 0.9 and a mutation probability of 0.15 are chosen. The population size and the maximum number of generation were set according to the problem difficulty.
- for PICPA, we set the population size to 1000 and the search effort to 0.2.

Notice that these settings lead to very close computing time for the two algorithms, ranging from some few seconds to about three minutes according to the test problems. For each test problem, NSGA-IIc was run ten times and the best run was taken for our comparisons. As PICPA doesn't use any random value, only one run is required to get the result.

4.1 The Tanaka Test Problem

We first used a test problem introduced by Tanaka [18]:

$$\text{TNK} \begin{cases} \text{Minimize} & f_1(x) = x_1 \\ \text{Minimize} & f_2(x) = x_2 \\ \text{s.t.} & c_1(x) \equiv x_1^2 + x_2^2 - 1 - 0.1\cos(16\arctan(\frac{x_1}{x_2})) \geq 0 \\ & c_2(x) \equiv (x_1 - 0.5)^2 + (x_2 - 0.5)^2 \leq 0.5 \\ \text{and} & x_1, x_2 \in [0..\pi] \end{cases}$$

In this problem, the feasible objective space is the same as the feasible decision variable space. In the TNK experiments, we used a population of size 150 and a maximum generation of 500 for NSGA-IIc. The optimal Pareto set as well as the bounds given by PICPA are presented in figure 5(a). Figure 5(b) shows the return set given by PICPA and NSGA-IIc.

From these figures, we notice that 1) the bounds of PICPA match almost perfectly the Pareto optimal front and 2) PICPA and NSGA-IIc give very similar solutions.

4.2 The Osyczka and Kundu Test Problem

For our second experiment, we chose a six variables test problem presented by Osyczka and Kundu [16]:

---

[1] downloadable at: http://www.iitk.ac.in/kangal/soft.htm

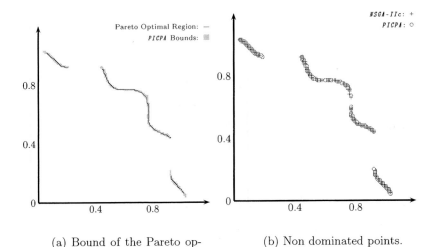

(a) Bound of the Pareto optimal front.

(b) Non dominated points.

Fig. 5. Simulation results on TNK

$$\text{OSY} \begin{cases} \text{Minimize} & f_1(x) = -(25(x_1 - 2)^2 + (x_2 - 2)^2 + (x_3 - 1)^2 + (x_4 - 4)^2 + \\ & \quad (x_5 - 1)^2) \\ \text{Minimize} & f_2(x) = x_1^2 + x_2^2 + x_3^2 + x_4^2 + x_5^2 + x_6^2 \\ \text{s.t.} & c_1(x) \equiv x_1 + x_2 - 2 \geq 0 \\ & c_2(x) \equiv 6 - x_1 - x_2 \geq 0 \\ & c_3(x) \equiv 2 - x_2 + x_1 \geq 0 \\ & c_4(x) \equiv 2 - x_1 + 3x_2 \geq 0 \\ & c_5(x) \equiv 4 - (x_3 - 3)^2 - x_4 \geq 0 \\ & c_6(x) \equiv (x_5 - 3)^2 + x_6 - 4 \geq 0 \\ \text{and} & x_1, x_2, x_6 \in [0..10] \\ & x_3, x5, \in [1..5] \\ & x_4, \in [0..6] \end{cases}$$

Like in the TNK experiment (see Section 4.1), we set a population size of 150 and a maximum generation of 500 for *NSGA-IIc*.

From figure 6(a), we observe that the bounds of *PICPA* are globally very close to the optimal front, with exceptions for some areas. These larger areas are due to the great percent of additive constraints. Figure 6(b) shows that the quality of the two non dominated sets found by *PICPA* and *NSGA-IIc* are very close, even if some points found by *NSGA-IIc* dominate some of *PICPA*. But we notice that on this problem, different runs of *NSGA-IIc* give mixed results.

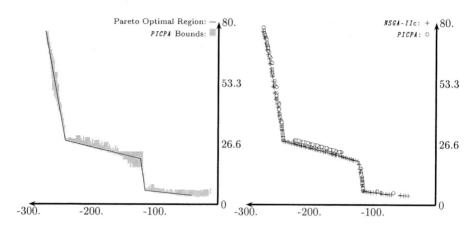

(a) Bound of the Pareto optimal front.

(b) Non dominated points.

Fig. 6. Simulation results on OSY

4.3 Constrained Test Problems (CTP)

For the last experiments, we used two test problems presented in [8] and [5]. As these problems are tunable, we use a functional $g()$ which is a Rastrigin's function-base:

$$
\text{CTP} \begin{cases}
\text{Minimize} & f_1(x) = x_1 \\
\text{Minimize} & f_2(x) = g(x) - f_1(x) \\
\text{s.t.} & g(x) = 41 + x_2^2 - 10\cos(4\pi x_2) + x_3^2 - 10\cos(4\pi x_3) + \\
& \quad x_4^2 - 10\cos(4\pi x_4) + x_5^2 - 10\cos(4\pi x_5) \\
& C_1(x) \equiv \cos(\theta)(f_2(x) - e) - \sin(\theta)f_1(x) \geq \\
& \quad a|\sin(b\pi(\sin(\theta)(f_2(x) - e) + \cos(\theta)f_1(x))^c)|^d \\
\text{and} & x_1 \in [0..1] \\
& x_{i,i>1} \in [-5..5]
\end{cases}
$$

For the following experiments, we set a population of size 300 and a maximum generation of 1000 for *NSGA-IIc*. The *PICPA* parameters remain the same as in the previous sections.

CTP7. The parameter values used to get CTP7 are as follows:

$$\theta = -0.05\pi, \quad a = 40, \quad b = 5, \quad c = 1, \quad d = 6, \quad e = 0$$

The feasible search space, the corresponding disconnected Pareto optimal regions and the *PICPA* bounds are shown in figure 7(a).

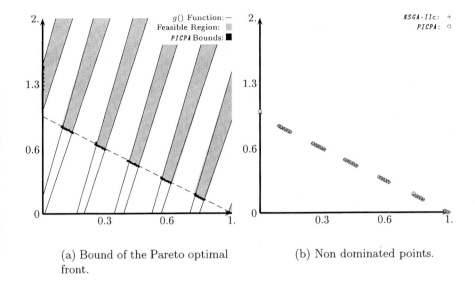

(a) Bound of the Pareto optimal front.

(b) Non dominated points.

Fig. 7. Simulation results on CTP7

Figure 7(a) shows the bounds found by `PICPA`, which are very tight with respect to the Pareto optimal front. On this more difficult test problem, the contraction procedures used by `PICPA` proved to be very useful. In figure 7(b), we see that `PICPA` and `NSGA-IIc` have comparable results.

CTP8. CTP8 is composed of many disconnected feasible regions. In CTP8, unlike CTP7, we have two constraints. Here are the parameters used to generate these constraints:

$$C_1 : \theta = 0.1\pi, \quad a = 40, \quad b = 0.5, \quad c = 1, \quad d = 2, \quad e = -2$$

$$C_2 : \theta = -0.05\pi, \quad a = 40, \quad b = 2, \quad c = 1, \quad d = 6, \quad e = 0$$

Figure 8(a) shows the feasible search space, the corresponding disconnected Pareto optimal regions and the `PICPA` bounds. Figure 8(b) shows the solution sets by `NSGA-IIc` and `PICPA`.

From figure 8(a), we see once again that the `PICPA` bounds are very precise. For this problem, which is the most difficult problem tested here, `NSGA-IIc` and `PICPA` found some solutions on the most of disconnected Pareto optimal regions.

5 Conclusions

In this paper, we presented `PICPA`, a new algorithm to solve continuous constrained multi-objective problems. `PICPA` combines interval constraint propagation with evolutionary concepts (population and selection). This algorithm

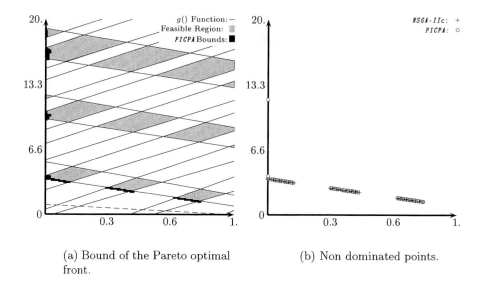

Fig. 8. Simulation results on CTP8

has the desirable property of bounding the Pareto optimal front. Experimental evaluation of *PICPA* on some well-known test problems show its practical efficiency to find high quality approximation solutions and very tight bounds of the Pareto optimal front. Also, a new dominance relation called PS-Dominance was proposed which allows to compare a point to a set of points. We think this work fills in a gap of existing population-based methods. *PICPA* strongly depends on the efficiency of the projection and on instantiation procedures. Currently, we are investigating these two issues.

Acknowledgments. We would like to thank the referees of the paper for their useful comments.

References

1. F. Benhamou, F. Goualard, L. Granvilliers, and J.F. Puget. Revising hull and box consistency. In *Proceedings of the International Conference on Logic Programming*, pages 230–244, 1999.
2. J.G. Cleary. Logical arithmetic. *Future Computing Systems*, 2(2):125–149, 1987.
3. D.W. Corne and J.D. Knowles. M-paes: a memetic algorithm for multiobjective optimization. In *Proceedings of the 2000 Congres on Evolutionary Computation*, pages 325–332, 2000.
4. E. Davis. Constraint propagation with interval labels. *Artificial Intelligence*, 32(3):281–331, 1987.
5. K. Deb. *Multi-objective optimization using evolutionary algorithms*. John Wiley, 2001.

6. K. Deb and R.B. Agrawal. Simulated binary crossover for continuous search space. *Complex Systems*, 9:115–148, 1995.
7. K. Deb and T. Goel. Controlled elitist non-dominated sorting genetic algorithms for better convergence. In *Proceedings of Evolutionary Multi-Criterion Optimization*, pages 67–81, 2001.
8. K. Deb, A. Pratap, and T. Meyarivan. Constrained test problems for multi-objective evolutionary optimization. In *Proceedings of Evolutionary Multi-Criterion Optimization*, pages 284–298, 2001.
9. C.M. Fonseca and P.J. Fleming. Genetic algorithms for multi-objective optimization: Formulation, discussion and generalization. In *Proceedings of The Fifth International Conference on Genetic Algorithms*, pages 416–423, 1993.
10. X. Gandibleux, N. Mezdaoui, and A. Freville. A multiobjective tabu search procedure to solve combinatorial optimization problems. In *Lecture Notes in Economics and Mathematical Systems*, volume 455, pages 291–300. Springer, 1997.
11. D.E. Goldberg. Genetic algorithms for search, optimization, and machine learning. *Reading, MA: Addison-Wesley*, 1989.
12. L. Jaulin, M. Kieffer, O. Didrit, and E. Walter. *Applied Interval Analysis, with Examples in Parameter and State Estimation, Robust Control and Robotics*. Springer-Verlag, London, 2001.
13. A.K. Mackworth. Consistency in networks of relations. *Artificial Intelligence*, 8:99–118, 1977.
14. R. Mohr and T.C. Henderson. Arc and path consistency revisited. *Artificial Intelligence*, 28:225–233, 1986.
15. R.E. Moore. Methods and applications of interval analysis. *SIAM, Philadelphia, PA*, 1979.
16. A. Osyczka and S. Kundu. A new method to solve generalized multicriteria optimization problems using the simple genetic algorithm. *Structural Optimization*, 10:94–99, 1995.
17. N. Srinivas and K. Deb. Multiobjective optimization using non dominated sorting in genetic algorithms. *Evolutionary Computation*, 2(3):221–248, 1994.
18. M. Tanaka. Ga-based decision support system for multi-criteria optimization. In *Proceedings of the International Conference on Systems, Man and Cybernetics-2*, pages 1556–1561, 1995.
19. E.L. Ulungu, J. Teghem, Ph. Fortemps, and D. Tuyttens. Mosa method: a tool for solving multiobjective combinatorial optimization problems. *Journal of Multi-Criteria Decision Analysis*, 8:221–336, 1999.
20. E. Zitzler and L. Thiele. Multiobjective evolutionary algorithms: a comparative case study and the strength pareto approach. *IEEE Transactions on Evolutionary Computation*, 3:257–271, 1999.

Multi-objective Binary Search Optimisation

Evan J. Hughes

Department of Aerospace, Power, and Sensors,
Cranfield University, Royal Military College of Science,
Shrivenham, Swindon, SN6 8LA, UK,
ejhughes@iee.org,
http://www.rmcs.cranfield.ac.uk/radar

Abstract. In complex engineering problems, often the objective functions can be very slow to evaluate. This paper introduces a new algorithm that aims to provide controllable exploration and exploitation of the decision space with a very limited number of function evaluations. The paper compares the performance of the algorithm to a typical evolutionary approach.

1 Introduction

Multi-Objective Evolutionary Algorithms [1] are becoming a well established technique for solving hard engineering problems. Like all optimisation algorithms they were first developed when processing resources and memory were scarce. As the use of optimisation algorithms migrates deeper into industry, and with more processing power available, the scale and characteristics of the problems being solved are changing. The objective functions are becoming more complex and consequently can take a long time to evaluate.

Problems such as aerodynamic optimisation and electromagnetic simulation often rely on finite element methods in order to simulate the systems of interest. These simulations can take from seconds to hours to run. The better the resolution and fidelity required, the longer the simulation time.

Many of the problems are highly multi-modal and so gradient based optimisers do not perform well. Unfortunately, with long simulation times, either small population sizes must be used within an evolutionary algorithm, or the algorithm must be run for a reduced number of generations in order to keep the total processing time within reasonable bounds. In both gradient based searches and evolutionary algorithms, every iteration or generation, the results of previous objective calculations are discarded to reduce processing and storage costs.

For example, a simple aerodynamic simulation takes 10 variables, produces two objectives and one constraint, and requires 1 hour per evaluation. An evolutionary algorithm with a population of 20 for 100 generations would take over 83 days to complete. With some multi-objective algorithms (e.g. NSGA and MOGA) only 20 points on the Pareto surface could be generated per generation. Algorithms such as SPEA with an external store for the Pareto set could retain more points. A Pareto surface would need to be generated from every point that was evaluated in the run of the evolutionary algorithm to make sure that no points had been lost. To confine the problem to a more realistic time scale, only about 500 points could be generated, taking nearly 21 days to complete.

Generating only 500 evaluations is equivalent to a population of 20 for 25 generations only, not much for many evolutionary multi-objective optimisation algorithms.

This paper proposes a new algorithm that is designed specifically to provide controllable exploration and exploitation, but with few objective calculations. The new algorithm uses many of the techniques developed for multi-objective evolutionary algorithms to guide the search process. The algorithm is not generational however and utilises all the objective calculations made when deciding where to place the next point in the hypercube that defines the search space.

This paper describes two approaches to implementing the idealised algorithm. The first uses Voronoi decomposition to locate the exact centre of the largest empty hypersphere in each case, but is computationally expensive for even moderate numbers of variables. The second algorithm is the prime focus of the paper and uses binary space subdivision to approximate the unexplored regions, producing a faster and more scalable algorithm.

Section 2 describes the ideal search algorithm, section 3 discusses multi-objective optimisation, section 4 details one approach to implementing the ideal algorithm using Voronoi diagrams and section 5 details an alternative implementation using a binary search. Section 6 describes the two multi-objective test functions used, section 7 presents results of the optimisation trials and a comparison with a typical evolutionary approach, and section 8 concludes.

2 The Ideal Algorithm

2.1 Introduction

The idealised algorithm is:

1. **Exploration:** Next point is the centre of the largest empty convex region.
2. **Exploitation:** Next point is the centre of the largest empty convex region that has a selected *good* point at one edge.

The aim of the idealised algorithm is to reduce the size of unexplored regions, resulting in uniform search coverage, while still being able to focus on the areas forming the Pareto surface. With only a limited number of function evaluations available, every evaluation must count.

2.2 Exploration

The exploration search step of the algorithm identifies the most unexplored region of the search hypercube, and places the next point at the centre of the region. The region could be described in a number of ways, the ideal being to find the largest convex region that will reside between the existing evaluation points. Sections 4 & 5 describe two alternative methods for approximating the most unexplored region.

2.3 Exploitation

The exploitation step involves first identifying a good point. In both algorithm implementations presented in this paper, tournament selection is used to identify a *good* point for a localised search to begin from. Once a point has been selected, the largest unexplored volume that contains the point at its edge is identified, and a new evaluation generated for the point corresponding to the centre of the volume.

2.4 Exploration versus Exploitation

The two phases of the algorithm, exploration and exploitation, must be controlled in order to provide effective coverage of the decision space. The algorithm must begin with an exploration phase to allow interesting regions to be identified, then the exploitation phase can be applied to refine the regions. As noted in most evolutionary algorithms, it is wise to always have a low level of exploration, even in the exploitation phase.

In evolutionary algorithms, the initial population provides pure exploration. The selective pressure and crossover in subsequent generations provide exploitation, with a low level mutation providing exploration of the decision space throughout the remaining optimisation process.

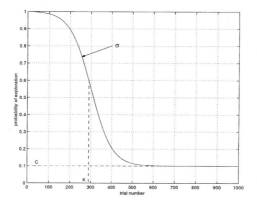

Fig. 1. Probability of performing exploration for C=0.1, K=0.3, $\sigma = 0.1$ and 1000 trials

In both the algorithms presented, at each iteration the decision of whether to perform exploration or exploitation is made based on a probability distribution, $P(n)$, that varies with the number of function evaluations, n. The distribution is detailed in (1) and illustrated graphically in Fig.1. In (1), C is the minimum probability of performing an exploration step, σ is the rate at which the probability of exploration decays (smaller σ gives faster decay), K is the mid point of the decay (midpoint of the range $[C,1]$) and n_x is the maximum number of trials that are to be performed. Note, in Fig.1, C is not zero and so the point marked for K does not occur at 300 as would be the case if $C = 0$.

$$P(n) = (C-1)\frac{\tanh\left(\frac{n/n_x - K}{\sigma}\right) - \tanh\left(\frac{-k}{\sigma}\right)}{\tanh\left(\frac{1-k}{\sigma}\right) - \tanh\left(\frac{-k}{\sigma}\right)} + 1 \qquad (1)$$

3 Multi-objective Optimisation

Much research has been performed on discriminating between members of a Pareto set to allow the entire Pareto set to be approximated in a single run of an evolutionary algorithm. In an evolutionary algorithm, the Pareto set may either be maintained within the base population (NSGA, MOGA, etc.) or externally (SPEA, PAES, etc.)[1]. For the methods that use the population to store the set, a large population size is required to provide sufficient sampling of the Pareto surface.

A Pareto ranking method is required to allow 'good' solutions to be chosen. In the algorithms described in this paper, with the small number of evaluations used, **all** the points generated in the run of the optimiser will be used to create the final Pareto surface. Methods that do not maintain an external store of the Pareto set will be used in the examples as at each iteration of the algorithm, all the points that satisfy the constraints created so far will be accounted for. If larger numbers of evaluations are to be performed, only the solutions in the current tournament set need to be ranked. This approach will lead to reduced performance, but the processing overhead of the algorithms will scale better with increasing function evaluations.

4 Voronoi Optimisation Algorithm

4.1 Largest Empty Convex Region

The idealised algorithm in section 2 relies on being able to identify the largest empty convex region either in the entire search space, or with a chosen point at its edge. Figure 2 shows a 2D Euclidean plane representation of the largest empty convex region between a set of points in the decision space. The region may be approximated by finding the largest empty hypersphere that can be placed between the existing points. The new point would then be generated at the centre of the hypersphere. Finding the centre of the largest empty hypersphere is still not a trivial problem to solve.

4.2 Voronoi Diagrams

The *Voronoi diagram* [2,3] can be used to identify the centre of the largest empty hypersphere. A typical Voronoi Diagram is shown in Fig. 3 with the largest empty circle indicated. The centre of the largest empty circle will always coincide with either a Voronoi vertex, or a vertex generated by the intersection of the Voronoi diagram with the convex hull of the set of points.

The Voronoi diagram divides a hyperspace containing points into regions, each region surrounding a single point. The space is divided so each point is associated with the region of space closest to it. If $P = p_1, p_2, \ldots, p_n$ is a set of points (or *sites*) in the 2D Euclidean

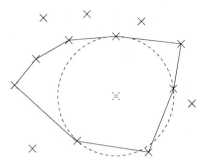

Fig. 2. Largest empty hypersphere

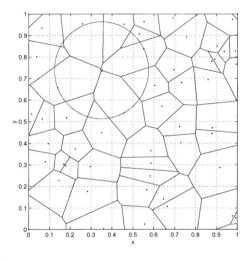

Fig. 3. Example Voronoi diagram showing how centre of largest empty hypersphere lies at a Voronoi vertex

plane, the plane is partitioned by assigning every point in the plane to its nearest site. All those points assigned to p_i form the Voronoi region $V(p_i)$.

$$V(p_i) = \{x : |p_i - x| \leq |p_j - x|, \forall j \neq i\} \qquad (2)$$

The Voronoi diagram is formed as the boundaries of the set of Voronoi regions. The Voronoi edges are points on the plane that lie on the boundaries of the Voronoi regions and will be by definition equidistant from two sites. A Voronoi vertex is formed at the junction of multiple Voronoi edges. The generation of Voronoi diagrams is computationally expensive and so direct use is only really possible for problems with low-dimensionality. Indirect calculation of the Voronoi diagram is still slow but can lead to useful optimisation systems [4].

In all the optimisation algorithms used in this paper, the full ranges of the parameter optimisation in each dimension are mapped into a hypercube with each side having the limits of [0,1]. This mapping scales each of the parameters with respect to their minimum and maximum values, allowing unbiased Euclidean distances to be calculated.

To simplify the processing for finding the largest empty hypersphere, a point is placed at each corner of the hypercube in the decision space, simplifying the calculation of the intersection of the Voronoi diagram with the convex hull of the points. The next point is then placed uniformly at random within the hypercube, allowing the Voronoi diagram to be generated and the optimisation process to begin. With a 10 dimensional problem, the hypercube has 1024 corners, therefore 1025 points would be required in the initial sampling of the decision space. In practice, for many engineering problems that are to be optimised on a single processor, the direct Voronoi approach is limited to problems with less than 10 dimensions due to a rapid expansion of computational complexity with increasing dimensionality. With multiple processors, the objective calculations and calculation of the Voronoi diagram can be 'farmed' out to the next free processor, or a slightly sub-optimal approach can be used of generating a small set of points, one for each processor, from each Voronoi diagram.

5 Binary Search Algorithm

Although the Voronoi approach gives a very neat and near optimal algorithm, the computational complexity for high dimensionality problems is immense. An alternative strategy has been developed that uses a binary search tree [5] to divide the search space into empty regions, allowing the largest empty region to be approximated. The search tree is constructed as shown in Fig. 4 by generating a point within the chosen hypercube, then dividing the hypercube along the dimension that yields the most 'cube-like' subspaces. The definition of 'cube-like' is the split that minimises (3), where $d_{\max}$ is the maximum side length of the sides of the two subspaces, and correspondingly $d_{\min}$ is the overall shortest side length.

$$C = \frac{d_{\max}}{d_{\min}} \qquad (3)$$

The ideal algorithm in section 2 has been modified for the binary search tree thus:

1. **Exploration:** Next point is generated within the largest empty region,
2. **Exploitation:** Next point is generated within the largest empty region that is within a small distance of a selected *good* point.

When the region with the largest area is identified for exploitation, a point is generated at random that lies within the bounds of the region. For this paper, the point is generated with a normal distribution about the mean of the region axes in each dimension. The normal distribution is scaled so that each dimension of the region is ±4 standard deviations wide.

The identification of a local region for exploitation is illustrated in Fig. 4. A small offset distance d_p is used to generate a hypercube of interest about the chosen point (chosen with tournament selection). The small hypercube is placed around the point of

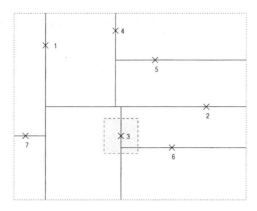

Fig. 4. Binary search process

interest simply to provide an efficient means of identifying neighbouring regions. A new point is then generated at random using a normal distribution in the largest region that intersects the hypercube. For a region A to intersect the hypercube about the point P, (4) &(5) must not be satisfied for any dimension i, where A_{i_L} and A_{i_H} are the lower and upper extents of region A in dimension i.

$$A_{i_L} > P_i + d_p \qquad (4)$$
$$A_{i_H} < P_i - d_p \qquad (5)$$

As the tree is traversed, the subspaces of the higher order nodes (say upper subspace of node 2, marked with a cross in Fig. 4 for example) are also tested. If the subspace fails, all the child nodes of the subspace can be ignored.

At each iteration, the tree search to find the largest empty region is at worst $O(mn)$, where n is the number of evaluation points so far and m is the number of dimensions. The tree pruning can lead to $O(mlog(n))$ performance for exploitation, and at worst $O(mn)$. Thus the computational explosion associated with the Voronoi approach is avoided. Overall, the computational complexity for the binary search is at worst $O(mn_x{}^4)$ if a Pareto ranking method of $O(n^2)$ is used to rank all the points. If only the members of the tournament are ranked, the computational complexity will be between $O(mlog(n)^2)$ and $O(mn^2)$, depending on how much exploitation is performed.

6 Test Functions

For the trials in this paper, two multi-objective test functions have been used. The test functions both have concave Pareto sets, with one also being discontinuous. Both test functions are detailed in [6] and are given in (6) and (7). It should be noted that for (7), the decision space and objective space are co-located, with the Pareto set being defined by the intersection of two constraint boundaries. Thus points can appear in the objective space on the lower side of the Pareto set, but these points violate the constraints and as such are not acceptable.

$$\mathcal{O}_1 = 1 - \exp\left(-\sum_{i=1}^{n}\left(x_i - \frac{1}{\sqrt{n}}\right)^2\right)$$
$$\mathcal{O}_2 = 1 - \exp\left(-\sum_{i=1}^{n}\left(x_i + \frac{1}{\sqrt{n}}\right)^2\right)$$
$$-2 \leq x_i \leq 2 \tag{6}$$

$$\mathcal{O}_1 = x$$
$$\mathcal{O}_2 = y$$
$$0 \geq -(x)^2 - (y)^2 + 1 + 0.1\cos\left(16\arctan\left(\frac{x}{y}\right)\right)$$
$$0.5 \geq (x-0.5)^2 + (y-0.5)^2$$
$$0 \leq x, y \leq 1 \tag{7}$$

7 Experimental Results

Both of the optimisation algorithms described in the paper have been trialed on the two test functions. For comparison, a typical multi-objective evolutionary strategy has been used, and constrained to generate the same number of sample points. In all three algorithms, NSGA [1] has been used at each iteration (and correspondingly generation for the evolutionary strategy) to perform the Pareto ranking of the objective values generated so far that meet all the constraints. Only the Pareto ranking elements of NSGA are required for the Voronoi and binary algorithms. The same $\sigma_{\text{share}} = 0.05$ has been used for consistency. The ranking algorithm of NSGA is well suited to the Voronoi and binary search techniques as the Pareto set is held within the 'population' being ranked. Alternative methods based on SPEA, PAES [1] are less well suited to this application as they rely on holding the Pareto set externally. However, although SPEA and PAES have been demonstrated to out perform NSGA on many optimisation problems when used in an evolutionary algorithm, they still search with objective calculation distributions very similar to NSGA. Thus in this trial, only the point generation process is different for each of the algorithms, demonstrating how the point distribution is controlled in the new algorithms to give a much wider search with fewer unexplored regions.

The evolutionary strategy had a population of 20 and ran for 24 generations, giving a total of 500 evaluations (including the initial population). A crossover rate of 0.7 was used with real-valued intermediate crossover. The initial standard deviation of the mutations was set to one eighth of the range of each parameter, and then allowed to adapt during the run.

For the Voronoi and binary search optimisation, a total of 500 points were generated, with the exploration and exploitation parameters set to $K = 0.04, C = 0.02 \ \& \ \sigma = 0.1$ and a tournament size of 10. For the binary search optimisation, the local exploitation hypercube was set to be ± 0.02 in each normalised dimension.

Figures 5 & 6 show the objective and decision space for (6), the first test function and Voronoi optimisation. In Fig. 6 the regular pattern outside of the region of the optima is very clear, demonstrating well controlled and uniform exploration. The tight clustering around the centre of the plot corresponds to the Pareto set and the exploitation phase of the algorithm.

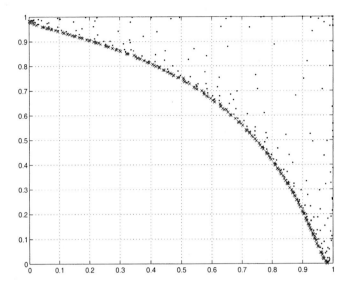

Fig. 5. Objective space for equation 6 and Voronoi Optimisation

Figures 7 & 8 show the objective and decision space for (7), the second test function and for Voronoi optimisation. Again, the regular pattern, created during the exploration phases, outside of the region of the optima is very clear in Fig. 8. The tight clustering in the lower left corner of the plot corresponds to the Pareto set and the exploitation phase of the algorithm. The non-convex and discontinuous Pareto set is clear in the plot. The light coloured points also indicate the region of the plot which is constrained. It is interesting to note that the boundary between the constrained and feasible regions that form the Pareto set has been explored in detail, providing an accurate approximation to the Pareto set.

Figures 9 & 10 show the objective and decision space for (6) and binary search optimisation. The regular pattern outside of the region of the optima is not quite as clear as with the Voronoi approach, but is still well defined and demonstrates a controlled and uniform exploration. Again we see a tight clustering around the centre of the plot which corresponds to the Pareto set and the exploitation phase of the algorithm.

Figures 11 & 12 show the objective and decision space for (7) and binary search optimisation. Again, the exploration pattern outside of the region of the optima is clear with most of the unexplored regions being of a similar size. The tight clustering in the lower left corner of the plot again corresponds to the Pareto set and the exploitation phase

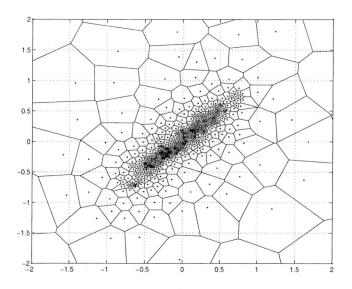

Fig. 6. Decision space for equation 6 and Voronoi Optimisation

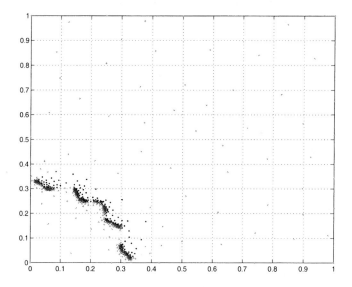

Fig. 7. Decision and Objective space for equation 7 and Voronoi Optimisation

of the algorithm. The non-convex and discontinuous Pareto set is clear in the plot. The light coloured points indicating the region of the plot which is constrained also show that the regions that form the Pareto set have been explored in detail.

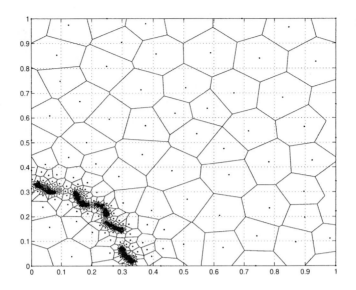

Fig. 8. Decision and Objective space for equation 7 and Voronoi Optimisation

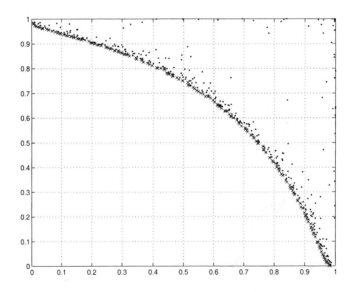

Fig. 9. Objective space for equation 6 and Binary Optimisation

Figures 13 & 14 show the objective and decision space for (6) and optimisation using a multi-objective evolutionary strategy. There is no regular pattern outside of the region of the optima and there are some large unexplored areas near the top of the plot. This demonstrates that the exploration is not as controlled or as uniform with the low number

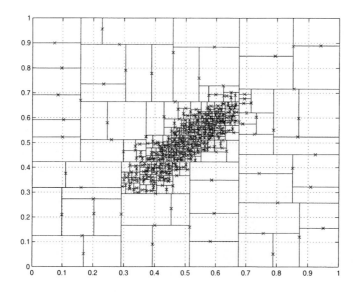

Fig. 10. Decision space for equation 6 and Binary Optimisation

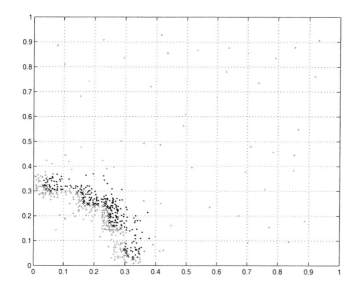

Fig. 11. Decision and Objective space for equation 7 and Binary Optimisation

of sample points used in the experiment. Again we see a clustering around the centre of the plot which corresponds to the Pareto set and the exploitation phase of the algorithm, but here the cluster is very loose and ill-defined.

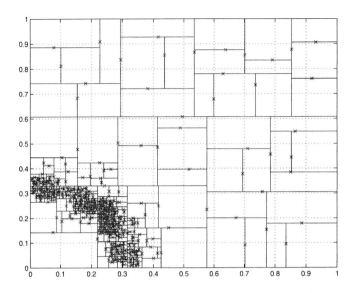

Fig. 12. Decision and Objective space for equation 7 and Binary Optimisation

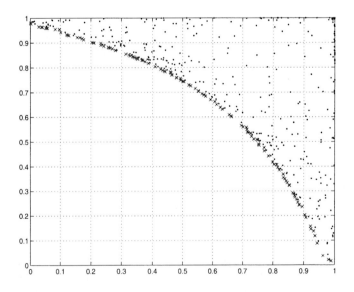

Fig. 13. Objective space for equation 6 and Evolutionary Strategy

Figure 15 shows the objective / decision space for (7) optimised with the evolutionary strategy. The exploration pattern away from the Pareto set is clear, but has large unexplored regions. The Pareto set is patchy and the constrained region in the lower left corner has been searched more heavily than the other constrained regions, but not in a

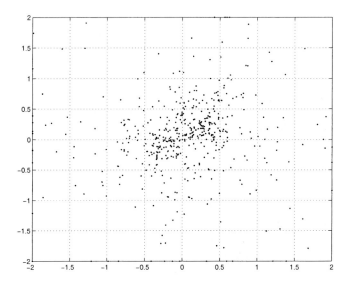

Fig. 14. Decision space for equation 6 and Evolutionary Strategy

controlled way near the Pareto set. This bias in part due to the performance of NSGA. The location of the Pareto set is not as accurate and well defined as with either the Voronoi or binary search optimisation algorithms.

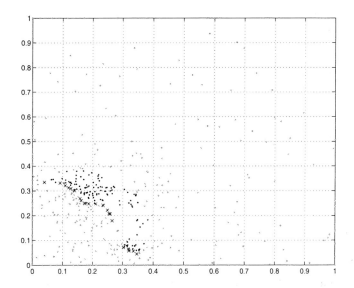

Fig. 15. Decision and Objective space for equation 7 and Evolutionary Strategy

Table 1. Number of distinct points on Pareto set over 100 trials

Algorithm	Function	Minimum	Mean	Maximum	σ
Voronoi	(6)	170	194.9	221	10.5
Binary	(6)	147	170.0	174	6.7
ES	(6)	74	98.4	122	11.3
Voronoi	(7)	56	65.9	75	3.6
Binary	(7)	13	28.8	37	4.2
ES	(7)	8	17.2	27	3.9

To quantify the ability of the algorithms to identify points on the Pareto set, 100 trials of each algorithm for each objective were performed and the number of distinct points on the final non-dominated set recorded. Table 1 shows that the Voronoi optimisation algorithm is by far the most effective at identifying non-dominated points, with the binary search algorithm a close second. Experiments have also shown that the algorithms are well suited to problems where up to $10,000$ points or more are to be generated.

8 Conclusions

This paper has introduced the concept of a new idealised search algorithm and two methods for approximating it, for problems where very few objective calculations can be performed. The results have shown that both the Voronoi and binary search approach allow full, independent control over the exploration and exploitation phases of the search. The exploration phase is designed to give uniform coverage of the search volume by targeting unexplored areas directly. The exploitation phase gives uniform coverage in 'good' areas by targeting unexplored regions close to points that perform well. The algorithms exploit multi-objective techniques developed for evolutionary algorithms and can handle multiple objectives and constraints easily for problems with multiple variables.

The research has also shown that for problems with very few parameters, the Voronoi search algorithm performs the best when compared to the binary search optimisation and a typical evolutionary solution. The computational complexity increases rapidly though for the Voronoi approach when the number of variables is increased. The binary approach scales linearly with an increasing number of dimensions, allowing large problems to be tackled. The research demonstrates how the two new algorithms exploit the information from all the evaluations performed to give much more structure to the location of trial points when compared to a typical evolutionary approach.

Acknowledgements. The author would like to acknowledge the use of the Department of Aerospace, Power, and Sensors DEC Alpha Beowulf cluster for the production of the results.

References

1. Kalyanmoy Deb. *Multi-objective optimization using evolutionary algorithms*. John Wiley & Sons, 2001. ISBN 0-471-87339-X.
2. Joseph O'Rourke. *Computational Geometry in C*. Cambridge University Press, 1993. ISBN 0-521-44592-2.
3. Franz Aurenhammer. Voronoi diagrams – a survey of a fundamental geometric data structure. *ACM Comput. Surveys*, 23:345–405, 1991.
4. Malcolm Sambridge. Geophysical inversion with a neighbourhood algorithm – I. Searching a parameter space. *International Journal of Geophysics*, 138:479–494, 1999.
5. Mark Allen Weiss. *Algorithms, data structures, and problem solving with C++*. Addison-Wesley Publishing Company, Inc., 1996. ISBN 0-8053-1666-3.
6. David A. Van Veldhuizen and Gary B. Lamont. Multiobjective evolutionary algorithm research: A history and analysis. Technical Report TR-98-03, Air Force Institute of Technology, 1 Dec 1998.

Covering Pareto Sets by Multilevel Evolutionary Subdivision Techniques

Oliver Schütze[1], Sanaz Mostaghim[2], Michael Dellnitz[1], and Jürgen Teich[2]

[1] Department of Computer Science, Electrical Engineering and Mathematics
http://math-www.uni-paderborn.de/~agdellnitz
[2] Department of Computer Science, Electrical Engineering and Mathematics
University of Paderborn, Paderborn, Germany
http://www-date.uni-paderborn.de

Abstract. We present new hierarchical set oriented methods for the numerical solution of multi-objective optimization problems. These methods are based on a generation of collections of subdomains (boxes) in parameter space which cover the entire set of Pareto points. In the course of the subdivision procedure these coverings get tighter until a desired granularity of the covering is reached. For the evaluation of these boxes we make use of evolutionary algorithms. We propose two particular strategies and discuss combinations of those which lead to a better algorithmic performance. Finally we illustrate the efficiency of our methods by several examples.

1 Introduction

In the optimization of technical devices or economical processes one frequently has the goal to minimize simultaneously several conflicting objectives. Such objectives can be, for instance, cost and energy. Thus, in applications one is typically confronted with *multi-objective optimization problems* (MOPs) and one has to find the set of optimal trade-offs, the so-called *Pareto set*.

In this paper we propose several new methods for the numerical computation of Pareto sets of MOPs. Similar to [7,4] we use multilevel subdivision techniques for the solution of these problems. However, in contrast to this previous work we now combine the hierarchical search with multi-objective evolutionary algorithms (MOEAs) which are used for the evaluation of subdomains of parameter space. In this way the robustness of our algorithms is significantly increased, and this is of particular interest for higher dimensional problems. Simultaneously the number of function calls can be reduced.
In a second step we discuss combinations of these algorithms in order to improve the total performance. We also describe in which way these methods can be coupled with "standard" MOEAs for the solution of MOPs. In the final section we illustrate the efficiency of these algorithms by a couple of examples.
An outline of the paper is as follows: in Section 2 we summarize the necessary background for the algorithms which are described in Section 3. The computational results are presented in Section 4. We conclude the paper with a summary in Section 5.

2 Background

In this section we briefly summarize the background for the algorithms which are described in Section 3.

2.1 Multi-objective Optimization

In an MOP several objective functions are to be minimized or maximized at the same time. In the following, we state the MOP in its general form:

minimize $(f_1(\boldsymbol{x}), f_2(\boldsymbol{x}), \cdots, f_m(\boldsymbol{x}))$
subject to $\boldsymbol{x} \in Q$.

Here $f_i : \mathbb{R}^n \to \mathbb{R}$, $i = 1, 2, \ldots, m$, are conflicting objective functions that we want to minimize simultaneously. The *decision vectors* $\boldsymbol{x} = (x_1, x_2, \cdots, x_n)^T$ belong to the feasible region $Q \subset \mathbb{R}^n$. We assume Q to be compact and that its boundary can be defined by constraint functions.
We denote the image of the feasible region by $Z \subset \mathbb{R}^m$ and call it the *feasible objective region*. That is,

$$Z = \{F(\boldsymbol{x}) = (f_1(\boldsymbol{x}), \ldots, f_m(\boldsymbol{x})) : \boldsymbol{x} \in Q\}.$$

The elements of Z are called *objective vectors*.

A decision vector $\boldsymbol{x}_1 \in Q$ is said to *dominate* a decision vector $\boldsymbol{x}_2 \in Q$ if the following two conditions are satisfied:

(i) $f_i(\boldsymbol{x}_1) \leq f_i(\boldsymbol{x}_2)$ for all $i = 1, \cdots, m$, and
(ii) $f_j(\boldsymbol{x}_1) < f_j(\boldsymbol{x}_2)$ for at least one $j = 1, \cdots, m$.

A decision vector $\boldsymbol{x}_1 \in Q$ is called *Pareto-optimal* (relative to Q) if there is no $\boldsymbol{x}_2 \in Q$ dominating $\boldsymbol{x}_1$. Finally, an objective vector is also called *Pareto-optimal* if a corresponding decision vector is Pareto-optimal.

2.2 The Subdivision Algorithm

We now briefly introduce set oriented hierarchical subdivision techniques. For the details we refer to [6],[5] and [7]. The common aim of these algorithms is the computation of invariant sets of dynamical systems of the form

$$\boldsymbol{x}_{j+1} = g(\boldsymbol{x}_j), \qquad j = 0, 1, \ldots \qquad (2.1)$$

where $g : \mathbb{R}^n \to \mathbb{R}^n$ is continuous. (Recall that a set $A \subset \mathbb{R}^n$ is *invariant* if $g(A) = A$.) The computation starts with a (big) compact set (box[1]) in parameter space. By a repeated bisection and selection of boxes the box coverings $\mathcal{B}_k$ of the invariant sets get tighter until the desired granularity of

---
[1] A n-dimensional box can be represented by a center $c \in \mathbb{R}^n$ and a radius $r \in \mathbb{R}^n$. Thus $B = B_{c,r} = \{x \in \mathbb{R}^n : c_i - r_i \leq x_i \leq c_i + r_i \; \forall i = 1, .., n\}$.

this outer approximation is reached. In computer implementations the selection process is based on an application of the underlying dynamical system (2.1) on a finite set of test points within each box.

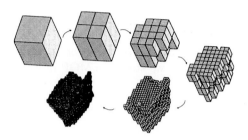

Fig. 1. Schematic illustration of the subdivision algorithm for the approximation of invariant sets

The general structure of the subdivision algorithm is as follows:

Algorithm DS-Subdivision
1.) **Subdivision** Construct from $\mathcal{B}_{k-1}$ a new system $\hat{\mathcal{B}}_k$ of subsets such that

$$\bigcup_{B \in \hat{\mathcal{B}}_k} B = \bigcup_{B \in \mathcal{B}_{k-1}} B$$

and

$$\mathrm{diam}^2(\hat{\mathcal{B}}_k) = \theta_k \, \mathrm{diam}(\mathcal{B}_{k-1}),$$

where $0 < \theta_{min} \leq \theta_k \leq \theta_{max} < 1$.
2.) **Selection**
Define the new collection $\mathcal{B}_k$ by

$$\mathcal{B}_k = \left\{ B \in \hat{\mathcal{B}}_k : \text{there exists } \hat{B} \in \hat{\mathcal{B}}_k \text{ such that } g^{-1}(B) \cap \hat{B} \neq \emptyset \right\}.$$

In [7] this algorithm was modified and adapted to the context of global zero finding, and in [4] these methods were applied to the context of multi-objective optimization. The common idea in these papers is to interpret iteration schemes as dynamical systems. For instance, using the descent direction for MOPs with differentiable objectives as proposed in [12] one may formulate a dynamical system which has as an invariant set all the points where the Kuhn-Tucker condition holds (see [10]). In other words, the set of points which are locally nondominated – including points on the boundary of Q – becomes an invariant set of this specific dynamical system. Thus, it remains to compare images of boxes in Z in order to find the (global) Pareto set. The so-called `sampling algorithm` reads as follows:

Sampling Algorithm
1.) **Subdivision**
 as in algorithm DS-Subdivision.
2.) **Selection**
 for all $B \in \hat{\mathcal{B}}_k$
  choose a set of test points P_B
  $N :=$ nondominated points of $\bigcup_{B \in \hat{\mathcal{B}}_k} P_B$
 $\mathcal{B}_k := \left\{ B \in \hat{\mathcal{B}}_k : \exists p \in P_B \cap N \right\}$

By using this algorithm – in some realizations in combination with DS-Subdivision – it is possible to detect the entire set of (global) Pareto points, including points on the boundary of the feasible region Q. Observe that these subdivision techniques possess naturally a "smoothing" property which is the basis for the combination with MOEAs: a box is kept if it contains *at least* one "good" point. This smoothing property is illustrated in Example A below.

DS-Subdivision works particularly well in the case when both the number of objective functions and the number of parameters is not too big. Otherwise the number of boxes created in the subdivision procedure is getting too large. In this article we propose the combination of DS-Subdivision with MOEAs in order to overcome this problem (see Section 3).

EXAMPLE A *We consider an MOP $F = (f_1, f_2) : \mathbb{R}^3 \to \mathbb{R}^2$ taken from [12].*

$$f_1(x) = \sum_{j=1}^{3} x_j, \quad f_2(x) = 1 - \prod_{j=1}^{3}(1 - p_j(x_j))$$

where

$$p_j(x) = \begin{cases} 0.01 \cdot exp(-(\frac{x_j}{20})^{2.5}) & \text{for } j = 1, 2 \\ 0.01 \cdot exp(-\frac{x_j}{15}) & \text{for } j = 3 \end{cases}$$

In Figure 2 we present box collections computed by the sampling algorithm. The smoothing property of the box approach becomes apparent because the coverings preserve the symmetry of the function in the first two parameters ($F(x_1, x_2, x_3) = F(x_2, x_1, x_3) \; \forall x \in \mathbb{R}^3$).

2.3 MOEAs

Evolutionary algorithms (EAs) are iterative stochastic search methods that are based on the two concepts of generate and evaluate [1]. Up to now, there are many multi-objective optimization methods that are based on this idea of EAs. MOEAs have demonstrated the advantage of using population-based search algorithms for solving multi-objective optimization problems. In all of these methods converging to the Pareto-optimal front and maintaining a spread of solutions (diversity) are the most important factors. MOEAs can be divided into two groups. The first group contains the MOEAs that always keep the best solutions (non-dominated solutions) of each generation in an *archive*, and they are called

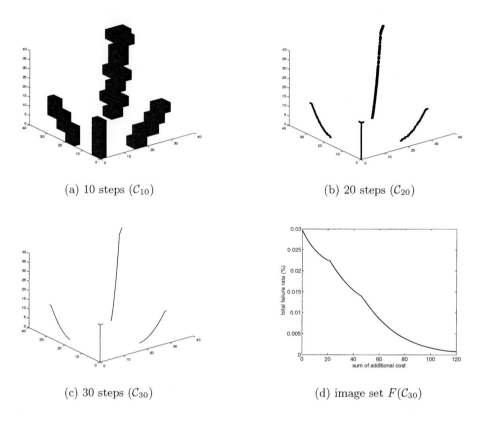

Fig. 2. Computation of the Pareto set of a model $F : \mathbb{R}^3 \to \mathbb{R}^2$ using subdivision techniques

MOEAs with elitism. It is proved by Rudolph [11] that in some cases elitism will provide convergence to the Pareto set. In the second group, there is no archive for keeping best solutions and MOEA may loose them during generations. MOEAs with elitism are studied in several methods like Rudolph and Agapie' Elitist GA, Elitist NSGA-II, SPEA, PAES (see e.g. [2] for all) and SPEA2 [13].

Figure 3 shows the typical structure of a MOEA with elitism, where t denotes the number of the generation, P_t the population, and A_t the archive at generation t. The aim of function $Generate$ is to generate new solutions in each iteration t which is done through selection, recombination and mutation. The function $Evaluate$ calculates the *fitness* value of each individual in the actual population P_t. Fitness assignment in MOEA is done in different ways such as by Pareto-ranking [8], non-dominated sorting [3], or by calculating Pareto-strengths [14]. Since only the superior solutions must be kept in the archive, it must be updated after each generation. The function $Update$ compares whether members of the current population P_t are non-dominated with respect to the members of the actual archive A_t and how and which of such candidates should be considered

```
BEGIN
    Step 1: t = 0;
    Step 2: Generate the initial population P₀ and initial archive A₀
    Step 3: Evaluate Pₜ
    Step 4: A_{t+1} := Update(Pₜ, Aₜ)
    Step 5: P_{t+1} := Generate(Pₜ, Aₜ)
    Step 6: t = t + 1
    Step 7: Unless a termination criterion is met, goto Step 3
END
```

Fig. 3. Typical structure of an archive-based MOEA

for insertion into the archive and which should be removed. Thereby, an archive is called *domination-free* if no two points in the archive do dominate each other. Obviously, during execution of the function $Update$, dominated points must be deleted in order to keep the archive domination-free. These three phases of an elitist MOEA are iteratively repeated until a termination criterion is met such as a maximum number of generations or when there has been no change in non-dominated solutions found for a given number of generations. The output of an elitist MOEA is the set of non-dominated solutions stored in the final archive. This set is an approximation of the Pareto-set and often called *quality set*. The above algorithm structure is common to most elitist MOEAs. In some of these methods (e.g., Rudolph and Agapie' Elitist GA, NSGA2), in the case of inadequate available space in the archive to store all of the non-dominated solutions, only those non-dominated solutions that are maximally apart from their neighbors are chosen. Therefore a crowding method is executed to select the solutions in less crowded areas. However, the true convergence property cannot be achieved, since an existent Pareto-optimal solution may get replaced by one which is not Pareto-optimal during the crowding selection operation. In some other methods (e.g., SPEA) clustering is done among the archive members, when the size of the archive exceeds. The use of clustering among the archive members guarantees spread among them. However, these algorithms lack a convergence proof, simply because of the same reason as in crowding methods: during the clustering procedure an existent Pareto-optimal archive member may get replaced by a non-Pareto-optimal.

3 Combination of Subdivision Techniques with MOEAs

3.1 Basic Idea

The basic idea behind the following algorithms is to view MOEAs as special (set oriented) dynamical systems which have to be combined properly with the subdivision techniques in order to increase the total performance. In particular the following property of MOEAs will be utilized for the combination of these methods: *MOEAs (typically) generate very quickly some very good approximations of Pareto points.*

We illustrate this by the following example:

$$f_1, f_2 : Q \subset \mathbb{R}^2 \to \mathbb{R}$$
$$f_1(x) = (x_1 - 1)^2 + (x_2 - 1)^4 \qquad (3.1)$$
$$f_2(x) = (x_1 + 1)^2 + (x_2 + 1)^2$$

In Figure 4 (a) we show a starting population consisting of 10 randomly chosen points in the domain $Q = [-3,3] \times [-3,3]$. The following two figures show the resulting populations after 5 and 10 generations using SPEA (see [2]). Here we observe that already after 5 generations there are some individuals close to the Pareto set. This property makes it possible to improve the sam-

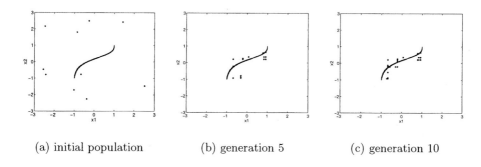

(a) initial population (b) generation 5 (c) generation 10

Fig. 4. One advantage of EAs is to find some good solutions quickly. The solid line indicates the actual Pareto set (in parameter space)

pling algorithm described above: instead of using many test points to evaluate a (high-dimensional) box, it is better to take just a few test points as the initial population of a "short" MOEA[3]. The EA only has to run for a short time because a box B is kept if it contains at least one point in N, namely the set of nondominated points of the total set of test points.

3.2 The Algorithms

Here we propose different algorithms with the desire to combine the advantages both of the subdivision techniques and the MOEAs. Practical combinations of these algorithms for the efficient solution of MOPs will be given in the last paragraph.

---

[3] A short MOEA is characterized by a short running time; that means small initial population and few generations.

EA-Subdivision. The discussion made above leads directly to the first algorithm: *use the sampling algorithm combined with a "short" MOEA for the evaluation of every box.* That is, we propose a particular choice of test points as follows:

$$P_B := \text{final population of "short" MOEA}$$

The only task of the MOEA is to find as fast as possible one good approximation of a Pareto point relative to the given domain. Therefore, diversity or even clustering are not needed. But special attention should be paid so that the MOEA does not get stuck on local minima. In particular "hill climbers" failed in some situations.

EXAMPLE B *In Figure 5 we show the coverings of the set of Pareto points after 4, 8 and 12 subdivision steps. The black "line" indicating the Pareto set is in fact the resulting box collection after 20 subdivision steps. In this example the population size and the number of generations were chosen to be 5.*

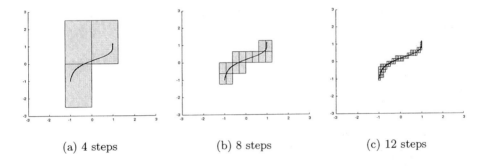

(a) 4 steps   (b) 8 steps   (c) 12 steps

Fig. 5. Application of EA-Subdivision

Recovering. It may be the case that in the course of the subdivision procedure boxes get lost although they contain part of the Pareto set. We now describe two algorithms containing a kind of "healing" process which allows to recover the Pareto set.

Before we can state the algorithms we give some details about the box collections: For theoretical purposes denote $\mathcal{P}_k$ a *complete* partition of the set $Q = B_{\hat{c},\hat{r}}$ into boxes of subdivision size - or $depth^4$ - k, which are generated by successive bisection of Q. Then there exists for every point $p \in Q$ and every depth k exactly one box $B(p, k) \in \mathcal{P}_k$ with center c and radius r such that
$c_i - r_i \leq p_i < c_i + r_i, \quad \forall i = 1, ..., n.$
Let us first consider the case where the covering is not complete but every box contains a part of the Pareto set (like box B_1 in Figure 6). The aim of the

---

[4] $\mathcal{P}_k$ and hence every box collection considered here can be identified with a set of leaves of a binary tree of depth k.

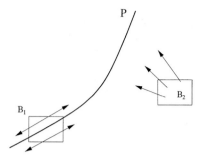

Fig. 6. Different problems for recovering: the algorithms have to cope with the fact that some boxes contain a part of the Pareto set (B_1) but others do not (B_2)

algorithm is to extend the given box collection step by step along the covered parts of the Pareto set until no more boxes are added. In order to find the corresponding neighboring boxes of a given box B with center c and radius r we run a MOEA in the extended box $\hat{B}$ given by center c and radius $\lambda \cdot r$ with $\lambda > 1$, say $\lambda = 3$. Afterwards the box collection is extended by the boxes $B \in \mathcal{P}_k$ which contain points from the resulting population (see Figure 8). In the first step this is done for all boxes from the box collection, for the following steps this local search has to be done only in the neighborhood of the boxes which were added in the preceding step. With a given box collection $\mathcal{B}_k$ the complete algorithm StaticRecover reads as follows:

Algorithm StaticRecover

1.) for all $B \in \mathcal{B}_k$
    $B.active := TRUE$
2.) for $i = 1,..,MaxStep$
    $\hat{\mathcal{B}}_k := \mathcal{B}_k$
    for all $\{B \in \mathcal{B}_k : B.active == TRUE\}$
        run MOEA in an extended universe $\hat{B} := (B.c, \lambda \cdot B.r)$
        $P :=$ final population
        $B.active = FALSE$
        for all $p \in P$:
            if $B(p,k) \notin \mathcal{B}_k$
                $\mathcal{B}_k := \mathcal{B}_k \cup B(p,k)$
                $B(p,k).active := TRUE$
    if $\hat{\mathcal{B}}_k == \mathcal{B}_k$ $STOP$

Hence StaticRecover only allows the addition of boxes into the given collection. The desired covering of the set of Pareto points cannot get worse, but will improve if the parameters of the algorithm are adjusted properly. On the other hand, StaticRecover does not treat adequately the case where a box does not contain some part of the Pareto set but is possibly far away (e.g. box B_2 in Figure 6). In this case the algorithm would extend the box covering by many undesired regions on their way towards the Pareto set (in particular in higher

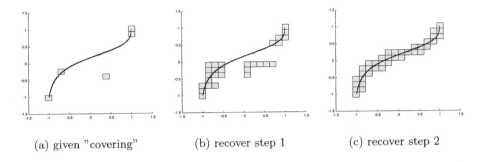

(a) given "covering"   (b) recover step 1   (c) recover step 2

Fig. 7. Application of `DynamicRecover` on a simple example

dimensions). Thus, when there are "good" and "bad" boxes like in Figure 6 we propose the following algorithm.

Algorithm `DynamicRecover`

1.) for all $B \in \mathcal{B}_k$
     $B.active := TRUE$
2.) for $i = 1,..,MaxStep$
     $\hat{\mathcal{B}}_k := \mathcal{B}_k$, $\mathcal{B}_k := \emptyset$
     for all $B \in \hat{\mathcal{B}}_k$: $B.active == TRUE$
         run MOEA in an extended universe $\hat{B} := (B.c, \lambda \cdot B.r)$
         $P_B :=$ final population
     $P :=$ nondominated points of $\bigcup_{B \in \hat{\mathcal{B}}_k} P_B$
     for all $p \in P$:
         $\mathcal{B}_k := \mathcal{B}_k \cup B(p,k)$
         if $B(p,k) \in \mathcal{B}_k$ $B(p,k).active := FALSE$
         else $B(p,k).active := TRUE$
     if $\hat{\mathcal{B}}_k == \mathcal{B}_k$ $STOP$

In contrast to `StaticRecover` this algorithm has again the disadvantage that good boxes can be deleted while they have been computed once[5]. But otherwise there would be no chance to sort out the boxes which contain no optimal solution. The speed of the algorithm depends - besides of the MOEA - on the choice of the extension factor λ. A larger value of λ yields faster convergence but lower robustness. In general, the number of generations and the size of the initial population should increase with λ. For this local covering of the part of the Pareto set the MOEA has to preserve diversity. Furthermore the convergence of the MOEA should be good enough in order not to insert too many superfluous boxes.

---

[5] The usage of an archive seems to be suitable and has to be tested in future work.

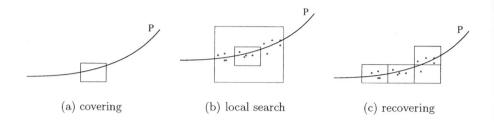

(a) covering   (b) local search   (c) recovering

Fig. 8. Working principle of StaticRecover

EXAMPLE C *We again consider the MOP (3.1). Algorithm DynamicRecover was applied to a chosen initial box collection (see Figure 7). The algorithm stops after 2 iterations with a total covering of the Pareto set.*

Combination of the Algorithms. Here we propose two possible combinations of the algorithms described above which turned out to be practical for the numerical solution of MOPs.

First, if the MOP is "moderate" dimensional we suggest the use of the algorithm EA-Subdivision in combination with StaticRecover. For most MOPs it is sufficient to use StaticRecover only once or twice during the computation, but for more complicated problems it can turn out that both algorithms have to be used in alternation: after some number of iteration steps of EA-Subdivision, the number of new boxes added to this box collection by StaticRecover gives a feedback on the quality of the computed "covering" and hence the adjustment of the MOEA can be adapted to the next subdivision steps. Eventually the algorithm stops if the desired granularity is reached and no more boxes are added by the recovery step. Here, the global optimization is done by the subdivision techniques while the MOEAs are used for local optimization.

For higher dimensional MOPs we recommend another strategy since the evaluation of the boxes by MOEAs is expensive even if one is only interested in just a few good solutions. Here we suggest to improve the result of a MOEA - where the granularity K can be low - by using DynamicRecover. Therefore, the recovery process should extend the box collection which is generated by the final population of the MOEA. Finally, this extended (hopefully complete) covering can be further refined using subdivision techniques. Important for this method is the proper choice of size of the boxes. We suggest to adjust the edge lengths of the boxes to be approximately κK, where κ is a safety factor, say $\kappa = 2$.

Also, in this method the stopping condition is given by the size of the boxes and the number of boxes which are added in a recovery step. But in contrast to the first combination in this method the global optimization is done by the MOEA while the subdivision techniques can only give local improvements.

4 Computational Results

4.1 Example 1

The following three objectives have to be minimized at once:

$$f_1, f_2, f_3 : Q = [-5, 5]^3 \to \mathbb{R},$$
$$f_1(x_1, x_2, x_3) = (x_1 - 1)^4 + (x_2 - 1)^2 + (x_3 - 1)^2,$$
$$f_2(x_1, x_2, x_3) = (x_1 + 1)^2 + (x_2 + 1)^4 + (x_3 + 1)^2,$$
$$f_3(x_1, x_2, x_3) = (x_1 - 1)^2 + (x_2 + 1)^2 + (x_3 - 1)^4.$$

Figure 9 shows the result of a combination of EA-Subdivision and StaticRecover. To achieve the uncomplete covering of Figure 9 (a) no EA techniques were necessary but only the center point was used to evaluate each box. Figure 9 (c) shows a tight and complete covering of the Pareto set. The picture was produced by the software package GRAPE[6].

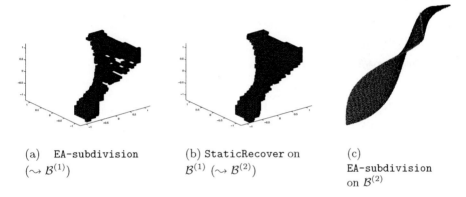

(a) EA-subdivision ($\rightsquigarrow \mathcal{B}^{(1)}$)

(b) StaticRecover on $\mathcal{B}^{(1)}$ ($\rightsquigarrow \mathcal{B}^{(2)}$)

(c) EA-subdivision on $\mathcal{B}^{(2)}$

Fig. 9. Combination of EA-subdivision and StaticRecover

4.2 Example 2

Now we consider the following MOP

$$f_1, f_2 : [-5.12, 5.12]^{10} \to \mathbb{R},$$
$$f_1(x) = \sum_{i=1}^{9} (-10 e^{-0.2\sqrt{x_i^2 + x_{i+1}^2}}),$$
$$f_2(x) = \sum_{i=1}^{10} (|x_i|^{0.8} + 5\sin(x_i)^3).$$

Figure 10 shows a final population using SPEA (size of initial population: 200; number of generations: 300) and the local improvement by an application of DynamicRecover on this result.

---
[6] http://www.iam.uni-bonn.de/sfb256/grape/

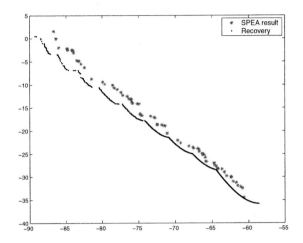

Fig. 10. Local improvement of a MOEA result by using `DynamicRecover`

4.3 Example 3

Finally we consider an MOP which arises in antenna design ([9]):

$$f_1(x_\nu, y_\nu) = -4\pi^2 \left| \sum_{\nu=-n}^{n} (-i)^\nu \mathcal{J}_\nu(\ell)(x_\nu + iy_\nu) \right|^2,$$

$$f_2(x_\nu, y_\nu) = \max_{\eta=0,\ldots,5} \left(4\pi^2 \left| \sum_{\nu=-n}^{n} (-i)^\nu \mathcal{J}_\nu(\ell)(x_\nu + iy_\nu) e^{i\nu s_\eta} \right|^2 \right)$$

subject to the constraints

$$x_\nu, y_\nu \in \mathbb{R} \quad (\nu \in \mathbb{Z}, |\nu| \leq n),$$

$$2\pi \sum_{\nu=-n}^{n} (x_\nu^2 + y_\nu^2) \leq 1$$

with the specific discretization points $s_\eta = \frac{3}{4}\pi + \eta\frac{\pi}{10}$. Here $\mathcal{J}_\nu$ denotes the Bessel function of ν-th order. We have tested the algorithms for $n = 5$ and $\ell = 10$. Since $\mathcal{J}_\nu(x) = (-1)^\nu \mathcal{J}_{-\nu}(x)$ and $\mathbb{C} \cong \mathbb{R}^2$ this leads to a model with 12 free parameters. First, we have applied the recovery techniques on the result of a SPEA with the following parameter settings: 1000 initial individuals coded to 13 bits, 500 generations and archive size 1000 (see Figure 11). As in Example 2 also here local improvements can be observed. The total running time was 7.5 hours for the SPEA result and 20 minutes for the application of `StaticRecover` on it. Furthermore, we have taken another SPEA result with 300 initial individuals coded to 7 bits, 300 generations and archive size 300 in order to decrease the computational time. Afterwards we have used the recovering techniques - with larger box sizes due to the lower granularity of the MOEA - and got similar results (see Figure 12). The total running time was 20 minutes for the SPEA result and another 15 minutes for the recovering. Therefore, at least here it is possible to use the recovering techniques to speed up the running time while the total performance of the combination of a MOEA and the subdivision techniques is kept.

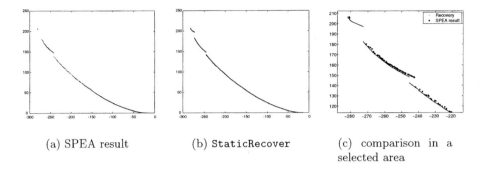

(a) SPEA result   (b) StaticRecover   (c) comparison in a selected area

Fig. 11. Application of StaticRecover on a SPEA result and a comparison in selected area

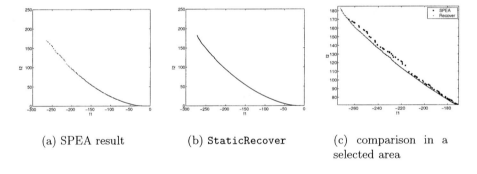

(a) SPEA result   (b) StaticRecover   (c) comparison in a selected area

Fig. 12. Application of StaticRecover on a SPEA result and a comparison in selected area

5   Conclusion and Future Work

We have presented robust set oriented algorithms for the numerical solution of MOPs. More precisely, we have presented two different methods. The first one is an improvement of the sampling algorithm described in Section 2. Due to the evaluation of the subdomains by MOEAs the speed of the computation of MOPs can be increased while its robustness is kept. This allows to apply the subdivision techniques to higher dimensional MOPs.

The second method is a technique for the local improvement of the diversity of MOEA results, but can also be used to speed up the computational time for the solution of an MOP.

The general advantages of our algorithms are certainly their robustness, the preservation (or improvement) of the diversity, and the "natural" stopping conditions. On the other hand so far the subdivision techniques are restricted to "moderate" dimensions. Thus, for higher dimensional MOPs recovering techniques should be applied.

In future work we would like to adjust the structure of the MOEAs for the special requirements of the different algorithms. Also other optimization metaheuristics like e.g. *particle swarm optimization* have to be tested with respect to a combination with subdivision strategies.

References

1. D. Corne, M. Dorigo, and F. Glover. *New Ideas in Optimization*. Mc Graw Hill, 1999.
2. K. Deb. *Multi-Objective Optimization using Evolutionary Algorithms*. John Wiley & Sons, 2001.
3. K. Deb, S. Agrawal, A. Pratap, and T. Meyarivan. A fast elitist non-dominated sorting genetic algorithm for multi-objective optimization: Nsga-ii. In *Parallel Problem Solving from Nature VI (PPSN-VI)*, pages 849–858, 2000.
4. M. Dellnitz, R. Elsässer, T. Hestermeyer, O. Schütze, and S. Sertl. Covering Pareto sets with multilevel subdivision techniques. to appear, 2002.
5. M. Dellnitz and A. Hohmann. The computation of unstable manifolds using subdivision and continuation. In H.W. Broer, S.A. van Gils, I. Hoveijn, and F. Takens, editors, *Nonlinear Dynamical Systems and Chaos*, pages 449–459. Birkhäuser, *PNLDE* 19, 1996.
6. M. Dellnitz and A. Hohmann. A subdivision algorithm for the computation of unstable manifolds and global attractors. *Numerische Mathematik*, 75:293–317, 1997.
7. M. Dellnitz, O. Schütze, and St. Sertl. Finding zeros by multilevel subdivision techniques. 2002.
8. D. E. Goldberg. *Genetic Algorithms in Search, Optimization and Machine Learning*. Addison-Wesley Publishing Company, Inc., 1989.
9. A. Jüschke and J. Jahn. A bicriterial optimization problem of antenna design. *Comp. Opt. Appl.*, 7:261–276, 1997.
10. H. Kuhn and A. Tucker. Nonlinear programming. *Proc. Berkeley Symp. Math. Statist. Probability, 2nd, (J. Neumann, ed.)*, pages 481–492, 1951.
11. G. Rudolph. On a Multi-Objective Evolutionary Algorithm and Its Convergence to the Pareto Set. In *5th IEEE Conference on Evolutionary Computation*, pages 511–516, 1998.
12. S. Schäffler, R. Schultz, and K. Weinzierl. A stochastic method for the solution of unconstrained vector optimization problems. To appear in J. Opt. Th. Appl., 2002.
13. E. Zitzler, M. Laumanns, and L. Thiele. SPEA2: Improving the strength Pareto evolutionary algorithm. In *Evolutionary Methods for Design, Optimisation and Control with Applications to Industrial Problems*, 2002.
14. E. Zitzler and L. Thiele. Multiobjective evolutionary algorithms: A comparative case study and and the strength Pareto approach. *IEEE Trans. on Evolutionary Computation*, 3(4):257–271, 1999.

An Adaptive Divide-and-Conquer Methodology for Evolutionary Multi-criterion Optimisation

Robin C. Purshouse and Peter J. Fleming

Department of Automatic Control and Systems Engineering,
University of Sheffield, Mappin Street, Sheffield, S1 3JD, UK.
{r.purshouse, p.fleming}@sheffield.ac.uk

Abstract. Improved sample-based trade-off surface representations for large numbers of performance criteria can be achieved by dividing the global problem into groups of independent, parallel sub-problems, where possible. This paper describes a progressive criterion-space decomposition methodology for evolutionary optimisers, which uses concepts from parallel evolutionary algorithms and nonparametric statistics. The method is evaluated both quantitatively and qualitatively using a rigorous experimental framework. Proof-of-principle results confirm the potential of the adaptive divide-and-conquer strategy.

1 Introduction

Treatments for the evolutionary optimisation of large numbers of criteria can potentially be discovered through analysis of the relationships between the criteria [1]. This paper considers *divide-and-conquer* strategies based on the concept of *independence* between a pair of criteria, in which performance in each criterion is entirely unrelated to performance in the other.

An *independent set* is herein defined as a set whose members are linked by dependencies, and for which no dependencies exist with elements external to the set. Consider a problem with n independent sets of criteria $[z_1,...,z_n]$ and associated independent sets of decision variables $[x_1,...,x_n]$. If knowledge of these sets is available then the global problem, p, can be decomposed into a group of parallel sub-problems $[p_1,...,p_n]$ that can be optimised independently of each other to ultimately yield n independent trade-off surfaces.

This paper demonstrates the benefit of using such a divide-and-conquer strategy when the correct decompositions are known in advance. It also proposes a general methodology for identifying, and subsequently exploiting, the decomposition during the optimisation process. An empirical framework is described in Sect. 2, which is then used to establish the case for divide-and-conquer in Sect. 3. An on-line adaptive strategy is proposed in Sect. 4 that exploits the iterative, population-based nature of the evolutionary computing paradigm. Independent sets of criteria are identified using nonparametric statistical methods of independence testing. Sub-populations are assigned to the optimisation of each set, with migration between these occurring as the decomposition is revised over the course of the optimisation. Proof-of-principle results are presented in Sect. 5, together with a discussion of issues raised by the study.

2 Experimental Methodology

2.1 Baseline Algorithm

The baseline evolutionary multi-criterion optimiser chosen in this work is an elitist multi-objective genetic algorithm (MOGA) [2]. An overview is shown in Table 1. Parameter settings are derived from the literature and tuning has not been attempted.

Table 1. Baseline multi-objective evolutionary algorithm (MOEA) used in the study

EMO component	Strategy
General	Total population = $100n$, Generations = 250
Elitism	Ceiling of 20%-of-population-size of non-dominated solutions preserved. Reduction using SPEA2 clustering [3].
Selection	Binary tournament selection using Pareto-based ranking [4].
Representation	Concatenation of real number decision variables. Accuracy bounded by machine precision.
Variation	Uniform SBX crossover with $\eta_c = 15$, exchange probability = 0.5 and crossover probability = 1 [5]. Element-wise polynomial mutation with $\eta_m = 20$ and mutation probability = (chromosome length)$^{-1}$ [6].

2.2 Test Functions

A simple way to create independent multi-criterion test functions is to concatenate existing test problems from the literature, within which dependencies exist between all criteria. In this proof-of-principle study, only the test function *ZDT-1* proposed in [7] is used. The recently proposed scalable problems [8] will be used in later studies.

Definition 1. Concatenated ZDT-1 test function.

$$\text{Min. } \mathbf{z} = \left[z_1(x_1), z_2(x_1,...,x_m),..., z_{2n-1}(x_{(n-1)m+1}), z_{2n}(x_{(n-1)m+1},...,x_{nm}) \right],$$

with $z_{2i-1} = x_{(i-1)m+1}$,

$$z_{2i} = 1 - \sqrt{z_{2i-1} \Big/ \left(1 + \left[9/(m-1) \right] \sum\nolimits_{j=(i-1)m+2}^{im} x_j \right)},$$

where $i = [1,...,n]$ is a particular sub-problem, $m = 30$ is the number of decision variables per sub-problem, and $x_j \in [0,1] \forall j$. An instance is denoted by *C-ZDT-1(i)*.

The global solution to this problem is a set of bi-criterion trade-off surfaces (z_1 versus z_2, z_3 versus z_4, and so forth). Each trade-off surface is a convex curve in the region $[0\ 1]^2$ (for which the summation in Definition 1 is zero). The *ideal vector* is [0 0]. The *anti-ideal vector* of worst possible performance in each criterion is [1 10].

2.3 Performance Metrics

Hypervolume. A quantitative measure of the quality of a trade-off surface is made using the hypervolume S unary performance metric [9]. The hypervolume metric measures the amount of criterion-space dominated by the obtained non-dominated front, and is one of the best unary measures currently available, although it has limitations [10][11]. The anti-ideal vector is taken as the reference point. The metric is normalised using the hypervolume of the ideal vector, as illustrated in Fig. 1a.

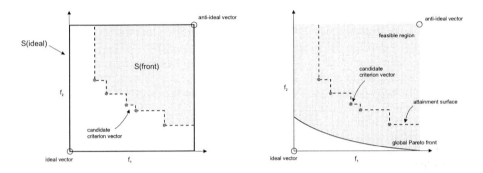

Fig. 1. (a) Hypervolume metric, (b) Attainment surface

Attainment surfaces. Given a set of non-dominated vectors produced by a single run of an MOEA, the attainment surface is the boundary in criterion-space that separates the region that is dominated by or equal to the set from the region that is non-dominated. An example attainment surface is shown in Fig. 1b. Performance across multiple runs can be described in terms of regions that are dominated in a given proportion of runs and can be interpreted probabilistically. For example, the 50%-attainment surface is similar to the median, whilst the 25%- and 75%-attainment surfaces are akin to the quartiles of a distribution. Univariate statistical tests can be performed on attainment surfaces using the concept of auxiliary lines [12][13]. However, in this paper, the technique is simply used to provide a qualitative comparison of the median performance of two algorithms, supported by quantitative hypervolume-based significance testing.

2.4 Analysis Methods

For the type of MOEA described in Table 1, the final population represents an appropriate data set upon which to measure performance. 35 runs of each algorithm configuration have been conducted in order to generate statistically reliable results.

Quantitative performance is then expressed in the distribution of obtained hypervolumes. A comparison between configurations is made via the difference between the means of the distributions.

The significance of the observed result is assessed using the simple, yet effective, nonparametric method of *randomisation testing* [14]. The central premise of the method is that, if the observed result has arisen by chance, then this value will not appear unusual in a distribution of results obtained through many random relabellings of the samples. Let *S1* be the distribution of hypervolume metrics for *algorithm_1*, and let *S2* be the corresponding distribution for *algorithm_2*. The randomisation method for *algorithm_1 versus algorithm_2* proceeds as follows:

- Subtract the mean of *S2* from the mean of *S1*: this is the observed difference.
- Randomly reallocate half of all samples to one algorithm and half to the other. Compute the difference between the means as before.
- Repeat Step 2 until 5000 randomised differences have been generated, and construct a distribution of these values.
- If the observed value is within the central 99% of the distribution, then accept the null hypothesis that there is no performance difference between the algorithms. Otherwise consider the alternative hypotheses. Since optimal performance is achieved by maximising hypervolume, if the observed value falls to the left of the distribution then there is strong evidence to suggest that *algorithm_2* has outperformed *algorithm_1*. If the observed result falls to the right, then superior performance is indicated for *algorithm_1*. This is a two-tailed test at the 1%-level.

2.5 Presentation of Results

Comparisons of *algorithm_1* versus *algorithm_2* for $n = [1,...,4]$ are summarised within a single figure such as Fig. 2, for which *algorithm_1* is the baseline algorithm of Table 1 (with no decomposition) and *algorithm_2* is a parallel version with an a priori decomposition of both criterion-space and decision-space. Region (a) shows the validation case of one independent set, *C-ZDT-1(1)*, whilst regions (b), (c), and (d) show two, three, and four sets respectively. Within each region, each row indicates a bi-criterion comparison. The left-hand column shows the results of the randomisation test on hypervolume (if the observed value, indicated by the filled circle, lies to the right of the distribution then this favours *algorithm_1*), whilst the right-hand column shows the median attainment surfaces (the unbroken line is *algorithm_1*).

3 The Effect of Independence

The potential of a divide-and-conquer strategy can be examined by comparing a global solution to the concatenated ZDT-1 problem to a priori correct decompositions in terms of decision-space, criterion-space, or both. Consider a scheme in which a sub-population of 100 individuals is evolved in isolation for each independent set. Each EA uses only the relevant criteria and decision variables. This is compared to a global approach, with a single population of size $100n$, using all criteria and decision variables in Fig. 2. Substantially improved performance is shown for the divide-and-conquer scheme.

An Adaptive Divide-and-Conquer Methodology   137

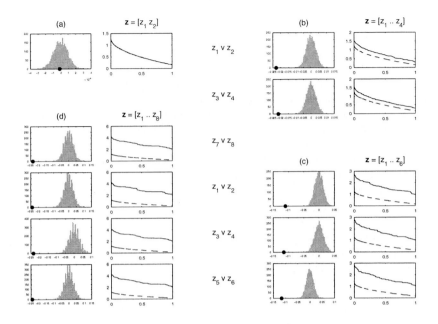

Fig. 2. Global model versus decomposition of both criterion-space and decision-space

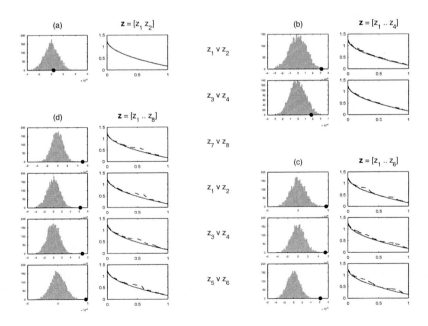

Fig. 3. Decomposition of both criterion-space and decision-space versus decision-space decomposition alone

To clarify which parts of the decomposition are important, sub-population schemes that decompose decision-space whilst treating criterion-space globally and vice versa are now considered. In the decision-space scheme, each 100 individual sub-population operates on the correct subset of decision variables. However, fitness values (and elitism) are globally determined. Elite solutions are reinserted into the most appropriate sub-population depending on their ranking on local criterion sets. Assignment is random in the case of a tie. Performance is compared to the ideal decomposition in Fig. 3. It is evident that decision-space decomposition alone is not responsible for the results in Fig. 2, and that the quality of the trade-off surfaces deteriorates with n. The attainment surfaces for cases (c) and (d) suggest that the global treatment of criteria may be affecting the shape of the identified surface.

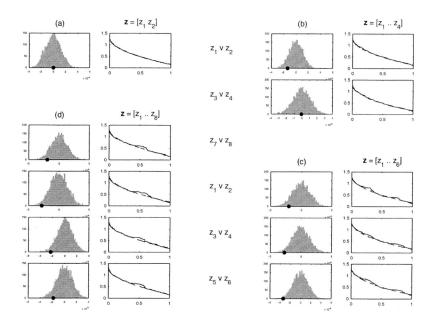

Fig. 4. Decision-space decomposition versus criterion-space decomposition

In the criterion-space scheme, each sub-population operates on the correct subset of criteria (fitness and elitism is isolated within the sub-population), but the EA operates on the global set of decision variables. A comparison with the decision-space method is shown in Fig. 4. No statistically significant performance difference is evident in any of the cases. Thus, criterion-space decomposition alone is also not responsible for the achievement in Fig. 2. Note that if single-point rather than uniform crossover had been used then results would have been much worse for the global treatment of decision-space since, for the former operator, the probability of affecting any single element of the chromosome (and thus the relevant section) decreases with chromosome length.

The above results indicate that a sub-population-based decomposition of either criterion-space or decision-space can significantly benefit performance. Best results are obtained when both domains are decomposed. Given that, in general, the correct decomposition for either domain is not known in advance, the choice of domain will depend on which is less demanding to analyse. Note that if criterion-space is decomposed then decision-space decomposition is also required at some point in order to synthesise a global solution. However, the converse is not the case. Decomposition may be a priori, progressive, or a posteriori with respect to the optimisation. A correct early decomposition in both spaces would be ideal but this may not be achievable.

4 Exploiting Independence via Criterion-Space Decomposition

4.1 Overview of the Methodology

A progressive decomposition of criteria, together with a retrospective decomposition of decision variables, is proposed in this paper. This is appropriate for problem domains where the number of criteria is significantly fewer than the number of decision variables. A sub-population approach is taken, in which the selection probability of an individual in a sub-population is determined using only the subset of criteria assigned to that sub-population. The topology of this parallel model can vary over the course of the optimisation.

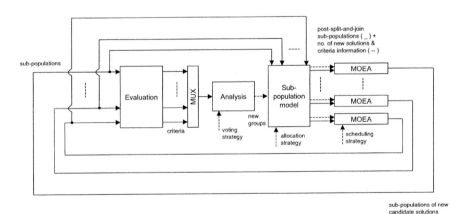

Fig. 5. Schematic overview of the progressive decomposition methodology

An overall schematic of the technique is given in Fig. 5. The process begins with a global population model. The multi-criterion performance of each candidate solution is then obtained. From the perspective of a single criterion, the population provides a set of observations for that criterion. Pair-wise statistical tests for independence are then performed for all possible pairs of criteria to determine between which criteria dependencies exist. Linkages are created for each dependent relationship. A sub-problem is then identified as a linked set of criteria. This concept is illustrated for an ideal decomposition of C-ZDT-1(2) in Fig. 6. Of all pair-wise dependency tests, significant dependencies have been identified between z_1 and z_2, and between z_3 and z_4.

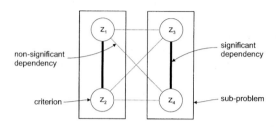

Fig. 6. Identification of sub-problems [z_1 z_2] and [z_3 z_4] via linkage

The new topology of the population model follows from the decomposition. *Split* and *join* operations are implemented to allow criteria (and associated candidate solutions) to migrate between sub-problems as appropriate.

When each new sub-population has been formed, selection probabilities and the identification (and management) of elites are determined using the current subset of criteria. Performance across all other criteria is ignored. Selection and variation operators are then applied within the boundaries of the sub-population. The size of the resulting new sub-population is pre-determined by the population management process.

All new solutions are evaluated across the complete set of criteria. This new data is then used to determine an updated decomposition. In this study, the update is performed at every generation, although in general it could be performed according to any other schedule. The process then continues in the fashion described above.

4.2 Population Management

The population topology is dependent on the identified decomposition. This can change during the course of the optimisation, thus requiring some sub-problem resources to be reallocated elsewhere. Operations are required to split some portion of the candidate solutions from a sub-population and subsequently join this portion on to another sub-population.

The size of the split is decided using an allocation strategy. Since the number of candidate solutions required to represent a trade-off surface grows exponentially with the number of criteria, k, it would seem a reasonable heuristic to use an exponential allocation strategy such as 2^k. As an example, consider a four-criterion problem, $k = 4$, with an initial decomposition of $\{[z_1\ z_2],[z_3\ z_4]\}$. The suggested new decomposition is $\{[z_1],[z_2\ z_3\ z_4]\}$. The situation is depicted in Fig. 7.

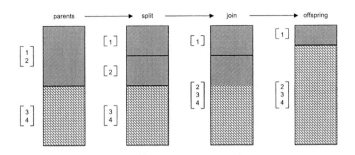

Fig. 7. Example of the split and join operations

Prior to reallocation, each sub-population should have a proportion $2^2/(2^2+2^2) = 1/2$ of the available resources. z_2 is now to be linked with z_3 and z_4, and must thus be split from its grouping with z_1. Both z_1 and z_2 will receive $2^1/(2^1+2^1) = 1/2$ of the resources in this sub-population. The actual candidate solutions to be assigned to each part of the split are determined randomly. The resources allocated to z_2 are then added to the $[z_3\ z_4]$ sub-population. Now z_1 has 1/4 of the resources, whilst $[z_2\ z_3\ z_4]$ has 3/4 of the resources. The selection and variation operations are then used to return to the required proportions of $2^1/(2^1+2^3) = 1/5$ for z_1 and $2^3/(2^1+2^3) = 4/5$ for $[z_2\ z_3\ z_4]$.

4.3 Tests for Independence

A sub-problem group is generated by collecting all criteria that are linked by observed pair-wise dependencies. In this sub-section, the tests used to determine if a connection should be made are introduced. Several tests for variable independence based on sample data exist in the statistics literature [15]. Two nonparametric procedures, the first based on the *Kendall K* statistic and the second on the *Blum-Kiefer-Rosenblatt D* statistic, are used in this work.

Both methods require special care for the handling of tied data. This is of concern in an evolutionary algorithm implementation since a particular solution may have more than a single copy in the current population. Large-sample approximations to each method have been implemented. This is possible because reasonably large population sizes have been used (100 individuals per independent bi-criterion set). All significance tests are two-tailed at the 1%-level, the null hypothesis being that the criteria are independent.

Kendall K. A distribution-free bivariate test for independence can be made using the Kendall sample correlation statistic, K. This statistic measures the level of *concordance* (as one variable increases/decreases, the other increases/decreases) against the level of *discordance* (as one variable increases/decreases, the other decreases/increases). This is somewhat analogous to the concepts of harmony and conflict in multi-criterion optimisation [1]. The standardised statistic can then be tested for significance using the normal distribution $N(0,1)$. Ties are handled using a modified paired sign statistic. A modified null variance is also used in the standardisation procedure. For further details, refer to [15, pp363-394].

The main concern with this method is that if $K = 0$ this does not necessarily imply that the two criteria are independent (although the converse is true). This restricts the applicability of the method beyond bi-criterion dependencies, where relationships may not be monotonous.

Blum-Kiefer-Rosenblatt D. As an alternative to the above test based on the sample correlation coefficient, Blum, Kiefer, and Rosenblatt's large-sample approximation to the *Hoeffding D* statistic has also been considered in this study. This test is able to detect a much broader class of alternatives to independence. For full details, refer to [15, pp408-413].

4.4 Decision-Space Decomposition: An Aside

Discussion. In the above methodology, and throughout the forthcoming empirical analysis of this method in Sect. 5, different sub-populations evolve solutions to different sets of criteria. Decision-space decomposition is not attempted. Thus, at the end of the optimisation process, complete candidate solutions exist for each criterion set. It is now unclear which decision variables relate to which criterion set. In order to finalise the global solution, solutions from each trade-off surface must be synthesised via partitioning of the decision variables.

An a posteriori decomposition, as described below, is simple to implement but has two clear disadvantages: (1) some reduction in EA efficiency will be incurred because the operators search over inactive areas of the chromosome (operators that are independent of chromosome length should be used), and (2) further analysis is required to obtain the global solution.

Method. A candidate solution should be selected at random from the overall final population. Each variable is then perturbed in turn and the effect on the criteria should be observed. The variable should be associated with whichever criteria are affected. Then, when the decision-maker selects a solution from the trade-off featuring a particular set of criteria, the corresponding decision variables are selected from the sub-population corresponding to this set.

This method requires as many extra candidate solution evaluations as there are decision variables in the problem. For the 4-set concatenated ZDT-1 test function, this is 120 evaluations or 30% of a single generation of the baseline algorithm.

Two special cases must be addressed: (1) If the perturbation of a decision variable affects criteria in more than one criterion subset, this indicates an invalid decomposition of criterion-space. Information of this kind could be used progressively to increase the robustness of the decomposition. (2) It is possible that no disturbance of criteria is seen when the decision variable is perturbed. Here, the alternatives are to consider another candidate solution or to consider more complicated variable interactions. This may also be an indication that the variable is globally redundant.

5 Preliminary Results

Proof-of-principle results for the adaptive divide-and-conquer strategy devised in Sect. 4 are presented herein for the concatenated ZDT-1 test function (Definition 1) with $n = [1,\ldots,4]$. A summary of the chosen strategy is given in Table 2. Both the Kendall K method and the Blum-Kiefer-Rosenblatt D method have been considered.

Table 2. Divide-and-conquer settings

EMO component	Strategy	
Independence test	(either)	Blum-Kiefer-Rosenblatt D
	(or)	Kendall K
Resource allocation	2^k	
Schedule	Revise the decomposition every generation	

5.1 Blum-Kiefer-Rosenblatt D Results

The performance of this strategy when compared to the baseline case of no decomposition is shown in Fig. 8. Both the hypervolume metric results and the attainment surfaces indicate that the divide-and-conquer strategy produces trade-off surfaces of higher quality in cases where independence exists. However, the attainment surfaces also show that the absolute performance of the method degrades as more independent sets are included.

The degradation can be partially explained by considering the percentage of correct decompositions made by the algorithm at each generation (measured over the 35 runs) shown in Fig. 9a. As the number of independent sets increases, the proportion of correctly identified decompositions decreases rapidly. Note that this does not necessarily mean that the algorithm is making invalid decompositions or no decomposition: other valid decompositions exist, for example $\{[z_1\ z_2], [z_3\ z_4\ z_5\ z_6]\}$ for $n = 3$, but these are not globally optimal (also the number of possible decompositions increases exponentially with n.). Indeed, on no occasion did the test produce an invalid decomposition (identified independence when dependency exists). This is evident from plots of the decomposition history over the course of the optimisation. A typical history is depicted in Fig. 9b. Each criterion is labelled on the vertical axis, whilst the horizontal axis depicts the current generation of the evolution. At a particular generation, criteria that have been identified as an independent set are associated with a unique colour. Thus, as shown in Fig. 9b, at the initial generation z_1 and z_2 have been identified as a cluster (white), as have $[z_3\ z_4]$ (black), $[z_5\ z_6]$ (light grey), and $[z_7\ z_8]$ (dark grey). At generation 200 all the criteria have been grouped together, as indicated by the complete whiteness at this point in the graph. Note that there is no association between the colours across the generations.

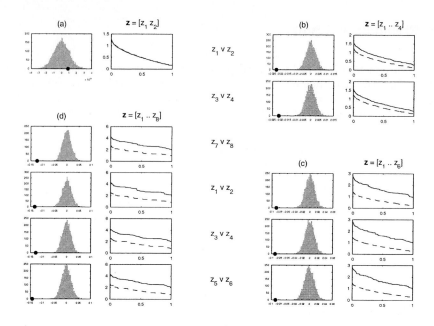

Fig. 8. No decomposition versus Blum-Kiefer-Rosenblatt D divide-and-conquer

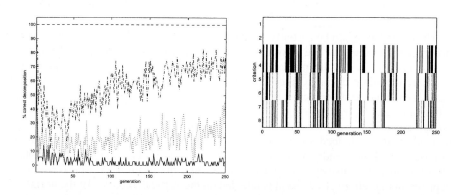

Fig. 9. Blum-Kiefer-Rosenblatt D: **(a)** Correct decompositions as a percentage of total runs over the course of the optimisation. ---- $[z_1\ z_2]$; -·-· $[z_1,\ldots,z_4]$; ···· $[z_1,\ldots,z_6]$; ──── $[z_1,\ldots,z_8]$; **(b)** Typical decomposition history for a single replication. Each identified criterion cluster is represented by a colour (white, light grey, dark grey, black)

5.2 Kendall K Results

The performance of the divide-and-conquer algorithm with the Kendall K test for independence is compared to Blum-Kieffer-Rosenblatt D in Fig. 10. The former test appears to offer superior performance as the number of independent sets increases. No significant performance difference can be found for $n = 2$, but such a difference can be seen for two of the surfaces for $n = 3$, and every surface for $n = 4$.

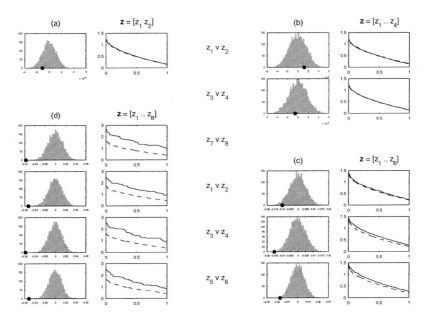

Fig. 10. Blum-Kiefer-Rosenblatt D divide-and-conquer versus Kendall K divide-and-conquer

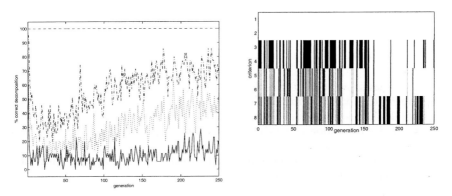

Fig. 11. Kendall K: **(a)** Correct decompositions as a percentage of total runs over the course of the optimisation. ---- $[z_1, z_2]$; -·-· $[z_1,...,z_4]$; ···· $[z_1,...,z_6]$; _____ $[z_1,...,z_8]$; **(b)** Typical decomposition history. Each identified criterion cluster is represented by a colour (white, light grey, dark grey, black)

This difference in performance can be explained by the plot of correct decompositions shown in Fig. 11a. Whilst the proportion of correct decompositions degrades as n increases, this degradation is not as severe as for Blum-Kiefer-Rosenblatt D (Fig. 9a). Also, from the typical decomposition history depicted in Fig. 11b, the valid decompositions tend to be of higher resolution than those developed by the alternative method (Fig. 9b).

5.3 Discussion

The obtained results show that the adaptive divide-and-conquer strategy offers substantially better performance than the global approach in terms of the quality of trade-off surfaces generated.

Of the two independence tests considered, Kendall K would appear more capable of finding good decompositions on the benchmark problem considered, especially as the number of independent sets increases. However, Kendall K may experience difficulties when the dimension of the trade-off surface increases, since it may incorrectly identify independence due to the variation in the nature of bi-criterion relationships over the surface (the relationship is not always conflicting, as it is for a bi-criterion problem). By contrast, Blum-Kiefer-Rosenblatt D offers a more robust search in these conditions, but is more conservative.

There is a clear need for the procedure to be robust (invalid decompositions should be avoided, although the progressive nature of the process may somewhat mitigate the damage from these), but conservativism should be minimised in order to increase the effectiveness of the methodology. Under these circumstances, it may be prudent to adopt a voting strategy, in which a decision is made based on the results from several tests for independence.

The adaptive divide-and-conquer strategy carries some overhead in terms of the test for independence and the sub-population management activity, which may be controlled using a scheduling strategy. This must be balanced against the improvements in the quality of the trade-off surfaces identified and the reduction in the complexity of the MOEA ranking and density estimation procedures.

6 Conclusion

This study has shown that, if feasible, a divide-and-conquer strategy can substantially improve MOEA performance. The decomposition may be made in either criterion-space or decision-space, with a joint decomposition proving the most effective. Criterion-space decomposition is particularly appealing because it reduces the complexity of the trade-off surfaces to be presented to the decision-maker. Furthermore, no loss of trade-off surface shape was observed for the ideal criterion-space decomposition as it was for the sole decomposition of decision variables.

An adaptive criterion-space decomposition methodology has been presented and proof-of-principle results on the concatenated ZDT-1 problem have been shown to be very encouraging. It should be noted that the approach is not confined to criterion-space: independence tests and identified linkages could equally well have been applied to decision variable data. In this case, the sub-populations would evolve different decision variables, whilst the evaluation would be global. It should also be possible to use the technique on both spaces simultaneously.

The main limitation of the methodology is the number of pair-wise comparisons that have to be conducted for high-dimensional spaces. In the concatenated ZDT-1 problem, analysis of the decision variables would be very compute-intensive. Further techniques for the progressive decomposition of high-dimensional spaces are required to complete the framework. Future work will also consider further concatenated problems, especially those of high dimension, with emphasis on real-world applications.

References

1. Purshouse, R.C., Fleming, P.J.: Conflict, Harmony, and Independence: Relationships in Evolutionary Multi-Criterion Optimisation. This volume. (2003)
2. Purshouse, R.C., Fleming, P.J.: Why use Elitism and Sharing in a Multi-Objective Genetic Algorithm? Proceedings of the Genetic and Evolutionary Computation Conference (GECCO 2002). (2002) 520–527
3. Zitzler, E., Laumanns, M., Thiele, L.: SPEA2: Improving the Strength Pareto Evolutionary Algorithm. TIK-Report 103, ETH Zürich. (2001)
4. Fonseca, C.M., Fleming, P.J.: Genetic algorithms for multiobjective optimization: Formulation, discussion and generalization. Proceedings of the Fifth International Conference on Genetic Algorithms. (1993) 416–423
5. Deb, K., Agrawal, R.B.: Simulated binary crossover for continuous search space. Compex Systems **9** (1995) 115–148
6. Deb, K., Goyal, M.: A combined genetic adaptive search (GeneAS) for engineering design. Computer Science and Informatics **26** (1996) 30–45
7. Zitzler, E., Deb, K., Thiele, L.: Comparison of Multiobjective Evolutionary Algorithms: Empirical Results. Evolutionary Computation **8** (2000) 173–195
8. Deb, K., Thiele, L., Laumanns, M., Zitzler, E.: Scalable Multi-Objective Optimization Test Problems. Proceedings of the 2002 IEEE Congress on Evolutionary Computation (CEC 2002), Vol. 1. (2002) 825–830
9. Zitzler, E.: Evolutionary Algorithms for Multiobjective Optimization: Methods and Applications. Doctoral dissertation, ETH Zürich. (1999)
10. Zitzler, E., Laumanns, M., Thiele, L., Fonseca, C.M., Grunert da Fonseca, V.: Why Quality Assessment Of Multiobjective Optimizers Is Difficult. Proceedings of the Genetic and Evolutionary Computation Conference (GECCO 2002). (2002) 666–674
11. Knowles, J.D., Corne, D.W.: On Metrics for Comparing Nondominated Sets. Proceedings of the 2002 IEEE Congress on Evolutionary Computation (CEC 2002), Vol. 1. (2002) 711–716
12. Fonseca, C.M., Fleming, P.J.: On the Performance Assessment and Comparison of Stochastic Multiobjective Optimizers. Parallel Problem Solving from Nature – PPSN IV. Lecture Notes in Computer Science, Vol. 1141 (1996) 584–593
13. Knowles, J.D., Corne, D.W.: Approximating the non-dominated front using the Pareto archived evolution strategy. Evolutionary Computation **8** (2000) 149–172
14. Manly, B.F.J.: Randomization and Monte Carlo Methods in Biology. Chapman and Hall, London New York Tokyo Melbourne Madras (1991)
15. Hollander, M., Wolfe, D.A.: Nonparametric Statistical Methods. 2nd edn. Wiley, New York Chichester Weinheim Brisbane Singapore Toronto (1999)

Multi-level Multi-objective Genetic Algorithm Using Entropy to Preserve Diversity

S. Gunawan, A. Farhang-Mehr, and S. Azarm

Department of Mechanical Engineering, University of Maryland
College Park, Maryland 20742, U.S.A.
{vector@wam, mehr@wam, azarm@eng}.umd.edu

Abstract. We present a new method for solving a multi-level multi-objective optimization problem that is hierarchically decomposed into several sub-problems. The method preserves diversity of Pareto solutions by maximizing an entropy metric, a quantitative measure of distribution quality of a set of solutions. The main idea behind the method is to optimize the sub-problems independently using a Multi-Objective Genetic Algorithm (MOGA) while systematically using the entropy values of intermediate solutions to guide the optimization of sub-problems to the overall Pareto solutions. As a demonstration, we applied the multi-level MOGA to a mechanical design example: the design of a speed reducer. We also solved the example in its equivalent single-level form by a MOGA. The results show that our proposed multi-level multi-objective optimization method obtains more Pareto solutions with a better diversity compared to those obtained by the single-level MOGA.

1 Introduction

A general multi-objective optimization problem can be formulated as follows:

$$\begin{aligned}
&\text{minimize} \quad f(x) = \{f_1,\ldots,f_M\}^t \quad (M \geq 2) \\
&\text{subject to:} \quad g_k(x) \leq 0; \quad k = 1,\ldots,K \\
&\qquad\qquad x_L \leq x \leq x_U
\end{aligned} \quad (1)$$

In this formulation, $x = \{x_1,\ldots,x_N\}^t$, is the design variable vector, f is the design objective vector, and g_k is the kth inequality constraint. x_L and x_U are the lower and upper bounds on the design variables, respectively. For notational simplicity, equality constraints are not shown in the formulation.

Eq. 1 formulates an optimization problem in its "single-level" form. However, because of their large size and multi-subsystem nature, real world engineering problems may have to be structured and solved in a "multi-level" form (Figure 1). In a multi-level form, an optimization problem is composed of several smaller sub-problems. Then, optimization of the sub-problems is coordinated in some prescribed manner in order to obtain the overall solution. The structure shown in Figure 1 is called hierarchical because there is no coupling between the same level sub-problems.

There is a large body of literature on multi-level (or "multi-disciplinary") optimization methods (e.g., Lasdon [1], Balling and Sobieski [2], Sobieski and Haftka [3], and Haimes et al. [4]). Most of these methods were developed for multi-level single-objective problems with continuous variables and differentiable (objective and/or constraint) functions. Nearly all of the methods that were developed for multi-level multi-objective optimization problems convert the problem to a single-objective form (e.g., Haimes et al. [4], Tappeta and Renaud [5]). Such methods can only obtain a single Pareto solution from each run of an optimizer. Gunawan et al. [6] reported initial progress on the development of a two-level MOGA in order to obtain the entire Pareto solutions.

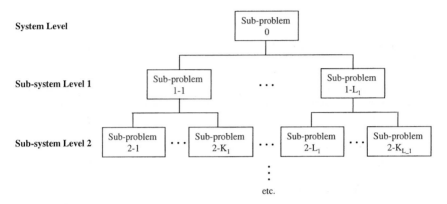

Fig. 1. Multi-level hierarchical structure of an optimization problem

For a multi-objective problem, however, just as it is important to obtain the entire Pareto solutions, it is equally important that the solutions are as diverse as possible [7]. When solving a multi-objective optimization problem in its multi-level form, it is generally difficult to maintain the diversity of the solutions. This is because the overall solutions are biased towards the solutions of the sub-problems. This difficulty in obtaining a diverse set of solutions is compounded further when using a MOGA. This is because MOGA solutions also tend to converge into clusters leaving the rest of the Pareto frontier empty or sparsely populated, a phenomenon known as *genetic drift* [8].

Many different diversity-preserving techniques have been proposed in the literature (e.g., Fonseca and Fleming [9], Srinivas and Deb [10], Horn et al. [11], Narayanan and Azarm [12]). An overview of different diversity-preserving techniques for MOGAs can be found in Deb [13]. The diversity-preserving techniques in the literature are based on meta-heuristic operators. These operators work by modifying the evolutionary pressure put on a MOGA population during an optimization run. Although it has been shown that these techniques can produce relatively diverse solutions, the diversity of the solutions is not optimal. Many of these techniques also require parameters (such as niche size, crowding distance, etc.) that are problem dependent.

In this paper, we present a new method to obtain the Pareto solutions of a multi-level multi-objective optimization problem using MOGA, while at the same time preserving the diversity of those solutions. In preserving the diversity, instead of using heuristics operators, the method explicitly optimizes the "quality" of the solutions as measured by the entropy metric [14] (see Appendix). The MOGA used in this paper is based on our implementation of the MOGA developed by Fonseca and Fleming [9] along with Kurapati et al's constraint handling technique [15].

The rest of the paper is organized as follows. Section 2 gives a formulation of the multi-objective problem of interest. Section 3 gives a detailed description of the method. Section 4 demonstrates an application of the method to a mechanical design example. Section 5 concludes the paper with a brief summary. For completeness, the appendix gives a brief introduction of the entropy metric developed by Farhang-Mehr and Azarm [14].

2   Problem Formulation

The structure of a multi-level multi-objective optimization problem is shown in Figure 2. The problem consists of $J+1$ sub-problems organized into two levels: one sub-problem at the system level, and J sub-problems at the sub-system level. (In this paper, we present our method only for a two level problem, but the method can be readily extended to more than two levels.) Each sub-system sub-problem has a design variable vector, objective vector, and constraint vector that are exclusive to that sub-problem. The system sub-problem has a design variable vector that is shared by the sub-system sub-problems, i.e., a shared variable vector. It also has a system constraint vector that must be satisfied by the shared variable vector.

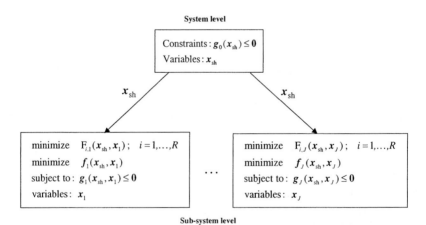

Fig. 2. Multi-level form of a multi-objective optimization problem

In this formulation, x_{sh} denotes the shared variable vector while x_j ($j=1,...,J$) denotes the variable vector that is exclusive to the jth sub-problem. For the constraints, g_0 denotes the system constraint vector while g_j ($j=1,...,J$) denotes the constraint vector that is exclusive to the jth sub-problem. The constraint vector g_0 is a function of x_{sh} only, while constraint vector g_j is a function of both x_{sh} and x_j.

In the sub-system sub-problem, there are two types of objectives: the functionally separable objective and the exclusive objective (denoted by $F_{i,j}$ and f_j, respectively, $j=1,...,J$). A functionally separable objective $F_{i,j}$ ($i=1,...,R$) is an objective function which is formed by combining the corresponding functionally separable objectives from all sub-problems to obtain the objective of the overall problem. Weight is an example of a functionally (i.e., additively) separable objective function since the weight of a product is equal to the sum of the weights of its components. The exclusive objective, on the other hand, is an objective that is unique to a particular sub-system sub-problem. The exclusive objective does not need to be combined with the objective functions from the other sub-problems.

The goal of the optimization approach is to obtain a set of solutions $(x_{sh}, x_1, \ldots, x_J)$ that minimizes all objectives (i.e., functionally separable objectives as well as exclusive objectives) while satisfying the constraints in all sub-problems. Eq. 2 shows the equivalent single-level form of the multi-level problem shown in Figure 2.

$$\begin{aligned} \text{minimize} \quad & \mathbf{F} = \{F_1(x), \ldots, F_i(x), \ldots, F_R(x)\}^{1t} \\ & F_i(x) = function\bigl(F_{i,1}(x_{sh}, x_1), \ldots, F_{i,J}(x_{sh}, x_J)\bigr) \quad ; \quad i = 1, \ldots, R \\ \text{minimize} \quad & f = \{f_1(x_{sh}, x_1), \ldots, f_J(x_{sh}, x_J)\}^{1t} \\ \text{subject to:} \quad & g(x) = \{g_0(x_{sh}), g_1(x_{sh}, x_1), \ldots, g_J(x_{sh}, x_J)\}^{1t} \le 0 \\ & x_L \le x \le x_U \end{aligned} \quad (2)$$

3 Approach

We now describe our method to solve the multi-level problem shown in Figure 2 in order to obtain the most diverse set of solutions. The main idea behind the method is to optimize the sub-problems independently, and to systematically use the entropy value of intermediate solutions to guide the optimization of sub-problems to a diverse set of solution. To solve the sub-problems, the method uses MOGA for each sub-problem. Since our method uses an entropy metric and solves a multi-level problem, we call our method: Entropy-based Multi-level Multi-Objective Genetic Algorithm (EM-MOGA).

In EM-MOGA, the system sub-problem is removed, and the shared variables and constraints (x_{sh} and g_0) are integrated into the sub-system sub-problems. Following this integration, the optimization at each sub-system sub-problem is performed with respect to $\{x_{sh}, x_j\}$ and subject to $\{g_0, g_j\}$, $j = 1,\ldots,J$. The objectives of the sub-problem remain the same. Eq. 3 shows the mathematical formulation of the modified sub-system sub-problem j.

$$\begin{aligned} \text{minimize} \quad & F_{i,j}(x_{sh}, x_j) \; ; \quad i = 1, \ldots, R \\ \text{minimize} \quad & f_j(x_{sh}, x_j) \\ \text{subject to:} \quad & \begin{cases} g_0(x_{sh}) \\ g_j(x_{sh}, x_j) \end{cases} \le 0 \end{aligned} \quad (3)$$

Figure 3 shows the overall schematic of EM-MOGA. Each MOGA at the sub-problems operates on its own population of $\{x_{sh}, x_j\}$. The population size, P, for each sub-problem is kept the same. In addition, EM-MOGA maintains two more populations external to the sub-problems: the grand population and the grand pool.

Both the grand population and the grand pool are populations of complete design variable vector: $\{x_{sh}, x_1, \ldots, x_J\}$. The grand population is an estimate of the solutions of the overall problem. It is EM-MOGA's equivalent of a MOGA population when solving the problem in its single-level form. The grand pool is an archive of the union of solutions generated by the sub-problems. It is a population on which the entropy metric will operate. The size of the grand population is the same as the sub-problem population size, P. The size of the grand pool is J times the size of the sub-system's population.

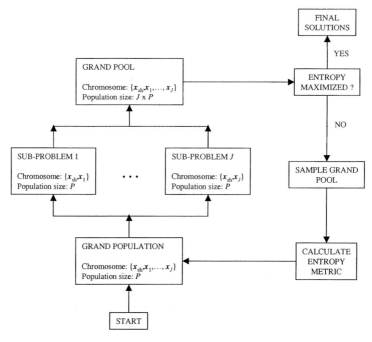

Fig. 3. Overall schematic of EM-MOGA

Initially, the grand population is generated randomly while the grand pool is empty. The chromosomes of the grand population are used as the initial population (or seed) to run the MOGA at each sub-problem. Since the sub-problem MOGA operates on its own variables $\{x_{sh}, x_j\}$, only the chromosomes corresponding to x_{sh} and x_j are used as the seed for the jth sub-problem. Each sub-problem MOGA takes these initial chromosomes and evolves them to optimize its own objectives subject to its own constraints. After each sub-problem MOGA is run for SR iterations, all the MOGAs are terminated. Since there are J independent MOGAs, after the termination, there will be J populations having P chromosomes each. Each of the J populations contains the chromosomes of only $\{x_{sh}, x_j\}$, $j=1,\ldots,J$. To obtain chromosomes of complete design variable vector $\{x_{sh}, x_1, \ldots, x_J\}$, the chromosomes obtained from the jth sub-problem MOGA are completed using the rest of the chromosome sequence $\{x_1, \ldots, x_{j-1}, x_{j+1}, \ldots, x_J\}$ from the grand population.

After the chromosomes in all J populations are re-sequenced to form the complete design variable vector, they are added to the grand pool. The addition of the

J populations to the grand pool marks the end of one EM-MOGA iteration. Before the next EM-MOGA iteration, the grand population is updated based on the chromosomes in the grand pool. The grand pool contains a total of $J \times P$ chromosomes of complete design variable vectors. Each of the chromosomes in the grand pool maps to a point in the objective space of the problem ($\mathbf{F}$ and f of Eq. 2). The goal of EM-MOGA is to obtain solutions that are as diverse as possible in this space. To do so, from the $J \times P$ chromosomes in the grand pool, EM-MOGA picks P chromosomes that maximize the entropy metric (i.e., the most diverse according to the entropy metric [14]) and substitutes the P chromosomes in the grand population with these new P chromosomes.

The problem of choosing P chromosomes that maximize the entropy metric can be posed as a separate optimization problem. However, since the size of the search space for this optimization problem can become very large (the total number of combinations to be searched is $\binom{J \times P}{P}$) it is recommended to perform a simple random sampling instead. For the EM-MOGA implementation, we use a simple random sampling of size $10P$ to search for the set of P chromosomes (or design points) that maximize the entropy metric.

The new chromosomes in the grand population serve as the new seed for the sub-problem MOGAs, and the EM-MOGA iteration continues. EM-MOGA is run until there is no further improvement in the entropy value (i.e., entropy is maximum), and then it is stopped. The solution of the overall problem is obtained by taking the non-dominated frontier of the points in the grand pool of the last EM-MOGA iteration.

The step-by-step algorithm of EM-MOGA is given next.

Step 1: Initialize EM-MOGA's iteration counter, $ITR = 0$. Initialize the entropy value, $ENTRP = 0$.

Step 2: Initialize the grand pool to empty. Create an initial grand population by generating P random chromosomes of complete variables: $\{x_{sh}, x_1, \ldots, x_J\}$.

Step 3: Use the chromosome of $\{x_{sh}, x_j\}$ from the grand population as the initial population of the jth sub-problem MOGA. Run each of the J sub-problem MOGAs for SR iterations.

Step 4: Re-sequence the evolved $\{x_{sh}, x_j\}$ chromosomes from the jth sub-problem MOGA with the $\{x_1, \ldots, x_{j-1}, x_{j+1}, \ldots, x_J\}$ chromosomes from the grand population.

Step 5: Remove all chromosomes from the grand pool, and add all the re-sequenced chromosomes from all sub-problems to the grand pool.

Step 6: If there is no improvement in the entropy value $ENTRP$ for the last GR EM-MOGA iterations, stop EM-MOGA. Obtain the solution of the problem by taking the non-dominated frontier of the points in the grand pool. Else, continue to Step 7.

Step 7: From the $J \times P$ points in the grand pool, randomly pick $10P$ sets of points (each set contains P points). Calculate the entropy of each set in the objective space of the overall problem. Identify S_{MAX}, the set with the highest entropy value.

Step 8: Delete the chromosomes in the grand population. Add the P chromosomes from S_{MAX} to the grand population.
Step 9: Set $ITR = ITR + 1$. Set $ENTRP$ = entropy (S_{MAX}). Go to Step 3.

4 Example

As a demonstration, we applied EM-MOGA to a modified formulation of a speed reducer problem (Figure 4), originally formulated by Golinski [16] as a single-objective problem.

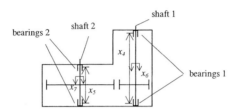

Fig. 4. A speed reducer

Figure 5 shows the speed reducer problem in its multi-level multi-objective optimization form. The problem consists of three sub-problems structured into two levels: one sub-problem at the system level, and two sub-problems at the sub-system level. The system sub-problem has three shared variables: the gear face width (x_1), the tooth module (x_2), and the number of teeth pinion (x_3 – discrete variable). The first sub-system sub-problem (Sub-problem-1) has two exclusive variables: the distance between bearings 1 (x_4) and the diameter of shaft 1 (x_6). The second sub-system sub-problem (Sub-problem-2) also has two exclusive variables: the distance between bearings 2 (x_5) and the diameter of shaft 2 (x_7). Upper and lower bounds are imposed on each of the design variables.

The system sub-problem has five constraints: an upper bound on the bending stress of the gear tooth (g_1), an upper bound on the contact stress of the gear tooth (g_2), and dimensional restrictions based on space and/or experience (g_3, g_4, g_5). The first sub-system sub-problem has three exclusive constraints: an upper bound on the transverse deflection of shaft 1 (g_6), design requirement on shaft 1 based on experience (g_7), and an upper bound on the stress in shaft 1 (g_8). The second sub-system sub-problem also has three exclusive constraints: an upper bound on the transverse deflection of shaft 2 (g_9), design requirement on shaft 2 based on experience (g_{10}), and an upper bound on the stress in shaft 2 (g_{11}).

Sub-problem-1 has two objectives: to minimize the volume of the gears and shaft 1 ($F_{1,1}$), and to minimize the stress on shaft 1 (f_1). Sub-problem-2 also has two objectives: to minimize the volume of shaft 2 ($F_{1,2}$), and to minimize the stress on shaft 2 (f_2). $F_{1,1}$ and $F_{1,2}$ are additively separable objectives. They must be added to obtain the objective of the overall problem, i.e., $F_1 = F_{1,1} + F_{1,2}$. On the other hand, f_1 and f_2 are exclusive objectives. So, in total, the overall problem has three objectives: $F_1, f_1,$ and f_2.

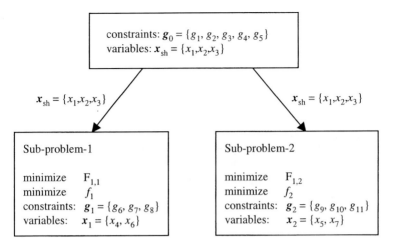

Fig. 5. Multi-level multi-objective speed reducer example

The detailed mathematical formulation of the system sub-problem, the sub-system sub-problem 1, and the sub-system sub-problem 2 are shown by Eqs. 4, 5, and 6, respectively.

SYSTEM SUB-PROBLEM:

$$g_1 : \frac{1}{\left(x_1 x_2^2 x_3\right)} - \frac{1}{27} \leq 0 \qquad g_2 : \frac{1}{\left(x_1 x_2^2 x_3^2\right)} - \frac{1}{397.5} \leq 0$$

$$g_3 : x_2 x_3 - 40 \leq 0 \qquad g_4 : \frac{x_1}{x_2} - 12 \leq 0$$

$$g_5 : 5 - \frac{x_1}{x_2} \leq 0$$

$$2.6 \leq x_1 \leq 3.6$$
$$0.7 \leq x_2 \leq 0.8$$
$$17 \leq x_3 \leq 28$$

(4)

SUB-PROBLEM 1:

$$\text{minimize } F_{1.1} = 0.7854 x_1 x_2^2 \left(\frac{10 x_3^2}{3} + 14.933 x_3 - 43.0934 \right)$$
$$- 1.508 x_1 x_6^2 + 7.477 x_6^3 + 0.7854 x_4 x_6^2$$

$$\text{minimize } f_1 = \frac{\sqrt{\left(\frac{745 x_4}{x_2 x_3}\right)^2 + 1.69 \times 10^7}}{0.1 x_6^3}$$

$$\text{subject to : } g_6 : \frac{x_4^3}{\left(x_2 x_3 x_6^4\right)} - \frac{1}{1.93} \leq 0$$

$$g_7 : 1.9 - x_4 + 1.5 x_6 \leq 0$$
$$g_8 : f_1 - 1300 \leq 0$$
$$7.3 \leq x_4 \leq 8.3$$
$$2.9 \leq x_6 \leq 3.9$$

(5)

SUB-PROBLEM 2:

$$\text{minimize} \quad F_{1,2} = -1.508 x_1 x_7^2 + 7.477 x_7^3 + 0.7854 x_5 x_7^2$$

$$\text{minimize} \quad f_2 = \frac{\sqrt{\left(\dfrac{745 x_5}{x_2 x_3}\right)^2 + 1.575 \times 10^8}}{0.1 x_7^3}$$

$$\text{subject to:} \quad g_9 = \frac{x_5^3}{\left(x_2 x_3 x_7^4\right)} - \frac{1}{1.93} \leq 0$$

$$g_{10} : 1.9 - x_5 + 1.1 x_7 \leq 0$$

$$g_{11} : f_3 - 1100 \leq 0$$

$$7.3 \leq x_5 \leq 8.3 \tag{6}$$

$$5.0 \leq x_7 \leq 5.5$$

Table 1 shows the parameters used in applying EM-MOGA to the above problem. For comparison purposes, in this example, we stop EM-MOGA after *GR* EM-MOGA iterations.

Table 1. EM-MOGA parameters for the speed reducer example

Parameter	Value
Chromosomes length (bits per variable)	
Continuous variable	10
Discrete variable	5
Population size (P)	100
MOGA population replacement	20
MOGA crossover probability	0.85
MOGA mutation probability	0.10
MOGA crossover type	Two-Point Crossover
MOGA selection type	Stochastic Universal Selection
Number of MOGA iterations (SR)	300
Number of EM-MOGA iterations (GR)	10

The Pareto points obtained by EM-MOGA are shown in Figure 6. We have shown the points projected into a two-dimensional plane. Since there are three objectives, F_1, f_1, and f_2, we have three projections: F_1-f_1, F_1-f_2, and f_1-f_2 projections. For comparison, we solved the same problem in its single-level form using MOGA. The single-level form of the speed reducer problem is shown in Eq. 7.

$$\begin{aligned}
\text{minimize} \quad & F_1(x) = F_{1,1}(x_1, x_2, x_3, x_4, x_6) + F_{1,2}(x_1, x_2, x_3, x_5, x_7) \\
\text{minimize} \quad & f_1(x_1, x_2, x_3, x_4, x_6) \\
\text{minimize} \quad & f_2(x_1, x_2, x_3, x_5, x_7) \\
\text{subject to:} \quad & \begin{Bmatrix} g_1, g_2, g_3, g_4, g_5, g_6 \\ g_7, g_8, g_9, g_{10}, g_{11} \end{Bmatrix}^t \leq \mathbf{0} \\
& x = \{x_1, x_2, x_3, x_4, x_5, x_6, x_7\}^t \\
& x_L \leq x \leq x_U
\end{aligned} \tag{7}$$

The parameters used in solving the single-level problem are the same as those shown in Table 1 except that MOGA is allowed to run for 3000 iterations (instead of

$SR=300$). This is done so that the number of function calls used in both EM-MOGA and single-level MOGA is the same. The Pareto points obtained by the single-level MOGA are also shown in Figure 6.

From Figure 6, we observe that while EM-MOGA estimates the Pareto frontier similar to MOGA, EM-MOGA produces more distinct solutions compared to MOGA. Using a population size of 100 points, MOGA found only 32 distinct Pareto points (6 points are inferior points, while the other 62 points are duplicates of the 32 distinct Pareto points - effect of genetic drift). On the other hand, using the same population size of 100 for the sub-problem MOGAs and the grand population, EM-MOGA found 114 distinct Pareto points (recall that since there are 2 sub-problems, the size of the grand pool is $2 \times 100 = 200$ points).

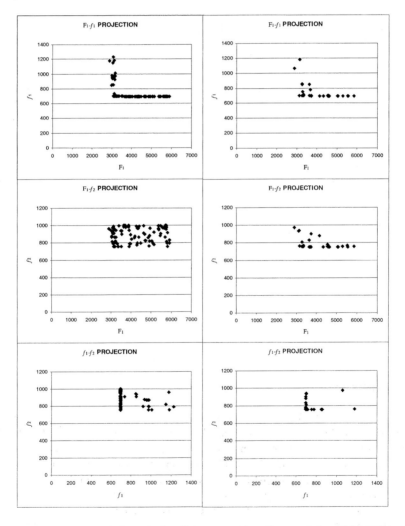

Fig. 6. Pareto frontier of the speed reducer example. Generated by EM-MOGA (*left*). Generated by MOGA (*right*)

Figure 7 shows the changes in the entropy value of both EM-MOGA and single level MOGA during the optimization run. The horizontal axis shows the progress of the optimization run (in percentage of process completed), while the vertical axis shows the entropy value of intermediate solutions. We observe from Figure 7 that at the completion of the process, EM-MOGA solutions have a higher entropy value than the MOGA solutions. Thus, from the entropy metric point of view, EM-MOGA solutions are more diverse and uniformly distributed than MOGA solutions. In addition, Figure 7 shows that the entropy value of MOGA saturates after the process is approximately 60% completed (MOGA can no longer improve diversity). The entropy value of EM-MOGA, on the other hand, continues to improve until the end of the process. The Figure also shows that the entropy values of both methods are fluctuating throughout the optimization run (i.e., the entropy value is not monotonically increasing). This is expected for the single-level MOGA since it is a stochastic method. Since EM-MOGA uses MOGAs in its sub-problems, EM-MOGA also exhibits this fluctuating property.

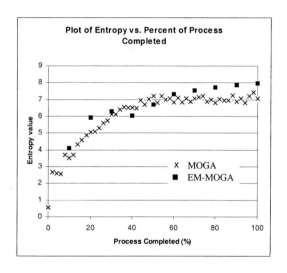

Fig. 7. Changes in entropy value

5 Summary

We have presented EM-MOGA, a new method for solving multi-level multi-objective optimization problems while preserving diversity of Pareto solutions. EM-MOGA preserves the diversity of solutions by explicitly maximizing an entropy quality metric. The main idea behind EM-MOGA is to optimize sub-problems independently using a MOGA, while systematically using the entropy value of intermediate solutions to guide the optimization of the sub-problems to an overall Pareto solution set. As a demonstration, we applied EM-MOGA to a mechanical design problem: the design of a speed reducer, decomposed into a two-level form. For comparison, we also solved the problem in its single-level form using MOGA. We observed from the results that EM-MOGA obtained more Pareto points compared to MOGA. In addition, EM-MOGA Pareto points are more diversified along the Pareto frontier.

Acknowledgments. The work presented in this paper was supported in part by a contract from Indian Head Division - NSWC. Indian Head Division was funded by the U.S. Office of Naval Research. The work was also supported in part by the U.S. National Science Foundation Grant 0112767.

References

1. Lasdon, L. S.: Optimization Theory for Large Systems. The Macmillan Company, London (1970)
2. Balling, R.J. and Sobieszczanski-Sobieski, J.: Optimization of Coupled Systems: A Critical Overview of Approaches. AIAA Journal, Vol. 34, No. 1 (1996) 6–17
3. Sobieszczanski-Sobieski, J. and Haftka, R.T.: Multidisciplinary Aerospace Design Optimization - Survey of Recent Developments. Structural Optimization, Vol. 14, No. 1 (1997) 1–23
4. Haimes, Y.Y., Tarvainen, K., Shima, T., and Thadathil, J.: Hierarchical Multiobjective Analysis of Large-Scale Systems. Hemisphere Publishing Corporation, New York (1990)
5. Tappeta, R.V. and Renaud, J.: Multiobjective Collaborative Optimization. Transactions of the ASME, Journal of Mechanical Design, Vol. 119 (1997) 403–411
6. Gunawan, S., Azarm, S., Wu, J., and Boyars, A.: Quality Assisted Multi-Objective Multi-Disciplinary Genetic Algorithms. CD-ROM Proceedings of the 9th AIAA/ISSMO Symposium on Multidisciplinary Analysis and Optimization, Atlanta, Georgia, September 4–6 (2002)
7. Deb, K.: Multi-Objective Genetic Algorithms: Problem Difficulties and Construction of Test Problems. IEEE Journal of Evolutionary Computation 7 (1999) 205–230
8. Goldberg, D.E.: Genetic Algorithms in Search, Optimization and Machine Learning. Addison-Wesley, Reading, Massachusetts (1989)
9. Fonseca, C. M., Fleming, P. J.: Genetic Algorithms for Multiobjective Optimization: Formulation, Discussion, and Generalization. Proceedings of the 5th International Conference on Genetic Algorithms (1993) 416–423
10. Srinivas, N., Deb, K.: Multi-Objective Function Optimization using Non-Dominated Sorting Genetic Algorithms. Evolutionary Computation, 2(3) (1994) 221–248
11. Horn, J., Nafploitis, N., Goldberg, D.E.: A Niched Pareto Genetic Algorithm for Multi-Objective Optimization. Proceedings of the 1st IEEE Conference on Evolutionary Computation (1994) 82–87
12. Narayanan, S., Azarm, S.: On Improving Multiobjective Genetic Algorithms for Design Optimization. Structural Optimization 18 (1999) 146–155
13. Deb, K.: Multi-Objective Optimization using Evolutionary Algorithms. Wiley-Interscience series in systems and optimization (2001)
14. Farhang-Mehr, A., and Azarm, S.: Diversity Assessment of Pareto Optimal Solutions: An Entropy Approach. CD-ROM Proceedings of the IEEE 2002 World Congress on Computational Intelligence, Honolulu, Hawaii, May 12–17, (2002)
15. Kurapati, A., Azarm, S., and Wu, J.: Constraint Handling Improvements for Multi-Objective Genetic Algorithms. Structural and Multidisciplinary Optimization, Vol. 23, No. 3 (2002) 204–213
16. Golinski, J.: Optimal Synthesis Problems Solved by Means of Nonlinear Programming and Random Methods. Journal of Mechanisms, Vol. 5 (1970) 287–309
17. Shannon, C. E.: A Mathematical Theory of Communication. Bell System Technical Journal, 27 (1948) 379–423 and 623–656 (July and October)

Appendix: Entropy Quality Metric

In a recent paper [14], Farhang-Mehr and Azarm developed an entropy-based metric that can be used to assess the quality of a set of solutions obtained for a multi-objective optimization problem. This metric quantifies the 'goodness' of a set of solutions in terms of its distribution quality over the Pareto frontier. The following is a brief description of the entropy metric (for more details see [14]).

The basic idea behind the entropy metric is that each solution point provides some information about its neighborhood that can be modeled as a function, called an influence function. Figure 8, for instance, shows several solution points along a line segment (one-dimensional). The impact of each solution point on its neighborhood is modeled by a Gaussian influence function. The influence function of the ith solution point is maximum at that point and decreases gradually with the distance. Now, the density function, D, is defined as the aggregation of the influence functions from all solution points, i.e.,

$$D(y) = \sum_{i=1}^{N} \Omega_i \left(r_{i \rightarrow y} \right) \qquad (8)$$

where y represents an arbitrary point, $r_{i \rightarrow y}$ is a scalar that represents the Euclidean distance of the point y and the ith solution point, and $\Omega_i(.)$ is the influence function of the ith solution point. The solution set contains N solution points.

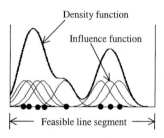

Fig. 8. A set of solution points in a one-dimensional feasible space with the corresponding influence and density functions

The density curve of Figure 8 (or density hyper-surface for 2 or more dimensions) consists of peaks and valleys. The peaks correspond to dense areas with many solution points in the vicinity and the valleys correspond to sparse areas with few adjacent points. Farhang-Mehr and Azarm [14] showed that there is a direct relationship between the flatness of the density function and the distribution uniformity of the solution set. A set of uniformly distributed solutions yields a relatively even surface without significant peaks and valleys. In contrast, if the solution points are grouped into one or more clusters, leaving the rest of the area sparsely populated, the density function contains sharp peaks and deep valleys.

So far, we have established a relationship between the goodness of a distribution of a solution set and the flatness of its corresponding density surface. In the following, the entropy metric is formulated to quantify this flatness.

Claude Shannon [17] introduced the information-theoretic entropy to measure the information content of a stochastic process. Assume a stochastic process with n possible outcomes where the probability of the ith outcome is pi. The probability distribution of this process can be shown as:

$$\mathbf{P} = [p_1,\ldots, p_i,\ldots, p_n]\,; \quad \sum_{i=1}^{n} p_i = 1\,; \quad p_i \geq 0 \quad (9)$$

This probability vector has an associated Shannon's entropy, H, of the form:

$$H(\mathbf{P}) = -\sum_{i=1}^{n} p_i \ln(p_i) \quad (10)$$

where $p_i \ln(p_i)$ approaches zero as p_i goes to zero asymptotically. This function is at its maximum, $H_{max}= \ln(n)$, when all probabilities have the same value, and at its minimum of zero when one component of P is 1 and the rest of entries are zero. In fact, the Shannon's entropy measures the uniformity of P, i.e., if the values of the entries in the vector are approximately the same then the entropy is high, but if the values are very different (uneven probability distribution), the corresponding entropy is low.

As mentioned before, a desirable solution set must have a 'flat' density surface. To quantify this flatness, one may take advantage of the formal similarities between this problem and the Shannon's entropy, which also measures the flatness of a distribution. This can by done by constructing a mesh in the solution hyper-plane and normalizing the values of density function, measured at the nodes:

$$\rho_i = \frac{D_i}{\sum_i D_i} \quad (11)$$

where D_i represents the density function at the ith node. Now we have:

$$\sum_i \rho_i = 1 \quad ; \rho_i \geq 0 \quad (12)$$

The entropy of such a distribution can then be defined as:

$$H = -\sum_i \rho_i \ln(\rho_i) \quad (13)$$

The entropy metric, as obtained from Equation (13), measures the distribution quality of a set of solution points. We arrive at a conclusion: *A solution set with a higher entropy is spread more evenly and covers a larger area on the solution hyper-plane* [14].

Solving Hierarchical Optimization Problems Using MOEAs*

Christian Haubelt[1], Sanaz Mostaghim[2], Jürgen Teich[1], and Ambrish Tyagi[2]

[1] Department of Computer Science 12
Hardware-Software-Co-Design
University of Erlangen-Nuremberg
{christian.haubelt, teich}@cs.fau.de
[2] Computer Engineering Laboratory (DATE)
Department of Electrical Engineering and Information Technology
University of Paderborn
{mostaghim, tyagi}@date.upb.de

Abstract. In this paper, we propose an approach for solving hierarchical multi-objective optimization problems (MOPs). In realistic MOPs, two main challenges have to be considered: (i) the complexity of the search space and (ii) the non-monotonicity of the objective-space. Here, we introduce a hierarchical problem description (chromosomes) to deal with the complexity of the search space. Since *Evolutionary Algorithms* have been proven to provide good solutions in non-monotonic objective-spaces, we apply genetic operators also on the structure of hierarchical chromosomes. This novel approach decreases exploration time substantially. The example of system synthesis is used as a case study to illustrate the necessity and the benefits of *hierarchical optimization*.

1 Introduction

The increasing complexity of typical search spaces demands new strategies in solving optimization problems. One possibility to overcome the large computation times is the hierarchical decomposition of the search as well as the objective space. The decomposition of optimization problems was already mentioned in [1] and formalized in [2]. While [1] only shows experimental results, Abraham et al. [2] discuss a very special kind of search space which possesses certain monotonicity properties that do not hold in SoC design.

Since Multi-Objective Evolutionary Algorithms (MOEAs) [3] provide good results in non-monotonic optimization problems, we propose an extension of MOEAs towards hierarchical chromosomes. By using hierarchical chromosomes, we capture the knowledge about the search space decomposition. The concept of hierarchical chromosomes is based on the idea of regulatory genes as described in [4] where the activation and deactivation of genes is used to adapt to non-stationary functions. In this paper, the structure of the chromosome itself affects

---

* This work was supported in part by the German Science Foundation (DFG), SPP 1040.

the fitness calculation, i.e., also the structure of the chromosomes is subject to the optimization. Therefore, two new genetic operators are introduced: (i) *composite mutation* and (ii) *composite crossover*.

In this paper, we consider the example of system synthesis as a case study which includes binding and allocation problems to illustrate the necessity and benefits of hierarchical optimization. When applying hierarchical optimization to system synthesis, we have to consider two extensions: a) *hierarchical problem decomposition* and b) *hierarchical design space exploration*. In previous approaches such as [5], [6], etc., both the application specification and the architecture are modeled non-hierarchically. Here, we introduce a hierarchical approach to model embedded systems. This hierarchical structure could be coded directly in hierarchical chromosomes. We will show by experiment that for our particular problem the computation time required for the optimization decreases substantially.

This paper is organized as follows: Section 2 introduces the problem of hierarchical optimization. Also first approaches to code these problems in hierarchical chromosomes including the genetic composite operators are presented. Afterwards, we apply this novel approach to the task of system synthesis. The benefits of hierarchical optimization in system synthesis are illustrated in Section 4.

2 Hierarchical Optimization Problems

This section describes the formalization of hierarchical optimization problems. Starting from (non-hierarchical) multi-objective optimization problems, we introduce the notion of *hierarchical decision and hierarchical objective spaces*.

2.1 Multi-objective Optimization and Pareto-Optimality

First, we give a formal notation of multi-objective optimization problems.

Definition 1 (Multi-Objective Optimization Problem (MOP)). *A multi-objective optimization problem (MOP) is given by:*

$$\text{minimize } o(x), \text{ subject to } c(x) \leq 0$$

where $x = (x_1, x_2, \ldots, x_m) \in X$ is the decision vector and X is called the search space. Furthermore, the constraints $c(x) \leq 0$ determine the set of feasible solutions, where c is k-dimensional, i.e., $c(k) = (c_1(x), c_2(x), \ldots, c_k(x))$.

The *objective function* o is n-dimensional, i.e., we optimize n objectives simultaneously. There are q constraints c_i, $i = 1, \ldots, q$. Only those *decision vectors* $x \in X$ that satisfy all constraints c_i are in the set of feasible solutions, or for short in the *feasible set* called $X_\text{f} \subseteq X$. The image of X is defined as $Y = o(X) \subset \mathbb{R}^n$, where the objective function o on the set X is given by $o(X) = \{o(x) \mid x \in X\}$. Analogously, the *objective space* is denoted by $Y_\text{f} = o(X_\text{f}) = \{o(x) \mid x \in X_\text{f}\}$.

Since we are dealing with multi-objective optimization problems, there is generally not only one global optimum, but a set of so-called *Pareto-points* [7]. A Pareto-optimal solution x_p is a decision vector which is not worse than any

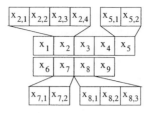

Fig. 1. Example of a hierarchical decision vector consisting of a non-hierarchical part x_1, x_3, x_4, x_6, x_9 and a hierarchical part x_2, x_5, x_7, x_8

other decision vector $\tilde{x} \in X$ in all objectives. The set of all Pareto-optimal solutions is called the *Pareto-optimal set*, or the *Pareto-set* X_p for short. An approximation of the Pareto-set X_p will be termed *quality set* X_q subsequently.

2.2 Hierarchical Search Spaces and Hierarchical Objective Spaces

In the following, we assume that the search space has a hierarchical structure, i.e., each element x_i of a decision vector x itself may be a vector $(x_{i1}, x_{i2}, \ldots, x_{ik_i})$. In general, the decision vector x consists of a non-hierarchical and a hierarchical part, i.e., $x = (x^n, x^h)$. Each element $x_j^h \in x^h$ itself may be a decision vector consisting of a non-hierarchical and a hierarchical part. Hence, this structure is not limited to only a single level of hierarchy.

Example 1. Fig. 1 shows the example of a hierarchical decision vector x. The non-hierarchical part x^n of the decision vector is given by the elements x_1, x_3, x_4, x_6, and x_9. The hierarchical part x^h of x consists of four elements x_2, x_5, x_7, and x_8.

By using hierarchical decision vectors, we must reconsider the objective functions, too. On the top-level, the objective function is given by: $o(x_1, x_2, \ldots, x_9)$. Since we do not assume monotonicity for the objective functions, we must introduce a decomposition operator $\otimes$ to construct the top-level objective function from deeper levels of hierarchy, i.e., $o(x_1, x_2, \ldots, x_9) = o^n(x_1, x_3, x_4, x_6, x_9) \otimes o^h(x_2, x_3, x_4, x_6, x_9)$, where o^n denotes the partial objective function for the non-hierarchical part of the decision vector. o^h denotes the partial objective function for the elements of the hierarchical part of the decision vector. In general, $o(x) = o^n(x^n) \otimes o^h(x^h)$ where $o^h(x^h) = \bigotimes_{x_i^h} o(x_i^h)$.

Abraham et al. name three advantages of hierarchical decomposition [2]:

1. The size of each search space of a decision vector $x_i^h \in x^h$ in the hierarchical part x^h of the decision vector is smaller than the top-level search space.
2. The evaluation effort for each decision vector $x_i^h \in x^h$ of the hierarchical part x^h is low because these elements are less complex.
3. The number of top-level decision vectors to be evaluated is a small fraction of the size of the original search space, when using a hierarchical search space.

The last advantage refers to the fact that a feasible decision vector must be composed of feasible decision vectors of the elements in the subsystem. Thus,

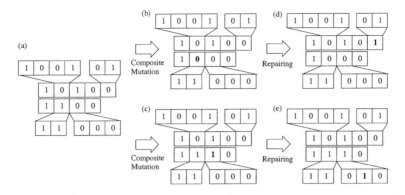

Fig. 2. Example of composite mutation in hierarchical chromosome

we are able to restrict our search space dramatically. Unfortunately, we cannot generally assume that a Pareto-optimal top-level decision vector is composed of Pareto-optimal decision vectors of the elements in the hierarchical part. Abraham et al. define necessary and sufficient conditions of the decomposition function of the objectives which guarantee Pareto-optimality for the top-level decision vector depending on the Pareto-optimality of elements in the hierarchical part of the decision vector [2]. Pareto-optimal solutions of a top-level decision vector are only composable of Pareto-optimal solutions of the decision vector $x_i^h \in x^h$ iff the decomposition function of each objective function is a monotonic function.

Although this observation is important and interesting, many optimization goals unfortunately do not possess these monotonicity properties. In fact, they depend on the decomposition operator $\otimes$.

2.3 Chromosome Structure for Hierarchical Decision Vectors

Due to the non-monotonicity property of the decomposition of the objective function, heuristic techniques are preferable for the optimization of hierarchically coded problems. Furthermore, Evolutionary Algorithms (EAs) seem to be good at exploring selection and assignment problems. Since we are dealing with multi-objective optimization problems, we propose a special chromosome structure particular to our problem for Multi-Objective Evolutionary Algorithms.

In this paper, we focus on problems where the hierarchical part x^h of the decision vector corresponds to a selection problem, i.e., the selection of an element $x_i^h \in x^h$ implies the selection of at least one subelement $x_{i,j}^h \in x_i^h$. For example, consider Fig. 1. The selection of element x$_2$ implies the selection of at least one of the elements x$_{2,1}$, x$_{2,2}$, x$_{2,3}$, or x$_{2,4}$. Many problems like the selection of scenarios supported by a network processor can be coded that way [8]. This results in a number of network processors optimized for different scenarios.

The advantage of such a strategy should be clear. As in the example of the network processors, we are not only interested in a single type of network processor. In fact, we would like to design a diversity of processors special to different

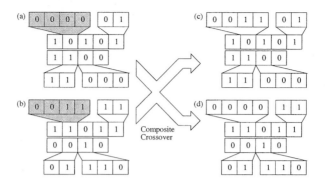

Fig. 3. Example of composite crossover applied to a hierarchical chromosome resulting in one valid and one invalid decision vector

packet streams in type and number. The knowledge about an optimal implementation of a simple network processor may help to design more sophisticated types.

The novelty of this approach is that not only parameters are optimized but also the structure of such a problem. Note, this differs from [4] where redundant chromosomes are provided for adapting to non-stationary environments. Thus, we need to define two new genetic operators regarding the structure of the chromosomes, namely (i) *composite mutation* and (ii) *composite crossover*.

Composite Mutation. The hierarchical nodes $x_i^h \in x^h$ in a hierarchical chromosome directly encode the use of associated elements $x_{i,j}^h \in x_i^h$ of the decomposition of the problem. The composite mutation of a hierarchical chromosome changes a selection bit in a selection list L_S from selected to deselected, or vice versa. As a result, leaves of the chromosome are selected or deselected. Fig. 2 shows the two cases that may occur during composite mutation.

Here, we assume that at least one element $x_i^h \in x^h$ has to be selected and if an element x_i^h is selected at least one associated subelement $x_{i,j}^h \in x_i^h$ must be selected, too. Fig. 2(a) shows one valid assignment to our decision vector.

In Fig. 2(b), we see the deselection of an element $x_i^h \in x^h$. Since no element $x_i^h \in x^h$ of the hierarchical part x^h is selected, the assignment is invalid. Hence, in Fig. 2(d) one of the hierarchical elements x_i^h is selected. In the second case (Fig. 2(c)), an element $x_i^h \in x^h$ is selected leading to an invalid assignment (no associated subelement is selected). Thus, we randomly choose one of the associated subelements $x_{i,j}^h \in x_i^h$ (Fig. 2(e)).

Composite Crossover. The second operator is called composite crossover. An example of how composite crossover works is shown in Fig. 3. Two individuals (Fig. 3(a) and (b)) are cut at the same hierarchical node in the chromosome. After that we interchange these two subvectors. This results in two new decision vectors as depicted in Fig. 3(c) and (d). This operation again may invalidate one

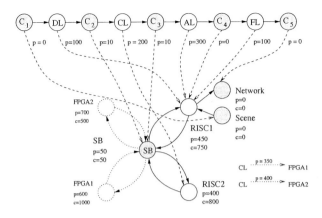

Fig. 4. Specification graph for an MPEG4 coder

or both assignments (as in Fig. 3(d)). We can repair this infeasibility by randomly choosing one of the associated subelements of a new selected hierarchical element.

3 Case Study: Hierarchical Optimization in System Synthesis

This section illustrates a way to code hierarchical optimization problems. As case study, we use the example of system synthesis. Since Evolutionary Algorithms have been proven to be a good optimization technique in system design [5], we extend this approach by proposing a hierarchical chromosome structure, the objective space decomposition, and the genetic composite operators.

3.1 Specification of Embedded Systems

Blickle et al. propose a graph-based approach for embedded system optimization and synthesis [5]. We introduce this model as a starting point and derive our enhanced hierarchical model subsequently based on this basic model.

The specification model [5] consists of two main components:

- A given functional specification that should be mapped onto a suitable architecture of hardware components as well as the class of possible architectures are described each by means of a universal directed graph $g(V, E)$.
- The user-defined mapping constraints E_m between tasks and architecture components are specified in a specification graph g_s. Additional parameters which are used for formulating the objective functions and further functional constraints may be assigned to either vertices or edges of g_s.

Example 2. The example used throughout this paper is an MPEG4 coder. The problem graph is shown in the upper part of Fig. 4. We start with a given scene

(C_1). This scene is decomposed into audio/visual objects (AVO) like images, video, speech, etc. using the decomposition layer (DL). Each AVO is coded by an appropriate coding algorithm (CL). In the next step (Access Unit Layer, AL), the data are provided with time stamps, data type, etc. The FlexMux Layer (Flexible Multiplexer, FL) allows to group streams with the same quality of service requirements and sends the data to the network (C_5). Between two data flow dependent operations, we insert an additional vertex in order to model the required communication.

The processes of the problem graph are mapped onto a target architecture shown in Fig. 4 (lower part), consisting of two RISC processors, two field programmable gate arrays (FPGAs), and a single shared bus (SB). Additionally, the processor RISC1 is equipped with two special ports.

The mapping edges (dashed edges) relate the vertices of the problem graph to vertices of the architecture graph. The edges represent user-defined mapping constraints in the form of a relation: "can be implemented by". For the purpose of better visibility, additional mapping edges are depicted in the lower right corner of Fig. 4. The mapping edges are annotated with additional power consumptions which arise when a mapping edge is used in the implementation. Furthermore, all resources in Fig. 4 are annotated with allocation cost and power consumptions. These values have to be taken into account if the corresponding resource is used in an implementation of the problem.

3.2 Implementation and the Task of System Synthesis

An implementation, being the result of a system synthesis, consists of two parts:

1. the *allocation* α that indicates which elements of the problem and architecture graph are used in the implementation and
2. the *binding* β, i.e., the set of mapping edges which define the binding of vertices in the problem graph to components of the architecture graph.

The term implementation will be used in the same sense as formally defined in [5]. It is useful to determine the set of feasible allocations and feasible bindings: Given a specification graph g_s and an allocation α, a *feasible binding* is a binding β that satisfies the following requirements:

1. Each mapping edge $e \in \beta$ starts and ends at an allocated vertex.
2. For each problem graph vertex being part of the allocation, exactly one outgoing mapping edge is part of the binding β.
3. For each allocated problem graph edge $e \in (v_i, v_j)$:
   – either both operations v_i and v_j are mapped onto the same vertex
   – or they are mapped on adjacent resources connected by an allocated edge to establish the required communication.

With these definitions, we can define the feasibility of an allocation: A *feasible allocation* is an allocation α that allows at least one feasible binding β.

We define an implementation by means of a feasible allocation and binding.

Definition 2 (Implementation). *Given a specification graph g_s, a (valid or feasible) implementation is a pair (α, β) where α is a feasible allocation and β is a corresponding feasible binding.*

Example 3. The dashed mapping edges shown in Fig. 4 indicate a binding for all processes: $\beta = \{(C_1, \text{Scene}), (\text{DL}, \text{RISC1}), (C_2, \text{SB}), (\text{CL}, \text{RISC2}), (C_3, \text{SB}), (\text{AL}, \text{RISC1}), (C_4, \text{RISC1}), (\text{FL}, \text{RISC1}), (C_5, \text{Network})\}$. The allocation of vertices in the architecture graph is: $\alpha = \{\text{SB}, \text{RISC1}, \text{RISC2}, \text{Network}, \text{Scene}\}$. Given this allocation and binding, one can see that our implementation is indeed feasible, i.e., α and β are feasible.

With the model introduced previously, the task of system synthesis can be formulated as a multi-objective optimization problem.

Definition 3 (System Synthesis). *The task of system synthesis is the following multi-objective optimization problem:*
$$\text{minimize } o(\alpha, \beta),$$
subject to:
$\quad \alpha$ is a feasible allocation,
$\quad \beta$ is a feasible binding,
$\quad c_i(\alpha, \beta) \leq 0, \ \forall i \in \{1, \ldots, q\}.$

The constraints on α and β define the set of valid implementations. Additionally, there are functions $c_i, i = 1, \ldots, q$, that determine the set of feasible solutions. All possible allocations α and bindings β span the design space X.

The task of system synthesis defined in Definition 3 is similar to the optimization problem defined in Definition 1 where α and β correspond to the decision vector x. In the following, we show how to solve this (non-hierarchical) optimization by using Evolutionary Algorithms as described in [5].

3.3 Chromosome Structure for System Synthesis

We start with the chromosome structure of an EA for non-hierarchical specification graphs as described in [5]: Each individual consists of two components, an *allocation* and a *binding*, as defined in the previous section. The mapping task as described in [5] can be divided in three steps:

1. The allocation of resources is decoded from the individual and repaired with a simple heuristic,
2. The binding is performed, and
3. The allocation is updated in order to eliminate unnecessary resources and all necessary edges in the architecture graph are added to the allocation.

One iteration of this loop results in a feasible allocation and binding of the vertices and edges of the problem graph g_p to the vertices and edges of the architecture graph g_a. If no feasible binding could be found, the whole decoding of the individual is aborted.

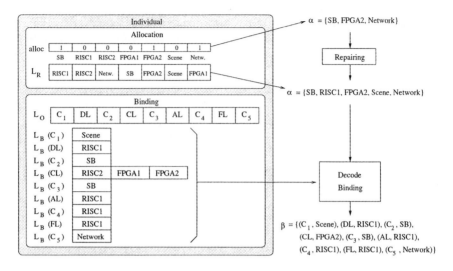

Fig. 5. An example of the coding of an allocation

The allocation of resources is directly encoded as bit string in the chromosome. This simple coding might result in many infeasible allocations. A simple repair heuristic only adds new resources to infeasible allocation until each process could be bound to at least one resource [5]. The order in which additional resources are added is automatically adapted by a *repair allocation priority list* L_R. This list also undergoes genetic operators.

The problem of coding the binding lies in the strong inter-dependence of the binding and the allocation. As crossover or mutation might change the allocation, a directly encoded binding could be meaningless for a different allocation. Hence, an allocation independent coding of the binding is used: A binding order list L_O determines the next problem graph vertex to be bound. For each problem graph vertex, a priority list L_B containing all adjacent mapping edges is coded. The first edge in L_B that gives a feasible binding is included in the binding. Note that it is possible that no feasible binding is specified by the individual. In this case, β is the empty set, and the individual will be given a penalty value as its fitness value.

Finally, resources that are not used will be removed from the allocation. Furthermore, all edges $e \in E_a$ in the architecture graph g_a are added to the allocation that are necessary to obtain a feasible binding.

3.4 Hierarchical Modeling

Many applications are composed of alternative functions and algorithms. For example, the coding layer of the MPEG4 coder could be refined by several different coding schemes. We can apply the same refinement also to the architecture graph in order to model different IP cores placed on an FPGA, etc. Thus, hierarchy

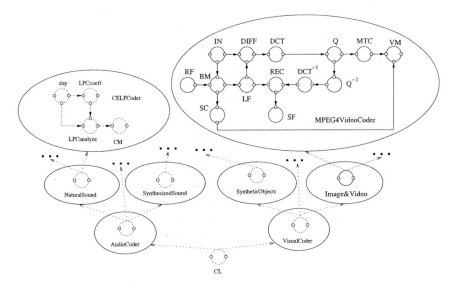

Fig. 6. Refinements of the coding layer of the MPEG4 coder shown in Fig. 4

is used to model the designer's knowledge about possible refinements and the decomposition of the system.

To model these refinements, we propose a hierarchical extension to the previously introduced model of a specification graph. This hierarchical model [9] is based on hierarchical graphs. Each vertex v in the architecture or problem graph can be refined by a set of subgraphs associated with v. If a subset of these subgraphs is selected as a refinement of a vertex v, we are able to flatten our model, i.e., to replace the vertex v by the selected subgraphs.

Example 4. Fig. 6 shows possible refinements of the coding layer of the MPEG4 coder shown in Fig. 4. There are two types of codings: audio and visual coding. The audio coder subgraph consists only of a single vertex and no edges. In the next level of the hierarchy, for example we can refine the audio coder v_{ac} by different subgraphs (NaturalSound, SynthesizedSound, ...). One of the coding schemes for natural sounds is the CELP algorithm (Code Excited Linear Prediction) depicted in the upper left corner of Fig. 6.

Thus, a hierarchical specification graph consists of three parts:

1. a hierarchical problem graph $g_p(V_p, E_p)$
2. a hierarchical architecture graph $g_a(V_a, E_a)$, and
3. a set of *mapping edges* E_m map leaf problem graph vertices to leaf architecture graph vertices.

3.5 Hierarchical Chromosomes

In Section 3.3, an Evolutionary Algorithm coding for allocations and bindings of non-hierarchical specification graphs was revised. Here, we want to present

EA-based technique for hierarchical graphs. Therefore, we consider an EA that exploits the hierarchical structure of the specification by encoding the structure in the chromosome itself. In order to extend the presented approaches, we have to capture the hierarchical structure of the specification in our chromosomes first. Fig. 7 gives an example for such a chromosome.

The depicted chromosome encodes an implementation of the problem graph first introduced in Fig. 6. The underlying architecture graph is the one shown in Fig. 4. The structure of the chromosome in Fig. 7 resembles the structure of the problem graph given in Fig. 6. The leaves of the chromosome are nearly identical to the non-hierarchical chromosome structure described in Section 3.3 except for the lack of the allocation and allocation repairing list. These have moved to the top-level node of the chromosome (see Fig. 7). Note that the allocation and allocation repairing list form the non-hierarchical part (x^n) of the chromosome.

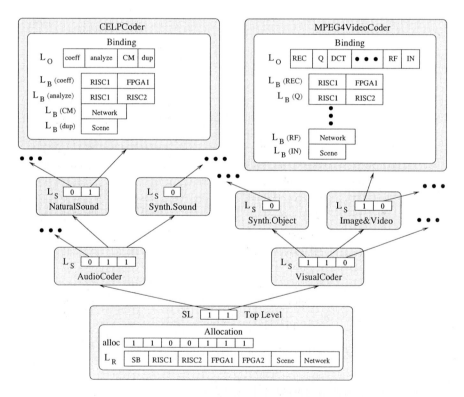

Fig. 7. Hierarchical chromosome structure for the problem graph given in Fig. 6 and the architecture graph given in Fig. 4

Each hierarchical node in the chromosome resembles a subgraph of the underlying problem graph. For each hierarchical vertex in the corresponding subgraph, the hierarchical node contains a selection list L_S. Each entry in this list describes the use of a subgraph in the implementation. Thus, the selection list L_S corresponds to the hierarchical part (x^h) of the chromosome. If we encounter a subgraph that does not include hierarchical vertices, we encode it by using a non-hierarchical chromosome as described above. The binding order list and the binding priority list form the non-hierarchical part (x^n) at this level of hierarchy.

Despite the modified internal structure, our hierarchical chromosome resembles exactly the non-hierarchical chromosome by still encoding an allocation and a binding. The main difference lies in the variable number of problem graph vertices allocated and bound in any given individual. We therefore still can use the same evolutionary operations, namely mutation and crossover. However, we propose two additional genetic operators making use of the hierarchical structure of the chromosome which have been introduced in Section 2.3.

In summary, composite mutation is used in order to explore the design space of different allocations of leaf graphs of the problem graph (flexibility). The second genetic operator, composite crossover, is used for the same purpose, allowing larger changes as when using the composite mutation operator only.

4 Experimental Results

This section presents first results obtained by using our new hierarchical chromosome structure with the multi-objective evolutionary algorithm SPEA2 [10].

As example, we use the specification of the MPEG4 coder layer. Due to space limitations, we omit a detailed description. A detailed description of this case study including possible mappings of all modules as well as the results can be found in [9]. The specification consists of six leaf graphs for the coding layer, each representing a different coding algorithm. Our goal is to implement at least one of these algorithms while minimizing cost, power, and maximizing flexibility. The underlying architecture consists of 15 resources. The search space for this example consists of more than 2^{200} points.

4.1 Objective Space

Here, we introduce the three most important objectives during system synthesis namely the *cost*, *power consumption*, and *flexibility* of an implementation.

Implementation Cost. The *implementation cost* $\text{cost}(i)$ for a given implementation $i = (\alpha, \beta)$ is given by the sum of cost of all allocated resources $v \in \alpha$. Note, due to the resource sharing, the decomposition function of the implementation cost is a non-monotonic function.

Power Consumption. Our second objective, the overall power consumption is the sum of the power consumptions of all allocated resources and the additional power consumptions originating by binding processes to resources.

Example 5. The implementation described in Example 3 possesses the overall power consumption $\text{power}(i) = 1620$.

Flexibility. The third objective, the reciprocal of a system's flexibility, captures the functional richness an implementation possesses. Without loss of generality, we only treat system specifications throughout this paper where the maximal flexibility equals the number of leaf graphs. For a comprehensive illustration of a system's flexibility, see [11].

4.2 Parameters of the Evolutionary Algorithm

All results were obtained by using the SPEA2 algorithm [10] with the following parameter: In our experiments, we have chosen a population size of 300 and an archive size of 70. The specific encoding of an individual makes special crossover and mutation schemes necessary. In particular, for the allocation α uniform crossover is used, that randomly swaps a bit between two parents with a probability of 0.25. For the lists (repair allocation priority lists L_R, binding order lists L_O and the binding priority lists $L_B(v)$), order based crossover (also names position-based crossover) is applied (see [12]).

For the composite crossover in hierarchical chromosomes, single-point crossover with a probability of 0.25 is chosen. Composite mutation is done on the selection list of each node of the hierarchical chromosome, and by swapping one element of the selection list with a probability of 0.2.

4.3 Exploration Results

As due to complexity reasons, we do not know the true Pareto-set, we compare the quality sets obtained by each approach against the quality set obtained by combining all these results and taking the Pareto-set of this union of optimal points. This set consists of 38 Pareto-optimal design points. A good measure of comparing two quality sets A and B is then to compute the so-called *coverage* $\mathcal{C}(A,B)$ as defined in [13], where $\mathcal{C}(A,B) = \frac{|\{b \in B | \exists a \in A : a \succeq b\}|}{|B|}$.

Hierarchical Chromosomes. By using hierarchical chromosomes, the coverage of the Pareto-set increases fast. After $t = 1100$ generations we achieved a coverage of $\mathcal{C}(o(X_{q,t}^{hc}), o(X_p)) \approx 0.83\%$ (average over the results from 5 different runs). As Fig. 8 shows, the hierarchical EA produces only a few Pareto-optimal points at the beginning. This is due to the fact, that the EA is also responsible for the exploration of allocations in the problem graph. The hierarchical EA could reach a full coverage of the Pareto-set when run sufficiently long.

Non-hierarchical EAs. Here, we compare our new approach against a non-hierarchical approach. In this non-hierarchical exploration algorithm, we explore the design spaces individually for all $2^k - 1$ possible combinations where k is the total number of leaf subgraphs in the problem graph. Therefore, we perform six different exploration runs for each individual leaf subgraph, $\binom{6}{2} = 15$ runs for combinations that select exactly two leaf subgraphs, etc. All in all, there are

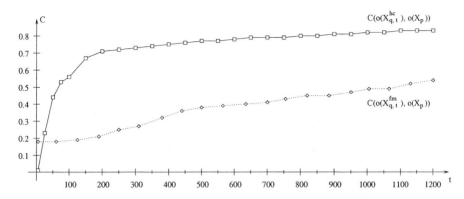

Fig. 8. Coverage of the Pareto-optimal implementations found after a given number of generations compared to the Pareto-set

$2^6 - 1 = 63$ combinations of leaf subgraphs, where at least one leaf subgraph has to be chosen. Note that this method is in general not a feasible way to go as the number of EA runs grows exponentially with the number of leaf graphs.

For each of these 63 cases we apply the EA for a certain number of generations for each combination k to obtain the quality set of the different leaf graph selections. Since we use the same number of generations for each combination, we simulate the case where each combination is selected with the same probability. With the given archives, we are able to construct the quality set of the top-level design, denoted by $X_{q,t}^{fm}$, by simply taking the union of all archives of the combinations and calculating the Pareto-optimal points in the union. Fig. 8 shows the result compared with the Pareto-set of our particular problem.

For our particular problem, we see that the hierarchical EAs is superior to the non-hierarchical exploration. At the beginning, the non-hierarchical EA constructs more Pareto-optimal solutions, since the hierarchical EA has to *compose* problem graphs with Pareto-optimal implementations first. By using the non-hierarchical EA, the maximum coverage of the Pareto-front is $C(X_{q,1200}^{fm}, X_p) \approx 52\%$. However, as in the case of the hierarchical chromosomes, it should be possible to find all Pareto-optimal solutions by using this non-hierarchical EA.

5 Conclusions

In this paper, we propose an approach for solving hierarchical multi-objective optimization problems. Two challenges have to be considered: (i) the complexity of the search space and (ii) the non-monotonicity of the objective-space. We propose hierarchical chromosomes in order to deal with the complexity of these search spaces. Furthermore, two new genetic operators, namely *composite mutation* and *composite crossover*, have been introduced in order to not only optimize parameters but also the structure of the chromosome. We applied our novel approach to the problem of synthesis of embedded systems. The results to our particular problem have shown the benefits of this new idea of *hierarchical optimization* by finding solutions with higher convergence than a non-hierarchical method in less number of generations.

In the future, we would like to show the relevance of this approach on other more general optimization problems.

References

1. Josephson, J.R., Chandrasekaran, B., Carroll, M., Iyer, N., Wasacz, B., Rizzoni, G., Li, Q., Erb, D.A.: An Architecture for Exploring Large Design Spaces. In: Proc. of the Nat. Conference of AI (AAAI-98), Madison, Wisconsin (1998) 143–150
2. Abraham, S.G., Rau, B.R., Schreiber, R.: Fast Design Space Exploration Through Validity and Quality Filtering of Subsystem Designs. Technical report, Hewlett Packard, Compiler and Architecture Research, HP Laboratories Palo Alto (2000)
3. Rudolph, G., Agapie, A.: Convergence Properties of Some Multi-Objective Evolutionary Algorithms. In: Proc. of the 2000 Congress on Evolutionary Computation, Piscataway, NJ, IEEE Service Center (2000) 1010–1016
4. Dasgupta, D., McGregor, D.R.: Nonstationary Function Optimization using the Structured Genetic Algorithm. In Männer, R., Manderick, B., eds.: Proceedings of Parallel Problem Solving from Nature (PPSN 2), Brussels, Belgium, Elsevier Science (1992) 145–154
5. Blickle, T., Teich, J., Thiele, L.: System-Level Synthesis Using Evolutionary Algorithms. In Gupta, R., ed.: Design Automation for Embedded Systems. 3. Kluwer Academic Publishers, Boston (1998) 23–62
6. Dick, R., Jha, N.: MOGAC: A Multiobjective Genetic Algorithm for Hardware-Software Cosynthesis of Distributed Embedded Systems. In: IEEE Transactions on Computer-Aided Design of Integrated Circuits and Systems 17(10). (1998) 920–935
7. Pareto, V.: Cours d'Économie Politique. Volume 1. F. Rouge & Cie., Lausanne, Switzerland (1896)
8. Thiele, L., Chakraborty, S., Gries, M., Künzli, S.: A Framework for Evaluating Design Tradeoffs in Packet Processing Architectures. In: Proceedings of the 39th Design Automation Conference (DAC 2002). (2002) 880–885
9. Teich, J., Haubelt, C., Mostaghim, S., Slomka, F., Tyagi, A.: Techniques for Hierarchical Design Space Exploration and their Application on System Synthesis. Technical Report 1/2002, Institute Date, Department of EE and IT, University of Paderborn, Paderborn, Germany (2002)
10. Zitzler, E., Laumanns, M., Thiele, L.: SPEA2: Improving the Strength Pareto Evolutionary Algorithm. Technical report, Swiss Federal Institute of Technology (ETH) Zurich (2001) TIK-Report 103. Department of Electrical Engineering.
11. Haubelt, C., Teich, J., Richter, K., Ernst, R.: Flexibility/Cost-Tradeoffs in Platform-Based Design. In Deprettere, E., Teich, J., Vassiliadis, S., eds.: Embedded Processor Design Challenges. Volume 2268 of Lecture Notes in Computer Science (LNCS)., Berlin, Heidelberg, Springer (2002) 38–56
12. Deb, K.: Multi-Objective Optimization Using Evolutionary Algorithms. John Wiley & Sons (2001)
13. Zitzler, E.: Evolutionary Algorithms for Multiobjective Optimization: Methods and Applications. PhD thesis, Department of Electrical Engineering, Swiss Federal Institute of Technology (ETH) Zurich (1999)

Multiobjective Meta Level Optimization of a Load Balancing Evolutionary Algorithm

David J. Caswell and Gary B. Lamont

Department of Electrical and Computer Engineering
Graduate School of Engineering and Management, Air Force Institute of Technology
Wright-Patterson AFB, OH 45433-7765
{david.caswell,gary.lamont}@afit.edu[*]

Abstract. For any optimization algorithm tuning the parameters is necessary for effective and efficient optimization. We use a meta-level evolutionary algorithm for optimizing the effectiveness and efficiency of a load-balancing evolutionary algorithm. We show that the generated parameters perform statistically better than a standard set of parameters and analyze the importance of selecting a good region on the Pareto Front for this type of optimization.

1 Introduction

The effectiveness of an Evolutionary Algorithm(EA) can be largely based on the operators and parameters chosen for implementation [1]. The analysis of parameters is itself a large combinatoric problem[16][2]. Thus, in order to further optimize the execution of an allocation EA a meta-EA is developed that tunes its parameters.

The objectives optimized by the meta-EA are the effectiveness and efficiency of the allocation EA. The representation of the chromosomes within the meta-EA is composed of the different operators and parameters available for tuning the allocation EA. Specifically the efficiency objective is based on number of evaluations executed by the allocation EA. This metric is chosen based on its direct correlation with the time required to execute the algorithm. The second meta-level objective (effectiveness) is measured by the evaluation of the solution obtained from the allocation EA execution using the chosen parameters.

Since the meta-EA has two objectives to evaluate a Pareto front *a posteriori* approach is used for finding solutions. With this approach we can evaluate the tradeoff between the two objectives and compare the optimized allocation EA results with results from a set of parameters found effective in literature.

In order to have an appreciation of what the meta-EA does we must first examine the algorithm that it is optimizing. Section 2 provides a brief description of the allocation EA and the objectives that it uses. Section 3 then discusses the meta-EA and it's approach to parameter optimization. A design of experiments is given in Section 4 with the results in Section 5 and conclusions in Section 6.

---

[*] The views expressed in this article are those of the authors and do not reflect the official policy of the United States Air Force, Department of Defense, or the U.S. Government.

2 HPC Load Balancing

The high level goal of processor allocation is to create a mapping of processes to processors in such a way as to get better performance and better utilization of the processors available than would have been possible otherwise. It can be viewed as a spatial balancing algorithm much like the temporal based flow shop scheduling problem[8]. The processor allocation problem is an NP-complete optimization problem[6,9][2], yet is one of such practical application that numerous researchers have examined the effectiveness of a variety of different heuristics for solving the problem including neural networks [9], Recursive Spectral Bisection [14], diffusion method[5], [4], Evolutionary Strategies [7], and a variety of others too numerous to cite.

For this analysis we design the allocation objectives so that the problem may be represented as a minimization problem with an assumed homogenous HPC cluster running p processors and n processes. The four objectives selected are: request cost ($C_{request}$), preemption cost ($C_{preempt}$), communication linkage cost (C_{link}), and rollover cost ($C_{rollover}$).

2.1 Constraints

Since some processes require upper and lower limits on the amount of processors that they can support we must define the linear constraints binding them to these limits. If we allow $Required(i)$ and $Requested(i)$ to define the amount of processors that process i must have to execute and the maximum allowed to execute respectively. Then to bound the amount of processors that can be used we must ensure that no more than $Requested(i)$ are used, and that no less than $Required(i)$ are used. We define S_k to be the state of the system at time $k \in \{o, n\}$ where S_o represents the current state of the system and S_n the next state of the system. Furthermore if we allow N_i, $i \in \{1 \ldots n\}$ to represent the specific tasks to be executed where n is the amount of processes, then $\mid S_k(N_i) \mid$ represents the amount of processors assigned to task i at state k allowing the constraint to be given as:

$$Required(i) \leq \mid S_n(N_i) \mid \leq Requested(i) \, \forall i \in \{1 \ldots n\} \qquad (1)$$

2.2 Request Cost

The first objective that is to be dealt with is the request cost, $C_{request}$. Each process has some upper and lower limit on the number of processors that it execute on. These upper and lower limits create bounded constraints on the amount of processors that can be allocated to each process. As shown in Equation 1 this constraint does not allow for any advantage to the process for using any more than the lower bound of processors. Thus the request cost objective exists in order to try to provide some benefit for using more than the lower bound of processors. For any given process the amount of extra processors used is found using Equation 2.

$$C_{request_i} = \mid S_n(N_i) \mid -Required(i) \qquad (2)$$

2.3 Preemption Cost

For the cost of preemption, $C_{preempt}$ we want to minimize the overall cost of preempting a process. There are two manners of preemption examined: the true preemption, that which occurs when a process is stopped so another process can use that processor, and the preemption that occurs when a process is moved to a different processor. Since not all processes can be preempted the feasibility of a solution based on it's being preemptive or not is dealt with as a feasibility constraint. For those processes that can be preempted we decided to examine the two types of preemption from the perspective of new processors added to a preexisting process, and currently allocated processors being removed from a preexisting process. In this manner we can examine not only the preempted processors, but also include in it any additional cost that occurs from adding extra processors to an already ongoing process.

For the first type of preemption, where processes are added to a preexisting processor, we use the cost function as given by Equation 3.

$$C_{preempt+_i} = \mid S_n(N_i) - S_o(N_i) \mid \qquad (3)$$

This equation enumerates, for some preemptible process i, the amount of processors that exists in the potential next-state of the system that did not exist in the previous-state of the system.

Using the same logic, Equation 4 calculates the amount of processors that existed in the old-state of the system that are not present in the potential next-state of the system for process i. In other words they determine the amount of processors given to and removed from task i respectively.

$$C_{preempt-_i} = \mid S_o(N_i) - S_n(N_i) \mid \qquad (4)$$

2.4 Link Cost

The link cost, C_{link} is a measure of the overall communication overhead experienced by each process. By using a normalized communication matrix, calculated a priori to the EA execution, we are able to take into account all communication costs that would be associated due to the relative node speeds, the backplane structure, etc. Thus for each process we would calculate

$$C_{link_i} = \sum_{j=1}^{(\mid S_n(N_i) \mid - 1)} \sum_{k=j}^{(\mid S_n(N_i) \mid)} l(S_n(N_i))_{jk} \qquad (5)$$

as the summation of all possible communication occurrences that could occur for process i.

2.5 Rollover Cost

The final cost chosen for examination is that of the rollover cost $C_{rollover}$. This is the cost associated with not including available processes for execution in the next-state of the system. We find the amount of rolled-over processes from Equation 6 where m represents the amount of processes available to be assigned and n again is the amount of processes being assigned. This is a critical element of the algorithm in that we want to ensure that as few processes are not executed as can be contained within the constraints as explained in section 2.1.

$$C_{rollover} = m - n \tag{6}$$

2.6 Normalization

Now that we have identified the objectives that are incorporated into the system we must adapt them so that they can be reasonably compared with each other. As can be seen by the left hand side of Table 1 most of the objectives only take into account a single process and its relative configuration. Thus, in order to be able to aggregate the objectives they must all account for the overall system. In this manner the right hand side of Table 1 defines these objectives in terms of the entire set of processes. It also normalizes the equations so that they can be easily compared against each other. For these equations n_n and n_o represent the amount of processes assigned in the new and old states of the system respectively.

Table 1. Standardized Objective Functions for the allocation EA

Standard Function	Normalized Function
$C_{request_i} = \mid S_n(N_i) \mid - Required(i)$	$C_{request} = \frac{\sum_{i=1}^{n_n}(\mid S_n(N_i)\mid - Required(i))}{\sum_{i=1}^{n_n}\mid S_n(N_i)\mid}$
$C_{preempt+_i} = \mid S_n(N_i) - S_o(N_i) \mid$	$C_{preempt+} = \frac{\sum_{i=1}^{n_o}\mid S_n(N_i) - S_o(N_i)\mid}{1 + \sum_{i=1}^{n_o}\mid S_n(N_i)\mid}$
$C_{preempt-_i} = \mid S_o(N_i) - S_n(N_i) \mid$	$C_{preempt-} = \frac{\sum_{i=1}^{n_o}\mid S_o(N_i) - S_n(N_i)\mid}{1 + \sum_{i=1}^{n_o}\mid S_n(N_i)\mid}$
$C_{link_i} = \sum_{j=1}^{\mid S_n(N_i)\mid -1} \sum_{k=j}^{\mid S_n(N_i)\mid} l(S_n(N_i))_{jk}$	$C_{link} = 1 - \frac{\sum_{i=1}^{n_n}\left(\sum_{j=1}^{(\mid S_n(N_i)\mid -1)} \sum_{k=j}^{(\mid S_n(N_i)\mid)} l(S_n(N_i))_{jk}\right)}{\sum_{j=1}^{p-1}\sum_{k=j}^{P} l(S_n)_{jk}}$
$C_{rollover} = m - n$	$C_{rollover} = \frac{m_n - n_n}{m_n}$

In order to evaluate the results of the meta-EA we aggregate the objectives of the allocation EA into a single function. Since they are now normalized we can combine them in an aggregate form with some weights to take into account the differing degrees of importance they are towards each other. Using $\alpha_1 \ldots \alpha_5$ for the weights then gives us the aggregate objective Equation 7.

$$min\left(\alpha_1 C_{request} + \alpha_2 * (C_{preempt-} - \frac{1}{10} C_{preempt+}) + \alpha_3 * C_{link} + \alpha_4 * C_{rollover}\right) \tag{7}$$

Of these objectives the only one that immediately stands out as needing some priority is the cost of rollover, $C_{rollover}$. We of course want to try to execute as many tasks as we can if it is at all possible. The other objectives are related to the system performance and while important, are not nearly as much so as that of the rollover cost. Thus we set the coefficient to $C_{rollover}$ to ten, and all of the other coefficients to one, making it one order of magnitude larger than the others objective coefficients.

3 Meta-level Research

Definition 1 *Coevolution of a population with itself occurs when individuals within the population are in symbiosis with other individuals in the same population. Coevolution of a pair of populations with each other occurs when individuals in one population are in symbiosis with individuals of another population.*

Using Definition 1 all a system needs in order to be coevolutionary is a fitness based on the evaluation of chromosomes from a population other than its own [12]. With this definition the meta-EA is coevolutionary and the interaction can be described as commensalism [13]. However, this relationship is different from most coevolutionary systems in that in a meta-EA the interactions only go one way. The meta-EA optimizes the operators and parameters for the sub-EA, and the sub-EA uses these parameters and operators to optimize the problem being examined.

3.1 Meta-EA

Our research builds on the ideas used with the multi-layer DAGA2 algorithm [16], [15], [17]. However, for this research a GENOCOP III algorithm is used, modified so that it is able to execute in a multiobjective and multi-layer manner. The multiobjective capability is included so that the algorithm can optimize not only effectiveness but also the efficiency of the allocation EA in a Pareto manner. The meta-EA can thus be described as a coevolutionary algorithm with a commensal relationship to the allocation EA[13]. With this relationship, the meta-EA's objective evaluation improves as it is able to improve the results from the allocation EA. The meta-EA works in the same manner as other evolutionary algorithms, the primary difference being that it's chromosomes determine the parameters used by the allocation EA and its evaluation thus entails an execution of the allocation EA. By doing this the algorithm evolves sets of parameters for the allocation EA so that it is able to more effectively and efficiently find solutions to whatever allocation request is given to it.

3.2 Parameter Representation

The set of candidate solutions are the parameters necessary to evolve the next execution of the allocation EA. Because GENOCOP III is used for both the meta-level and allocation algorithms, these parameters are based directly off the input required to execute one iteration of the GENOCOP III system.

Since GENOCOP was designed as an easily configurable system the basic parameters are input via an input file. Their are 33 available parameters total for any given execution. Since some of these parameters represent constant based on the representation of the problem itself only 21 of these parameters are actually allowed to be altered by the meta-EA. Those fields that can be altered must be bounded so as to avoid infeasible executions. Table 2 lists those elements that are allowed to be modified by the meta-EA. Table 2 also displays the bounds for each of the different allocation EA parameters. The domain constraints for this algorithm are selected to test a large expanse of the search space.

Table 2. GENOCOP III Available Parameters and their limits for the meta-EA execution

Parameter	Lower Bound	Upper Bound
Reference Population Size	1	1000
Search Population Size	1	1000
Total Number of Evaluations	1	5000
Reference Population Evolution Period	1	2000
Number of Offspring for Ref. Population	1	1000
Selection of Ref point for repair	0	1
Selection of Repair Method for Search Pop	0	1
Search Point Replacement Ratio	0.0	1.0
Reference Point Initialization Method	0	1
Search Point Initialization Method	0	1
Frequency of Operators	0	100

3.3 Objectives

There are two objectives that the meta-EA tries to accomplish. The first objective deals with the effectiveness of the program that it is controlling. In this respect the meta-EA wants to minimize the objective solution produced by the aggregate allocation EA. The second objective deals with the efficiency of the program it is controlling, for this the meta-EA works to minimize the amount of calculations necessary to produce a solution, which for this case is directly proportional to the number of evaluations. Since the choice of which objective we want to use depends on requirements of the system a Pareto multiobjective approach is used for this algorithm, as seen in Equation 8 where f_1 represents the number of evaluations that the meta-EA assigns to the allocation EA and f_2 represents the evaluation of the allocation EA as given in Equation 7. With this type of approach we can examine what the tradeoffs between the efficiency and effectiveness of this algorithm are. Because the results are given in a Pareto nondominated fashion the decision maker is able to choose the chromosome that most suits the research area for implementation of this algorithm.

$$PF_{known} = (min(f_1), min(f_2)) \qquad (8)$$

4 Design of Experiments

While the meta-EA controls the parameters of the allocation algorithm, it itself need only to operate effectively. The overall execution time of the algorithm is important only with respect to the amount of time available for experimentation. In this manner the overall parameters must be constrained based on the lack of infinite time to run this algorithm. Thus the meta-EA itself is set to run using the parameters as given by Table 3. These parameters are chosen based on the length of time necessary to execute the multi-layer algorithm. The Reference Evolution Period parameter is also set atypically low so as to allow for migration between the parallel populations based on a coarse grain parallelization of the meta-EA algorithm.

Table 3. Meta-EA parameters for execution

Parameter	Value
Reference Population Size	20
Search Population Size	20
Total Number of Evaluations	200
Reference Population Evolution Period	20
Number of Offspring for Ref. Population	20

With this in mind we generate three test cases as given in Table 4 for the meta parameter development. These test cases were chosen because they provide three different sizes of problems for the allocation EA to use. The 16 processor, 4 process problem is the easiest, and the 128 processor 16 process is the most complex with respect to the amount of processors/processes that must be examined by allocation EA. Due to the necessary generality of the program their is no constraint to the size of the problem that can be generated by changing either the amount of processors or the amount of processes (or both). The problems selected in Table 4 are chosen based on their varied combination of processor sizes with amounts of processes, this allows for a comparison on the scalability of the algorithm.

Table 4. Amount of Processors and Processes for Testing

Test Case Number	Processors	Number Processes
1	16	4
2	64	16
3	128	16

The goal of the meta-EA is to generate a set of parameters tuned for the allocation EA in order to optimize its effectiveness and efficiency. This should help to optimize the allocation EA so that it can run well over a variety of different potential inputs. In order to ensure that the parameters are not being tuned for any specific problem we provide a set of simulated load balancing challenges for which the evaluation of the allocation EA can be measured against. Thus, each meta-EA chromosome evaluation is based on the summation of the allocation EA results on test cases comprised of different original states of the processing system. These test cases are standardized across all of the different combinations of processors and processes so that the results can be compared by using the following

- Empty original state- such that none of the processes has yet been allocated
- Missing one process- such that all but one process has been allocated
- Full state but one, with preemption- such that all of the processors but one have been allocated with all of the processes

There are two experiments that must be done using the meta-EA. The first tests the ability of the meta-EA to provide effective and efficient parameters for the allocation EA algorithm. The second test is to examine the ability of the meta-EA for generating a set of results that is effective across a range of problem sizes. With these two tests we can validate the effectiveness of the meta-EA and generate good experimental results that can be used as parameters for the allocation EA.

4.1 Parameter Generation Experiment

If we define $\Phi(meta[i])$ to be the evaluation result of the allocation EA based on the meta generated parameter for test number i, and $\Phi(generic[j])$ to be the allocation EA results based on the parameter set from literature for some test number j, then we can express our first statistical hypothesis as:

Hypothesis 1 *The meta-EA can generate effective and efficient parameters that are statistically better than a set of parameters based on previous experiments, given some static p amount of processors and N tasks.*

$$H_0 : \mu_{\Phi(meta[1])..\Phi(meta[m])} = \mu_{\Phi(generic[1])..\Phi(generic[m])}$$
$$H_1 : \mu_{\Phi(meta[1])..\Phi(meta[m])} < \mu_{\Phi(generic[1])..\Phi(generic[m])}$$

In order to test this hypothesis we must have a set of parameters, based on generic experiments and literature suggestions, that the meta-EA results can be compared against when executed on the allocation EA [11]. We selected the parameters as given in Table 5 [3][10][11]. In order to make the result comparison between the generic and the meta developed parameters as fair as possible solutions are compared from the meta generated Pareto front that reside near the evaluation amount that the generic parameter uses.

The experiment is run through the allocation EA 10 times each over the same system designs as given in Table 4 for an empty processor problem, changing the random seed each time. We then examine the max, min, median, mean, and standard deviation of the results. With these results we also use a statistical student T-test to test Hypothesis 1.

Table 5. GENOCOP III Parameters From Generic Experiments

Parameter	Literature Setting
Reference Population Size	100
Search Population Size	100
Total Number of Evaluations	2000
Reference Population Evolution Period	100
Number of Offspring for Ref. Population	100
Selection of Ref point for repair	0 (Random)
Selection of Repair Method for Search Pop	0 (Random)
Search Point Replacement Ratio	0.2
Reference Point Initialization Method	1 (Multiple)
Search Point Initialization Method	1 (Multiple)
Frequency of Operators	10 (For All)

4.2 Meta Problem Size Experiment

Given that the meta-EA validates itself as being statistically more effective than generic experimental parameters, the next step is testing what scale of the parameters to select. It is not prudent to assume that what works on a small 16 processors 4 process system would necessarily work on a larger 128 processor system. Simply because of the simplicity of the system, a 16 processor 4 process system would not necessarily need as many evaluations as the larger experiment. However, the converse may not be true as shown in the following hypothesis.

Hypothesis 2 *Given the parameters on a large system(with respect to the amount of evaluations necessary to optimize the system), these parameters are just as effective (if not more so) on a small system.*

$H_0 : \mu_{\Phi(largeMeta[1])..\Phi(largeMeta[m])} = \mu_{\Phi(smallMeta[1])..\Phi(smallMeta[m])}$
$H_1 : \mu_{\Phi(largeMeta[1])..\Phi(largeMeta[m])} < \mu_{\Phi(smallMeta[1])..\Phi(smallMeta[m])}$

For this hypothesis, we let $\Phi(largeMeta[i])$ represent the allocation EA result from the parameters gained from a meta execution using a large problem size and $\Phi(smallMeta[i])$ represents the parameters gained from a meta execution using a small problem size. In order to test this hypothesis the allocation EA is executed 10 times each on the afore mentioned 16/4, 64/16, 128/16 sized problems. The results are then statistically compared, again using the standard student T-test for hypothesis testing. The results are also compared for each of the parameters for each of the problem sizes to see if they are statistically equivalent (again using the student T-test for each). This allows a more narrow selection of parameters that are effective/efficient for the allocation EA.

5 Results

As shown from Figure 1, Figure 2, and Figure 3 the Pareto Front generated by the meta-EA has a definite tradeoff between effectiveness and efficiency for the

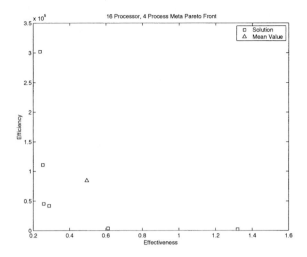

Fig. 1. Known Pareto Front for meta-EA Evaluation on 16 Processor, 4 Process Experiment

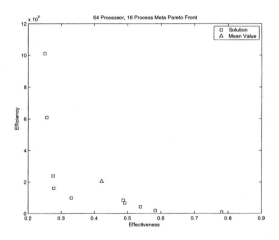

Fig. 2. Known Pareto Front for meta-EA Evaluation on 64 Processor, 16 Process Experiment

allocation EA. With this Pareto front we can then select solutions with a range of tradeoffs between the two objectives. The triangle in each of these figures represents the mean value of the Pareto solutions.

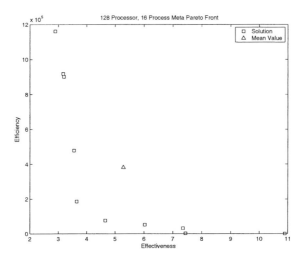

Fig. 3. Known Pareto Front for meta-EA Evaluation on 128 Processor, 16 Process Experiment

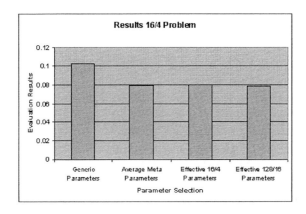

Fig. 4. Allocation EA results comparison for 16 Processor, 4 Process Experiment run with different sets of parameters

5.1 Parameter Generation Experiment

The first hypothesis that needs to be examined with regards to the meta-EA is Hypothesis 1. In this hypothesis we are testing the meta-EA's ability to find effective parameters that are statistically better than those generated with a known good set of parameters. For this experiment we examine each problem size independent of the others.

For the 16 processor, 4 process experiment all of the student T-test P-values were well below our *alpha* of 0.05, thus we reject H_0 and must conclude that the

Table 6. P-value for one-tailed student T-test (alpha=0.05) comparing each parameter set against the generic parameters chosen from previous works

Parameter Type	16/4 Problem	64/16 Problem	128/16 Problem
Average Meta	1.03 E-4	1.10 E-3	8.79 E-6
Effective Meta (Problem Size)	4.55 E-3	9.11 E-3	3.70 E-2
Effective 128/16	7.58 E-5	2.07 E-1	3.70 E-2

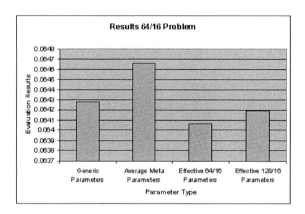

Fig. 5. Allocation EA results comparison for 64 Processor, 16 Process Experiment run with different sets of parameters

meta-generated results (even the average ones) are statistically greater than the results generated by our selected generic parameters. This is what we expected to see based on the ability of the meta-EA to fine tune the parameters so that the allocation EA can more effectively search the search space.

For the 64 processor, 16 process problem the results are more varied. The T-test P-values for the mean comparison between the generic parameters and the effective 128/16 parameters is clearly greater than 0.05, thus we would fail to reject H_0 and conclude that the 128/16 parameters generate statistically equivalent results as the generic parameters. When we examine the P-value for the average meta parameters the results are much less than 0.05 and thus must reject H_0 in favor of H_1 for this problem. This rejection leans towards the statistical conclusion that the generic parameters are better than the mean meta parameters. This is not surprising as these parameters are selected as being average points in the population and are thus not feasible members of the Pareto front. As can be seen be Figures 1, 2, and 3 the mean of the Pareto front solutions is clearly not a member of the Pareto front itself, the same is true for the phenotypic level. This being the case we do not expect to have the most effective nor efficient solutions found using these averaged parameters. As for the parameters deemed effective for the 64 processor 16 process problem specific the P-value is also less than 0.05 allowing us to once again reject H_0 and conclude that these parameters are statistically more effective than the generic parameters.

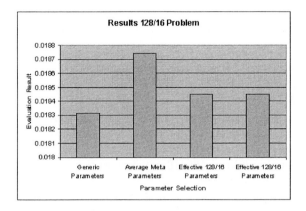

Fig. 6. Allocation EA results comparison for 128 Processor, 16 Process Experiment run with different sets of parameters

With the 128 processor 16 process problem we again see that the average meta results clearly have a P-value that must reject H_0 and conclude that the generic parameters are statistically more effective. The set of effective parameters used for this experiment also do not have a P-value greater than alpha and therefore rejects our hypothesis. These results too lend themselves to concluding that the results of the generic parameters are statistically better than the meta generated solutions. However, if we move to a two-tailed test to determine whether or not the effective 128/16 parameters are statistically equivalent to the generic parameters we find a P-value of 0.074086, which would mean that our results are statistically equivalent (from a two-sided perspective).

Thus, overall the meta generated solutions selected for their effectiveness statistically prove themselves to be better than the generic parameters for the easier problems and statistically equivalent for the harder problems. The mean Pareto front parameters show themselves to be worse than the generic parameters in two out of three trials, but this is easily understood when we consider that they are based on the cumulative parameters for an entire Pareto front.

5.2 Meta Problem Size Experiment

Table 7 displays the student T-test probabilities based on a comparison of the results of using the effective 128/16 parameters on the allocation EA against using the meta generated parameters specific to the problem size. This experiment relates again to the Figures 4, 5, and 6. For all of these experiments we fail to reject H_0 in favor of H_1 for all of the experiments except the average meta

Table 7. P-value for one-tailed student T-test (alpha=0.05) comparing each parameter set against the effective 128/16 parameters

Parameter Type	16/4 Problem	64/16 Problem	128/16 Problem
Average Meta Parameters	1.96 E-1	9.95 E-6	1.13 E-4
Effective Parameters (Problem Size)	1.38 E-1	1.20 E-1	n/a

parameters of the 64/16 problems and the average meta results of the 128/16 problem. Meaning that statistically the results found by the effective 128/16 parameters were statistically equivalent to the results found by these other parameters sets. When comparing the effective 128/16 parameters to the average parameters for the 64/16 and the 128/16 problems we must reject H_0 for H_1 and conclude that the effective 128/16 parameters are statistically more effective than these two sets of parameters.

This validates that the effective parameters for a larger problem are effective on smaller sized problems. Thus in real-world application of the allocation EA, using these larger parameters would allow for optimization of the smaller problem sizes.

6 Conclusions

We have used a meta-level algorithm for optimizing the parameters of a sub-level allocation EA. We have shown that the results provided by the high level algorithm provide a concave Pareto front with a distinctive tradeoff between the effectiveness and efficiency objectives. Using this knowledge we have been able to select Pareto front members that were effective and have compared them against an averaged parameter set, a parameter set derived from previous works, and a parameter set evolved from the larger problem size. We have then validated that the meta generated parameters are more effective than those found from previous work. What's more we have shown that the parameters generated for a large allocation problem are effective on the smaller problems, thus allowing us to use these parameters for a range of problem sizes.

References

1. Thomas Bäck. *Evolutionary Algorithms in Theory and Practice*. Oxford University Press, New York, 1996.
2. David Caswell. Active processor scheduling using evolutionary algorithms. Master's thesis, Wright-Patterson Air Force Base, Ohio.
3. David J. Caswell and Gary B. Lamont. Wire-antenna geometry design with multiobjective genetic algorithms, 2001.
4. Y. Chan, S. Dandamudi, and S. Majumdar. Performance comparison of processor scheduling strategies in a distributed-memory multicomputer system. In *Proc. Int. Parallel Processing Symp (IPPS)*, pages 139–145, 1997.

5. Hluch'y Dobrovodsk'y Dobruck'y. Static mapping methods for processor networks.
6. Paul C. Messina Geoffrey C. Fox, Roy D. Williams. *Parallel Computing Works*. Morgan Kaufmann Publishers, Inc., San Francisco, 1994.
7. G. Greenwood, A. Gupta, and K. McSweeney. Scheduling tasks in multiprocessor systems using evolutionary strategies, 1994.
8. Maciej Hapke, Andrzej Jaszkiewicz, and Krzysztof Kurowski. Multi-objective genetic local search methods for the flow shop problem. In *Proceedings of the Evolutionary Multiobjective Optimizations Conference*, 2002.
9. Tracy Braun Howard, Howard Jay Siegel, and Noah Beck. A comparison of eleven static heuristics for mapping a class of independent tasks onto heterogeneous distributed computing systems, 2001.
10. Zbigniew Michalewicz. *Genetic Algorithms + Data Structures = Evolution Programs*. Springer-Verlag, New York, 2nd edition, 1994.
11. Zbigniew Michalewicz and David B. Fogel. *How to Solve It: Modern Heuristics*. Springer-Verlag, New York, 2000.
12. Jason Morrison. Co-evolution and genetic algorithms. Master's thesis, Carleton University, Ottawa, Ontario, 1998.
13. Jason Morrison and Franz Oppacher. A general model of co-evolution for genetic algorithms. In *Int. Conf. on Artificial Neural Networks and Genetic Algorithms ICANNGA 99*, ?, 1999.
14. Horst D. Simon. Partitioning of unstructured problems for parallel processing. *Computing Systems in Engineering*, 2:135–148, 1991.
15. G. Wang, T. Dexter, E. Goodman, and W. Punch. Optimization of a ga and within the ga for a 2-dimensional layout problem, 1996.
16. G. Wang, E. Goodman, and W. Punch. Simultaneous multi-level evolution, 1996.
17. Gang Wang, Erik D. Goodman, and William F. Punch. On the optimization of a class of blackbox optimization algorithms, 1997.

Schemata-Driven Multi-objective Optimization

Skander Kort

Concordia University, ECE Department,
1455 De Maisonneuve Blvd. W.,
Montreal, PQ, Canada, H3G 1M8
kort@ece.concordia.ca

Abstract. This paper investigates the exploitation of non-dominated sets' schemata in guiding multi-objective optimization. Schemata capture the similarities between solutions in the non-dominated set. They also reflect the knowledge acquired by multi-objective evolutionary algorithms. A schemata-driven genetic algorithm as well as a schemata-driven local search algorithm are described. An experimental study to evaluate the suggested approach is then conducted.

1 Introduction

Recent years witnessed a tremendous growth in the application of metaheuristics to solve multi-objective optimization problems. This growth was motivated by the lack of a deep knowledge about the aforementioned problems. Such a lack makes it more difficult to apply conventional optimization techniques such as branch and bound or dynamic programming. This paper focuses on a specific class of metaheuristics, namely evolutionary algorithms. In such a class, a population of solutions evolves as the consequence of applying some unary and n-ary operators.This class of metaheuristics includes genetic algorithms, some variants of local search algorithms as well as memetic algorithms.

Despite their success, evolutionary algorithms do not generally scale up very well with large problem instances. This is theoretically justified by the no-free-lunch theorems [16]. Practically speaking, larger problem instances imply larger search spaces. Moreover, the likelihood of getting good solutions as the result of applying evolutionary operators becomes lower. This leads the evolutionary algorithm to spend a lot of time performing useless evaluations. This also leads the algorithm to prematurely converge toward poor-quality solutions.

The outcome of a multi-objective optimization algorithm is a set of non-dominated solutions. Since, many multi-objective optimization problems are NP-hard, this set is rather just an approximation of the optimal non-dominated set. Assessing the quality of such an approximation has been covered by many papers (e.g. [8,5]). Intuitively, an approximation is good if it is very close to the optimal non-dominated set and if it provides a more or less uniform sampling of this set.

The aim of this paper is to exploit the information conveyed by schemata in order to improve the outcome of multi-objective evolutionary algorithms.

The concept of schemata was first introduced by Holland [6] to theoretically justify the power of genetic algorithms (GAs). The suggested approach extracts periodically the schema describing the non-dominated set found so far. This schema is then used to guide the evolutionary algorithm toward better solutions. The remaining of the paper is organized as follows. Next section introduces some basic concepts and notations as well as the multi-objective knapsack problem ($MOKP$). This problem will be used to evaluate the suggested approach. Section 3, motivates the use of schemata in multi-objective optimization. It also describes how can schemata be used in multi-objective genetic and local search algorithms. The suggested approach is evaluated in Sec. 4. Finally, some guidelines for future research are given in Sec. 5

2 Basic Concepts and Notations

Let f be a vector function defined over some domain D. The multi-objective optimization problem is defined as follows [18] :

$$\max f(x) = (f_1(x), \ldots, f_k(x)) \text{ s.t. } x \in D.$$

$x = (x_1, \ldots, x_l)$ is called the decision vector. Set D is called the decision variables space. Set $f(D)$ is called the objectives space.

The dominance relation on set D is defined as follows. Let x and y be two decision vectors in D. Vector x is said to dominate vector y (we write $x \succ y$) iff

$$\forall i \in \{1, \ldots, m\}, \ f_i(x) \geq f_i(y) \text{ and } \exists j \in \{1, \ldots, m\} \text{ s.t. } f_j(x) > f_j(y).$$

Let S be a subset of D. The set of non-dominated decision vectors in S is defined by :

$$ND(S) = \{x | x \in S \text{ and } \not\exists y \in S \text{ s.t. } y \succ x\}$$

In particular, set $ND(D)$ is called the pareto optimal set of the multi-objective optimization problem. We will rather use notation ND^* to refer to this set in the future. The image of ND^* by f is called the pareto front.

Let EA be an evolutionary algorithm for solving the multi-objective optimization problem described above. Let $P(EA, t)$ be the population of solutions produced by algorithm EA after iteration t. Conventially, the initial population is denoted by $P(EA, 0)$. The set of non-dominated solutions at iteration t is denoted by $ND(EA, t)$. More formally :

$$ND(EA, t) = ND(P(EA, t)).$$

We denote by $\overline{ND}(EA, t)$ the set of non-dominated solutions found by algorithm EA up to iteration t. More specifically :

$$\overline{ND}(EA, t) = \cup_{g=0}^{t} ND(EA, g)$$

where $\cup$ is a union operator that preserves the non-dominance property[1]. The set of all the non-dominated solutions found by algorithm EA, after it terminates, is denoted by $\overline{ND}(EA)$.

We assume that every decision vector x in D is encoded as a binary string $enc(x)$ of length n. A schema is defined to be a string made up of symbols 0, 1 and $*$. Let S be a subset of D. Let r be in $[0,1]$. Similarities between the elements of S may be described by schema $SCH(S,r)$. The ith symbol of this schema (denoted by $SCH(S,r)_i$) is defined as follows :

- $SCH(S,r)_i = 1$ iff there are at least $r * |S|$ vectors having their i^{th} symbol (or gene) set to 1.
- $SCH(S,r)_i = 0$ iff there are at least $r * |S|$ vectors having their i^{th} gene set to 0.
- $SCH(S,r)_i = *$ otherwise.

Let $sch1$ and $sch2$ be two schemata of length n. $sch1$ and $sch2$ are said to be *contradictory* iff there exists $i \in \{1,\ldots,n\}$ s.t. $sch1_i, sch2_i \in \{0,1\}$ and $sch1_i \neq sch2_i$. Such situation is called a *contradiction*. Two schemata that are not contradictory are said to be *compatible*. Let sch be a schema. Let sch_i denote the ith symbol of sch. Symbol sch_i is said to be *set* iff $sch_i \neq *$.

The multi-objective knapsack problem ($MOKP$) will be used to evaluate the approach suggested in this paper. $MOKP$ can be formulated as follows:

$$\begin{cases} Max(f_k(x)) = \sum_{j=1}^{n} c_j^k x_j,\ k \in \{1,\ldots,K\} \\ x \in \{0,1\}^n \\ \sum_{j=1}^{n} w_j x_j \leq W \end{cases}$$

All factors in the objectives and in the constraint are natural numbers. In this formulation, w_j denotes the weight of item j, and c_j^k denotes the utility of the same item according to objective k. Item j is packed in the knapscack iff $x_j = 1$, $j \in \{1,\ldots,n\}$. The formulation above assumes a single constraint as in [15]. This is opposed to the definition given in [18] where k objectives as well as k constraints are considered.

3 Schemata-Driven Multi-objective Optimization

3.1 Framework

Early work on multi-objective optimization recognized the importance of archiving the non-dominated solutions found so far by an evolutionary algorithm [7]. This was mainly motivated by two reasons . Some non-dominated solutions might be lost after applying the evolutionary operators. Moreover, these operators are not guaranteed to always provide better solutions. Second, the population's size might not be large enough to accomodate all the pareto optimal solutions.

---

[1] This operator eliminates all the dominated solutions as well as all the duplicates.

Later on, evolutionary algorithms started to exploit the archive in order to guide the search toward better solutions. For instance, elitism was applied by many multi-objective genetic and memetic algorithms. The most obvious way to implement elitism is to select some individuals randomly from the archive and to inject them in the mating pool. Though, many evolutionary algorithms use more sophisticated selection techniques. For example, Tamaski et. al. [13] used a parallel selection method and a fitness sharing mechanism. Parallel selection chooses the best individuals according to every objective, independently [11].

Zitzler and Thiele promote the selection of non-dominated solutions with a lower strength [18]. The strength of a non-dominated solution x is equal to the number of individuals in the current population who are covered by x. An individual is said to be covered by x iff it is equal to or dominated by x. A recent paper described an improved version of this algorithm taking into account a density measure [17]. Moreover, this new version builds the next generation exclusively from the non-dominated solutions archive.

Some population-based local search methods also use the archive to explore many regions in the search space. For instance, Knowles and Corne [9] described an algorithm that decides when is the mutant individual[2] included in the archive and when does it become the current solution. The mutant becomes the current solution if it either dominates the current solution or if it is not dominated by any solution in the archive and if it is located in a less crowded region than the current solution.

Besides applying the well-established elitism technique, our approach exploits the similarities between individuals in set $\overline{ND}(EA,t)$. These similarities are captured by schema $SCH(\overline{ND}(EA,t),r)$, where r is a number in $[0,1]$. The motivation of using schemata to guide multi-objective optimization is shown by the following example.

Consider two knapsack problem instances denoted by $mokp\_100$ and $mokp\_500$. Both instances have 2 objectives and 1 constraint. $mokp\_100$ is a small problem instance with 100 variables. On the other hand, $mokp\_500$ is a large problem instance with 500 variables. Let $MOGA$ be a genetic algorithm for solving the multi-objective knapsack problem. This algorithm uses the non-dominated sort ranking suggested by Goldberg [3] and implemented in [12]. Selection is performed according to the stochastic universal sampling method [1]. This algorithm keeps track of the non-dominated solutions found so far in set $\overline{ND}(MOGA,t)$. Furthermore, a proportion of this set contributes in building the next generation.

MOGA is run 10 times on each problem instance. A run is stopped after 100 generations. Population size is set to 100 for $mokp\_100$ and to 300 for $mokp\_500$. Schema $SCH(\overline{ND}(MOGA,100),1)$ is built after each run. Fig. 1 shows two schemata derived from two different runs of $MOGA$ on $mokp\_100$. We start by calculating the percentage of set symbols in every schema. Notice

---

[2] The mutant individual is obtained by randomly mutating the current solution.

that, when a symbol is set in some schema this implies that the corresponding gene has the same value for all vectors in $\overline{ND}(MOGA, 100)$. This also means that $MOGA$ has made the same decision for the corresponding knapsack item. On average, MOGA sets 65% and 29% of the genes after 100 iterations for $mokp\_100$ and $mokp\_500$ respectively. As expected, $MOGA$ sets more genes in the case of smaller problem instances.

run 1 : 1 ∗ 111 ∗ 01 ∗ 111 ∗ 111 ∗ 101 ∗ 110 ∗ ∗ ∗ ∗0100 ∗ 1 ∗ ∗00 ∗ 11 ∗ ∗01 ∗ 100011 ∗ 11 ∗ ∗00101 ∗ 1 ∗ 0 ∗ ∗1111 ∗ 11 ∗ 01111 ∗ ∗10 ∗ ∗ ∗ ∗ ∗ 1 ∗ ∗101 ∗ ∗01
run 2 : 1 ∗ 11110 ∗ ∗111 ∗ 111 ∗ 1 ∗ 111 ∗ 01 ∗ ∗00100 ∗ 11 ∗ 00111 ∗ 101 ∗ ∗000 ∗ 1 ∗ 110100 ∗ ∗111000 ∗ 1 ∗ ∗ ∗ ∗1 ∗ ∗0111 ∗ 1 ∗ 1 ∗ 101 ∗ ∗1 ∗ ∗101 ∗ ∗01

Fig. 1. Schemata derived from 2 runs of MOGA on $mokp\_100$

Now, we determine how stable are the decisions made by $MOGA$ over the 10 runs. For this purpose, shemata pairs derived from different runs are compared. These pairs are selected randomly. The comparison is based on two criteria. The first criterion is the number of set symbols who are common to both schemata. The second criterion is the number of contradictions between both schemata. The results of the comparisons are shown in Tab. 1. The percentage of contradictions is very low for both the small and the large problem instances. Moreover, the number of common set symbols is quite high. This means that, different runs of MOGA agree on a significant number of decisions and that they scarcely contradict the decisions made by each other[3].

Table 1. Common symbols and contradictions between schemata derived from different runs of MOGA

pair	$mokp\_100$		$mokp\_500$	
	common (%)	contradictions (%)	common (%)	contradictions (%)
(1, 2)	50%	0%	16%	0%
(2, 3)	50%	1%	16%	0%
(3, 4)	44%	1%	16%	1%
(4, 5)	44%	1%	17%	3%
(5, 6)	32%	0%	19%	4%
(6, 7)	37%	0%	23%	6%
(7, 8)	65%	5%	23%	2%
(8, 9)	54%	0%	19%	2%
(9, 10)	50%	0%	12%	1%

To summarize these results, we mention that when MOGA is run for some iterations, the algorithm converges to a set of non-dominated solutions hav-

---

[3] Though this result might not be generalizable to any optimization problem.

ing many genes in common. Moreover, many decisions made by the algorithm remain unchanged over different runs. This fact could be interepreted in two different ways. The first interpretation stems from the realm of automatic learning. According to this interpretation, MOGA learned some interesting features about trade-off solutions. Therefore, this acquired knowledge should be used to guide the algorithm toward better individuals. The second interpretation is that MOGA converged, may be prematurely, to some region in the search space. Ignoring this convergence would lead to many useless evaluations and to a slight improvement in the results' quality. This "premature" convergence could be avoided by applying some diversification techniques [4]. An alternative solution, is to focus on exploring this region of the search space. This latter solution will be detailed in the following paragraphs to improve the results produced by genetic and local algorithms. Some ideas to maintain the diversity of the non-dominated set will be overviewed in Sec. 5.

3.2 A Schemata-Driven GA (SMOGA)

This section describes how schemata could be used to improve the outcome of a multi-objective genetic algorithm. As mentioned earlier, genetic algorithms perform usually well on small problem instances. On the other hand, these algorithms could produce poor solutions when confronted with larger instances. Therefore, one would expect to get better results if the size of the problem instance is iteratively reduced during the execution of the genetic algorithm. Implementing this idea involves dealing with the three following issues. First, when should the problem instance be reduced? Second, how and on which basis is the reduction performed? Finally, how are the reduced-instances' solutions used to derive solutions for the original problem?

Static and dynamic criteria could be used to decide when a problem instance is reduced. The simplest static criterion is to perform instance reduction every g generations, where g is a constant. A dynamic criterion is to perform instance reduction when the *quality* of set $\overline{ND}(MOGA, t)$ does not improve too much over a given number of generations. For example, when no new individuals are added to $\overline{ND}(MOGA, t)$ for some number of successive generations.

We deal with the aforementioned second and third issues by expoiting non-dominated-sets' schemata. This is detailed by algorithm $SMOGA$ shown below.

The algorithm deals with a different problem instance (denoted by *currentInstance*) on every iteration of the *for* loop. At the start, *currentInstance* is the initial problem instance. A multi-objective genetic algorithm (called $MOGA$) is run on the current instance for a fixed number of generations. This algorithm produces a set of non-dominated solutions (denoted by $ndSet$) for this instance. Schema $SCH(ndSet, 1)$ is then extracted. The set symbols in this schema correspond to the genes common to all the individuals in $ndSet$. That is to the genes for which a decision was made by the previous run of $MOGA$. On the other hand, all the $*$ symbols correspond to the genes

Algorithm 1 SMOGA

1: currentInstance := initInstance
2: **for** $i = 1$ to $step$ **do**
3:   MOGA(currentInstance, ndSet)
4:   SchemaOfArchive(ndSet,SCH(ndSet,1))
5:   ReduceInstance(currentInstance, SCH(ndSet,1), redInstance)
6:   BuildArchive(ndSet, $SCH(\overline{ND},1)$, initInstance, tempSet)
7:   JoinNonDominatedSets($\overline{ND}$, tempSet)
8:   UpdateSchema($SCH(\overline{ND},1)$, SCH(ndSet,1))
9:   currentInstance := redInstance
10: **end for**

for which $MOGA$ did not make a decision. In the case of $MOKP$, these are the items for which a decision whether they should or should not be taken, was not made yet. Schema $SCH(ndSet, 1)$ is then used to reduce the size of the current instance yielding $redInstance$. The instance reduction is illustrated by the following example in the case of $MOKP$.

Consider an $MOKP$ instance with 8 items, 2 objectives and 1 constraint. The utilities and the weight of each item are shown in Tab. 2. Furthermore, we assume that the knapsack capacity is $C = 20$.

Table 2. An instance of $MOKP$ before reduction

item	w_i	c_i^1	c_i^2
1	4	5	4
2	3	2	1
3	2	1	2
4	1	3	1
5	5	4	4
6	3	3	5
7	3	2	3
8	1	1	1

Now, consider the following schema :

$$1 * * * 1 1 * 0$$

Reducing the previous instance according to this schema yields a new instance consisting of items $2, 3, 4$ and 7. The new knapsack capacity is calculated from the original one by subtracting the weights of the items whose corresponding symbol in the schema is set to 1. Thus, the new knapsack capacity is equal to 8 in this example.

The individuals in $ndSet$ are then extended to build solutions for the initial problem instance. This is performed according to schema $SCH(\overline{ND}, 1)$, as shown in Fig. 2. Notice that the individuals in $ndSet$ could be thought of as partial solutions of the original problem instance. The so obtained solutions are used to update set $\overline{ND}$. Finally, schema $SCH(\overline{ND}, 1)$ is incrementally updated using the information conveyed by $SCH(ndSet, 1)$. This is illustrated by Fig. 2.

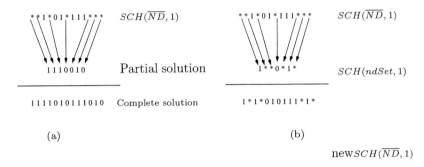

Fig. 2. (a) Building a complete solution from a schema and a partial solution. (b) Incremental update of $SCH(\overline{ND}, 1)$

3.3 A Schemata-Driven Local Search Algorithm (SLS)

Early multi-objective local search algorithms applied the aggregation technique to decide whether a neighbor solution will become the current solution [15]. This technique transforms a multi-objective optimization problem into a single-objective one by combining the objectives all together. Recent multi-objective local search algorithms maintained a population of non-dominated solutions [9]. This population is used to determine the next current solution. We describe in this section a population-based local search algorithm (called SLS) to solve MOKP. This algorithm exploits the schema of the maintained non-dominated set to guide the search procedure. An outline of SLS is given below.

Algorithm 2 SLS

1: RandomNDSet($\overline{ND}$)
2: **for** $i = 1$ to $maxGen$ **do**
3:    **if** $i \bmod period = 0$ **then**
4:       SchemaOfArchive($\overline{ND}, SCH(\overline{ND}, 1)$)
5:    **end if**
6:    GetArchiveNeighbor($\overline{ND}$, ndNeighbor)
7:    JoinNeighbor($\overline{ND}$, ndNeighbor)
8: **end for**

The algorithm starts from a randomly-generated non-dominated set (denoted by $\overline{ND}$). Such a set consists of the non-dominated individuals of a randomly-generated population. Schema $SCH(\overline{ND}, 1)$ is periodically extracted from $\overline{ND}$. A neighbor of the current non-dominated set is built on every iteration of the *for* loop. This is a set consisting of neighbors of all the individuals in $\overline{ND}$. Finally, sets $\overline{ND}$ and its neighbor are joined according to some specific rules that will be detailed in the sequent.

The neighbor of an individual in $\overline{ND}$ is determined with respect to schema $SCH(\overline{ND}, 1)$. More specifically, neighbors have always to be instances of $SCH(\overline{ND}, 1)$. An individual x is said to be an instance of some schema sch iff both sch and x have the same length and every set symbol in sch is equal to the corresponding gene in x. More formally,

$$sch_i = x_i, \; \forall i \in \{1, \ldots, n\} \text{ s.t. } sch_i \neq *$$

where n is the length of sch and x_i is the ith gene of x. Notice that by the definition of $SCH(\overline{ND}, 1)$, all the individuals in $\overline{ND}$ are instances of $SCH(\overline{ND}, 1)$. Therefore, SLS focuses on exploring a specific region of the search space. This region gets norrower as the algorithm evolves. Algorithm $SNGBR$ is used to determine the neighbor of a given individual in $\overline{ND}$. This algorithm is an adaptation of the one described in [15]. An outline of $SNGBR$ is given below.

Algorithm 3 SNGBR

1: GetRandomUnpacked(individual, $SCH(\overline{ND}, 1)$, i)
2: **while** not IsPackable(MOKPInstance, i) **do**
3:    GetRandomPacked(individual, $SCH(\overline{ND}, 1)$, j)
4:    Unpack(individual, j)
5: **end while**

An unpacked item i is randomly selected. The algorithm checks that packing i in the knapsack does not contradict schema $SCH(\overline{ND}, 1)$. In other words, symbol $SCH(\overline{ND}, 1)_i$ is not equal to 0. Some randomly-selected items are then unpacked to make enough space for i. The algorithm checks that unpacking an item does not contradict $SCH(\overline{ND}, 1)$. That is, the corresponding symbol in $SCH(\overline{ND}, 1)$ is not equal to 1.

Local search algorithms generally accept worse solutions under some control mechanism. The rationale is to help the algorithm escape local optima. We applied this idea in our algorithm by accepting dominated individuals in the new $\overline{ND}$ set. The rationale of this idea is better explained by the following example. Consider sets $\overline{ND}$ and $ndNeighbor$ shown in Fig. 3. Individual Z from $ndNeighbor$ is dominated by A from $\overline{ND}$. Nevertheless, the distance between both individuals is quite large. Therefore dismissing Z would prevent exploring a potentially less crowded region in the search space. On the other hand, individual Y could be dismissed since it is dominated by a near individual, namely B. In other words, B is a better approximation of some pareto optimal solution(s) than Y.

More formally, an individual x is accepted in the new $\overline{ND}$ set iff one of the following conditions is fulfilled :

1. x is not dominated by any individual in $(\overline{ND} \cup ndNeighbor) \setminus \{x\}$.
2. $rank(x) > minRank$ and all the individuals y dominating x verify :

$$distance(x,y) \geq \sigma$$

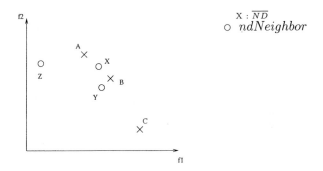

Fig. 3. Joining $\overline{ND}$ and its neighbor set

4 Experimental Study

4.1 Experimentation Protocol

The aim of this experimental study is to evaluate the impact of using the non-dominated set schemata in a genetic algorithm as well as in a local search algorithm. The multi-objective knapsack problem was considered. A set of 10 problem instances from [15] were used. All these instances have 2 objectives and a single constraint. The number of items ranges from 50 to 500 by a step of 50. The pareto optimal set ND^* is already known for each of these instances.

Two performance indicators defined in [15], namely WD and MD, were considered. Indicator WD is the worst distance between any solution in ND^* and set $\overline{ND}$. More formally :

$$WD = \max\{d(\overline{ND}, y) \text{ s.t. } y \in ND^*\}$$

Lower values of WD imply that $\overline{ND}$ is closer to ND^* and/or is more dispersed. Indicator MD is the average distance between any solution in ND^* and set $\overline{ND}$. More formally :

$$MD = \frac{\sum_{y \in ND^*} d(\overline{ND}, y)}{|ND^*|}$$

This indicator tells how good are the pareto optimal solutions approximated. We introduce another performance indicator denoted by MD' and defined as follows :

$$MD' = \frac{\sum_{y \in \overline{ND}} d(ND^*, y)}{|\overline{ND}|}$$

This indicator tells how close are the solutions produced by an EA to the optimal set. In other words, this metric emphasizes on set $\overline{ND}$ instead of set ND^*.

In a first step, a schemata-driven GA (called $SMOGA$) was compared to a more conventional GA (called $MOGA$). Both algorithms applied NSGA ranking and the stochastic universal sampling selection method. Moreover, in both algorithms, 30% of population $P(EA, t+1)$ is built using individuals in set $\overline{ND}(EA, t)$. A greedy repair algorithm suggested in [10] for the single objective case and generalized in [18] to the multi-objective framework was used to cope with infeasible individuals. Also, crossover and mutation rates were set to 0.8 and 0.01 respectively. In a second step, a schemata-driven local search algorithm (SLS) was compared to a more conventional one (called LS). Both algorithms use the neighborhood described in section 3.3. The *joinNeighbor* procedure in SLS applies $NSGA$ ranking. Parameter σ in this procedure is calculated using the estimators described in [2].

Each algorithm is run 10 times on every problem instance. The arithmetic means of performance indicators over the 10 runs are then calculated. The maximum number of generations was set to 1000 for all the runs. The population size was set to 100, 200 and 300 depending on the problem instance. Parameter *step* in $SMOGA$ was set to 4. Whereas, parameter *period* in SLS was set to 50. These values were set after analyzing some preliminary runs.

4.2 Results

The results of comparing $SMOGA$ and $MOGA$ are shown in Tab. 3. These results show that $SMOGA$ produces more non-dominated solutions than $MOGA$, except for $n = 50$. This is due to the fact that $SMOGA$ explores restricteed regions in the search space. Moreover, $MOGA$ yields better values of performance indicator WD for all problem instances. This suggests that the non-dominated sets produced by $SMOGA$ are more compact than the ones produced by $MOGA$. On the other hand, $SMOGA$ provides better values of performance indicator MD except for $n = 50$. Moreover, the difference between the values provided by both algorithms becomes more important for larger problem instances. The same remarks apply to performance indicator MD'[4]. This suggests that algorithm $SMOGA$ provides better approximations of the pareto optimal set than

---

[4] In fact, $SMOGA$ outperforms $MOGA$ on all problem instances according to this indicator.

$MOGA$. This is especially true for large problem instances. The result is predictable since the iterative instance reduction process applied in $SMOGA$ is likely to be more effective for large instances. Fig. 4 shows the non-dominated sets produced by both algorithms as well as the pareto optimal set when $n = 500$. Notice that the set produced by $SMOGA$ is much closer to the pareto optimal set than the one produced by $MOGA$.

Table 3. Comparison of $MOGA$ and $SMOGA$

	MOGA				SMOGA			
n	$|ND|$	WD	MD	MD'	$|ND|$	WD	MD	MD'
50	15	0.077	0.017	0.007	13	0.108	0.022	0.004
100	17	0.161	0.053	0.037	26	0.190	0.048	0.016
150	16	0.164	0.059	0.038	23	0.184	0.053	0.018
200	20	0.157	0.061	0.051	30	0.179	0.045	0.019
250	20	0.203	0.085	0.071	28	0.210	0.057	0.022
300	13	0.175	0.096	0.084	33	0.193	0.056	0.029
350	12	0.180	0.097	0.082	28	0.195	0.057	0.030
400	16	0.166	0.113	0.100	30	0.182	0.057	0.030
450	15	0.190	0.127	0.111	30	0.200	0.069	0.044
500	15	0.186	0.137	0.116	31	0.199	0.070	0.042

The results of comparing SLS and LS are shown in Tab. 4. As in the case of genetic algorithms, SLS yields more non-dominated solutions than LS. Furthermore, LS leads to better values of performance indicator WD than SLS. Though, the difference between the values provided by both algorithms becomes less important for larger problem instances. On the contrary to the genetic algorithms, both local search algorithms provide comparable values of performance indicator MD. On the other hand, SLS outperforms LS according to indicator MD' and for larger problem instances. The previous results suggest that SLS produces better approximations of the pareto optimal set for large problem instances. Moreover, the non-dominated sets produced by SLS are more compact than those produced by LS. Fig. 5 shows the non-dominated sets produced by LS and SLS for $n = 500$. Almost all the solutions produced by LS in this case are dominated by solutions produced by SLS. Finally, notice that SLS outperforms $SMOGA$ according to all performance indicators.

5 Conclusions

This paper dealt with the exploitation of non-dominated sets' schemata in order to improve the outcome of multi-objective evolutionary algorithms. A schemata-driven genetic algorithm (SMOGA) as well as a schemata-driven local search algorithm (SLS) were described. Both algorithms produced better approximations

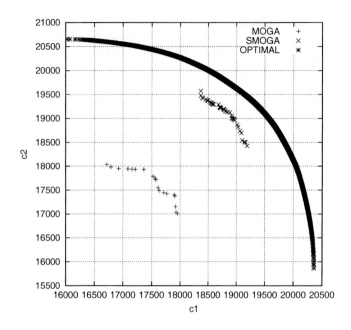

Fig. 4. The pareto fronts produced by MOGA and SMOGA for $n = 500$. Optimal solutions are produced by a branch and bound algorithm [14]

Table 4. Comparison of LS and SLS

	LS				SLS			
n	$|ND|$	WD	MD	MD'	$|ND|$	WD	MD	MD'
50	29	0.035	0.003	0.001	28	0.053	0.004	0.000
100	97	0.044	0.005	0.003	125	0.064	0.004	0.001
150	89	0.071	0.008	0.004	134	0.091	0.008	0.001
200	100	0.087	0.011	0.009	193	0.112	0.009	0.001
250	115	0.102	0.013	0.008	220	0.132	0.014	0.001
300	104	0.107	0.016	0.010	202	0.124	0.014	0.001
350	84	0.115	0.022	0.015	185	0.139	0.021	0.003
400	92	0.113	0.021	0.013	219	0.126	0.018	0.002
450	106	0.113	0.026	0.021	243	0.130	0.018	0.004
500	102	0.125	0.029	0.019	219	0.149	0.026	0.005

than their equivalent more conventional versions. The observed improvements are more important in the case of larger problem instances.

The sets produced by the schemata-driven algorithms tend to be compact. Some techniques could be used to maintain the diversity of these sets. For instance, in our implementation of $SMOGA$, the algorithm systematically took

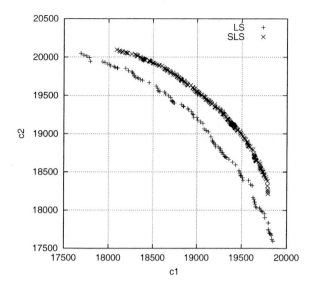

Fig. 5. Pareto fronts produced by LS and SLS for $n = 500$

into account the decisions made by the previous run of $MOGA$ in reducing the problem instance. Instead, the algorithm could take into account these decisions with some probability p_{rs}. This is comparable to the mutation probability used by genetic algorithms to hillclimb local optima neighborhoods. On the other hand, the decisions made by SLS are always compliant with the current schema. The algorithm could accept instead to contradict the schema with some probability p_{cs}. These contradictions will help SLS escape from the search-space region it is currently exploring. Parameter p_{cs} may have a high value at the start of the algorithm. This value could then be lowered like in simulated annealing. In other words, the algorithm would accept contradictions with the current schema according to an over-time decreasing probability.

References

1. J. E. Baker. Reducing Bias and Inefficiency in the Selection Algorithm. In *Second Intl. Conference on Genetic Algorithms, ICGA'2*, pages 14–21, Pittsburg, 1987.
2. C. M. Fonseca and P. J. Fleming. An Overview of Evolutionary Algorithms in Multiobjective Optimization. *Evolutionary Computation*, 3(1):1–16, 1995.
3. D. E. Goldberg. *Genetic Algorithms in Search, Optimization and Learning*. Addison-Wesley, 1989.
4. D. E. Goldberg and J. Richardson. Genetic Algorithms with Sharing for Multimodal Function Optimization. In *Second Int. Conf. on Genetic Algorithms ICGA'2*, pages 41–49, NJ, 1987. Lawrence Erlbaum.

5. M. P. Hansen and A. Jaskiewicz. Evaluating the Quality of Approximations to the Non-dominated Set. Technical Report IMM-REP-1998-7, Technical University of Denmark, March 1998.
6. J. H. Holland. *Adaptation in Natural Artificial Systems*. University of Michigan Press, 1975.
7. H. Ishibuchi and T. Murata. A Multi-objective Genetic Local Search Algorithm and its Application to Flowshop Scheduling. *IEEE Transactions on Systems, Man, and Cybernetics – Part C: Applications and Reviews*, 28(3):392–403, Aug 1998.
8. J. Knowles and D. Corne. On Metrics for Comparing Non-dominated Sets, 2002.
9. J. D. Knowles and D. Corne. Approximating the Nondominated Front Using the Pareto Archived Evolution Strategy. *Evolutionary Computation*, 8(2):149–172, 2000.
10. Z. Michalewicz and J. Arabas. Genetic Algorithms for the 0/1 Knapsack Problem. In Zbigniew W. Raś and Maria Zemankova, editors, *Proceedings of the 8th International Symposium on Methodologies for Intelligent Systems*, volume 869 of *LNAI*, pages 134–143, Berlin, October 1994. Springer.
11. J. D. Schaffer. Multiple Objective Optimization with Vector Evaluated Genetic Algorithms. In J. J. Grefenstette, editor, *ICGA Int. Conf. on Genetic Algorithms*, pages 93–100. Lawrence Erlbaum, 1985.
12. N. Srinivas and K. Deb. Multiobjective Optimisation Using Non-dominated Sorting in Genetic Algorithms. *Evolutionary Computation*, 2(8):221–248, 1995.
13. H. Tamaki, H. Kita, and S. Kobayashi. Multi-objective Optimization by Genetic Algorithms: A Review. In *IEEE Int. Conf. on Evolutionary Computation ICEC'96*, pages 517–522, 1996.
14. E. L. Ulungu and J. Teghem. The Two-phase Method: An Efficient Procedure to Solve Bi-objective Combinatorial Optimization Problems. In *Foundations of Computing and Decision Sciences*, volume 20, pages 149–165. 1995.
15. E. L. Ulungu, J. Teghem, Ph. Fortemps, and D. Tuyttens. MOSA Method: A Tool for Solving Multiobjective Combinatorial Optimization Problems. *Journal of Multi-Criteria Decision Analysis*, 8(4):221–236, 1999.
16. D. H. Wolpert and W. G. Macready. No Free Lunch Theorems for Optimization. *IEEE Transactions on Evolutionary Computation*, 1(1):67–82, April 1997.
17. E. Zitzler, M. Laumanns, and L. Thiele. SPEA2: Improving the Strength Pareto Evolutionary Algorithm for Multiobjective Optimization. In K. Giannakoglou, D. Tsahalis, J. Periaux, K. Papailiou, and T. Fogarty, editors, *Evolutionary Methods for Design, Optimisation and Control*. CIMNE, Barcelona, Spain, 2002.
18. E. Zitzler and L. Thiele. Multiobjective Evolutionary Algorithms: A Comparative Case Study and the Strength Pareto Approach. *IEEE Transactions on Evolutionary Computation*, 3(4):257–271, 1999.

A Real-Coded Predator-Prey Genetic Algorithm for Multiobjective Optimization

Xiaodong Li

School of Computer Science and Information Technology
RMIT University, GPO Box 2476v
Melbourne, VIC 3001, Australia
xiaodong@cs.rmit.edu.au

Abstract. This paper proposes a real-coded predator-prey GA for multiobjective optimization (RCPPGA). The model takes its inspiration from the spatial predator-prey dynamics observed in nature. RCPPGA differs itself from previous similar work by placing a specific emphasis on introducing a dynamic spatial structure to the predator-prey population. RCPPGA allows dynamic changes of the prey population size depending on available space and employs a BLX-α crossover operator that encourages a more self-adaptive search. Experiments using two different fitness assignment methods have been carried out, and the results are compared with previous related work. Although RCPPGA does not employ elitism explicitly (such as using an external archive), it has been demonstrated that given a sufficiently large lattice size, RCPPGA can consistently produce and maintain a diverse distribution of nondominated optimal solutions along the Pareto-optimal front even after many generations.

1 Introduction

In recent years, evolutionary algorithms have gained much attention for solving multiobjective optimization problems. In contrast to conventional multicriteria analysis models, evolutionary algorithms are population based, hence they are capable of evolving multiple solutions simultaneously approaching the non-dominated Pareto front in a single run [1]. In order to find a good set of Pareto-optimal solutions, an efficient evolutionary algorithm for multiobjective optimization must achieve two goals - to ensure a good convergence to Pareto-optimal front, and to maintain a set of solutions as diverse as possible along the Pareto front [1]. The benefit of such approach is that it gives the decision-maker the freedom to choose from many alternative solutions.

This second goal of maintaining a diverse set of solutions is unique to multiobjective optimization. As only when we have a diverse set of solutions, we can provide a good set of trade-off solutions for the decision-maker to choose from. To preserve solution diversity, several niching and non-niching techniques have been proposed [2]. For example fitness sharing has been adopted as a niching method in a number of multiobjective optimization EAs [3],[4]. Among the non-niching techniques, Laumanns et al. proposed a spatial Predator-Prey Evolutionary Strategy (PPES) for multiobjective optimization [5]. Their preliminary study demonstrated that their predator-prey approach is feasible for multiobjective optimization, though there are some serious drawbacks associated with it. Deb also examined Laumanns' model and introduced a modified predator-prey model using different weighted vectors associated with each predator [1]. Deb showed that the modified model produced a better distribution of Pareto optimal solution than the original PPES.

Along the same line of thought, in this paper a real-coded predator-prey genetic algorithm for multiobjective optimization (RCPPGA) is developed. Although RCPPGA share some of its similarity with PPES as proposed by Laumanns et al. [5], RCPPGA makes special emphasis on the use of a dynamic spatial structure of the predator-prey populations (which is lacking in the original PPES). By doing this, we can introduce to our model the kind of predator-prey dynamics more analogous to nature. The main objective of this research is to investigate RCPPGA's ability to approximate Pareto-optimal front, and more importantly to see if the model can produce well-distributed nondominated solutions along the Pareto front. This is of particular interest to us because the PPES proposed by Laumanns et al. seems to have difficulty in doing so when it is applied to a more difficult two-objective optimization function [5], in which case, after many iterations, PPES converges to solutions that are concentrated only partially on the Pareto front.

2 Background and Related Work

Studies in ecology and in particular population dynamics have shown that populations have unique spatial characteristics such as density and distribution [12]. Individuals that make up a population affect one another in various ways. For example, interactions among individuals of a population are often confined within an individual's immediate neighbourhood. Individuals interact in space and time not only with its own species, but also with competitors, predators, and the environment. These properties play a critical role in the evolutionary process of the individuals in an evolving population.

Many evolutionary algorithms have been developed explicitly exploring the use of spatial structure in a GA population. It has been found that the use of spatial structure is especially effective in maintaining a better population diversity, which is critical in improving the performance of many evolutionary algorithms on difficult optimization problems [6], [7]. Cantu-Paz has provided a very good survey in this area [7]. However, most of these algorithms have been developed for single objective optimization, and they commonly use a static structure, whether it is fine-grained or coarse-grained, that is the spatial structure of the population remains unchanged throughout a GA run. A few exceptions are work done by Kirley et al [8], Kirely [9], and Li and Kirley [10], where the authors examined the effects of introducing dynamic ecological features (e.g., disturbances and varying population density) into a fine-grained parallel GA model. It was found that such ecologically inspired parallel GA models using a dynamic spatial population structure are comparable or better in performance than those without such features, especially when dealing with difficult multi-modal function optimization tasks. Since multi-modal function optimization shares its similarity with multiobjective optimization in many aspects, such observation led to our current investigation of a recently proposed predator-prey GA [11], more specifically its ability in handling multiobjective optimization problems. For definitions and key concepts used in multiobjective optimization, the readers can refer to [1], [2].

Laumanns proposed a spatial predator-prey evolutionary strategy (PPES) for multiobjective optimization [5]. In this model, prey representing potential solutions are mapped onto the vertices of a 2-dimensional lattice. Predators are also introduced to the 2d lattice. Each predator is associated with a particular objective function. Each predator performs random walk on the lattice, and chases and kills the weak prey in its nearest neighbourhood according to its associated objective. A detailed description of PPES can be found in [5]. Laumanns et al. suggested that an adaptive step size for the EA is mandatory in order to obtain a reasonably good distribution of solutions on the Pareto front. However, even with this adaptive step size, in this case a decreasing step size $\sigma_{k+1} = 0.99\sigma_k$, their model produced a rather disappointing result. Only some subsets of the Pareto front were obtained. In fact, in Deb's reproduction of the experiment, as the model was run for more iterations, the nondominated solutions produced by the PPES became less and less diverse, eventually even converging to a single solution on the nondominated front [1]. To overcome this difficulty of maintaining a diverse set of non-dominated solutions, Deb suggested a modified version of PPES [1]. Instead of assigning one objective to each predator, a different weighted vector is associated with each predator. Each predator evaluates its neighbouring prey with respect to the weighted sum of the objectives. For instance, for a two-objective optimization problem, each predator is assigned with a weighted vector (w_1, w_2). If 9 predators are used, then the first predator takes a weight vector (1, 0), the 2nd predator takes (0.875, 0.125), and so on, until the 9th predator takes (0, 1). The least-fit (according to the weighted sum) prey individual is eliminated at each iteration. By using a different weighted vector for each predator, this method allows each predator to emphasize solutions on different part of the Pareto-front. Deb's result on the second function of the original study of Lanmanns et al. gave a much improved performance. Even after many iterations, the diversity of prey solutions was still well maintained in the Pareto-front region.

In this study, a real-coded predator-prey GA (RCPPGA) is developed to handle multiobjective optimization problems. The two above described fitness assignment methods, used by Laumanns et al. [5] and Deb [1] respectively, are adopted in RCPPGA. In the following sections, we will refer Laumanns' method as **method 1**, and Deb's method as **method 2**. By employing these two fitness assignment methods, we can then look at whether RCPPGA would have the same kind of difficulty in obtaining well distributed optimal solutions, as described by Laumanns et al., and verify whether Deb's weighted vector approach is effective in this model.

3 A Real-Coded Predator-Prey GA for Multiobjective Optimization – RCPPGA

We emphasize the spatial dynamics of predator-prey interaction by using a two-dimensional lattice where the prey and predator populations reside and interact. The 2d lattice has its boundaries wrapped around to the opposite edge, therefore eliminating any boundary conditions. The initial populations of prey and predator individuals are randomly generated and distributed across the lattice. As illustrated in Fig. 1, we often start with a large number of prey and a relatively small number of predators in the initial populations. Each prey presents a possible solution, whereas each predator does not represent a solution but is able to roam around in order to keep the prey in check (ie., its task is to kill the least-fit prey in its vicinity).

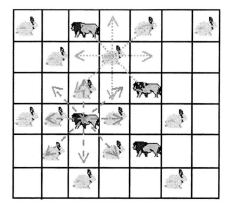

Fig. 1. Predators and prey are randomly distributed across the 2d lattice at the beginning of a run

After the above initialization, the predator-prey model proceeds in the following steps:

1. Prey and predators are allowed to move from one cell to another on the lattice according to a *randomMoveProbability*, which is normally set to 0.5, so that half the prey would attempt to move one step on the lattice whereas the other half would remain where they were. If the prey were allowed to move, they could choose a random direction, i.e., one of the eight cells in a 8-cell Moore neighbourhood (north, south, ast, west, and plus the four diagonal neighbours), to move into. They then attempt to move. If the cells they are attempting to move into are occupied by other prey or predators, then they try again. Each prey is allowed to try 10 times. If the prey is still unable to find a place to move, it remains where it is.
2. After the prey have moved they are then allowed to breed. Space plays a critical role in this model as each prey can only breed with another prey within its neighbourhood (excluding itself). If the prey has no neighbours it is not allowed to breed. Otherwise the prey is allowed to breed with another

randomly selected neighbour to produce an offspring using real crossover and mutation operators (see Section 3.2). The offspring is randomly placed over the entire lattice, which can be seen as migration among different clusters of prey across the lattice. 10 attempts are made to place the child on the lattice. If all the attempted cells are occupied, the child is not generated. Note that in this step, the creation of a new child is essentially one function evaluation.
3. The prey population is under constant threat from the predators, which are initially allocated at random across the lattice. Selection pressure is exerted upon the prey population through the predator-prey interaction, that is, predators are given the task of killing the least-fit prey in their vicinity. The predators first look around their neighbourhood to see if there are any prey. If so, the predator selects the least-fit prey and kills it. The predator then moves onto the cell held by that prey. If a predator has no neighbouring prey, it moves in exactly the same way as prey. However it is possible to allow the predators to move more than once per prey step (refer to equation (1)).
4. Go back to step 1), if the number of required evaluations is not reached.

In order to prevent predators from completely wiping out the entire prey population, the following formula is adopted to keep the prey population at an acceptable level:

$$iterations = \lfloor \frac{numPrey_{actual} - numPrey_{preferred}}{num_{predators}} \rfloor \quad (1)$$

where *iterations* is the number of moves the predators may take before the prey can make their moves. A predator can kill at most one prey per iteration. Basically equation (1) is used to keep the actual number of prey ($numPrey_{actual}$) to a number similar to the preferred number of prey ($numPrey_{preferred}$). The predators are encouraged to minimize the difference between these two values. The floor operator ensures that the predators do not wipe out the prey population entirely. For example, if there are 250 prey, the preferred number of prey is 120, and the number of predators is 20, then the predators would iterate 6 times before the prey have a chance to move and breed again. This is also analogous to the fact that predators are often faster than prey in speed. Another merit of using equation (1) is that as the minimum number of prey (the floor value) is reached, or in another word, the predators become 'slower' in speed, new born prey would have a better chance of survival than otherwise. As a result, the number of prey individuals would start to increase, rather than to continue its decline. This trend would continue until it gets to a point, where the effect of applying equation (1) is once again tipped to be in favour of predators. Using equation (1) in a way provides a mechanism of varying the prey population size dynamically.

One distinct feature of RCPPGA is its explicit implementation of a dynamic spatial structure of the predator-prey populations. In addition to the mating restriction feature as seen in PPES, the predators and prey in RCPPGA can interact via dynamically changing population structure, as predators and prey are capable of moving around on the 2d lattice. Over time, prey clusters of various sizes can be formed naturally by the roaming prey themselves in parallel across the lattice. These different clusters tend to form niches of their own and therefore help preserve prey population diversity over the successive generations. By randomly placing new-born offspring over the entire lattice, we hope to stop potential 'inbreeding' or 'genetic drift' from occurring [12], and meanwhile help continue to maintain a diverse set of solutions until the end of many iterations of a simulation run.

3.1 Selection Mechanism

In RCPPGA, selection pressure is dynamically exerted upon the prey population through the killing and removal of the least-fit prey by the roaming predators. We do not use any direct replacement, for example, fitter offspring directly replacing the less fit ones in the population, as often seen in a typical GA. RCPPGA does not use any explicit elitist mechanism such as an external archive to keep the best-fit prey at each generation. It only adopts a weak "mating" selection method, that is, at each generation, a prey simply breeds with another prey randomly chosen from its neighbourhood.

3.2 Real-Coded Crossover and Mutation Operators

In this real-coded predator-prey GA model, each prey individual represents a chromosome that is a vector of genes, where each gene is a floating point number [13]. For example, a parameter vector corresponding to a GA individual can be represented as $x = (x_1, x_2, \ldots, x_n)$ $(x_i \in [a_i, b_i] \subset \Re, i = 1, \ldots, n)$. The GA works in exactly the same way as the binary counterpart except that the crossover and mutation operations are slightly different.

We follow the suggestion of Wright that a mixed real-coded crossover seems to give better results. This mixed real-coded crossover involves two operations [13]. The first operation uses a real crossover operator, which behaves similarly to a standard crossover. The difference is that instead of swapping binary values, the values in the slots of floating point array (i.e., a chromosome consisting of genes each representing a real-number variable) are swapped. For example, if we have two parents, $x = (x_1, x_2, \ldots, x_n)$ and $y = (y_1, y_2, \ldots, y_n)$, and the crossover point is between x_i and x_{i+1}, then one child corresponds to $c_1 = (x_1, x_2, \ldots, x_i, y_{i+1}, \ldots, y_n)$ and the other $c_2 = (y_1, y_2, \ldots, y_i, x_{i+1}, \ldots, x_n)$. We apply this operator to two parents to produce 50% of gene values of a prey offspring. The second crossover operator is so called blend crossover operator (BLX-α), first introduced by Eshelman and Schaffer [14]. BLX-α uses two parent solutions x_1^t and x_2^t at generation t, to define a range $[x_1^t - \alpha(x_2^t - x_1^t), x_2^t + \alpha(x_2^t - x_1^t)]$ (assuming $x_1^t < x_2^t$), within which a child solution can be randomly picked. If μ is a random number between 0 and 1, then a child can be generated as follows:

$$x_1^{t+1} = (1 - \gamma)x_1^t + \gamma x_2^t \qquad (2)$$

where $\gamma = (1 + 2\alpha)\mu - \alpha$. The above equation can be rewritten as:

$$(x_1^{t+1} - x_1^t) = \gamma(x_2^t - x_1^t). \tag{3}$$

From equation (3) we can observe that if the difference between the two parent solutions x_1^t and x_2^t is small, then the difference between the child x_1^{t+1} and parent x_1^t is also small. This interesting property of BLX-α can be seen as a self-adaptive mechanism for the real-coded GAs [15], because the spread of the population at generation t dictates the spread of the solutions in the population at generation $t + 1$. Eshelman and Schaffer reported BLX-0.5 (with α=0.5) gives better results than BLX with other α values. It seems that with α=0.5, BLX provides a nice balance between exploration and exploitation (convergence), therefore we choose to use α=0.5 in this model. We apply BLX-0.5 crossover to the two parents to produce the remaining 50% of the gene values of the prey offspring.

Mutation is applied with a probability to the entire prey population. The mutation operator simply replaces a gene (i.e., a real parameter value) in a chromosome with another floating point number randomly chosen within the bounds of the parameter values.

4 Experiments

We chose the same two test functions used in Laumanns' PPES [5], and a test function adopted by Deb in his modified PPES [1].

Test function F1:

$$\begin{aligned}
& \text{minimize} & & f_1(x) = x_1^2 + x_2^2, \\
& \text{minimize} & & f_2(x) = (x_1 + 2)^2 + x_2^2, \\
& \text{where} & & -50 \leq x_1 \leq 50, \text{ and } -50 \leq x_2 \leq 50.
\end{aligned}$$

Test function F2:

$$\begin{aligned}
& \text{minimize} & & f_1(x) = -10 exp(-0.2\sqrt{x_1^2 + x_2^2}), \\
& \text{minimize } f_2(x) = & & \mid x_1 \mid^{4/5} + \mid x_2 \mid^{4/5} + 5(sin^3 x_1 + sin^3 x_2), \\
& \text{where} & & -50 \leq x_1 \leq 50, \text{ and } -50 \leq x_2 \leq 50.
\end{aligned}$$

Test function F3:

$$\begin{aligned}
& \text{minimize} & & f_1(x) = x_1^2, \\
& \text{minimize} & & f_2(x) = \frac{1+x_2^2}{x_1^2}, \\
& \text{where} & & \sqrt{0.1} \leq x_1 \leq 1, \text{ and } 0 \leq x_2 \leq \sqrt{5}.
\end{aligned}$$

In order to compare RCPPGA with the PPES proposed by Laumanns et al and the subsequently modified PPES by Deb [5], [1], experiments were conducted on RCPPGA using the two fitness assignment methods as suggested by them - **method 1**: A different type of predators associated with a different objective [5]; **method 2**: Each predator associated with a different weighted vector [1]. We are

214   X. Li

interested in finding out whether RCPPGA is able to approximate the Pareto front using the above two fitness assignment methods. In particular, we would like to verify if RCPPGA also has the difficulty in obtaining a good distribution of the nondominated solutions, like the findings on PPES obtained by Laumanns and Deb.

To study the effectiveness of using the dynamic spatial structure of the predator-prey populations, we run RCPPGA in the following experiments using different lattice sizes. A smaller lattice size would restrain the prey population from increasing its size, because the new-born offspring are always competing for the limited vacant space (see equation (1)). On the other hand, if we use the same number of preys in the initial prey population on a larger lattice, the prey population would enjoy a rapid increase in its size, until the lattice is occupied to a threshold governed by equation (1).

RCPPGA is given the following parameter configurations: the lattice size is varied in 3 different sizes, 20x20, 30x30, and 50x50; the number of prey in the initial prey population is 240; the number of predators is 20 and it remains constant throughout a simulation run, however for the fitness assignment method 1, there are two objectives, hence we divide these 20 predators into two types, 10 for each, optimizing according to the two objectives respectively; the *randomMoveProbability*, which is a variable specifying the probability of a prey moving to another cell or remaining stationary, is assigned with a value of 0.5; mutation rate is 0.01; the number of evaluations required is set to 30000 (note that each time a new born is created, it is essentially one function evaluation).

In the following figures, we use a filled circle '•' to indicate the nondominated solutions found at the end of 30000 evaluations, and the plus sign '+' for the dominated solutions. Note that we do not use an external archive to extract the dominated solution in the course of a RCPPGA run. The nondominated solutions shown in the figures are exactly those left in the final prey population when the 30000 evaluations are completed.

5 Results and Analysis

5.1 Effects of Using Different Lattice Sizes

From Fig. 2 - 5, we can see that lattice size has a significant impact on the number of different prey solutions obtained in the final iteration. For a lattice size of 20x20, the RCPPGA converged to only very few nondominated solutions. When the lattice size is increased to 30x30, method 1 gives a much improved performance in terms of the number of different nondominated solutions found. However, surprisingly, method 2, the weighted vector approach proposed by Deb for the modified PPES, did not perform as well as method 1 for all the 3 test functions. This suggests the inter-dependency of the model parameters, and the fact that parameters have to be chosen carefully in order to obtain satisfactory results.

When the lattice size is increased again to 50x50, both methods 1 and 2 give very good distributions of nondominated solutions for all the 3 test functions. Using a much larger lattice size such as 50x50 in RCPPGA means more prey individuals are allowed to survive. This is because there are more vacant cells available, but the same constant number of predators has to cover a much larger lattice (see Section 3).

One important observation from the above figures is that given a sufficiently large lattice size, RCPPGA seems to be able to obtain a very good distribution of nondominated solutions, even without a direct use of adaptive variation operators such as a decreasing step size suggested in PPES. However RCPPGA does use a BLX-α crossover operator, which has the effect of self-adaptation (see Section 3.2). This crossover operator, along with the use of a migration mechanism (i.e., a random distribution of prey offspring over the entire lattice), seems to be able to produce and maintain a diverse prey population. Implicit niching is carried out via the natural formation of prey clusters of various sizes across the lattice, which allows a diverse set of nondominated solutions obtained and maintained until the last iteration.

Fig. 5 shows that even when a very large lattice of size 80x80 is used, RCPPGA still managed to get a fairly good solution distribution on the Pareto front for F3 within 30000 evaluations, especially for method 1. Note that this time there are many more individuals that are not quite converged to the nondominated front, though there are already good approximation and distribution of optimal solutions on the nondominated front. Another interesting observation is that method 1 gives a much more uniform distribution of both nondominated and dominated solutions in the final iteration than those of method 2. Fig. 5b) shows that method 2 converged heavily towards one end of the Pareto front, but missing out converging to some solutions on the other end. However, as shown in Fig. 5 a), method 1 does not seem to have such a problem.

If we examine closely over Fig. 2 - 5, it can be noted that method 2 seems to be more sensitive to different lattice sizes than method 1. As far as solution distribution is concerned, method 2 performed particularly worse than method 1 on both of a smaller lattice size of 30x30 and a larger lattice size of 80x80 (for F3).

5.2 Dynamic Changes of the Prey Population Size

Fig. 6 shows the dynamic changes of the prey population size when different lattice sizes are used. The larger the lattice size is, the larger the prey population would become. Since the maximum number of evaluations is fixed at 30000, a larger prey population would run for fewer generations than a smaller one. For example, for a lattice size of 50x50, only 67 generations are run as shown in Figure 6, whereas a lattice of size 30x30 needs 165, and a lattice of size 20x20 needs 225 generations. The lattice of size 50x50 gives the most diverse nondominated solution set however. It can be observed that for the lattice size of 50x50, the number of prey is increased very rapidly at the beginning of a run. Within only a few generations, the prey population has gone from an initial 240 up to around 500, and then fluctuates around this number. It appears that this larger diverse prey population in the early stage of the run provides a basis for RCPPGA to obtain good diverse nondominated solutions towards the end of the run.

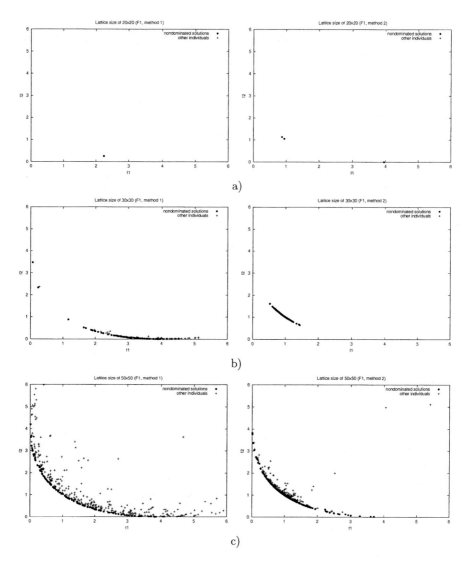

Fig. 2. Prey solutions for F1 using method 1 (left) and method 2 (right) - a) for a lattice of size 20x20, b) 30x30, and c) 50x50

5.3 Sensitivity of Predator-Prey Ratio

For method 2, Deb's weighted vector approach, it may appear that by increasing the number of predators, one would hope a better approximation and distribution of the nondominated solutions can be obtained along the Pareto front, however selection pressure depends on the predator-prey ratio. Fig. 7 shows the effects of changing the predator-prey ratio when we used the same set of parameter values as for Fig. 4 c), but doubled the number of predators. It can

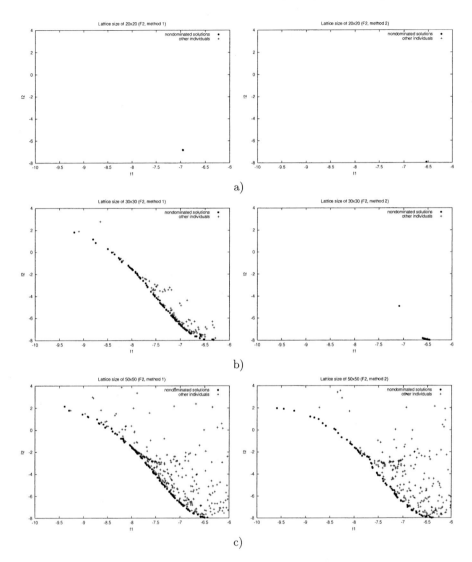

Fig. 3. Prey solutions for F2 using method 1 (left) and method 2 (right) - a) for a lattice of size 20x20, b) 30x30, and c) 50x50

be noted that method 2 is more sensitive to the increased number of predators than method 1. The dramatically increased selection pressure occurred when using method 2 resulted in a poor convergence, which in fact further reduced the diversity of the nondominated solutions on the Pareto front. In contrast, method 1 still mananged to obtain well-distributed solutions consistently.

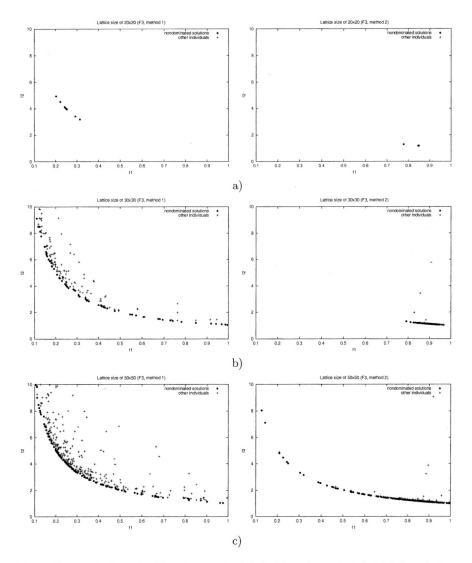

Fig. 4. Prey solutions for F3 using method 1 (left) and method 2 (right) - a) for a lattice of size 20x20, b) 30x30, and c) 50x50

6 Conclusion and Future Work

In this study a real-coded predator-prey GA (RCPPGA) for multiobjective optimization has been developed as an extension of the original predator-prey model proposed by Laumanns et al. [5]. From the experiments carried out over the 3 test functions, it has been shown that RCPPGA is able to produce a good set of diverse nondominated solutions along the Pareto front. The RCPPGA's performance when using two different fitness assignment methods in conjunction

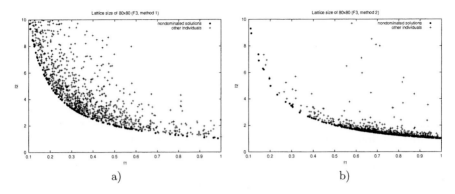

Fig. 5. Prey solutions for F3 when a very large lattice size of 80x80 is used - a) method 1, and b) method 2

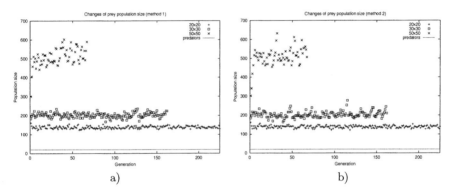

Fig. 6. The dynamic changes of the prey population size over generations - a) method 1, and b) method 2

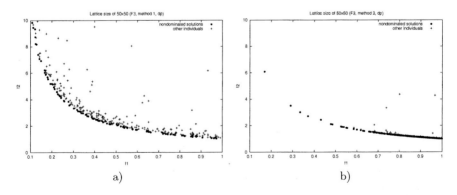

Fig. 7. The effects of changing the predator-prey ratio (doubling the number of predators in Fig. 4 c)) - a) method 1, and b) method 2

with serval lattice sizes have been studied in detail. It has been found that when using different types of predators associated with different objectives, given a sufficiently large lattice size, RCPPGA can consistently produce and maintain a diverse distribution of nondominated optimal solutions along the Pareto front even after many generations. This method is also empirically shown to be less sensitive to the predator-prey ratio than the weighted vector approach.

Many current evolutionary algorithms place emphasis on using an elitist mechanism such as an external archive in order to keep a good optimal solution distribution [1]. By contrast, RCPPGA does not use any explicit elitist mechanism, but still manages to obtain a diverse set of optimal solutions for all the 3 test functions. There is no separate archive used in RCPPGA to store nondominated solutions during a run. The nondominated and dominated solutions are always in the same population until the last iteration step of a run. One possible explanation for RCPPGA's good performance is that using the self-adaptive BLX-α crossover operator together with the random allocation migration method (see Section 3) is effective and rather non-detrimental as compared with the mutation operators used in the previous studies [5], [1]. In future we will test RCPPGA over more difficult multiobjective optimization problems, especially problems with more than just two objectives. We will also need to carry out more formal performance measurements on the algorithm, for example using the Mann-Whitney rank-sum test [16].

References

1. Deb, K.: Multi-Objective Optimization using Evolutionary Algorithms. John Wiley & Sons, Chichester, UK (2001)
2. Zitzler, E. and Thiele, L.: Multiobjective Evolutionary Algorithms: A Comparative Case Study and the Strength Pareto Approach, IEEE Transactions on Evolutionary Computation, 3(4): 257–271, (1999)
3. Hajela, P. and Lin, C.-Y.: Genetic search strategies in multicriterion optimal design, Structural Optimization, vol.4. New York: Springer, (1992) 99–107
4. Horn, J. and Nafpliotis, N.: Multiobjective optimization using niched pareto genetic algorithm, IllGAL Report 93005, Illinois Genetic Algorithm Lab., Univ. Illinois, Urbana-Champaign (1993)
5. Laumanns, M., Rudolph, G. and Schwefel, H. P.: A Spatial Predator-Prey Approach to Multi-Objective Optimization: A Preliminary Study, In: Eiben, A. E., Schoenauer, M. and Schwefel, H.P (eds.): Proceedings of the Parallel Problem Solving From Nature – PPSN V, Springer-Verlag, Amsterdam, Holland, (1998) 241–249
6. Manderick, B. and Spiessens, P.: Fine-grained parallel genetic algorithms. In Proceedings of the Third International Conference on Genetic Algorithms, Morgan Kaufmann (1989) 428–433
7. Cantu-Paz, E.: A Survey of Parallel Genetic Algorithms. Technical Report IlliGAL 97003, University of Illinois at Urbana-Champaign (1997)
8. Kirley, M., Li, X., and Green, D.: Investigation of a cellular genetic algorithm that mimics landscape ecology. In: McKay, R., et al. (eds.): Simulated Evolution and Learning – SEAL98, volume 1585 Lecture Notes in Artificial Intelligence, Springer-Verlag (1998) 90–97

9. Kirley, M. A.: A Cellular Genetic Algorithm with Disturbances: Optimisation Using Dynamic Spatial Interactions. In: Journal of Heuristics. 8(3):321–342 (2002)
10. Li, X. and Kirley, M.: The Effects of Varying Population Density in a Fine-grained Parallel Genetic Algorithm. In: The Proceedings of the 2002 Congress on Evolutionary Computation (CEC'02). Vol: 2, 1709–1714 (2002)
11. Li, X. and Sutherland, S.: A Cellular Genetic Algorithm Simulating Predator-Prey Interactions. In: Proceeding of the 4th Asia-Pacific Conference on Simulated Evolution And Learning (SEAL'02), edited by Wang, L.,Tan, K.C.,Furuhashi,T., Kim, J-H and Yao, X., (2002) 76–80
12. Smith, R.L. and Smith, T.M.: Elements of Ecology – fourth edition, The Benjamin/Cummings Publishing Company, Inc., Menlo Park, CA 94025 (1998)
13. Wright, A.: Genetic Algorithms for Real Parameter Optimization. In: Rawlin, G.J.E. (eds.): Foundations of Genetic Algorithms 1. Morgan Kaufmann, San Mateo (1991) 205–218
14. Eshelman, L. J. and Schaffer, J.: Realcoded genetic algorithms and interval-schemata. In: Foundation of Genetic Algorithms. (1991) 187–202
15. Deb, K. and Beyer, H.: Self-Adaptive Genetic Algorithms with Simulated Binary Crossover, Evolutionary Computation 9(2): 197–221, MIT Press (2001)
16. Knowles, J.D. and Corne, D.W.: Approximating the nondonimated frond using the Pareto Archived Evolution Strategy. Evolutionary Computation, 7(3): 1–26 (2000)

Towards a Quick Computation of Well-Spread Pareto-Optimal Solutions

Kalyanmoy Deb, Manikanth Mohan, and Shikhar Mishra

Kanpur Genetic Algorithms Laboratory (KanGAL)
Indian Institute of Technology Kanpur
Kanpur, PIN 208016, India
deb@iitk.ac.in

Abstract. The trade-off between obtaining a good distribution of Pareto-optimal solutions and obtaining them in a small computational time is an important issue in evolutionary multi-objective optimization (EMO). It has been well established in the EMO literature that although SPEA produces a better distribution compared to NSGA-II, the computational time needed to run SPEA is much larger. In this paper, we suggest a clustered NSGA-II which uses an identical clustering technique to that used in SPEA for obtaining a better distribution. Moreover, we propose a steady-state MOEA based on ϵ-dominance concept and efficient parent and archive update strategies. Based on a comparative study on a number of two and three objective test problems, it is observed that the steady-state MOEA achieves a comparable distribution to the clustered NSGA-II with a much less computational time.

1 Introduction

It has been experienced in the recent past that the computation of a well-diverse set of Pareto-optimal solutions is usually time-consuming [4,8]. Some MOEAs use a quick-and-dirty diversity preservation operator, thereby finding a reasonably good distribution quickly, whereas some MOEAs use a more computationally expensive diversity preservation operator for obtaining a better distribution of solutions. For example, NSGA-II [2] uses a crowding approach which has a computational complexity of $O(N \log N)$, where N is the population size. On the other hand, SPEA [11] uses a clustering approach which has a computational complexity of $O(N^3)$ involving Euclidean distance computations. Although in two-objective problems, the difference in obtained diversity between these two MOEAs was not reported to be significant [2], the difference was clear while solving three-objective problems [4]. SPEA produced a much better distribution at the expense of a large computational effort.

In this paper, we address this trade-off and suggest a steady-state MOEA based on the ϵ-dominance concept suggested elsewhere [8]. The ϵ-dominance does not allow two solutions with a difference ϵ_i in the i-th objective to be non-dominated to each other, thereby allowing a good diversity to be maintained in a population. Besides, the method is quite pragmatic, because it allows the user to choose a suitable ϵ_i depending on the desired resolution in the i-th objective. In the proposed ϵ-MOEA, two populations (EA and archive) are evolved simultaneously. Using one solution each from both populations, two offspring solutions are created. Each offspring is then used to update both parent and archive populations. The archive population is updated based on the ϵ-dominance concept, whereas an usual domination concept is used to update the parent population. Since the ϵ-dominance concept reduces the cardinality of the Pareto-optimal set and since a steady-state EA is proposed, the maintenance of a diverse set of solutions is possible in a small computational time.

In the reminder of the paper, we briefly discuss a clustered version of NSGA-II and then present the ϵ-MOEA approach in details. Thereafter, these two MOEAs are compared along with the original NSGA-II on a number of two and three objective test problems. Finally, some useful conclusions are drawn from the study.

2 Distribution versus Computation Time

In addition to the convergence to the Pareto-optimal front, one of the equally important aspects of multi-objective optimization is to find a widely distributed set of solutions. Since the Pareto-optimal front can be a convex, non-convex, disconnected, piece-wise continuous hyper-surfaces, there are differences in opinion as to how to define a diversity measure denoting the true spread of a finite set of solutions on the Pareto-optimal front. Although the task is easier for a two-objective objective space, the difficulty arises in the case of higher-dimensional objective spaces. This is the reason why researchers have developed different diversity measures, such as the hyper-volume measure [11], the spread measure [9], the chi-square deviation measure [10], the R-measures [6], and others. In maintaining diversity among population (or archive) members, several researchers have used different diversity-preserving operators, such as clustering [11], crowding [2], pre-specified archiving [7], and others. Interestingly, these diversity-preserving operators produce a trade-off between the achievable diversity and the computational time.

The clustering approach of SPEA forms N clusters (where N is the archive size) from $N'(> N)$ population members by first assuming each of N' members forming a separate cluster. Thereafter, all $\binom{N'}{2}$ Euclidean distances in the objective space are computed. Then, two clusters with the smallest distance are merged together to form one cluster. This process reduces the number of clusters to $N' - 1$. The inter-cluster distances are computed again[1] and another merging is done. This process is repeated until the number of clusters is reduced to N. With multiple population members occupying two clusters, the average distance of all pair-wise distances between solutions of two clusters is used. If $(N' - N)$ is of the order of N (the archive size), the procedure requires $O(N^3)$ computations

---

[1] A special book-keeping procedure can be used to eliminate further computation of pair-wise Euclidean distances.

in each iteration. Since this procedure is repeated in every iteration of SPEA, the computational overhead as well as storage requirements for implementing the clustering concept are large.

On the other hand, NSGA-II used a crowding operator, in which N' (as large as $2N$, where N is the population size) solutions are processed objective-wise. In each objective direction, solutions are first sorted in ascending order of objective value. Thereafter, for each solution an objective-wise crowding distance is assigned equal to the difference between normalized objective value of the neighboring solutions. The overall crowding distance is equal to the sum of the crowding distances from all objectives. Once all distance computations are achieved, solutions are sorted in descending order of crowding distance and the first N solutions are chosen. This procedure requires $O(N \log N)$ computations. It is clear that although NSGA-II is faster than SPEA, the diversity in solutions achievable by NSGA-II is not expected to be as good as that achievable with SPEA.

3 Two Approaches for a Better Spread

Here, we propose a modification to the NSGA-II procedure and an ϵ-dominance based MOEA in order to achieve a better distribution of solutions.

3.1 Clustered NSGA-II

The first approach is a straightforward replacement of NSGA-II's crowding routine by the clustering approach used in SPEA. After the parent and offspring population are combined into a bigger population of size $2N$ and this combined population is sorted into different non-domination levels, only N good solutions are required to be chosen based on their non-domination levels and *nearness* to each other. This is where the original NSGA-II used a computationally effective crowding procedure. In the clustering NSGA-II approach suggested here, we replace the crowding procedure with the clustering approach. The solutions in the last permissible non-dominated level are used for clustering. Let us say that the remaining population slots are N' and the solutions in the last permissible non-domination level from the combined population is n'. By definition, $n' \geq N'$. To choose N' solutions from n', we form N' clusters from n' solutions and choose one representative solution from each cluster. The clustering algorithm used in this study is exactly the same as that used in SPEA [11]. Although this requires a larger computational time, the clustered NSGA-II is expected to find a better distributed set of Pareto-optimal solutions than the original NSGA-II.

3.2 A Steady-State ϵ-MOEA

Next, we suggest a steady-state MOEA based on the ϵ-dominance concept introduced in [8]. The search space is divided into a number of grids (or hyper-boxes) and diversity is maintained by ensuring that a grid or hyper-box can be occupied by only one solution. Although, PAES and its variants [7] are developed with the similar idea, ϵ-dominance is a more general concept.

Here, we suggest a steady-state MOEA based on the ϵ-dominance concept. In the proposed MOEA, we have two co-evolving populations: (i) an EA population $P(t)$ and (ii) an archive population $E(t)$ (where t is the iteration counter). The MOEA is started with an initial population $P(0)$. To start with, $E(0)$ is assigned with the ϵ-non-dominated solutions of $P(0)$. Thereafter, one solution each from $P(t)$ and $E(t)$ are chosen for mating. To choose a solution from $P(t)$, two population members from $P(t)$ are picked at random and a domination check (in the usual sense) is made. If one solution dominates the other, the former is chosen. Otherwise, it indicates that two picked solutions are non-dominated to each other and we simply choose one of the two solutions at random. Let us denote the chosen solution by p. To choose a solution e from $E(t)$, several strategies involving a certain relationship with p can be made. However, here we randomly pick a solution from $E(t)$. After this selection phase, solutions p and e are mated to create λ offspring solutions (c_i, $i = 1, 2, \ldots, \lambda$). Now, each of these offspring solutions are compared with the archive and the EA population for their possible inclusion. Different strategies are used in each case.

For its inclusion in the archive, the offspring c_i is compared with each member in the archive for ϵ-dominance. Every solution in the archive is assigned an identification array (**B**) as follows:

$$B_j(\mathbf{f}) = \begin{cases} \lfloor (f_j - f_j^{\min})/\epsilon_j \rfloor, & \text{for minimizing } f_j, \\ \lceil (f_j - f_j^{\min})/\epsilon_j \rceil, & \text{for maximizing } f_j. \end{cases} \quad (1)$$

where $f_j^{\min}$ is the minimum possible value of the j-th objective and ϵ_j is the allowable tolerance in the j-th objective beyond which two values are significant to the user. This ϵ_j value is similar to the ϵ used in the ϵ-dominance definition [8]. The identification arrays make the whole search space into grids having ϵ_j size in the j-th objective. Figure 1 illustrates that the solution P ϵ-dominates the entire region ABCDA (in the minimization sense), whereas the original dominance definition allows P to dominate only the region PECFP. For brevity, the rest

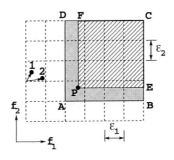

Fig. 1. The ϵ-dominance concept is illustrated (for minimizing f_1 and f_2)

of the discussion is made here for minimization cases only. However, a similar analysis can be followed for maximization cases as well. The identification array of P is the coordinate of point A in the objective space. With the identification arrays calculated for the offspring c_i and each archive member a, we use the following procedure. If the $\mathbf{B}_a$ of any archive member a dominates that of the offspring c_i, it means that the offspring is ϵ-dominated by at least one archive member and the offspring is not accepted. Otherwise, if $\mathbf{B}_c$ of the offspring dominates $\mathbf{B}_a$ of any archive member a, those archive members are deleted and the offspring is accepted. If neither of the above two cases occur, this means that the offspring is ϵ-non-dominated with the archive members. We separate this case into two. If the offspring shares the same $\mathbf{B}$ vector with an archive member (meaning that they belong to the same hyper-box), then they are checked for the usual non-domination. If the offspring dominates the archive member or the offspring is non-dominated to the archive member but is closer to the $\mathbf{B}$ vector (in terms of the Euclidean distance) than the archive member, the offspring is retained. The latter case is illustrated in Figure 1 with solutions 1 and 2. They occupy the same hyper-box (or have the same $\mathbf{B}$ vector) and they are non-dominated according to the usual definition. Since solution 1 has a smaller distance to the $\mathbf{B}$ vector, it is retained and solution 2 is deleted. In the event of an offspring not sharing the same $\mathbf{B}$ vector with any archive member, the offspring is accepted. It is interesting to note that the former condition ensures that only one solution with a distinct $\mathbf{B}$ vector exists in each hyper-box. This means that each hyper-box on the Pareto-optimal front can be occupied by exactly one solution, thereby providing two properties: (i) solutions will be well distributed and (ii) the archive size will be bounded. For this reason, no specific upper limit on the archive size needs to be pre-fixed. The archive will get bounded according to the chosen ϵ-vector.

The decision whether an offspring will replace any population member can be made using different strategies. Here, we compare each offspring with all population members. If the offspring dominates any population member, the offspring replaces that member. Otherwise, if any population member dominates the offspring, it is not accepted. When both the above tests fail, the offspring replaces one randomly chosen population member. This way the EA population size remains unchanged.

The above procedure is continued for a specified number of iterations and the final archive members are reported as the obtained solutions. A careful thought will reveal the following properties of the proposed algorithm:

1. It is a steady-state MOEA.
2. It emphasizes non-dominated solutions.
3. It maintains diversity in the archive by allowing only one solution to be present in each pre-assigned hyper-box on the Pareto-optimal front.
4. It is an elitist approach.

3.3 Differences with Past Studies

Although the above steady-state ϵ-MOEA may look similar to the multi-parent PAES [7], there is a subtle difference. The ϵ-dominance concept implemented here with the **B**-vector domination-check does not allow two non-dominated solutions with a difference less than ϵ_i in the i-th objective to be both present in the final archive. Figure 1 can be used to realize that fewer non-dominated solutions will be obtained with the proposed approach than PAES. The procedure will not only allow a reduction in the size of the Pareto-optimal solutions, but also has a practical significance. Since a user is not interested in obtaining solutions within a difference ϵ_i in the i-th objective, the above procedure allows the user to find solutions according to his/her desire. However, the actual number of solutions to be obtained by the ϵ-dominance procedure is unknown, but it is bounded. Because of this reason, the overall computation is expected to be faster.

The proposed archive update strategy is also similar to that proposed in [8], except in the case when two solutions have the same **B** vector. Here, a non-dominated solution to an existing archive member can still be chosen if it is closer to the **B** vector. The earlier study only accepted an offspring if it dominated an existing member. Moreover, here we have proposed a steady-state MOEA procedure with an EA population update strategy, an archive update strategy, and a recombination plan.

4 Simulation Results

In this section, we compare the three MOEAs discussed above on a number of two and three objective test problems. We have used a convergence measure for exclusively computing the extent of convergence to the Pareto-optimal front and a sparsity measure for exclusively computing the diversity in solutions. The hyper-volume measure [11] for computing a combined convergence and diversity estimate is also used. Since all problems considered in this paper are test problems, the exact knowledge of the Pareto-optimal front is available. We calculate H uniformly distributed (on the f_1-f_2-$\cdots$-f_{M-1}-plane) solutions P^* on the Pareto-optimal front. For each such point in the $(M-1)$-dimensional plane, f_M is calculated from the known Pareto-optimal front description. Then, the Euclidean distance of each obtained solution from the nearest solution in P^* is computed. The average of the distance value of all obtained solutions is defined as the convergence measure here. The sparsity measure is described in detail while discussing the three-objective optimization results. We also present the computational time needed to run each MOEA on the same computer.

4.1 Two-Objective Test Problems

ZDT1 Test Problem: The $n = 30$-variable ZDT1 problem has a convex Pareto-optimal front. We use a population size of 100 and real-parameter SBX recombination operator with $p_c = 1$ and $\eta_c = 15$ and a polynomial mutation operator with $p_m = 1/n$ and $\eta_m = 20$ [1]. In order to investigate the effect of ϵ used in the ϵ-MOEA, we use different ϵ values and count the number of solutions found in the archive in each case after 20,000 solution evaluations. Figure 2 shows that as ϵ increases, the number of obtained solutions varies proportional to $1/\epsilon$. For an equally-sloped Pareto-optimal straight line in the range $f_1 \in [0,1]$, we would expect $1/\epsilon$ solutions to appear in the archive. But with a non-linear Pareto-optimal front, we would expect smaller number of archive members with the ϵ-dominance concept. This will be clear from Figure 3, which shows the distribution of solutions obtained with $\epsilon = 0.05$. The boxes within which a solution lies are also shown in the figure. It is interesting to note that all solutions are ϵ-nondominated with respect to each other and in each of the expected boxes only one solution is obtained. In the boxes with $f_1 \in [0, 0.05]$, one Pareto-optimal

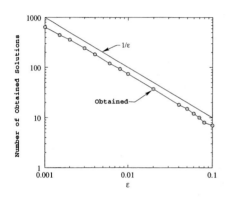

Fig. 2. The number of solutions versus ϵ on ZDT1

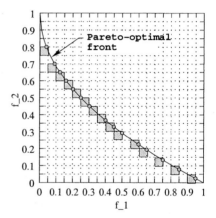

Fig. 3. ϵ-MOEA distribution on ZDT1 with $\epsilon = 0.05$

solution in the minimum f_2 box ($f_2 \in [0.75, 0.80]$) is obtained. Other four boxes on top of this box (with larger f_2) are ϵ-dominated and hence not retained in the final archive.

To compare the three MOEAs, we have used the convergence metric discussed above, the hyper-volume metric [11], and a sparsity measure. We use $\epsilon_i = 0.008$ in order to get roughly 100 solutions in the archive after 20,000 solution evaluations. The convergence metric is computed with $H = 1,000$ equi-spaced solutions on the Pareto-optimal front. Table 1 shows that the convergence of solutions achieved with ϵ-MOEA is the best. For the hyper-volume measure, we use the reference point at $(1.1, 1.1)^T$. Although the convergence of the ϵ-MOEA is better than that of the C-NSGA-II in this problem, the hyper-volume of ϵ-MOEA is worse. This is mainly due to the absence of extreme solutions in the Pareto-optimal front (refer Figure 3). Since the ϵ-dominance concept is used in ϵ-MOEA,

Table 1. Performance comparison of three MOEAs for ZDT1, ZDT4, and ZDT6. Best metric values are shown in a slanted font

Test Problem	MOEA	Convergence measure Avg.	Convergence measure Std. Dev.	Hyper-volume Measure	Sparsity Measure	Time (sec)
ZDT1	NSGA-II	0.00060	9.607e-05	0.87060	0.857	7.25
	C-NSGA-II	0.00054	1.098e-04	*0.87124*	0.997	1984.83
	ϵ-MOEA	*0.00041*	2.818e-05	0.87031	*1.000*	1.20
ZDT4	NSGA-II	0.01356	2.955e-04	0.85039	0.937	5.64
	C-NSGA-II	0.00884	2.314e-04	*0.85747*	*1.000*	99.01
	ϵ-MOEA	*0.00227*	6.019e-05	0.85723	0.943	*0.57*
ZDT6	NSGA-II	0.07104	1.186e-04	0.40593	0.803	3.93
	C-NSGA-II	*0.06507*	6.943e-04	*0.41582*	*1.000*	2365.95
	ϵ-MOEA	0.07280	1.777e-04	0.40174	0.997	*0.91*

the extreme solutions usually get dominated by solutions within ϵ and which are better in other objectives. Figure 3 illustrates this aspect. The presence or absence of the extreme solutions makes a significant difference in the hyper-volume metric, a matter we discuss further later. Thus, the hyper-volume measure may not be an ideal metric for measuring diversity and convergence of a set of solutions. We suggest and use a different diversity measure (which we have described in Section 4.2 in detail for any number of objectives). In short, the sparsity measure first projects the obtained solutions on a suitable hyper-plane (a unit vector $(1/\sqrt{2}, 1/\sqrt{2})^T$ is used here) and then computes the non-overlapping area occupied by the solutions on that plane. The higher the measure value, the better is the distribution. This measure is also normalized so that the maximum possible non-overlapping area is one, meaning that there is no overlap (or 100% non-overlap) among projected solutions. The measure involves a size parameter, which is adjusted in such a way that one of the competing MOEAs achieve a 100% non-overlapping area. For ZDT1, ϵ-dominance achieves the best sparsity measure.

Finally, the table presents the actual computational time needed by each MOEA. It is clear that the ϵ-MOEA obtained a better convergence and sparsity measure in a much smaller computational time than the other two MOEAs.

ZDT4 Test Problem: Table 1 also shows the performance measures on the 10-variable ZDT4. This problem has a number of local Pareto-optimal fronts. All chosen parameters used here are identical to that used in ZDT1, except that $\epsilon_i = 0.006$ are used to get about 100 solutions in the final archive (after 20,000 solution evaluations). The table shows that ϵ-MOEA is better than the other two MOEAs in terms of convergence and computational time, and is the second-best in terms of diversity among obtained solutions.

ZDT6 Test Problem: The 10-variable ZDT6 problem has a non-uniform density of solutions on the Pareto-optimal front. Here, we use $\epsilon_i = 0.0067$ to get about 100 solutions in the archive after 20,000 solution evaluations. Here, the convergence measures of C-NSGA-II and NSGA-II are slightly better than that of ϵ-MOEA. However, the diversity measures (Table 1) obtained in ϵ-MOEA and C-NSGA-II are comparable. But, ϵ-MOEA is superior in terms of the computational time.

Thus, from the two-objective problems studied above, we can conclude that ϵ-MOEA produces a good convergence and diversity with a smaller (at least an order of magnitude smaller) computational time than the other two MOEAs.

4.2 Three-Objective Test Problems

Now, we consider a few three-objective test problems developed elsewhere [4].

DTLZ1 Test Problem: First, we consider the $n = 7$ variable, three-objective DTLZ1 test problem. The Pareto-optimal solutions lie on a three-dimensional plane satisfying: $f_3 = 0.5 - f_1 - f_2$ in the range $f_1, f_2 \in [0, 0.5]$. For each algorithm, we use a population size of 100, real-parameter representation with the SBX recombination (with $\eta_c = 15$ and $p_c = 1$), the polynomial mutation operator (with $\eta_m = 20$ and $p_m = 1/n$) [1], and a maximum of 30,000 function evaluations. For ϵ-MOEA, we have chosen $\epsilon = [0.02, 0.02, 0.05]^T$. The solutions obtained in each case are shown on the objective space in Figures 4, 5, and 6, respectively. It is clear from the figures that the distribution of solutions

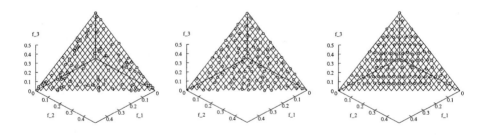

Fig. 4. NSGA-II distribution on DTLZ1   **Fig. 5.** C-NSGA-II distribution on DTLZ1   **Fig. 6.** ϵ-MOEA distribution on DTLZ1

with the original NSGA-II is poor compared to the other two MOEAs. Using $H = 5,000$ solutions, we tabulate the average and standard deviation of the convergence measure in columns 2 and 3 in Table 2, respectively. It is observed that the convergence of the ϵ-MOEA is relatively better than that of the other two MOEAs with identical function evaluations.

Column 4 of the table shows the hyper-volume measure calculated with a reference solution $f_1 = f_2 = f_3 = 0.7$. Since for minimization problems, a larger hyper-volume is better, the table indicates that the C-NSGA-II is best,

Table 2. Comparison of three MOEAs in terms of their convergence and diversity measures on DTLZ1

MOEA	Convergence measure		Hyper-volume Measure	Sparsity Measure	Time (sec)
	Avg.	Std. Dev.			
NSGA-II	0.00267	1.041e-04	0.313752	0.867	8.46
C-NSGA-II	0.00338	8.990e-05	0.314005	1.000	3032.38
ϵ-MOEA	0.00245	9.519e-05	0.298487	0.995	2.22

followed by the NSGA-II. The ϵ-MOEA performs the worst. However, by visually investigating solutions displayed in Figures 4 and 6 for NSGA-II and ϵ-MOEA, respectively, it can be seen that the distribution of ϵ-MOEA is better. The hyper-volume measure fails to capture this aspect, despite a better convergence and sparsity of ϵ-MOEA solutions. The extent of solutions obtained by these two MOEAs are shown below:

	f_1	f_2	f_3
NSGA-II:	0–0.500525	0–0.501001	0–0.500763
ϵ-MOEA:	0.000273–0.435053	0.000102–0.434851	0.044771–0.499745

Since the ϵ-dominance does not allow two solutions with a difference of ϵ_i in the i-th objective to be mutually non-dominated to each other, it will be usually not possible to obtain the extreme corners of the Pareto-optimal front. However, the diversity of solutions elsewhere on the Pareto-optimal front is ensured by the archive update procedure of ϵ-MOEA. It is then trivial to realize that without the extreme solutions the hyper-volume measure fails to add a large portion of hyper-volume calculated from the extreme points. The advantage in diversity metric obtained from the well-distributed solutions on the interior of the Pareto-optimal front was not enough to account for the loss in diversity due to the absence of boundary solutions. We argue that in this sense the hyper-volume measure is biased towards the boundary solutions. Thus, we do not use this measure for the rest of the test problems used in this paper.

Instead, we define a *sparsity* measure, which is similar to the entropy measure [5] or the grid diversity measure [3]. The Pareto-optimal solutions are first projected on a suitable hyper-plane (with a unit vector η). Figure 7 illustrates the calculation procedure of this measure. A hyper-box of certain size d is centered around each projected solution. The total hyper-volume covered by these hyper-boxes is used as the measure of sparsity of solutions. If a solution set has many clustered points, their hyper-boxes will overlap with each other and the obtained sparsity measure will be a small number. On the other hand, if solutions are well distributed, the hyper-boxes do not overlap and a large overall measure will be obtained. To normalize the measure, we divide the total hyper-volume by the total expected hyper-volume calculated with a same-sized solution set having no overlap between the hyper-boxes. Thus, the maximum sparsity achievable is one and the larger the sparsity measure the better is the distribution. However, the choice of the parameter d is important here. A too small value of d will make any

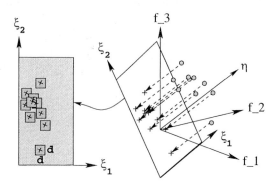

Fig. 7. The sparsity measure is illustrated

distribution to achieve the maximum sparsity measure of one, whereas a very large value of d will make every distribution to have a small sparsity measure. We solve this difficulty of choosing a suitable d value by finding the smallest possible value which will make one of the competing distributions to achieve the maximum sparsity value of one and use the same d for computing sparsity for other distributions. In all case studies here, points are projected on an equally-inclined plane to coordinate axes ($\boldsymbol{\eta} = (1/\sqrt{3}, 1/\sqrt{3}, 1/\sqrt{3})^T$). The column 5 of Table 2 shows that C-NSGA-II attains the best distribution, followed by ϵ-MOEA, and then NSGA-II. The distribution of solutions plotted in Figures 4 to 6 also support this observation visually.

Although the C-NSGA-II achieves the best distribution, it is also computationally the slowest of the three MOEAs. Since the clustering algorithm requires comparison of each population member with each other for computing adequate number of clusters in every generation of C-NSGA-II, the table shows that it takes three orders of magnitude more time than the other two MOEAs. Based on both convergence and diversity measures and the required computational time tabulated in the table, the ϵ-MOEA emerges out as a good compromised algorithm.

DTLZ2 Test Problem: Next, we consider the 12-variable DTLZ2 test problem having a spherical Pareto-optimal front satisfying $f_1^2 + f_2^2 + f_3^2 = 1$ in the range $f_1, f_2 \in [0, 1]$. Identical parameters to those used in DTLZ1 are used here. About $H = 8,000$ Pareto-optimal solutions are considered as P^* for the convergence metric computation. For the ϵ-MOEA, we have used $\epsilon = [0.06, 0.06, 0.066]^T$. This produces about 100 solutions on the Pareto-optimal front. Table 3 shows the comparison of performance measures of three MOEAs.

Figures 8 and 9 shows the distribution of solutions obtained by C-NSGA-II and ϵ-MOEA, respectively. Although both approaches produced a very similar sparsity measure, they produced them in different ways. In the C-NSGA-II, since

Table 3. Performance comparison of three MOEAs for DTLZ2

	Convergence measure		Sparsity	Time
MOEA	Avg.	Std. Dev.	Measure	(sec)
NSGA-II	0.01606	0.00112	0.821	6.87
C-NSGA-II	*0.00986*	0.00088	0.998	7200.51
ε-MOEA	0.01160	0.00119	*1.000*	*1.86*

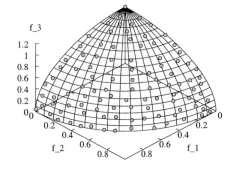

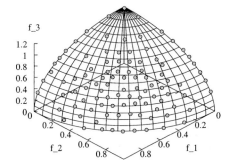

Fig. 8. C-NSGA-II distribution on DTLZ2

Fig. 9. ε-MOEA distribution on DTLZ2

the Euclidean distance measure is used in the clustering approach for maintaining diversity, a uniform spread of solutions on the front is observed. In the case of ε-MOEA, there seems to be a considerable amount of gap observed between the boundary solutions and their nearest neighbors. This happens because of the fact that there is a gentle slope near the boundary solutions on a spherical surface and the ε-dominance consideration does not allow any solution to be non-dominated within an ϵ_i in the i-th objective. But, wherever there is a considerable change of slope, more crowded solutions are found. It is argued earlier that such a set of solutions with a minimum pre-specified difference in objective values has a practical significance and hence we advocate the use of ε-dominance in this paper.

It is clear from the Table 3 that ε-MOEA achieves a good convergence and diversity measure with a much less computational time than C-NSGA-II.

DTLZ4 Test Problem: The 12-variable DTLZ4 test problem introduces a non-uniform density of solution on the Pareto-optimal front. The first two variables are raised to the power of 100, thereby making the $x_1 = 0$ and $x_2 = 0$ solutions to have a larger probability to be discovered. In such a problem, a uniform distribution of Pareto-optimal solutions is difficult to obtain. Table 4 shows the performance of three MOEAs. For ε-MOEA, we have used $\epsilon_i = 0.062$ to obtain

Table 4. Performance comparison of three MOEAs for DTLZ4

MOEA	Convergence measure Avg.	Std. Dev.	Sparsity Measure	Time (sec)
NSGA-II	0.03658	0.00172	0.339	10.61
C-NSGA-II	0.01235	0.00115	0.999	5041.85
ϵ-MOEA	*0.00938*	*0.00094*	*1.000*	*3.86*

100 solutions on the Pareto-optimal front. Once again, the sparsity measure of C-NSGA-II and ϵ-MOEA are much better than that of NSGA-II. However, the ϵ-MOEA achieves this diversity with a much smaller computational time.

DTLZ5 Test Problem: The DTLZ5 is a 12-variable problem having a Pareto-optimal curve: $f_3^2 = 1 - f_1^2 - f_2^2$ with $f_1 = f_2 \in [0, 1]$. Table 5 shows the performance measures. Here, we use $\epsilon_i = 0.005$ for the ϵ-MOEA. It is also clear

Table 5. Performance comparison of three MOEAs for DTLZ5

MOEA	Convergence measure Avg.	Std. Dev.	Sparsity Measure	Time (sec)
NSGA-II	0.00208	0.00038	0.970	5.23
C-NSGA-II	0.00268	0.00034	0.990	1636.96
ϵ-MOEA	*0.00091*	*0.00018*	*1.000*	*1.58*

from the table that ϵ-MOEA is the quickest and best in terms of achieving convergence and diversity.

DTLZ8 Test Problem: The 30-variable DTLZ8 test problem is chosen next. The overall Pareto-optimal region is a combination of two fronts: (i) a line and (ii) a plane. Figure 10 shows the distribution of points obtained using the C-NSGA-II after 100,000 evaluations. As discussed elsewhere [4], the domination-based MOEAs suffers from the so-called 'redundancy problem' in this test problem. For two distinct solutions on the line portion of the Pareto-optimal front, many other non-Pareto-optimal solutions appear as non-dominated. In the figure, solutions on the adjoining sides (shown shaded) of the Pareto-optimal line are these redundant solutions and the clustered NSGA-II is unable to get rid of these solutions, because these solutions are non-dominated to some Pareto-optimal solutions. But with ϵ-dominance, many of these redundant solutions get ϵ-dominated by the Pareto-optimal solutions. Figure 11 shows the solutions obtained with ϵ-MOEA having $\epsilon = [0.02, 0.02, 0.04]^T$. With a 30-variable decision

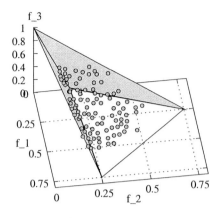
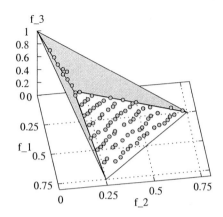

Fig. 10. C-NSGA-II distribution on DTLZ8

Fig. 11. ϵ-MOEA distribution on DTLZ8

space, the density of solutions near the Pareto-optimal line and near the $f_3 = 0$ part of the Pareto-optimal plane are very small. Thus, it may be in general difficult to find solutions at these portions in the Pareto-optimal front. For ϵ-MOEA, we have used $\eta_c = 2$ and $\eta_m = 5$, however for C-NSGA-II $\eta_c = 15$ and $\eta_m = 20$ produced better results. It is clear from the plot that the ϵ-MOEA is able to find a reasonable distribution of solutions on the line and the plane. Although ϵ-MOEA is able to get rid of most redundant solutions, the problem is not entirely eradicated by this procedure. However, the number of redundant solutions is much smaller than a procedure which uses the original dominance criterion.

5 Conclusions

Achieving a good distribution of solutions in a small computational time is a dream to an MOEA researcher or practitioner. Although past studies have either demonstrated a good distribution with a large computational overhead or a not-so-good distribution quickly, in this paper we have suggested an ϵ-dominance based steady-state MOEA to achieve a good distribution with a computationally fast procedure. Careful update strategies for archive and parent populations ensure a continuous progress towards the Pareto-optimal front and maintenance of a good diversity. The use of ϵ-dominance criterion has been found to have two advantages: (i) it helps in reducing the cardinality of Pareto-optimal region and (ii) it ensures that no two obtained solutions are within an ϵ_i from each other in the i-th objective. The first aspect is useful in using the proposed ϵ-MOEA to higher-objective problems and to somewhat lessen the 'redundancy' problem [4]. The second aspect makes the approach highly suitable for practical problem solving, particularly in making the MOEA approach interactive with a decision-maker.

On a number of two and three-objective test problems, the proposed ϵ-MOEA has been successful in finding well-converged and well-distributed solutions with a much smaller computational effort than the original NSGA-II and a clustered yet computationally expensive NSGA-II. The consistency in the solution accuracy and the requirement of a fraction of computational effort needed in other MOEAs suggest the use of the proposed ϵ-MOEA to more complex and real-world problems. The study recommends a more rigorous comparison of ϵ-MOEA with other competing MOEAs, such as the new versions of SPEA and PAES. By choosing ϵ_i as a function of f_i, the method can also be used to find a biased distribution of solutions on the Pareto-optimal front, if desired.

References

1. K. Deb. *Multi-objective optimization using evolutionary algorithms*. Chichester, UK: Wiley, 2001.
2. K. Deb, S. Agrawal, A. Pratap, and T. Meyarivan. A fast and elitist multi-objective genetic algorithm: NSGA-II. *IEEE Transactions on Evolutionary Computation*, 6(2):182–197, 2002.
3. K. Deb and S. Jain. Running performance metrics for evolutionary multi-objective optimization. In *Simulated Evolution and Learning (SEAL-02)*, in press.
4. K. Deb, L. Thiele, M. Laumanns, and E. Zitzler. Scalable multi-objective optimization test problems. In *Proceedings of the Congress on Evolutionary Computation (CEC-2002)*, pages 825–830, 2002.
5. A. Farhang-Mehr and S. Azarm. Diversity assessment of pareto-optimal solution sets: An entropy approach. In *Proceedings of the World Congress on Computational Intelligence*, pages 723–728, 2002.
6. M. P. Hansen and A. Jaskiewicz. Evaluating the quality of approximations to the non-dominated set. Technical Report IMM-REP-1998-7, Lyngby: Institute of Mathematical Modelling, Technical University of Denmark, 1998.
7. Joshua D. Knowles and David W. Corne. Approximating the non-dominated front using the Pareto archived evolution strategy. *Evolutionary Computation Journal*, 8(2):149–172, 2000.
8. M. Laumanns, L. Thiele, K. Deb, and Eckart Zitzler. Combining convergence and diversity in evolutionary multi-objective optimization. *Evolutionary Computation*, 10(3):263–282, 2002.
9. J. R. Schott. Fault tolerant design using single and multi-criteria genetic algorithms. Master's thesis, Boston, MA: Department of Aeronautics and Astronautics, Massachusetts Institute of Technology, 1995.
10. N. Srinivas and K. Deb. Multi-objective function optimization using non-dominated sorting genetic algorithms. *Evolutionary Computation Journal*, 2(3):221–248, 1994.
11. E. Zitzler and L. Thiele. Multiobjective evolutionary algorithms: A comparative case study and the strength pareto approach. *IEEE Transactions on Evolutionary Computation*, 3(4):257–271, 1999.

Trade-Off between Performance and Robustness: An Evolutionary Multiobjective Approach

Yaochu Jin and Bernhard Sendhoff

Honda Research Institute Europe GmbH
63073 Offenbach/Main, Germany
{yaochu.jin,bs}@honda-ri.de

Abstract. In real-world applications, it is often desired that a solution is not only of high performance, but also of high robustness. In this context, a solution is usually called robust, if its performance only gradually decreases when design variables or environmental parameters are varied within a certain range. In evolutionary optimization, robust optimal solutions are usually obtained by averaging the fitness over such variations. Frequently, maximization of the performance and increase of the robustness are two conflicting objectives, which means that a trade-off exists between robustness and performance. Using the existing methods to search for robust solutions, this trade-off is hidden and predefined in the averaging rules. Thus, only one solution can be obtained. In this paper, we treat the problem explicitly as a multiobjective optimization task, thereby clearly identifying the trade-off between performance and robustness in the form of the obtained Pareto front. We suggest two methods for estimating the robustness of a solution by exploiting the information available in the current population of the evolutionary algorithm, without any additional fitness evaluations. The estimated robustness is then used as an additional objective in optimization. Finally, the possibility of using this method for detecting multiple optima of multimodal functions is briefly discussed.

1 Motivation

The search for robust optimal solutions is of great significance in real-world applications. Robustness of an optimal solution can usually be discussed from the following two perspectives:

- The optimal solution is insensitive to small variations of the design variables.
- The optimal solution is insensitive to small variations of environmental parameters. In some special cases, it can also happen that a solution should be optimal or near-optimal around more than one design point. These different points do not necessarily lie in one neighborhood.

Mostly, two methods have been used to increase the robustness of a solution [1,2].

- Optimization of the expectation of the objective function in a neighborhood around the design point. If the neighborhood is defined using a probability distribution $\phi(z)$ of a variation parameter z, an effective evaluation [3] function can be defined using the original evaluation function f as

$$f_{eff} = \int_{-\infty}^{\infty} f(\boldsymbol{x}, \boldsymbol{z})\, \phi(\boldsymbol{z})\, d\boldsymbol{z}, \qquad (1)$$

where $\boldsymbol{x}$ is the design variable.
- Optimization of the second order moment or higher order moments of the evaluation function. For example, minimization of the variance of the evaluation function over the neighborhood around the design point has been used to maximize the robustness of a solution.

Unfortunately, the expectation based measure does not sufficiently take care of fluctuations of the evaluation function as long as these fluctuations are symmetric around the average value. At the same time, a purely variance based measure does not take the absolute performance of the solution into account. Thus, it can only be employed in combination with the original quality function or with the expectation based measure. Different combinations of objectives are possible:

- maximizing the expectation and maximizing the original function, for example in [1];
- maximizing the expectation and minimizing the variance, for example in [4];
- maximizing the original function and minimizing the variance.

Since robustness and performance (even if its measure is expectation based), are often exclusive objectives, see Figure 1 for an example, it makes sense to analyze this problem in the framework of multicriteria optimization. In this way, the user can get a better understanding of the relation between robustness and performance for the optimization problem at hand. Besides, the Pareto front can provide the user with valuable information about the stability of the solutions.

In this paper, we employ two variance based measures which are outlined in Section 3 as robustness objectives. Both measures use the information which is already available within the population to estimate the variance. Thus, no additional fitness evaluations are necessary, which is very important when fitness evaluation is computationally expensive, such as in aerodynamic design optimization problems [5]. In Section two, we will briefly review some of the expectation based approaches to searching for robust optimal solutions. The multiobjective optimization algorithm used in this paper, the dynamic weighted aggregation method proposed in [6,7], is described in Section 4. Simulation results on two test problems are presented in Section 5 to demonstrate the effectiveness of the proposed method. A summary of the method and a brief discussion of future work conclude the paper, where a simple example of detecting multiple optima using the proposed method is also provided.

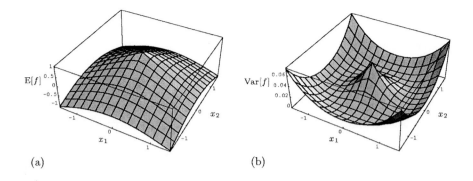

Fig. 1. Example for the trade-off between average and variance. Figure (a) shows the average and (b) the variance of function $f(\mathbf{x}) = a - (z+1)\|\mathbf{x}\|^\alpha + z, z \sim \mathcal{N}(0, \epsilon_z^2), a = 5, \alpha = 1, \epsilon_z = 0.25$. The maximum of the average is given for $\mathbf{x} = (0,0)$, whereas the variance is minimized for $\|\mathbf{x}\| = 1$

2 Expectation-Based Search for Robust Solutions

In evolutionary optimization, efforts have been made to obtain optimal solutions that are insensitive to small changes in the design variables. Most of these approaches are mainly based on the optimization of the expectation of the fitness function. The calculation of the expected performance is usually not trivial in many real-world applications.

To estimate the expected performance, one straightforward way is to calculate the fitness of a solution ($\boldsymbol{x}$) by averaging several points in its neighborhood [8,9,10,11,12]:

$$\tilde{f}(\boldsymbol{x}) = \frac{\sum_{i=1}^{N} w_i f(\boldsymbol{x} + \boldsymbol{\Delta x}_i)}{\sum_{i=1}^{N} w_i}, \qquad (2)$$

where $\boldsymbol{x}$ denotes a vector of design variables and possibly some environmental parameters, $i = 1, 2, ..., N$ is the number of points to be evaluated. $\boldsymbol{\Delta x}_i$ is a vector of small numbers that can be generated deterministically or stochastically and w_i is the weight for each evaluation. In the simplest case, all the weights are set equally to 1. If the $\boldsymbol{\Delta x}_i$ are random variables $\boldsymbol{z}$ and are drawn according to a probability distribution $\phi(\boldsymbol{z})$, we obtain in the limit $N \to \infty$, the effective evaluation function f_{eff}, equation 1.

One problem of the averaging method for estimating the expected performance is the increased computational cost. To alleviate this problem, several ideas have been proposed to use the information in the current or in previous populations [12] to avoid additional fitness evaluations. Note that throughout this paper, the terminology *population* is used as defined in evolutionary algorithms[1]. An alternative is to construct a statistical model for the estimation of

---

[1] In statistics, a population is defined as any entire collection of elements under investigation, while a sample is a collection of elements selected from the population.

the points in the neighborhood using the historical data [13]. Statistical tests have been used to estimate how many samples are needed to decide which solutions should be selected for the next generation [14].

Besides the averaging methods, it has been showed in [3] that the "perturbation" of design variables in each fitness evaluation leads to the maximization of the effective fitness function, equation 1, under the assumption of linear selection (in [3] the schema theorem is used as a basis for the mathematical proof, however, it can be shown that the important assumption is the linearity of the selection operator and an infinite population. Note that the "perturbation" method is equivalent to the averaging method for $N = 1$ and stochastic $\boldsymbol{\Delta x}$.

Whereas most methods in evolutionary optimization consider the robustness with respect to the variations of design variables, the search for robust solutions that are insensitive to environmental parameters has also been investigated [15]. An additional objective function has been defined at two deterministic points symmetrically located around the design point. Let a define an environmental parameter, and the design point is given by $a = a_1$. Thus, the first objective (performance) is defined by $f(\boldsymbol{x}, a_1)$. As a second objective f_2, the following deterministic function is used:

$$f_2(\boldsymbol{x}) = f(\boldsymbol{x}, a_1 + \Delta a) + f(\boldsymbol{x}, a_1 - \Delta a). \tag{3}$$

Note that deterministic formulations of a robustness measure like in equation 3 seems only sensible in the special case when a fixed number of design conditions can be determined. In most cases where parameters can vary within an interval, a stochastic approach seems to be more sensible.

A general drawback of the expectation based methods is that (with the exception of the last example) only one objective has been used. As discussed in Section 1, it is necessary to combine two objectives to search for robust solutions, where a trade-off between performance and robustness exists, as often occurs in many real-world applications. In this case, it is able to present a human user with a set of solutions trading off between the robustness and the optimality, from which the user has to make a choice according to the need of the application. A method for achieving multiple robust solutions has been suggested in [3] using the sharing method suggested in [16]. However, no information on the relative robustness increase and performance decrease of the solutions is available and thus, no trade-off decisions can be made on the obtained solutions.

3 Variance-Based Measures for Robustness

The search for robust optimal solutions has been widely investigated in the field of engineering design [17]. Consider the following unconstrained minimization problem:

$$\text{minimize } f = f(\boldsymbol{a}, \boldsymbol{x}), \tag{4}$$

where $\boldsymbol{a}$ and $\boldsymbol{x}$ are vectors of environmental parameters and design variables. For convenience, we will not distinguish between environmental parameters and design variables and hereafter, both are called design variables denoted uniformly with x.

Now consider the function $f(x) = f(x_1, x_2, ..., x_n)$, where the x_i's are n design variables and function f is approximated using its first-order Taylor expansion about the point $(\mu_{x_1}, \mu_{x_2}, ..., \mu_{x_n})$:

$$f \approx f(\mu_{x_1}, \mu_{x_2}, ..., \mu_{x_n}) + \sum_{i=1}^{n} \left[\frac{\partial f}{\partial x_i}(\mu_{x_1}, \mu_{x_2}, ..., \mu_{x_n}) \right] \cdot (x_i - \mu_{x_i}), \quad (5)$$

where $\mu_{x_i}, i = 1, 2, ..., n$ is the mean of x_i. Thus, the variance of the function can be derived as follows:

$$\sigma_f^2 = \sum_{i=1}^{n} \left(\frac{\partial f}{\partial x_i} \right)^2 \sigma_{x_i}^2 + \sum_{i=1}^{n} \sum_{j=1, i \neq j}^{n} \left(\frac{\partial f}{\partial x_i} \right) \left(\frac{\partial f}{\partial x_j} \right) \sigma_{x_i x_j}, \quad (6)$$

where $\sigma_{x_i}^2$ is the variance of x_i and $\sigma_{x_i x_j}$ is the covariance between x_i and x_j. Recall that the function has to be evaluated using the mean value of the variables. If the design variables are independent of each other, the resulting approximated variance is

$$\sigma_f^2 = \sum_{i=1}^{n} \left(\frac{\partial f}{\partial x_i} \right)^2 \sigma_{x_i}^2. \quad (7)$$

A measure for robustness of a solution can be defined using the standard deviation of the function and that of the design variables as:

$$f^R = \frac{1}{n} \sum_{i=1}^{n} \frac{\sigma_f}{\sigma_{x_i}}. \quad (8)$$

It should be pointed out that with this definition of robustness, the smaller the robustness measure, the more robust the solution is. In other words, the search for robust optimal solutions can now be formulated as a multiobjective optimization problem where both the fitness function and the robustness measure are to be minimized.

In robust design, the variation of the objective function in the presence of small variations in the design variables is the major concern. Therefore, it is reasonable to discuss the variance of the function defined in equation (7) in a local sense. Take a one-dimensional function $f(x)$ for example, as shown in Fig. 2. If the robustness of a target point x_j is considered, the function is then expanded in a Taylor series about $x = \mu_{x_j} = x_j$, which assumes that the variations of the design variable are zero-mean. Similarly, if the robustness of x_k is to be evaluated, the function will be expanded about $x = \mu_{x_k} = x_k$, refer to Fig. 2. In the figure, $\mu_{f,j}$ and $\mu_{f,k}$ denote the mean of the function calculated around the point x_j and x_k, respectively.

In the following, an estimation of the robustness measure based on the fitness evaluations in the current population will be proposed. Suppose the population size is λ, and N_j ($1 \leq N_j \leq \lambda$) individuals are located in the neighborhood of

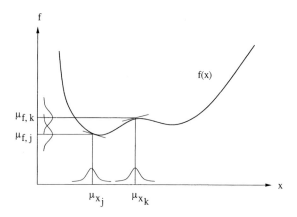

Fig. 2. Illustration of the local variations in the design variables

the j-th individual. The robustness of the j-th individual can be approximated by

$$\text{Robust measure 1: } f_j^R = \frac{1}{n} \sum_{i=1}^{n} \frac{\bar{\sigma}_{f,j}}{\sigma_{x_i}}, \quad (9)$$

where $\bar{\sigma}_{f,j}$ is an estimation of the variance of the j-th individual according to equation (7):

$$\bar{\sigma}_{f,j}^2 = \sum_{i=1}^{n} \left(\frac{\partial f}{\partial x_i}\right)^2 \sigma_{x_i}^2$$

$$\approx \sum_{i=1}^{n} \left(\frac{1}{N_j} \sum_{k \in D_j} \frac{f_j - f_k}{x_{i,j} - x_{i,k}} \right)^2 \sigma_{x_i}^2, \text{k} \neq j, \quad (10)$$

where $x_{i,j}$ and $x_{i,k}$ denote the i-th element of x of the j-th and k-th individuals, and D_j denotes a set of the individuals that belong to the neighborhood of the j-th individual. The neighborhood of j-th individual D_j is defined using the Euclidean distance between the individual $x_k, k = 1, 2, ..., \lambda$ and the j-th individual x_j:

$$D_j : k \in D_j, \text{ if } d_{jk} \leq d^2, \quad 1 \leq k \leq \lambda, \quad d_{jk} = \sqrt{\frac{1}{n} \sum_{i=1}^{n} (x_{i,j} - x_{i,k})^2}, \quad (11)$$

where $k = 1, 2, ..., \lambda$ is the index for the k-th individual, λ is the population size of the evolutionary algorithm, d_{jk} is the Euclidean distance between individual j and k, and d is a threshold to be specified by the user according to the requirements in real applications. This constant should be the same for all individuals.

Actually, a more direct method for estimating the robustness measure can be used. Using the current population and the definition of the neighborhood, the robustness measure of the j-th individual can be estimated by dividing the local standard deviation of the function by the average local standard deviation of the variables. Assume N_j ($1 \leq N_j \leq \lambda$) is the number of individuals in the neighborhood of the j-th individual in the current population, then the local variance of the function corresponding to the j-th individual in the population can be estimated as follows:

$$\mu_{f,j} = \frac{1}{N_j} \sum_{k \in D_j} f_k, \tag{12}$$

$$\sigma_{f,j}^2 = \frac{1}{N_j - 1} \sum_{k \in D_j} (f_k - \mu_{f,j})^2, \tag{13}$$

where $\mu_{f,j}$ and $\sigma_{f,j}^2$ are the local mean and variance of the function calculated from the individuals in the neighborhood of the j-th individual. Thus, the robustness of the j-th individual can be estimated in the following way:

$$\text{Robustness measure 2: } f_j^R = \frac{\sigma_{f,j}}{\bar{\sigma}_{x,j}}, \tag{14}$$

where $\bar{\sigma}_{x,j}$ is the average of the standard deviation of x_i estimated in the j-th neighborhood:

$$\bar{\sigma}_{x,j} = \frac{1}{n} \sum_{i=1}^{n} \sigma_{x_i,j}. \tag{15}$$

The calculation of the mean and variance of x_i in the j-th neighborhood is similar to the calculation of the local mean and variance of the j-th individual as follows:

$$\mu_{x_i,j} = \frac{1}{N_j} \sum_{k \in D_j} x_{i,k}, \tag{16}$$

$$\sigma_{x_i,j}^2 = \frac{1}{N_j - 1} \sum_{k \in D_j} (x_{i,k} - \mu_{x_i,j})^2. \tag{17}$$

Note that the individuals in the neighborhood can be seen as a small sample of a probability distribution around the concerned point, i.e., the current solution. If this probability distribution would coincide with $\phi(z)$ in equation 1, the approximation of the variance would be exact in the sense of the given "variation rule" $\phi(z)$. For example, if we assume that manufactoring tolerance of the final solution leads to a "noise" term which is normally distributed with a given standard deviation, the estimation of the robustness is exact if the sub-sample of the population represents the same normal distribution. Of course, this will not be the case in general. In other words, the sample will usually not be able to reproduce exactly the given distribution $\phi(z)$. Nevertheless, the results obtained in the simulations in the next section demonstrate that the estimations seem to be sufficient for a qualitative search for robust solutions.

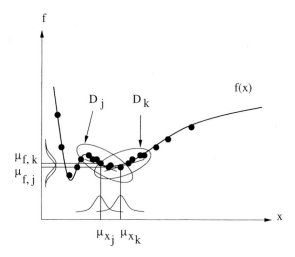

Fig. 3. Samples of the local statistics of the objective function on the basis of the current population of the evolutionary algorithm. The black dots represent the individuals in the current population

With the robustness measures defined above, it is then possible to explicitly treat the search for robust optimal solutions as a multiobjective optimization problem.

Some remarks can be made on the robustness measures defined by equations (9) and (14). The former definition is based on an approximation of the partial derivative of the function with respect to each variable. Theoretically, the smaller the neighborhood, the more exact the estimation will be. However, the estimation may fail if two individuals are too close in the design space due to numerical errors. In this method, neither the variance of the function nor the variance of the variables needs to be estimated. In contrast, the latter definition directly estimates the local variance of the variables and the function using the individuals in the neighborhood.

In this section, we have discussed possible ways to estimate a variance based second criterion for the integration of robustness in the evaluation of solutions. These criteria can now be used in a multi-objective evolutionary algorithm to visualize the trade-off between performance and robustness with the help of the Pareto front. We will employ the Dynamic Weighted Aggregation (DWA) method due to its simplicity and ease of use with evolution strategies.

4 Dynamic Weighted Aggregation for Multiobjective Optimization

4.1 Evolution Strategies

In the standard evolution strategy (ES), the mutation of the object parameters is carried out by adding an $N(0, \sigma_i^2)$ distributed random number. The standard deviations, σ_i's, usually known as the step sizes, are encoded in the genotype together with the object parameters and are subject to mutations. The standard ES can be described as follows:

$$\boldsymbol{x}(t) = \boldsymbol{x}(t-1) + \tilde{\boldsymbol{z}}, \qquad (18)$$

$$\sigma_i(t) = \sigma_i(t-1)\exp(\tau' z)\exp(\tau z_i); i = 1, ..., n, \qquad (19)$$

where $\boldsymbol{x}$ is an n-dimensional parameter vector to be optimized, $\tilde{\boldsymbol{z}}$ is an n-dimensional random number vector with $\tilde{\boldsymbol{z}} \sim N(\boldsymbol{0}, \boldsymbol{\sigma}(t)^2)$, z and z_i are normally distributed random numbers with $z, z_i \sim N(0,1)$. Parameters τ, τ' and σ_i are the strategy parameters, where σ_i is mutated as in equation (19) and τ, τ' are constants as follows:

$$\tau = \left(\sqrt{2\sqrt{n}}\right)^{-1}; \ \tau' = \left(\sqrt{2n}\right)^{-1} \qquad (20)$$

4.2 Dynamic Weighted Aggregation

The classical approach to multiobjective optimization using weighted aggregation of objectives has often been criticized. However, it has been shown [6,7] through a number of test functions as well as several real-world applications that the shortcomings of the weighted aggregation method can be addressed by changing the weights dynamically during optimization using evolutionary algorithms. Two methods for changing the weights have been proposed. The first method is to change the weights gradually from generation to generation. For a bi-objective problem, an example for the periodical gradual weight change is illustrated in Fig. 4(a). The first period of the function can be described by:

$$w_1(t) = \begin{cases} \frac{t}{T}, & 0 \le t \le T, \\ -\frac{t}{T} + 2, & T \le t \le 2T. \end{cases} \qquad (21)$$

$$w_2(t) = 1 - w_1(t), \qquad (22)$$

where T is a constant that controls the speed of the weight change.

A special case of the gradual weight change method described above is to switch the weights between 0 and 1, which has been termed the bang-bang weighted aggregation (BWA) method, as shown in Fig. 4(b). The BWA has shown to be very effective in approximating concave Pareto fronts [7]. A combination of the two methods will also be very practical, as shown in Fig. 4(c).

5 Simulation Studies

5.1 Test Problem 1

The first test problem is constructed in such a way that it exhibits a clear trade-off between the performance and robustness. The function can be described as follows, which is illustrated in Fig. 5.

$$f(x) = 2.0 \sin(10 \exp(-0.08x)x) \exp(-0.25x), \qquad (23)$$

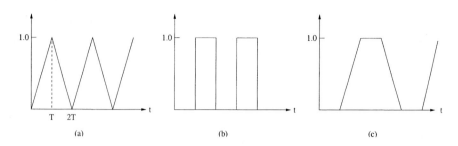

Fig. 4. Patterns of dynamic weight change. (a) Gradual change; (b) Bang-bang switching; (c) Combined

where $0 \leq x \leq 10$. From Fig. 5, it is seen that there is one global minimum together with six local minima in the feasible region. Furthermore, the higher the performance of a minimum, the less robust it is. That is, there is a trade-off between the performance and robustness and the Pareto front should consist of seven separated points.

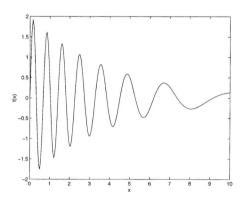

Fig. 5. The one-dimensional function of test problem 2

At first, robustness measure 1 defined by equation (9) is used. That is to say, the individuals in the neighborhood are used to estimate the partial derivatives. The obtained Pareto front is given in Fig. 6(a). It can be seen that an obvious trade-off between the performance and the robustness of the minima has been correctly reflected. Thus, it is straightforward for a user to make a choice among the trade-off solutions according to the problem at hand.

The result using the robustness measure defined by equation (14) is presented in Fig. 6(b). The Pareto fronts of both Figures 6(a) and 6(b) are qualitatively the same and the robustness values at the corners of the Pareto fronts share very similar values even quantitatively.

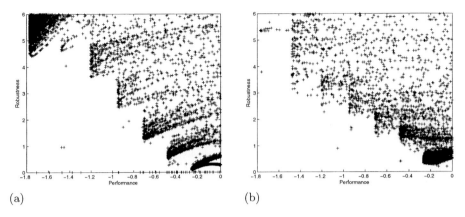

Fig. 6. The trade-off between performance and robustness of test problem 1 based on (a) robustness measure 1, eq. (9) and (b) robustness measure 2, eq. (14)

In the following, we extend the test function in equation (23) to a two-dimensional one. The two-dimensional test function is shown in Fig. 7(a). It can be seen that there are a large number of minima with a different degree of robustness.

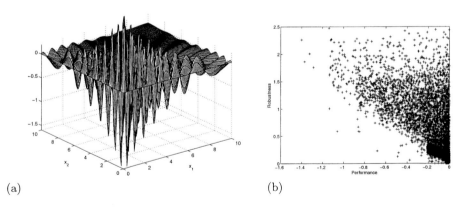

Fig. 7. (a) The 2-dimensional function of the test problem 1. (b) The Pareto front obtained using robust measure 2, eq. (14)

The trade-off between performance and robustness is shown in Fig. 7(b) using the robustness measure 2. It can be seen that the Pareto front seems to be continuous due to the large number of minima and the small robustness difference between the neighboring minima. Nevertheless, the result provides a qualitative picture about the trade-off between performance and robustness, from which a user can make a decision and choose a preferred solution.

5.2 Test Problem 2

The second test problem is taken from reference [18]. The original objective function to minimize is as follows:

$$f(\boldsymbol{x}) = (x_1 - 4.0)^3 + (x_1 - 3.0)^4 + (x_2 - 5.0)^2 + 10.0, \tag{24}$$

subject to

$$g(\boldsymbol{x}) = -x_1 - x_2 + 6.45 \leq 0, \tag{25}$$
$$1 \leq x_1 \leq 10, \tag{26}$$
$$1 \leq x_2 \leq 10. \tag{27}$$

The standard deviation of the function can be derived as follows, assuming the standard deviation of x_1 and x_2 are the same:

$$\sigma_f(\boldsymbol{x}) = \sigma_x \sqrt{(3.0(x_1 - 4.0)^2 + 4.0(x_1 - 3.0)^3)^2 + (2.0(x_2 - 5.0))^2}, \tag{28}$$

where σ_x is the standard deviation of both x_1 and x_2, which is set to:

$$\sigma_x = \frac{1}{3}\Delta x, \tag{29}$$

where Δx is the maximal variation of x_1 and x_2. According to [18], the search for robust optimal solutions can be formulated as follows, assuming the maximal deviation of both variables is 1:

$$\text{minimize } f_1 = f, \tag{30}$$
$$f_2 = \sigma_f, \tag{31}$$
$$\text{subject to } g(\boldsymbol{x}) = -x_1 - x_2 + 8.45, \tag{32}$$
$$2 \leq x_1 \leq 9, \tag{33}$$
$$2 \leq x_2 \leq 9. \tag{34}$$

We call the objective for robustness in equation (31) the theoretical robustness measure, which is explicitly derived from the original fitness function.

The dynamic weighted aggregation method with a $(15, 100)$-ES is used to solve the multiobjective optimization problem. The obtained Pareto front is shown in Fig. 8(a), which is obviously concave. Note, that no archive of the non-dominated solutions has been used in the optimization, which also indicates that the success of the dynamic weighted aggregation method for multiobjective optimization has nothing to do with the archive that has been used in [6,7].

An estimated local standard deviation is used as the robustness measure so that the obtained Pareto front is comparable to the one in Fig. 8(a).

The optimization result is provided in Fig. 8(b). It is seen that although the Pareto front is quite "noisy", it does provide a qualitative approximation of the theoretical trade-off between performance and robustness.

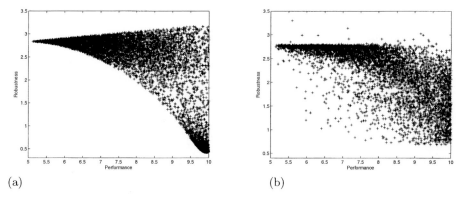

Fig. 8. (a) The Pareto front of test problem 2 using the theoretical robustness measure. (b) The approximated Pareto front using the estimated standard deviation as the robustness measure

6 Conclusion and Discussions

In this paper, we discussed the important property of robustness of solutions in the light of multiobjective optimization. Robustness measures were used as additional objectives together with the usual performance measure. With the help of the obtained Pareto fronts the relation between robustness and performance becomes visible and the decision by the application engineer for a particular solution will be more grounded.

In order to minimize the computational cost involved in estimating the robustness of solutions, we suggested two robustness measures based on the "local" variance of the evaluation function have been introduced. For both methods only information available within the current population has been used, thus, no additional fitness evaluations were necessary. In the case of computationally expensive evaluation functions this property can be essential. The basic idea is to define a neighborhood of a solution and thus to estimate the local mean and variance of a solution. The methods have been applied to two test problems and encouraging results have been obtained.

Although the proposed method is originally targeted at achieving trade-off optimal solutions between performance and robustness, it is straightforward to imagine that the method can also be used in detecting multiple optima of multimodal functions [16]. To show this capability, we consider the central two peak trap function studied in [19]. The function is modified to be a minimization problem and rescaled as shown in Fig. 9(a)

$$f(x) = \begin{cases} -0.16x & \text{if } x < 10, \\ -0.4(20 - x) & \text{if } x > 15, \\ -0.32(15 - x) & \text{otherwise.} \end{cases} \quad (35)$$

The function has two minima and is believed to be deceptive because values of x between 0 and 15 lead toward the local minima.

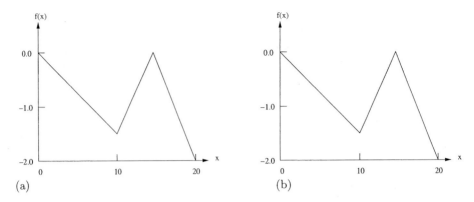

Fig. 9. (a) The trap function. (b) The detected minima

The proposed method is employed to detect the two minima of the function and the result is shown in Fig. 9(b). It can be seen that both minima have successfully been detected.

Of course, it will be difficult to distinguish different optima using the proposed method either if the function values of the optima are very similar or if the robustness values of the optima are very similar.

Several issues still deserve further research efforts. For example, how to improve the quality of the robustness estimation. Currently, the robustness estimation is quite noisy, which to some extent, degrades the performance of the algorithms. Meanwhile, it may be desirable to use the information not only in the current generation, but also from previous generations. Finally, the current algorithm is based on evolution strategies. It will be interesting to extend the method to genetic algorithms.

Acknowledgments. The authors would like to thank E. Körner for his support and M. Olhofer for discussions on robustness.

References

1. I. Das. Robustness optimization for constrained nonlinear programming problems. *Engineering Optimization*, 32(5):585–618, 2000.
2. H.-G. Beyer, M. Olhofer, and B. Sendhoff. On the behavior of $(\mu/\mu_i, \lambda)$-es optimizing functions disturbed by generalized noises. In *Foundations of Genetic Algorithms*, 2002.
3. S. Tustsui and A. Ghosh. Genetic algorithms with a robust solution searching scheme. *IEEE Transactions on Evolutionary Computation*, 1(3):201–208, 1997.
4. W. Chen, J. K. Allen, K.-L. Tsui, and F. Mistree. A procedure for robust design: Minimizing variations caused by noise factors and control factors. *ASME Journal of Mechanical Design*, 118:478–485, 1996.
5. Y. Jin, M. Olhofer, and B. Sendhoff. Managing approximate models in evolutionary aerodynamic design optimization. In *IEEE Congress on Evolutionary Computation*, volume 1, pages 592–599, Seoul, South Korea, 2001.

6. Y. Jin, T. Okabe, and B. Sendhoff. Adapting weighted aggregation for multi-objective evolution strategies. In *First International Conference on Evolutionary Multi-Criterion Optimization*, pages 96–110, Zurich, March 2001. Springer.
7. Y. Jin, M. Olhofer, and B. Sendhoff. Evolutionary dynamic weighted aggregation for multiobjective optimization: Why does it work and how? In *Genetic and Evolutionary Computation Conference*, pages 1042–1049, San Francisco, CA, 2001.
8. A.V. Sebald and D.B. Fogel. Design of fault tolerant neural networks for pattern classification. In D.B. Fogel and W. Atmar, editors, *Proceedings of 1st Annual Conference on Evolutionary Programming*, pages 90–99, 1992.
9. H. Greiner. Robust optical coating design with evolution strategies. *Applied Optics*, 35(28):5477–5483, 1996.
10. D. Wiesmann, U. Hammel, and T. Bäck. Robust design of multilayer optical coatings by means of evolutionary algorithms. *IEEE Transactions on Evolutionary Computation*, 2(4):162–167, 1998.
11. A. Thompson. On the automatic design of robust electronics through artificial evolution. In *Proceedings on 2nd International Conference on Evolvable Systems*, pages 13–24. Springer, 1998.
12. J. Branke. Creating robust solutions be means of evolutionary algorithms. In *Parallel Problem Solving from Nature*, volume V of *Lecture Notes in Computer Science*, pages 119–128. Springer, Berlin, 1998.
13. J. Branke, C. Schmidt, and H. Schmeck. Efficient fitness estimation in noisy environment. In *Proceedings of Genetic and Evolutionary Computation Conference*, pages 243–250, San Francisco, 2001. Morgan Kaufmann.
14. Peter Stagge. Averaging efficiently in the presence of noise. In A. E. Eiben, T. Bäck, M. Schoenauer, and H.-P. Schwefel, editors, *Parallel Problem Solving from Nature - PPSN V*, pages 188–197. Springer Verlag, 1998.
15. Y. Yamaguchi and T. Arima. Multiobjective optimization for transonic compressor stator blade. In *8th AIAA/USAF/NASA/ISSMO Symposium on Multidisciplinary Analysis and Optimization*, 2000.
16. D.E. Goldberg and J. Richardson. Genetic algorithms with sharing for multimodal function optimization. In J.J. Grefenstette, editor, *Proceedings of the 2nd International Conference on Genetic Algorithms*, pages 41–49, 1987.
17. K.-L. Tsui. An overview of Taguchi method and newly developed statistical methods for robust design. *IIE Transactions*, 24(5):44–57, 1992.
18. W. Chen, M. Wiecek, and J. Zhang. Quality utility: A compromise programming approach to robust design. *Journal of Mechanical Design*, 121(2):179–187, 1999.
19. D. Beasley, D.B. Bull, and R.R. Martin. A sequential niche technique for multimodal function optimization. *Evolutionary Computation*, 1(2):101–125, 1993.

The Micro Genetic Algorithm 2: Towards Online Adaptation in Evolutionary Multiobjective Optimization

Gregorio Toscano Pulido and Carlos A. Coello Coello

CINVESTAV-IPN
Evolutionary Computation Group
Depto. de Ingeniería Eléctrica
Sección de Computación
Av. Instituto Politécnico Nacional No. 2508
Col. San Pedro Zacatenco
México, D. F. 07300
gtoscano@computacion.cs.cinvestav.mx
ccoello@cs.cinvestav.mx

Abstract. In this paper, we deal with an important issue generally omitted in the current literature on evolutionary multiobjective optimization: on-line adaptation. We propose a revised version of our micro-GA for multiobjective optimization which does not require any parameter fine-tuning. Furthermore, we introduce in this paper a dynamic selection scheme through which our algorithm decides which is the "best" crossover operator to be used at any given time. Such a scheme has helped to improve the performance of the new version of the algorithm which is called the micro-GA2 (μGA^2). The new approach is validated using several test function and metrics taken from the specialized literature and it is compared to the NSGA-II and PAES.

1 Introduction

One of the research topics that has been only scarcely covered in the current literature on evolutionary multiobjective optimization has been (on-line or self-) adaptation [4]. This is an important issue since evolutionary multiobjective optimization techniques normally require more parameters than a traditional evolutionary algorithm (e.g., a niche radius or sharing threshold). In some of our recent research, we have emphasized the importance of designing computationally efficient evolutionary multiobjective optimization approaches. One of our proposals in this direction was the micro-GA for multiobjective optimization, which uses a very small internal population (four individuals) combined with three forms of elitism and a reinitialization process [3]. The micro-GA has been found to be highly competitive with respect to other techniques that are representative of the state-of-the-art in the area (namely, the Nondominated Sorting Genetic Algorithm or NSGA-II [6] and the Pareto Archived Evolution Strategy or PAES [8]). However, its main drawback is that the micro-GA requires several parameters whose fine-tuning may not be intuitive for a newcomer.

This paper presents a revised version of the micro-GA for multiobjective optimization [3] which does not require any parameters fine-tuning from the user. The new approach, called the micro-GA2 (μGA^2), tends to perform better than our original micro-GA without needing any non-intuitive parameters. To validate our proposal, we compare it against PAES [8], the NSGA-II [6] and our original micro-GA [3] using several test functions of high degree of difficulty which have been recently proposed in the specialized literature.

2 Related Work

There have been very few attempts in the literature to produce an evolutionary multiobjective optimization technique that adapts its parameters during the evolutionary process and that, therefore, does not require any fine-tuning from the user. One of the earliest attempts to incorporate self-adaptive mechanisms in evolutionary multiobjective optimization was Kursawe's proposal of providing individuals with a set of step sizes for each objective function such that his multiobjective evolution strategy could deal with a dynamic environment [10]. Laumanns et al. [11] showed that a standard self-adaptive evolution strategy had problems to converge to the true Pareto set of a multiobjective optimization problem and proposed alternative self-adaptation mechanisms that, however, were applied only to an aggregating fitness function. Tan et al. [13] proposed the incrementing multiobjective evolutionary algorithm (IMOEA), which uses a dynamic population size that adapts based on the tradeoffs produced so far and the desired population distribution density. The IMOEA relies on a measure of convergence based on population domination and progress ratio [14]. The IMOEA also uses dynamic niches (i.e., no sharing factor needs to be defined). Another interesting proposal is the idea of Büche et al. [2] of using self-organizing maps of Kohonen [9] to adapt the mutation step size of an evolutionary multiobjective optimization algorithm. The authors also define a recombination operator using self-organizing maps (something similar to intermediate recombination). Abbass [1] recently proposed a differential evolution algorithm used to solve multiobjective problems that self-adapts its crossover and mutation rates. Zhu and Leung [16] proposed an asynchronous self-adjustable island genetic algorithm in which certain information about the current search status of each island in a parallel evolutionary algorithm is used to focus the search effort into non-overlapping regions.

The approach proposed in this paper introduces on-line adaptation into a genetic algorithm (which uses Pareto ranking and an external memory) in order to make unncessary to fine tune its parameters. Unlike the proposals previously discussed, our approach is mainly focused on performing an appropriate exploration and exploitation of the search space relying on very simple information and statistical measures obtained from the evolutionary process itself.

3 The μGA^2

Since this paper presents a variation of the micro-GA reported in [3], we consider convenient to describe our original proposal first. The way in which our micro-GA works is the following: First, a random population is generated. This random population feeds the population memory, which is divided in two parts: a replaceable and a non-replaceable portion. The non-replaceable portion of the population memory will never

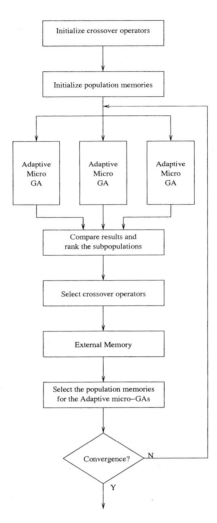

Fig. 1. Diagram that illustrates the way in which our μGA^2 works

change during the entire run and is meant to provide the required diversity for the algorithm. In contrast, the replaceable portion will experience changes after each cycle of the micro-GA.

The population of the micro-GA at the beginning of each of its cycles is taken (with a certain probability) from both portions of the population memory so that we can have a mixture of randomly generated individuals (non-replaceable portion) and evolved individuals (replaceable portion). During each cycle, the micro-GA undergoes conventional genetic operators. After the micro-GA finishes one cycle, we choose two non-dominated vectors from the final population and compare them with the contents of the external memory (this memory is initially empty). If either of them (or both) remains as non-dominated after comparing it against the vectors in this external memory, then they are included there (i.e., in the external memory). This is our historical archive of non-dominated vectors. All dominated vectors contained in the external memory are eliminated. The same two vectors previously mentioned are also compared against two elements from the replaceable portion of the population memory. If either of these vectors dominates to its match in the population memory, then it replaces it. Otherwise, the vector is discarded. Over time, the replaceable part of the population memory will tend to have more non-dominated vectors, some of which will be used in some of the initial populations of the micro-GA.

The main difference between the μGA^2 and its ancestor is that we now provide on-line adaptation mechanisms. The way of which μGA^2 works is illustrated in Figure 1. Since one of the main features of the new approach is the use of a parallel strategy to adapt the crossover operator (i.e., we have several micro-GAs which are executed in parallel), we will start by describing this mechanism. First, the initial crossover operator to be used by each micro-GA is selected. The three crossover operators available in our approach are: 1) SBX, 2) two-point crossover, and 3) a crossover operator proposed by us. The behavior of the crossover operator that we proposed depends on the distance between each variable of the corresponding parents: if the variables are closer than the mean variance of each variable, then intermediate crossover is performed; otherwise, a recombination that emphasizes solutions around the parents is applied. These crossover operators were selected because they exhibited the best overall performance in an extensive set of experiments that we conducted.

Once the crossover operator has been selected, the population memories of the internal micro-GAs are randomly generated. Then, all the internal micro-GAs are executed, each one using one of the crossover operators available (this is a deterministic process). The nondominated vectors found by each micro-GA are compared against each other and we rank the contribution of each crossover operator with respect to its effectiveness to produce nondominated vectors. At this point, the crossover operator which exhibits the worst performance is replaced by the one with the best performance. The external memory stores the globally nondominated solutions, and we fill the new population memories (of every internal micro-GA) using this external memory. The new external memories of the micro-GA are identical to this external memory.

When all these processes are completed, we check if the algorithm has converged. For this sake, we assume that convergence has been reached when none of the internal micro-GAs can improve the solutions previously reached (see below for details). The rationale here is that if we have not found new solutions in a certain (reasonably large) amount of time, it is fruitless to continue the search.

The μGA^2 works in two stages: the first one starts with a conventional evolutionary process and it concludes when the external memory of each slave process is full or when at least one slave has reached convergence (as assumed in the previous paragraph). We finish the second stage when global convergence (i.e., when all of the slaves have converged) is reached.

An interesting aspect of the μGA^2 is that it attempts to balance between exploration and exploitation by changing the priorities of the genetic operators. This is done during each of the two stages previously described. During the first stage, we emphasize exploration and during the second we emphasize exploitation, which are performed in the following way:

- **Exploration stage**: At this stage, mutation has more importance than crossover so that we can locate the most promising regions of the search space. We use at this point a low crossover rate and the mutation operator is the main responsible of directing the search. We also decrease the nominal convergence (i.e., the internal cycle of the micro-GA), since we are not interested in recombining solutions at this point.
- **Exploitation stage**: At this stage, the crossover operator has more importance and therefore nominal convergence is increased to reach better results.

Since the main drawback of the micro-GA is that it requires several additional parameters which have to be fine-tuned by hand [3], the main goal of the μGA^2 was the elimination of these parameters. With this goal in mind, we divided the parameters of the micro-GA into two groups: parameters which cannot be adapted on-line and parameters that can be adapted. The first class is composed by those parameters that depend on the problem characteristics and it includes: bounds for the decision variables, number of decision variables, number of objectives, and number of nondominated vectors that we aim to find (this defines the size of the external memory which will be called T_{pareto}).

For those parameters that can be adapted, we followed a rationale based on their possible dependences. Certain parameters such as the mutation rate can be easily fixed to one divided by the number of genes (or decision variables). After a careful analysis of the parameters of the micro-GA susceptible of adaptation, we decided the following:

- **Crossover rate**: It is important that it behaves in such a way that we can explore more at the beginning of the evolutionary process and we can exploit more at later stages of the search. In the first stage of the μGA^2 we use only a 50% for the crossover rate, while during the exploitation stage, we use a 100% for the crossover rate. As indicated before, we adapt the crossover operator using a parallel strategy.
- **Size of the population memory**: The size of the population was set to $T_{pareto} \div 2$.
- **Percentage of non-replaceable memory**: Since the μGA^2 is a parallel algorithm, we decided to decrease the percentage of the non-replaceable memory to a 10% (with respect to the 30% used in the original micro-GA [3]) of the size of the population memory.
- **Total number of iterations**: This refers to the external cycle of the micro-GA [3]. This parameter is set such that we finish after the external memory has been replaced $T_{pareto} \times 2$ times without having any dominated individual nor any replacement of the limits of the adaptive grid.
- **Replacement cycle**: We replace the replaceable memory whenever T_{pareto} individuals had been evaluated. Also note that when the replaceable memory is refreshed, we also refresh the non-replaceable memory, but using randomly generated individuals (as it was done at the begining of the execution of the algorithm).
- **Number of subdivisions of the adaptive grid**: The number of subdivisions is treated with respect to the number of individuals desired per region. This value is set such that it never exceeds (on average) three individuals per region and is

never less than 1.5 individuals per region. Thus, the number of subdivisions (and therefore the number of individuals per hypercube) is either increased or decreased in consequence.

The constraint-handling approach of the original micro-GA was kept intact (see [3] for details).

4 Metrics Adopted

In order to give a numerical comparison of our approach, we adopted three metrics: generational distance [15], error ratio [14] and spacing [12]. The description and mathematical representation of each metric are shown below.

1. **Generational Distance (GD)**: The concept of generational distance was introduced by Van Veldhuizen & Lamont [15] as a way of estimating how far are the elements in the set of nondominated vectors found so far from those in the Pareto optimal set and is defined as:

$$GD = \frac{\sqrt{\sum_{i=1}^{n} d_i^2}}{n} \qquad (1)$$

   where n is the number of vectors in the set of nondominated solutions found so far and d_i is the Euclidean distance (measured in objective space) between each of these and the nearest member of the Pareto optimal set. It should be clear that a value of $GD = 0$ indicates that all the elements generated are in the Pareto optimal set. Therefore, any other value will indicate how "far" we are from the global Pareto front of our problem.

2. **Error Ratio (ER)**: This metric was proposed by Van Veldhuizen [14] to indicate the percentage of solutions (from the nondominated vectors found so far) that are not members of the true Pareto optimal set:

$$ER = \frac{\sum_{i=1}^{n} e_i}{n}, \qquad (2)$$

   where n is the number of vectors in the current set of nondominated vectors available; $e_i = 0$ if vector i is a member of the Pareto optimal set, and $e_i = 1$ otherwise. It should then be clear that $ER = 0$ indicates an ideal behavior, since it would mean that all the vectors generated by our algorithm belong to the Pareto optimal set of the problem.

3. **Spacing (SP)**: Here, one desires to measure the spread (distribution) of vectors throughout the nondominated vectors found so far. Since the "beginning" and "end" of the current Pareto front found are known, a suitably defined metric judges how well the solutions in such front are distributed. Schott [12] proposed such a metric measuring the range (distance) variance of neighboring vectors in the nondominated vectors found so far. This metric is defined as:

$$S \triangleq \sqrt{\frac{1}{n-1}\sum_{i=1}^{n}(\overline{d}-d_i)^2}\,, \qquad (3)$$

where $d_i = \min_j(\mid f_1^i(\boldsymbol{x}) - f_1^j(\boldsymbol{x}) \mid + \mid f_2^i(\boldsymbol{x}) - f_2^j(\boldsymbol{x}) \mid)$, $i,j = 1,\ldots,n$, $\overline{d}$ is the mean of all d_i, and n is the number of nondominated vectors found so far. A value of zero for this metric indicates all members of the Pareto front currently available are equidistantly spaced.

5 Test Functions and Numerical Results

Several test functions were taken from the specialized literature to compare our approach. In all cases, we generated the true Pareto fronts of the problems using exhaustive enumeration (with a certain granularity) so that we could make a graphical and metric-based comparison of the quality of the solutions produced by the μGA^2. We also compared our results with respect to the NSGA-II [6], with respect to the PAES [8] and with respect to our original micro-GA [3]. In the following examples, the NSGA-II was run using a population size of 100, a crossover rate of 0.8 (using SBX), tournament selection, and a mutation rate of 1/vars, where vars = number of decision variables of the problem. PAES was run using a depth of 5, a size of the archive of 100, and a mutation rate of 1/bits, where bits refers to the length of the chromosomic string that encodes the decision variables. Our micro-GA used a crossover rate of 0.7, an external memory of 100 individuals, a number of iterations to achieve nominal convergence of two, a population memory of 50 individuals, a percentage of non-replaceable memory of 0.3, a population size (for the micro-GA itself) of four individuals, 25 subdivisions of the adaptive grid, and a mutation rate of $1/L$ (L = length of the chromosomic string).

The number of fitness function evaluations for the original micro-GA, the NSGA-II and PAES was the closest value to the average number of fitness function evaluations obtained from performing 20 runs with the μGA^2.

5.1 Test Function 1

Our first example is a n-objective, n-variable test function proposed by *Deb et al.* [7]:

$$\begin{aligned}
&\text{Minimize } f_1(\mathbf{x}) = x_1, \\
&\quad\vdots \\
&\text{Minimize } f_{M-1}(\mathbf{x}) = x_{M-1}, \\
&\text{Minimize } f_M(\mathbf{x}) = (1 + g(\mathbf{x}_M))h(f_1, f_2, \ldots, F_{M-1}, g), \\
&\text{where } g(\mathbf{x}_M) = 1 + \frac{9}{|\mathbf{x}_M|}\sum_{x_i \in \mathbf{x}_M} xi, \\
&\qquad h = M - \sum_{i=1}^{M-1}\left[\frac{f_i}{1+g}(1+\sin(3\pi f_i))\right], \\
&\text{subject to } 0 \le x_i \le,\ \text{for } i = 1, 2.., n,
\end{aligned} \qquad (4)$$

This test function has 2^{M-1} disconnected Pareto-optimal regions in the search space. Deb et al. [7] propose to use 22 variables to make it more challenging. That is precisely the number of decision variables that were adopted for our experiments. Results for the first test function are summarized in Table 1. Figures 2 and 3 show the average behavior of each algorithm with respect to the generational distance metric. From the results shown in Table 1, it can be seen that in the first test function, the μGA^2 had the best performance with respect to generational distance and spacing and it placed second (after the NSGA-II) with respect to the error ratio.

Table 1. Results obtained in the first test function (DTLZ6) by the μGA^2, the NSGA-II, PAES and the original micro-GA. We show in **boldface** the best values for each metric

Perf. measure	Iterations	μGA^2		
		GD	ER	SP
Average	20382	**0.003561016**	0.171	**0.07382801**
Best	16954	0.00304658	0.1	0.0598198
Worst	24394	0.00440405	0.25	0.0886338
Std. Dev.	2019.793840	0.000372	0.04290	0.007385
		PAES		
Average	20382	0.0161938745	0.49492855	0.125067925
Best	20382	0.00260934	0.2	0.0770419
Worst	20382	0.109795	0.75	0.258494
Std. Dev.	0	0.023217	0.1603101	0.049333
		NSGA-II		
Average	20100	0.003606146	**0.115**	0.077738055
Best	20100	0.00281355	0.07	0.039322
Worst	20100	0.0052915	0.16	0.0940669
Std. Dev.	0	0.000634	0.030174	0.012038
		micro-GA		
Average	20376	0.8760464	1.0425015	0.97022395
Best	20376	0.381188	1.025	0.232188
Worst	20376	1.66206	1.07143	3.4051
Std. Dev.	0	0.3524874	0.01302171	1.0298174

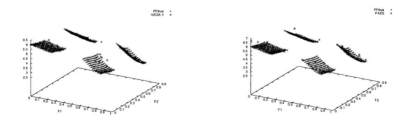

Fig. 2. Pareto fronts produced by the NSGA II (left) and PAES (right) for the first test function

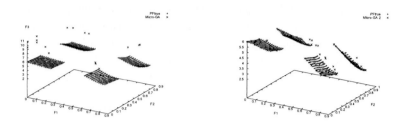

Fig. 3. Pareto fronts produced by the micro-GA (left) and the μGA^2 (right) for the first test function

5.2 Test Function 2

Our second example is an n-objective, n-variable test function proposed by *Deb et al.* [7]:

$$\begin{aligned}
&\text{Minimize } f_1(\mathbf{x}) = (1 + g(\mathbf{x}_M))\cos(x_1\pi/2)\ldots\cos(x_{M-1}\pi/2),\\
&\text{Minimize } f_2(\mathbf{x}) = (1 + g(\mathbf{x}_M))\cos(x_1\pi/2)\ldots\sin(x_{M-1}\pi/2),\\
&\quad\vdots\qquad\vdots\\
&\text{Minimize } f_M(\mathbf{x}) = (1 + g(\mathbf{x}_M))\sin(x_1\pi/2),\\
&\text{subject to } 0 \le x_i \le,\ \text{for } i = 1, 2.., n,\\
&\text{where } g(\mathbf{x}_M) = \sum_{xi \in \mathbf{x}_M}(x_i - 0.5)^2.\\
&\text{subject to } 0 \le xi \le 1,\ for\ i = 1, 2, \ldots, n.
\end{aligned} \tag{5}$$

Results for the second test function are summarized in Table 2. Figures 4 and 5 show the average behavior of each algorithm with respect to the generational distance metric. It is interesting to observe in Table 2, that for the second test function, PAES had the best performance with respect to generational distance and spacing, and that the original micro-GA produced better results than the μGA^2 for all of the metrics considered. In fact, the μGA^2 exhibited the worst performance with respect to spacing although it was not the worst of all with respect to error ratio and generational distance. This poor performance is mainly due to the fact that the μGA^2 could not get rid of some points that did not lie on the true Pareto front of the problem.

5.3 Test Function 3

The third example is a bi-objective optimization problem proposed by Deb [5]:

$$\begin{aligned}
&\text{Minimize } f_1(x_1, x_2) = x_1\\
&\text{Minimize } f_2(x_1, x_2) = g(x_1, x_2) \cdot h(x_1, x_2)
\end{aligned} \tag{6}$$

where:

$$g(x_1, x_2) = 11 + x_2^2 - 10 \cdot \cos(2\pi x_2) \tag{7}$$

Table 2. Results obtained in the second test function (DTLZ2) by the μGA^2, the NSGA-II, PAES and the original micro-GA. We show in **boldface** the best values for each metric

Perf. measure	Iterations	μGA^2 GD	ER	SP
Average	28035.6	0.06768955	0.4782749	0.12874763
Best	22662	0.0466557	0.07	0.0864405
Worst	38626	0.0808061	0.87	0.258394
Std. Dev.	3935.705252	0.009093	0.307199	0.036003
		PAES		
Average	28035	**0.00188515945**	0.661	**0.07808585**
Best	28035	0.000960599	0.49	0.0645913
Worst	28035	0.00372801	0.83	0.0969163
Std. Dev.	0	0.000729	0.099994	0.007436
		NSGA-II		
Average	28100	0.071364695	0.595	0.086525355
Best	28100	0.0012885	0.37	0.0600044
Worst	28100	0.146352	0.88	0.151025
Std. Dev.	0	0.039513	0.12279	0.026498
		micro-GA		
Average	28032	0.0079807725	**0.43616665**	0.07967385
Best	28032	0.00196516	0.25	0.0469776
Worst	28032	0.0139376	0.82	0.1319
Std. Dev.	0	0.0035755	0.1511332	0.0271390

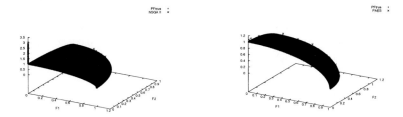

Fig. 4. Pareto fronts produced by the NSGA II (left) and PAES (right) for the second test function

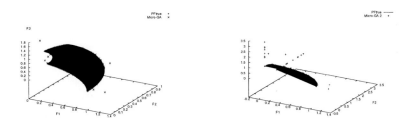

Fig. 5. Pareto fronts produced by the micro-GA (left) and the μGA^2 (right) for the second test function

$$h(x_1, x_2) = \begin{cases} 1 - \sqrt{\frac{f_1(x_1,x_2)}{g(x_1,x_2)}} & \text{if } f_1(x_1, x_2) \leq g(x_1, x_2) \\ 0 & \text{in other case} \end{cases} \quad (8)$$

and $0 \leq x_1 \leq 1, -30 \leq x_2 \leq 30$.

Results for the third test function are summarized in Table 3. Figures 6 and 7 show the average behavior of each algorithm with respect to the generational distance metric. Table 3 shows that in the third test function, the μGA^2 produced the best results for both the generational distance and the error ratio metrics. With respect to spacing, it placed second (after the NSGA-II).

Table 3. Results obtained in the third test function (Deb) by the μGA^2, the NSGA-II, PAES and the original micro-GA. We show in **boldface** the best values for each metric

Perf. measure	Iterations	μGA^2 GD	ER	SP
Average	9171.8	**0.00016127085**	**0.115**	0.0088751215
Best	6186	0.000102157	0	0.00721023
Worst	13826	0.000218467	0.32	0.0100066
Std. Dev.	0.081917	1956.912487	4.252223-05	0.000822
		PAES		
Average	9171	0.4651514	0.70408545	5.46232964
Best	9171	0.242424	0.0252054	0.0829736
Worst	9171	1	7.97044	64.8108
Std. Dev.	0	0.180424	2.012568	16.406210
		NSGA-II		
Average	9100	0.0002118179	0.2105	**0.0079981215**
Best	9100	0.000155758	0.01	0.00646298
Worst	9100	0.000282185	0.74	0.0089998
Std. Dev.	0	3.577123-05	0.224252	0.000594
		micro-GA		
Average	91068	0.0556739552	0.162	0.281928729
Best	91068	0.000159071	0.05	0.00637886
Worst	91068	0.465348	0.31	1.22778
Std. Dev.	0	0.1079727	0.0796439	0.3647516

5.4 Test Function 4

Our fourth example is a bi-objective test function proposed by *Kursawe* [10]:

$$\text{Minimize } f_1(x) = \sum_{i=1}^{n-1} \left(-10 \exp\left(-0.2\sqrt{x_i^2 + x_{i+1}^2} \right) \right) \quad (9)$$

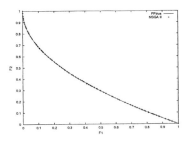
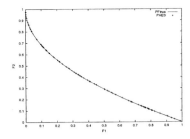

Fig. 6. Pareto fronts produced by the NSGA II (left) and PAES (right) for the third test function

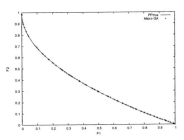
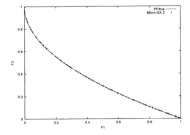

Fig. 7. Pareto fronts produced by the micro-GA (left) and the μGA^2 (right) for the third test function

$$\text{Minimize } f_2(x) = \sum_{i=1}^{n} \left(|x_i|^{0.8} + 5\sin(x_i)^3\right) \quad (10)$$

where:

$$-5 \leq x_1, x_2, x_3 \leq 5 \quad (11)$$

Our fourth test function has a discontinuous Pareto optimal set and a discontinuous Pareto front. The main difficulty of this test function resides in the fact that it has a fairly large search space.

Results for our fourth test function are summarized in Table 4. Figures 8 and 9 show the average behavior of each algorithm with respect to the generational distance metric. Table 4 shows that in the fourth test function, the μGA^2 produced better results than our original micro-GA for all the metrics, but it had poorer values than the NSGA-II, which beats all of the other approaches in this problem with respect to all the metrics considered. Note, however, that the NSGA-II does not completely cover the true Pareto Front (see Figure 8, left graph). In contrast, the μGA^2 has a wider spread of nondominated solutions, but with a poorer distribution than the NSGA-II.

Table 4. Results obtained in the fourth test function (Kursawe) by the μGA^2, the NSGA-II, PAES and the original micro-GA. We show in **boldface** the best values for each metric

Perf. measure	Iterations	μGA^2		
		GD	ER	SP
Average	12521.2	0.005006659	0.3505	0.11070785
Best	9350	0.00326891	0.2	0.103748
Worst	16262	0.0176805	0.51	0.131396
Std. Dev.	2075.483192	0.003133	0.080031	0.005867
		PAES		
Average	12521	0.1963858	1.01001	0.30378948
Best	12521	0.139281	1.01	**0.00477854**
Worst	12521	0.383218	1.0102	1.24913
Std. Dev.	0	0.061842	4.4721-05	0.355380
		NSGA-II		
Average	12100	**0.0036544415**	**0.2765**	**0.06236726**
Best	12100	**0.00311988**	**0.21**	0.0536878
Worst	12100	**0.00541468**	**0.44**	**0.0815969**
Std. Dev.	0	0.000506	0.052342	0.007113
		micro-GA		
Average	12520	0.0065217035	0.59144695	0.130839305
Best	12520	0.00357395	0.39	0.0636691
Worst	12520	0.00969116	0.73	0.180439
Std. Dev.	0	0.00171691	0.08471597	0.02968868

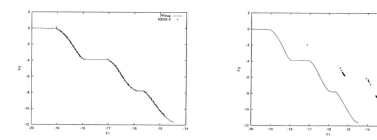

Fig. 8. Pareto fronts produced by the NSGA-II (left) and PAES (right) for the fourth test function

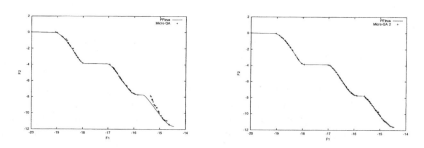

Fig. 9. Pareto fronts produced by the micro-GA (left) and the μGA^2 (right) for the fourth test function

6 Conclusions and Future Work

We have proposed an on-line adaptation scheme that allows the use of a micro-GA designed for multiobjective optimization without requiring the definition of any non-intuitive parameters. The proposed approach obtains information from the evolutionary process to guide the search efficiently. Among other things, our scheme decides on the most appropriate crossover operator and it switches between an exploration and an exploitation stage by changing the importance of the crossover and the mutation operators. We also define a criterion that allows to stop the execution of the algorithm when the search seems to be fruitless and therefore, the definition of a maximum number of generations is no longer necessary.

The approach has been validated with several test functions from which four were included in this paper. We compared our results with respect to our original micro-GA and with respect to PAES and the NSGA-II using three metrics: generational distance, error ratio and spacing. Our preliminary results indicate that although our approach does not always beat the other approaches, it remains competitive, and it normally improves on the results generated by the micro-GA without the need of defining any parameters.

As part of our future work, we want to experiment with spatial data structures to make more efficient the storage and retrieval of nondominated vectors from the external memory. We are also working on the development of a mechanism that allows to reduce the number of fitness function evaluations required to approximate the true Pareto front of a problem.

Acknowledgments. This paper is representative of the research performed by the Evolutionary Computation Group at CINVESTAV-IPN (EVOCINV). The first author acknowledges support from the mexican Consejo Nacional de Ciencia y Tecnología (CONACyT) through project number 34201-A. The second author acknowledges support from CONACyT through a scholarship to pursue graduate studies at the Computer Science Section of the Electrical Engineering Department at CINVESTAV-IPN.

References

1. Hussein A. Abbass. The Self-Adaptive Pareto Differential Evolution Algorithm. In *Congress on Evolutionary Computation (CEC'2002)*, volume 1, pages 831–836, Piscataway, New Jersey, May 2002. IEEE Service Center.
2. Dirk Büche, Gianfranco Guidati, Peter Stoll, and Petros Kourmoursakos. Self-Organizing Maps for Pareto Optimization of Airfoils. In Juan Julián Merelo Guervós et al., editor, *Parallel Problem Solving from Nature—PPSN VII*, pages 122–131, Granada, Spain, September 2002. Springer-Verlag. Lecture Notes in Computer Science No. 2439.
3. Carlos A. Coello Coello and Gregorio Toscano Pulido. Multiobjective Optimization using a Micro-Genetic Algorithm. In Lee Spector et al., editor, *Proceedings of the Genetic and Evolutionary Computation Conference (GECCO'2001)*, pages 274–282, San Francisco, California, 2001. Morgan Kaufmann Publishers.
4. Carlos A. Coello Coello, David A. Van Veldhuizen, and Gary B. Lamont. *Evolutionary Algorithms for Solving Multi-Objective Problems*. Kluwer Academic Publishers, New York, May 2002. ISBN 0-3064-6762-3.

5. Kalyanmoy Deb. Multi-Objective Genetic Algorithms: Problem Difficulties and Construction of Test Problems. *Evolutionary Computation*, 7(3):205–230, Fall 1999.
6. Kalyanmoy Deb, Amrit Pratap, Sameer Agarwal, and T. Meyarivan. A Fast and Elitist Multiobjective Genetic Algorithm: NSGA–II. *IEEE Transactions on Evolutionary Computation*, 6(2):182–197, April 2002.
7. Kalyanmoy Deb, Lothar Thiele, Marco Laumanns, and Eckart Zitzler. Scalable Multi-Objective Optimization Test Problems. In *Congress on Evolutionary Computation (CEC'2002)*, volume 1, pages 825–830, Piscataway, New Jersey, May 2002. IEEE Service Center.
8. Joshua D. Knowles and David W. Corne. Approximating the Nondominated Front Using the Pareto Archived Evolution Strategy. *Evolutionary Computation*, 8(2):149–172, 2000.
9. Teuvo Kohonen, T.S. Huang, and M.R. Schroeder, editors. *Self-Organizing Maps*. Springer-Verlag, 2001.
10. Frank Kursawe. A Variant of Evolution Strategies for Vector Optimization. In H. P. Schwefel and R. Männer, editors, *Parallel Problem Solving from Nature. 1st Workshop, PPSN I*, volume 496 of *Lecture Notes in Computer Science*, pages 193–197, Berlin, Germany, Oct 1991. Springer-Verlag.
11. Marco Laumanns, Günter Rudolph, and Hans-Paul Schwefel. Mutation Control and Convergence in Evolutionary Multi-Objective Optimization. In *Proceedings of the 7th International Mendel Conference on Soft Computing (MENDEL 2001)*, Brno, Czech Republic, June 2001.
12. Jason R. Schott. Fault Tolerant Design Using Single and Multicriteria Genetic Algorithm Optimization. Master's thesis, Department of Aeronautics and Astronautics, Massachusetts Institute of Technology, Cambridge, Massachusetts, May 1995.
13. K.C. Tan, T.H. Lee, and E.F. Khor. Evolutionary Algorithms with Dynamic Population Size and Local Exploration for Multiobjective Optimization. *IEEE Transactions on Evolutionary Computation*, 5(6):565–588, December 2001.
14. David A. Van Veldhuizen. *Multiobjective Evolutionary Algorithms: Classifications, Analyses, and New Innovations*. PhD thesis, Department of Electrical and Computer Engineering. Graduate School of Engineering. Air Force Institute of Technology, Wright-Patterson AFB, Ohio, May 1999.
15. David A. Van Veldhuizen and Gary B. Lamont. Multiobjective Evolutionary Algorithm Research: A History and Analysis. Technical Report TR-98-03, Department of Electrical and Computer Engineering, Graduate School of Engineering, Air Force Institute of Technology, Wright-Patterson AFB, Ohio, 1998.
16. Zhong-Yao Zhu and Kwong-Sak Leung. Asynchronous Self-Adjustable Island Genetic Algorithm for Multi-Objective Optimization Problems. In *Congress on Evolutionary Computation (CEC'2002)*, volume 1, pages 837–842, Piscataway, New Jersey, May 2002. IEEE Service Center.

Self-Adaptation for Multi-objective Evolutionary Algorithms

Dirk Büche, Sibylle Müller, and Petros Koumoutsakos

Institute of Computational Science, Swiss Federal Institute of Technology (ETH),
CH - 8092 Zürich, Switzerland
{bueche,muellers,petros}@inf.ethz.ch
http://www.icos.ethz.ch

Abstract. Evolutionary Algorithms are a standard tool for multi-objective optimization that are able to approximate the Pareto front in a single optimization run. However, for some selection operators, the algorithm stagnates at a certain distance from the Pareto front without convergence for further iterations.
We analyze this observation for different multi-objective selection operators. We derive a simple analytical estimate of the stagnation distance for several selection operators, that use the dominance criterion for the fitness assignment. Two of the examined operators are shown to converge with arbitrary precision to the Pareto front. We exploit this property and propose a novel algorithm to increase their convergence speed by introducing suitable self-adaptive mutation. This adaptive mutation takes into account the distance to the Pareto front. All algorithms are analyzed on a 2- and 3-objective test function.

1 Introduction

Real-world optimization problems often include multiple and conflicting objectives. A solution to such a problem is often a compromise between the objectives, since no solution can be found that is ideal to all objectives. The set of the best compromise solutions is referred to as the Pareto-ideal set, characterized by the fact that starting from a solution within the set, one objective can only be improved at the expense of at least one other objective.
Evolutionary Algorithms (EAs) are a standard tool for Pareto optimization, since their population-based search allows approximating the Pareto front in a single optimization run. EAs operate by evolving the population in a cooperative search towards the Pareto front. They incorporate biologically inspired operators such as mutation, recombination, and fitness-based selection.
In Pareto optimization, recent research has focused on multi-objective selection operators and in particular fitness assignment techniques. Various selection operators are compared in the literature [1,2,3] and according to Van Veldhuizen and Lamont [4], the dominance criterion in combination with niching techniques is one of the most efficient techniques for the fitness assignment. This group of algorithms is referred to by Horn [5] as "Cooperative Population Searches

with Dominance Criterion" and two prominent representatives are SPEA [1] and NSGA-II [6].

While these algorithms perform well in a number of test problems, we observe a stagnation in the convergence of these algorithms at a certain distance from the Pareto front. The distance is dependent on the selection operator as usually these algorithms do not employ any mutation or recombination operators. In this paper we estimate this stagnation distance by deriving an analytical solution for a simplified Pareto front. This raises the question, which selection operators are able to converge to the Pareto front and in addition, which operators converge efficiently?

Two alternatives to the dominance criterion are the Constraint Method-based Evolutionary Algorithm (CMEA) [2] and Subdivision Method (SDM) [7]. These algorithms perform selection by optimizing one objective, while the other objectives are treated as constraints. They are able to converge to the Pareto front for certain test cases.

In conjunction with the selection operator, the mutation and recombination operators are important for an efficient convergence and should adapt while converging towards the Pareto front. In recent years, some efforts have been made in order to apply adaptation in multi-objective optimization. Kursawe [8] and Laumanns *et al.* [9] developed two implementations of self-adaptation [10]. Kursawe performs selection based on a randomly chosen objective. In his work each individual contains a separate vector of design variables and step sizes for each objective (polyploid individuals). Laumanns *et al.* assign a single step size to each individual, which yields an isotropic mutation distributions.

Sbalzarini et al. [11] use a simple self-adaptation scheme for mutating step sizes. Each individual in the population is assigned an individual step size for each design variable. Abbass [12] implemented a Pareto optimization algorithm with recombination and mutation based on the Differential Evolution of Storn and Price [13]. He used self-adaptation in order to find appropriate crossover and mutation rates (probabilities). Büche *et al.* [14] trained self-organizing maps (SOMs) [15] on the currently nondominated solutions, and recombination was performed within the network. The mutation strength was related to the distance between the neurons in the SOM.

Compared to single objective optimization, applying self-adaptive mutation to multi-objective selection operators reveals an additional difficulty. Self-adaptation contains several strategy parameters, which describe the mutation distribution. These parameters benefits from the recombination (interpolation) of several parent solutions. However, in multi-objective optimization, the individuals of the population converges towards different areas of the Pareto front and efficient strategy parameters differ between the individuals. Thus, recombination will also be discussed for the different selection operators.

In the next section, the multi-objective optimization problem is introduced and we briefly outline the basic concepts of multi-objective evolutionary algorithms, which will also be used later on for comparison. Self-adaptation is presented for multi-objective optimization and the key properties of suitable selection and recombination operators are discussed. Different multi-objective algorithms are analyzed in terms of their ability to converge to the Pareto front. In addition, the implementation of self-adaptive mutation into these algorithms is discussed. Finally, the performance of the proposed algorithms is analyzed on a 2- and 3-objective test function. In the performance comparison, the number of resulting nondominated solutions of the different algorithms is allowed to be small, since in real-world applications, analyzing these solutions is often an expensive process.

2 Multi-objective Optimization

2.1 Definition of Multi-objective Optimization

A multi-objective optimization problem can be formulated by an objective vector $\mathbf{f}$ and a corresponding set of design variables $\mathbf{x}$. Without loss of generality minimization of all objectives is considered:

$$\begin{aligned} \text{find } \min \mathbf{f}(\mathbf{x}) &= (f_1(\mathbf{x}), f_2(\mathbf{x}), \ldots, f_m(\mathbf{x})) \in F \\ \text{where } \mathbf{x} &= (x_1, x_2, \ldots, x_n) \in X, \end{aligned} \quad (1)$$

where $X \in \mathcal{R}^n$ is the n-dimensional design space, $F \in \mathcal{R}^m$ is the m-dimensional objective space. A partial ordering can be applied to solutions to the problem by the dominance criterion. A solution a in X is said to dominate a solution b in X ($a \succ b$) if it is superior or equal in all objectives and at least superior in one objective. This is expressed as:

$$\begin{aligned} a \succ b, \text{ if } \forall\, i \in \{1, 2, \ldots, m\}: \quad & f_i(a) \leq f_i(b)\ \wedge \\ \exists\, j \in \{1, 2, \ldots, m\}: \quad & f_j(a) < f_j(b) \end{aligned} \quad (2)$$

The solution a is said to be indifferent to a solution c, if neither solution is dominating the other one. When no a priori preference is defined among the objectives, dominance is the only way to determine, if one solution performs better than the other [16]. The best solutions to a multi-objective problem are the Pareto ideal solutions, which represent the nondominated subset among all feasible solutions. In other words, starting from a Pareto solution, one objective can only be improved at the expense of at least one other objective.

2.2 Self-Adaptation in Multi-objective Optimization

Self-adaptation [10] is associated with mutation or recombination operators and has been mainly used in Evolution Strategies (ES) and Evolutionary Programming (EP) for single objective optimization. In the following we outline the basic principles of self-adapting the mutation distribution of ES as described in [17]. These concepts are subsequently implemented into multi-objective optimization. The optimization of objective functions $\mathbf{f}(\mathbf{x})$, which depend on a set of design variables $\mathbf{x} \in \mathcal{R}^n$ is considered. In ES, mutation is performed by adding a normally distributed random vector to the design variables $\mathbf{x}$:

$$\mathbf{x}' = \mathbf{x} + \mathbf{z}, \quad \mathbf{z} \sim \mathbf{N}(0, \mathbf{C}), \tag{3}$$

where $\mathbf{z}$ is a realization of a normally distributed random vector with zero mean and covariance matrix $\mathbf{C}$. Choosing a constant covariance matrix might be efficient in the beginning of the optimization, but can become inefficient close to the optimum. Adaptation of the mutation distribution has been show to be necessary for efficient optimization algorithms [17].

The elements c_{ij} of the covariance matrix are strategy parameters, which can be build by a set of n standard deviations σ_i and $n(n-1)/2$ rotation angles α_k where:

$$\sigma_i^2 = c_{ii} \tag{4}$$

$$\tan(2\alpha_k) = \frac{2c_{ij}}{\sigma_i^2 - \sigma_j^2}, \quad \text{with } k = \frac{1}{2}(2n-i)(i+1) - 2n + j \tag{5}$$

The mutation of the strategy parameters is performed similarly to the mutation of the design variables by:

$$\sigma_i' = \sigma_i \exp(\tau_0 \mathbf{N}(0,1) + \tau \mathbf{N}_i(0,1)) \tag{6}$$

$$\alpha_k' = \alpha_k + \beta \mathbf{N}_k(0,1), \tag{7}$$

where τ_0, τ and β are the learning rates and recommended values are:

$$\tau_0 = \frac{1}{\sqrt{2n}}, \quad \tau = \frac{1}{\sqrt{2\sqrt{n}}}, \quad \beta = 5° \tag{8}$$

This mutation is referred to as *correlated mutation* or *rotation angle mutation*. Simplifications of the covariance matrix can be obtained in a first step by removing the correlation (i. e., $c_{ij,j\neq i} \equiv 0$) and in a second step by reducing all standard deviations to a single one, i. e., $\sigma_i \equiv \sigma$.

To promote the adaptation of the strategy parameters, the following procedure is recommended [17]:

1. **Non-elitist selection operators should be preferred:**
   Although an elitist strategy (($\mu + \lambda$)-strategy) guarantees a continuous improvement of the objective value of the parent population, it inhibits the risk of getting stuck in parents with inappropriate strategy parameters with a low chance of generating better offspring. Thus, non-elitist (μ, κ, λ)-strategies with a limited lifetime κ are preferred. A typical population size for parents and offspring are $\mu = 15$ and $\lambda = 100$, respectively [17].
2. **Recombination of a parent population is necessary:**.
   A further improvement is obtained by recombination: The design variables are usually recombined by discrete or intermediate recombination of always two parents and the standard deviations are recombined by computing the mean of all parents (global intermediate recombination). No recombination is usually applied to the rotation angles.

2.3 Multi-objective Evolutionary Algorithms

We consider multi-objective evolutionary algorithms, performing a population-based search in order to find a set of approximately Pareto-ideal solutions along the Pareto front. Promising methods have been proposed and evaluated by several researchers [1,4,18]. The various multi-objective evolutionary algorithms mainly employ a selection operator. In the following we classify different approaches as proposed by Horn [5] and discuss the applicability of these algorithms to self-adaptation. In particular, we focus on the ability of an algorithm to converge with arbitrary precision to the Pareto front.

Independent Sampling. An approximation of the Pareto front can be obtained by performing several independent runs with different aggregation of the objectives by e.g. a weighted sum or a constraint approach. This leads to a discrete approximation of the Pareto front, with each optimization run converging to a different point of the Pareto front. Independent sampling is an ideal candidate for self-adaptation as the multi-objective problem is transformed to a set of single objective problems, thus self-adaptation is directly applicable.

Ranjithan et al.[2] proposed to use a constraint method-based evolutionary algorithm (CMEA) for aggregating the objectives. One objective f_h is selected for optimization, while all other objectives $f_{i, i \neq h}$ are treated as constraints:

$$\min f_h, \text{ while } f_i < u_i^t \ \forall \ i = 1, \ldots, m \ ; i \neq h, \tag{9}$$

where u_i^t are the constraint values. For varying the constraint values, different Pareto solutions are obtained.

In order to find appropriate constraint values, the algorithm searches for the extreme corners of the Pareto front by separately optimizing all objectives $f_i, i \neq h$ [5]. Then, for each objective, a certain number of different constraint values u_i^t is chosen uniformly within the extreme corners. For each possible combination of one constraint value per objective, an optimization run is performed and the Pareto front is approximated.

Some knowledge of previous runs can be exploited by using the best solution(s) obtained so far as initial solution(s) for the next run [2].

Cooperative Population Searches with Dominance Criterion. Cooperative population searches converge in a single run towards the Pareto front. The dominance criterion in combination with niching techniques is used in order to select on average the less dominated solutions and preserve diversity in the population, respectively. According to Van Veldhuizen and Lamont [4] this class of fitness assignment is most efficient. Two recent performance comparisons [7] [2] show however that other classes of optimization approaches can also lead to comparable results. One of the most prominent representative is the Strength Pareto Evolutionary Algorithm (SPEA) [1]. SPEA uses the nondominated solutions for the fitness assignment. First, the fitness of each nondominated solution is computed as the fraction of the population, which it dominates. The fitness of a dominated individual is equal to one plus the fitness of each nondominated solution by which it is dominated. This fitness assignment promotes solutions in sparse areas.

Elitism is a key element in this class of algorithms and represents a technique of preserving always the best solutions obtained so far. It is often performed by preserving the nondominated solutions in an archive. In order to preserve diversity in the archive and to keep its size limited, clustering algorithms are applied. SPEA does not employ a mutation or recombination operator. To compare the performance on continuous problems, Zitzler and Thiele [3] use the polynomial distributed mutation and the simulated binary crossover proposed by Deb et al. [19]. Both methods do not implement any adaptation process. So they do not exploit explicitly any knowledge available from the evolution of the population. We discuss these methods in order to analyze their limited convergence towards the Pareto front:

In Fig. 1, a simple example for a 2-objective minimization problem is given with the Pareto front being a straight line between $\mathbf{f} = \{1,0\}$ and $\mathbf{f} = \{0,1\}$. For a limited archive size (or number of parents) s, the objective space cannot be completely dominated. For the ideal case, that all archive solutions are uniformly distributed along the Pareto front, the nondominated area is described by the Pareto front and the dashed lines. This area contains, from the aspect of dominance, indifferent solutions, i. e. solutions of the same quality as the considered archive solutions, and the maximal distance of a nondominated solution to the Pareto front can be calculated as $\frac{1}{\sqrt{2}(s-1)}$. Thus, the minimal number of archive solutions in order to dominate a solution with a certain distance to the Pareto front scales with s^{-1} and in order to converge to the Pareto front, an infinite number of archive solutions is necessary. For 3 objectives, it can be shown that the maximal distance of a nondominated solution scales with s^{-2} and the dominance criterion as selection criterion becomes less efficient. The dominance criterion works well for the approximation of the Pareto front, but fails in the final convergence since the archive size of this class of algorithms is usually limited. This is a general problem of selection operators, using the dominance criterion, e. g., SPEA, SPEA2 [3] and NSGA-II [6], as it will be experimentally shown in Section 3.

Cooperative Population Searches without Dominance Criterion. This class of optimization approaches converges towards the Pareto front in a single optimization run, but does not use the dominance criterion within the selection operator. Moreover, it could be considered as performing several Independent Sampling optimizations within a single optimization. Several selections are performed from one population with different aggregation of the objectives. This might be beneficial compared to Independent Sampling, since information can be exchanged while converging as one individual can be preferred by several different aggregations.

One example is the Subdivision Method (SDM) [7], an optimization approach with some similarities to the CMEA. In the following, the algorithm is briefly

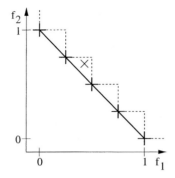

Fig. 1. Disadvantage of the dominance as selection criterion: for a limited number of archive solutions [+ *symbols*], which may be on the Pareto front [*solid line*], the objective space cannot be completely dominated, e.g. the solution [*x symbol*] is not dominated, even though it is still far from the Pareto front

described and an example for a 2-objective minimization problem is given in Fig. 2. In the objective space, the SDM performs several local (μ, κ, λ) selections and then unifies all selected solutions to the parent population. Similar to the CMEA one objective f_h is selected for optimization, while all other objectives $f_{i,i \neq h}$ are treated as constraints. However, a lower and upper constraint value is set:

$$\min f_h, \quad \text{while } l_i^t \leq f_i \leq u_i^t, \ \forall \ i = 1, \ldots, m \ ; i \neq h. \tag{10}$$

The constraints l_i^t and u_i^t are set such that they divide the objective axis of f_i in k intervals $t = 1, \ldots, k$ of equal width, where the upper constraint u_i^t of an interval t is always the lower constraint l_i^{t+1} of the adjacent interval $t+1$, i. e., $l_i^{t+1} = u_i^t$. The lower constraint value l_i^1 of the first interval and the upper value of the kth interval u_i^k are set equal to the minimal and maximal value for objective i of the current nondominated front, respectively. Thus, the constraints change along the optimization process as the current nondominated front changes. For each possible combination of choosing one interval t for each of the objectives $f_{i,i\neq h}$, a separate selection is performed with respect to f_h, where the constraints are hard, i. e., a solution which violates the constraints is not considered. Then, this process is repeated until each objective is chosen once as a selection criterion, in order to avoid a preference between the objectives. In total $m \cdot k^{m-1}$ local selections are performed.

Self-adaptation can be implemented in this selection operator by the following procedure: The selection process can be considered as performing several local selections in a "subdivided" objective space. The mean distance of the selected individuals to the Pareto front may differ between the local selections, resulting in different sets of efficient strategy parameters. Thus, recombination as described in Sec. 2.2 is always performed within a local selection. Finally, self-adaptive mutation is applied to the recombined individuals.

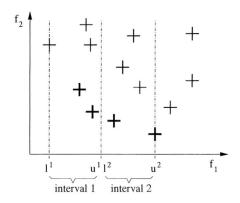

Fig. 2. Selection by SDM for 2 objectives: The objective space is divided along f_1 into two intervals by specifying a hard lower and upper constraint value l and u for f_1, respectively [dash-dotted lines]. From all solutions [+ symbols] in an interval, always the μ best solutions [bold + symbols] with respect to f_2 are selected. Then the procedure is repeated by dividing the space along f_2 and considering f_1 as selection criterion

3 Experimental Analysis

A simple test function for an arbitrary number of objectives is considered, which is a generalization of the sphere model to multiple objectives [9]. It allows to analyze the convergence of optimization algorithms as a function of the number of objectives:

$$f_i = (x_i - 1)^2 + \sum_{j=1, j \neq i}^{n} x_j^2, \quad i = 1, \ldots, m \ \wedge \ m \leq n \tag{11}$$

with $x_{1,\ldots,n} \in [-2.0; 2.0]$, i is the index of the objective and the number of variables is set to $n = 10$. For two objectives the Pareto front is given by:

$$x_1 + x_2 = 1, \ x_{3,\ldots,n} = 0 \wedge x_{1,2} \geq 0, \tag{12}$$

and for 3 objectives by:

$$x_1 + x_2 + x_3 = 1, \ x_{4,\ldots,n} = 0 \wedge x_{1,2,3} \geq 0. \tag{13}$$

In the design space, the Pareto fronts of the 2- and 3-objective problem describes a straight line or a plane, respectively.

3.1 Performance Measures

The definition of the quality of the results from a Pareto optimization considers two aspects. The first aspect is the convergence of the solutions towards the Pareto front, and will be addressed by the mean distance D of solutions to the Pareto front. The second aspect reflects the distribution of the solutions, whereas a uniform distribution along the Pareto front is desired. This will be addressed by plotting the objective space.

3.2 Results

In the following we analyze the performance of the 3 different classes of multi-objective algorithms. For simplicity, just one representative of each class is considered in order to focus on the general convergence properties and not on the exact convergence speed.

Independent Sampling. The CMEA is analyzed as a representative of Independent Sampling on a 2-objective sphere. In total 6 independent optimization runs are performed with a $(15, 3, 100)$ strategy, implementing correlated mutation. All initial step sizes are set to 0.1 and 17.000 solutions are evaluated for each run, leading in total to about 100.000 evaluated solutions.

First, 2 single objective optimization runs are started for f_1 and f_2 from random initial solutions. Then, 4 constraint optimization runs are performed, where f_2 is optimized and f_1 is treated as constraint. Since the f_1 value of the best solution of the two single objective runs is about 0 and 2, respectively, the constraints for f_1 for the 4 remaining runs are set to 0.4, 0.8, 1.2, and 1.6.

For the constraint optimization runs, some knowledge of previous runs is exploited by using the best solutions obtained so far as initial solutions for the next run [2]. We consider a hard constraint: If solutions violate the constraint, they are not considered for selection. This constraint ensures a convergence to a point on the Pareto front. A soft penalty with a gradient smaller than the gradient of f_2 may converge to a point in the neighborhood of the Pareto front. Fig. 3a shows the performance measure for all 6 optimization runs, which represents the mean distance of the parent population at each generation to the Pareto front. The figure shows large differences in the convergence between the different runs. While the two single objective optimizations converge linearly, the convergence of the constraint optimizations slows down at a distance of about $D = 0.01$ to the Pareto front. Fig. 4 addresses this aspect and shows the contour lines for f_1 and f_2 and the Pareto front in the (x_1, x_2) space. Each contour line of f_1 represents a different constraint setting. The optimum for each constraint optimization is located in the intersection of a f_1 contour line with the Pareto front. In the vicinity of the optimum, the topology is badly scaled and oriented such that it is difficult to optimize it with the given self-adaptation scheme (compare Hansen et al. [20], Fig. 2).

Fig. 7a shows the best solution of each optimization run and the Pareto front. While the solutions are equally spaced in f_1 direction, the distribution is nonuniform along the Pareto front, with a concentration on the part of the Pareto front, which is almost parallel to the f_1 axis. This results from treating f_1 and f_2 differently as constraint and objective, respectively.

Cooperative Population Searches with Dominance Criterion. For SPEA, different archive sizes of 30, 100, and 300 are analyzed for the 2- and 3-objective sphere. Here, the focus is on the maximal convergence of a limited archive size to the Pareto front and not on the actual convergence speed of

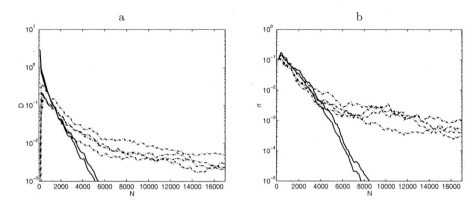

Fig. 3. Mean distance D of the parent population (a) and mean standard deviation σ of the mutation distribution (b) for single objective optimizations [*solid lines*] and constraint optimizations [*dash-dotted lines*] for CMEA

SPEA. The number of parents and offspring is set equal to the archive size and in total 100.000 solutions are evaluated. Discrete and intermediate recombination of 2 parents is considered with 33% probability each. In order to analyze just the effect of the archive, a normally distributed mutation with zero mean and standard deviation of 0.0005 and 0.002 is used with a mutation probability of 15% per variable, which lead to the best results.

Fig. 5 shows the convergence measure D, which represents the mean distance of the solutions in the archive to the Pareto front. It can clearly be seen that the convergence stagnates at a certain value of D and this value is decreasing with an increase of the archive size. Comparing the 2- and 3-objective optimization results, a second observation can be made. The final mean distance D for the 3-objective problem is significantly larger than for the 2-objective problem and the increase in the archive size leads to a smaller relative decrease in D. This underlines the previously stated rule that for a limited archive size an evolutionary algorithm, which considers the dominance criterion for selection in a similar way than SPEA, cannot converge to the Pareto front. In addition, in order to obtain the same mean distance D from the Pareto front, the necessary number of archive solutions rises with the number of objectives. The distribution of the solutions of the final archive is shown for the 2-objective problem with an archive size of 300 in Fig. 7b and shows the uniform approximation of the Pareto front with a large number of archive solutions.

Cooperative Population Searches without Dominance Criterion. The SDM is now analyzed on the multi-objective sphere. Always 8 solutions are selected in each local selection and a maximal lifetime $\kappa = 3$ is assigned to each solution. The convergence measure D for the SDM is set to the mean distance of all selected parents of a generation to the Pareto front. For the 2-objective problem, 3 intervals are chosen along each objective axis, leading in total to 6 lo-

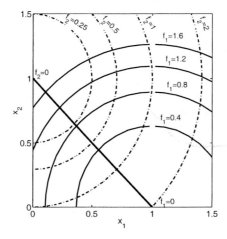

Fig. 4. Convergence difficulty in the CMEA for optimizing f_2 and setting f_1 as constraint. The optimum for a specific constraint value for f_1 is located at the intersection of the Pareto front [*bold solid line*], with the contour line of f_1 [*solid line*]. In the vicinity of the optimum, the topology is badly scaled and oriented as shown above

cal selections and a maximal number of 48 parents. For the 3-objective problem, each objective axis is divided into 2 intervals, leading to a total number of 12 local selections and a maximal number of 96 parents. The bounds of the intervals are obtained from the current nondominated front of the optimization run and thus need no user specification. For self-adaptation, a ratio of 7 offspring per parent is recommended, leading to a total number of 336 and 772 offspring for the 2- and 3-objective problem, respectively. Similar to CMEA, a global intermediate recombination of the parents from one local selection is performed for the variables and step sizes. No recombination is applied to the rotation angles. In total 100.000 solutions are evaluated and the convergence measure is plotted in Fig. 6. In addition, the mean step size of the mutation distribution is given. The convergence speed decreases at a value of about 10^{-2} for the same reason as for the CMEA: The convergence becomes difficult due to the constraint optimization. Here, the convergence speed for the 3-objective problem is about a factor of 2 smaller than for the 2-objective problem. This results mainly due to the doubled number of local selections. The parents of the final population are plotted in Fig. 7c and are uniformly distributed along the Pareto front.

4 Conclusions

Evolutionary Algorithms for multi-objective optimization should implement efficient techniques in order to improve convergence towards the Pareto front, while maintaining diversity spanning the front. We study three different classes of multi-objective algorithms and compare a representative of each class with

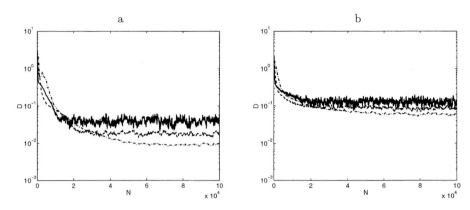

Fig. 5. Mean distance D of the archive solutions of SPEA for the 2-objective (a) and 3-objective (b) sphere problem. The archive sizes is varied between 30 [*solid line*], 100 [*dashed line*] and 300 [*dash-dotted line*]

each other. The question is which of these algorithms is able to converge to the Pareto front with an arbitrary precision.

We found that cooperative population searches like SPEA, which use the dominance criterion in the fitness assignment, cannot approximate the Pareto front with arbitrary precision. For a 2-objective optimization problem, the necessary number of archive solutions to assign fitness by dominance scales inversely with the distance of solutions to the Pareto front. For 3 objectives, convergence becomes even more difficult, since the necessary number of archive solutions scales inversely with the square of the distance. This result holds for other algorithms using the dominance criterion and a limited population (e. g. NSGA-II, SPEA2).

The algorithms CMEA and SDM do not use dominance. In these algorithms, one objective is selected for optimization, while the other objectives are treated as constraints. Both algorithms converge to a fixed number of discrete points on the Pareto font, which can be reached in arbitrary precision. For the considered test functions, this number is significantly smaller than the number of nondominated solutions of SPEA. However, the limited number of converged solutions of CMEA and SDM is often sufficient in real-world applications, especially if the analysis of these solutions is expensive. Thus, algorithms like CMEA and SDM are interesting alternatives to the well established algorithms based on the dominance criterion.

CMEA finds one optimal point in each optimization run and thus needs to be run for several times in order to find an approximation of the Pareto front. In contrast, SDM finds an approximate Pareto front in a single optimization run. It is shown that self-adaptation can easily be applied to CMEA and SDM and that both algorithms converge successfully to the Pareto front and outperform SPEA in terms of the final distance to the Pareto front. Comparing CMEA and SDM in terms of convergence speed, CMEA is faster on the considered test function,

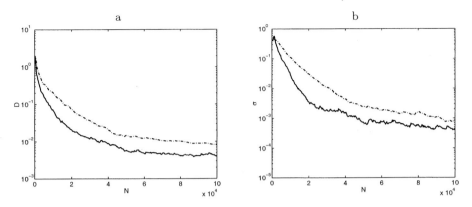

Fig. 6. Convergence of the SDM for the 2-objective [*solid line*] and 3-objective [*dash-dotted line*] sphere problem. The mean distance of the parents to the Pareto front is given (a) as well as the mean standard deviation for the mutation (b)

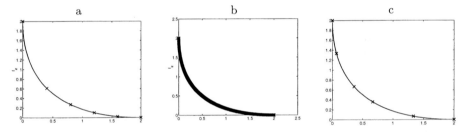

Fig. 7. Location of the best solution from each independent run of CMEA (a), the final archive of SPEA (b) and final parent population of SDM (c) [x] on the Pareto front [*bold line*]

although it is not known if this generalizes to other functions and to other adaptation schemes. For CMEA, one has to decide before optimization, which objective is optimized subject to the other objectives, which are then treated as constraints. This is in contrast to SDM that gives no a-priori preference to any of the objectives.

Some difficulty in converging has been found for CMEA and SDM. CMEA and SDM transfer the multi-objective sphere problem into a difficult constraint optimization problem: While optimizing a single objective of the sphere problem converges linearly (see Fig. 3a, solid lines), the convergence speed of the constraint optimization problem decreases over the number of function evaluations (see Fig. 3a, dash-dotted lines). In general the question arises if objectives could be aggregated by a different method, leading to a linear convergence also in the constraint case. In addition, Hansen *et al.* [20] stated some difficulty of self-adaptation in generating arbitrary correlation distributions and thus methods like the Covariance Matrix Adaptation [20] might perform better on the constraint problem.

Acknowledgments. We would like to acknowledge support from the Commission for Technology and Innovation, Switzerland (Project 4817.1) and Alstom Power Technology Center Dättwil, Switzerland.

References

1. Zitzler, E., Thiele, L.: Multiobjective evolutionary algorithms: A comparative case study and the strength Pareto approach. IEEE Transactions on Evolutionary Computation **3** (1999) 257–271
2. Ranjithan, S.R., Chetan, S.K., Dakshima, H.K.: Constraint Method-Based Evolutionary Algorithm (CMEA) for Multiobjective Optimization. In Zitzler, E., Deb, K., Thiele, L., Coello Coello, C.A., Corne, D., eds.: First International Conference on Evolutionary Multi-Criterion Optimization, Springer-Verlag. Lecture Notes in Computer Science No. 1993 (2001) 299–313
3. Zitzler, E., Laumanns, M., Thiele, L.: SPEA2: Improving the Strength Pareto Evolutionary Algorithm. Technical Report 103, Computer Engineering and Networks Laboratory (TIK), Swiss Federal Institute of Technology (ETH) Zurich, Gloriastrasse 35, CH-8092 Zurich, Switzerland (2001)
4. Van Veldhuizen, D.A., Lamont, G.B.: Multiobjective Evolutionary Algorithms: Analyzing the State-of-the-Art. Evolutionary Computation **8** (2000) 125–147
5. Horn, J.: Multicriterion decision making. In Bäck, T., Fogel, D.B., Michalewicz, Z., eds.: Handbook of Evolutionary Computation, Sec. F1.9: pp. 1–15. (1997)
6. Deb, K., Agrawal, S., Pratab, A., Meyarivan, T.: A Fast Elitist Non-Dominated Sorting Genetic Algorithm for Multi-Objective Optimization: NSGA-II. In Schoenauer, M., Deb, K., Rudolph, G., Yao, X., Lutton, E., Merelo, J.J., Schwefel, H.P., eds.: Proceedings of the Parallel Problem Solving from Nature VI Conference, Paris, France, Springer. Lecture Notes in Computer Science No. 1917 (2000) 849–858
7. Büche, D., Dornberger, R.: New evolutionary algorithm for multi-objective optimization and the application to engineering design problems. In: Proceedings of the Fourth World Congress of Structural and Multidisciplinary Optimization, Dalian, China, International Society of Structural and Multidisciplinary Optimization (ISSMO) (2001)
8. Kursawe, F.: A Variant of Evolution Strategies for Vector Optimization. In Schwefel, H.P., Männer, R., eds.: Parallel Problem Solving from Nature. 1st Workshop, PPSN I. Volume 496 of Lecture Notes in Computer Science., Berlin, Germany, Springer-Verlag (1991) 193–197
9. Laumanns, M., Rudolph, G., Schwefel, H.P.: Mutation Control and Convergence in Evolutionary Multi-Objective Optimization. In: Proceedings of the 7th International Mendel Conference on Soft Computing (MENDEL 2001), Brno, Czech Republic (2001)
10. Schwefel, H.P.: Numerical Optimization of Computer Models. Wiley, Chichester (1981)
11. Sbalzarini, I.F., Müller, S., Koumoutsakos, P.: Microchannel optimization using multiobjective evolution strategies. In Zitzler, E., Deb, K., Thiele, L., Coello Coello, C.A., Corne, D., eds.: Proceedings of the First International Conference on Evolutionary Multi-Criterion Optimization (EMO), Zürich, Switzerland, Springer Lecture Notes in Computer Science (2001) 516–530

12. Abbass, H.A.: The Self-Adaptive Pareto Differential Evolution Algorithm. In: Congress on Evolutionary Computation (CEC'2002). Volume 1., Piscataway, New Jersey, IEEE Service Center (2002) 831–836
13. Storn, R., Price, K.: Minimizing the real functions of the icec'96 contest by differential evolution. In: IEEE Conference on Evolutionary Computation. (1996) 842–844
14. Büche, D., Milano, M., Koumoutsakos, P.: Self-Organizing Maps for Multi-Objective Optimization. In Barry, A.M., ed.: GECCO 2002: Proceedings of the Bird of a Feather Workshops, Genetic and Evolutionary Computation Conference, New York, AAAI (2002) 152–155
15. Kohonen, T.: Self-Organizing Maps. 3 edn. Springer series in information sciences (2001)
16. Fonseca, M.C., Fleming, P.J.: Multi-objective genetic algorithms made easy: Selection, sharing and mating restrictions. In: Proceedings of the 1st International Conference on Genetic Algorithms in Engineering Systems: Innovations and Application. (1995) 45–52
17. Bäck, T.: Self-adaptation. In Bäck, T., Fogel, D.B., Michalewicz, Z., eds.: Handbook of Evolutionary Computation, Sec. C7.1: pp. 1–15. (1997)
18. Coello Coello, C.A.: An updated survey of evolutionary multiobjective optimization techniques: State of the art and future trends. In: Congress on Evolutionary Computation. (1999) 3–13
19. Deb, K., Agrawal, R.B.: Simulated binary crossover for continuous search space. Complex Systems **9** (1995) 115–148
20. Hansen, N., Ostermeier, A.: Completely derandomized self-adaptation in evolution strategies. Evolutionary Computation **9** (2001) 159–195

MOPED: A Multi-objective Parzen-Based Estimation of Distribution Algorithm for Continuous Problems

Mario Costa[1] and Edmondo Minisci[2]

[1] Department of Electronics, Polytechnic of Turin, C.so Duca degli Abruzzi, 24, 10129 Turin, ITALY
mario.costa@polito.it
[2] Department of Aerospace Engineering, Polytechnic of Turin, C.so Duca degli Abruzzi, 24, 10129, Turin, ITALY
edmondo_minisci@yahoo.it

Abstract. An evolutionary multi-objective optimization tool based on an estimation of distribution algorithm is proposed. The algorithm uses the ranking method of non-dominated sorting genetic algorithm-II and the Parzen estimator to approximate the probability density of solutions lying on the Pareto front. The proposed algorithm has been applied to different types of test case problems and results show good performance of the overall optimization procedure in terms of the number of function evaluations. An alternative spreading technique that uses the Parzen estimator in the objective function space is proposed as well. When this technique is used, achieved results appear to be qualitatively equivalent to those previously obtained by adopting the crowding distance described in non-dominated sorting genetic algorithm-II.

1 Introduction

The extensive use of evolutionary algorithms in the last decade demonstrated that an optimization process can be obtained by combining effects of interactive operators such as selection - whose task is mainly to identify the best individuals in the current population - and crossover and mutation, which try to generate new and better solutions starting from the selected ones. But, if the mimicking of natural evolution in living species has been a source of inspiration of new strategies, the attempt to copy natural techniques as they are sometimes introduces a great complexity without a corresponding improvement of algorithms performance. Moreover standard evolutionary algorithms can be ineffective when problems exhibit a high level of interaction among variables. This is mainly due to the fact that recombination operators are likely to disrupt promising sub-structures of optimal solutions.

Alternatively, in order to make a rational use of the evolutionary metaphor and/or to create optimization tools that are able to handle very hard problems (with several parameters, with difficulties in linkage learning, deceptive), some algorithms have been proposed that automatically learn the structure of the search space. Following this way, several works, based on explicit probabilistic-statistic tools, have been carried out.

Generally, these methods, starting from results of current populations, try to identify a probabilistic model of the search space, and crossover and mutation operators are replaced with sampling. Those methods have been named Estimation of Distribution Algorithms (EDAs).

Most EDAs have been developed to manage optimization processes for mono-objective, combinatorial problems, but several works regarding problems in continuous domains have been proposed.

We can distinguish three types of EDAs depending on the way the probabilistic model is built: a) without dependences among variables ([1] - [4]); with bivariate dependences among variables ([5] - [7]); c) with multivariate dependences ([8] - [10]).

Recently, EDAs handling multi-objective optimizations have been proposed. References [11] and [12] respectively extend the mono-objective version in [10] and [8]. They describe the algorithms and present some results when applied to well known test problems.

In this paper we propose a multi-objective optimization algorithm for continuous problems that uses the Parzen method to build a probabilistic representation of Pareto solutions, with multivariate dependences among variables.

Notwithstanding main features can be used for single-objective optimization, because of future practical applications in our case the algorithm has been directly developed for multi-objective problems. Similarly to what was done in [12] for multi-objective Bayesian Optimization Algorithm (BOA), the already known and implemented techniques of Non Dominated Sorting Genetic Algorithm II (NSGA-II) [13] are used to classify promising solutions, while new individuals are obtained by sampling from the Parzen model.

The Parzen method, as introduced in the next section, can appear analogous to the normal kernel method described and used in [10]. Actually, the two methods are different and, even if both put kernels on each sampled point, our method uses classical Parzen dictates to set terms of the covariance matrix (non-diagonal) of kernels in order to directly approximate the joint Probability Density Function (PDF).

A brief introduction on the general problem of building probabilistic models is followed by a description of the main characteristics of the Parzen method. In section 3 the structure of the algorithm and the practical implementation are discussed; results of application to test cases are detailed and an alternative spreading technique is briefly described. A final section of concluding remarks summarizes the present work and indicates future developments.

2 Parzen Method

When dealing with continuous-valued random variables, most statistical inferences rely on the estimation of PDFs and/or associated functionals from a finite-sized sample. Whenever something is known in advance about the PDF to be estimated, it is worth exploiting that knowledge as much as we can in order to shape a special-purpose estimator. In fact any additional information we are able to implement in the estimator as a built-in feature is equivalent to some effective increase in the sample size. Otherwise stated, in so doing we improve the estimator's efficiency.

In the statistician's wildest dream some prime principles emerge and dictate that the true PDF must belong to a certain parametric family of model PDFs. This restricts the set of admissible solutions to a finite-dimensional space, and cuts the problem down to the identification of the parameters thereby introduced. In fact parametric estimation is so appealing that few popular families of model PDFs are applied almost everywhere even in lack of any guiding principle, and often little effort is made to check their actual faithfulness. On the other hand, a serious check has to rely on composite hypothesis tests that are like to be computationally very expensive.

While designing an EDA for general-purpose multi-objective optimization there is really no hint on how the true PDF should look like. For instance, that PDF could well have several modes, whereas most popular models are uni-modal. The possible occurrence of multiple modes is usually handled through *mixtures* of uni-modal kernel PDFs. Since the "correct" number of kernels is not known in advance, the size of the mixture is optimized (e.g. by data clustering) just like any other parameter: that is, the weight and the inner parameters of each kernel.

The usage of mixtures does however not alleviate us from worrying about faithfulness. Otherwise stated, the choice of the parametric family the kernels belong to still matters. In fact the overall number of parameters (and therefore the number of kernels) must grow sub-linearly with the sample size n, or else the variance of the resulting estimator would not vanish everywhere as $n \to \infty$, thus precluding ubiquitous converge to the true PDF in the mean square sense. But if that condition is met, then even a single "wrong" kernel can spoil convergence wherever it injects some bias. This is nothing but another form of the well-known bias-variance dilemma.

The Parzen method [14] pursues a non-parametric approach to kernel density estimation. It gives rise to an estimator that converges everywhere to the true PDF in the mean square sense. Should the true PDF be uniformly continuous, the Parzen estimator can also be made uniformly consistent. In short, the method allocates exactly n identical kernels, each one "centered" on a different element of the sample. In contrast with parametric mixtures, here no experimental evidence is spent to identify parameters. This is the reason why the presence of so many kernels does not inflate the asymptotic variance of the estimator. As a consequence, the detailed shape of the kernels is irrelevant, and the faithfulness problem is successfully circumvented. Of course some restrictions are in order: here is a brief explanation.

Let z be a real-valued random variable. Let $p^z(\cdot):\Re \to \Re_+ \cup \{0\}$ be the associated PDF. Let $D_n = \{z_1,...z_n\}$ be a collection of n independent replicas of z. The empirical estimator $\hat{p}_n^E(\cdot)$ of $p^z(\cdot)$ based on D_n is defined as follows:

$$\forall z \in \Re \quad \hat{p}_n^E(z) = \frac{1}{n}\sum_{i=1}^{n} \delta(z - z_i) \cdot \tag{1}$$

The estimator just defined is unbiased everywhere but it converges nowhere to $p^z(\cdot)$ in the mean square sense because $Var[\hat{p}_n^E(z)] = \infty$ irrespective of both n and z. This last result is not surprising, since the Dirac's delta is not squared integrable.

The Parzen estimator $\hat{p}_n^S(\cdot)$ of $p^z(\cdot)$ based on D_n is obtained by convolving the empirical estimator with some squared integrable kernel PDF $g_s(\cdot)$:

$$\forall z \in \Re \quad \hat{p}_n^S(z) = \int_{-\infty}^{\infty} \hat{p}_n^E(x) \frac{1}{h_n} g_S\left(\frac{z-x}{h_n}\right) dx = \frac{1}{n} \sum_{i=1}^{n} \frac{1}{h_n} g_S\left(\frac{z-z_i}{h_n}\right). \quad (2)$$

The kernel acts as a low-pass filter whose "bandwidth" is regulated by the scale factor $h_n \in \Re_+$. It exerts a "smoothing" action that lowers the sensitivity of $\hat{p}_n^S(z)$ w.r.t. D_n so as to make $Var[\hat{p}_n^S(z)] < \infty \; \forall z \in \Re$. Thus for any given sample size the larger is the scale factor, the smaller is the variance of the estimator. But the converse is also true: since $\hat{p}_n^S(z)$ is nothing but a mean, then for any given scale factor the larger is the sample size, the smaller is the variance of the estimator (indeed it is inversely proportional to the sample size). Both statements are in fact special cases of the following property:

$$\forall z \in \Re \quad \lim_{n \to \infty} nd_n = \infty \Rightarrow \lim_{n \to \infty} Var[\hat{p}_n^S(z)] = 0. \quad (3)$$

On the other hand, the same smoothing action produces an unwanted "blurring" effect that limits the resolution of the approximation. Intuitively the scale factor should therefore vanish as $n \to \infty$ in order to let the estimator closely follow finer and finer details of the true PDF. Also this last remark finds a precise mathematical rendering in the following property:

$$\forall z \in \Re \quad \lim_{n \to \infty} d_n = \infty \Rightarrow \lim_{n \to \infty} E[\hat{p}_n^S(z)] = p^z(z). \quad (4)$$

To summarize, the conflicting constraints dictated by the bias-variance dilemma can still be jointly satisfied by letting the scale factor decrease slowly enough as the sample size grows. The resulting estimator converges everywhere to the true PDF in the mean square sense irrespective of the kernel employed, provided that it is squared integrable.

The above results were later extended to the multi-variate case by Cacoullos [15].

3 Parzen EDA

The main idea of the work is the use of the Parzen method to build a probabilistic model and to sample from the estimated PDF in order to obtain new promising solutions. A detailed description of the Multi-Objective Parzen EDa (MOPED algorithm) follows, and some results are presented in order to show capabilities and potentialities of the algorithm.

Moreover, an extensive use of the Parzen method could lead to simplify the overall optimization procedure towards a parameter-less tool. As a first step in this direction, at the end of section we introduce a different spreading technique for solutions in the Pareto front.

3.1 General Algorithm

As summarized in figure 1, the general optimization procedure can be described as follows:
1. Starting: N_{ind} individuals are sampled from a uniform m-dimensional PDF.
2. Classification & Fitness evaluation: by using NSGA-II techniques [13], individuals of current population are ranked and ordered in terms of dominance criterion and crowding distance in the objective function. A fitness value, linearly varying from $2-\alpha$ (best individual) to α (worst individual), with $0 < \alpha < 1$, is assigned to each individual.

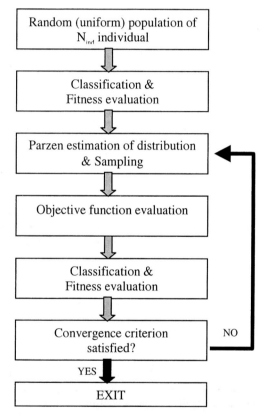

Fig. 1. General structure of the algorithm

3. Building model & sampling: on the basis of information given by N_{ind} individuals, by means of the Parzen method a probabilistic model of promising search space portion is built. For generic processes can be useful adopting different kernels alternatively from a generation to the other in order to obtain an effective exploration. In this work Gauss and Cauchy distributions are used. Actually, these types of kernel, for their intrinsic characteristics, are complementary and results will show that the use of only one of them could be inefficient for some problems.

From the probabilistic model so determined, τN_{ind} new individuals are sampled. Fitness values are used to calculate variance of kernels (the fitness values are related to the scale factors introduced in section 2) and to favor sampling from most important kernels.
4. Evaluation: New τN_{ind} individuals are evaluated in terms of objective functions.
5. Classification & Fitness evaluation: following NSGA-II criteria, individuals of intermediate population, of which dimension is $(1+\tau) N_{ind}$, are ordered. A fitness value, linearly varying from $2-\alpha$ (best individual) to α (worst individual), with $0 < \alpha < 1$, is assigned to each individual.
6. New population: best N_{ind} individuals are selected to be next generation.
7. EXIT or NEXT ITER: if convergence criteria are achieved the algorithm stops, otherwise it restarts from point 3.

The algorithm presented above demonstrated satisfactory performance in solving several test cases, when performance is measured in terms of objective function evaluations to obtain a good approximation of the Pareto front. Some results will be shown in the next paragraph.

The still open question is finding an efficient convergence criterion that could be adopted for a generic optimization. That is finding a convergence criterion that guaranties an optimal approximation of Pareto front (efficacy) and requires a number of objective function evaluations as low as possible (efficiency).

Following results show that neither the maximum generation number nor all of the individuals in first class are without gaps. The former because of an extremely low efficiency if a too high maximum number of generations is used, the latter because of premature convergence on a local, non-optimal, front.

Consequently, the maximum generation number is always used. An upper limit for iteration is imposed as suggested from literature results, even if this kind of stopping criterion makes the algorithm inefficient.

3.2 Test Cases Results

In order to have some ideas regarding effectiveness and efficiency of the method, the proposed algorithm has been applied to some well-known test problems taken from literature [16].

For all of test cases 10 independent runs have been performed and results in terms of number of function evaluations are given as average values.

As said in the previous description of the algorithm, in absence of an effective and efficient criterion a maximum number generation criterion has been adopted. In order to allow comparison with obtained results in literature, our results are presented in terms of effective number of iteration, or better, in terms of number of functions evaluations required to obtain the approximation of the optimal front as well. Mainly, results will be compared with those achieved in [11] by using the Multi-objective Mixture-based Iterated Density Estimation Evolutionary Algorithm (**MIDEA**).

All tests have been run with the same values of the following parameters: a) the number of individuals (N_{ind} = 100), the sampling parameter ($\tau = 2$), and the fitness parameter ($\alpha = 0.2$).

For the sake of simplicity, considered problems are reported in Table 1, which for each problem shows type of problem (minimization or maximization) and number of parameters, objective functions and limits of parameters.

Table 1. Test cases problems

Type & m		Objective functions	Limits
min, $m = 3$	MOP2	$f_1(x) = 1 - \exp\left(-\sum_{i=1}^{m}\left(x_i - \frac{1}{\sqrt{m}}\right)^2\right)$ $f_2(x) = 1 - \exp\left(-\sum_{i=1}^{m}\left(x_i + \frac{1}{\sqrt{m}}\right)^2\right)$	$-4 \leq x_i \leq 4$, $i = 1, ..., m$
min, $m = 3$	MOP4	$f_1(x) = \sum_{i=1}^{m-1}\left(-10\exp\left(-0.2\sqrt{x_i^2 + x_{i+1}^2}\right)\right)$ $f_2(x) = \sum_{i=1}^{m}\left(|x_i|^{0.8} + 5\sin(x_i)^3\right)$	$-5 \leq x_i \leq 5$, $i = 1, ..., m$
min, $m = 10$	EC4	$f_1(x) = x_1$ $f_2(x) = g\left(1 - \sqrt{\frac{x_1}{g}}\right)$ $g(x) = 91 + \sum_{i=2}^{m}\left(x_i^2 - 10\cos(4\pi x_i)\right)$	$0 \leq x_1 \leq 1$; $-5 \leq x_i \leq 5$, $i = 2, ..., m$
min, $m = 10$	EC6	$f_1(x) = 1 - \exp(-4x_1)\sin(6\pi x_1)^6$ $f_2(x) = g\left(1 - \left(\frac{f_1}{g}\right)^2\right)$ $g(x) = 1 + 9\left(\sum_{i=2}^{m}\frac{x_i}{9}\right)^{0.25}$	$0 \leq x_i \leq 1$, $i = 1, ..., m$

Figure 2 shows one of the fronts obtained for the MOP2 problem. The process stops after 15 iterations, that means 3,100 function evaluations, but from graphical results displayed during the run, it is possible to see that solutions remain the same after 9.75 (average value) iterations, or 2,062 function evaluations. For this problem MIDEA gives an approximation of the optimal front after 3,754 function evaluations.

In figure 3 one of the fronts obtained for the MOP4 problem is shown in the upper left corner. The other three parts of the figure represent the marginal bivariate PDFs of variables when normal kernels are used. The triple structure of the approximated front can be identified from every marginal PDF, even if for this run it is more evident in the $x_1 - x_2$ PDF.

For this problem a maximum number of iterations is set equal to 55 (11,100 function evaluations), but still in this case in order to have a stable configuration of the solutions a less number of iterations is needed, which is 44.9 (9,090 function evaluations). In [11] **MIDEA** requires 10,762 function evaluations in the best case when objective clustering is adopted.

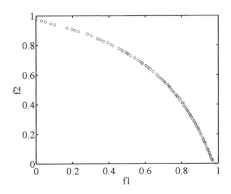

Fig. 2. Obtained non-dominated solutions on MOP2 problem

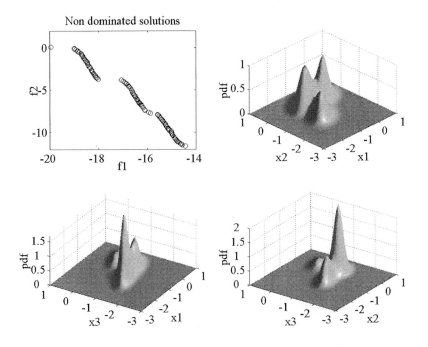

Fig. 3. MOP4 problem. In the left upper corner non-dominated solutions in the objective plain are shown. The other three parts of the figure show the marginal bivariate PDFs of problem's variables when normal kernels are used

Problems EC4 and EC6 are more complex and a presentation of relative results allows a deeper discussion of advantages and gaps of the proposed algorithm.

EC6 is presented as a problem that tests the ability of algorithms to spread solutions on the whole front. MOPED demonstrates to be able to cover the entire optimal range, even if most of runs produce one or two sub-optimal solutions on the left part of the Pareto front. What happens is similar to the results of the Strength Pareto Evolutionary Algorithm (SPEA) when applied to the same problem as reported in [13].

For both EC4 and EC6 we know that achievement of optimal front corresponds to $g(x)=1$. Therefore, for these problems we adopted the following exit criterion: when the $g(x)$ value averaged on the whole population is ≤ 1.01, this allows to have an error less that 1%.

For EC6 problem we imposed 15,000 maximum function evaluations, but the convergence criterion related to $g(x)$ function has been reached after approximately 8,300 evaluations. For this problem **MIDEA** needs 8,426 function evaluations to obtain the front with no clustering, but better results (2,284 evaluations) can be obtained if conditionals dependencies among variables are not learned.

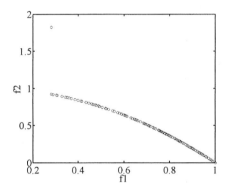

Fig. 4. Obtained non-dominated solutions on EC6 problem. Most of obtained fronts display some sub-optimal solutions

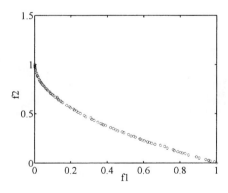

Fig. 5. Obtained non-dominated solutions on EC4 problem

Problem EC4 is the hardest in terms of number of function evaluations needed to reach the true optimal front. Because of the form of the functions to optimize, process tends to get stuck on sub-optimal fronts.

Results demonstrate that the optimal solution (figure 5 shows one of the fronts) can be obtained after 153,710 function evaluations, with a minimum value of 66,100 in one of the ten runs, and a maximum of 244,100 , when the upper limit of function evaluations is set to 300,000. In this case **MIDEA** needs at the best 1,019,330 function evaluations when objective clustering is used. For this problem ignoring conditional dependences among variables can be useful for **MIDEA** as well; with a mixture of univariate factorizations caching the front needs only 209,635 evaluations.

For EC4 problem NSGA-II acts much better than both the algorithms (MOPED and **MIDEA**) [13]. It is able to bring the population to the front with less then 25,000 evaluations if an ad hoc parameter setting is used.

From figure 6, which shows a general trend of $g(x)$ function, it is possible to see how the process goes on. Ranges with null slope mean a transitorily convergence on a local Pareto-optimal front. In order to allow some comparison we monitored the $g(x)$ function and it reaches the value of 3 after 40,862 objective function evaluations.

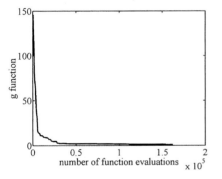 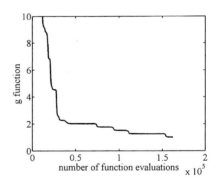

Fig. 6. General trend of $g(x)$ function for EC4 problem. On the left the trend during the whole process is shown. On the right side, last part of the process allows a deeper comprehension of difficulties in terms of function evaluation to jump from a local front to a better one

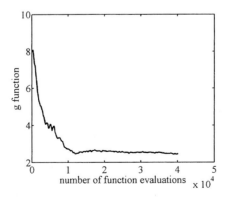

Fig. 7. General trend of $g(x)$ function for EC6 problem when only Cauchy PDF is used

As anticipated in the previous paragraph describing the algorithm, use of only one type of kernel (with other parameters and structures fixed) could make the procedure ineffective or, at the most, inefficient. Figure 7 represents the $g(x)$ function trend on EC6 problem, when in the structure algorithm Cauchy PDF is the only used kernel. Clearly, in this case the optimization process has difficulties to converge to the optimal front, even if we allow a higher number of function evaluations.

On the other hand some results allow thinking that a different scheduling sequence of kernels could provide results better than presented ones. It will be subject of future investigations.

3.3 Alternative Spreading Technique

In the original version of the algorithm, in order to spread solutions on the whole Pareto optimal front the crowding distance technique is used as described in reference [13]. Alternatively it is possible to use Parzen itself to measure the degree of crowding of solutions in order to favour most isolated individuals.

In section 2, where Parzen method is described, we can see that probability density in a point of the m-variable space is related to density of kernels around the point. In this way, if we estimate the PDF. of solutions in the space of objective functions, instead that in the space of m variables, and we evaluate the probability density for each solution, we actually measure the crowding. This technique replaces the local crowding distance used in NSGA-II by a global one. Studies of theoretical main differences between local and global crowding measure are still in progress. However, the proposed method demonstrates to be able to spread solutions in the front for EC6 problem (figure 8 shows one of the obtained fronts).

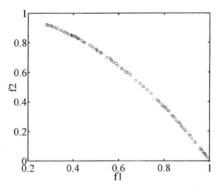

Fig. 8. Obtained non-dominated solutions on EC6 problem when Parzen is used to spread solutions in the Pareto front

4 Conclusions

Here we have presented a new estimation of distribution algorithm that is able to manage multi-objective problems following Pareto criterion. The algorithm uses the

Parzen method in order to create a non-parametric, model-independent probabilistic representation of promising solutions in the search space. Results obtained when the algorithm is applied to well-known test cases show good performance of the optimization process in terms of the number of objective function evaluations and in the spreading of solutions on the whole front.

Showed results allow some comparisons with the best ones obtained in [11] by using MIDEA which appears to be the most similar algorithm. Except for the EC4 problem the reported values are fairly similar. But both works lack a measure of the degree of approximation of the true Pareto front. This could allow deeper considerations.

In this paper an alternative method to spread solutions is also proposed. It uses the Parzen method in the objective space to identify crowding level of individuals. Results are comparable with those obtained through NSGA-II crowding measure.

Contrary to previous works, in this paper we do not attempt to identify a conditionally independent structure in the genome. We know that this may increase the efficiency of the Parzen estimator. It is our intention to address this important point in the future along a frequentist approach with minor changes in the underlying philosophy. In fact the hypothesis testing inherent in the frequentist approach allows the user to impose a known error of the first kind.

For that said above, this work can be seen as the first step towards a more complex and efficient algorithm able to manage multi-objective optimization problems, with constraints too.

References

1. Mühlenbein, H., The equation for the response to selection and its use for prediction, Evolutionary Computation 5(3), pp. 303–346, 1998.
2. Baluja, S., Population based incremental learning: A method for integrating genetic search based function optimization and competitive learning (Tech. Rep. No. CMU-CS-94-163). Pittsburgh, PA: Carnegie Mellon University, 1994.
3. Harik, G. R., Lobo, F. G., and Goldberg, D. E., The compact genetic algorithm, In Proceedings of the International Conference on Evolutionary Computation 1998, pp. 523–528, Piscataway, NJ: IEEE Service Center, 1998.
4. Sebag, M., Ducoulombier, A., Extending population-based incremental learning to continuous search spaces, In Parallel Problem Solving from Nature PPSN V, pp. 418–427, Berlin: Springer Verlag, 1998
5. De Bonet, J. S., Isbell, C. L., and Viola, P., MIMIC: Finding optima by estimating probability densities. In Mozer, M. C., Jordan, M. I., & Petsche, T. (Eds.), Advances in Neural Information Processing Systems, Vol. 9, pp. 424, The MIT Press, Cambridge, 1997.
6. Baluja, S., Davies, S., Using optimal dependency-trees for combinatorial optimization: Learning the structure of the search space. In Proceedings of the 14th International Conference on Machine Learning, pp. 30–38, Morgan Kaufmann, 1997.
7. Pelikan, M., Mühlenbein, H., The bivariate marginal distribution algorithm. In Roy, R., Furuhashi, T., & Chawdhry, P. K. (Eds.), Advances in Soft Computing Engineering Design and Manufacturing, pp. 521–535, London: Springer-Verlag, 1999.
8. Mühlenbein, H., Mahnig, T., The Factorized Distribution Algorithm for Additively Decomposed Functions, Proceedings of the 1999 Congress on Evolutionary Computation, pp. 752–759, 1999,

9. Pelikan, M., Goldberg, D. E., and Cant'u-Paz, E. (1999). BOA: The Bayesian optimization algorithm. In Banzhaf, W., Daida, J., Eiben, A. E., Garzon, M. H., Honavar, V., Jakiela, M., & Smith, R. E. (Eds.), Proceedings of the Genetic and Evolutionary Computation Conference GECCO-99, Vol. I, pp. 525–532. Orlando, FL, Morgan Kaufmann Publishers, San Francisco, CA, 1999.
10. Bosman, P.A.N., Thierens, D., Expanding from discrete to continuous estimation of distribution algorithms: The IDEA, in M. Schoenauer, K. Deb, G. Rudolph, X. Yao, E. Lutton, J.J. Merelo, and H.-P. Schwefel, eds., Parallel Problem Solving from Nature, pp 767–776, Springer, 2000.
11. Thierens, D., Bosman, P.A.N., Multi-Objective Mixture-based Iterated Density Estimation Evolutionary Algorithms L. Spector, E.D. Goodman, A. Wu, W.B. Langdon, H.-M. Voigt, M. Gen, S. Sen, M. Dorigo, S. Pezeshk, M.H. Garzon and E. Burke, editors, Proceedings of the Genetic and Evolutionary Computation Conference – GECCO-2001, pages 663–670, Morgan Kaufmann Publishers, 2001.
12. Khan, N., Goldberg, D.E., and Pelikan, M., Multi-Objective Bayesian Optimization Algorithm, IlliGAL Report No. 2002009, March 2002.
13. Deb, K., Pratap, A., Agarwal, S., and Meyarivan, T., A fast and elitist Multiobjective Genetic Algorithm: NSGA-II, IEEE Transactions on Evolutionary Computation, Vol. 6, No. 2, April 2002.
14. Parzen, E., On Estimation of a Probability Density Function and Mode, Ann. Math. Stat., Vol. 33, pp. 1065–1076, 1962.
15. Cacoullos, T., Estimation of a Multivariate Density, Ann. Inst. Stat. Math., Vol. 18, pp. 179–189, 1966.
16. Deb, K., Multi-Objective Genetic Algorithm: Problem Difficulties and Construction of Test Problems, Evolutionary Computation, Vol. 7, No. 3, pp. 205–230, The MIT Press, 1999.

Instance Generators and Test Suites for the Multiobjective Quadratic Assignment Problem

Joshua Knowles[1]* and David Corne[2]

[1] IRIDIA - CP 194/6, Université Libre de Bruxelles, Avenue F D Roosevelt, 50, 1050 Brussels, Belgium. jknowles@ulb.ac.be
[2] School of Systems Engineering, University of Reading, Reading RG6 6AY, UK. d.w.corne@reading.ac.uk

Abstract. We describe, and make publicly available, two problem instance generators for a multiobjective version of the well-known quadratic assignment problem (QAP). The generators allow a number of instance parameters to be set, including those controlling epistasis and inter-objective correlations. Based on these generators, several initial test suites are provided and described. For each test instance we measure some global properties and, for the smallest ones, make some initial observations of the Pareto optimal sets/fronts. Our purpose in providing these tools is to facilitate the ongoing study of problem structure in multiobjective (combinatorial) optimization, and its effects on search landscape and algorithm performance.

1 Introduction

Configuring a search metaheuristic to work effectively on a specific problem class or instance depends upon being able to relate measurable properties of the problem to performance predictions of the metaheuristic and its configuration. In single objective optimization, some fairly general, measurable properties of problems and/or search 'landscapes', have already been proposed and analysed in the literature (e.g. see [15]), and progress towards a science of heuristic search is under way [7]. In *multiobjective* optimization (MOO), however, problem and landscape structures have additional degrees of freedom, necessitating the consideration of other unique factors. For example, in MOO, fitness landscapes have multiple 'vertical' dimensions and the desired optima form a Pareto front (PF) that can vary in dimension, cardinality, extent, connectedness, and convexity. Consequently, search algorithms operating on these landscapes have more potential modes of operation: when trying to obtain the PF, many search paths are possible but some will be far more efficient than others. It is therefore, perhaps, even more important to understand how to configure MOO metaheuristics than in the case of single objective optimization.

To facilitate empirical studies of problems/landscapes and their relationships to algorithm/configuration performance, test suites and problem generators are

* http://iridia.ulb.ac.be/~jknowles/

useful tools. Arguably, generators are better (particularly those with many controllable parameters) because they enable the effects of different properties to be investigated in isolation or in groups, whereas test suites (alone) can often be too limited in the instances, which can lead to a false impression of progress while, in fact, important problem properties remain uncharted.

In the field of evolutionary multiobjective optimization (EMOO), some good generators and test suites already exist. The frameworks proposed by Deb et al. [4,5,6] have been widely appreciated, although the test suite proposed in [21] had some drawbacks that have, arguably, somewhat constrained progress. The knapsack problems, originally used in [22] have also become a popular choice for benchmarking algorithms. Once again, there are advantages and difficulties with the popularity of this suite, however. It is advantageous for researchers to have a common standard of comparison and it is encouraging to see multiobjective combinatorial optimization (MOCO) problems being tackled in the EMOO literature, but the knapsack problems are limited for two main reasons: 1. a generator is not publicly available and the instances provided do not have many varying parameters, and 2. because, being a constrained problem, heuristic repair or other mechanisms must be introduced, affecting the landscape 'seen' by a metaheuristic, and clouding the important issue of measuring algorithm performance. Other MOCO problems exist in the literature, including our own mc-MST problems [12], but few have become useful benchmarks for investigating general problem characteristics.

Much more extensive study of problems and algorithm performance has been carried out in relation to single objective combinatorial problems. After the traveling salesman problem, perhaps the most studied of all is the quadratic assignment problem (QAP). QAP is both practically important and very difficult, making it particularly relevant for approximate search. In a recent paper [11] we proposed a multiobjective version of the QAP, where $m \geq 2$ distinct QAPs must be minimized simultaneously over the same permutation space. We believe the problem to have practical applications but our main purpose in proposing it is the opportunity it may provide for more general understanding of multiobjective combinatorial optimization (MOCO). There are several advantages to the QAP as a candidate to become a useful test-bed in multiobjective optimization. It has a very simple formulation, solutions being permutations of the integers from $1..n$, so that specialized heuristics and/or repair mechanisms are not needed (or used) to tackle the problem as, for example, they are in knapsack and graph-based problems. The objective function is fast to compute and it can also be delta-evaluated, enabling local search to be efficiently applied [19]. Furthermore, much of the knowledge about the problem, including global measures and landscape analysis tools can be adapted from the vast QAP literature, and 'imported' into the multiobjective domain.

Our aim here is to encourage study of the multiobjective QAP (mQAP), and of MOCO problems in general, through the provision of two mQAP instance generators which allow some important parameters to be controlled and investigated. Some initial test suites derived from these are also given to illustrate the effects of some parameters and to facilitate algorithm performance comparisons. Properties and measures for QAP instances are briefly discussed and, for the test suites, some simple measures are applied. For the smallest instances we compute the entire search space and present some observations of the Pareto front structures. It is our aim in future work to make much further investigations of this problem, and we have already begun this work in [11]. But our focus here is on providing what we hope is a useful new MOCO problem for the EMOO field in a form that makes it easy to use.

The rest of this paper is organized as follows. In section 2 we briefly review some of the literature related to the QAP, focusing on available instances, global properties and landscape measures. Section 3 describes the mQAP and reviews some other MOCO problems and what is known about their problem/landscape structures. Section 4 introduces our generators, while section 5 presents some test suites and provides some initial measures on some instances. Section 6 concludes.

2 The Scalar QAP

Many diverse planning tasks of practical importance can be formulated as instances of the *quadratic assignment problem* (QAP), an NP hard [17] combinatorial optimization problem that dates back to the early sixties. The QAP is a broad problem class embracing both the *graph partitioning problem* and *travelling salesman problem* as special cases [14]. It is also an unusually difficult problem, where even relatively small ($n \geq 20$) general instances cannot be solved to optimality, and it has thus been important in stimulating research in approximate methods. The great research effort attracted by the problem means that a relatively large amount is now known about QAP instance structures and how this relates to global and local statistical measures of their fitness landscapes. This knowledge has been put to effective use in designing and tuning metaheuristics for this problem, e.g. [19,18,14].

The quadratic assignment problem (QAP) entails the assignment of n facilities to n locations so as to minimize a sum of flow/distance products. It may be formulated as:

$$\text{Minimize } C(\pi) = \sum_{i=1}^{n} \sum_{j=1}^{n} a_{ij} b_{\pi_i \pi_j} \qquad (1)$$

where n is the number of facilities/locations, a_{ij} is the distance between location i and location j, b_{ij} is the flow from facility i to facility j, and π_i gives the location of facility i in permutation $\pi \in P(n)$ where $P(n)$ is the QAP search space: the set of all permutations of $\{1, 2, \ldots, n\}$.

2.1 Instances and Generators

A number of instances of the QAP are publicly accessible from QAPlib [3], and come in several categories: those with uniformly random distance and flow matrix entries; those that derive from real applications, which are generally more structured and where typically some of the off-diagonal flow matrix entries are zero; and, because the latter are quite small, random 'real-like' instances [19] that have been artificially constructed using a parameterized generator. The real-like instances are more practically interesting than the uniformly random instances and although it is easier to find the optimum (or best known solution) of them on small instances, they are generally more difficult in terms of reaching a given % above this value. The reason for this effect is that in the uniform instances there are far more solutions at a given % cost above the best-known (easy to find), but it is then difficult to find which of these is close (in permutation space) to the best-known solution, making it difficult to direct the search. These properties can be related to the 'dominance' of the flow and distance matrices, defined below.

2.2 Measures of QAP Instances and Landscapes

Several papers on QAP have conjectured that certain (measurable) characteristics of QAP instances can be related to measurable properties of their fitness landscapes. To merely *characterize* an instance it is sufficient to describe some properties of the distance and flow matrices, a computationally tractable task. To measure some property of the fitness landscape, however, it is necessary (for all but very small instances) to sample the search space. Sampling approaches can be broadly divided into global and local methods. The former operate by taking arbitrary solutions (e.g. uniformly at random) and measuring some properties such as parameter correlations between these. These global methods have been criticized, however, because when performing optimization, little time should be spent sampling these 'average' points [15]. Therefore, it may be better to use local measures that are based on biased sampling.

Vollmann and Buffa [20] introduced flow dominance as a basic means of characterizing QAP instances. Flow dominance is a measure of the flow matrix, A, given by

$$fd(A) = 100\frac{\alpha}{\beta}, \quad \text{where } \alpha = \sqrt{\frac{1}{n^2}\sum_{i=1}^{n}\sum_{j=1}^{n}(a_{ij} - \beta)^2} \quad \text{and } \beta = \frac{1}{n^2}\sum_{i=1}^{n}\sum_{j=1}^{n}a_{ij}. \tag{2}$$

When there is high flow dominance there is low epistasis, in general [14]. The distance dominance dd can be defined on the distance matrix in an analogous fashion.

Measures of the search landscape (whether global or local) often depend on a measure of distance between solutions in the parameter space. Bachelet [1] measures distance $\text{dist}(\pi, \mu)$ between two QAP solutions π and μ as the smallest number of 2-swaps that must be performed to transform one solution into the other (this can be computed in $O(n)$ time); this distance measure has a range of $0..n-1$.

From this, other measures can be easily defined. Bachelet gives the diameter of a population of solutions P as

$$\mathrm{dmm}(P) = \frac{\sum_{\pi \in P} \sum_{\mu \in P} \mathrm{dist}(\pi, \mu)}{|P|^2}. \tag{3}$$

The entropy [9], which is a further measure of the dispersion of solutions, is given by

$$\mathrm{ent}(P) = \frac{-1}{n \log n} \sum_{i=1}^{n} \sum_{j=1}^{n} \left(\frac{n_{ij}}{|P|} \log \frac{n_{ij}}{|P|} \right) \tag{4}$$

where n_{ij} is the number of times facility i is assigned to location j in the population. Low values of entropy indicate highly clustered solutions; high values that they are more randomly distributed.

These measures can be used on a random sample of points (i.e. as global measures) but it is also possible to measure these properties on local optima, or on, say, the fittest $p\%$ of points found during an optimization. The latter can give a better picture of how desirable solutions are distributed in the search space.

The fitness-distance correlation [10] has also been widely used. Often a plot is shown, giving the fitness against distance from the nearest global optimum or best known solution.

3 The Multiobjective QAP Model

The multiobjective QAP (mQAP), with multiple flow matrices (defined below) naturally models any facility layout problem where we are concerned with the flow of more than one type of item or agent. For example, in a hospital layout problem we may be concerned with simultaneously minimizing the flows of doctors on their rounds, of patients, of hospital visitors, and of pharmaceuticals and other equipment.

3.1 Problem Definition

The mQAP may be formulated as follows:

$$\text{`minimize'} \ \boldsymbol{C}(\pi) = \{C^1(\pi), C^2(\pi), \ldots, C^m(\pi)\} \tag{5}$$

where

$$C^k(\pi) = \sum_{i=1}^{n} \sum_{j=1}^{n} a_{ij} b^k_{\pi_i \pi_j}, \ k \in 1..m \tag{6}$$

and where n is the number of facilities/locations, a_{ij} is the distance between location i and location j, b^k_{ij} is the kth flow from facility i to facility j, π_i gives the location of facility i in permutation $\pi \in P(n)$, and finally, 'minimize' means to obtain the Pareto front, or an approximation of it.

3.2 Fitness Landscapes in MOCO

Although a fairly wide variety of MOCO problems have been defined and tackled in the literature (see [8] for a survey), little work has attempted to characterize the search landscapes of these problems. A rare and valuable foray into this area is [2], where the property of 'global convexity' in bi-objective TSPs was investigated; the distances in objective space between solutions near the PF were found to correlate with parameter space distances. One of the implications of this study was that applying wholly separate runs of a single-objective metaheuristic may not be as effective as making use of information from 'nearby' points in the objective or weight space. With the mQAP we expect global convexity to be less marked but similar techniques could be used to test this hypothesis, and to observe how global convexity varies with instance type.

The knapsack problems introduced in [22] have been used as benchmarks for studying the performance of various metaheuristics. Generalizing over a number of results, it now appears clear that population-based methods using only Pareto selection have the tendency to concentrate their solutions on the centre of the PF. This is the case, even when using advanced strategies such as the BMOA [13] that also incorporate modern archiving methods. This 'central tendency' may be due to a clustering in the parameter space of the central PF solutions, with the extremes more isolated. Further investigations of this tendency and its relation to MOCO problem structure is needed.

Finally, we have observed previously [12] that correlations between the weights making up different objectives in a multiobjective MST problem can strongly affect the PF shape, and this, in turn, has an effect on the success of weighted-sum based approaches. The introduction of the mQAP generators (described next) is intended to facilitate further study of the above relationships, and to this end we have also outlined some methods [11], making use of a fast local search for the QAP, aimed eventually at answering the following question: what problem features affect the relative difficulty of approaching the PF vs moving along it? This, we believe impacts strongly on the most appropriate choice of search strategy.

4 Instance Generators for the mQAP

One approach to obtaining instances of the mQAP (or any other MOCO problem) is simply to use and concatenate available single-objective instances. This approach has the advantage of allowing comparison between single-objective and multiobjective algorithm performance on the problem. The extremes of the Pareto front—or best known single-objective solutions—would also be known, although *only* the extremes. We do not criticize this approach and would encourage more researchers to use it (being explicit about where the instances came from). However, some aspects of multiobjective problems cannot be easily controlled by concatenating single-objective instances. In particular, we would like to control the correlations between corresponding components (flows in our case) of the different objectives of a problem instance. This is desirable because

Algorithm 1 Real-like mQAP instance generator with correlated flow elements and overlap parameter

1: **Input:** $n \in \mathbb{Z}^+$, $m \in \mathbb{Z}^+$, $c[1]..c[m] \in [-1, 1]$, $\eta \in [0, 1]$, $A \in \mathbb{Z} < B \in \mathbb{Z}^+$, $r_{max} \in \mathbb{Z}^+$, $R_{max} \in \mathbb{Z}^+$, $N_{max} \in \mathbb{Z}^+$, $seed \in \mathbb{Z}^+$
2: $i \leftarrow 1$
3: **while** $i < n$ **do**
4:    $\Theta \leftarrow \mathcal{U}[0, 2.\pi)$, $R \leftarrow \mathcal{U}[0, R_{max})$, $N \leftarrow \lfloor \mathcal{U}[1, N_{max} + 1) \rfloor$
5:    **for** $j \leftarrow 1..N$ **do**
6:       $\theta \leftarrow \mathcal{U}[0, 2.\pi)$, $r \leftarrow \mathcal{U}[0, r_{max})$
7:       **if** $i < n$ **then**
8:          $loc[i] \leftarrow (R \cos \Theta + r \cos \theta, R \sin \Theta + r \sin \theta)$
9:          $i \leftarrow i + 1$
10:      **end if**
11:    **end for**
12: **end while**
13: **for** $i \leftarrow 1..n$ **do**
14:    **for** $j \leftarrow 1..n$ **do**
15:       $dmatrix[i][j] \leftarrow$ Euclid$(loc[i], loc[j])$
16:    **end for**
17: **end for**
18: print2d($dmatrix$)
19: **for** $i \leftarrow 1..n$ **do**
20:    **for** $k \leftarrow 1..m$ **do**
21:       $fmatrix[k][i][i] \leftarrow 0$
22:    **end for**
23: **end for**
24: **for** $i \leftarrow 1..n - 1$ **do**
25:    **for** $j \leftarrow i + 1..n$ **do**
26:       **for** $k \leftarrow 1..m$ **do**
27:          **if** $k = 1$ **then**
28:             $R1 \leftarrow \mathcal{U}[0, 1)$
29:             $fmatrix[k][i][j] \leftarrow fmatrix[k][j][i] \leftarrow \lfloor 10^{(B-A).R1+A} \rfloor$
30:          **else**
31:             $R2 \leftarrow \mathcal{U}[0, 1)$
32:             **if** $R2 > \eta$ **then**
33:                **if** $fmatrix[1][i][j] = 0$ **then**
34:                   $R2 \leftarrow \mathcal{U}[0, 1)$
35:                   $fmatrix[k][i][j] \leftarrow fmatrix[k][j][i] \leftarrow \lfloor 10^{(B.R2)} \rfloor$
36:                **else**
37:                   $fmatrix[k][i][j] \leftarrow fmatrix[k][j][i] \leftarrow 0$
38:                **end if**
39:             **else**
40:                $V \leftarrow$ correl_val$(R1, c[k])$ /* see equation (7) */
41:                $fmatrix[k][i][j] \leftarrow fmatrix[k][j][i] \leftarrow \lfloor 10^{(B-A).V+A} \rfloor$
42:             **end if**
43:          **end if**
44:       **end for**
45:    **end for**
46: **end for**
47: **for** $k = 1$ **to** m **do**
48:    print2d($fmatrix[k]$)
49: **end for**

realistic problems are likely to exhibit such inter-objective correlations, and describing how these affect the search landscape should be a key element of any useful exposition on MOCO problem/landscape relationships.

4.1 Uniformly Random Instance Generator

In the first of our generators, makeQAPuni.cc[1], only one inter-objective correlation can be controlled. The generator makes symmetric[2] QAP instances with one distance and multiple flow matrices, and its basic parameters are for the instance size n and number of objectives m. All flows and distances are integers in $1..f_{max}$ and $1..d_{max}$, respectively where f_{max} and d_{max} are two further parameters. The desired correlation between corresponding entries in the first and all other flow matrices, is set using the parameter c. Correlated random variables are generated using:

$$p(r^k) = \mathcal{N}(r^1, 1-\sqrt{c})/((1-\sqrt{c}).\sqrt{2\pi}), \qquad k \in 2..m \qquad (7)$$

where $p(r^k)$ is the probability of accepting a random variable $r^k \in [0,1)$ given the value of a uniform random variable, $r^1 \in [0,1)$, and $\mathcal{N}(\overline{x}, \sigma^2)$ is a normal distribution with mean $\overline{x}$ and variance σ^2. The actual flows are made from the random variables using:

$$b^k = 1 + \lfloor r^k.b \rfloor, \qquad k \in 1..m \qquad (8)$$

The generator makes the flow entries in all flow matrices, one at a time using the above procedure. Pseudocode for the generator is not given here but follows a similar structure as that for the real-like generator described next.

4.2 Real-Like Instance Generator

The real-like instance generator, makeQAPrl.cc[1], makes instances where the distance and flow matrices have structured entries. The generator follows procedures for making the non-uniformly random QAP problems given the appellation TaiXXb in the literature, outlined in [19]. Pseudocode for the generator is presented in Algorithm 1.

The distance matrix entries generated are the Euclidean distances between points in the plane. The points are randomly distributed in small circular regions, with these regions distributed in a larger circle. The size and number of the small and larger circles can be controlled by the parameters $r_{max} \in \mathbb{Z}^+$, $R_{max} \in \mathbb{Z}^+$, and $N_{max} \in \mathbb{Z}^+$.

The flow entries are non-uniform random values, controlled by two parameters, $A \in \mathbb{Z}$ and $B \in \mathbb{Z}^+$, with $A < B$. Let X be a random variable uniformly distributed in $[0,1)$. Then a flow entry is given by

$$\lfloor 10^{((B-A)*X+A)} \rfloor. \qquad (9)$$

With negative values of A the flow matrix is sparse, i.e. it contains a number of off-diagonal zero entries. The non-zero entries have non-uniformly distributed

---

[1] Available from http://iridia.ulb.ac.be/~jknowles/mQAP/. C code for reading instances is also provided

[2] Without loss of generality, since an asymmetric matrix can always be transformed into a symmetric one [14]

values. Different values of A and B cannot be set for each different flow matrix but this is an easy extension that might later be added.

The entries in the kth flow matrix ($2 \le k \le m$) are generated using (9) but the random variable X is correlated with the value of X that was used in the corresponding entry in the first flow matrix. Here, correlations (in $[-1,1]$) can be set between the first and *each* of the additional flow matrices using $m - 1$ further parameters to the generator.

A degree of 'overlap' between the matrices can also be specified using a parameter $\eta \in [0,1]$. It controls the fraction of entries in the jth flow matrix that are correlated with the corresponding entries in the 1st flow matrix. With the overlap parameter $\eta = 0$, a random un-correlated value, calculated using

$$\lfloor 10^{(B*X)} \rfloor. \tag{10}$$

will be placed in each entry of the jth flow matrix that corresponds to a zero entry in the first flow matrix. Using this, (and not (9)) ensures that the flow is non-zero. Conversely, a zero will be placed in each entry of the jth flow matrix that corresponds to a non-zero value in the first flow matrix. Thus there is no overlap between the flows of the first and jth matrix when $\eta = 0$. With the $\eta = 1$ all the flows overlap and are correlated. With the overlap set to intermediate values some of the flows will overlap and others will not.

5 Test Suites

Table 1 describes three 10 facility, 2 flow matrix instances, produced by the uniform instance generator. All off-diagonal distances and flows in the matrices are from the set 1..100. The instances differ in the correlations between corresponding elements in the first and second flow matrix. All instances have very similar flow and distance dominance values so the epistasis present (when considering just one flow matrix at a time) should be about the same. However, the different correlations in the instances may affect difficulty for two reasons. First, the shape of the Pareto front may be changed. For example, in [12] it was found that positive correlations between weights in mc-MST problems led to more convex and smaller Pareto fronts, whereas negative correlations led to large, flat Pareto fronts. The latter were more difficult for weighted sum-based approaches in the sense that many weight vectors (in a naïve approach where these are uniformly randomly generated) tend to push the search to either extreme of the PF, making it difficult to find the intermediate optima. The mQAP instances here should allow these correlation effects to be isolated and studied further. The difficulty of moving toward the Pareto front may also be affected by the correlation. One might guess that if the correlation between objectives is strongly positive then the problem reduces to the single objective one. Whereas if there is little, or even negative correlation, then there should exist more optima and they are more likely to be spread out in the space. Therefore the search to find at least one (but not a specific one) may be easier. The correlation parameter used to

Table 1. Test suite of three 10 node, 2 flow matrix, uniformly random instances with different flow correlations and other parameters. The first figure in an instance name is the number of nodes, the second is the number of flow matrices and the third is just an index. The global parameters/properties are: $\max(d)$, the maximum distance in the distance matrix; $\max(f)$, the maximum flow in any of the flow matrices; $\text{corr}(f^i, f^j)$, the correlation parameter affecting corresponding flow matrix entries of the ith and j flow; Spearman r, the measured sample rank correlation between (off-diagonal) corresponding entries in the first and jth flow matrix; fd^j, the flow dominance of the jth flow matrix; dd, the distance dominance; #PO, the number of Pareto optima; #supported the number of supported Pareto optima; diam(PO), the diameter of the Pareto optimal set; ent(PO), the entropy of the Pareto optimal set. The random seed to the generator is also given for reference

Param/Property	Parameter/Property Values by Instance		
	KC10-2fl-1uni	KC10-2fl-2uni	KC10-2fl-3uni
$\max(d)$	100	100	100
$\max(f)$	100	100	100
$\text{corr}(f^1, f^2)$	0.0	0.8	-0.8
Spearman r	0.18	0.92	-0.75
fd^1, fd^2	76.7, 69.9	66.1, 63.2	64.8, 69.8
dd	57.9	622.3	69.8
#PO	27	4	135
#supported	7	3	14
diam(PO)	7	6	8
ent(PO)	0.71	0.39	0.78
seed	68203720	289073914	73892083

set the desired correlation may be compared with the value, given in the table, of the Spearman rank correlation measure [16] of the actual flow entries.

Some of the correlation effects can also be appreciated by observation of the values of the other measures given. The number of Pareto optima (found by exhaustive search) is smaller when the correlation is large and larger when it is negative. Similarly, the diameter and entropy of the Pareto optima vary with inter-objective correlation. From the values we see that the optima are in fewer, larger clusters (in parameter space) when the correlation is positive and more spread out, and less clustered when the correlation is negative. The number of supported solutions (those lying on the convex hull of the Pareto front) is small in these instances, indicating that it may be difficult for a weighted sum approach to find all the Pareto optima. However, this is an artifact of the problems being very small; with larger instances the number of unsupported Pareto optima would generally be smaller as the PF assumes a smoother shape due to the 'central-limit' effect from summing larger numbers of flow/distance products.

Five small, two objective instances of the real-like problems are presented in Table 2. Similar observations apply to these problems. The Spearman r values correlate well with the correlation parameter setting for the first three instances. However, when the overlap parameter is set to 0.6, in instances 4 and 5, the r

Table 2. Test suite of five 10 node, 2 flow matrix, real-like instances with different flow correlations and other parameters. A and B control the distribution of flow values, in particular the fraction of off-diagonal zero entries, thus they influence the flow dominance and epistasis. The overlap parameter η indicates the fraction of entries in the first flow matrix that are correlated with the corresponding entries in the other flow matrices. The distance matrices of all these instances are the Euclidean distances between random points in a single circle of radius 100. Other parameters are as in Table 1

Param/Prop	Parameter/Property Values by Instance				
	KC10-2fl-1rl	KC10-2fl-2rl	KC10-2fl-3rl	KC10-2fl-4rl	KC10-2fl-5rl
max(d)	155	111	145	136	138
max(f)	9445	9405	9732	9419	95476
A	-2	-2	-2	-2	-5
B	4	4	4	-4	5
corr(f^1, f^2)	0.0	0.7	-0.7	0.7	0.7
overlap η	1.0	1.0	1.0	0.6	0.6
fd^1, fd^2	243.0, 194.1	230.8, 242.8	248.9, 219.6	304.7, 272.1	405.3, 561.99
dd	69.5	60.7	65.6	58.6	58.6
Spearman r	0.15	0.83	-0.77	0.39	-0.10
#PO	38	17	58	33	48
#supported	13	3	25	12	14
diam(PO)	8	7	8	8	8
ent(PO)	0.68	0.49	0.62	0.58	0.63
seed	35243298	18178290	7810398	48972324	2129704715
Parameter/Property Values Common to All Instances					
$N_{\max} = 1$, $r_{\max} = 100$, $R_{\max} = 0$					

values drop significantly, as one would expect. Looking at the number of Pareto optima, and their entropy and diameter values, we see that the r value explains these quite well. For the fifth instance, which has the largest flow dominance (lowest epistasis), the number, distance and entropy of Pareto optima is somewhat surprising. We might expect these values to be lower, but because the r value is negative they lie between those of the first and third instances, underlining the importance of the inter-objective correlation in determining the distribution of Pareto optima. Of course, many further observations are needed to verify this effect.

In Figure 1 we compare the shape of the Pareto front in instances 4 and 5, which differ in parameters only in the value of A and B. It is seen that instance 5 (despite having a negative r correlation) has a much more convex-shaped PF.

In Table 3 we present the first of our two publicly available test suites, produced using the uniform instance generator. The first three instances are of size $n = 20$ and have two objectives. The second three instances are of size $n = 30$ and have three objectives. For these instances it is not practical to perform exhaustive search, so we are unable to give measures of the Pareto front. However,

Table 3. Test suite of three 20 node, 2 flow matrix, and three 30 node, 3 flow matrix, uniformly random instances with different flow correlations and other parameters. Parameters and properties are as for previous tables, except that nondom replaces PO here, and refers to an internally nondominated set from a total of 100 000 local optima

Param/Property	Parameter/Property Values by Instance		
	KC20-2fl-1uni	KC20-2fl-2uni	KC20-2fl-3uni
$\max(d)$	100	100	100
$\max(f)$	100	100	100
$\text{corr}(f^1, f^2)$	0.0	0.7	-0.7
Spearman r	-0.14	0.79	-0.82
fd^1, fd^2	58.5, 64.6	58.6, 56.9	60.3, 61.8
dd	60.7	61.9	62.2
# nondom	80	19	178
$\text{diam}(nd)$	15	14	16
$\text{ent}(nd)$	0.828	0.43	0.90
seed	89749235	20983533	48927232

Param/Property	KC30-3fl-1uni	KC30-3fl-2uni	KC30-3fl-3uni
$\max(d)$	100	100	100
$\max(f)$	100	100	100
$\text{corr}(f^1, f^{k>1})$	0.0	0.4	-0.4
Spearman r	-0.02, -0.02	0.43, 0.50	-0.43, -0.42
fd^1, fd^2, fd^3	59.4, 61.3, 64.8	60.3, 57.2, 58.3	61.0, 59.2, 61.8
dd	59.9	58.7	60.1
# nondom	705	168	1257
$\text{diam}(nd)$	24	22	24
$\text{ent}(nd)$	0.97	0.92	0.96
seed	549852743	394908123	121928193

we are able to repeatedly (100 000 times) apply a deterministic local search to these instances, and can measure the number of internally nondominated optima found, as well as their entropy and diameter. For more information on our procedure for doing this see [11]. We can observe that on these instances, where the fd and dd levels are fairly constant, the number of nondominated optima and their diameter and entropy follow closely the value of r.

Table 4 presents the second of our test suites and gives identical measures for the local search samples. Figure 2 summarises the effects of inter-objective correlations on the number of PO/nondominated solutions found, over otherwise homogeneous sets of instances.

6 Conclusion

We have presented instance generators and test suites for the mQAP, with the aim of providing new benchmarks for EMOO algorithms and to facilitate studies of the relationships between problem characteristics, search landscapes, and

Table 4. Test suite of five 20 node, 2 flow matrix, and three 30 node, 3 flow matrix, real-like instances with different flow correlations and other parameters. Parameters and properties as in previous tables

Param/Prop	Parameter/Property Values by Instance			
	KC20-2fl-1rl	KC20-2fl-2rl	KC20-2fl-3rl	KC20-2fl-4rl
$\max(d)$	196	173	185	164
$\max(f)$	9954	9644	9326	96626
A	-2	-2	-2	-5
B	4	4	4	5
$\text{corr}(f^1, f^2)$	0.0	0.4	-0.4	0.4
Spearman r	0.08	0.43	-0.49	0.32
overlap η	1.0	1.0	1.0	1.0
fd^1, fd^2	206.8, 230.1	243.8, 367.0	246.2, 350.0	305.9, 477.5
dd	54.9	57.9	56.3	56.5
# nondom	541	842	1587	1217
diam(nd)	15	14	15	15
ent(nd)	0.63	0.6	0.66	0.51
seed	1213563	45767	8347234	1121932

Param/Prop	KC20-2fl-5rl	KC30-3fl-1rl	KC30-3fl-2rl	KC30-3fl-3rl
$\max(d)$	185	172	180	172
$\max(f)$	98776	9929	9968	9852
A	-5	-2	-2	-2
B	5	4	4	4
$\text{corr}(f^1, f^2)$	0.4	0.4	0.7	-0.4
$\text{corr}(f^1, f^3)$	–	0.0	-0.5	-0.4
$\text{Sp.}r(f^1, f^2)$	-0.25	0.10	0.46	-0.75
$\text{Sp.}r(f^1, f^3)$	–	-0.29	-0.67	-0.70
overlap η	0.5	0.7	0.7	0.2
fd^1, fd^2, fd^3	351.5, 391.1,–	235.3, 320.7, 267.5	233.7, 337.5, 341.8	251.7, 359.5, 328.3
dd	57.3	56.1	58.6	56.8
# nondom	966	1329	1924	1906
diam(nd)	15	24	24	24
ent(nd)	0.56	0.83	0.86	0.86
seed	3894673	20983533	34096837	9346873

Parameter/Property Values Common to All Instances
$N_{\max} = 1$, $r_{\max} = 100$, $R_{\max} = 0$

algorithm performance in MOCO. In the previous section we have observed some simple relationships between two problem characteristics, namely flow-dominance and inter-objective correlations, and the resulting search landscapes. In [11], we also defined some methods for measuring properties of *local* landscape features. Of particular interest to us, is obtaining some measure of the relative difficulty of moving towards the Pareto front versus moving along it. This, we think, should bear strongly on the overall search strategy that is most effective on a given problem. Our future work will focus on this issue.

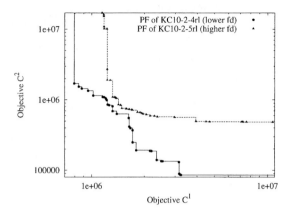

Fig. 1. The effect of different flow dominance (fd) on the Pareto front shape. Notice the logarithmic scale; with high flow dominance the PF is strongly convex

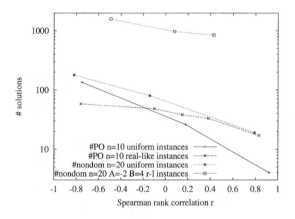

Fig. 2. The effect of inter-objective correlations on the number of nondominated solutions

Acknowledgments. Joshua Knowles gratefully acknowledges the support of a Marie Curie research fellowship of the European Commission, contract number HPMF-CT-2000-00992.

References

1. V. Bachelet. *Métaheuristiques Parallèles Hybrides: Application au Problème D'affectation Quadratique.* PhD thesis, Université des Sciences et Technologies de Lille, December 1999.
2. P. C. Borges and M. P. Hansen. A basis for future successes in multiobjective combinatorial optimization. Technical Report IMM-REP-1998-8, Institute of Mathematical Modelling, Technical University of Denmark, March 1998.

3. R. E. Burkard, S. Karisch, and F. Rendl. QAPLIB—A quadratic assignment problem library. *European Journal of Operational Research*, pages 115–119, 1991.
4. K. Deb. Multi-Objective Genetic Algorithms: Problem Difficulties and Construction of Test Problems. *Evolutionary Computation*, 7(3):205–230, Fall 1999.
5. K. Deb, A. Pratap, and T. Meyarivan. Constrained Test Problems for Multi-objective Evolutionary Optimization. In E. Zitzler et al,. editors, *First International Conference on Evolutionary Multi-Criterion Optimization*, pages 284–298. Springer-Verlag. LNCS 1993, 2001.
6. K. Deb, L. Thiele, M. Laumanns, and E. Zitzler. Scalable Test Problems for Evolutionary Multi-Objective Optimization. Technical Report 112, Computer Engineering and Networks Laboratory (TIK), Swiss Federal Institute of Technology (ETH), Zurich, Switzerland, 2001.
7. S. Droste, T. Jansen, K. Tinnefeld, and I. Wegener. A new framework for the valuation of algorithms for black-box optimization. In *FOGA 2002, Proceedings of the Seventh Foundations of Genetic Algorithms Workshop*, pages 197–214, 2002.
8. M. Ehrgott and X. Gandibleux. A Survey and Annotated Bibliography of Multi-objective Combinatorial Optimization. *OR Spektrum*, 22:425–460, 2000.
9. C. Fleurent and J.A. Ferland. Genetic hybrids for the quadratic assignment problem. *DIMACS Series in Discrete Mathematics and Theoretical Computer Science*, 16:173–188, 1994.
10. T. Jones and S. Forrest. Fitness distance correlation as a measure of problem difficulty for genetic algorithms. In L. J. Eshelman, editor, *Proceedings of the 6th International Conference on Genetic Algorithms*, pages 184–192. Morgan Kaufmann, 1995.
11. J. Knowles and D. Corne. Towards Landscape Analyses to Inform the Design of Hybrid Local Search for the Multiobjective Quadratic Assignment Problem. In A. Abraham et al., editors, *Soft Computing Systems: Design, Management and Applications*, pages 271–279, Amsterdam, 2002. IOS Press.
12. J. D. Knowles and D. W. Corne. A Comparison of Encodings and Algorithms for Multiobjective Minimum Spanning Tree Problems. In *Proceedings of the Congress on Evolutionary Computation 2001 (CEC'2001)*, volume 1, pages 544–551, Piscataway, New Jersey, May 2001. IEEE Service Center.
13. M. Laumanns and J. Ocenasek. Bayesian Optimization Algorithms for Multi-objective Optimization. In J. J. Merelo Guervós et al., editors, *Parallel Problem Solving from Nature—PPSN VII*, pages 298–307, 2002. Springer-Verlag. LNCS 2439.
14. P. Merz and B. Freisleben. Fitness landscape analysis and memetic algorithms for the quadratic assignment problem. *IEEE Transactions on Evolutionary Computation*, 4(4):337–352, 2000.
15. B. Naudts and L. Kallel. A comparison of predictive measures of problem difficulty in evolutionary algorithms. *IEEE Transactions on Evolutionary Computation*, 4(1):1–15, April 2000.
16. W. Press, S. Teukolsky, W. Vetterling, and B. Flannery. *Numerical Recipes in C: The Art of Scientific Computing*. Cambridge University Press, 1992.
17. S. Sahni and T. Gonzalez. P-complete approximation problems. *Journal of the ACM*, 23:555–565, 1976.
18. T. Stützle and M. Dorigo. Available ACO algorithms for the QAP. In D. Corne et al., editors, *New Ideas in Optimization*, chapter 3, pages 33–50. McGraw Hill, 2000.
19. E. D. Taillard. A comparison of iterative searches for the quadratic assignment problem. *Location Science*, 3:87–105, 1995.

20. T. E. Vollmann and E. S. Buffa. The facilities layout problem in perspactive. *Management Science*, 12(10):450–468, 1966.
21. E. Zitzler, K. Deb, and L. Thiele. Comparison of Multiobjective Evolutionary Algorithms: Empirical Results. *Evolutionary Computation*, 8(2):173–195, Summer 2000.
22. E. Zitzler and L. Thiele. Multiobjective Evolutionary Algorithms: A Comparative Case Study and the Strength Pareto Approach. *IEEE Transactions on Evolutionary Computation*, 3(4):257–271, November 1999.

Dynamic Multiobjective Optimization Problems: Test Cases, Approximation, and Applications

M. Farina[1], K. Deb[2], and P. Amato[1]

[1] STMicroelectronics, Via C. Olivetti, 2, 20041, Agrate (MI), Italy
{marco.farina,paolo.amato}@st.com
[2] Kanpur Genetic Algorithms Laboratory (KanGAL),
Dept. of Mechanical Engg. IITK Kanpur, India, deb@iitk.ac.in

Abstract. Parametric and dynamic multiobjective optimization problems for adaptive optimal control are carefully defined; some test problems are introduced for both continuous and discrete design spaces. A simple example of a dynamic multiobjective optimization problems arising from a dynamic control loop is given and an extension for dynamic situation of a previously proposed search direction based method is proposed and tested on the proposed test problems.

1 Introduction

The EMO literature contains a large number of test problems covering different difficulties which may be encountered by an EMO algorithm when converging towards the Pareto- optimal front. All these problems correspond to a static optimization, in which the objective functions or associated problem parameters are static and the task is to find a set of design variables to optimize the objective functions. But several other important applications require a time-dependent (on-line) multiobjective optimization in which either the objective function or the problem parameters or both vary with time. In handling such problems, there exist not many algorithms and certainly there is a lack of test problems to adequately test a dynamic multi-objective evolutionary algorithm. In this study, we refer to adaptive control (eventually on-line) of time-varying systems where the optimal controller is time-dependent because the system's properties are time-dependent. Moreover we consider several objectives for the controller dynamic optimization and we give a full formulation of the resulting multiobjective dynamic nonlinear optimization problem. We formulate some continuous and discrete test problems where the time dependent Pareto-optimal solutions are known analytically. We also give an example of a simple control problem. Finally, an extension of a search direction based method is proposed and tested on one of the aforementioned analytical test problems. Optimal design of controllers is a classical field of application for evolutionary computation and evolutionary multiobjective optimization. Once closed loop stability is assured, several additional criteria for performances improvement can be considered such as maximum overshooting minimization, settling time minimization and rise time minimization, in order to design stable and powerful controllers. Several examples of such optimization procedure are available in literature in case of static design problems, that is when the optimization is to be performed off-line and when the model of the system (the plant or the device) is not time dependent. Two early examples can be found in [7] where some controllers (among which an $\mathcal{H}_2/\mathcal{H}_\infty$ one) are optimized with an EMO algorithm. Another classical application of EMO for static controllers optimization consider fuzzy rule set optimization for fuzzy controllers, some examples can be found in [1,4]. When considering dynamic single-objective optimization problems, the use of genetic algorithms for time dependent fitness landscape have been considered for single-objective problems and several studies are available in the literature [6,14,5,12]. Major modifications in the operators are required for a prompt reaction to time dependent changing. Moreover, several non-GA strategies for dynamic optimization procedure for single objective problems are also proposed in the literature. An example can be found in [13], where an artificial-life inspired algorithm was developed. But when dynamic multi-objective optimization is concerned, very few studies are available in literature [3,15] and a complete formulation of the problem together with proper test problems is still missing. In this paper, we make an attempt to fill this gap and suggest a test suite of five problems testing various aspects of tracking the optima whenever there is a change in the problems. Hopefully, this study will motivate researchers interested in dynamic optimization to develop and test their algorithms towards creating efficient multi-objective EAs.

2 Problem Setting

From the most general point of view any dynamic multiobjective optimal control problem can be included in the following general extension of non-linear multi-criteria optimization problem:

Definition 1. *Let* $\mathbf{V_O}$, $\mathbf{V_F}$ *and* $\mathbf{W}$ *be* n_O-*dimensional,* n_F-*dimensional and* M-*dimensional continuous or discrete vector spaces,* $\mathbf{g}$ *and* $\mathbf{h}$ *be two functions defining inequalities and equalities constraints and* $\mathbf{f}$ *be a function from* $\mathbf{V_O} \times V_F$ *to* $\mathbf{W}$. *A parametrical non-linear multi-criteria (minimum) optimization problem with* M *objectives is defined as:*

$$\min_{v_O \in V_O} \mathbf{f} = \{f_1(\mathbf{v_O}, \mathbf{v_F}), ..., f_M(\mathbf{v_O}, \mathbf{v_F})\} \quad \text{s. t.} \quad \mathbf{g}(\mathbf{v_O}, \mathbf{v_F}) \leq 0, \ \mathbf{h}(\mathbf{v_O}, \mathbf{v_F}) = 0.$$

In problem 1 some variables are available for optimization (v_O) and some other (v_F) are imposed parameters being independent from optimization variables; both objective functions and constraints are parameter-dependent. As evident the degree of complexity of such problems (even when only one parameter is considered) is much higher than classical multiobjective optimization problems in terms of formulation, computation and visualization. All problem's features depends on v_F variables, search space shape, utopia point location and even degree of contrast among objectives; in other words the problem may move from a cooperative objectives situation to a truly multiobjective situation when parameters v_F are changed. On the other hand, when v_F variables are fixed standard multiobjective optimization theory can be applied. A special case of problem 1 is the following, where only one parameter is considered, the time t:

Definition 2. *Let t be the time variable,* $\mathbf{V}$ *and* $\mathbf{W}$ *be n-dimensional and M-dimensional continuous or discrete vector spaces,* $\mathbf{g}$ *and* $\mathbf{h}$ *be two functions defining inequalities and equalities constraints and* $\mathbf{f}$ *be a function from* $\mathbf{V} \times t$ *to* $\mathbf{W}$. *A dynamic non-linear multi-criteria (minimum) optimization problem* with M *objectives is defined as:*

$$\min_{\mathbf{v}\in\mathbf{V}} \mathbf{f} = \{f_1(\mathbf{v},t), ..., f_M(\mathbf{v},t)\} \quad \text{s.t.} \quad \mathbf{g}(\mathbf{v},t) \leq 0, \; \mathbf{h}(\mathbf{v},t) = 0.$$

Definition 3. *We call* Pareto-optimal set at time t $(\mathcal{S}_\mathrm{P}(t))$ *and* Pareto-optimal front at time t $(\mathcal{F}_\mathrm{P}(t))$ *the set of Pareto-optimal solutions at time t in design domain and objective domain, respectively.*

3 Test Problems

In this section, we suggest some a number of test problems for dynamic multi-objective optimization for continuous and discrete search spaces. Unlike in the single-objective optimization problems, here we are dealing with two different search spaces: decision variable space and objective space. Therefore, the following are the four possible ways a problem can dynamically change:

Type I: The Pareto-optimal set (optimal decision variables) $\mathcal{S}_P$ changes, whereas the pareto-optimal front (optimal objective values) $\mathcal{F}_P$ does not change.

Type II: Both $\mathcal{S}_P$ and $\mathcal{F}_P$ change.

Type III: $\mathcal{S}_P$ does not change, whereas $\mathcal{F}_P$ changes.

Type IV: Both $\mathcal{S}_P$ and $\mathcal{F}_P$ do not change, although the problem can dynamically change.

There is, of course, a possibility that while the problem is changing one or more types of the above changes can occur in the time scale. Here, we concentrate on the first three types of changes, as Type IV change is not that interesting from the point of testing an optimization algorithm. One of the requirements for a test problem is that the resulting Pareto-optimal solutions must be known exactly. In the following, we describe some such test problems.

3.1 Continuous Search Space Test Problems

A straightforward extension of ZDT and DTLZ test problems developed earlier ([8,9]) for two and higher objectives can be considered in order to insert time dependance factors into multiobjective optimization test cases. As it is well known these test problems provide different difficulties which may be encountered when considering real-life multiobjective optimization problems: non-concavity, discontinuity, deceptiveness, presence of local fronts, etc. When solving dynamically changed problems, such difficulties may transform themselves from one of the above features to another with random sudden jumps or with a gradual change. A generic test problem for such a dynamic situation is presented in the following equation:

$$\min_{\mathbf{x}}(f_1(\mathbf{x}), f_2(\mathbf{x})) = (f1(\mathbf{x}_I), g(\mathbf{x}_{II}) \cdot h(\mathbf{x}_{III}, f_1, g)) \tag{1}$$

In the above test problem, there are three functions $f1$, g, and h. In the original paper, the following functions were suggested:

$$f1(\mathbf{x}_I) = x_1, \quad g(\mathbf{x}_{II}) = \sum_{x_i \in \mathbf{X}_{II}} x_i^2, \quad h(f_1, g) = 1 - \left(\frac{f_1}{g}\right)^2. \tag{2}$$

Each of them can change dynamically or in combination. Let us consider the following scenarios.

• The function g changes by varying $G(t)$ as follows:

$$g(\mathbf{x}_{II}) = \sum_{x_i \in \mathbf{X}_{II}} (x_i - G(t))^2, \quad x_{II}^{\min} \leq G(t) \leq x_{II}^{\max}, \tag{3}$$

but $f1$ and h are kept fixed as above. The above equation makes $x_i = G(t)$ as the optimal solution for all $x_i \in \mathbf{x}_{II}$. Here each variable $x_i \in \mathbf{x}_{II}$ lies in $[x_{II}^{\min}, x_{II}^{\max}]$. Since $G(t)$ changes with time, S_P changes with time. However, the resulting Pareto-optimal front $\mathcal{F}_P$ does not change. Thus, the above change in ZDT functions will cause a Type I test problem. However, if the g function is changed as follows: $g(\mathbf{x}_{II}) = G(t) + \sum_{x_i \in \mathbf{X}_{II}}(x_i - G(t))^2$, $x_{II}^{\min} \leq G(t) \leq x_{II}^{\max}$, the Pareto-optimal front also changes and the resulting problem is a Type II test problem.

• Next, the h function can be changed as follows:

$$h(\mathbf{x}_{III}, f_1, g) = 1 - \left(\frac{f_1}{g}\right)\left(H(t) + \sum_{x_i \in \mathbf{X}_{III}}(\mathbf{x}_i - H(t))^2\right)^{-1}, x_{III}^{\min} \leq H(t) \leq x_{III}^{\max}. \tag{4}$$

Here, a third set of variables $\mathbf{x}_{III}$ (with $x_i \in [x_{III}^{\min}, x_{III}^{\max}]$) is introduced to change h dynamically. Here, we suggest $x_{III}^{\min} = 0.75$ and $x_{III}^{\max} = 1.5$. Functions $f1$ and g are not changed. The resulting $\mathcal{F}_P$ changes through a change in $\mathbf{x}_{III}$ only. This will make a Type III test problem; an example is FDA2 in next subsection. Interestingly, if g and $f1$ functions are not changed and h is changed as follows:

$$h = 1 - \left(\frac{f_1}{g}\right)^{H'(t)}, \quad H'(t) > 0, \tag{5}$$

the Pareto-optimal set S_P does not change, but the Pareto-optimal front $\mathcal{F}_P$ changes. This makes the corresponding problem a Type-III problem as well. It is intuitive that such a problem is of not much use for testing an MOEA.

- The $f1$ function can be changed as follows:

$$f1(\mathbf{x}_I) = \sum_{x_i \in \mathbf{X}_I} x_i^{F(t)}, \quad F(t) > 0. \tag{6}$$

By varying $F(t)$, the density of solutions on the Pareto-optimal front can be varied. Since the objective of an MOEA would still be to find a well distributed set of Pareto-optimal solutions, the resulting $\mathbf{x}_I$ set (of finite size) would be different from before. Since a change in $F(t)$ does not change the location of the Pareto-optimal front, or $\mathcal{F}_P$, the resulting problem is a Type I problem.

- Finally, a more complex problem can be formed by changing all three functions by simply varying $G(t)$, $H(t)$, and $F(t)$. This will result in a Type II test problem. In each case, a systematic set of time-varying functions ($G(t)$, $F(t)$, and $H(t)$) can be chosen. Since both $\mathcal{S}_P$ and $\mathcal{F}_P$ will be known in each case, it will be easier to test the performance of an MOEA. The functions suggested above can also be changed with more complex functions to make the test problem harder.

In a similar fashion, the DTLZ functions [9] can also be converted into dynamic test problems.

- The $g(x_M)$ function can be changed as follows:

$$g(\mathbf{x}_M) = G(t) + \sum_{x_i \in \mathbf{X}_M} (x_i - G(t))^2, \quad 0 \leq G(t) \leq 1. \tag{7}$$

This will change both $\mathcal{S}_P$ and $\mathcal{F}_P$, thereby making the modified problem a Type II problem. Wherever a multimodal g function was used (DTLZ1 and DTLZ3), it can also be changed accordingly. With the above change in g function, the DTLZ5 and DTLZ6 problems will get changed in an interesting manner. Since variables x_2 to x_{M-1} are dependent on the g function, the resulting Pareto-optimal front may not lead to a curve, instead for $G(t) > 0$ it will lead to a surface.

- The change of x_i to $x_i^{F(t)}$ (where $F(t) \in (0, 1]$) would be another interesting modification. Such a problem but with a fixed modification ($F(t) = 0.1$) was used in constructing the DTLZ4 function. This will give rise to a Type I test problem. The h function in DTLZ7 function can be modified as described in ZDT functions above.

- The spherical shape of the Pareto-optimal front in DTLZ2 to DTLZ6, can be changed to an ellipsoidal shape by changing the $(1 + g(\mathbf{x}_M))$ term in every f_i expression by $(1 + g(\mathbf{x}_M) + K_i(t))$, where $0 \leq K_i(t) \leq 1$ and $g(\mathbf{x}_M)$ is defined in equation 7. Such a change will lead to a Type II test problem. The Pareto-optimal front will then be defined by $\sum_{i=1}^{M} [f_i/(1 + K_i(t))]^2 = 1$.

The following test problems set is proposed:

Definition 4 (FDA1). *Type I, convex POFs*

$$\begin{cases} f1(\mathbf{x}_I) = x_1, \quad g(\mathbf{x}_{II}) = 1 + \sum_{x_i \in \mathbf{X}_{II}} (x_i - G(t))^2, \quad h(f_1, g) = 1 - \sqrt{\frac{f_1}{g}}, \\ G(t) = \sin(0.5\pi t), \quad t = \frac{1}{n_t} \lfloor \frac{\tau}{\tau_T} \rfloor, \\ \mathbf{x}_I = (x_1) \in [0,1], \quad \mathbf{x}_{II} = (x_2, ..., x_n) \in [-1, 1]. \end{cases} \quad (8)$$

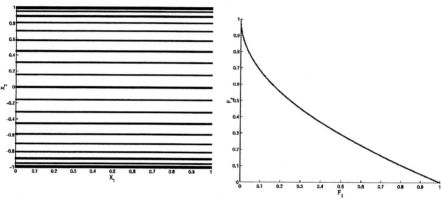

Fig. 1. $\mathcal{S}_P(t)$ for FDA1, first two decision variables, 24 time steps

Fig. 2. $\mathcal{F}_P(t)$ for FDA1, 24 time steps

Here, τ is the generation counter, τ_T is the number of generation for which t remains fixed, and n_t is the number of distinct steps in t. The suggested number of variables is $n = 20$, $\tau_T = 5$, and $n_T = 10$. The task of a dynamic MOEA would be to find the same Pareto-optimal front $f_2 = 1 - \sqrt{f_1}$ every time there is a change in t.

Definition 5 (FDA2). *Type II, convex to non-convex POFs*

$$\begin{cases} f1(\mathbf{x}_I) = x_1, \quad g(\mathbf{x}_{II}) = 1 + \sum_{x_i \in \mathbf{X}_{II}} x_i^2, \\ h(\mathbf{x}_{III}, f_1, g) = 1 - \left(\frac{f_1}{g}\right)^{\left(H(t) + \sum_{x_i \in \mathbf{X}_{III}} (x_i - H(t))^2\right)^{-1}}, \\ H(t) = 0.75 + 0.7 \sin(0.5\pi t), \quad t = \frac{1}{n_t} \lfloor \frac{\tau}{\tau_T} \rfloor, \\ \mathbf{x}_I = (x_1) \in [0,1], \quad \mathbf{x}_{II}, \mathbf{x}_{III} \in [-1, 1]. \end{cases} \quad (9)$$

It is recommended to use $|\mathbf{x}_{II}| = |\mathbf{x}_{III}| = 15$. Other parameters are the same as in FDA1. Here, the Pareto-optimal front will swing from convex non-convex shapes.

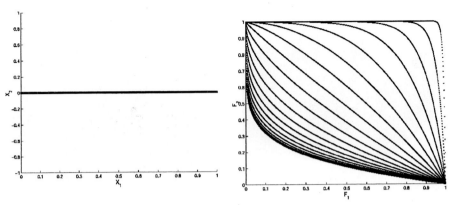

Fig. 3. $\mathcal{S}_P(t)$ for FDA2, first two decision variables, 24 time steps

Fig. 4. $\mathcal{F}_P(t)$ for FDA2, 24 time steps

Definition 6 (FDA3). *Type II, convex POFs*

$$\begin{cases} f1(\mathbf{x}_I) = \sum\limits_{x_i \in \mathbf{X}_I} x_i^{F(t)}, \ g(\mathbf{x}_{II}) = 1 + G(t) + \sum\limits_{x_i \in \mathbf{X}_{II}} (x_i - G(t))^2, \ h(f_1, g) = 1 - \sqrt{\frac{f_1}{g}}, \\ G(t) = \sin(0.5\pi t), \quad F(t) = 10^{2\sin(0.5\pi t)}, \quad t = \frac{1}{n_t}\lfloor\frac{\tau}{T_T}\rfloor, \\ \mathbf{x}_I \in [0,1], \quad \mathbf{x}_{II} \in [-1,1]. \end{cases} \quad (10)$$

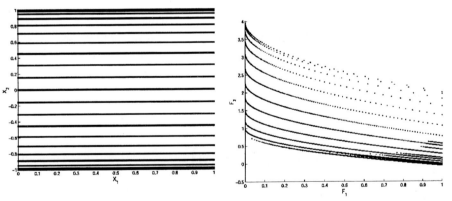

Fig. 5. $\mathcal{F}_P(t)$ for FDA3, first two decision variables, 24 time steps

Fig. 6. $\mathcal{F}_P(t)$ for FDA3, 24 time steps

Here, we recommend $|\mathbf{x}_I| = 5$ and $\mathbf{x}_{II} = 25$. In this test problem, the density of solutions on the Pareto-optimal front varies with t. The task of an MOEA in FDA3 would be to find a widely distributed set of solutions every time there is a change in t

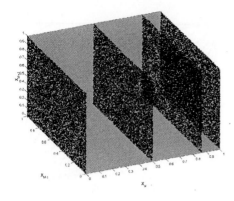
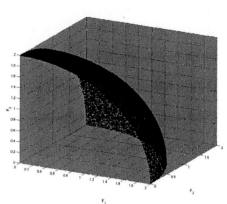

Fig. 7. $\mathcal{S}_P(t)$ for FDA4, M-th, (M+1)-th and (M-1)-th decision variables, 4 time steps

Fig. 8. $\mathcal{F}_P(t)$ for FDA4, 4 time steps

Definition 7 (FDA4). *Type I, non-convex POFs*

$$\begin{cases} \min_{\mathbf{x}} \quad f_1(\mathbf{x}) = (1 + g(\mathbf{x}_{II})) \prod_{i=1}^{M-1} \cos(x_i \pi/2), \\ \min_{\mathbf{x}} \quad f_k(\mathbf{x}) = (1 + g(\mathbf{x}_{II})) \left(\prod_{i=1}^{M-k} \cos(x_i \pi/2) \right) \sin(x_{M-k+1} \pi/2) \quad k = 2 : M-1, \\ \min_{\mathbf{x}} \quad f_M(\mathbf{x}) = (1 + g(\mathbf{x}_{II})) \sin(x_1 \pi/2), \\ \text{where } g(\mathbf{x}_{II}) = 1 + \sum_{x_i \in \mathbf{x}_{II}} (x_i - G(t))^2, \\ G(t) = |\sin(0.5\pi t)|, \quad t = \frac{1}{n_t} \lfloor \frac{\tau}{\tau_T} \rfloor, \quad \mathbf{x}_{II} = (x_M, \ldots, x_n), \quad x_i \in [0,1], \ i=1:n. \end{cases} \tag{11}$$

A recommended value of n is 12. Here the task of a dynamic MOEA would be to find the same spherical surface (with radius one) whenever there is a change t.

Definition 8 (FDA5). *Type II, non-convex POFs*

$$\begin{cases} \min_{\mathbf{x}} \quad f_1(\mathbf{x}) = (1 + g(\mathbf{x}_{II})) \prod_{i=1}^{M-1} \cos(y_i \pi/2), \\ \min_{\mathbf{x}} \quad f_k(\mathbf{x}) = (1 + g(\mathbf{x}_{II})) \left(\prod_{i=1}^{M-k} \cos(y_i \pi/2) \right) \sin(y_{M-k+1} \pi/2) \quad k = 2 : M-1, \\ \min_{\mathbf{x}} \quad f_M(\mathbf{x}) = (1 + g(\mathbf{x}_{II}) \sin(y_1 \pi/2), \\ \text{where } g(\mathbf{x}_{II}) = 1 + G(t) + \sum_{x_i \in \mathbf{x}_{II}} (x_i - 0.5)^2, \\ y_i = x_i^{F(t)} \quad \text{for } i = 1, \ldots, M, \\ G(t) = |\sin(0.5\pi t)|, \quad F(t) = 1 + 100 \sin^4(0.5\pi t), \quad t = \frac{1}{n_t} \lfloor \frac{\tau}{\tau_T} \rfloor, \\ \mathbf{x}_{II} = (x_M, \ldots, x_n), \quad x_i \in [0,1], \ i = 1 : n. \end{cases} \tag{12}$$

Identical parameter values as those in FDA4 can be used here. Here, the density of solutions on the Pareto-optimal front changes with t. The distribution

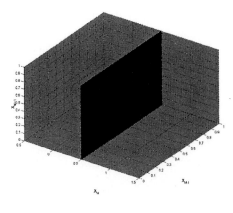
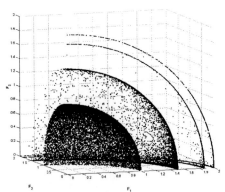

Fig. 9. $S_P(t)$ for FDA5, M-th, (M+1)-th and (M-1)-th decision variables, 4 time steps

Fig. 10. $\mathcal{F}_P(t)$ for FDA5, 4 time steps

achieved with a previously found good distribution will no more be a good one. The task of an dynamic MOEA would be to find a good distribution every time there is a change in the density of solutions.

3.2 Discrete Search Space Test Problems

Although discrete version of the above test problems can be chosen here, we mainly suggest combinatorial test problems. As it is well known, the travelling salesman problem (TSP) can be considered as the prototype of optimal routing problems and it is widely used for testing evolutionary algorithms and heuristic search strategies. In real-life problems, optimal routing may require several criteria to be considered and may be required in a dynamic situation where features of the search space changes during optimization. We thus consider an extension of the classical TSP as a benchmark for dynamic multiobjective optimization in discrete search spaces. The most general dynamic multiobjective n-city travelling salesman problem (DMTSP) can be given as follows:

$$\begin{cases} \min_{\mathbf{p} \in \mathbf{P}^n} \mathbf{f} = \{f_1(\mathbf{p}, t), ..., f_M(\mathbf{p}, t)\} \\ f_j = \sum_{i=1}^{n} w_{i,j}(t) \|\mathbf{x}(p_i, t) - \mathbf{x}(p_{i+1}, t)\| \quad p_{n+1} = p_1 \end{cases} \quad (13)$$

As can be seen from equation 13, both coefficients in weighted sum and coordinates of cities can be time-dependent; the first case may express different traffic conditions on the path, while the second may represent changing in cities.

As a specific test case, we suggest a problem in which n cities are arranged equi-spaced on the circumference of a circle of radius one. The following fixed weight vectors can be used:

$$w_{i1}(t) = 1.0, \qquad w_{i2}(t) \frac{1}{\|\mathbf{x}(p_i, t) - \mathbf{x}(p_{i+1}, t)\|^2}, \qquad (14)$$

but the cities get interchanged dynamically. It is clear that the above two weight vectors will give rise to a set of Pareto-optimal solutions on which a successive arrangements of the cities and the diametric arrangement of cities are two extreme solutions (as shown in Figure 11).

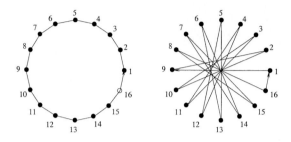

Fig. 11. Two extreme Pareto-optimal solutions are shown

The initial arrangement of cities may be in any order. Thereafter, some cities in the first quadrant of the circle can get interchanged with the cities from the third quadrant. The number of cities and the choice of cities to be exchanged can be varied dynamically. Since the weight vectors do not change, clearly the Pareto-optimal front ($\mathcal{F}_P$ also does not change. So, this is a Type I DMTSP problem. However, the solution sequence changes with time and it would be the task of an MOEA to track the modified sequence of cities every time there is a change in arrangement of the cities. Since the Pareto-optimal front does not change with time, it would be easier to test the performance of the underlying MOEA.

3.3 A Simple Dynamic Multiobjective Optimal Control Problem

Static controller optimization requires the system to be known and fixed (not time-dependent) during the entire optimal control procedure. This is quite often not true and a simple example of this can be the aging of systems or the intrinsic randomness of some plants. In both cases, because the system changes during time, the controller is to be adaptive in order the closed loop system performances to be satisfactory. We point out here that the knowledge of multiple controllers allows to shift to different controllers for changing condition. Moreover, knowledge of multiple controller often provides additional insight about the underlying problem. As an example consider the control of combustion in a rubbish burner [2] where the income properties (the rubbish to be burned) is typically random. We consider here a simplified test case of such situation, that is the control of a randomly variable system via PID controller. The system's $S(s)$ and controller's $C(s)$ transfer functions are the following:

$$S(s) = \frac{1.5}{50s^3 + a_2(t)s^2 + a_1(t)s + 1} \qquad C(s) = K_p(t) + K_i(t)\frac{1}{s} + K_d(t)s \qquad (15)$$

where $a_1(t)$ and $a_2(t)$ are time-dependent parameters that can be varied in order to simulate the aforementioned aging or intrinsic random changes in the system. Moreover, in order to introduce some non-linearity, two additional limit and rate blocks are inserted. The derivative coefficient in the PID controller has been fixed to $K_{\rm d} = 8.3317$ and the other two coefficients, $K_{\rm p}(t), K_{\rm i}(t)$ are to be varied in order the closed loop performances of the controlled system to be as similar as possible to a reference response in terms of small rising time $R(t)$, small maximum overshooting $O(t)$ and small settling time $S(t)$. The multiobjective problem is thus the following:

$$\min_{(K_p(t),K_d(t))\in(.5,5)\times(.1,1)} \{R(K_p(t),K_d(t)), O(K_p(t),K_d(t)), S(K_p(t),K_d(t))\} \tag{16}$$

Figure 12 shows the system and controller loop; $a_1(t)$ and $a_2(t)$ in the system transfer function can be varied in the following way:

$$a_1 = 3 + 30f(t) \quad a_2 = 43 + 30f(t) \tag{17}$$

where the function $-1 < f(t) < 1$ can be tuned for different time dependence simulation. A search space sampling when only $O(t)$ and $R(t)$ are considered is shown in figure 13 together with the corresponding set of Pareto optimal solutions for nine time steps with $f(t) = sin(\pi * t/18)$.

When considering the single-objective dynamic optimization of controllers the control strategy is dynamically changed based on the the dynamic optimization algorithm; in the multiobjective case an additional a-posteriori decision making procedure is to be considered for a proper choice among dynamic Pareto optimal controllers. This block is inserted in figure 12 between the DMOEA block and the PID block.

4 Dynamic MOEAs: Extending Search Direction Based Methods

When EMO algorithms are considered for the solution of time dependent problems and thus for the approximation of time dependent POF and/or POS, the following two situations or a combination of the two may happen:
- the time dependence is seldom but sudden and random (type B),
- the time dependence is slow but continuous (type A).

In the second case a full evolution of an evolutionary algorithms is possible during the rest time and a time change monitoring strategy together with an update of the starting population may be sufficient for a proper approximation of the moving front and set. An example of such a strategy is shown in subsection 4 where the static method described in [10] is extended for a dynamic situation.

If on the other hand the time dependence is of type A an extension to the multiobjective case of single objective evolutionary algorithms for variable fitness landscape is to be considered [6,14,5,12,11] and some examples can be found

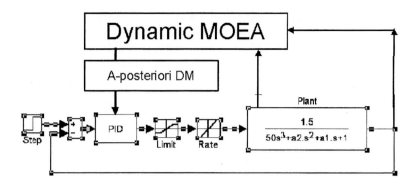

Fig. 12. Schema of the dynamic PID controller optimization test problem: four blocks are considered: the plant, the PID controller, the dynamic multiobjective optimizer and the a-posteriori decision making procedure

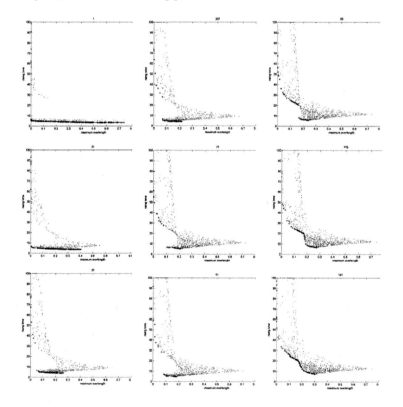

Fig. 13. Dynamic-POF (black (o)) and sampling (gray (·)) for the dynamic PID controller optimization problem, snapshots at iteration 1, 21, 27, 38, 77, 81, 89, 113, 141 . The maximum overlenght $O(K_p(t), K_d(t))$ and the rising time $R(K_p(t), K_d(t))$ are shown on the x and y axes respectively. The shape of the POF changes quite significantly moving from a single-objective problem (step 1) to a disconnected multiobjective problem (step 4-5-6) up to a non convex multiobjective one (step 8-9)

```
BEGIN
Start time t
Compute random starting pop. pst(0).
WHILE t=STOP TIME
    IF ε(t) > ε̃
        FOR I=1:M
            min. f_i(t) with (1+1)ES + GDA or NMA with pst(t) as st. pop.
        END
        compute U(t) and M̃(t)
        FOR J=1:n_s
            compute P_j(t) from P_{k=1:j-1}(t) and S_{k=1:j-1}(t)
            min. f̃_j(t) with (1+1)ES + GBA or NMA with pst(t) as st. pop.
            obtain S_j(t)
        END
    END
    S_j(t) = pst(t)
    END
    Compute ε(t)
END
```

Fig. 14. Pseudo-code of the proposed strategy. $\tilde{f}_j(t)$: scalar objective functions corresponding to different directions, $P_j(t), \widetilde{M}(t)$ and $S_j(t)$: values in design space necessary for building the $\tilde{f}_j(t)$, GDA: Gradient based single-objective optimization algorithm, NMA: Nelder Mead simplex single-objective optimization algorithm, $pst(t)$: Starting population for the $npop$ (1+1)ES searches

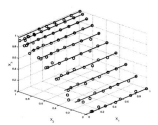

Fig. 15. Exact $\mathcal{S}_P$(dots) and approximated (o) with a search direction based method on FDA1 (10 solutions)

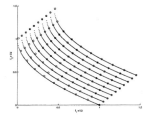

Fig. 16. Exact $\mathcal{F}_P$ (dots) and approximated (o) with the proposed search direction based method on FDA1 (10 solutions)

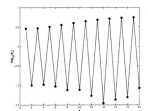

Fig. 17. Log of POS approximation error ($log(e_x(t))$, equation 19) versus time

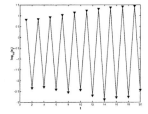

Fig. 18. Log of POF approximation error ($log(e_f(t))$, equation 19), versus time

in [3,15]. The proposed strategy is an immediate extension of the static search direction based method described in [10]. We consider the case of sudden random time-dependent changes among which the system behaves as a static one. A full fast multiobjective search algorithm can thus be run in the time between one changing and the other. In order the strategy to be fully automated, a time changes checking is inserted through the computation of the following value:

$$\varepsilon(t) = \frac{1}{n_\varepsilon} \sum_{j=1}^{n_\varepsilon} \left\| \frac{(\mathbf{f}_j(X,t) - \mathbf{f}_j(X,t-1))}{\mathbf{R} - \mathbf{U}} \right\| \qquad (18)$$

where n_ε points in search space are chosen randomly for test problems or intentionally in case of real devices where some working point may be particularly sensitive to time changes; if $\varepsilon(t) > \tilde{\varepsilon}$ it means that a significant time changes has happened and a new search is to be done. A pseudo-code of the whole strategy is shown in figure 14 (for more details see [10]). From a general point of view any EMO algorithm may be run in type B problems and no particular modification for operators is necessary. The advantage of the proposed strategy is that any number of solution ($n_s \geq 2$) can be chosen and the cost C_O in terms of objective function number is $C_O = n_s \times (ni_{(1+1)ES} + ni_L)$ where n_{iter} is the iteration number of each $(1+1)ES$ global search and ni_L is the iteration number of each local search algorithm.

In figure 15 the solution of FDA1 test problem (type I problem) is shown in the design domain for the first three variable (20 decision variables are considered) and in objective domain in figure 16; note that for better visualization purposes $f_1 + t/2$ and $f_2 + t/2$ are shown on the x and y axes respectively. The method encounters some difficulties when the $\mathcal{S}_P$ get close to the bounding box region. Moreover in figures 17 and 18 the log of approximation errors ($log(e_x(t))$ and $log(e_f(t))$) for both $\mathcal{F}_P$ and $\mathcal{S}_P$ are shown. Ten time steps have been considered and the approximation errors are computed before and after each jump; as can be seen after each time step corresponding to a changing of the problem a new solution is found with very low values of approximation errors in both design and objective space. The following two formulas where used for error evaluation.

$$e_x(t) = \frac{1}{np} \sum_{j=1}^{np} \min_{i=1:ns} \left\| \frac{\mathcal{F}_{P,i}(t) - \mathbf{F}_j^{sol}(t)}{\mathbf{R}(t) - \mathbf{U}(t)} \right\| \qquad (19)$$

$$e_f(t) = \frac{1}{np} \sum_{j=1}^{np} \min_{i=1:ns} \left\| \mathcal{S}_{P,i}(t) - \mathbf{X}_j^{sol}(t) \right\| \qquad (20)$$

where ns is the number of sampling points for the analytical $\mathcal{F}_P$ and $\mathcal{S}_P$, np is the solution number, $f^{sol}(t)$ and $\mathbf{X}^{sol}(t)$ are computed solutions in design space and objective space respectively and where $\|\cdot\|$ is the euclidean distance in design and objective spaces.

5 Conclusion

The extension to the dynamic case of some multiobjective optimization test problems available for the static case leads to a wide typology of problems where the problem features typical of static cases (convexity, disconnectedness, deceptiveness...) can be time dependent. The dynamic EMO algorithm is asked to perform well as time goes on. As for single-objective optimization the case of random sudden changes and slow continuous changes can be considered as extreme cases where different methods may behave in different way. When considering the first case (random sudden changes) an extension to the dynamic case of a search direction based method is proposed and tested on some of the introduced test problems with satisfactory results. In such cases the key point of the strategy is the time change monitoring and the consequent proper restart of the algorithm.

References

1. Jessica M. Anderson, Tessa M. Sayers, and M. G. H. Bell. Optimization of a Fuzzy Logic Traffic Signal Controller by a Multiobjective Genetic Algorithm. In *Proceedings of the Ninth International Conference on Road Transport Information and Control*, pages 186–190, London, April 1998. IEE.
2. M. Annunziato. http://erg055.casaccia.enea.it/.
3. Zafer Bingul, Ali Sekmen, and Saleh Zein-Sabatto. Adaptive Genetic Algorithms Applied to Dynamic Multi-Objective Problems. In Cihan H. Dagli, Anna L. Buczak, Joydeep Ghosh, Mark Embrechts, Okan Ersoy, and Stephen Kercel, editors, *Proceedings of the Artificial Neural Networks in Engineering Conference (ANNIE'2000)*, pages 273–278, New York, 2000. ASME Press.
4. Anna L. Blumel, Evan J. Hughes, and Brian A. White. Fuzzy Autopilot Design using a Multiobjective Evolutionary Algorithm. In *2000 Congress on Evolutionary Computation*, volume 1, pages 54–61, Piscataway, New Jersey, July 2000. IEEE Service Center.
5. J. Branke. Evolutionary approaches to dynamic optimization problems - A survey. *Juergen Branke and Thomas Baeck editors: Evolutionary Algorithms for Dynamic Optimization Problems*, 13:134–137, 1999.
6. C. O. Wilke C. Ronnewinkel and T. Martinetz. Genetic algorithms in time-dependent environments. In L. Kallel, B. Naudts, and A. Rogers, editors, *Theoretical Aspects of Evolutionary Computing*, pages 263–288, Berlin, 2000. Springer.
7. Carlos Manuel Mira de Fonseca. *Multiobjective Genetic Algorithms with Applications to Control Engineering Problems*. PhD thesis, Department of Automatic Control and Systems Engineering, University of Sheffield, Sheffield, UK, September 1995.
8. Kalyanmoy Deb. Multi-Objective Genetic Algorithms: Problem Difficulties and Construction of Test Problems. *Evolutionary Computation*, 7(3):205–230, Fall 1999.
9. Kalyanmoy Deb, Lothar Thiele, Marco Laumanns, and Eckart Zitzler. Scalable Test Problems for Evolutionary Multi-Objective Optimization. Technical Report 112, Computer Engineering and Networks Laboratory (TIK), Swiss Federal Institute of Technology (ETH), Zurich, Switzerland, 2001.

10. M. Farina. A minimal cost hybrid strategy for pareto optimal front approximation. *Evolutionary Optimization*, 3(1):41–52, 2001.
11. J.J. Grefenstette. Genetic algorithms for changing environments. *Proc. 2nd International Conference On Parallel problem Solving from Nature, Brussels*, 1992.
12. J.J. Grefenstette. Evolvability in dynamic fitness landscapes: A genetic algorithm approach. *Proc. Congress on Evolutionary Computation (CEC99) Washington DC IEEE press*, pages 2031–2038, 1999.
13. P.Amato, M.Farina, G.Palma, and D.Porto. An alife-inspired evolutionary algorithm for adaptive control of time-varying systems. In *Proceedings of the EUROGEN2001 Conference, Athens, Greece, September 19-21, 2001*, pages 227–222. International Center for Numerical Methods in Engineering (CIMNE), Barcelona, Spain, March 2002.
14. F. Vavak, K. A. Jukes, and T. C. Fogarty. Performance of a genetic algorithm with variable local search range relative to frequency of the environmental changes. *Genetic Programming 1998: Proceedings of the Third Annual Conference*, 1998.
15. Kazuo Yamasaki. Dynamic Pareto Optimum GA against the changing environments. In *2001 Genetic and Evolutionary Computation Conference. Workshop Program*, pages 47–50, San Francisco, California, July 2001.

No Free Lunch and Free Leftovers Theorems for Multiobjective Optimisation Problems

David W. Corne[1], Joshua D. Knowles[2]

[1]Department of Computer Science, University of Reading, UK
d.w.corne@reading.ac.uk
[2]IRIDIA, ULB, Belgium
jknowles@ulb.ac.be

Abstract. The classic NFL theorems are invariably cast in terms of single objective optimization problems. We confirm that the classic NFL theorem holds for general multiobjective fitness spaces, and show how this follows from a 'single-objective' NFL theorem. We also show that, given any particular Pareto Front, an NFL theorem holds for the set of all multiobjective problems which have that Pareto Front. It follows that, given any 'shape' or class of Pareto fronts, an NFL theorem holds for the set of all multiobjective problems in that class. These findings have salience in test function design. Such NFL results are cast in the typical context of *absolute* performance, assuming a performance metric which returns a value based on the result produced by a single algorithm. But, in multiobjective search we commonly use *comparative* metrics, which return performance measures based non-trivially on the results from two (or more) algorithms. Closely related to but extending the observations in the seminal NFL work concerning *minimax distinctions* between algorithms, we provide a 'Free Leftovers' theorem for comparative performance of algorithms over permutation functions; in words: over the space of permutation problems, *every* algorithm has some companion algorithm(s) which it outperforms, according to a certain well-behaved metric, when comparative performance is summed over all problems in the space.

1 Introduction

The so-called 'No Free Lunch' theorem (NFL) is a class of theorems concerning the average behaviour of 'black-box' optimization algorithms over given spaces of optimisation problems. The primary such theorems were proved in [15], and disseminated to the evolutionary computation community in [16], showing that, when averaged over all possible optimization problems defined over some search space X, no algorithm has a performance advantage over any other. A considerable debate has grown over the importance and applicability of these theorems, and the consequences for the field of optimization in general are much discussed in [16], and other work reviewed briefly below. Perhaps the main consequence for the majority of researchers in optimization, particularly those investigating generic *algorithms* rather than those focused on particular *problems*, is that the NFL result denies any claim of the form 'algorithm A is better than algorithm B', for *any* pair of algorithms A and B. For example, the statement 'Genetic Programming outperforms random search' is false. NFL points out that any such statement must be carefully qualified in terms of the set of problems under consideration. As we see later in a brief review, it has been shown that NFL does not hold over certain subsets of problems, and proofs in this vein are a growing area. In fact, of course, this is easy to show (at least empirically), when the set of problems under consideration is small, such as, perhaps, three particular test problems on which well-designed comparative experiments are performed. However, generalization of any such 'A is better than B' statement to all problems in a large *class* of problems is always unwise without convincing argumentation.

In this paper we are concerned with NFL on spaces of *multiobjective problems*. We will confirm that NFL results hold over such spaces, and in particular we will show that NFL holds over spaces of problems which share a particular (or particular class of) Pareto front(s). We also prove a 'Free Leftovers' theorem for *comparative* metrics; these are technically distinct from the straightforward 'absolute' metrics typically assumed in NFL work, and are the most commonly applied metrics when comparing the performance of multiobjective algorithms..

The remainder is set out as follows. In section 2 we briefly review NFL-related research in (largely) the evolutionary computation community to date, and then start to discuss the question of NFL in the context of multiobjective search. Section 3 then provides some standard preliminaries and notation, and in section 4 we confirm the standard NFL result in multiobjective problem spaces, via showing how it would follow from a 'single-objective' NFL theorem. Section 5 then proves NFL results for interesting subsets of multiobjective spaces: those with particular (or particular classes of) Pareto fronts; this section also notes that a more general (than shown in section 4) NFL result holds, where we allow the range of fitnesses in each objective to be different. Section 6 brings up the issue of *comparative metrics*, which are of particular interest in multiobjective optimization. We show, with our 'Free Leftovers' theorem, that the spirit of NFL does *not* hold in general when we use comparative metrics; this is consistent and closely linked to the observations concerning 'minimax distinctions' in the seminal NFL work. We briefly discuss and conclude in section 7.

2 About NFL and the Multiobjective Case

Work on NFL since the seminal papers has broadly concerned two themes; one theme is new proofs of NFL/and or interesting subclasses of problems for which NFL holds; the other theme concerns classes of problems for which NFL doesn't hold (i.e. proofs of 'Free Lunch' theorems). We first briefly review some work on NFL since [15], covering first theme one and then theme two.

Regarding theme one, quick on the heels of [15], [10] provided another proof of NFL, and reviewed the implications of NFL with a focus on representation issues. In particular, suggestions for representation/operator performance measures were proposed, given any well-defined set over which NFL does not hold. More recently, [14] showed that a general NFL result holds over the set of permutation functions, that is, problems in which every point in the search space has a unique objective function value. In [14] it is shown that these functions are, on average, 'unusually' complex (uncompressible) – more so than the NP-complete class – implying that this set may not be of practical interest. Furthermore, it is shown that all functions in this set have identical summary statistics, implying that the latter cannot be used to guide search.

A 'sharpened' NFL result is proved in [11], which shows that the NFL holds over a set of functions F, if and only if F is closed under permutation (c.u.p.). They also show that the compressibility of a set of functions *cannot* be appealed to, as a means of 'escaping' NFL: thus, a highly compressible function such as a needle-in-a-haystack function, when closed under permutation, forms a set over which NFL still holds. Expoiting the result in [11], Igel and Toussaint [7] consider its implications. They prove that nearly all subsets of a typical space of problems are not c.u.p, and, in particular, some practically interesting problem classes are not c.u.p. Hence, certain practically interesting classes of problems are not subject to NFL.

Theme two concerns identifying and focusing on particular, usually 'small', subsets of problems for which NFL does not hold. For example, in [4] it was shown that for a certain very simple set of objective functions, and a particular restriction on this set, NFL does not hold, and adaptive algorithms have provably better performance than non-adaptive ones on this set. In [13] it is proven that Gray encoding is worse than binary encoding over a set of functions with $2^L - 1$ optima, from which it follows that Gray is better than binary over all *other* functions. In [1], a free lunch is proved for a simple search algorithm, applied over a set of multivariate polynomial functions of bounded degree d. This is of general interest because it is shown that performance is better than random search up to a very large value of d, such that many functions of practical interest could be approximated by the polynomial. Finally, in [5] it is proved that the time to optimize a specific, realistic, and simply defined instance of MAXSAT is exponential for a heuristic designed specifically for the MAX-SAT class, and it is further hypothesised that all 'reasonable' heuristic algorithms will exhibit this performance. This shows that even in complexity-restricted problem classes, in which NFL does not hold, it still 'almost' holds since expected worst-case performance of an optimizer may be very poor.

It so happens that, in the prominent published NFL work so far, the *single-objective* case has been assumed, at least implicitly. For example, [16] talks of the space of fitnesses Y as a space of 'values' and indeed cast the set of problems as 'cost functions', and give examples of performance measures which relate to finding minima in that space; when specific examples of functions are given in [11], they are numerically valued; similar can be said of [1], [5] and so forth. Meanwhile, universally, NFL theorems are notated in terms of a space of problems (functions) $f : X - Y$, clearly implying single-objective optimization. One exception we have found is the early work of [10], which once notes that the domain of the problems concerned could be multiobjective.

That NFL holds over multiobjective optimization is hence not clear from an abrupt reading of the literature in the literature. Since Y only needs to be a finite set, we could glibly and simply say 'it does', however this ignores one key difference between multiobjective and single objective optimization: in the multiobjective case, algorithm performance is commonly done via a *comparative metric*, which is non-trivially distinct from 'absolute' performance metrics. As it turns out, NFL does *not* in hold in general in such a scenario. As we'll see, the fact that fitnesses are vectors rather than values is immaterial with regard to NFL in the context of absolute performance metrics (where performance of an algorithm is a measure of the result produced by that algorithm alone), however as we just indicated, the greater emphasis in multiobjective search on comparative metrics renders this less applicable, and in general there *is* a free lunch (or, rather, 'freee leftovers') when certain well-behaved comparative metrics are involved.

Regarding the embracing of multiobjective optimization by the 'absolute performance' NFL theorem, one may still feel uneasy about this, since various aspects and elements of the various 'single-objective' proofs, which at least implicitly assume single-objective optimization, need to be revisited. So, rather than simply say 'since Y only needs to be a finite space, NFL holds for multiobjective optimization', we also provide a proof which shows how this follows from the 'single-objective' case; i.e. if we assume only that NFL for absolute performance holds on single objective problems, we can derive its truth for multiobjective problems in a more interesting way. This is done in section 4, following preliminaries and notation in section 3. As indicated earlier, in section 5 we then look at certain salient subsets of multiobjective problems, in particular those which share a certain Pareto front, and in section 6 we look at the issue of comparative metrics.

3 Preliminaries and Notation

Using similar notation to that in [16], we have a search space X and a set of 'fitnesses' Y, and we will generally assume maximization. A single-objective optimisation problem f is identified with a a mapping $f : X - Y$ and $F = Y^X$ is the space of all problems. Its size is clearly $|Y|^{|X|}$. A black-box optimisation algorithm a generates a time-ordered sequence of points in the search space, associated with their fitnesses, called a *sample*. A sample of size m is denoted $d_m \equiv \{(d_m^X(1), d_m^Y(1)), ..., (d_m^X(m), d_m^Y(m))\}$, where, in general, $d_m^X(i)$ is the ith distinct point in the search space X visited by the algorithm, and $d_m^Y(i)$ is the fitness of this point. We will also use the term d_m^X to denote the vector containing only the $d_m^X(i)$, and similarly d_m^Y will indicate the time-ordered set of fitness values in the sample. The space of all samples of size m is denoted $D_m = (X \times Y)^m$, and the space of all samples is denoted D, and is simply the union of D_m for all positive integers m.

An optimisation algorithm a is characterised as a mapping: $a : D - X$ with the restriction that $a(d_m) \notin d_m^X$. That is, it does not revisit previously visited points. This restriction, plus that of the algorithm being *deterministic*, is imposed for simplicity but ultimately does not affect the generality of the resulting theorems.

The (absolute) performance a after m iterations on problem f is measured using $P(d_m^y | f, m, a)$; this is the probability of a particular sample being obtained after iterating a for m iterations on problem f. The NFL theorem (for static cost functions) is this [15, 16]: for any pair of algorithms a_1 and a_2,

$$\sum_f P(d_m^Y \mid f, m, a_1) = \sum_f P(d_m^Y \mid f, m, a_2). \quad (1)$$

Hence, when summed over all cost functions, the probabilities of obtaining a particular sample using a_1 amount to the same as that of using a_2. The key message is that, since *performance* of an algorithm is a function only of the sample it generates, when performance is summed over the whole space of problems, the result is the same for any pair of algorithms. We can define a performance measure as a function $\Phi: d_m^Y \to \Re$ (with the choice of $\Re$ being natural, and relatively arbitrary), and a corollary of equation (1) is:

$$\sum_f P(\Phi(d_m^Y) \mid f, m, a_1) = \sum_f P(\Phi(d_m^Y) \mid f, m, a_2). \quad (2)$$

Now, we need to establish notation that will be helpful in the multiobjective case. A multiobjective problem with k objectives can be cast as a function $g: X \to Y^k$, in which, for simplicity at this stage, we assume that the space of fitness values in each objective is the same (Y). We now need new notation for a multiobjective sample, which we will denote as follows

$$v_m = \{(d_1^X(1), v_1^Y(1)), \ldots, (d_m^X(m), v_m^Y(m))\},$$

where $v_j^Y(j) \in Y^k$; i.e. $v_j^Y(j)$ is the k-objective vector of fitnesses resulting from evaluation of $d_j^X(j)$. Finally, using V to denote the set of all multiobjective samples, we define a multiobjective algorithm q as a function $q: V - X$ with the restriction that $q(v_m) \notin v_m^X$, with notation assuming the obvious meaning.

4 No Free Lunch on Multiobjective Problem Spaces

What we wish to confirm is that, for any two multiobjective algorithms q_1 and q_2,

$$\sum_g P(v_m^Y \mid g, m, q_1) = \sum_g P(v_m^Y \mid g, m, q_2). \quad (3)$$

As indicated, we choose to do so by showing how it follows from a 'single-objective only' NFL theorem. The sketch of the proof is as follows – the idea is straightforward, but the book-keeping issues will take up a little space. We will first assume that equation (3) is false, and hence there is some pair of multiobjective algorithms q_1 and q_2 whose overall performance differs over the space of all k-objective problems. Via showing that a $1 \leftrightarrow 1$ mapping exists between the space of k-objective problems and a certain space of single objective problems, we can then show that this implies that there are two single-objective algorithms who differ in performance over this single objective space. This must be false (by equation (1)), and hence equation (3) follows by *reductio ad absurdum*. So, we now formally state and prove an NFL theorem for static multiobjective problems.

Theorem 1 Equation (3) holds for any two multiobjective algorithms q_1 and q_2.

Proof of Theorem 1:
Assume that equation (3) does not hold, thus we can state that, for a particular pair of multiobjective algorithms q_1 and q_2,

$$\sum_g P(v_m^Y \mid g, m, q_1) > \sum_g P(v_m^Y \mid g, m, q_2), \qquad (4)$$

in which, without loss of generality, we cast q_1 as the algorithm which performs best overall (when the performance measure correlates with achieving the sample v_m^Y).

Now consider the mapping $s : Y^k \to W$, where $W = \{1, 2 ..., |Y|^k\}$, defined as follows, where $z \in Y^k = \{z_1, z_2, ..., z_k\}$:

$$s(z) = \sum_{i=1}^{k} z_i \mid Y \mid^{i-1}.$$

This is clearly a one-to-one mapping; we omit a proof, but simply note that this map is between a number $s(z)$, and a unique representation of that number in base $|Y|$. The set W simply enumerates, or uniquely labels, all possible multiobjective vectors in Y^k.

Now consider the mapping $s \circ g : X - W$. This is a single objective function on X. Further, from any multiobjective algorithm $q : V - X$ we can construct a corresponding single objective algorithm $q' : D - W$ as follows:

$$q'(v_m') \equiv s(q(v_m)),$$

where $v_m' = \{(d_1^X(1), s(v_1^Y(1))), ..., (d_m^X(m), s(v_m^Y(m)))\}$. Hence, every k-objective algorithm q defined on the search space X and involving the space of fitnesses Y^k corresponds to a unique single objective algorithm defined on the search space X and involving the single-objective space of fitnesses W, and *vice versa*.

Recall our assumption (equation (4)); now, we can say

$$P(v_m^Y \mid g, m, q) = P(v_m' \mid s \circ g, m, q'),$$

since there are one-to-one correspondences between v_m^Y and v_m', g and $s \circ g$, and q and q'. It trivially follows (from this and equation (4)) that:

$$\sum_{s \circ g} P(v_m' \mid s \circ g, m, q_1') > \sum_{s \circ g} P(v_m' \mid s \circ g, m, q_2').$$

However, we notice that the space of all functions $s \circ g$ corresponds precisely to the space of all functions $f : X - W$, that both q_1' and q_2' are single objective algorithms from D to X (where fitnesses in d_m^Y are from W), and that any sample v_m' corresponds uniquely to a sample d_m^Y from D. Hence, we can write

$$\sum_f P(d_m^W \mid f, m, q_1') > \sum_f P(d_m^W \mid f, m, q_2').$$

But this essentially expresses a free lunch for single-objective algorithm q_1' at the expense of single-objective algorithm q_2', over the space of all single-objective problems $f : X - W$. By equation (3), this is false, and hence we conclude our original assumption is wrong, and so the following holds:

$$\sum_g P(v_m^Y \mid g, m, q_1) = \sum_g P(v_m^Y \mid g, m, q_2); \tag{5}$$

and hence

$$\sum_g P(\Psi(v_m^Y) \mid g, m, q_1) = \sum_g P(\Psi(v_m^Y) \mid g, m, q_2),$$

in which $\Psi : V - \Re$ is an arbitrary absolute performance metric for multiobjective samples. So, for any k, any search space X, and any space of fitnesses Y, an NFL result holds for the complete space of k-objective problems $g : X \to Y^k$.

5 NFL on Certain Subsets of Multiobjective Problem Spaces

A further (and perhaps more interesting) issue concerns the shape of the Pareto Front; this is a discussion which we will now motivate. The fact that there is no free lunch over the set of all k-objective problems means that any multiobjective algorithm (such as SPEA [17]) has no absolute performance advantage over any other algorithm (e.g. NSGA-II [3]), over the (understood) space of all multiobjective problems. This result, however, concerns averaging performance over all possible problems, and hence including all possible shapes of Pareto front. It could be argued, however, that particular algorithms have particular advantages correlated with the shape of the Pareto Front. E.g. we may imagine that SPEA with a large internal population size will do better on problems with a highly discontinuous Pareto front than would SPEA with a small internal population size. We can show, however, that for a particular shape of Pareto front, an NFL result holds over the space of all multiobjective problems which share that shape of Pareto front. We proceed towards proving this, and then note the case of 'general' multiobjective problems.

Consider again the space of all k-objective problems $g : X \to Y^k$, and the set $W = \{1, 2, ..., |Y|^k\}$ which labels the space of k-objective fitness vectors. It will help to illustrate a small example of the correspondence between W and Y^k as follows. Let $Y = \{1,2,3\}$ and $k = 2$, and therefore $W = \{1,...,9\}$. The correspondence between Y^k and W is given by Table 1. We will refer to this Table in illustrations next.

Table 1: A simple correspondence between a space of multiobjective fitness vectors Y^k (in this case 2-objective, where $k = 2$ and each objective has 3 possible values) and a single-objective space W of the appropriate size (in this case 9). Four points are bold, representing a particular subset of problems, with the underlined representing a particular Pareto front

Vector from Y^k	(1,1)	**(1,2)**	(1,3)	**(2,1)**	(2,2)	(2,3)	**(3,1)**	(3,2)	(3,3)
'Label' from W	1	2	3	4	5	6	7	8	9

Consider a subset $B \subset Y^k$, and the corresponding labels $L(B) \subset W$. E.g., in table 1, the bold entries identify $B = \{(1,1),(1,2),(2,1),(3,1)\}$ and $L(B) = \{1,2,4,7\}$. Any such subset has its own non-dominated 'front' $N(B)$, in this case $N(B) = \{(1,2),(3,1)\}$ and $L(N(B)) = \{2,7\}$.

If we are interested in a *particular* Pareto front C which exists for at least some multiobjective problems in $g: X \to Y^k$, we can pick out the points from Y^k which constitute this front, and then easily construct the maximal subset $Q \subseteq Y^k$ such that $N(Q) = C$ and the remainder of Q is made up of all points in Y^k which are *dominated* by at least one point in C. This corresponds to, in table 1, choosing the front represented by $C = \{(1,2),(3,1)\}$, in which case $Q = \{(1,1),(1,2),(2,1),(3,1)\}$.

Clearly we can construct such a Q for any given Pareto front C in the space of problems $g: X \to Y^k$, for any X, Y, and k. Given C and Q, we can ask: does NFL hold over the space H of all multiobjective problems $h: X - Q$, where a restriction on $h \in H$ is that the domain of h contains C? The restriction ensures that every function $h \in H$ has the same Pareto front C, but in general h may only contain additionally in its domain a subset of the remainder of Q. Hence H is the space of all possible multiobjective problems $g: X \to Y^k$ whose Pareto front is C. The anwer is affirmative, and we prove it simply, by using this result from [11]:

Theorem 2:
For any two algorithms a and b, equation (1) holds iff the set of functions F concerned is closed under permutation (c.u.p.).

This is (effectively) stated and proved in [11]. Recall that a function $f: X - Y$ is an assignment of a $y \in Y$ to each $x \in X$, and can hence be viewed as a vector of $|X|$ elements from Y, where element $f_i = y$ iff $f(x_i) = y$. If a set of functions F is 'closed under permutation', this simply means that we can arbitrarily permute the elements of f, i.e. we apply a bijective mapping $\pi: X \leftrightarrow X$ and the resulting new function f', with $f'(\pi(x)) = f(x)$, remains a member of F. We can now state and prove the following:

Theorem 3:
Given the set of multiobjective functions $g: X \to Y^k$, and a subset H, where $h \in H \Leftrightarrow h: X - Q$ and $\forall c \in C, \exists x \in X, h(x) = c$, where C is the Pareto subset of $Q \subseteq Y^k$, equation (5) holds over the H. That is, given a particular Pareto front, NFL holds over the space of multiobjective problems which share that Pareto front.

Proof of Theorem 3:

Given H, C and Q as defined in the theorem, we can identify a corresponding subset J of single-objective functions, and ask if that set is closed under permutation. This set is $j \in J \Leftrightarrow j : X - L(Q)$ and $\forall c \in C, \exists x \in X, j(x) = L(c)$. Consider any $j \in J$, and any point x such that $j(x) = L(c)$ for some $c \in C$. A permutation $\pi : X \leftrightarrow X$ leads to a function j' such that $j'(\pi(x)) = j(x)$, so clearly there is a point in the new function ($\pi(x) \in X$) whose fitness value is $L(c)$. Thus, every point in $L(c)$ is in the domain of j'. Next, suppose the domain of j' contains some point $u \notin L(Q)$; this means there is some point u such that $j'(x) = u$. But, $j'(x) = j'(\pi(z)) = j(z)$ for some $z \in X$, and the domain of j is $L(Q)$; this is a contradiction, so j' cannot contain a point outside $L(Q)$. This proves that $j' \in J$, and hence the set J is closed under permutation.

We know that NFL holds for the set of single objective problems J. By appeal to the one-to-one correspondences between H and J, Q and $L(Q)$, C and $L(C)$, and by using machinery developed earlier to produce one-to-one correspondences between multiobjective samples and single objective samples, and multiobjective and single objective algorithms, we can conclude that

$$\sum_{h \in H} P(v_m^y | h, m, q_1) = \sum_{h \in H} P(v_m^y | h, m, q_2),$$

where q_1 and q_2 are multiobjective algorithms $q_1, q_2 : V - X$ and H is as described.

This has shown that, over all k-objective problems with a particular Pareto front C, there is no free lunch in terms of absolute metrics. We can still ask about the set of k-objective problems with a particular *shape* of Pareto front C. We do not define how to characterize a 'shape' here, but would expect that any such definition would amount to enumerating a particular finite set of k-objective Pareto fronts $S = \{C_1, C_2, ..., C_n\}$ where $C_i \subset Y^k$ for each i. Given such a set S, we can easily prove:

Theorem 4:

Given any particular *shape* (or other characterisation) of Pareto front, corresponding to a set of specific fronts $S = \{C_1, C_2, ..., C_n\}$ where $C_i \subseteq Y^k$ for each i, there is no free lunch over the set of all k-objective problems whose Pareto front is in S.

Proof of Theorem 4:

We know that, for each individual C_i, the corresponding space of multiobjective problems H is closed under permutation. Since a union of such sets is also closed under permutation, it follows that the set S is closed under permutation.

Finally, we have confirmed that the absolute performance of black-box multiobjective algorithms is subject to NFL over the space of all problems $g : X \rightarrow Y^k$; however, in general, a multiobjective problem will typically have a different range of fitnesses for each objective. That is, a given k-objective problem will be of the form $g : X - Y_1 \times Y_2 \times ... \times Y_k$, where $|Y_i| > 1$ for each i, but with no other constraint (such as $|Y_i| = |Y_j|$ for different i and j). One way to confirm that NFL holds over this more general multiobjective problem space is by appeal to the results about Pareto fronts. Any Pareto front defined in the space $Y_1 \times Y_2 \times ... \times Y_k$ corresponds to a Pareto front for some set of problems in Y_M^k, where Y_M has the largest cardinality of the $Y_i, i \in \{1,...,k\}$. So, that NFL holds over this general space can be shown as a special case of Theorem 4, where the set S includes all Pareto fronts of Y_M^k which are also Pareto fronts of $Y_1 \times Y_2 \times ... \times Y_k$.

6 Comparative Metrics

The basis of NFL theorems published to date concerns there being ultimately no performance advantage for one black box algorithm over another, when averaged over a suitable collection of problems. In NFL work published thus far, it tends to be assumed that the measure of performance of algorithm a is purely a function of the sample yielded by algorithm a when run on a problem. So far, we have confirmed that NFL holds over certain spaces of multiobjective problems, and what follows from this, in terms of statements about algorithm *performance*, can only be stated with respect to 'absolute' performance metrics – i.e. performance measures which are based only on the sample produced by a single algorithm. Several such metrics are occasionally used in multiobjective optimisation studies (such as hypervolume [17], or others which can be found in [12]). However, a more common and favoured way to measure the performance of a multiobjective algorithm is via a metric: $\Psi : V^d \to \Re$, where $d > 1$ typically $d = 2$), where the metric takes in the results of two or more algorithms and produces a measure of relative performance; Zitzler's 'Cover' metric is a commonly employed example [17], while more sophisticated but currently less used examples can be found in [6].

Before we consider NFL in relation to such metrics, it is worth mentioning that they are far from trivial counterparts to 'absolute' performance metrics. E.g., some such metrics yield intransitive performance relationships between algorithms. See [8] for a full treatment of multiobjective comparison metrics and their properties in this sense, as well as [9] and [18].

It turns out that there *is* a potential 'free lunch' when comparative performance is concerned, at least for certain comparative metrics $\Psi : V \times V - \Re$. This arises owing to an asymmetry which occurs when we look at the performance of a pair of algorithms on the same problem. Every algorithm will generate a given sample on some problem in the set, but the *pair* of samples generated by two algorithms on some problem may not be generated by the same pair of algorithms on any other problem.

Preliminaries to what we call a 'Free Leftovers' theorem are as follows, with the earlier preliminaries (section 2) already in place. Let the set of algorithms be A, let the set of all comparative metrics be C, and let the set of problems be F (we will use the notation of [16] for simplicity, hence often oriented towards the single objective case but not ruling out the multiobjective case). A comparative 'experiment', E, is a function $E : A \times A \times F \times N \times C - \Re$ which separately runs two algorithms, $a, b \in A$ on a function $f \in F$ for $m \in N$ distinct function evaluations each, and then compares the resulting two samples using a comparative metric $\Psi \in C$, where $\Psi : D \times D \to \Re$. Hence, given two algorithms a and b, and a problem f, running experiment $E(a,b,f,m,\Psi)$ produces the result $\Psi(v_m^Y, w_m^Y)$, where v_m^Y is the sample produced by a on f, and w_m^Y is the result produced by b on f.

Importantly, we will be concerned here only with comparative metrics which are well-behaved in the following reasonable sense. Where v and w are samples, $v = w \Rightarrow \Psi(v,w) = 0$, and $\Psi(v,w) + \Psi(v,w) = 0$. Hence, when samples are the same, the 'performance difference' is 0, and in particular the metric is symmetric. That is, assuming without loss of generality that $\Psi(v,w) = r > 0$ indicates that (the algorithm which produced) sample v is better than (the algorithm which produced) sample w), the same metric will indicate that $\Psi(w,v) = -r$; hence, that w is worse than v, to the same extent that v is better than w. Now, familiarity with the NFL theorems and intuition may suggest that, when summed over all problems f, $E(a,b,f,m,\Psi)$ will amount to zero, since the performance advantage for a on some problem will presumably be countered by advantages for b on others. However, this is not so in general. In fact, [16] proves by example that

$$\exists a,b \in A, \sum_f P(d_m^Y | f,m,a) \cdot P(e_m^Y | f,m,b) \neq \sum_f P(d_m^Y | f,m,b) \cdot P(e_m^Y | f,m,a). \qquad (6)$$

In [16] it is called a *minimax distinction* when this type of asymmetry exists between algorithms, and [16] was concerned with the consequences as regards whether a large performance advantage for a over b on one function is matched by a similar advantage for b on another function, or instead by several small advantages for b on other functions. Here, however, we see how this type of result relates directly to the issue of comparative performance, which is especially salient in multiobjective optimisation. In the following, we build on this result and express it in our 'Experiment' notation, and derive a result which guarantees that, at least on a certain large space of problems, every algorithm has a 'straw man' – some other algorithm b which it outperforms on the whole over all functions in a space according to some metric. However, we only show here that this holds over spaces of permutation functions. Such a space can be defined as: $\Pi = \{f : X \leftrightarrow Y \,|\, \forall x,y \in X, f(x) = f(y) \Rightarrow x = y\}$, i.e. every point in the search space has a unique fitness. Our 'Free Leftovers in Permutation Spaces' theorem shows that *every* algorithm has a comparative performance advantage in a permutation space, (over some other algorithm, and given a particular metric).

Theorem 5 (Free Leftovers in Permutation Spaces)
Given any algorithm a, there are other algorithms b and well-behaved comparative metrics Ψ such that, when summed over all problems in a permutation space, a outperforms b. Formally:

$$\forall a \in A, \exists b \in A, \Psi \in C : \sum_{f \in \Pi} E(a,b,f,m,\Psi) > 0. \qquad (7)$$

Proof of Theorem 5

We start by showing a proof of the following statement:

$$\exists a,b \in A, \Psi \in C : \sum_f E(a,b,f,m,\Psi) \neq 0, \qquad (8)$$

which proposes the existence of at least one pair of algorithms whose summed comparative performance, for some well behaved metric, is nonzero. This is not confined to permutation spaces, and a theorem from which this follows is already proved by example in [15], however we also give a proof by example, since in so doing we rehearse and clarify some concepts in the proof of Theorem 6.

Towards proving equation (8), let $X = \{1,2,...,p\}$ and $Y = X$, with $p \in N$, and let problem $f_1 : X \to Y$ simply be defined by: $f_1(x) = x$. Further, let algorithm a be the algorithm which simply enumerates X from point 1 to point p. So, when we apply a to f_1 for m ($\leq p$) iterations, the sample v_m^Y obtained is simply $\{1,2,...,p\}$. Now consider algorithm b, which enumerates X in a different order as follows: the first point b visits is point 2, and thereon the nth point visited is point $n+1$, for $n < p$, and the final point visited is point 1. So, on problem f_1, algorithm b generates the sample $w_m^Y = \{2,3,...,p,1\}$. Now, we will construct a problem f_2 on which algorithm a produces sample v_m^Y. This is easily done – a visits points in a particular order, and so we simply arrange f_2 such that the appropriate fitnesses are at the points visited so as to generate w_m^Y. That is, we define f_2 as follows: $f_2(x) = x+1$ for $1 \leq x < p$, and $f_2(p) = 1$. Algorithm a will now generate w_m^Y on this function (and note that this is the *only* function on which algorithm a will generate this sample), but algorithm b does *not* generate v_m^Y; in fact algorithm b generates: $\{3,4,...,p,1,2\}$.

So far, to couch it in the more familiar notation of the NFL theorems, this has shown, by example, that equation (6) is true, which was proved in [15, 16]; we next derive equation (8) from this, which indicates what this means explicitly in terms of well-behaved comparative performance metrics, of the type commonly used in multiobjective optimisation.

Let algorithm a generate sample $v*_m^Y$ on problem f, and let algorithm b generate $w*_m^Y$ on the same problem. Further, let a and b be a pair, whose existence is guaranteed above, such that for any function g on which a generates sample $w*_m^Y$, b does *not* generate $v*_m^Y$. Now, define a well-behaved comparative performance metric Ψ by setting $\Psi(v*_m^Y, w*_m^Y) = 1$ (entailing $\Psi(w*_m^Y, v*_m^Y) = -1$), and $\Psi(v,w) = 0$ for all other distinct pairs of samples. When we run experiment $E(a,b,f,m,\Psi)$, we get the result $E(a,b,f,m,\Psi) = 1$, since this experiment results in a pair of samples for which the metric will provide this result. However, given the preceding, there no function g for which $E(a,b,g,m,\Psi) = -1$, which leads to and proves equation (8).

Now, on to Theorem 6. We will illustrate with a simple example as we go along. Let $a \in A$ and $f \in \Pi$; a generates a sample vaf_m on f. For example, f may be the function $f(x) = x, x \in \{1,...5\}$, with $m = 3$, and so $vaf_m = \{(1,1),(2,2),(3,3)\}$, and therefore $vaf_m^Y = (1,2,3)$. We will show that another algorithm b can always be constructed with the following properties: b produces a sample $vbf_m \neq vaf_m$ on f, (which, given $f \in \Pi$, guarantees that $vbf_m^Y \neq vaf_m^Y$), however there is no function $g \in \Pi$ such that both a produces sample vbf_m^Y on g and b produces sample vaf_m^Y on g. To show this, we start with the samples vaf_m and vbf_m. We first construct b so as to ensure that these samples are different in terms of performance (i.e. $vbf_m^Y \neq vaf_m^Y$), but that they 'overlap' (by which we mean $\exists x \in X, x \in vaf_m^X \wedge x \in vbf_m^X$ – i.e. the same point in the search space occurs at some position (not necessarily the same) in each sample). We can handle these conditions at one stroke by just arranging that $b(\phi) = vaf_2^X(2)$. I.e., the first point visited by b is the second point visited by a when a runs on f. Continuing our simple example, we have arranged b such that $vbf_m = \{(2,2),...\}$ – i.e. the first point visited by b is the second point visited by a, which ensures that, on problem f, different samples are generated by a and b, however some point in the search space (2) is visited by both. Now, assume that we have a function g on which b generates a sample vbg_m, where $vbg_m^Y = vaf_m^Y$, and on which a generates a sample vag_m where $vag_m^Y = vbf_m^Y$. So, in our example, we are assuming that we have some mapping $g := \{1,...,5\} \leftrightarrow \{1,...,5\}$ on which algorithm a generates vbf_m^Y (which so far we know starts with 2), and b generates (1,2,3) First, since a is constrained to first visit $a(\phi) = vaf_1^X(1)$, we need $g(vaf_1^X(1)) = vbf_1^Y(1)$. In our example, since a visits point 1 first, we must have $g(1)=2$. Similarly, since $b(\phi) = vbf_1^X(1)$, we need $g(vbf_1^X(1)) = vaf_1^Y(1)$. In our example, we must have $g(2)=1$, However, now recall that $b(\phi) = vaf_2^X(2)$, which is the second point visited by a on f. The second point visited by b on g must have the same fitness as this, in order to ensure $vbg_m^Y = vaf_m^Y$. In our example, this means that we have to arrange g such that $g(z)=2$, where z is the second point visited by b on g. But, we are free to construct b so that this is violated. So far our construction of b does not preclude any particular assignment to $b(\{(vaf_1^X(1), vbf_1^Y(1))\})$ (in our example, $b(\{(2,1)\})$) must in fact be 1, if we are to ensure $vbg_m^Y = vaf_m^Y$, since $g(1)=2$. However, we are free to construct b such that $b(\{(2,1)\})$ maps to any point we wish other than 2, which has already been visited by b. We will impose that $b(\{(vaf_1^X(1), vbf_1^Y(1))\}) = vaf_1^X(3)$, ensuring that no function g can exist with the assumed properties. In the example, we have made b visit point 3, which cannot have fitness 2 (since we already needed to ensure that point 1 had fitness 2, and each point must have a unique fitness), and so our attempt to construct a function g which allowed a and b to 'swap' their performances on f was doomed to failure.

7 Concluding Summary

We have confirmed that the NFL holds for absolute performance metrics in the realm of multiobjective optimization. Interestingly, we have shown that NFL holds over certain salient subsets of multiobjective problems, particularly: over the space of all problems which share the same type of Pareto front (subject to a reasonable way to define 'type' of front as an enumeration of examples). This is of interest since it speaks to the notion of test-function design. For example, Deb [2] has given the community an excellent framework for test function design, which explicitly differentiates between problems on the basis of the shape and nature of their Pareto fronts. The NFL result concerning Pareto fronts essentially means that we ultimately learn nothing from the finding 'algorithm A outperforms algorithm B on a problem with Pareto front P'.

This is because the NFL result guarantees that this will be counterbalanced by B outperforming A on some other problem or set of problems which share the same Pareto front P. However, inasmuch as the shape of the Pareto front (for problems in this or similar test suites) correlates with or corresponds to certain regularities of the *landscapes* in these test suites, there is certainly value in such test suites; the point is that any statements about comparative multiobjective algorithm performance should be carefully qualified. We have also shown that, in general, there are 'free leftovers' available for every algorithm, when compared with at least some other algorithm using a particular well-behaved performance metric. Much needs to be done to indicate how applicable this overall result is. For example, it is currently unknown which families of comparison metrics will yield such asymmetry and which won't, or what proportion of algorithms are thus summarily outperformed by a given algorithm. This and related work are the topic of further research.

References

1. Christensen, S. and Oppacher, F. (2001) What can we learn from No Free Lunch? A First Attempt to Characterize the Concept of a Searchable Function, in L. Spector et al (eds), *Proc. of GECCO 2001*, Morgan Kaufmann, pp. 1219–1226.
2. Deb, K. (1999) Multi-objective Genetic Algorithms: Problem Difficculties and Construction of Test Problems, *Evolutionary Computation* 7(3): 205–230.
3. Deb, K, Agrawal, S., Pratab, A. and Meyarivan, T. (2000) *A fast elitist non-dominated sorting genetic algorithm for multiobjectie optimization: NSGA-II*, KanGAL Technical Report 200001, Indian Institute of Technology, Kanpur, India.
4. Droste, S., Jansen, T. and Wegener, I. (1999), Perhaps Not a Free Lunch But At Least a Free Appetizer, in W. Banzhaf, J. Daida, A. E. Eiben, M. H. Garzon, V. Honavar, M. Jakiela and R. E. Smith (eds.) *Proc. of GECCO 9)*, Morgan Kaufmann Publishers, Inc., pp. 833–839.
5. Droste, S., Jansen, T. and Wegener, I. (2002) Optimization with Randomized Search Heuristics – The (A)NFL Theorem, Realistic Scenarios, and Difficult Functions", Theoretical *Computer Science*, to appear, and citeseer.nj.nec.com/droste97optimization.html .
6. Hansen, M.P. and Jaszkiewicz, A. (1998) *Evaluating the quality of approximations to the non-dominated set*, Tech. Report IMM-REP-1998-7, Technical University of Denmark.
7. Igel, C. and Toussaint, M. (2001) On Classes of Functions for which No Free Lunch Results Hold, see http://citeseer.nj.nec.com/528857.html.
8. Knowles, J.D. (2002) *Local search and hybrid evolutionary algorithms for Pareto optimization*, PhD Thesis, Department of Computer Science, University of Reading, UK.
9. Knowles, J.D. and Corne, D.W. (2002) On metrics for comparing non-dominated sets, in *Proc. 2002 Congress on Evolutionary Comp.*, IEEE Service Center, Piscataway, NJ.
10. Radcliffe, N.J. and Surry, P.D. (1995) Fundamental Limitations on Search Algorithms: Evolutionary Computing in Perspective, in *Computer Science Today*, pp. 275–291
11. Schumacher, C., Vose, M.D. and Whitley, L.D. (2001) The No Free Lunch Theorem and Problem Description Length, in L. Spector et al (eds), Proceedings of the Genetic and Evolutionary Computation Conference (GECCO 2001), Morgan Kaufmann, pp. 565–570.
12. Van Veldhuizen (1999) *Multiobjective Evolutionary Algorithms: Classifications, Analyses, and New Innovations*, PhD Thesis, Dept. of Electrical and Computer Eng., Graduate School of Engineering, Air Force Institute of Technology, Wright-Paterson AFB, Ohio.

13. Whitley, L.D. (1999) A Free Lunch Proof for Gray versus Binary Encodings, in Wolfgang Banzhaf and Jason Daida and Agoston E. Eiben and Max H. Garzon and Vasant Honavar and Mark Jakiela and Robert E. Smith (eds.) *Proceedings of the Genetic and Evolutionary Computation Conference, volume 1*, Morgan Kaufmann, pp. 726–733.
14. Whitley, L.D. (2000) Functions as Permutations: Implications for No Free Lunch, Walsh Analysis and Statistics, in Schoenauer, M., Deb, K., Rudolph, G., Yao, X., Lutton, E., Merelo, J.J. and Schwefel, H.-P. (eds.) *Proc. of the Sixth International Conference on Parallel Problem Solving from Nature (PPSN VI)*, Springer Verlag, Berlin, pp. 169–178.
15. Wolpert, D.H. and Macready, W.G. (1995) *No Free Lunch Theorems for Search*, Santa Fe Institute Technical Report SFI-TR-05-010, Santa Fe Institute, Santa Fe, NM.
16. Wolpert, D.H. and Macready, W.G. (1997) No Free Lunch Theorems for Optimization, *IEEE Transactions on Evolutionary Computation* **1**(1): 67–82.
17. Zitzler, E. (1999) *Evolutionary Algoirthms for Multiobjective Optimization: Methods and Applications*, PhD Thesis, Swiss Federal Institute of Technology (ETH), Zurich, Switzerland, November.
18. Zitzler, E., Thiele, L., Laumanns, M., Fonseca, C.M., Fonseca, V.G. (2002) *Performance assessment of multiobjective optimizers: an analysis and review*, available from the url: http://citeseer.nj.nec.com/zitzler02performance.html.

A New MOEA for Multi-objective TSP and Its Convergence Property Analysis

Zhenyu Yan[1], Linghai Zhang[2], Lishan Kang[1], and Guangming Lin[3,4]

[1] State Key Laboratory of Software Engineering, Wuhan University,
Wuhan 430072, P.R. China
{Zyyan,Kang}@Whu.edu.cn
[2] Department of Computer Science, York University,
Toronto, Ontario, M3J 1P3, Canada
lhzhang@cs.yorku.ca
[3] School of Computer Science, UC, UNSW Australian Defence Force Academy
Northcott Drive, Canberra, ACT 2600 Australia
[4] Capital Bridge Securities Co.,Ltd, Floor 42, Jinmao Tower,
Shi Ji avenue No.88, Shanghai, P.R.China
Lgm@Chinacbs.com

Abstract. Evolutionary Multi-objective Optimization(EMO) is becoming a hot research area and quite a few aspects of Multi-objective Evolutionary algorithms(MOEAs) have been studied and discussed. However there are still few literatures discussing the roles of search and selection operators in MOEAs. This paper studied their roles by a representative combinatorial Multi-objective Problem(MOP): Multi-objective TSP. In the new MOEA, We adopt an efficient search operator, which has the properties of both crossover and mutation, to generate the new individuals and chose two kinds of selection operators: Family Competition and Population Competition with probabilities to realize selection. The simulation experiments showed that this new MOEA could get good uniform solutions representing the Pareto Front and outperformed SPEA in almost every simulation run on this problem. Furthermore, we analyzed its convergence property using Finite Markov Chain and proved that it could converge to Pareto Front with probability 1.We also find that the convergence property of MOEAs has much relationship with search and selection operators.

1 Introduction

Evolutionary Multi-objective Optimization(EMO) is becoming a hot research area and quite a few aspects of MOEAs have been studied and discussed. However there are still few literatures discussing the roles of search and selection operators in MOEAs. Then one question emerges: do search operator and selection operator play important roles in MOEAs? To answer this question, we try to solve a representative combinatorial Multi-objective Problem(MOP): Multi-objective TSP with the search and selection operator we designed specially for this problem and compare our algorithm with the other current MOEAs.

TSP is one of well-known and wide-studied NP hard problems. In short, TSP is, given a finite number of "cities" along with the cost of travel between each pair of them, to find the cheapest way of visiting all the cities and returning to the starting point.

In many real-world problems, we should consider more than one kind of costs simultaneously. For example, if we arrange the routes for traveling, we may hope to not only minimize the traveling expense but also save the traveling time. This is a two-objective problem. We even often confront with the optimization problem with more than two objectives. We call this kind of TSP(including two objectives) multi-objective TSP.

Generally speaking, multi-objective problems(MOPs) are harder than single-objective problems. The goal of single-objective optimization is to find the optimal solution. But in MOPs, there are several conflicting objectives, so there is usually not a solution that is optimal to all objectives but a group of solutions (Pareto Optimal Set) that are not dominated by each other. With their Domain knowledge, decision makers (DM) can choose their preferable solutions from the Pareto Optimal Set. So our goal of solving MOP is to find a group of "best" solutions for the decision maker to choose.

Evolutionary algorithms (EAs) are becoming an important alternative to solve MOP. Besides the common merits of traditional EAs, MOEAs has their unique advantage, namely finding a group of solutions to represent the Pareto Front rather than one solution in a run because of their inherent parallel character and group strategy. In this paper, We will use this technology to solve Multi-objective TSP.

The organization of the paper is as follows. In section 2, we give some definitions and conceptions which we will use in the later sections. In section 3, we present the new algorithm for multi-objective TSP. In section 4, we illustrate the power of the new MOEA by some experiments and comparisons. In section 5, we analyze its convergence property using Finite Markov Chain and prove that it can converge to Pareto Front with probability 1.

2 Some Definitions and Conceptions

Definition 1 (Single-objective TSP). *Given n cities and an n∗n distance matrix $D=(d_{i,j})$, to minimize $f(\pi) = (\sum_{j=1}^{n-1} d_{\pi(j),\pi(j+1)}) + d_{\pi(n),\pi(1)}$, where π is a permutation over the set(1,2,...,n).*

Definition 2 (Multi-objective TSP). *Given n cities and a set $\{D_1, D_2, ..., D_3\}$ of n∗n matrices with $D_h = (d_{i,j}^h)$, to minimize $f(\pi) = (f_1(\pi), f_2(\pi), ..., f_k(\pi))$ with $f_i(\pi) = (\sum_{j=1}^{n-1} d_{\pi(j),\pi(j+1)}^i) + d_{\pi(n),\pi(1)}^i$, where π is a permutation over the set(1,2,...,n).*

Definition 3 (Pareto Dominance).

Definition 4 (Pareto Optimality).

Definition 5 (Pareto Optimality Set).

Definition 6 (Pareto Front).

Please refer to [5] for Definition 3–6.

3 A New MOEA for Multi-objective TSP

We use the conceptions and definitions in Section 2 to design our algorithm, which is based on the two aims: (1) Preference to nondominated solutions in a population. (2)Maintenance of diversity among nondominated solutions.

3.1 The Flow of the Algorithm

In the following subsections, we will give the explanations of **Reproduction Operator**, **better**, **Population Competition** and **Family Competition** respectively.

Algorithm 1 The flow of the new MOEA

t=0
Initialize the population P_t(population size =N)
Locate S_{best} and S_{worst} by sorting P_t using **better**
while not termination condition do
  for each individual $s_i \in P_t$ do
    Generate s'_i, which is the offspring of s_i, with **Reproduction Operator**
    if rand()$> \tau$ then
      Population Competition
    else
      Family Competition
    end if
  end for
  Update S_{best} and S_{worst}
  t=t+1
end while
Output the set S_{best} as the solutions

3.2 Reproduction Operator

Concerning the encoding, an individual is a vector of n integers ranging from 1 to n, each of which appears exactly once. Here we use the inver-over operator (*Algorithm 2*) [2],[3], instead of traditional crossover and mutation operators to generate new individuals. By integrating both crossover and mutation, the inver-over operator can make full use of the heuristic information in the population.

Algorithm 2 Reproduction operator (input s_i, output s'_i)

$s = s_i$
Select a city c randomly from the cities in s
loop
  if rand()$< \delta$ **then**
    Select a city c' randomly from the remaining cities in s
  else
    Select an individual x_k randomly from P_t
    Assign to c' the next city to the city c in s_k
  end if
  if the next or the previous city of city c in individual s is c' **then**
    Jumps out loop and return s
  else
    Assign to c'' the next city to the city c in s
    Inverse s from city c'' to city c'
    $c=c''$
  end if
end loop

To illustrate algorithm 2, we give an example. Suppose rand()>probability δ, Let n=8, s=(8,3,2,6,5,1,4,7), c =3, s_k=(4,6,3,1,8,7,5). Then the city next to city 3 in s_k is city 1 and the next city to city 3 in s is city 2, so we inverse s from city 2 to city 1. The result is (8,3,1,5,6,2,4,7). The edge 3-1 is from s_k and 2-4 is a new edge. we can go on with the inversion of s_i until the termination criterion is satisfied. It is obvious that the new edges of s_i are almost obtained from other individuals in the population, which can guarantee the full use of the heuristic information in the population. On the other hand, if rand()<probability δ, the algorithm can generate two new edges by random inverse, which can prevent the algorithm from premature convergence.

3.3 Pareto Niching

We use niches to preserve the diversity of the population in order to find a uniform distribution of vectors. Here we use phenotypic space[6]. We define the normalized distance between any two solutions i and j as follows:

$$d(i,j) = [\sum_{k=1}^{M} \frac{(f_k^{(i)} - f_k^{(j)})^2}{(f_k^{max} - f_k^{min})^2}]^{\frac{1}{2}} \tag{1}$$

Where f_k^{max} and f_k^{min} are the maximum and minimum objective function values of the k-th objective respectively. These values are pre-determined before run by our estimation. $f_k^{(i)}$ and $f_k^{(j)}$ are the objective function values of i and j. For the solution i, $d(i,j)$ is computed for each solution j. The sharing function value is computed as follows:

$$Sh(d) = \begin{cases} 1 - (\frac{d}{\sigma_{share}})^a, & \text{if } d < \sigma_{share}; \\ 0, & \text{otherwise.} \end{cases}$$

σ_{share} is a maximum phenotypic distance allowed between any two individuals to become a member of a niche. a is a constant and we let $a=2$ here. Thereafter, the niche count of i is calculated by summing the sharing function values:

$$nc_i = \sum_{j=1}^{H} Sh(d_{i,j}) \qquad (2)$$

H is the size of the non-dominated front to which solution i belongs

3.4 Better

We compare individuals using **better** rather than assign fitness to them in traditional EAs, so that **better** is the basic motivator of the evolution.

The return value of $better(x_1, x_2)$ is Boolean. The flow of the algorithm is as follows:

Algorithm 3 $better(x_1, x_2)$

if x_1 dominates (refer to **def3**.) x_2 then
    Return *true*
else
    if x_2 dominates x_1 then
        Return *false*
    else
        if $x_1.nichecount < x_2.nichecount$ then
            Return *true*
        else
            Return *false*
        end if
    end if
end if

We sort the individuals in the population by **better** and try to find S_{best} and S_{worst}. However, there are still some individuals that can not be compared with each other by **better**(eg.if x_1 is not dominated by x_2, x_2 is not dominated by x_1 and $x_1.nichecount = x_2.nichecount = 0$, then we can not tell which is the better). Therefore the results of S_{best} and S_{worst}, after we sort the population using **better**, are two sets, which may contain more than one individuals.

3.5 Population Competition(PC) and Family Competition(FC)

FC means the competition occurs only between the father s_i and the son s'_i, i.e., if better (s'_i, s_i), then s'_i replaces s_i ,otherwise no replacement happens. PC happens in the whole population, we realize it by *algorithm 4*.

Algorithm 4 Population competition (input s'_i)

$S = S_{worst}$
while $S \neq \Phi$ **do**
  Select an individual a randomly from S and delete a from S
  if $\mathbf{better}(s'_i, a)$ **then**
    $P_t = P_t \cup s'_i$, $P_t = P_t - a$, update S_{worst}
    Algorithm terminates
  else
    if $\mathbf{better}(a, s'_i)$ **then**
      Discard s'_i
      Algorithm terminates
    end if
  end if
end while
$P_t = P_t \cup s'_i$, $S_{worst} = S_{worst} \cup s'_i$
Select b randomly from S_{worst}
$P_t = P_t - b$, $S_{worst} = S_{worst} - b$, update S_{worst}

Where the set S_{worst} is obtained by sorting the population with **better**. In PC, we first compare the new individual s'_i with the individuals in set S_{worst} using **better**. If the new one is **better** than some individual in set S_{worst}, we add the former into the population and eliminate the latter from the population, if some individual in set S_{worst} is **better** than the new one, then the new will be discarded. Otherwise, we add s'_i to S_{worst} and to the population and eliminate an individual in S_{worst} randomly from the population to keep the population size.

Because FC only happens between the father and the son, it can preserve the diversity of the families (or individuals). But a nondominated solution which can dominate other individuals in the population, may be discarded, if it is no **better** than its father. So the PC is also indispensable. By PC, a nondominated solution will be preserved if it at least dominates another solution in S_{worst} and a solution dominated by it will be eliminated. In order to take advantage of the two strategies, we choose PC and FC randomly basing on probability τ.

4 Simulation Experiments

4.1 Test Data

The problem instances were considered with 100, 250 and 500 cities, each having two objectives. As in [1], the distance matrices were generated at random, where each was assigned a random number in the interval[1,100].For the the convenience of comparison, different algorithms used the same distance matrices, once the matrices were generated at random.

4.2 Parameter Settings

The experiments were carried out using the following parameter settings for the new MOEA:

    Number of generations T : 500
    Population size N   : equal to n (number of cities)
    Random inverse rate δ : 0.02
    Inver-over rate 1-δ : 0.98
    PC rate   : 0.1
    FC rate   : 0.9
    σ_{share} : 0.001

The parameter settings for SPEA are from [1].
Both T and N are same in the parameter settings of the two algorithms for the convenience of comparison.

4.3 Experimental Results

Under the same parameter settings (T and N), the results we got by the new algorithm and SPEA during the first 5 runs were shown in figure 1-3. It is obvious that the new MOEA outperforms SPEA. After several runs, we found that the fronts achieved by the new MOEA completely dominate the fronts produced by SPEA almost every time. It seems a likely supposition that search operators (*inver-over*) and selection operators (*FC* and *PC*) play an important role here. We will discuss the properties of these operators in the next section. In order to probe the potential of this algorithm, we increased the number of generations T to 10000, and the results of 100 cities we got are shown in figure 4. The results are obviously better than the results of 100 cities in figure 1, which illustrates that this algorithm not only has high convergence speed, but can avoid premature and we can get better results by increasing iteration rounds in finite steps. We will discuss the related theoretical issues in the next section.

5 Convergence Property Analysis

The Convergence property analysis of MOEA is more difficult than traditional EA [4]. Only a little work has been done in this area and further research work is urgently needed. We analyze the new MOEA's convergence property and prove that it can converge to Pareto Front with probability 1. We also try to provide a general framework for the convergence property analysis of MOEA.

Definition 7 (Finite Markov Chain). *If S is a finite set and $\{X_t : t \in N_0\}$ an S-valued random sequence with the property*
$P\{X_{t+1} = j | X_t = i, X_{t-1} = i_{t-1}, \ldots, X_0 = i_0\} = P\{X_{t+1} = j | X_t = i\} = p_{ij}$
for all $t \geq 0$ and for all pairs $(i,j) \in S \times S$ then the sequence $\{X_t : t \in N_0\}$ is called a homogeneous finite Markov chain with state space S.

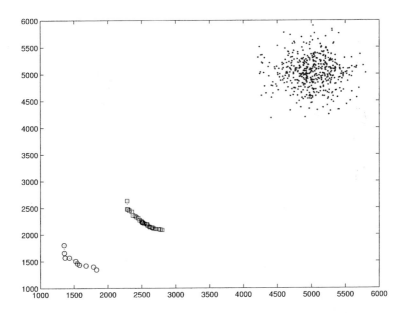

Fig. 1. Trade-off Fronts for the multi-objective TSP with 100 cities. The circles was the fronts achieved by the new MOEA, the squares was the fronts achieved by SPEA and the red dots represent the 5 initial populations

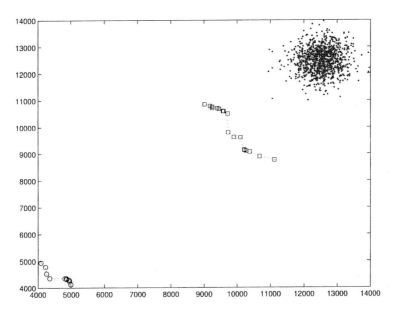

Fig. 2. Trade-off Fronts for the multi-objective TSP with 250 cities. The circles was the fronts achieved by the new MOEA, the squares was the fronts achieved by SPEA and the red dots represent the 5 initial populations

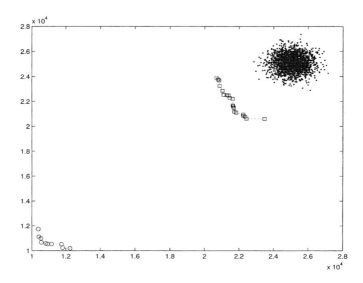

Fig. 3. Trade-off Fronts for the multi-objective TSP with 500 cities. The circles was the fronts achieved by the new MOEA, the squares was the fronts achieved by SPEA and the red dots represent the 5 initial populations

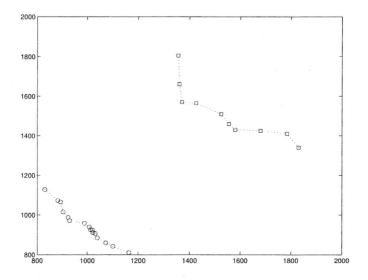

Fig. 4. Trade-off Fronts for the multi-objective TSP with 100 cities. The circles was the fronts achieved by the new MOEA with 10000 generations and the squares was the fronts achieved by the new MOEA with 500 generations. The former was much improved comparing with the latter

Since S is finite the transition probabilities can be gathered in the transition matrix $P = (p_{ij})_{i,j \in S}$. The row vector $\pi(t)$ with $\pi_i(t) = \text{P}\{X_t = i\}$ denotes the distribution of the Markov chain at step $t \geq 0$. Since

$$\pi(t) = \pi(t-1)P = \pi(0)P_t$$

for all $t \geq 1$, a homogeneous finite Markov chain is completely specified by its initial distribution $\pi(0)$ and its transition matrix P. the kth step transition probabilities are

$$p_{ij}(k) = \text{P}\{X_k = j | X_0 = i\} = e_i P^k e'_j,$$

where e_i is the ith unit vector, such that

$$\pi_j(t) = \sum_{i \in S} \pi_j(0) \cdot p_{ij}(t)$$

A matrix $A : n * m$ is termed *nonnegative* if $a_{ij} \geq 0$ and *positive* if $a_{ij} > 0$ for all $i = 1, \ldots, n$ and $j = 1, \ldots, m$. A nonnegative matrix is called *stochastic* if each row sum equals one. Thus, transition matrices are stochastic. A stochastic square matrix P is *irreducible* if

$$\forall i, j \in S : \exists k \in N : p_{ij}(k) > 0$$

and it is *primitive* or *regular* if

$$\exists k \in N : \forall i, j \in S : p_{ij}(k) > 0.$$

Therefore, every positive matrix P is regular and every regular matrix P is irreducible.

Lemma 1. [7] A homogeneous Markov chain with finite state space and irreducible transition matrix visits every state infinitely often with probability one regardless of the initial distribution.

Lemma 2. [8],[9] Let I,D,C,P,A be stochastic matrices where I is irreducible, D is diagonal-positive, C is column-allowable, P is positive, A is arbitrary. Then the products:
(a) AP and PC are positive,
(b) ID and DI are irreducible.

A stochastic matrix is termed diagonal-positive if every diagonal entry is nonzero, whereas it is called column-allowable if each column contains at least one positive entry. Thus, every diagonal-positive matrix is column-allowable.

Definition 8. *An element $x^* \in F$ is called a minimal element of the partial order set $(F, \preceq)$ if there is no $x \in F$ such that $x \prec x^*$. The set of all minimal elements, denoted $M(F, \preceq)$ is said to be complete if for each $x \in F$ there is at least one $x^* \in M(F, \preceq)$ such that $x^* \preceq x$.*

Let Ω be the MOP's variable space, then the Pareto Optimality Set (**Def5.**) is a minimal element set of the partial order set $(\Omega, \textbf{better})$.

Definition 9. *Let A_t be the population of some evolutionary algorithm at iteration $t \geq 0$ and $F_t = f(A_t)$ its associated image set. The algorithm is said to converge with probability 1 to the entire set of minimal elements if*

$$\lim_{t \to \infty} d(F_t, \Gamma^*) \to 0 \text{ with probability 1 as } t \to \infty$$

whereas it is said to converge with probability 1 to the set of minimal element if

$$\lim_{t \to \infty} \delta_{\Gamma^*}(F_t) \to 0 \text{ with probability 1 as } t \to \infty$$

Here, $d(A,B) = |A \cup B| - |A \cap B|$, $\delta_B A = |A| - |A \cap B|$, Γ^* denotes the set of minimal elements.

In the first case, the population size will eventually grow at least to the size of the set Γ^*. In the single-objective problems, $|\Gamma^*|=1$. But in MOPs, $|\Gamma^*|$ may be finite or infinite even as large as the search space.

Proposition 1. *The population sequence $(P_t)_t \geq 0$ of the new MOEA is a homogeneous finite Markov chain with positive transition matrix.*

Proof: It is obvious that the state space S is finite because TSP is discrete. The transition probabilities matrix P is irrelevant with t, because the parameter in the algorithm is constant during the run. So $P\{P_{t+1} = j | P_t = i, P_{t-1} = i_{t-1}, \ldots, P_0 = i_0\} = P\{P_{t+1} = j | P_t = i\} = p_{ij}$. Def 7. ensures that the sequence is a homogeneous finite Markov chain.

Now we need to prove the transition matrix is positive. Since the transition matrix, as it appears in this algorithm, is usually a product of several other transition matrices (describing e.g. mutation, crossover, selection cetera), we first discuss the properties of these matrices in the algorithm. As in section 3.2, we don't use traditional mutation and crossover operator to generate new individuals, but employ the *inver-over operator*. This operator inverses the cities between two random cities with rate δ, and inverses the cities between two "pre-determined" cities *(Algorithm 2)* with rate 1-δ. Both of the two actions are chosen stochastically and won't terminate until the two cities are next to each other. The latter action enables the algorithm to make full use of the heuristic information (the edges which have been in the population) in the population. The former action can generate new edges that the population doesn't have. With the two actions happening alternately, an individual can be changed into an arbitrary legal individual with probability more than zero. So the transition matrix of generate operator is positive.

It remains to characterize the column-allowable selection operator. Because the selection is realized by PC and FC, we discuss them respectively. If FC is chosen, and **better**$(s'_i, s_i) = false$, then s'_i will not replace s_i and no change of population will happen. If PC *(Algorithm 4)* is chosen, and some individual in S_{worst} is **better** than the new individual s'_i, then s'_i will be discarded and still no changes happen. In all, there is a positive probability that selection leaves the population unaltered. As a consequence, the transition matrix is diagonal-positive and therefore column-allowable.

Therefore, **Lemma 2.** ensures that the product of transition matrices of inver-over operator and selection operator is positive.

From the discussion above, we can conclude that population sequence $(P_t)_t \geq 0$ of the new MOEA is a homogeneous finite Markov chain with positive transition matrix.

Proposition 2. If the population sequence $(P_t)_t \geq 0$ is a homogeneous finite Markov chain with positive transition matrix then, this new MOEA converges to the set of minimal elements (Pareto Optimality Set) with probability 1.

Proof: Let N be the population size, Ω be the variable space and $\Gamma^* = M(F, \preceq)$ be minimal optimal set(Pareto Optimality)

i) Suppose N$\leq |\Gamma^*|$ and not all members of (P_t) are optimal. **Def4.** ensures that there exists at least one element in Ω whose image dominates the images of one or more than one elements in P_t. Since the transition matrix of $(P_t)_t \geq 0$ is positive, there exists a positive minimum probability that an arbitrary element of Ω is created by the **Reproduction Operator** (*Algorithm 2*) in one step. Owing to the Borel-Cantelli Lemma it is guaranteed that this arbitrary (and therefore every) element will be generated infinitely often and that the waiting time for the first occurrence as well as for the second, third, and so forth will be finite with probability 1. Consequently, a dominated(non-Pareto optimal) element can reproduce a son that dominates the father in finite time with probability one, and according to FC (3.5), it will enter the set P_t. According to the selection approach (PC, FC in 3.5), this element will stay in P_t forever *iff.* it is optimal. If it is not optimal, then it will be replaced by an optimal one after finite time by a repetition of the arguments given so far. Summing up: Although P_t is initialized with arbitrary elements, non-optimal elements in P_t will be replaced by optimal elements in finite time and P_t contains only optimal elements in the end. So $f(P_t) \subseteq \Gamma^*$ when $t \to \infty$.

ii) Let N$> |\Gamma^*|$. As previous discussion, the Pareto optimal will enter P_t in finite time. But there are still some non-optimal elements in P_t, because N$> |\Gamma^*|$. After we discard the non-optimal elements, $f(P_t) = \Gamma^*$

On the condition of i) and ii), $\delta_{\Gamma^*}(f(P_t)) = |f(P_t)| - |f(P_t) \cap \Gamma^*| = |f(P_t)| - |f(P_t)| = 0$ (see the definition of $\delta_B A$ in **Def 9.**), when $t \to \infty$. So **Def 9.** ensures that the new MOEA converges to the set of minimal elements (Pareto Optimality Set) with probability 1.

Proposition 3. The new MOEA converges to the set of minimal elements (Pareto Optimality Set) with probability 1.

Proof: Proposition 1 and proposition 2 lead to proposition 3.

6 Conclusion

Recently many literatures emphasized the importance of elitism, subpopulation, diversity preservation, niche and Pareto rank in MOEAs. As this paper shows, the search (Reproduction) operators and selection operators also play the importance roles in MOEAs. The inver-over operator exhibits its power in searching the good solutions, and the PC and FC also perform effectively in selection. Although SPEA outperform most of current MOEAs in some test problems because it combines established and new techniques in a unique manner [1], this algorithm obviously outperforms SPEA on multi-objective TSP. We consider the search and selection operators are the key reasons for that the new MOEA outperforms the other MOEAs in solving this problem. By the theoretical analysis, we can find that search (Reproduction) operators and selection operators are also the important factors to the convergence property of MOEA. so we propose that more attention should be paid to the search and selection operators and their influence to the efficiency and the convergence property of the MOEAs. And we need to design the appropriate and efficient search and selection operators for specified problems. In addition, convergence property analysis is also urgently needed to provide MOEAs with more theoretical foundations.

Acknowledgement. This paper was supported by National Natural Science Foundation of China (NO.70071042, NO.60073043 and NO.60133010).

References

1. Eckart Zitzler: Evolutionary Algorithm for Multi-Objective Optimization: Methods and Applications. dissertation of Swiss Federal Institute of Technology Zurich for the degree of Doctor of Technical Sciences, November 11, 1999.
2. Guo Tao: Evolutionary Computation and Optimization. dissertation of Wuhan University for the degree of Doctor of Technical Sciences, May, 2000.
3. Michalewicz, Z.: How to Solve It: Modern heuristics. Springer-Verlag, Berlin Heidelberg New York 2000
4. G.Rudolph and A.Agapie: Convergence Properties of Some Multi-Objective Evolutionary Algorithms. Proceedings of the 2000 IEEE Congress on Evolutionary Computation, pp.1010–1016.
5. David A.VanVeldhuizen and Gary B.Lamont: Multiobjective Evolutionary Algorithms: Analyzing the State-of-the-Art. Evolutionary Computation, 2000, 8(2) 125–147
6. N.Srinivas and Kalyanmoy Deb: Multiobjective Optimization Using Nondominated Sorting In Genetic Algorithms. Evolutionary Computation, 1995, 2(3) 221–248
7. M.Iosifescu: Finite Markov Processes and Their Applications. Chichester: Wiley, 1980.
8. G.Rudolph: Convergence properties of canonical genetic algorithms. IEEE Transactions on Neural Networks 1994, 5(1) 96–101.
9. A.Agapie: Genetic algorithms: Minimal conditions for convergence. Artificial Evolution: Third European Conference; select papers/AE'97,1998, pp.183–193. Berlin: Springer.

Convergence Time Analysis for the Multi-objective Counting Ones Problem

Dirk Thierens

Institute of Information and Computing Sciences, Utrecht University,
P.O. Box 80.089, 3508 TB Utrecht, The Netherlands
dirk.thierens@cs.uu.nl

Abstract. We propose a multi-objective generalisation for the well known Counting Ones problem, called the Multi-objective Counting Ones (MOCO) function. It is shown that the problem has four qualitative different regions. We have constructed a convergence time model for the Simple Evolutionary Multi-objective Optimiser (SEMO) algorithm. The analysis gives insight in the convergence behaviour in each region of the MOCO problem. The model predicts a $\ell^2 \ln \ell$ running time, which is confirmed by the experimental runs.

1 Introduction

In recent years, the evolutionary algorithms community has witnessed a vast increase in the design and application of multi-objective evolutionary algorithms (MOEAs) [2]. The popularity of MOEAs is largely due to the fact that, contrary to more traditional search techniques, population based search algorithms have the potential to convergence to the Pareto optimal front in a single optimisation run. At the moment, a number of different state-of-the-art MOEAs can be found in the literature. All these algorithms have their own characteristics, and are typically compared to each other in an experimental way. The formal analysis of their performance is still very limited. When evaluating the performance of a MOEA, two key questions need to be answered:

- first, how quickly and reliably does the algorithm converge ?
- second, how well does it find and maintain a diverse set of solutions ?

In this paper we address the convergence issue. A discussion on the diversity issue can be found elsewhere ([7] [1]). Existing work mainly focuses on an asymptotic analysis ([5] [6] [11] [12]). A notable exception can be found in [8] where a running time analysis for simple MOEAs on a specific problem - the Leading Ones Trailing Zeroes (LOTZ) function - is made. This study provides key insights complimentary to the asymptotic analyses. The specific choice of problem function is of course very important. It can be argued though that the LOTZ function is less suitable for investigating the performance of recombinative, population based, evolutionary algorithms. The convergence has an outspoken domino-like behaviour ([14]), and the first leading ones (resp. last trailing zeroes) take over the entire population very rapidly, thus preventing any recombination operator to contribute to the search process.

In this paper we also focus on a single problem function, and build a convergence model for a simple MOEA. The problem we propose is a multi-objective generalisation of the well known Counting Ones function (also known as One-Max). The study of the Counting Ones has provided key insights in the convergence behaviour of many single-objective evolutionary algorithms ([3] [4] [10] [9] [13]). By introducing the Multi-objective Counting Ones (MOCO) function, we hope to obtain similar insights in the performance of different MOEAs.

The next section formally defines the MOCO function. In Section 3 we derive the convergence model for a simple MOEA. Section 4 shows an experimental validation. A discussion and possible extensions are presented in Section 5. Finally, we conclude in Section 6.

2 The Multi-objective Counting Ones Problem

The Counting Ones function $CO(X) : \{0,1\}^\ell \to \mathbb{N}$ is defined as:

$$CO(X) = CO(x_1 x_2 \ldots x_\ell) = \sum_{i=1}^{\ell} x_i.$$

We transform the Counting Ones problem into the Multi-objective Counting Ones function $MOCO(X) : \{0,1\}^\ell \to \mathbb{R}^2$ by mapping the number of bits 1 to the unit circle ($[0\ldots\ell] \to [0\ldots 2\pi]$), and taking the cosine and sine:

$$MOCO(X) = (\cos\theta(X), \sin\theta(X)),$$

with

$$\theta(X) = \frac{2\pi}{\ell} CO(X).$$

The multi-objective optimisation problem is specified by searching for the maximal value of both $\cos\theta(X)$ and $\sin\theta(X)$. Figure 1 shows a plot of the MOCO function. It is easy to see that there are four qualitatively different regions to discern:

1. **Region I** is defined by $0 \le \theta \le \frac{\pi}{2}$, or equivalently, $0 \le CO(X) \le \frac{\ell}{4}$. None of the points on this part of the circle is Pareto dominated by any other point, therefore the entire region belongs to the Pareto optimal front.
2. **Region II** is defined by $\frac{\pi}{2} < \theta \le \pi$, or equivalently, $\frac{\ell}{4} < CO(X) \le \frac{\ell}{2}$. Every point in this region with $\theta = \theta_i$ Pareto dominates all points in region II and region III with $\theta = \theta_j$ and $\cos\theta_i > \cos\theta_j$. In addition it also Pareto dominates the point in region III with $\theta_j = 2\pi - \theta_i$.
3. **Region III** is defined by $\pi < \theta < \frac{3\pi}{2}$, or equivalently, $\frac{\ell}{2} < CO(X) < \frac{3\ell}{4}$. All points in this region do not Pareto dominate any other point.
4. **Region IV** is defined by $\frac{3\pi}{2} \le \theta \le 2\pi$, or equivalently, $\frac{3\ell}{4} \le CO(X) \le \ell$. Every point in this region with $\theta = \theta_i$ Pareto dominates points in region III and region IV with $\theta = \theta_j$ and $\sin\theta_j < \sin\theta_i$.

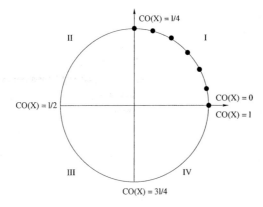

Fig. 1. The Multi-objective Counting Ones problem MOCO(X)

Formally, the Pareto optimal front $\mathcal{PF}$ and the corresponding Pareto optimal set $\mathcal{PS}$ are given by:

$$\mathcal{PF} = \{(\cos\theta, \sin\theta) | (0 \leq \theta \leq \frac{\pi}{2}) \vee (\theta = 2\pi)\},$$

and

$$\mathcal{PS} = \{X | (0 \leq CO(X) \leq \frac{\ell}{4}) \vee (CO(X) = \ell)\}.$$

3 Convergence Model

3.1 Simple Evolutionary Multi-objective Optimiser

The single-objective counting ones problem has been extensively studied in the EA literature, and simple closed-form convergence models have been obtained. Unfortunately, the multi-objective counting ones problem does not lend itself to such a straightforward approach. The difficulty is caused by a number of factors not present in the single-objective counting ones analysis, for instance the substantially different convergence characteristics in the four regions, the influence of the archive, the elitist mechanism, the need to keep diversity, ... In addition, the diverse set of efficient Multi-objective Evolutionary Algorithm proposed in the literature (eg. NSGA, SPEA, PAES, MIDEA), all have a different mechanism to construct the set of Pareto optimal solutions, and to encourage the search for diverse solutions. To facilitate the building of a convergence model, we restrict the analysis in this paper to a simple MOEA. The algorithm we consider here is the Simple Evolutionary Multi-objective Optimiser (SEMO) proposed by Laumanns et al. [8]. These authors investigated the running time complexity of SEMO on the multi-objective Leading-Ones-Trailing-Zeroes (LOTZ) problem. Although SEMO is much simpler than the full-fledged MOEAs, it possesses enough characteristics of these more complicated algorithms to make it a useful first step in the road towards a fully analytical understanding of the convergence behaviour of the state-of-the-art MOEAs. The SEMO algorithm is basically a stochastic local search algorithm, that keeps a population of all Pareto optimal solutions encountered so far. To explore the search space a randomly chosen solution from the archive is selected and mutated. Note that there is no use of recombination. Using the notation $X \succ Z$ for X Pareto dominating Z, the algorithm is specified as:

SEMO(ℓ)
1    $X \leftarrow$ GENERATE-RANDOM-BITSTRING(ℓ)
2    $P \leftarrow \{X\}$
3   repeat
4          $\leftarrow$ RANDOMLY-CHOOSE-STRING(P)
5          $X' \leftarrow$ FLIP-RANDOM-BIT(X)
6          $P \leftarrow P \setminus \{Z \in P | X' \succ Z\}$
7   if $\nexists Z \in P$ such that $(Z \succ X' \vee MOCO(Z) = MOCO(X'))$
8   then $P \leftarrow P \cup \{X'\}$
9   until TERMINATED-P(P)
10 return P

3.2 Convergence Time Analysis

Having defined the problem and the algorithm used, we can now proceed with the construction of the convergence model. Given the significant different behaviour in the four regions, we build the models for each region separately, ignoring for now the effects of the search process taking place in the other regions.

1. **Region I.** The exploration step in the SEMO algorithm consists of flipping exactly one randomly chosen bit. For any reasonable sized stringlength - say, $\ell \geq 20$ - the probability of starting the search process from within region I can be ignored. Therefore, the entire region I is generated in a fixed ordering: first, a string with $\frac{\ell}{4}$ bits-1 is generated, and since it Pareto dominates all points in the regions II and III, all points in the population P from those regions are removed (step 6 in the SEMO-algorithm). For the time being, we assume that no points from region IV have been added to P. Once a string with $\frac{\ell}{4} - 1$ bits-1 is generated, the Pareto set P contains 2 solutions, and further progress can only be obtained when the solution with the lowest number of bits equal to 1 is selected for exploration (step 4 in the SEMO-algorithm).

   Formally, call i the number of bits-1 of the solution with the lowest number of bits-1 in the current population P. Call $Pr(i)$ the probability of an improvement - this is, generating a string with one bit-1 less. Recognising that an improvement is only possible when the solution with the lowest number of bits-1 in P is chosen, and one of its bits-1 is flipped, we obtain:

$$Pr(i) = \frac{1}{\frac{\ell}{4} - i + 1} \frac{i}{\ell}.$$

The expected number of trials needed to generate the entire region I, $E[T^I]$, is the sum of the expected number of trials to go from a string with i bits-1 to a string with $i-1$ bits-1, starting from a string with $\ell/4$ bits-1 until the all zeroes string is reached. $E[T^I]$ can thus be calculated as:

$$E[T^I] = \sum_{i=1}^{\frac{\ell}{4}} \frac{1}{Pr(i)}$$

$$= \sum_{i=1}^{\frac{\ell}{4}} \frac{\ell}{i}(\frac{\ell}{4} - i + 1)$$

$$= \ell(\frac{\ell}{4} + 1) \sum_{i=1}^{\frac{\ell}{4}} \frac{1}{i} - \frac{\ell^2}{4}$$

$$\leq \ell(1 + (\frac{\ell}{4} + 1)\ln\frac{\ell}{4})$$

$$= O(\ell^2 \ln \ell)$$

The inequality is obtained by taking the upper boundary approximation for the harmonic series: $\sum_{i=1}^{n} \frac{1}{i} \leq \ln n + 1$.

2. **Region II.** Solutions in region II Pareto dominate all other solutions in this region that have a larger number of bits-1. Let us assume the search starts at a string with $\frac{\ell}{2}$ and continues into region II. Whenever a new string is generated with one bit-1 less than the current string, the new string replaces the current one in the population P. The entire region II is therefore traversed with a population size P equal to one. The probability of an improvement $Pr(i)$ when the current string in P has i bits-1 is simply:

$$Pr(i) = \frac{i}{\ell}.$$

The expected number of trials needed to traverse the entire region II, $E[T^{II}]$ is given by:

$$E[T^{II}] = \sum_{i=\frac{\ell}{4}-1}^{\frac{\ell}{2}} \frac{1}{Pr(i)}$$

$$= \sum_{i=\frac{\ell}{4}-1}^{\frac{\ell}{2}} \frac{\ell}{i}$$

$$= \sum_{i=1}^{\ell/2} \frac{\ell}{i} - \sum_{i=1}^{\ell/4-2} \frac{\ell}{i}$$

$$\leq \ell \ln \frac{2\ell}{\ell - 8}$$

$$\to \ell \ln 2$$

3. **Region III.** Points in region III do not dominate each other. When the search traverses the region from the string with $\frac{\ell}{2}$ bits-1 towards the string with $\frac{3\ell}{4}$ bits-1 all points are kept in the population. The probability of improvement $Pr(i)$ is similar to the one in region I, only now the number of bits-1 needs to increase:

$$Pr(i) = \frac{1}{i - \frac{\ell}{2} + 1} \frac{\ell - i}{\ell}.$$

The expected number of trials needed to traverse the entire region III, $E[T^{III}]$, can be calculated as:

$$E[T^{III}] = \sum_{i=\frac{\ell}{2}}^{\frac{3\ell}{4}-1} \frac{1}{Pr(i)}$$

$$= \sum_{i=\frac{\ell}{2}}^{\frac{3\ell}{4}-1} \frac{\ell}{\ell - i}(i - \frac{\ell}{2} + 1)$$

$$= \sum_{j=\frac{\ell}{4}+1}^{\frac{\ell}{2}} \frac{\ell}{j}(\ell - j - \frac{\ell}{2} + 1)$$

$$= \ell(\frac{\ell}{2} + 1) \sum_{j=\frac{\ell}{4}+1}^{\frac{\ell}{2}} \frac{1}{j} - \ell(\frac{\ell}{4} - 1)$$

$$\leq \ell(1 - \frac{\ell}{4} + (\frac{\ell}{2} + 1) \ln \frac{2\ell}{\ell + 4})$$

where we have used the substitution $l - i = j$.

4. **Region IV.** Points in region IV Pareto dominate all other points in this region that have a lower number of bits-1. The region is entered at the point with $\frac{3\ell}{4}$ bits-1 and moves towards the point with ℓ bits-1, which belongs to the overall Pareto optimal set of the MOCO problem. As in region II, region IV is traversed with a population size P equal to one. The probability of an improvement $Pr(i)$ when the current string in P has i bits-1 is:

$$Pr(i) = \frac{\ell - i}{\ell}.$$

The expected number of trials needed to traverse the entire region IV, $E[T^{IV}]$ is given by:

$$E[T^{IV}] = \sum_{i=\frac{3\ell}{4}}^{\ell-1} \frac{1}{Pr(i)}$$

$$= \sum_{i=\frac{3\ell}{4}}^{\ell-1} \frac{\ell}{\ell - i}$$

$$= \sum_{j=1}^{\frac{\ell}{4}} \frac{\ell}{j}$$

$$\leq \ell(\ln \frac{\ell}{4} + 1).$$

Until now we have modelled the convergence of the SEMO algorithm for the MOCO problem in each region separately. For the transition from region II to region I the two running times do not influence each other: when region I is entered - this is, when the string with $\frac{\ell}{4}$ bits-1 is generated - no points from region II (and III) can be in the population P.

The interaction between region II and III however is more significant. During the first generations of the search the population P expands in both regions simultaneously as long as the string with the largest number of bits-1 remains to have a higher value for $\cos \theta$ than the string with the lowest number of bits 1. If at any time in the search process this condition is not fulfilled, all points found so far in region III are Pareto dominated by the point in region II, and the population P is reduced to this single point. Once this happens it is impossible for the search to return to region III. The expected speed at which the search expands in region II and III is the same, causing most of the points in region III that have a smaller number of bits 1 than the current rightmost point to be dominated by the point in region II. Therefore the size of the population P remains small, and the time spend in this phase of the search is small compared to the $O(\ell^2 \ln \ell)$ running time of region I.

As a first order approximation we assume that the overall running time of the SEMO algorithm on the MOCO problem can be approximated by the sum of the independent expected running times in region I and region II:

$$E[T] \approx E[T^I] + E[T^{II}]$$

$$= \ell(\frac{\ell}{4} + 1) \sum_{i=1}^{\frac{\ell}{4}} \frac{1}{i} - \frac{\ell^2}{4} + \sum_{i=1}^{\ell/2} \frac{\ell}{i} - \sum_{i=1}^{\ell/4-2} \frac{\ell}{i}$$

$$\leq \ell(1 + \ln 2 + (\frac{\ell}{4} + 1) \ln \frac{\ell}{4})$$

3.3 Bitwise Mutation

Before we experimentally test the approximated expected running times, we consider the effect of the bitwise exploration step (step 5 in the SEMO-algorithm). The convergence model assumes a 1 bit-flip stochastic search - this is, each exploration step a single, randomly chosen bit is flipped. In the standard genetic algorithm approach every bit in the string is flipped with a small, constant probability p_m. For the Counting Ones problem it has been shown that the most difficult step is getting the last bit correct, and the optimal mutation probability for this to happen is $p_m = \frac{1}{\ell}$. The exact modelling of the constant mutation probability search is rather involved. We can however get an easy approximate model for the MOCO problem when calculating the probability of the number of bits-1 flipped in one particular string with $p_m = \frac{1}{\ell}$. The probability $Pr(k|i)$ that exactly k bits-1 are flipped when there are i bits-1 in a string of length ℓ is given by the probability that exactly k bits are flipped, times the probability that these k bits are all bits-1:

$$Pr(k|i) = \binom{\ell}{k} \frac{1}{\ell^k}(1-\frac{1}{\ell})^{l-k} \prod_{j=0}^{k-1} \frac{i-j}{l-j}.$$

It is instructive to compute the ratio between the expected waiting time for two bits 1 to be flipped simultaneously in one offspring, and in two consecutive 1 bit flip mutations:

$$\frac{\frac{1}{Pr(2|i)}}{\frac{2}{Pr(1|i)}} = \frac{\ell\frac{1}{\ell}(1-\frac{1}{\ell})^{\ell-1}\frac{i}{\ell}}{\frac{\ell(\ell-1)}{2}\frac{1}{\ell^2}(1-\frac{1}{\ell})^{\ell-2}\frac{i}{\ell}\frac{i-1}{\ell-1}}$$
$$= \frac{\ell-1}{i-1}$$

Considering that most time is spend when i is close to ℓ, it is clear that the vast majority of improvements are caused by single bit-flip mutations. Recognising that the probability of a single bit-flip mutation equals $(1-\frac{1}{\ell})^{\ell-1} \to e^{-1}$, we can approximate the expected running time $E[T_{p_m}]$ for the bitwise mutation algorithm ($p_m = 1/\ell$) with:

$$E[T_{p_m}] \approx e.E[T].$$

4 Experimental Validation

We compare the analytical results of the previous section with experimental results for strings with length $\ell \in \{20, 40, 60, 80, 100\}$, averaged over 100 runs. We calculate the expected running time $E[T]$ using the exact calculation for the harmonic series, and the approximated value ($\tilde{E}[T]$). The tables show the average, standard deviation, minimum, and maximum number of trials needed to generate the entire region I. The first table shows the results for the 1 bit-flip stochastic local search, while the second shows the results for the bitwise mutation algorithm. It is clear that the convergence model represents a good approximation of the average running time.

ℓ	$E[T]$	$\tilde{E}[T]$	T_{avg}	T_{std}	T_{min}	T_{max}
20	196	227	210	125	59	666
40	924	1081	976	477	310	2969
60	2334	2701	2402	1042	818	6899
80	4507	5168	4593	1717	1902	8911
100	7498	8538	8137	3466	2171	22360

ℓ	$E[T_{p_m}]$	$\hat{E}[T_{p_m}]$	T_{avg}	T_{std}	T_{min}	T_{max}
20	532	617	483	311	89	1695
40	2511	2938	2126	994	637	5171
60	6345	7343	5965	3566	1632	27457
80	12251	14049	11567	5642	3153	36431
100	20382	23210	18386	8201	5746	47023

5 Discussion & Extensions

The model build in this paper gives an insight in the convergence behaviour of a simple MOEA for the MOCO problem. This work can be extended in various ways:

- The SEMO algorithm spends a lot of time expanding the Pareto set in region I. Most of these trials cannot be successful since the population P can only be expanded when the point with the lowest number of bits-1 is selected for exploration. A similar problem occurred for the LOTZ problem, and there the FEMO algorithm was proposed to solve this problem. It would be interesting to build a convergence model for the FEMO algorithm on the MOCO problem. One difficulty is that the interacting search in region II and III is now more influential in comparison with the search in region I.
- We have mapped the number of bits-1 to the interval $[0\ldots 2\pi]$. We could however also map it to the interval $[0\ldots 4\pi]$, thus creating a discontinuous Pareto optimal front.
- The analysis has been limited to a non-recombinative MOEA. It is important that similar convergence models are constructed for recombinative, population based MOEAs.
- The MOCO problem considered has two objectives. An interesting question would be to build a convergence model for the 3-dimensional MOCO, where we map the number of bits-1 to the unit sphere as opposed to the unit circle.

6 Conclusion

We have proposed a multi-objective generalisation for the well known Counting Ones problem, called the Multi-objective Counting Ones (MOCO) function. We have constructed a convergence time model for this problem when running the Simple Evolutionary Multi-objective Optimiser (SEMO) algorithm. The analysis gives insight in the convergence behaviour in the four qualitative different regions of the problem. The model predicts a $\ell^2 \ln \ell$ running time, which is confirmed by experimental runs.

References

1. P.A.N Bosman and D. Thierens. The balance between proximity and diversity in multi–objective evolutionary algorithms. *IEEE Transactions on Evolutionary Computation*, 2003. Accepted for publication.

2. K. Deb. *Multi-objective optimization using evolutionary algorithms*. Wiley, Chichester, UK, 2001.
3. S. Droste, T. Jansen, and I. Wegener. On the analysis of the (1+1) evolutionary algorithm for separable functions with Boolean inputs. *Evolutionary Computation*, 6(2):185–196, 1998.
4. S. Droste, T. Jansen, and I. Wegener. On the analysis of the (1+1) evolutionary algorithm. *Theoretical Computer Science*, 276(1–2):51–81, 2002.
5. T. Hanne. On the convergence of multiobjective evolutioanry algorithms. *European Journal of Operations Research*, 117(3):553–564, 1999.
6. T. Hanne. Global multiobjective optimization with evolutionary algorithms: selection mechanisms and mutation control. In *Proceedings of the International Conference on Multi-Criterion Optimization*, pages 197–212. Springer, 2001.
7. M. Laumanns, L. Thiele, K. Deb, and E. Zitzler. Combining convergence and diversity in evolutionary computation. *Evolutionary Computation*, 10(3):263–282, 2002.
8. M. Laumanns, L. Thiele, E. Zitzler, E. Welzl, and K. Deb. Running time analysis of mulit-objective evolutionary algorithms on a simple discrete optimization problem. In J.J. Merelo Guervós et al., editor, *Proceedings of the 7th International Conference on Parallel Problem Solving from Nature*, pages 44–53. Springer, 2002.
9. H. Muehlenbein. How genetic algorithms really work: Mutation and hill-climbing. In R. Manner et B. Manderick, editor, *Proceedings of the International Conference on Parallel Problem Solving from Nature*, pages 15–26. North-Holland, 1992.
10. H. Muhlenbein and D. Schlierkamp-Voosen. Predictive models for the breeder genetic algorithm: I. continuous parameter optimization. *Evolutionary Computation*, 1(1):25–49, 1993.
11. G. Rudolph. On a multi-objective evolutionary algorithm and its convergence to the Pareto set. In *Proceedings of the International Conference on Evolutionary Computation*, pages 511–516. IEEE Press, 1998.
12. G. Rudolph and A. Agapie. Convergence properties of some multi-objective evolutionary algorithms. In *Proceedings of the International Congress on Evolutionary Computation*, pages 1010–1016. IEEE Press, 2000.
13. D. Thierens and D.E. Goldberg. Convergence models of genetic algorithm selection schemes. In *International Conference on Evolutionary Computation: The Third Conference on Parallel Problem Solving from Nature*, pages 119–129. Springer-Verlag, 1994.
14. D. Thierens, D.E. Goldberg, and A.G. Pereira. Domino convergence, drift, and the temporal-salience structure of problems. In *Proceedings of the 1998 IEEE World Congress on Computational Intelligence*, pages 535–540. IEEE Press, 1998.

Niche Distributions on the Pareto Optimal Front

Jeffrey Horn

Department of Mathematics and Computer Science
Northern Michigan University
Marquette, Michigan, 49855 USA
jhorn@nmu.edu
http://cs.nmu.edu/~jeffhorn

Abstract. This paper examines the use of fitness sharing in evolutionary multi-objective optimization (EMO) algorithms to form a uniform distribution of niches along the non-dominated frontier. A long-standing, implicit assumption is that fitness sharing within an equivalence class, such as the Pareto optimal set, can form dynamically stable (under selection) subpopulations evenly spaced along the front. We show that this behavior can occur, but that it is highly unlikely. Rather, it is much more likely that a steady-state will be reached in which stable niches are maintained, but at inter-niche distances much less than the specified niche radius, with several times more niches than previously predicted, and with non-uniform sub-population sizes. These results might have implications for EMO population sizing, and perhaps even for EMO algorithm design itself.

1 Introduction

Maintaining diversity in the evolving population of an evolutionary multi-objective optimization (EMO) algorithm has become widely recognized as essential to its success [2]. One of the most popular diversity promotion mechanisms is fitness sharing, first introduced by Goldberg and Richardson [4] for function optimization. While the use of fitness sharing for niching (i.e., niche formation) in single objective (function) optimization has been well studied [1,5,9], its use in the multi-objective domain has received little analysis. In particular, it seems that many of us EMO algorithmists assume that fitness sharing among members of the on-line Pareto optimal set will lead to formation of stable niches evenly spaced out along the Pareto front in objective space, with the use of a fixed niche radius controlling the inter-niche distance. In this paper we will investigate the validity of this assumption.

2 Background

Sharing methods [4], in which the objective fitness of an individual in the population is reduced because of the need to share limited resources with competitors (including copies of itself), are typically incorporated into artificial evolution in order to promote diversity, to find multiple solutions [4], or to evolve a group of "cooperative" individuals or species [5]. In the field of EMO, fitness sharing has been used in an at tempt to "cover" an entire tradeoff surface (i.e., Pareto optimal frontier) by sampling the surface at regular intervals. It is certainly clear from much empirical data that the combination of Pareto selection pressure and niching-based diversity pressure has proven successful [2], and some controlled experiments indicate the superiority of an EMO algorithm with niching over the same algorithm without niching [2,7,8].

2.1 Fitness Sharing

Goldberg and Richardson introduced fitness sharing in 1987 [4]. Under fitness sharing, the *objective fitness* f_i of an individual i is degraded by the presence of nearby individuals, j, in the (current) population *pop*:

$$f_{sh,i} = \frac{f_i}{\sum_{j \in pop} Sh(i,j)}. \qquad (1)$$

Here $f_{sh,i}$ is the "shared fitness" of individual i. The sharing function, $Sh(i,j)$, is a decreasing function of the "distance" $d(i,j)$ between individuals i and j, as typically measured in genotypic or phenotypic space. A widely used sharing function is the linearly decreasing *triangular sharing function*:

$$Sh(i,j) = \begin{cases} 1 - \frac{d(i,j)}{\sigma_{share}} & \text{for } d(i,j) < \sigma_{share} \\ 0 & \text{otherwise.} \end{cases} \qquad (2)$$

The "niche radius" σ_{share} is usually fixed, at the start of a run, to a value estimated to be the minimum desired distance between *niches*.

2.2 Fitness Sharing in EMO Algorithms

Sharing has been implemented in different ways in various EMO algorithms. In this paper we focus on the use of fitness sharing within the Pareto optimal set (that is, the locally non-dominated set, which is drawn from the current population). The use of sharing to maintain diversity within a locally non-dominated set occurs in many EMO algorithms, including some of the earliest Pareto-based approaches, such as MOGA [3], NSGA [10], and NPGA [7,8]. In all of these, shared fitnesses are calculated within an equivalence class, such as the Pareto optimal set, or any set of "equally dominated" individuals. That is, shared fitness is calculated for individuals whose objective (unshared) fitnesses are all equal. Therefore, shared fitnesses are determined solely by the niche counts of individuals, in other words by how crowded they are in the current population. Horn and Nafpliotis [8] have called this *equivalence class sharing*, and *flat fitness sharing*. David Goldberg noted (personal communication, 1994) that no one really knows how selection with fitness sharing will behave on a flat fitness landscape. What equilibrium population distributions are possible over an equivalence class? And how should we set our parameters, such as σ_{share}, to achieve desirable distributions?

In [3], Fonseca and Fleming suggest a method for sizing σ_{share} to "allow a number of points to sample the trade-off surface only tangentially interfering with one another..." [3, p. 418]. Given a problem with q conflicting objectives, they estimate the maximum possible area of the Pareto optimal front[1], divide the associated hypervolume by the hypervolume of a niche, namely $(\sigma_{share})^q$, set the resulting expression equal to the desired number of "points", and then solve for σ_{share}. Deb [2, p. 198] elaborates on this method with figures and calculations showing "how the hypervolume [of the Pareto front] is divided into ... equal squares.... [and] each niche occupies a hypervolume of $(\sigma_{share})^q$ and there are N such hypervolumes in [the hypervolume of the Pareto front]." In [8], Horn and Nafpliotis also propose to estimate σ_{share} by dividing the area of the Pareto front by the area of a niche of radius σ_{share}. They further suggest the use of various Holder metrics (for $d(i,j)$ in Equation 2 above) to achieve different niche shapes, other than spheres, to allow, for example, "niches along diagonal lines [to] be more densely packed than niches along lines parallel to an attribute axis" [8, p.12]. All of the discussions cited in this paragraph evoke images of "points" (individuals, or subpopulations of individuals) packed closely, but uniformly, across the Pareto optimal surface. But is such an image a realistic expectation?

3 Two Objective Case

We concentrate on the most simple, and the most easily visualized case: a two objective problem. The Pareto frontier then is a one-dimensional surface. This surface may be quite convoluted. It may be discrete. The achievable Pareto optimal solutions might be irregularly spaced. For our initial explorations of niching distributions on the front, we will first assume a Pareto front (or simply an isolated portion of the Pareto front) in which Pareto solutions can be found uniformly spaced between two extrema. This situation can be seen in the upper left portion of Figure 1. The function $P(x)$ is the Pareto optimal set membership function over the set of integers from 0 to 255 inclusive. Thus $P(x) = 1$ for $40 \leq x \leq 200$, and $P(x) = 0$ otherwise, where x is an integer between 0 and 255. We assume any encoding of the single decision variable x as a finite length chromosome, such as an eight bit binary coded integer. We next examine two cases: one in which an integer number of niches exactly span the length of the front, and one in which no such perfect fit exists.

3.1 Perfect Fit

In Figure 1 we show the results of a typical run of a GA, with selection only, on the Pareto membership function P. The population size is 2560, giving an expected number of 10 copies of each possible species in the initial random population. Only selection (binary tournament) is applied each generation, using the shared fitness values, with $\sigma_{share} = 20$. So a maximum of nine of niches can be fit into the Pareto front.

---

[1] Fonseca and Fleming actually suggest re-calculating these volumes, and hence σ_{share}, every generation.

The four graphs in Figure 1 show the species counts (where each integer value of x is a separate species) over the run. By generation 5 the dominated solutions have been eliminated from the population. We also see an "edge effect", where the niches at the very edges of the front accumulate population share more quickly than other niches. "Internal niches" are in turn affected by the growth of the edge niches. Those niches exactly σ_{share} units distant from the edges grow almost as quickly as the edge niches. Similarly, niches σ_{share} units in from these niches grow almost as quickly, and so on. By generation 180, the only niches remaining in the population are those that are exactly a multiple of σ_{share} distant from both edges. Thus the steady-state population consists of non-overlapping (non-competing) species.

Together, these nine niches might be said to collectively (cooperatively) "represent" the globally optimal plateau, by uniformly sampling the surface. Clearly what is happening is that the "edge effect" propagates in toward the middle of the plateau. There is no preference among any of the globals. However, the niches at the edges of the plateau benefit from reduced competition. Since all neighbors to one side have objective fitness of zero, the edge niches find themselves "neighborless" on one side within five generations. Having to share less (i.e., having a lower niche count) than other niches centered on "internal" global optima, the edges quickly recruit members. Their increased niche populations then degrade the shared fitnesses of neighboring niches less than σ_{share} distant from the edges. Niches exactly σ_{share} distant from the edges are not degraded by the edge populations, and instead benefit from the suppression of niches between them and the nearest edge. A chain of such alternating competition and cooperation is eventually established across the plateau, until the nine non-overlapping niches, anchored at the edges by the edge niches, are all that survive.

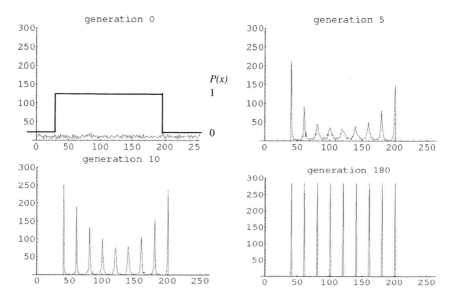

Fig. 1. Performance of fitness sharing (with selection only) when niche radius perfectly fits the Pareto front

The implicit cooperation at work in this example can be visualized more dramatically. In Figure 2, attention is called to groups of cooperating (non-competing) niches by drawing a line to connect the plotted points of niches that are separated by exactly σ_{share}. Thus there are 19 possible "lines" of eight niches, and one line of nine niches (the group containing the two edge niches). A few of the possible 20 cooperative groups are depicted in 2. We can see that the one "best" group, containing the edge niches, wins out over the other competing groups, being "pulled up" by the ends[2]. This figure illustrates how simple competition (for limited resources, namely "Pareto frontier real estate") among individuals can lead to complex cooperation within a group of individuals, which then leads to competition between groups of cooperating individuals.

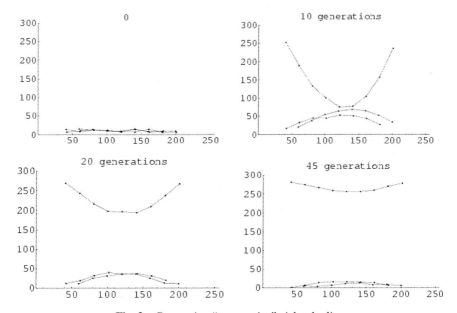

Fig. 2. Connecting "cooperative" niches by lines

We see that fitness sharing within the Pareto optimal equivalence class can indeed generate and maintain the uniformly spaced, stable distributions of niche populations that we might expect. That is, if we estimate the Pareto front length L_P correctly, and set our niche radius to some factor of that, say $\sigma_{share} = L_P / k$, we can expect the population to be divided evenly into $k+1$ niches, with approximately $N / (k + 1)$ individual in each niche (where N is the population size).

3.2 Imperfect Fit

When one considers the large number of Pareto optimal solutions in the initial population, and the much larger number of candidate subsets of solutions, it seems remark-

---

[2] This edge effect seems to be evident in all three plots in Figure 256 of [2], although the author does not mention it.

able that the GA selection (with fitness sharing) is able to so quickly select and promote the optimal subset of nine niches. To what extent can we count on this effect? How general is it?

In this section we examine the case of an imperfect fit of niche radius to Pareto front length. That is, we assume that the length of the "found" Pareto front is NOT an exact multiple of σ_{share}. We argue later that this is the more general case.

In this experiment, we use the same parameters (e.g., same GA, same population size) as in the previous section, with the exception of σ_{share} which is now set to 25 rather than 20. Since the total length of the Pareto front remains at 160, it is no longer possible to fit an integer number of niches perfectly into the front. The maximum number of non-overlapping niches that will fit into a front of length 160 is seven, and there exist many such sets of seven non-overlapping niches.

In Figure 3 we show some results of re-running the experiment of the previous section with the larger setting of σ_{share}. Again, we see the edge effect, but this time the two edge effects do not reinforce each other. They each promote a different subset of seven niches. By generation 20, selection seems to have reached a steady-state standoff between the two edge effects. This standoff seems capable of lasting, as evidenced by the generation 400 graph in the lower right of Figure 3.

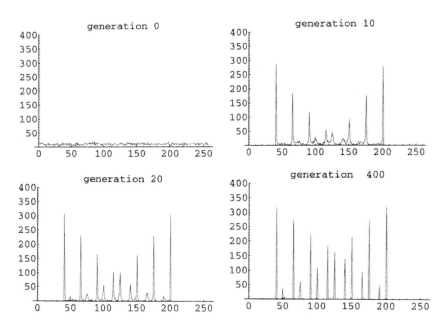

Fig. 3. An imperfect fit of niche radius to Pareto front length

Figure 4 reinforces the mental image of two opposing cooperative groups reaching a stalemate and maintaining it, while beating out all alternative groups of seven non-overlapping niches. We note that the apparent equilibrium distribution maintains twice as many niches (fourteen) as the maximum that should fit on the given front with the given niche radius. We further note that the inter-niche distance is not uniform. Before discussing the implications of these observations, it might be a good idea to explore further the nature of this apparent steady-state distribution.

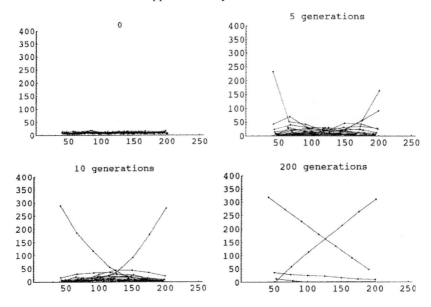

Fig. 4. Imperfect fits (of niche radius to Pareto front) lead to opposing teams of cooperating niches, each anchored by edge niches

3.3 Analysis of Sharing Equilibrium on a One-Dimensional Pareto Front

We first look at a small case. We assume a small Pareto optimal front, with respect to niche radius. As a matter of fact, we assume that the Pareto front (or at least this isolated portion) is exactly σ_{share} long. We then gradually increase the length of the front (or, equivalently, decrease σ_{share}).

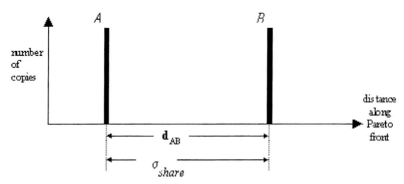

Fig. 5. Two niches with perfect fit (no overlap)

In Figure 5 the ideal two-niche case is illustrated, with the distance between the two end niches, **A** and **B**, being exactly equal to the niche radius: $d_{AB} = \sigma_{share}$. The equilibrium distribution is described by the simple equation:

$$f_{A,sh} = f_{B,sh} \Rightarrow \frac{f_A}{n_A} = \frac{f_B}{n_B} \Rightarrow n_A = n_B \text{ (since } f_A = f_B \text{)} \tag{3}$$

In the equation above, the denominators in the shared fitness formulae (i.e., the niche counts) are simply the number of copies of the respective individuals.

If we increase the distance between **A** and **B** so that it is greater than the niche radius but not equal to a multiple of the niche radius (i.e., $\sigma_{share} < d_{AB} < 2\sigma_{share}$), then we have a more general case, as shown in Figure 6. Here we see that two additional niches, **C** and **D**, have formed at a distance of σ_{share} from **B** and **A** respectively. While the end niches **A** and **B** overlap with only one intermediate niche (**C** and **D** respectively), the intermediate niches **C** and **D** overlap with two other niches: with each other and with one end niche. We note the symmetry in this scenario: $d_{AC} = d_{BD}$. And we note that $d_{AC} + d_{CD} = \sigma_{share}$.

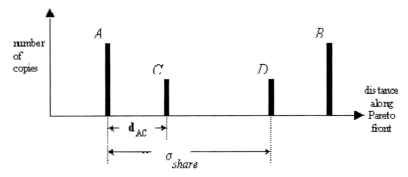

Fig. 6. Imperfect fit: $d_{AB} > \sigma_{share}$, allowing niches **C** and **D** to form

We now turn to the sharing equilibrium equations, to calculate the population distribution among niches **A**, **B**, **C**, and **D** at equilibrium. Noting that because of symmetry, $n_{A,eq} = n_{B,eq}$ and $n_{C,eq} = n_{D,eq}$, we choose to look only at $n_{A,eq}$ and $n_{C,eq}$ (i.e., the left half of the scenario in Figure 6):

$$f_{A,sh} = f_{C,sh} \tag{4}$$

$$\frac{f_A}{n_A + n_C(1 - \frac{d_{AC}}{\sigma_{share}})} = \frac{f_C}{n_C + n_A(1 - \frac{d_{AC}}{\sigma_{share}}) + n_D(1 - \frac{d_{CD}}{\sigma_{share}})}.$$

We define the ratio $d' = d_{AC} / \sigma_{share}$, so that $0 \leq d' \leq 1$. (When $d' = 0$ then the niches **A** and **C** coincide, as well as **B** and **D**, and we are back to the situation in Figure 5. When $d' = 1$, niches **C** and **D** coincide, and we are again in a "perfect niching" situation, but with three niches instead of two.) Furthermore, since $\mathbf{d}_{AC} + \mathbf{d}_{CD} = \sigma_{share}$, we can derive

$$\frac{d_{CD}}{\sigma_{share}} = \frac{\sigma_{share} - d_{AC}}{\sigma_{share}} = 1 - d'. \tag{5}$$

This result, plus the fact that $f_A = f_C$ under equivalence class sharing, means we can rewrite the equilibrium equation, in Equation 4 above, as

$$n_A + n_C(1-d') = n_C + n_A(1-d') + n_D(1-(1-d')) \Rightarrow \tag{6}$$
$$n_A + n_C(1-d') = n_C + n_A(1-d') + n_C d',$$

as we recall that by symmetry $n_C = n_D$ at equilibrium. We can manipulate this to yield a ratio:

$$\frac{n_C}{n_A} = \frac{d'}{2d'}. \tag{7}$$

Before completing the last and obvious step of canceling $\mathbf{d'}$ from the ratio, we note that for the degenerate case of d' = 0, corresponding to the merging of niches A and C, the ratio is undefined. For all other values of $\mathbf{d'}$, with $0 < \mathbf{d'} \leq 1$,

$$\frac{n_{C,eq}}{n_{A,eq}} = \frac{1}{2}, \tag{8}$$

which might seem surprising: the equilibrium ratio is independent of the distance between the niches. At $\mathbf{d'} = 1$ the ratio makes sense, as the two niches **C** and **D** would then coincide and the total of their counts should equal those of **A** and **B**: $n_A = n_B = n_C + n_D$. But for values of $\mathbf{d'}$ between 0 and 1, even for values very close to 0 or 1, in which the niche **C** is nearly indistinguishable from **A** or **D** respectively, the ratio is constant.

While the ratio n_C / n_A is constant, the actual numbers n_C and n_A are not; they can vary with $\mathbf{d'}$. Consider another niche on the Pareto front, call it **Z**, that does not overlap with any other niche. Then $f_{Z,sh} = 1/n_Z$. At equilibrium, the niche counts of all species present must be equal, so that $n_Z = n_A + n_C(1-d')$. Therefore, $n_A < n_Z$ and $n_C < n_Z$. Thus the closer niches **A** and **C**, the more overlapped they become, and the less carrying capacity they have, relative to niche **Z**.

We should also point out that this independence of the ratio n_C / n_A from the inter-niche distance $\mathbf{d'}$ holds only for the triangular sharing function, that is, when $\alpha_{share} = 1$. Recalling the complete form of the sharing function from [4]:

$$Sh(i,j) = \begin{cases} 1 - \left(\dfrac{d(i,j)}{\sigma_{share}}\right)^{\alpha_{share}} & \text{for } d(i,j) < \sigma_{sh} \\ 0 & \text{otherwise.} \end{cases} \quad (9)$$

Here the "power sharing parameter" α_{share} can be used to control the shape of the niche or the relative importance of niche overlap. However, α_{share} is usually set to one.

Re-deriving Equation 8 from Equation 4, with α_{share} left undefined, yields

$$\frac{n_{C,eq}}{n_{A,eq}} = \frac{1 - (1-d')^{\alpha_{share}}}{1 + d'^{\alpha_{share}} - (1-d')^{\alpha_{share}}}. \quad (10)$$

For example, with $\alpha_{share} = 2$, we get $n_{C,eq}/n_{A,eq} = 1 - d'/2$. Thus α_{share} can be used to vary the sensitivity of the ratio $n_{C,eq}/n_{A,eq}$ to d', if that is desired.

4 Discussion

The most salient point of the above analysis is that a stable equilibrium distribution exists for the general case of overlapping niches on the Pareto optimal front, shown in Figure 6. Furthermore this distribution, like the empirically derived ones in Figures 3 and 4, allow equilibrium distributions with the following, perhaps unexpected, characteristics:

1. niches can be spaced non-uniformly,
2. niches can be separated by considerably LESS than the niche radius,
3. there can be twice as many niches as predicted by the "usual" calculations, and
4. the niches can have widely varying sub-population counts.

So much, perhaps, for our intuitions. But what are the practical effects?

First, it should be argued that while the "perfect fit" situation in Figures 1 and 2 provide evidence of the elegance and power of fitness sharing, the more general case is the imperfect fit of Figures 3 and 4. In general, we cannot know the shape of the Pareto surface, nor can we know which particular Pareto solutions will be represented in any given generation. With all of the possible convolutions and gaps, it seems most likely that the imperfect fit case will describe most sections of any found Pareto front.

Next we note that the number of "additional" (unpredicted) niches should increase with the number of objectives. Since each additional objective adds a dimension to the surface that represents the Pareto front, then each new objective adds a new "edge" to the simplex and therefore defines (anchors) its own subset of non-overlapping niches. Thus for three objectives, the imperfect fit could mean triple the number of niches predicted for the perfect fit case. (This is an area for further investigation.)

How these nuances in Pareto set distribution affect the final outcome of an EMO algorithm is hard to say. We know little about how niching affects EMO search, although we have much empirical evidence that it helps in general. But it may be that we have to use larger population sizes, or larger niche radius settings, to achieve the kind of Pareto front coverage we expected under an assumption of "perfect fit". At the very least, we know now that our image of "perfect fit niching" is an illusive one.

References

1. Deb, K.: *Genetic Algorithms in Multi-Modal Function Optimization*. Masters thesis, University of Alabama at Tuscaloosa, AL (1989)
2. Deb, K.: *Multi-Objective Optimization using Evolutionary Algorithms*. John Wiley & Sons, Ltd., Chichester (2001)
3. Fonseca, C. M., & Fleming, P. J.: Genetic algorithms for multiobjective optimization: Formulation, discussion, and generalization. *Proceedings of the Fifth International Conference on Genetic Algorithms*. Morgan Kaufmann, San Mateo, CA (1993) 416–423
4. Goldberg, D. E., Richardson, J.: Genetic Algorithms with Sharing for Multimodal Function Optimization. In: Grefenstette, J. (ed.): *Proceedings of the 2nd International Conference on Genetic Algorithms*. Lawrence Erlbaum Associates, Hillsdale, New Jersey (1987) 1–8
5. Horn, J.: *The Nature of Niching: Genetic Algorithms and the Evolution of Optimal, Cooperative Populations*. Ph.D. thesis, University of Illinois at Urbana-Champaign, (UMI Dissertation Services, No. 9812622) (1997)
6. Horn, J.: Multicriterion Decision Making. In: Bäck, T., Fogel, D. (ed.s): *The Handbook of Evolutionary Computation*. Oxford University Press, New York (1997) F1.9:1–15
7. Horn J., Nafpliotis, N., Goldberg, D. E.: A niched Pareto genetic algorithm for multiobjective optimization. *Proceedings of 1st IEEE International Conference on Evolutionary Computation*, Volume 1. IEEE Service Center, Piscataway, New Jersey (1994) 82–87
8. Horn J., & Nafpliotis: Multi-objective optimization using the niched Pareto genetic algorithm. *IlliGAL Report Number 93005*. Illinois Genetic Algorithms Laboratory, University of Illinois at Urbana-Champaign (1993)
9. Mahfoud, S. W.: *Niching Methods for Genetic Algorithms*. Ph.D. thesis, University of Illinois at Urbana-Champaign (1995)
10. Srinivas, N. & Deb, K.: Multi-objective function optimization using non-dominated sorting genetic algorithms. *The Journal of Evolutionary Computation* 2(3) (1994) 221–248

Performance Scaling of Multi-objective Evolutionary Algorithms

V. Khare[1], X. Yao[1], and K. Deb[2]*

[1] School of Computer Science
The University of Birmingham
Edgbaston, Birmingham B15 2TT, UK
{V.R.Khare, X.Yao}@cs.bham.ac.uk
[2] Kanpur Genetic Algorithms Laboratory (KanGAL)
Indian Institute of Technology Kanpur
Kanpur 208016, INDIA
deb@iitk.ac.in

Abstract. MOEAs are getting immense popularity in the recent past, mainly because of their ability to find a wide spread of Pareto-optimal solutions in a single simulation run. Various evolutionary approaches to multi-objective optimization have been proposed since 1985. Some of fairly recent ones are NSGA-II, SPEA2, PESA (which are included in this study) and others. They all have been mainly applied to two to three objectives. In order to establish their superiority over classical methods and demonstrate their abilities for convergence and maintenance of diversity, they need to be tested on higher number of objectives. In this study, these state-of-the-art MOEAs have been investigated for their scalability with respect to the number of objectives (2 to 8). They have also been compared on the basis of -(1) Their ability to converge to Pareto front, (2) Diversity of obtained non-dominated solutions and (3) Their running time. Four scalable test problems (DTLZ1, 2, 3 and 6) are used for the comparative study.

1 Introduction

Evolutionary algorithms are often praised for their ability to search multiple solutions in parallel and to handle complicated tasks such as discontinuities, multi-modality and noisy function evaluations. Their population based approach, which enables them to find multiple optimal solutions in one single run, is especially useful in the conetxt of multi-objective optimization which involves the task of finding more than one optimal solution. Though there exists a number of efficient algorithms, most of the work that has been done in evolutionary multi-objective optimization is restricted to 2 and 3 objectives. The main motivation

---

* This work is partly supported by The Europian Commission through its grant AS1/B7-301/97/0126-08 and School of Computer Science, The University of Birmingham.

behind this work is to investigate how these algorithms behave when tested on problems with higher number of objectives.

Scalability of some recent algorithms (section 3) is to be investigated using an experimental study (section 5). This study would involve experiments with chosen algorithms (section 3) on four scalable (in terms of objectives) test problems (section 2) for 2 to 8 objectives. For comparison purposes, three performance metrics are to be used (section 4). Results are presented and discussed in section 6.

2 Test Problems

Most earlier studies on MOEAs introduced test problems which were either simple or not scalable [16,14,10]. Few others were too complicated to visualize the exact shape and location of the resulting Pareto-optimal (PO) front. In this study four scalable test problems, namely DTLZ1, DTLZ2, DTLZ3 and DTLZ6 [8], are chosen on the basis of the following properties- (1) They are easy to construct (Bottom-up approach [8]), (2) They can be scaled to any number of decision variables and objectives, (3) Exact shape and location of the resulting PO front for these problems are known and (4) Difficulties in both converging to the true PO front and maintaining a widely distributed set of solutions can be controlled. In these test problems, the total number of variables are $n = M+k-1$. Where M is the number of objectives and k can be set by the user giving him the freedom to scale to any number of variables. All these test problem require a functional $g(\mathbf{x}_M)$ to be set, where the dimension of $|\mathbf{x}_M|$ is k. Choices of $g(\mathbf{x}_M)$ and k were made according to the suggestions in [8].

For DTLZ1, DTLZ2 and DTLZ3, the PO solutions correspond to $\mathbf{x}_M = \{0.5, 0.5, \ldots, 0.5\}^T$ and the objective function values corresponding to the PO front lie on $\sum_{m=1}^{M} f_m = 0.5$ for DTLZ1 and on $\sum_{m=1}^{M} f_i^2 = 1$ for DTLZ2 and DTLZ3. The g function in DTLZ3 introduces $(3^k - 1)$ local PO fronts and one global PO front. All local PO fronts are parallel to the global PO front and an MOEA can get stuck to any of these local PO fronts, before converging to the global PO front (at $g^* = 0$). On the other hand, DTLZ6 tests an MOEA's ability to converge to a curve. In this case there is only one independent variable describing the PO front.

3 Algorithms Used

Earlier MOEAs (MOGA [9], NSGA [17] and NPGA [11]) were criticised for their dependence on sharing parameter [7] and lack of elitism [15,18]. Different algorithms that overcome these shortcomings have been proposed. Few such algorithms are - PAES [13], SPEA [21] and NSGA-II [7]. In these algorithms, elitism maintains the knowledge acquired during the algorithm execution by conserving the individuals with best fitness in the population or in an auxiliary population. For the maintenance of spread of solutions grid based techniques (PAES), clustering (SPEA) or crowding (NSGA-II) were used. Further improvements to these algorithms have also been proposed [1,19,5].

NSGA-II (Non-dominated Sorting Genetic Algorithm-II) with controlled elitism [5] limits the maximum number of individuals belonging to each front by a geometrically decreasing function (governed by the reduction rate r) to introduce more diversity into the population. NSGA-II outperformed PAES in preserving the spread of non-dominated front on five 2-objective test problems [7]. PESA (Pareto Enveloped-based Selection Algorithm) [1], an improvement of PAES, uses the hyper-cubes grid division not only for crowding as in PAES, but also for selection process. PESA was compared with PAES and SPEA on six test functions $\mathcal{T}_1$ to $\mathcal{T}_6$ [2] (each of which is a 2-objective problem defined on m parameters) and was reported to outperform SPEA and PAES on these test functions.

SPEA2 (Strength Pareto Evolutionary Algorithm 2) [19] was proposed as an improvement of SPEA and incorporated a revised fitness assignment strategy, a density estimation technique and an enhanced archive truncation method. Performance of SPEA2 was compared with SPEA, NSGA-II and PESA on some combinatorial and continuous problems. Similar performances of NSGA-II and SPEA2 were reported along with the fast convergence properties of PESA on these problems. All the test problems used here (except for the 3 and 4-objective *knapsack problem* [21]) were 2-objective problems.

In this study we have taken three of these most recent MOEAs (PESA, SPEA2 and NSGA-II) and compared their performances on specially designed scalable test problems (section 2) over 2 to 8 objectives. For a detailed description of these algorithm we refer the readers to the original papers. Since the test problems that we are dealing with have a continuous space, real encoding should be preferred to avoid problems related to hamming cliffs and to achieve arbitrary precision in the optimal solution. For this reason, in all the algorithms real-coded parameters were used and crossover (*Simulated Binary Crossover* or SBX [3]) and mutation (*polynomial mutation* [4]) operators were applied directly to real parameter values.

4 Performance Metrics

Zitzler [20] showed that for an $M-$objective optimization problem, at least M performance metrics are needed to compare two or more solutions and an infinite number of metrics to compare two or more sets of solutions. To make such a comparison possible Deb [6] suggested the use of a *functionally independent* set of variables, which would of course make it theoretically inaccurate but practically feasible. He also suggested two new *running* performance metrics - one for measuring the convergence to the reference set and other for measuring the diversity in population members at every generation of an MOEA run. In this study these two metrics (with slight variation) have been used in addition to a third one, which simply measures the running time of the MOEA. All three

Table 1. Parameter settings for calculating the diversity metric

Reference plane	M−th objective function $f_M = 0$
Number of grid cells (G_i)	Population size
Target (or reference) set of points (P^*)	One assumed solution in each grid cell

metrics, which were applied to only the final non-dominated set obtained by an MOEA to evaluate its performance, have been discussed in detail below.

The following metric represents the distance between the set of converged non-dominated solutions and the global PO front; hence lower values of convergence metric represent good convergence ability. Let P^* be the reference or target set of points on the PO front and let $\mathcal{F}$ be the final non-dominated set obtained by an MOEA. Then from each point i in $\mathcal{F}$ the smallest normalized Euclidean distance to P^* will be:

$$d_i = \min_{j=1}^{|P^*|} \sqrt{\sum_{k=1}^{M} \left(\frac{f_k(i) - f_k(j)}{f_k^{max} - f_k^{min}} \right)^2} \quad (1)$$

Here, f_k^{max} and f_k^{min} are the maximum and minimum function values of k-th objective function in P^*. However in this study, no target points were chosen because equations of PO fronts were known for all the four test problems. The orthogonal distance of a point A in the non-dominated set from the PO front was calculated directly from the equation of PO front. E.g. in DTLZ2 or DTLZ3

$$d_i = \|\boldsymbol{r}_A\| - 1 \quad (2)$$

Once these distances are known the convergence metric can be obtained by averaging the normalized distance for all points in $\mathcal{F}$.

Diversity metric is a number in the range [0, 1], where 1 corresponds to the best and 0 corresponds to the worst possible diversity. In calculating the diversity metric, the obtained non-dominated points are projected on a hyper-plane, thereby losing a dimension of the points. The plane is divided into a number of small grid cells (or $M-1$ dimensional boxes). Depending on whether each grid cell contains a non-dominated point or not, a diversity metric is defined. If all grid cells are represented with at least one point, the best possible (with respect to the chosen number of grids) diversity measure is achieved. If some grid cells are not represented by a non-dominated point, the diversity is poor. Various parameters required to calculate this metric and their chosen values are given in table 1.

Calculating the diversity metric. Following steps are involved in calculating the diversity metric:

Table 2. Lookup table for calculating diversity metric

h(... j-1 ...)	h(... j ...)	h(... j+1 ...)	m(h(... j ...))
0	0	0	0.00
0	0	1	0.50
1	0	0	0.50
0	1	1	0.67
1	1	0	0.67
0	1	0	0.75
1	1	1	1.00

1. For each grid cell indexed by $(i, j, \ldots)$ calculate following two arrays:

$$H(i,j,\ldots) = \begin{cases} 1, \text{ if the grid cell has a representative point in } P^*; \\ 0, \text{ otherwise.} \end{cases}$$

$$= 1, \text{ for the chosen reference set } P^* \quad (3)$$

$$h(i,j,\ldots) = \begin{cases} 1, \text{ if H(i, j, } \ldots \text{)} = 1 \text{ and the grid cell has a representative point in } \mathcal{F}; \\ 0, \text{ otherwise.} \end{cases}$$

$$= \begin{cases} 1, \text{ if the grid cell has a representative point in } \mathcal{F} \text{ for the chosen } P^*; \\ 0, \text{ otherwise.} \end{cases} \quad (4)$$

2. Using table 2, assign a value $m(H(i,j,\ldots))$ to each grid cell depending on its and its neighbor's $h()$. Similarly, calculate $m(H(i,j,\ldots))$ using $H()$ for reference points.
3. Calculate the diversity metric (of the population P of non-dominated solutions produced by an MOEA) by averaging the individual $m()$ values for $h()$ with respect to that for $H()$:

$$D(P) = \frac{\sum_{\substack{i,j,\ldots \\ H(i,j,\ldots) \neq 0}} m(h(i,j,\ldots))}{\sum_{\substack{i,j,\ldots \\ H(i,j,\ldots) \neq 0}} m(H(i,j,\ldots))} \quad (5)$$

Value function $m()$ for the grid was calculated by using its $h()$ and two neighboring $h()$ dimension-wise. With a set of three consecutive binary $h()$ values, there are a total of 8 possibilities. Suggested [6] values of $m()$ were used in this study (Table 2). Two or more dimensional hyper-planes are handled by calculating the above metric dimension-wise.

Figure 1 shows a sample calculation of diversity metric in case of 2-objective DTLZ2 or 2-objective DTLZ3 problem. Here circles represent the reference or target points, in every partition there is one such point and hence the number of partions is same as population size. Boxes represent the set of non-dominated points produced by an MOEA. The $f_2 = 0$ is used as the reference plane here and the complete range on f_1 values are divided into $G_1 = 10$ grid cells. This complete range depends on the PO front the algorithm has converged to and

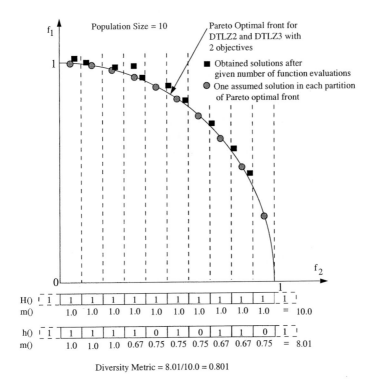

Fig. 1. Calculating the diversity metric

the resulting diversity metric will also be different. For boundary grid cells, an imaginary neighboring grid cell with a $h()$ and $H()$ value of one is always assumed. In figure 1, these grid cells are shown with dashed lines. Even if we have more than one points in a grid cell, $h()$ still remains as one. Based on moving window containing three consecutive grid cells, the $m()$ values are computed in the figure (using table 2). To avoid the boundary effects of using the imaginary grid cell, the metric value is normalized as follows:

$$\overline{D}(P) = \frac{\sum_{\substack{i,j,\ldots \\ H(i,j,\ldots) \neq 0}} m(h(i,j,\ldots)) - \sum_{\substack{i,j,\ldots \\ H(i,j,\ldots) \neq 0}} m(\mathbf{0})}{\sum_{\substack{i,j,\ldots \\ H(i,j,\ldots) \neq 0}} m(H(i,j,\ldots)) - \sum_{\substack{i,j,\ldots \\ H(i,j,\ldots) \neq 0}} m(\mathbf{0})} \quad (6)$$

where $\mathbf{0}$ is a zero-valued array. Though in our study $H(i,j,\ldots)$ is always equal to 1 (each grid cell contains one reference point), $H(i,j,\ldots) \neq 0$ consideration in computing the $\overline{D}(P)$ term and the boundary grid adjustment was suggested to allow a generic way to handle disconnected PO fronts in the original work [6].

Ideally we want our algorithm to give us non-dominated solutions on the global PO front and if we divide our objective space corresponding to this global

PO front into grid cells equal to the population size, then one point in each grid cell would be the best possible diversity ($\overline{D}(P) = 1$). We will call this diversity metric (obtained by splitting the global PO region into grid cells) *diversity metric1*. But if the algorithm isn't able to converge to the global PO front (figure 2) then the above metric would not be able to measure the diversity of non-dominated solutions produced by the MOEA. In such cases we should calculate our diversity metric based on the actual converged front (obtained after a fixed number of function evaluations) instead of global PO front. We will call this diversity metric (obtained by splitting the converged PO region into grid cells) *diversity metric2*.

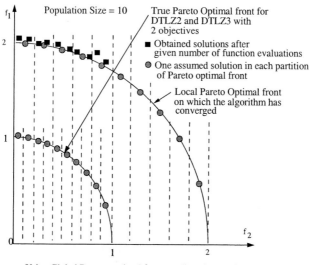

Fig. 2. Calculating diversity metrics - using global and local (converged) PO fronts

To compare two algorithms (for diversity) the use of diversity metric2 should be preferred because even if one of them has converged to true PO front and the other hasn't, the diversity metric2 of the former will be almost equal to its diversity metric1 value, which would not be the case with latter.

The third metric used in the study is simply the running time of an algorithm (in seconds) for the particular settings. It has been included in the performance metric set to evaluate how an MOEA scales in terms of time with increase in number of objectives. A linear or polynomial increase in running time is acceptable but an exponential increase is undesirable.

Table 3. Tuned parameter values

Parameter	PESA	SPEA 2	NSGA-II
Crossover probability p_c	0.8	0.7	0.7
Distribution index (DI) for SBX η_c	15	15	15
Mutation probability p_m (if $n = \#$ of variables)	1/n	1/n	1/n
DI for polynomial mutation η_m	15	15	20
Ratio of internal population size to archive size	1:1	1:1	1:1
# of grid cells per dimension (PESA)	10	-	-

5 Experimental Study

To make the comparisons fair the number of function evaluations were kept constant for all three algorithms on a particular setting. Before the actual experimentation some tuning of the parameters involved was required. Finding values for which an algorithm works best is, in itself, an optimization problem and if we are judging the performance on the three metrics it becomes a Muti-objective Optimization Problem (MOOP). A very simplistic approach was adopted to tune (instead of optimizing) these parameters. Experiments with only two objectives were used and the purpose of this tuning was to find a set of values for which an MOEA performs well. Table 3 gives the tuned values used for all the experimentation. Parameter tuning was carried out on problems DTLZ2 and DTLZ3 and only the convergence and diversity metrics were used to evaluate the performance of the algorithm, running time was not considered.

Population size plays a crucial rule in the performance of an MOEA. As the number of objective functions (M) increases, more and more solutions tend to lie in the first non-dominated front. Most MOEAs assign the similar fitness to all solutions in the first non-dominated front. So as the number of objective functions increase, there is no (very little) selection advantage to any of these solutions. In absence of any selection pressure for better solutions, the task of recombination and mutation operators to find better solutions may be difficult in general. It has been shown empirically [3, pages 404–405] that for a particular M, the proportion of non-dominated solutions decreases with population size. If we require a population with a user-specified maximum proportion of non-dominated solutions, then these empirical results can be used to estimate what would be a reasonable population size. This requirement on population size increases exponentially with M. Ideally to investigate the scaling of an MOEA we should present it with a population having equal proportion of non-dominated solutions, for all M, to start with, but this is practically impossible because of exponential increase in population size. The population scheme used in this study is given in table 4. This scheme is quadratic with correlation coefficient squared value, $R^2 = 0.9916$.

The number of generations used for different problems and different numbers of objectives (M) are listed in table 5. More number of function evaluations were used for DTLZ3 and DTLZ6 because they can introduce more difficulties

Table 4. Population scheme - for Internal or main population (which is same as external population as the result of parameter tuning)

M	Population Size	Maximum proportion of non-dominated solutions
2	20	0.2
3	50	0.22
4	100	0.28
5	150	0.36
6	250	0.45
7	400	0.52
8	600	0.60
9	850	0.68
10	1150	∼0.75

Table 5. # Of generations for different problems and # of objectives (M)

	# of generations	
	For $M = 2, 3$ and 4	For $M = 6$ and 8
DTLZ1 & DTLZ2	300	600
DTLZ3 & DTLZ6	500	1000

to a MOEA in converging to PO front and in finding a diverse set of solutions (section 2). DTLZ3 tests the ability of an MOEA by introducing local PO fronts and DTLZ6 tests them for their ability to converge to a curve.

The number of generations from 6-objectives onwards was doubled because none of the algorithms was able to converge to the global PO front in these many generations. Converging to a PO front means having a convergence metric less than a threshold (say ϵ). Any appropriate value of ϵ can be chosen. But here, instead of choosing some such threshold, the algorithms were compared (for convergence) solely on the basis of the convergence metric that they can achieve in given number of function evaluations.

6 Results

Figures 3 and 4 give the convergence metric, diversity metric(1 & 2) and running times for all three algorithms over 2 to 8 objectives. For a listing of these plotted values please refer to [12]. These values are averaged over 30 runs for 2, 3 and 4 objectives and 10 runs for 6 and 8 objectives. Corresponding standard deviation values for these experiments are listed in table 6. Following is the discussion over the results obtained. Few other points, which have been observed during this extensive comparative study, are also discussed.

1. ***Scalability.*** Each algorithm scale differently in terms of the performance metrics chosen.

- PESA scale very well in terms of convergence but poorly in terms of diversity maintenance and running time.
 - SPEA2 scales well in terms of diversity maintenance but suffers in converging to the global PO front and in running time.
 - NSGA-II scales well in terms of running time and diversity maintenance but suffers in converging to the global PO front.
2. **Convergence to PO front.** Ability to converge to the PO front was found best in PESA, though it cannot produce a very diverse solution on the final front at the end of the run.
 SPEA2 has better convergence than NSGA-II for small number of objectives but for higher number of objectives both of them have comparable performances, which is inferior to PESA. Both SPEA2 and NSGA-II had difficulties in dealing with local PO fronts (DTLZ3, figure 4) especially for higher number of objectives.
3. **Diversity in obtained solutions.** Even in terms of diversity of solutions in the converged front SPEA2 and NSGA-II have similar performances, which is much better than PESA.
4. **Running Time.** NSGA-II was the fastest of all three algorithms, primarily because it doesn't involve expensive grid based calculations (as in PESA) or archive truncation procedure (as in SPEA2). Exponential increase in running time for PESA makes it impractical for higher objectives.
5. **Grid size in PESA.** Grid size in PESA is a crucial factor. If we choose very fine grids we can hope to get a good performance in terms of diversity but that would make the algorithm even more expensive (in terms of running time).

7 Conclusion

All of the work that has been done in MOEAs is mostly limited to 2 and 3 objectives. In this paper scalability issues, in regards to the number of objectives, related to three of the state-of-the-art algorithms (PESA, SPEA2 and NSGA-II) were explored. These algorithms were tested for their scalability with respect to number of objectives (2 to 8). These algorithms were tested for their performance, on four scalable test problems, namely - DTLZ1, DTLZ2, DTLZ3 and DTLZ6. According to our study, it is clear that conclusions drawn from experimental comparisons on 2 or 3 objectives cannot be generalised to a higher number of objectives. More work needs to be done to fully understand the behaviour of MOEAs on different number of objectives.

In this study three performance metrics were used, in terms of which the scalability (in regards to the number of objectives) of these algorithms were assessed. First metric measures closeness of obtained non-dominated solution to the global PO front. The second metric indicates the diversity of solutions in the obtained non-dominated set. Running time was also included as one of the metric to evaluate how an MOEA scales in terms of time with increase in number of objectives.

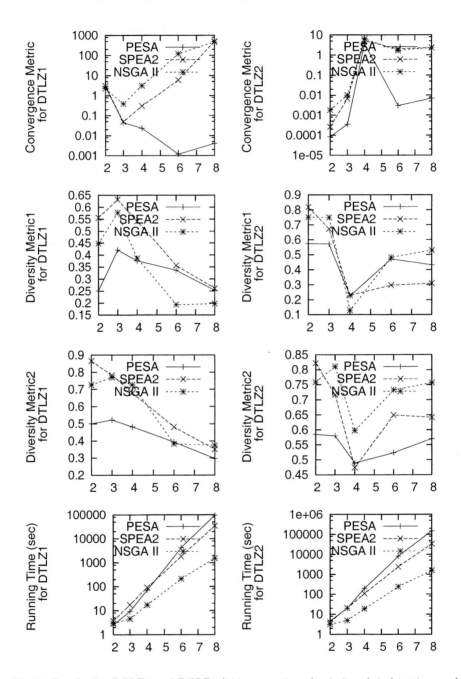

Fig. 3. Results for DTLZ1 and DTLZ2 (300 generations for 2, 3 and 4 objectives and 600 generations for 6 and 8 objectives). Results are averaged over 30 runs for 2,3 and 4 objectives and 10 runs for 6 and 8 objectives. Corresponding standard deviation values are given in table 6

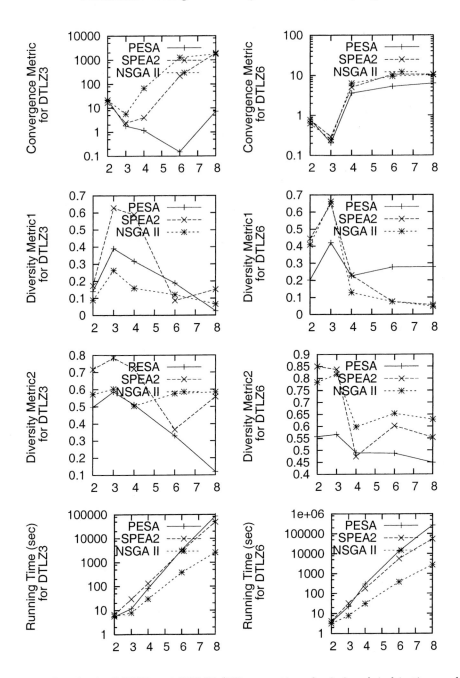

Fig. 4. Results for DTLZ3 and DTLZ6 (500 generations for 2, 3 and 4 objectives and 1000 generations for 6 and 8 objectives). Results are averaged over 30 runs for 2,3 and 4 objectives and 10 runs for 6 and 8 objectives. Corresponding standard deviation values are given in table 6

Table 6. Standard deviation values for the experiments (30 runs for 2, 3 and 4 objectives and 10 runs for 6 and 8 objectives)

Problem	Objectives	PESA	SPEA2	NSGA-II	PESA	SPEA2	NSGA-II
		Convergence Metric			Running Time		
DTLZ1	2	5.93164	5.35433	5.43593	0.51	0.47	0.55
	3	0.12320	0.05331	0.50094	1.56	0.49	0.80
	4	0.09059	0.66360	4.08272	5.53	1.04	0.39
	6	0.00089	7.67166	101.19802	184.11	251.08	4.53
	8	0.00015	13.38934	14.79745	4879.89	2121.46	51.05
DTLZ2	2	0.00019	0.00029	0.00082	0.50	0.26	0.59
	3	0.00013	0.00224	0.00234	3.69	0.59	0.83
	4	0.00039	0.00846	0.01373	17.41	1.28	2.42
	6	0.00040	0.07843	0.10533	127.77	315.43	30.78
	8	0.00109	0.03523	0.04242	1793.73	1283.36	61.32
DTLZ3	2	22.90480	16.72477	11.15397	0.74	0.49	0.79
	3	5.78546	4.72212	6.26729	1.50	0.63	1.13
	4	3.50522	4.00594	39.06815	8.91	1.38	2.84
	6	0.12692	76.62720	64.21416	189.63	357.13	44.93
	8	2.25611	11.09337	62.63447	2826.43	514.14	114.42
DTLZ6	2	0.32237	0.23794	0.29986	0.51	0.44	0.56
	3	0.21199	0.23631	0.22849	4.38	0.78	0.45
	4	0.38084	0.22360	0.36229	22.07	7.23	0.39
	6	0.31227	0.00819	0.27429	167.72	681.69	2.47
	8	0.10668	0.04585	0.05803	16819.82	5404.04	176.50
		Diversity Metric1			Diversity Metric2		
DTLZ1	2	0.14059	0.34527	0.23329	0.17016	0.11721	0.13887
	3	0.07563	0.04565	0.15660	0.10693	0.06054	0.09305
	4	0.07125	0.02007	0.11091	0.09175	0.04362	0.09545
	6	0.04046	0.15384	0.04365	0.05970	0.11778	0.01930
	8	0.00764	0.01021	0.01238	0.02745	0.02485	0.01172
DTLZ2	2	0.09135	0.01766	0.03891	0.09700	0.01992	0.04302
	3	0.04344	0.03255	0.02064	0.04528	0.03629	0.02452
	4	0.03692	0.01773	0.01881	0.04528	0.03629	0.02452
	6	0.02660	0.01939	0.01613	0.03503	0.01423	0.01139
	8	0.04908	0.00628	0.01019	0.02060	0.00327	0.01022
DTLZ3	2	0.14497	0.19218	0.06994	0.14339	0.15166	0.15533
	3	0.13220	0.14088	0.19989	0.09515	0.11143	0.18633
	4	0.09393	0.02637	0.03146	0.07063	0.05674	0.04957
	6	0.06554	0.01392	0.01290	0.06817	0.02543	0.01418
	8	0.00247	0.01223	0.01244	0.01179	0.00873	0.01278
DTLZ6	2	0.14198	0.25537	0.19784	0.12064	0.06017	0.08507
	3	0.06423	0.14591	0.13731	0.14311	0.03616	0.06102
	4	0.02790	0.01438	0.01845	0.03777	0.02237	0.02803
	6	0.02356	0.00710	0.01760	0.02286	0.01100	0.01746
	8	0.00488	0.00991	0.00391	0.03112	0.02598	0.00585

As the result of the study on four test problems, PESA was found to be best in terms of converging to the PO front, but it lacks good diversity maintenance. Also the algorithm was found to be slow because of expensive grid based calculations. Exponential increase in running time makes the algorithm infeasible for higher number of objectives. SPEA2 and NSGA-II performed equally well on convergence and diversity maintenance. Their convergence level was inferior to that of PESA but diversity maintenance was better. NSGA-II was found to be much faster than SPEA2 because of the expensive archive truncation procedure in SPEA2. Running times for NSGA-II were found to be an order of magnitude less than that of SPEA2 for higher objectives.

Acknowledgments. Authors would like to thank Joshua D. Knowles and Eckart Zitzler for sharing PESA and SPEA2 source codes, respectively.

References

1. D. W. Corne, J. D. Knowles, and M. J. Oates. The Pareto Envelope-based Selection Algorithm for Multiobjective Optimization. In M. Schoenauer, K. Deb, G. Rudolph, X. Yao, E. Lutton, J. J. Merelo, and H.-P. Schwefel, editors, *Proceedings of the Parallel Problem Solving from Nature VI Conference*, pages 839–848, Paris, France, 2000. Springer. Lecture Notes in Computer Science No. 1917.
2. K. Deb. Multi-Objective Genetic Algorithms: Problem Difficulties and Construction of Test Problems. Technical Report CI-49/98, Dortmund: Department of Computer Science/LS11, University of Dortmund, Germany, 1998.
3. K. Deb. *Multi-Objective Optimization using Evolutionary Algorithms*. John Wiley & Sons, Chichester, UK, 2001. ISBN 0-471-87339-X.
4. K. Deb and M. Goyal. A Combined Genetic Adaptive Search (geneAS) for Engineering Design. *Computer Science and Informatics*, 26(4):30–45, 1996.
5. K. Deb and T. Goyal. Controlled Elitist Non-dominated Sorting Genetic Algorithms for Better Convergence. KanGAL report 200004, Indian Institute of Technology, Kanpur, India, 2000.
6. K. Deb and S. Jain. Running Performance Metrics for Evolutionary Multi-objective Optimization. Technical Report 2002004, KanGAL, Indian Institute of Technology, Kanpur 208016, India, 2002.
7. K. Deb, A. Pratap, S. Agarwal, and T. Meyarivan. A Fast and Elitist Multiobjective Genetic Algorithm: NSGA-II. *IEEE Transactions on Evolutionary Computation*, 6(2):182–197, April 2002.
8. K. Deb, L. Thiele, M. Laumanns, and E. Zitzler. Scalable Test Problems for Evolutionary Multi-Objective Optimization. Technical Report 112, Computer Engineering and Networks Laboratory (TIK), Swiss Federal Institute of Technology (ETH), Zurich, Switzerland, 2001.
9. C. M. Fonseca and P. J. Fleming. Genetic Algorithms for Multiobjective Optimization: Formulation, Discussion and Generalization. In S. Forrest, editor, *Proceedings of the Fifth International Conference on Genetic Algorithms*, pages 416–423, San Mateo, California, 1993. University of Illinois at Urbana-Champaign, Morgan Kauffman Publishers.
10. C. M. Fonseca and P. J. Fleming. An Overview of Evolutionary Algorithms in Multiobjective Optimization. *Evolutionary Computation*, 3(1):1–16, Spring 1995.

11. J. Horn and N. Nafpliotis. Multiobjective Optimization using the Niched Pareto Genetic Algorithm. Technical Report IlliGAl Report 93005, University of Illinois at Urbana-Champaign, Urbana, Illinois, USA, 1993.
12. V. Khare. Performance Scaling of Multi-Objective Evolutionary Algorithms. Master's thesis, School of Computer Science, The University of Birmingham, Edgbaston, Birmingham B15 2TT, UK, September 2002.
13. J. D. Knowles and D. W. Corne. The Pareto Archived Evolution Strategy: A New Baseline Algorithm for Multiobjective Optimisation. In *1999 Congress on Evolutionary Computation*, pages 98–105, Washington, D.C., July 1999. IEEE Service Center.
14. F. Kursawe. A Variant of Evolution Strategies for Vector Optimization. In H. P. Schwefel and R. Männer, editors, *Parallel Problem Solving from Nature. 1st Workshop, PPSN I*, volume 496 of *Lecture Notes in Computer Science*, pages 193–197, Berlin, Germany, oct 1991. Springer-Verlag.
15. G. T. Parks and I. Miller. Selective Breeding in a Multiobjective Genetic Algorithm. In A. E. Eiben, M. Schoenauer, and H.-P. Schwefel, editors, *Parallel Problem Solving From Nature — PPSN V*, pages 250–259, Amsterdam, Holland, 1998. Springer-Verlag.
16. J. D. Schaffer. *Multiple Objective Optimization with Vector Evaluated Genetic Algorithms*. PhD thesis, Vanderbilt University, 1984.
17. N. Srinivas and K. Deb. Multiobjective Optimization Using Nondominated Sorting in Genetic Algorithms. *Evolutionary Computation*, 2(3):221–248, Fall 1994.
18. E. Zitzler, K. Deb, and L. Thiele. Comparison of Multiobjective Evolutionary Algorithms: Empirical Results. *Evolutionary Computation*, 8(2):173–195, Summer 2000.
19. E. Zitzler, M. Laumanns, and L. Thiele. SPEA2: Improving the Strength Pareto Evolutionary Algorithm. Technical Report 103, Computer Engineering and Networks Laboratory (TIK), Swiss Federal Institute of Technology (ETH) Zurich, Gloriastrasse 35, CH-8092 Zurich, Switzerland, May 2001.
20. E. Zitzler, M. Laumanns, L. Thiele, C. M. Fonseca, and V. Grunert da Fonseca. Why Quality Assessment of Multiobjective Optimizers Is Difficult. In W. Langdon, E. Cantú-Paz, K. Mathias, R. Roy, D. Davis, R. Poli, K. Balakrishnan, V. Honavar, G. Rudolph, J. Wegener, L. Bull, M. Potter, A. Schultz, J. Miller, E. Burke, and N. Jonoska, editors, *Proceedings of the Genetic and Evolutionary Computation Conference (GECCO'2002)*, pages 666–673, San Francisco, California, July 2002. Morgan Kaufmann Publishers.
21. E. Zitzler and L. Thiele. Multiobjective Evolutionary Algorithms: A Comparative Case Study and the Strength Pareto Approach. *IEEE Transactions on Evolutionary Computation*, 3(4):257–271, November 1999.

Searching under Multi-evolutionary Pressures

Hussein A. Abbass[1] and Kalyanmoy Deb[2]

[1] Artificial Life and Adaptive Robotics (A.L.A.R.) Lab, School of Computer Science, University of New South Wales, Australian Defence Force Academy Campus, Canberra, Australia. h.abbass@adfa.edu.au
[2] Mechanical Engineering Department, Indian Institute of Technology, Kanpur, Kanpur, PIN 208 016, India. deb@iitk.ac.in

Abstract. A number of authors made the claim that a multiobjective approach preserves genetic diversity better than a single objective approach. Sofar, none of these claims presented a thorough analysis to the effect of multiobjective approaches. In this paper, we provide such analysis and show that a multiobjective approach does preserve reproductive diversity. We make our case by comparing a pareto multiobjective approach against a single objective approach for solving single objective global optimization problems in the absence of mutation. We show that the fitness landscape is different in both cases and the multiobjective approach scales faster and produces better solutions than the single objective approach.

1 Introduction

In biological systems, natural selection is carried out on multi-traits that are usually in conflict. In real life animal breeding, a farmer selects those animals to cull or mate based on a multi-trait selection index. Yet, in applying artificial evolutionary systems to solve real life problems, researchers deviate from the biological phenomena claiming that a biological characteristic may not be the idle choice for a computational problem. Although this assertion contains some truth, it is somehow misused. An example for this is the use of very high mutation rates in evolutionary optimization algorithms to overcome the premature convergence problem due to loss of genetic diversity. As such, it seems that using very high mutation rates is becoming a common practice that sometimes makes us wonder whether the process is evolutionary or merely a random search? Mutation rates in biological systems such as human is extremely low (varies between 1 in 10,000 and 1 in 1,000,000 genes within a gamete). The largest mutation rate is 1 in 10,000 and is associated with neurofibromatosis type I, a cancer of the nervous system. A natural question which arises here is: why premature convergence did not occur in biological systems? Is it because of Niching, Speciation, or something else?

Another similar problem exists in genetic programming, where some papers use large populations in the order of 500-1000 individuals and the artificial evolution is carried out only for a few generations 30-50 to overcome premature convergence. A debate here is whether 30-50 generations are enough for evolution to find its way in a population size of 1000 and in very large search spaces. An obvious question is whether the final result is actually because of evolutionary forces or because of luck and random factors.

In an attempt to address some of these questions, we present a multiobjective evolutionary approach for solving single-objective global optimization problems. Our objective is not to invent a better method, although we will see that the proposed method is performing well. Rather, we will show that searching under multiple evolutionary pressures alone, in the absence of mutation, maintains a considerable amount of genetic diversity evidenced by the performance of the proposed approach on three global optimization hard functions. We will also show that the proposed approach exhibits better scalability. Our analysis will also cover a comparison to the fitness landscape as well as the evolutionary dynamics for multiple and single evolutionary pressures.

The research questions that this paper attempts to answer some of them are:

1. Is multiobjectivity a force that balances the high selection pressure in evolutionary computations?
2. Is multiobjectivity a diversity preserving and/or promoting mechanism?
3. How does the evolutionary dynamics compares between multiple and single selection pressures?
4. How does the fitness landscape compares between multiple and single selection pressures?
5. Should the analysis for neutral paths be within a multi-selection pressure framework in the absence of mutation?

In [5] a multiobjective approach is used for maintaining diversity in genetic programming. The second objective in this approach was an explicit diversity measure; therefore it is difficult to judge whether the good results are because of the multiobjective component (that evolution is searching in two dimensional objective space) or because the explicit inclusion of a diversity objective. In [4], a multiobjective approach is used once more for genetic programming. The second objective was to minimize the program size, which is an implicit diversity preserving mechanism. However, as the authors compare a number of different techniques, it is difficult to isolate the contribution of the multiobjective component from other evolutionary operators in the evolutionary method.

Knowles and Corne [8] also proposed the use of multiobjective optimization to solve single objective problems. They have shown that the multiobjective approach reduces the number of local optimum. However, their approach requires either decomposing the original objective or the decision space. Apart from the fact that the objective of this paper is different, the proposed approach in this paper does not require decomposing neither the objective nor the decision space.

In our experiments, we avoided using mutation neither comparing our results against traditional evolutionary algorithms with mutation operator. To this end, it is important that we differentiate between a diversity promoting mechanism and a diversity preserving or maintaining mechanism. The former, such as the use of mutation, adds variations to the population. The latter, does not add variations to the population, but slows down the selection pressure and maintains the diversity which already exists in the population.

Diversity preserving mechanisms include low selection pressure, large populations and as we present in this paper, multi–evolutionary pressures. In the absence of a diversity promoting mechanism, the only source of diversity is the initial population. In other words, these methods slow down the side–effect of the selection pressure but they do not add new variations. By using a good diversity preserving mechanism, we may not need high mutation.

In this paper, we present a first attempt to scrutinize the effect of a multiobjective approach on the fitness landscape of a problem as well as its contribution to the diversity in the population. The rest of this paper is divided as follows: in Section 2, the methods are introduced followed by the experimental setup and results in Sections 3 and 4 respectively. Conclusions are drawn in Section 5.

2 Methods

A variation of the Self–adaptive Pareto–frontier Differential Evolution (SPDE) algorithm [1] is used. We present this variation here followed by a similar algorithm using a *Breeder Genetic Algorithm* (BGA) [9,10] selection strategy for single objective evolutionary optimization.

2.1 The Multiobjective Algorithm: The SPDE Algorithm

SPDE is a variation of PDE [2,3] where both crossover and mutation rates self–adapt. In the current paper, we do not allow for mutation. The crossover rate is inherited from the parents in the same way crossover is undertaken for the decision variables. Here, a child is generated from three parents as follows

$$x^{child} \leftarrow x^{\alpha_1} + F \times (x^{\alpha_2} - x^{\alpha_3}) \tag{1}$$

where, $x^{\alpha_1}, x^{\alpha_2}, x^{\alpha_3}$ are three different genotypes and F is a step length generated from a Gaussian distribution with mean 0 and standard deviation 1.

If the maximum number of non–dominated solutions in a generation is greater than the user specified maximum, the following nearest neighbor distance function is adopted:

$$D(x) = \frac{(min||x - x^i|| + min||x - x^j||)}{2}, \tag{2}$$

where $x \neq x^i \neq x^j$. That is, the nearest neighbor distance is the average Euclidean distance between the closest two points. The non–dominated solution with the smallest neighbor distance is removed from the population until the total number of non–dominated solutions is retained to the user specified maximum.

If some variables in the child fall outside their range, a repair rule is used. The rule is simply to multiply the variable by a random number from a uniform distribution in the range 0–1 till the variable retains its original bound. A generic version of the SPDE algorithm used in this paper follows:

1. Create a random initial population of potential solutions. Each variable is assigned a random value according to a Gaussian distribution $N(\mu, \sigma)$, where μ is the mean of the variable's range and σ is usually taken to be one sixth of the variable possible range.
2. Repeat
   a) Evaluate the individuals in the population and label those who are non–dominated.
   b) If the number of non–dominated individuals in the population is less than 3, repeat the following until the number of non–dominated individuals in the population is greater than or equal to 3.
      i. Find a non–dominated solution among those who are not labelled.
      ii. Label the solution as non–dominated.
   c) If the number of non–dominated individuals in the population is greater than the allowed maximum, apply the neighborhood distance function (Equation 2) until the number of non–dominated individuals in the population is less than the allowed maximum.
   d) Delete all dominated solutions from the population.
   e) Repeat
      i. Select at random an individual as the main parent α_1, and two individuals, α_2 and α_3 as supporting parents.
      ii. Select at random a variable j.
      iii. **Crossover rate:** Let the crossover rate for the child be

      $$x_c^{child} \leftarrow x_c^{\alpha_1} + F \times (x_c^{\alpha_2} - x_c^{\alpha_3}) \quad (3)$$

      If the crossover rate is negative, make it positive. If the crossover rate is greater than 1, let it be $0.5 * (x_c^{child} - 1)$.
      iv. **Crossover:** For each variable i
      With some probability $Uniform(0, 1) > x_c^{child}$ or if $i = j$, do

      $$x_i^{child} \leftarrow x_i^{\alpha_1} + F \times (x_i^{\alpha_2} - x_i^{\alpha_3}) \quad (4)$$

      otherwise

      $$x_i^{child} \leftarrow x_i^{\alpha_1} \quad (5)$$

      where each variable i in the main parent, $x_i^{\alpha_1}$, is perturbed by adding to it a ratio, $F \in Gaussian(0, 1)$, of the difference between the two values of this variable in the two supporting parents. At least one variable must be changed.
      v. If the child dominates the main parent, place the child into the population.
   f) Until the population size is M
3. Until the maximum number of objective evaluations allowed is reached.

Sofar, we did not mention what the second objective is for the multi–pressure case. In designing our investigation into this problem, we were faced with the question of how can we transform a single objective unconstrained global optimization problem into a multiobjective one? A number of measures which may preserve diversity explicitly and are relative to each population is:

1. Entropy of the population (*ie.* Shannon's entropy term)
2. Mean variance of variables in the population
3. Information content (as introduced in Section 4.4)

However, we need to find a natural objective which may preserve diversity implicitly but not a direct measure for diversity. Also, we need to find a measure on the level of the individual rather than the population level. Assuming a global minimization problem, a number of options include:

1. Maximizing the reverse of the original objective.
2. Maximizing/minimizing a random value assigned to each chromosome as its second fitness.
3. Maximizing the age of the chromosome; analogue to reality where the age reflects the experience.

From a feasibility study to investigate the three options, the first option was the worst in terms of computational time because there was a large number of pareto–optimal solutions at each generation. Therefore, the neighborhood function needs to be applied a large number of times to cut down the number of pareto–optimal solutions to the user predefined number. The second was performing well but the third gave the best performance in terms of the quality of solutions and processing time. In this paper, we report the results obtained by the third option of maximizing the age.

The age of a chromosome is a fixed value assigned to the chromosome at the time it is created. For the first generation, the year of birth index starts with 1. Each time a chromosome is generated in the initial population, the year of birth index is incremented with 1. By the time the initial population is initialized, the year of birth ranges from 1 to the population size. In subsequent generations, the year of birth is increased by one; that is, all solutions to be generated in a subsequent generation will share the same year of birth, which is population size + generation number.

A possible explanation for the good performance of the age over other objective functions is, the age allows old solutions (presumably not so good) to be maintained in the population, thereby maintaining diversity and slowing down the selection pressure.

2.2 The Single Objective Algorithm

A similar algorithm to SPDE is used with two variations; the first is instead of maintaining at most P pareto–optimal solutions - which we do not have any more - we maintain the top P best solutions similar to the BGA strategy, where children are generated from these P solutions only. The second variation is we dropped the condition of placing the child into the population if it dominates the parent. We simply place the child into the population. If we place the child into the population only when it is better than the parent, we got very bad results because it increased the evolutionary pressures.

3 Test Suite and Experimental Setup

We tested the performance using three global optimization problems; these are: De Jong, Griewangk, and Rastrigin functions [6,7]. The three functions are given below.

1. De Jong's F1 function:

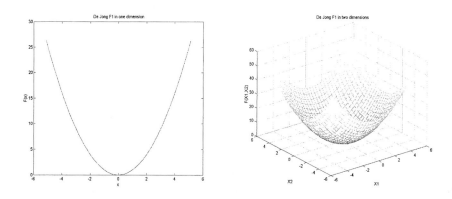

Fig. 1. De Jong function in one and two dimensions

$$f(x) = \sum x_i^2, i = 1\ldots n, -5.12 \le x_i \le 5.12$$

De Jong's F1 function is a unimodal function with a unique minimum at $x_1 = x_2 = 0$. It should be simple for a genetic algorithm as well as any hill–climbing algorithm with a proper neighborhood operator.

2. Griewangk's function:

$$f(x) = \sum \frac{x_i^2}{4000} - \prod \cos \frac{x_i}{\sqrt{i}} + 1, i = 1\ldots n, -600 \le x_i \le 600$$

In this function, the local optima are above the parabola level, which is produced by the summation term. The more we increase the search range, the flatter the function. This function is supposed to be difficult for a genetic algorithm because the product term causes the variables to be highly interdependent.

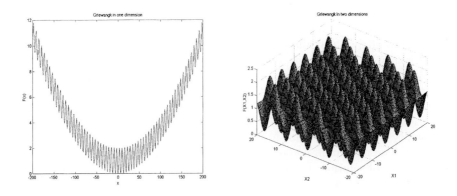

Fig. 2. Griewangk function in one and two dimensions

3. Rastrigin's function:

$$f(x) = A \times n + \sum(x_i^2 - A \times \cos(2 \times \pi \times x_i)), i = 1\ldots n, -5.12 \leq x_i \leq 5.12$$

This function is based on De Jong's F1 with the addition of the cosine term which creates a large number of local minima. The amplitude of the function's surface is controlled by the parameter A. When A is 10 (as in our experiments) the domain is highly multimodal. The local minima are located at a rectangular grid with size 1. With increasing distance to the global minimum the fitness values of local minima become larger.

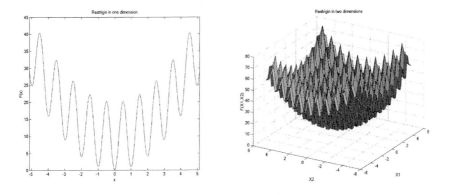

Fig. 3. Rastrigin function in one and two dimensions

In the rest of this paper, we will refer to the single and multiobjective algorithms as algorithm1 and algorithm2 respectively. The objective of our experiments is to analyze the differences between searching under single and multiple evolutionary pressures. We would also like to find if multiobjectivity preserves reproductive diversity. Therefore, in all of our experiments, we use the same initial populations for both algortihm1 and algorithm2. Mutation is not used in any experiment to avoid any source of diversity.

Each problem is run with 10, 20, 30, 40, and 50 variables to measure the scalability of each algorithm. A population size of 100 is used and the step size F is 0.1. The maximum number of pareto–optimal solutions for algorithm2 and the number of solutions cloned for algorithm1 are 50. The maximum number of objective evaluations is 1 million and the evolutionary run is terminated once this maximum is reached. Each experiment is run 10 times and the same ten seeds were used for all experiments.

4 Results

In Figure 4, the average and standard deviation of the best solutions found in the ten runs are reported for the three problems with the different number of variables. It is clear from the figure that the quality of solutions obtained by algorithm2 is much better than those obtained by algorithm1. The scalability of algorithm2 is notably better than algorithm1. The figure demonstrates that searching under multiple evolutionary pressures produced better solutions than searching under single evolutionary pressures.

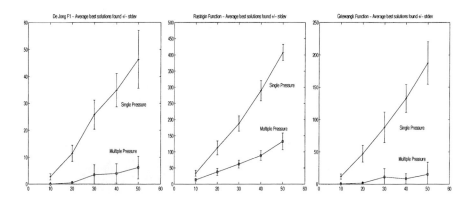

Fig. 4. The average best solution found in ten different runs for the three differen problems with dimensions of 10, 20, 30, 40, and 50

To scrutinize the effect of multiobjective evolutionary search and contrast it with the single–pressure evolutionary search, we will present a detailed analysis in the following sections using the Griewangk function. The results are consistent on the other problems.

4.1 Convergence

Figure 5 presents the best and average fitness over 300 generations for both algorithm1 and algorithm2. Premature convergence is very clear for algorithm1, where the best and average fitness stagnated very early in the search. The very bad solution quality is also clear from the figure. For algorithm2, we can see that some runs reached the global optimum. It is also apparent that the average fitness is not decaying over time, but fluctuating; therefore, this entails that the algorithm is exploring the search space better caused by the multiobjective pareto selection method. Convergence in algorithm2 occurred between 150 and 250 generations.

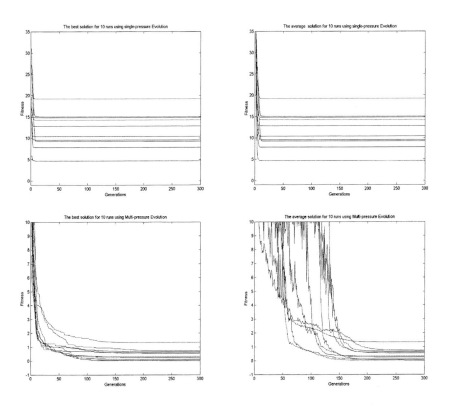

Fig. 5. The best (on left) and average (on right) solutions found in ten different runs for the Griewangk's problem with 10 variables for single (top graphs) and multiple (bottom graphs) pressures

4.2 Population Trajectories and Diversity

In Figure 6, we present the progress of genotype diversity in the evolution. We define genotype diversity as the average distance between all genotypes in a population and the population centroid. We also present the population trajectories represented by the change of the population centroid overtime.

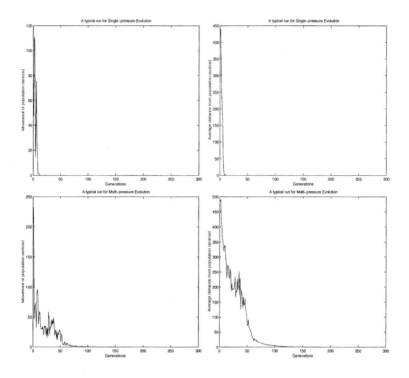

Fig. 6. The population trajectories (on left) and diversity measure (on right) for a typical run using the Griewangk's function for single (top graphs) and multiple (bottom graphs) pressures

We measured the average correlation coefficient between the diversity and the movement of the population over the ten runs. For the multiple pressure algorithm, the correlation was 0.79 for the first 200 generations and 0.78 for the single pressure algorithm over the first 50 generations. This may entail that the more the population shifts in the space, the more diversity it maintains.

The previous conclusion may not be entirely true. The population movement - measured by the change of the population centroid - can simply take place by deleting or injecting an outlier which will also affect the diversity measure. Therefore, we measured the average correlation coefficient between the population diversity and the best and worst solution for each algorithm. For algorithm1, the population diversity was highly correlated with the best and worst solutions (0.95 for each case). For algorithm2, however, the population diversity was highly correlated with worst solution (0.88) and less with the best (0.72). These findings can also be seen in Figure 7, which depicts a typical run. It is clear that the average fitness is closer to the worst solution than to the best for algorithm2. However, since the worst solution is not fluctuating and is monotonically decreasing, it cannot cause the oscillation that we see with the diversity and centroid's movement.

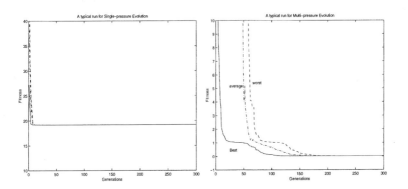

Fig. 7. A typical evolutionary run showing the best, average, and worst fitness for single pressure (left graph), and multiple pressures (right graph)

4.3 Fitness Landscape: Fitness Histograms

A fitness landscape is defined by the genotype representation, the fitness function, and the operators used to generate solutions in the neighborhood. In Figure 8, we present the fitness histogram for a hill–climbing algorithm, algorithm1, and algorithm2. In hill–climbing, the algorithm starts with a random solution, mutate it by adding a Gaussian random value $N(0, 10)$, if the resultant solution is better than the old one, it replaces it; otherwise a new solution is generated. The average ± standard deviation for the best solutions obtained by the hill–climbing algorithm in 10 runs are 0.550 ± 0.089. The corresponding numbers for algorithm1 and algorithm2 are 11.678 ± 3.978 and 0.365 ± 0.326 respectively. It is clear that algorithm1 was much worse than the hill–climbing algorithm while algorithm2 is better than both.

4.4 Fitness Landscape: Information Contents

A rugged landscape is expected to have large fitness values between neighborhood points while a smooth landscape will have small differences in the fitness values. In this section, we employ [11] information theoretic measures for analyzing the fitness landscape. We use three measures defined in [11]; the information

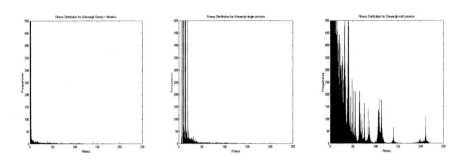

Fig. 8. The fitness histogram for the Griewangk's function using hill climbing (left graph), single pressure (middle graph), and multiple pressures (right graph)

content, the partial information content, and the information stability. The first measure provides an insight into the variety of shapes on, or local neighborhood of, the landscape. The second measure gives an insight into the modality of the path. The third measure captures the upper bound of the magnitude of the landscape optima.

In Table 1, we provide the information contents, partial information contents, and expected number of optima for single pressure and multiple pressures evolutionary algorithms. We considered the evolutionary algorithm as a black box which generates a sequence of solutions. This makes it easy to apply these measures.

Table 1. The information contents and partial information contents for the Griewangk function using single pressure and multiple pressures methods

Method	ϵ	Information Contents	Partial Information Contents	Expected Number of Optima
Single pressure	0	0.0034	0.0006	3302
	0.10	0.0017	0.0002	1398
	0.50	0.0016	0.0002	1216
	10.0	0.0008	0.0001	450
	90.0	0.0001	0.0	40
	100.0	0.0000	0.0	0
Multiple pressure	0	0.4261	0.2171	1085560
	0.10	0.0629	0.0182	91162
	0.50	0.0592	0.0141	70633
	10.0	0.0321	0.0061	30361
	100.0	0.0051	0.0006	2957
	210.0	0.0002	0.0000	42
	220.0	0.0000	0.0000	0

The expected number of local optima is simply half the partial information content multiplied by the size of the sample. When ϵ was zero, algorithm2 has the highest information content and partial information content.

It is worth mentioning that ϵ is a tolerance factor which flattens the landscape. When ϵ is zero, the measure is very sensitive to any variations and when ϵ is high, most of the landscape is flat. Information stability is the value of ϵ where the information content measure in the third column of the table reaches 0. It is clear from the table that the highest information stability is achieved with the multiple pressure algorithm, where ϵ is around 220.

4.5 A Fair Comparison

It may sound that the previous comparison was not that fair since the single pressure algorithm did not have obvious diversity preserving mechanism. To show that this is not entirely true, we did experiments with the single pressure algorithm while varying the selection pressure. The selection pressure is varied as follows. We used the breeder genetic algorithm strategy in the previous experiments by cloning the best 50 individuals and generating additional 50 children by breeding among them. Here, to reduce the selection pressure, we define a parameter ξ which represents the best ξ individuals. When ξ is 10, we clone 50 individuals from current population. These 50 are formed from the best 10 individuals and the other 40 are chosen at random without replacement. When selecting parents from these 50, we simply select parents at random regardless of their fitness.

Table 2. The results under different selection pressures using the single pressure algorithm

ξ	Best solution
1	10.365 ± 5.770
2	10.564 ± 4.456
3	9.780 ± 5.759
4	9.880 ± 5.793
5	8.090 ± 2.470
6	9.998 ± 5.151
7	11.353 ± 5.520
8	10.181 ± 3.580
9	9.960 ± 5.234
10	11.528 ± 3.745
20	12.484 ± 4.426
30	12.473 ± 4.441
40	12.218 ± 4.951
50	11.987 ± 5.536

We varied ξ between 1 and 10 in a step of 1 then up to 50 in a step of 10. The results are given in Table 2. Surprisingly, the best result obtained when ξ was 5;

that is, select the best five solutions to clone as parents then select the rest of the parents at random. This corresponds to a very low selection pressure, which entails that searching under single pressure is very sensitive to the selection pressure. Recalling from Section 4.3, the multi–pressure algorithm is still much better than the single pressure algorithm.

5 Conclusion

In this paper, we presented a multiobjective approach for solving single objective optimization problems. We have shown that the multiobjective approach preserves genetic diversity better than the single objective approach in the absence of mutation. The former also gave the best results. When analyzing the behavior of both algorithms, we found that the fitness landscapes for both algorithms are different. The multiobjective approach also gave the highest information contents.

References

1. H.A. Abbass. Self-adaptive pareto differential evolution. In *The IEEE Congress on Evolutionary Computation, Honolulu, USA*, pages 831–836. IEEE Press, 2002.
2. H.A. Abbass. A memetic pareto evolutionary approach to artificial neural networks. In M. Stumptner, D. Corbett, and M. Brooks, editors, *Proceedings of the 14th Australian Joint Conference on Artificial Intelligence (AI'01)*, pages 1–12, Berlin, 2001. Springer-Verlag.
3. H.A. Abbass. An evolutionary artificial neural network approach for breast cancer diagnosis. *Artificial Intelligence in Medicine*, 25(3):265–281, 2002.
4. S. Bleuler, M. Brack, L. Thiele, and E. Zitzler. Multiobjective genetic programming: Reducing bloat using SPEA2. In *Proceedings of the 2001 Congress on Evolutionary Computation CEC2001*, pages 536–543. IEEE Press, 2001.
5. E. De Jong, R. Watson, and J. Pollack. Reducing bloat and promoting diversity using multi-objective methods. In *Proceedings of the Genetic and Evolutionary Computation Conference*, pages 11–18, 2001.
6. K De Jong. *An analysis of the behavior of a class of genetic adaptive systems*. PhD thesis, 1975.
7. V.S. Gordon and D. Whitley. Serial and parallel genetic algorithms as function optimization. In *The 5th International Conference on Genetic Algortihms*, pages 177–183. Morgan Kaufmann, 1993.
8. J. Knowles and D. Corne. Reducing local optima in single-objective problems by multi-objectivization. In *First International Conference on Evolutionary Multi-criterion Optimization (EMO'01)*, pages 269–283. Springer-Verlag, 2001.
9. H. Mühlenbein and D. Schlierkamp-Voosen. Predictive models for the breeder genetic algorithms: continuous parameter optimization. *Evolutionary Computation*, 1(1):25–49, 1993.
10. H. Mühlenbein and D. Schlierkamp-Voosen. The science of breeding and its application to the breeder genetic algorithm bga. *Evolutionary Computation*, 1(4):335–360, 1994.
11. V.K. Vassilev, T.C. Fogarty, and J.F. Miller. Information characteristics and the structure of landscapes. *Evolutionary Computation*, 8(1):31–61, 2000.

Minimal Sets of Quality Metrics

A. Farhang-Mehr and S. Azarm

Department of Mechanical Engineering, University of Maryland
College Park, Maryland 20742, U.S.A.
{mehr@wam, azarm@eng}.umd.edu

Abstract. Numerous quality assessment metrics have been developed by researchers to compare the performance of different multi-objective evolutionary algorithms. These metrics show different properties and address various aspects of solution set quality. In this paper, we propose a conceptual framework for selection of a handful of these metrics such that all desired aspects of quality are addressed with minimum or no redundancy. Indeed, we prove that such sets of metrics, referred to as 'minimal sets', must be constructed based on a one-to-one correspondence with those aspects of quality that are desirable to a decision-maker.

1 Introduction

There are various multi-objective heuristic search techniques in the literature among which Multi-Objective Evolutionary Algorithms (MOEAs) have received significant attention [1][2][3][4]. These techniques usually generate a finite set of solutions to approximate the Pareto frontier of a multi-objective optimization problem. However, obtaining the 'best possible' set to represent the entire Pareto frontier is not always a trivial (or even an objectively-defined) task. Indeed, researchers have developed a myriad of techniques over the last few years to improve the quality of such solution sets in one way or another. Naturally, performance assessment and comparison study of such techniques have also gained much attention. One obvious way to compare MOEAs is to simply visualize the final sets of solutions and rely on intuitive judgments to decide on superiority of one technique to another. However, as discussed by Van Valdhuizen and Lamont [2], intuitive and visual assessment is not a reliable tool for comparison of different multi-objective optimization techniques. Especially, for problems with more than three-dimensions, visual judgment is either impossible or quite misleading yet it is the prevailing tool used by the researchers in the field.

More recently, there has been an emerging theme in the literature to quantitatively assess and compare the quality of non-dominated solution sets via *quality* (or *performance*) *metrics*. (A Non-Dominated Set, abbreviated as NDS, is defined as a set of mutually non-dominated solutions obtained from a MOEA to approximate the Pareto frontier.) These quality metrics generally assign an absolute or relative value (or a set of values) to a NDS to determine whether it is a 'good' representation of the actual Pareto frontier. For instance, Zitzler and Thiele [5] performed a comparative study of several multi-objective optimization methods using two metrics: "size of the dominated space" and "fraction of solutions dominated by the other set". Van Veldhuizen [6] introduced several quality metrics, such as: 'error ratio', 'generational distance', 'maximum Pareto frontier error' and 'overall non-dominated vector generation ratio' to assess different aspects of a solution set quality. Sayin [7] defined metrics for coverage, uniformity and cardinality to determine how 'good' a set of discrete solution points represents the true Pareto frontier. (For a recent review of the quality metrics and their shortcomings, see [8].)

Having many different quality metrics in the literature poses a new question to researchers: which metric or host of metrics must be used for an exhaustive (but not redundant) comparison study of different MOEAs? In fact, many of these metrics are coupled in the sense that they address a common aspect of quality. For example, there are several metrics in the literature that are claimed to assess 'diversity of solutions' in one way or another, including: 'spacing metric' [9]; 'overall non-dominated vector generation', 'overall non-dominated vector generation ratio' [6]; 'coverage', 'uniformity', 'cardinality' [7]; 'number of distinct choices', 'Pareto spread', and 'cluster' [10]. In a similar fashion, researchers developed numerous metrics to assess the closeness of solution sets to the Pareto frontier (see [4], for examples of these metrics). Obviously, many of these overlapping metrics are correlated, introducing redundancy in the comparison study of MOEAs. On the other hand, selecting too few of these quality metrics does not guarantee an exhaustive comparison with respect to all aspects of quality. Indeed, a desirable collection of quality metrics must be *minimal*, in the sense that: 1) There is at least one metric for every aspect of the solution set quality to guarantee an exhaustive performance assessment; 2) There exists minimum (or no) correlation among quality metrics to avoid redundancies (see Section 4 for a formal definition). Due to the subjective nature of quality metrics, and quality of a set of solutions in general, the above-mentioned properties may not be noticeable from the formulation of these metrics and require a more thorough investigation of the underlying concepts.

In this paper, we assume that the decision-maker's specific preferences are not known a priori, while the general aspects of interest are known. In other words, we would like to determine a minimal set of quality metrics that exhaustively addresses all desired aspects of quality without redundancy. Hansen and Jaszkiewicz [11] performed a similar study in which a family of outperformance relations was defined to compare Pareto-optimality of non-dominated solutions sets (see Section 2 for more on this). These relations account for pairs of NDSs where one solution set is objectively better (based only on the notion of dominance) than the other set and thus, they establish *strict partial orders* among NDSs. (A strict partial order in the set of all possible NDSs is defined as an irreflexive, antisymmetric, and transitive relation that compares some and not every given pair of NDSs. For details see almost any book on set theory, e.g. [12].) In other words, not all solution sets are comparable in this way, but at least one can verify the validity of a quality metric by examining its compatibility with these relations. Based on this idea, Zitzler et al. [13] proposed a theoretical framework to investigate the compatibility and completeness of different comparison methods and derived a set of theoretical restrictions for the existence of compatible and complete *unary* quality metrics. (A unary quality metric measures the *absolute* goodness of a NDS.) In fact, they prove that a finite combination of unary metrics that is compatible and complete at the same time does not exist in general. They also mention that this limitation does not apply to *binary* quality metrics (i.e., those that quantify only the *relative* goodness of two NDSs), wherein one can construct compatible and complete metrics with respect to any dominance relations.

The focus of this paper is on *binary* quality metrics and their correspondence with outperformance relations. More specifically, the contributions of this paper are as follows.

- While outperformance relations in the previous works address only one aspect of the quality, namely closeness to the Pareto frontier, in this paper, we propose the more general notion of *Excellence Relations*. These relations establish strict partial orders in the set of all NDSs with respect to different aspects of quality.

- Unlike previous studies by Hansen and Jaszkiewicz [11] and Zitzler et al. [13] who considered several outperformance relations (e.g., weak/strong/ outperformance) to address the closeness of a NDS to the Pareto frontier; here we assume that a decision-maker provides a *combination* of excellence relations, each addressing a *distinct* aspect of quality (e.g., a dominance relation for closeness to the Pareto frontier, a coverage relation for the coverage of the Pareto frontier, and so on). Then we define minimal sets of binary quality metrics, and investigate their ability to address all of these excellence relations at the same time (e.g., whether a set of quality metrics can address both closeness and coverage of the Pareto frontier.)

- We extend the definition of compatibility with outperformance relations [11][13] to excellence relations and show that: Given two uncorrelated binary quality metrics, they cannot be both compatible with the same excellence relation.

- Finally, one may find several minimal combinations of quality metrics, however, all of them possess the following important property (referred to as Minimality Lemma in Section 4): if one assumes n excellence relations with certain properties, a minimal set of binary quality metrics contains n and only n metrics, each compatible with exactly one excellence relation, i.e., there is a one-to-one correspondence. This property significantly narrows the search for minimal sets.

Although there are relatively fewer binary quality metrics in the literature as compared to unary metrics [13], the above property makes them very attractive for a comparison study of multiobjective optimization algorithms, mainly because one could select a minimal set of binary metrics to address all desired aspects of quality exhaustively and distinctly. (This is not possible in general with unary metrics, see [13].)

2 Excellence Relations

As mentioned in Section 1, Hansen and Jaszkiewicz [11] defined the outperformance relations to establish a strict partial order among NDSs, where some pairs of solution sets are objectively comparable in terms of Pareto optimality (or dominance). Here, the definition of a strong outperformance relation is given for the completeness of this paper. (In addition to the strong relation, Hansen and Jaszkiewicz [11] also defined weak and complete outperformance relations.)

Definition 1. A non-dominated set A strongly outperforms a non-dominated set B, denoted by AR_OB, iff (i.e., if and only if) $A \neq B$ and (in the objective space) for each $y \in B$, there exists $x \in A$ such that $x \succeq y$.

The notation '$\succeq$' in $x \succeq y$ indicates that a point x is either equal to or dominates a point y with respect to all objectives. Fig. 1 demonstrates two non-dominated solution sets generated for a 2-objective maximization problem. According to Definition 1, we observe that: AR_OB.

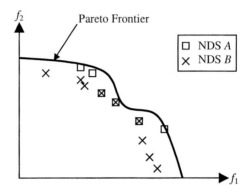

Fig. 1. AR_OB: Set A strongly outperforms Set B

As mentioned before, not every two sets are comparable in this way. If we denote the set of all possible non-dominated sets by U, then the above relation constructs a strict partial order in U. Here we define the *partially ordered domain* of this comparison in $U \times U$, shown by Λ_{Ro}, as the set of all 2-tuples of NDSs that are comparable via R_O, i.e.,

$$\Lambda_{Ro} = \{(A,B) \in U \times U \mid \text{either } AR_OB \text{ or } BR_OA\}$$

This relation by itself does not provide a tool to compare any given pair of NDSs, however, it can be used to verify the validity of quality metrics. In fact, this relation is based only on the concept of dominance (or closeness to the Pareto frontier), and therefore, if a quality metric aims at comparing NDSs in terms of dominance, it must be compatible with this relation in the first place. (The formal definition of compatibility is given later in the paper.)

We mentioned that the outperformance relation accounts for the Pareto optimality of solution sets. However, there are other aspects of quality that are especially important in the assessment of solution sets obtained from MOEAs, e.g., diversity of the solution sets, extent of the Pareto frontier that is covered by the solutions. Very similar to the outperformance relation, one can collect all 2-tuples of NDSs that are objectively comparable with respect to any aspect of quality and construct a strict partial order accordingly. This prompts for the definition of a more general concept, i.e., *excellence relations*, as proposed in this paper.

Definition 2. An excellence relation, denoted by R, is defined as a strict partial order in U that relates all non-dominated sets that are objectively comparable with respect to a common aspect of quality. The partially ordered domain of R in $U \times U$, denoted by Λ_R, is defined as: $\Lambda_R = \{(A,B) \in U \times U \mid \text{either } ARB \text{ or } BRA\}$.

For example, an outperformance relation is an excellence relation with respect to dominance. As another example, in the following we define a new excellence relation (i.e., *coverage relation*) to address a different aspect of quality: *coverage* (i.e., the span of the solution set over the Pareto frontier). In this example, it is assumed that all objective functions are positive.

Definition 3. A non-dominated set, B, is strictly superior to another non-dominated set, A, in terms of *coverage*, denoted by BR_CA, iff all solution points of Set A are contained in a convex cone generated by Set B, while there exists at least one solution in Set B that is not contained in a convex cone generated by Set A.

Here the convex cone generated by a solution set $A=\{\mathbf{a}_1, \mathbf{a}_2, ...\}$ is defined as all nonnegative linear combinations of $\mathbf{a}_i$'s, i.e., $\{\mathbf{v} \mid \mathbf{v} = \Sigma w_i \mathbf{a}_i ; \ w_i \geq 0\}$ [14]. The shaded area in Fig. 2(b) demonstrates the convex cone generated by Set B. This cone clearly contains all solution points of Set A. In contrast, the convex cone of Set A (Fig. 2(a)) does not include all solution points of Set B. Thus, according to Definition 3, we have: BR_CA. Finally, note that in this definition it is assumed that all objectives are to be maximized. Also, the nadir point (i.e., the lower bound of the Pareto frontier) is assumed to be located at the origin of the Cartesian objective space. If these assumptions do not hold for a given problem, one could always transform the objectives to meet these assumptions.

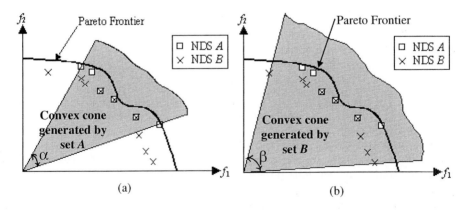

Fig. 2. Convex cones generated by: (a) NDS A; and (b) NDS B. According to Definition 3: BR_CA.

As expected, the excellence relation for this example establishes a strict partial order in U, i.e., objectively compares a pair of NDSs in terms of coverage.

Note that the first step in a comparison study of NDSs is to determine different aspects of quality that are of interest to the decision maker (e.g., Pareto optimality, coverage, diversity, and so on). Then, we collect all pairs of NDSs that are objectively comparable with respect to any of these quality aspects. These collections establish strict partial orders in U that we referred to as excellence relations. However, not all solution sets are comparable using these relations, and one must formulate quality metrics that enable an exhaustive comparison of all non-dominated solution sets. Each binary quality metric constructs a *total order* that quantitatively compares all NDSs pairs in U. (A total order in the set of all possible NDSs is defined as an irreflexive, antisymmetric, and transitive relation that compares all pairs of NDSs in the set U.) The correspondence of these quality metrics (i.e., total order) with excellence relations (i.e., partial orders) is the subject of the next section. We prove two key theorems that will be used later in Section 4 to investigate the properties of a minimal set of quality metrics.

3 Compatibility, Concordance, and Correlation

As briefly explained in Section 1, a binary quality metric compares the quality of two NDSs and returns a relative value, while a unary metric computes an absolute value for quality. The focus of this paper is only on binary metrics, i.e., if A and B are two NDSs, then $Q(A,B)$ returns a scalar that reflects how much better set A is when compared to set B. (The arguments of this paper also applies to other metrics that can be transformed into the above binary format.) Moreover, Q is assumed to be symmetric and homogeneous, meaning that: $Q(B,A) = -Q(A,B)$. Note that a symmetric metric as defined by Knowles and Corne [4] is: $Q'(B,A) = C - Q'(A,B)$. This latter format can be easily transformed into a homogenous symmetric metric by assuming: $Q(A,B) = C/2 - Q'(A,B)$. (See also [13] for a discussion on the properties of such metrics.) Without loss of generality, it is assumed that $Q(A,B) > 0$ iff A is strictly better than B. A binary quality metric (hereafter, 'metric' implies 'binary metric' unless otherwise stated) as defined above constructs a total order in U and compares any two non-dominated sets on a quantitative basis.

Now, assume a symmetric and homogeneous quality metric, Q, that compares any two given NDSs in terms of a certain aspect of quality. If R is an excellence relation that addresses the same aspect of quality, it is natural to expect Q to be compatible with R, as defined formally in the following. (This is very similar to the definition of compatibility with outperformance relation given in [11] and [13]; tailored for symmetric metrics, and generalized for excellence relations.)

Definition 4. A symmetric homogeneous binary metric, Q, is *compatible* with an excellence relation, R, iff: for any given pair of non-dominated sets A and B such that ARB, we also have $Q(A,B) > 0$, which implies set A has a better quality than set B. (The compatibility of metric Q with relation R is denoted as $Q \sim R$ in this paper.)

Knowles [8] studied the compatibility of several unary and binary quality metrics with respect to an outperformance relation. The same study can be carried out to determine the compatibility of those metrics with respect to any other excellence relations, such as the coverage relation, R_C. Obviously, if a metric is not intended to compare two NDSs in terms of a certain aspect of quality (e.g., a diversity assessment metric is not intended to account for closeness to the Pareto frontier) it does not need to be compatible with that excellence relation. In fact, we will prove that each quality metric in a minimal set must be compatible with one and only one excellence relation. Nevertheless, this compatibility is dependant on the definition of the excellence relation itself. Before formally defining minimal sets of quality metrics and their correspondence with excellence relations, in the following we introduce the notion of *concordance* among excellence relations.

Definition 5. Two excellence relations R and R' are *concordant* iff for each $(A, B) \in \Lambda_R \cap \Lambda_{R'}$ such that ARB, we also have $AR'B$.

Put another way, R and R' are concordant iff there do not exist two non-dominated sets A and B such that: ARB and $BR'A$. Concordance basically implies that the two excellence relations *cannot* work against each other. If two excellence relations are referring to different aspects of quality (e.g., diversity and Pareto optimality), there are always examples of NDSs that are better with respect to one aspect of quality and worse with respect to another, and therefore, those relations are not concordant (or are *non-concordant*). In contrast, if being better in terms of one relation always implies better with respect to another, then it implies that the two relations have essentially the same nature (i.e., refer to the same aspect or notion of quality) and thus are concordant. The two excellence relations of Section 2, i.e., R_C and R_O, are non-concordant because set A in Figures 1 and 2 is better than set B in terms of the outperformance relation (AR_OB), but worse in terms of coverage (BR_CA). Moreover, from the above definition two relations R and R', such that $R \subset R'$, are always concordant. Therefore, the family of outperformance relations defined by Hansen and Jaszkiewicz [11] are concordant because: complete outperfromance is a subset of strong outperformance, which in turn is a subset of weak outperformance. Therefore, although these relations are different, they are concordant according to Definition 5.

Concordance is a very strong assumption in the sense that if $(A,B) \in U \times U$ is comparable via two given concordant excellence relations, the outcome of the comparison from the first relation is always the same as that of the second one. On the other hand, non-concordance is a weak assumption in the sense that: two excellence relations are non-concordant even if there exists only one pair of non-dominated solution sets, (A,B), such that A is better than B with respect to one relation and worse with respect to another. Finally, the following theorem demonstrates an important property of non-concordant relations.

Theorem 1. There does not exist a symmetric and homogeneous quality metric that is compatible with two (or more) non-concordant excellence relations.

Proof. For the sake of contradiction, suppose there exists a quality metric, Q, which is compatible with two non-concordant excellence relations, namely R and R'. Since R and R' are non-concordant, there exists a pair of non-dominated sets, namely $A, B \in U$, $(A \neq B)$, such that ARB and $BR'A$. But since Q is compatible with R, from ARB we conclude $Q(A, B) > 0$. Similarly, Q is compatible with R' and from $BR'A$ we have $Q(B, A) > 0$, which is a contradiction because $Q(A, B) = -Q(A, B)$.

In fact, this theorem is somewhat intuitive from the definition of concordance and compatibility: a metric cannot simultaneously be compatible with two excellence relations that work against each other. For example a symmetric and homogeneous coverage metric, which is compatible with R_C, is necessarily incompatible with R_O (i.e., recall that R_O and R_C are non-concordant according to Definition 5 and Figures 1 and 2). The above theorem indicates that there must be at least one separate metric in a minimal set to individually address each aspect of quality, e.g., at least one metric compatible with diversity, another one compatible with Pareto optimality, and so on. Moreover, in the following we show that any two given metrics that address the same aspect of quality are necessarily correlated.

Theorem 2. If two symmetric homogeneous metrics are both compatible with an excellence relation, R, they are positively correlated within Λ_R.

Proof. Assume two symmetric and homogeneous metrics, Q and Q', are both compatible with R. Then the covariance of Q and Q' within Λ_R can be written as:

$$\text{Cov}\,[Q(A,B), Q'(A,B)] = <Q(A,B)Q'(A,B)> - <Q(A,B)> <Q'(A,B)> \;; (A,B) \in \Lambda_R$$

where the expected value of Q within Λ_R, i.e., $<Q(A,B)>$, is zero, because Q is symmetric, and therefore, for each $(A, B) \in \Lambda_R$, we also have $(B, A) \in \Lambda_R$, and $Q(A,B) = -Q(B,A)$. Similarly: $<Q'(A,B)> = 0$. On the other hand, Q and Q' are both compatible with R, and therefore, for each $(A, B) \in \Lambda_R$, $Q(A,B)$ and $Q'(A,B)$ have the same sign (both negative or both positive). Thus, $<Q(A,B)Q'(A,B)>$ is strictly positive, and the theorem follows.

Note that being 'positively correlated' is a necessary and not sufficient condition for 'compatibility with the same relation', i.e., two metrics that are not compatible with the same relation are not guaranteed to be uncorrelated. In the following section, we take advantage of the above theorems to investigate the properties of minimal sets of quality metrics.

4 Minimal Sets of Quality Metrics

In the following, we formally state the minimality conditions for a set of quality metrics.

Definition 6. A set of quality metrics, namely Γ, is said to be *minimal* with respect to a given set of non-concordant excellence relations, Φ, iff:

1. Each quality metric, $Q \in \Gamma$, is compatible with at least one excellence relation in Φ. Formally, $\forall Q \in \Gamma : \exists R \in \Phi$ such that $Q \sim R$.

2. For each excellence relation in Φ, there is at least one compatible quality metric in Γ. Formally, $\forall R \in \Phi : \exists Q \in \Gamma$ such that $Q \sim R$.

3. There is minimum (or no) correlation among quality metrics of Γ within the partially ordered domain of excellence relations.

The first property rejects unnecessary metrics that are not compatible with any of the excellence relations. The second property guarantees that Γ is exhaustive, in the sense that it addresses all aspects of quality that are of any interest to the decision-maker (i.e., expressed via excellence relations in Φ). The last property eliminates or minimizes the redundancy within the set, i.e., the selected quality metrics should have minimum or no correlation. From this definition and Theorems 1 and 2 we observe the following.

Minimality Lemma. Given a set of n non-concordant excellence relations, Φ, a corresponding minimal set of symmetric and homogeneous metrics, Γ, contains n and only n quality metrics. (Also, there is a one-to-one correspondence between Γ and Φ.)

Proof. From Theorem 1, a metric in Γ cannot be compatible with more than one excellence relations in Φ (because the excellence relations in Φ are non-concordant). Therefore, following the first property of Definition 6, each metric is compatible with exactly one excellence relation. Also, Theorem 2 indicates that two uncorrelated metrics cannot be compatible with the same excellence relation (because otherwise they would be positively correlated according to Theorem 2). Therefore, following the second property in Definition 6, there is a one-to-one correspondence between Γ and Φ.

Minimality Lemma suggests a recipe with a set of steps to be followed for the selection of a minimal set of quality metrics, as given next.

Step 1. The general aspects of performance that are of interest to the decision-maker are determined or presumed (e.g., closeness to the Pareto frontier, coverage, diversity, etc.)

Step 2. For a given aspect of quality, a strict partial order is established in U. In other words, an excellence relation is constructed that accounts for of all pairs of NDSs that are objectively comparable with respect to that given aspect of performance. For instance, outperformance relation addresses the closeness to the Pareto frontier; coverage relation of Definition 3 addresses coverage of the set, and so on. If the aspects of quality are defined properly in Step 1, these excellence relations are non-concordant (because if they are indeed referring to different aspects of quality, there exists a non-dominated sets that is better than another set with respect to one excellence relation and worse with respect to another). These non-concordant excellence relations constitute Φ.

Step 3. Suppose Φ consists of n excellence relations. To construct a minimal set, we select one and only one quality metric, compatible with each excellence relation in Φ (recall Minimality Lemma). Note that since Theorem 2 provides only a necessary condition for being uncorrelated, establishing a one-to-one compatibility correspondence does not guarantee a minimal set. However, it rules out many of non-minimal collections of metrics and significantly narrows the search for minimum correlations. The result is a set of size n of performance assessment metrics, i.e., Γ.

Γ constructs exactly n total orders in U, and can be used to compare any two given NDSs in terms of the quality aspects expressed in Step 1, and formulated as excellence relations (i.e., strict partial orders) in Step 2. It exhaustively and distinctly covers all desired aspects of quality, without unnecessary correlation among metrics. According to the Minimality Lemma, collections of more than n metrics are necessarily correlated, while less than n metrics cannot distinctly address all desired aspects of quality.

As an example, if the decision-maker desires only two aspects of quality: 1) Pareto optimality, and 2) Coverage, Φ consists of exactly two relations: $\Phi = \{R_O, R_C\}$ (recall that these two relations are non-concordant). A minimal set of size two of quality metrics should then be selected from the pool of existing metrics, $\Gamma=\{Q_1, Q_2\}$, such that $Q_1 \sim R_O$, and $Q_2 \sim R_C$. No other combination can address both of these quality aspects without redundancy. According to the Minimality Lemma, the same argument holds for any number of non-concordant excellence relations in Φ. Finally, the above guideline is only an abstraction of the notion of quality metrics and their desired properties, and therefore, it does not define or formulate new metrics by itself.

5   Practicality of the Minimality Lemma

Although the theoretical framework in this paper provides an approach for an objective selection of minimal sets of binary quality metrics, its real-world application may be hindered by several factors:

- A decision-maker may not be able to state his/her idea of 'quality of a solution set' explicitly in the form of excellence relations.

- Even if the decision-maker is able to state a set of excellence relations, Φ, as a basis for quality assessment, it is not always possible to find a corresponding set of minimal binary and symmetric quality metrics, Γ, such that each metric in Γ is compatible with an excellence relation in Φ.

For example consider a case where Φ consists of the two non-concordant excellence relations of Section 2, i.e., $\Phi = \{R_O, R_C\}$. Table 1 shows the compatibility of several quality metrics in the literature with these relations.

Table 1. Compatibility of quality metrics with $\Phi = \{R_O, R_C\}$. ('Y' indicates compatibility)

	Strong Outperformance Relation (R_O)	Coverage Relation (R_C, Definition 3)
C metric [4]	Y	N
Inferiority Index (*InfI*) [15]	Y	N
k-th Objective Pareto Spread (OS_k) [10]	N	N
Entropy [16]	N	N

(See [8] for a comprehensive compatibility analysis of common quality metrics with outperformance relations.) Note that Wu and Azarm's k-th Objective Pareto Spread (OS_k) is a unary metric [10]. So, in this paper, we revise this metric, i.e., compute the difference in the value of OS_k between two given NDSs to create a binary metric: $OS_k(A,B)=OS_k(A)-OS_k(B)$, which constructs a cardinal total order in U. Table 1 shows that only a portion of the examined metrics are compatible with R_O, while we were not able to find any quality metric in the literature to be compatible with R_C. Indeed, as shown by Knowles [8], a relatively small portion of the existing quality metrics is compatible with one of the outperformance relations introduced by Hansen and Jaszkiewicz [11]. Does this mean that we cannot obtain a set of quality metrics to address these two aspects of quality simultaneously? Zitzler et al. [13] show that this is in fact impossible in general with a finite number of unary quality metrics. As shown in this paper, however, this is in theory possible for the case of binary quality metrics, although it may not be practical because such metrics may not actually exist in the literature. In a case like this, one may go back to the initial set of the excellence relations, Φ, and try to compromise these relations such that a corresponding minimal set of quality metrics can be found. Alternatively, one may try to create new metrics to match the compatibility criterion with a given excellence relation. In the following, for example, we propose a new binary quality metric to quantify the difference between the spans of two non-dominated sets as a measure of extent of coverage. Later we show that this metric is compatible with R_c.

Definition 7. *Binary coverage metric*, denoted by $Q_c(A,B)$, is defined as:

$$Q_c(A,B)= \inf \{(\mathbf{b}_1.\mathbf{b}_2)/(\|\mathbf{b}_1\| \|\mathbf{b}_2\|) \text{ s.t. } \mathbf{b}_1,\mathbf{b}_2 \in B\} - \inf \{(\mathbf{a}_1.\mathbf{a}_2)/(\|\mathbf{a}_1\| \|\mathbf{a}_2\|) \text{ s.t. } a_1,a_2 \in A\}$$

where A and B are the convex cones generated by solution sets A and B. The term: $\inf \{(\mathbf{a}_1.\mathbf{a}_2)/(\|\mathbf{a}_1\| \|\mathbf{a}_2\|) \text{ s.t. } a_1,a_2 \in A\}$ measures the cosine of the largest possible angle between two vectors in the convex cone generated by the solution set A. This for example corresponds to $cos(\alpha)$ in Fig. 2, and $Q_c(A,B)= cos(\beta) - cos(\alpha) < 0$. We use this as a measure of maximum span of the solution sets on the Pareto frontier. Q_c is compatible with R_c because: if BR_cA, we have $A \subset B$, and therefore, the second term in the above equation is greater than the first term. Therefore, $Q_c(A,B)<0$. Similarly, if AR_cB we obtain $Q_c(A,B)>0$, and compatibility follows.

From Table 1, Definition 7, and minimality lemma we observe that $\Gamma = \{Infl; Q_c\}$ is a candidate minimal set of binary quality metrics with respect to $\Phi = \{R_O, R_C\}$.

Other than defining new compatible metrics, one could also modify the definition of excellence relations in Φ to make the existing quality metrics compatible. If the decision-maker modifies the definition of the coverage excellence relation (Definition 3) for example, Γ may or may not remain minimal. In the following, for example, we introduce a modified definition for coverage excellence relation that makes Wu and Azarm's OS_k metric compatible.

Definition 8 (Modified Coverage Relation; Compare to Definition 3). In a normalized multi-objective maximization, a non-dominated set $B=\{\mathbf{b}_1, \mathbf{b}_2, ...\}$ is strictly superior to another non-dominated set $A=\{\mathbf{a}_1, \mathbf{a}_2, ...\}$ in terms of a *modified coverage*, denoted by BR'_CA, iff: $\max_{i,j}(b_i^k - b_j^k)$ is strictly greater than $\max_{i,j}(a_i^k - a_j^k)$ for all k's (where b_i^k refers to the k-th objective value of the i-th solution in B).

OS_k is compatible with R'_C, and therefore, a combination of this quality metric and Inferiority Index (*Infl*) [15], i.e., $\Gamma' = \{Infl, OS_k\}$, is a candidate minimal set with respect to $\Phi' = \{R_O, R'_C\}$. In contrast, $\Gamma = \{Infl, Q_c\}$ which is minimal with respect to $\Phi = \{R_O, R_C\}$, is not minimal with respect to Φ', because Q_c is not compatible with R'_C (See Definition 7 for Q_C). Note that the process of defining and redefining Φ becomes increasingly difficult as more excellence relations are included. Nonetheless, it provides a formal platform and an objective starting point for selection of binary quality metrics.

6 Concluding Remarks

In this paper, we presented a theoretical framework for selection of minimal sets of quality metrics that can be used for a comparison study of different MOEAs. These metrics exhaustively account for all desired aspects of quality in non-dominated solution sets (obtained by MOEAs) without redundancy. In this framework, once the decision-maker's desired aspects of performance are determined, it is necessary to find all pairs of non-dominated sets that are objectively comparable. This in turn constructs partial orders in the set of all possible non-dominated sets, referred to as excellence relations in this paper. We proved that there is a one-to-one compatibility correspondence between these excellence relations (partial orders) and a minimal set of quality metrics (total orders), i.e., for each excellence relation there is one and only one compatible quality metric in a minimal set. This important result (referred to as Minimality Lemma) helps the decision-maker select a minimal set among the existing quality metrics in the literature, and thus, enables a quantitative and objective comparison of the solution sets obtained from different MOEAs.

Acknowledgments. The work presented in this paper was supported in part by the U.S. National Science Foundation, Grant number 0112767. This support is gratefully acknowledged.

References

1. Zitzler, E.: 1999, Evolutionary Algorithms for Multiobjective Optimization: Methods and Applications, Ph.D. Dissertation, Swiss Federal Institute of Technology (ETH), Zurich, Switzerland (1999)
2. Van Valdhuizen, D. A., Lamont G. B.: Multiobjective Evolutionary Algorithms: Analyzing the State-of-the-Art," Evolutionary Computation, 8(2), (2000) 125–147
3. Deb, K.: Multi-objective Optimization Using Evolutionary Algorithms. John Wiley and Sons, Chichester, UK (2001)
4. Knowles, J. D., Corne, D. W.: On Metrics for Comparing Non-Dominated Sets, Proceedings of the 2002 Congress on Evolutionary Computation Conference, (2002) 711–716
5. Zitzler, E., Thiele, L., 1998, "Multiobjective Optimization Using Evolutionary Algorithms – A Comparative Study," Proc. 5th International Conference: Parallel Problem Solving from Nature – PPSN V, Springer, Amsterdam, The Netherlands, (1998) 292–301
6. Van Veldhuizen, D. A.: Multiobjective Evolutionary Algorithm: Classification, Analyses and New Innovations, Ph.D. Dissertation, Dept. of Electrical and Computer Engineering, Air Force Institute of Technology, Wright-Patterson AFB, OH (1999)
7. Sayin, S.: Measuring the Quality of Discrete Representations of Efficient Sets in Multiple Objective Mathematical Programming, Mathematical Programming, 87, (2000) 543–560
8. Knowles, J. D.: Local-Search and Hybrid Evolutionary Algorithms for Pareto Optimization, Ph.D. Dissertation, University of Reading, Department of Computer Science, Reading, U.K. (2002)
9. Schott, J. R.: Fault Tolerant Design Using Single and Multicriteria Genetic Algorithm Optimization, M.S. Thesis, Department of Aeronautics and Astronautics, Massachusetts Institute of Technology, Cambridge, MA (1995)
10. Wu, J., Azarm, S.: Metrics for Quality Assessment of a Multiobjective Design Optimization Solution Set," Transactions of the ASME, Journal of Mechanical Design, 123, (2001) 18–25
11. Hansen, M. P., Jaszkiewicz, A.: Evaluating the Quality of Approximations to the Non-dominated Sets, Technical Report IMM-REP-1998-7, Technical University of Denmark, Lyngby, Denmark (1998)
12. Enderton, H. B.: Elements of Set Theory. Academic Press, Orlando, FL (1977)
13. Zitzler, E., Laumanns, M., Thiele, L., Fonseca, C. M., Grunert da Fonseca V.: Why Quality Assessment of Multiobjective Optimizers Is Difficult. Proceedings of the Genetic and Evolutionary Computation Conference (GECCO 2002) (2002) 666–674
14. Steuer, R. E.: Multiple Criteria Optimization: Theory, Computation, and Application, John Wiley & Sons, NY (1986)
15. Farhang-Mehr, A., Wu, J., Azarm, S.: Some Preliminary Results on the Development and Comparison of a New Multi-Objective Genetic Algorithm. Paper No. DETC2001/DAC-21021, CD-ROM Proceedings of the ASME Design Automation Conference, Pittsburgh, PA (2001)
16. Farhang-Mehr, A., Azarm, S.: Diversity Assessment of Pareto Optimal Solutions: An Entropy Approach. IEEE 2002 World Congress on Evolutionary Computation, Honolulu, Hawaii (2002)

A Comparative Study of Selective Breeding Strategies in a Multiobjective Genetic Algorithm

Andrew Wildman and Geoff Parks

Cambridge University Engineering Department,
Trumpington Street, Cambridge CB2 1PZ, UK
a.j.wildman.98@cantab.net, gtp@eng.cam.ac.uk

Abstract. The design of Pressurized Water Reactor (PWR) reload cores is a difficult combinatorial optimization problem with multiple competing objectives. This paper describes the use of a Genetic Algorithm (GA) to perform true multiobjective optimization on the PWR reload core design problem and improvements made to its performance in identifying nondominated solutions to represent the trade-off surface between competing objectives. The use of different pairing strategies for combining parents is investigated and found to produce promising results in some cases.

1 Introduction

Evolutionary Algorithms are well suited to multiobjective optimization and a number of different multiobjective GAs have been developed [1,2]. Multiobjective GAs provide the means to expose the trade-off surface between competing objectives in a single optimization and are therefore a very attractive tool for the designer.

A typical PWR core contains 193 fuel assemblies arranged with quarter-core symmetry. At each refuelling one third or one quarter of these may be replaced. It is common practice for fresh fuel assemblies to carry a number of burnable poisons (BP) pins (control material). It is also usual to rearrange old fuel in order to improve the characteristics of the new core. This shuffling can entail the exchange of corresponding assemblies between core quadrants, which is equivalent to changing the assembly 'orientations', or the exchange of different assemblies, which changes their locations and possibly their orientations also.

Thus, a candidate loading pattern (LP) of predetermined symmetry must specify:

- the fuel assembly to be loaded in each core location,
- the BP loading with each fresh fuel assembly, and
- the orientation of each assembly.

The PWR reload core design problem has been tackled in many different ways [3] and one interesting point to emerge from a review of past work is the diversity in objective functions chosen. It is clear that the PWR reload core design problem is in reality a multiobjective optimization problem, where an improvement in one objective is often only gained at the cost of deteriorations in others.

Thus, the search for the best LP is a formidable optimization problem characterized by:

- high combinatorial dimensionality,
- multiple nonlinear objectives and constraints,
- multimodality,
- computationally intensive objective and constraint function evaluations, and
- lack of direct derivative information

Parks [4] first demonstrated that the PWR reload core design problem could be tackled successfully using a multiobjective GA. More recently Parks and Miller [5] investigated the efficacy of various elitist selection strategies with parents being selected both from the current population and from the archive record of nondominated solutions encountered during search. They concluded that it was possible to improve the performance of their multiobjective GA in this rather specialized application domain through the use of strong elitism and high selection pressures. These conclusions were subsequently found to also hold true in another very different application domain [6].

In the algorithm used by Parks and Miller parents were paired off randomly prior to crossover. This paper presents the results of an investigation into the effectiveness of various pairing strategies in the same application domain.

2 A Multiobjective GA for PWR Reload Core Design

The structure of our GA is shown in Fig. 1 and is fairly standard. The details of the implementation are, however, problem-specific.

2.1 Coding

Each solution is represented by three two-dimensional arrays, corresponding to the physical layout of the fuel assemblies, their BP loadings and their orientations respectively.

2.2 Crossover Operator

For this application Poon's Heuristic Tie-Breaking Crossover (HTBX) operator [7] is used. HTBX maps the parent fuel arrays to reactivity-ranked arrays based on the assemblies' reactivities (reactivity being a parameter related to the amount of fissile material in the fuel and thus a convenient indicator of other attributes of interest). HTBX then combines randomly selected complementary parts of these arrays through a 'cut and paste' operation, and uses a simple tie-breaking algorithm to produce valid offspring reactivity-ranked arrays. Finally, the assembly-ranking mapping is reversed to produce the offspring assembly LPs.

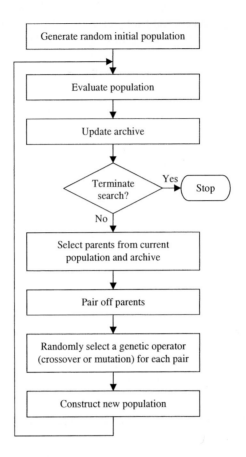

Fig. 1. The structure of the multiobjective GA

The BP loadings and assembly orientations are all inherited from one or other parent. Thus, the reactivity distribution (and it is hoped, in consequence, other attributes) of an offspring LP resembles, but is not necessarily identical to, parts of both parents.

2.3 Mutation Operator

For this application the mutation operator performs one, two or three fuel assembly shuffles, randomly allocating allowed BP loadings and orientations to the assemblies affected. It is used as an alternative to crossover, i.e. offspring are produced using either mutation or crossover but not both, the choice between operators being made randomly. The relative frequencies with which these operators are chosen are approximately 25% and 75% respectively, this ratio having been determined (by extensive testing) to give good performance in this application [4].

2.4 Population Evaluation

In the examples which follow the Generalized Perturbation Theory based reactor model employed in FORMOSA-P [8] was used to evaluate LPs, but, in principle, the evaluation of objectives and constraints can be performed using any appropriate reactor physics code.

2.5 Archiving

While the GA is running an archive of nondominated solutions is maintained. After each LP has been evaluated it is compared with existing members of the archive. If it dominates any members of the archive, those are removed and the new solution is added. If the new solution is dominated by any members of the archive, it is not archived. If it neither dominates nor is dominated by any members of the archive, it is archived if it is sufficiently 'dissimilar' to existing archive members, where, for this application, the degree of dissimilarity between two LPs is defined in terms of their reactivity distributions [9]. This dissimilarity requirement helps to maintain diversity in the archive.

The archiving logic is such that the solutions are arrayed by age, i.e. the first solution in the archive is earliest nondominated solution found and the last solution in the archive is the most recent nondominated solution found.

2.6 Selection

The current population is ranked by sorting through to identify all nondominated solutions, ranking these appropriately, then removing them from consideration, and repeating the procedure until all solutions have been ranked – a procedure first proposed by Goldberg [10] and implemented by Srinivas and Deb [11]. A selection probability is then assigned to each solution based on its ranking in a manner similar to Baker's single criterion ranking selection procedure [12]. The probability of a rank n member of the current population being selected is given by:

$$p_n = \frac{S(N+1-R_n) + R_n - 2}{N(N-1)} \qquad (1)$$

in which N is the population size, S is a selection pressure, and

$$R_n = 1 + r_n + 2\sum_{i=1}^{n-1} r_i \qquad (2)$$

where r_i is the number of solutions in rank i. If a value of S greater than 2.0 is used, some values of p_n may be negative. In this case, negative values of p_n are set to zero, and the positive values of p_n are rescaled so that they sum to unity. Parents are then selected by stochastic remainder selection without replacement.

In addition, some of the parents for each new generation can be chosen from the archive of nondominated solutions, thus introducing multiobjective elitism to the selection process. When this is done equation (1) is modified such that:

$$p_n = \frac{N-A}{N}\left(\frac{S(N+1-R_n) + R_n - 2}{N(N-1)}\right) \qquad (3)$$

where A is the number of parents selected from the archive.

Parks and Miller [5] observed the following elitist selection strategy to be effective:

- Selecting up to 50% of the parents of the next generation from the archive of nondominated solutions found;
- If there are insufficient solutions in the archive to meet the demand for parents, to make multiple (two or three) copies of the archive solutions;
- If there are more than sufficient solutions in the archive to meet the demand for parents, to select the solutions most recently added to the archive;
- To use a high selection pressure in selecting parents from the current population.

In the studies that follow, this selection strategy is employed with up to 2 copies being made of archive solutions and a selection pressure $S = 3.0$.

3 Pairing Strategies

The basic hypothesis of our research is that pairing the parents of the new population according to some information regarding their characteristics will lead to better algorithm performance. The method chosen was to maximize or minimize various measures of dissimilarity between the solutions and also to introduce pairing restrictions.

The following dissimilarity measures and pairing restrictions were implemented:

3.1 Genotypic Dissimilarity

The method chosen to represent genotypic dissimilarity (the difference in genetic composition between two solutions) was the difference in their reactivity distributions [9] – the same dissimilarity measure used to maintain diversity in the archive of nondominated solutions (see Sect. 2.5).

3.2 Phenotypic Dissimilarity

Phenotypic dissimilarity describes the difference between two solutions in terms of the representation (or result) of their genetic coding in the solution space.

The phenotypic dissimilarity between two solutions i and j can therefore be formulated in terms of the normalized distance between them in the solution space. Hence,

$$d_{ij}^2 = \sum_{k=1}^{M} \left(\hat{f}_i^k - \hat{f}_j^k \right)^2 \tag{4}$$

with M being the number of objectives and

$$\hat{f}_i^k = \frac{f_i^k - \mu^k}{\sigma^k} \tag{5}$$

where f_i^k is the value of the kth objective for solution i, and μ^k and σ^k are respectively the mean and standard deviation of the kth objective for the solutions selected to be parents.

3.3 Ranking Dissimilarity

Ranking dissimilarity represents a different measure of two solutions' comparative performance in the solution space. The formulation for ranking dissimilarity is simply the difference in their rankings, where a solution's rank is determined by nondominated sorting (as described in Sect. 2.6).

3.4 Pairing Restrictions

The use of pairing restrictions as a means of improving algorithm peformance was also investigated. The hypothesis was that either pairing archived solutions with 'new' solutions (solutions from the current population) (AN-NA), or archived with archived and new with new (AA-NN), might prove beneficial.

3.5 Pairing Strategies Implementation

Fig. 2 shows the flowchart for the subroutine implementing these pairing strategies. As pairs of parents are subject to either crossover or mutation, but not both, pairs which are to be mutated are always selected randomly. Pair selection on the basis of dissimilarity is done as follows (the same procedure is used whichever dissimilarity measure is used):

- If dissimilarity is to be maximized, then the remaining (allowed) parent with the greatest dissimilarity to the first parent chosen (parent x) is chosen.
- If dissimilarity is to be minimized, then the remaining (allowed) parent with the least dissimilarity to the first parent chosen (parent x) is chosen.

This procedure maximizes or minimizes the dissimilarity between pairs of parents (based on the availability of parents at the time of selection), but does not, of course, maximize or minimize the total dissimilarity across the complete set of parent pairs. The latter could be done, but would require that more difficult optimization problem to be solved every generation, thus increasing the overhead associated with running the GA.

4 Performance Measures

4.1 Pairing Visualization

Fig. 3(a) shows the distribution of points representing pairs (x, y) in two dimensions when the AA-NN pairing scheme is selected. The axes represent the order in which solutions are selected to become parents. Hence, the points in the top left corner represent pairs of parents from the archive and those in the bottom right corner represent pairs of parents from the current population. The plot clearly shows the desired bias towards pairing like with like. The random outliers are caused when the mutation operator has been selected and hence no pairing restrictions are imposed. The dotted lines represent the transition point from parents chosen from the archive to parents chosen from the current population.

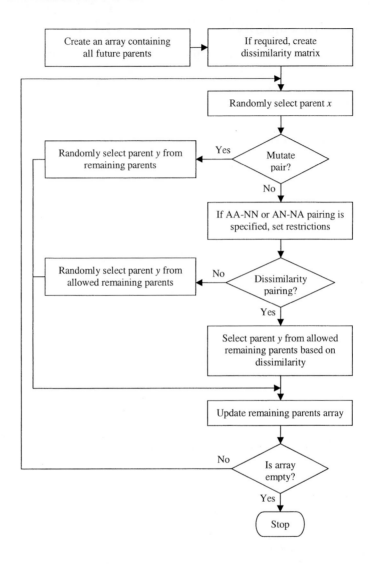

Fig. 2. Flowchart of the pairing subroutine

Fig. 3(b) shows the situation when AN-NA pairing has been selected. In this example, the number of parents chosen from the archive is less than the number of parents from the current population, so a degree of NN-pairing is found. These plots provide visual verification that the pairing restrictions are working correctly.

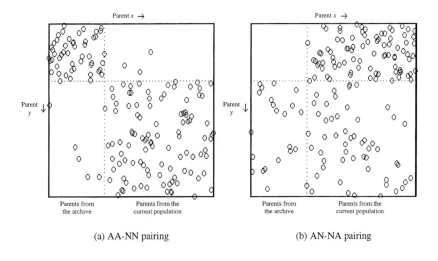

Fig. 3. Visual representations of pairing restrictions

4.2 Nondominated Set Comparison

Nondominated set comparison (NSC) of two final archives provides a direct means of quantifying the relative performance of two runs. NSC entails comparing every solution in archive A against every solution in archive B, and *vice-versa*. The resulting metrics are the *nondominated fractions*: the proportion of solutions in archive A that are not dominated by any solutions in archive B ($F_{A<B}$), and the proportion of solutions in archive B that are not dominated by any solutions in archive A ($F_{B<A}$).

In the studies that follow 30 runs were executed for each GA configuration. This allows two types of comparison:

- *Direct Comparison*: Individual runs using the same random number seed (and thus the same initial population) for each GA configuration can be compared (giving 30 different comparisons);
- *Complete Comparison*: Each of the 30 runs for one GA configuration can be compared with each of the 30 runs for the other (giving 900 different comparisons).

For each type of comparison, each pair of nondominated fractions ($F_{A<B}, F_{B<A}$) can be plotted against each other, and the distribution of points gives a measure of the two algorithm configurations' relative performance.

If the nondominated fraction for pairing strategy A is plotted on the vertical (y) axis and the nondominated fraction for pairing strategy B is plotted on the horizontal (x) axis, then a point lying above the line $y = x$ indicates that, for the two runs being compared, the GA using pairing strategy A outperformed the GA using pairing strategy B. Conversely, a point lying below the line $y = x$ indicates that the performance of the strategy B GA was better.

The difference $\Delta F_{AB} = F_{A<B} - F_{B<A}$ can be calculated for each of the comparisons available and then averaged, giving a measure $\overline{\Delta F}_{AB}$ of the superiority (if $\overline{\Delta F}_{AB} > 0$) or inferiority (if $\overline{\Delta F}_{AB} < 0$) of pairing strategy A compared to strategy B.

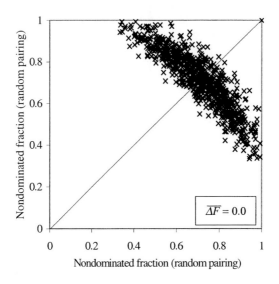

Fig. 4. Nondominated fraction plot for random pairing vs random pairing

Figure 4 shows such a plot for a complete comparison between the 30 runs obtained using the original random pairing strategy of Parks and Miller's algorithm [5] and themselves. This figure is presented in order to give an example of the variation in algorithm performance observed solely as a result of using different random number seeds. The cluster of points at $(1, 1)$ is due to the 30 self-comparisons generated by comparing a set of runs with itself. For this comparison $\overline{\Delta F} = 0$, of course.

5 Results

5.1 Test Problem Description

The performance of the GA using different pairing strategies was investigated on a representative PWR reload core design problem with three objectives:

- minimization of the enrichment of the fresh fuel (the feed enrichment),
- maximization of the burnup of the fuel to be discharged (the discharge burnup),
- minimization of the ratio of the peak to average assembly power (the power peaking).

In each run a population of 204 was evolved for 51 generations – the GA having been found to give satisfactory performance with these parameter settings [4]. The same random number seeds were used for each set of 30 runs, and the selection strategy described in Sect. 2.6 was employed.

5.2 Pairing Strategies

In the first set of investigations nine different pairing strategies were compared:

- *Rand*: random pairing;
- *AA-NN*: AA-NN pairing restriction (see Sects. 3.4 and 3.5);
- *AN-NA*: AN-NA pairing restriction (see Sects. 3.4 and 3.5);
- *Gmax*: maximization of genotypic dissimilarity (see Sects. 3.1 and 3.5);
- *Gmin*: minimization of genotypic dissimilarity (see Sects. 3.1 and 3.5);
- *Pmax*: maximization of phenotypic dissimilarity (see Sects. 3.2 and 3.5);
- *Pmin*: minimization of phenotypic dissimilarity (see Sects. 3.2 and 3.5);
- *Nmax*: maximization of ranking dissimilarity (see Sects. 3.3 and 3.5);
- *Nmin*: minimization of ranking dissimilarity (see Sects. 3.3 and 3.5).

Table 1 shows the average difference in the nondominated fractions $\overline{\Delta F}$ for direct NSC comparisons (each run for one strategy being compared with its immediate counterpart for the other – that using the same random number seed) between each pairing strategy. In this table a positive number indicates that the pairing strategy associated with that row performed better than the pairing strategy associated with that column. For instance, when comparing AN-NA and random pairing, on average 0.7599 (76.0%) of the solutions in the final archives for AN-NA pairing runs were not dominated by their counterparts for random pairing, while 0.6487 (64.9%) of the solutions in the final archives for random pairing runs were not dominated by their counterparts for AN-NA pairing, giving the difference of 0.1112 (11.1%) shown in Table 1.

Table 2 shows the same information as Table 1 but for complete comparisons between pairing strategies (each run for one strategy being compared with its 30 counterparts for the other, not just its immediate counterpart).

Analysing the data in Table 1, the ranking of the pairing strategies in terms of their effectiveness (from best to worst) is:

- AN-NA, Nmin, Gmax, Nmax, Pmax, AA-NN, Rand, Gmin, Pmin;

while, for the data in Table 2, the ranking is:

- AN-NA, Gmax, Nmax, Pmax, Nmin, Rand, AA-NN, Pmin, Gmin.

Table 1. $\overline{\Delta F}$ for direct NSC comparisons of pairing strategies

	Rand	AA-NN	AN-NA	Gmax	Gmin	Pmax	Pmin	Nmax	Nmin
Rand	0.0000	-0.0118	-0.1112	-0.0737	0.0294	-0.0686	0.0178	-0.0587	-0.0667
AA-NN	0.0118	0.0000	-0.0903	-0.1057	0.0045	-0.0850	0.0049	-0.0610	-0.0398
AN-NA	0.1112	0.0903	0.0000	0.0418	0.1223	0.0415	0.0835	0.0437	0.0294
Gmax	0.0737	0.1057	-0.0418	0.0000	0.1181	0.0302	0.0713	0.0415	-0.0077
Gmin	-0.0294	-0.0045	-0.1223	-0.1181	0.0000	-0.1202	0.0010	-0.0960	-0.0951
Pmax	0.0686	0.0850	-0.0415	-0.0302	0.1202	0.0000	0.0791	-0.0109	-0.0221
Pmin	-0.0178	-0.0049	-0.0835	-0.0713	-0.0010	-0.0791	0.0000	-0.0692	-0.0681
Nmax	0.0587	0.0610	-0.0437	-0.0415	0.0960	0.0109	0.0692	0.0000	-0.0188
Nmin	0.0667	0.0398	-0.0294	0.0077	0.0951	0.0221	0.0681	0.0188	0.0000

Table 2. $\overline{\Delta F}$ for complete NSC comparisons of pairing strategies

	Rand	AA-NN	AN-NA	Gmax	Gmin	Pmax	Pmin	Nmax	Nmin
Rand	0.0000	0.0197	-0.0928	-0.0784	0.0354	-0.0617	0.0135	-0.0617	-0.0487
AA-NN	-0.0197	0.0000	-0.1008	-0.0891	0.0167	-0.0707	0.0031	-0.0697	-0.0606
AN-NA	0.0928	0.1008	0.0000	0.0259	0.1197	0.0432	0.0866	0.0400	0.0424
Gmax	0.0784	0.0891	-0.0259	0.0000	0.1048	0.0215	0.0736	0.0091	0.0242
Gmin	-0.0354	-0.0167	-0.1197	-0.1048	0.0000	-0.0914	-0.0199	-0.0823	-0.0764
Pmax	0.0617	0.0707	-0.0432	-0.0215	0.0914	0.0000	0.0572	-0.0040	0.0030
Pmin	-0.0135	-0.0031	-0.0866	-0.0736	0.0199	-0.0572	0.0000	-0.0534	-0.0558
Nmax	0.0617	0.0697	-0.0400	-0.0091	0.0823	0.0040	0.0534	0.0000	0.0064
Nmin	0.0487	0.0606	-0.0424	-0.0242	0.0764	-0.0030	0.0558	-0.0064	0.0000

It can be seen that these orders of merit are broadly similar. Pairing strategies that seek to maintain diversity (by pairing 'unalike', in some sense, parents) are generally more successful than those that do not.

The AN-NA pairing restriction strategy, which seeks to pair solutions from the archive (of nondominated solutions found) with solutions selected from the current population, is the most successful. The converse strategy of pairing archive solutions with other archive solutions (AA-NN) does not seem to be significantly more or less successful than the strategy of purely random pairing.

The three strategies (Gmax, Nmax and Pmax) that seek to maximize various measures of dissimilarity (genotypic, nondominated ranking and phenotypic, respectively) between parents all give better performance than purely random pairing, with genotypic dissimilarity maximization being the most successful and phenotypic dissimilarity maximization the least. The benefit of such dissimilarity maximization strategies is reinforced by the fact that the strategies that seek to minimize the genotypic and phenotypic dissimilarity between parents (Gmin and Pmin) perform worse than purely random pairing.

An apparently anomalous result is the fact that the strategy that seeks to minimize the nondominated ranking dissimilarity between parents (Nmin) performs better than purely random pairing, and, indeed, on the basis of the direct comparison between individual runs with the same initial populations (Table

1), it is the second most successful strategy. This result is puzzling and merits further investigation. One comment, of partial explanation, that can be made now is that two solutions with the same nondominated ranking may be very significantly dissimilar genotypically and phenotypically.

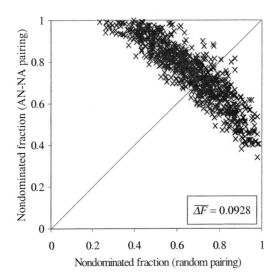

Fig. 5. Nondominated fraction plot for random pairing vs AN-NA pairing

To provide the reader with a feel for the relative gains in performance achieved, Fig. 5 shows a plot for a complete comparison between the 30 runs obtained using the random pairing strategy and the most successful strategy tested (AN-NA pairing). When compared to Fig. 4, the greater preponderance of points above the line $y = x$ is apparent, but nevertheless it is clear that individual random pairing runs do still outperform individual AN-NA runs.

5.3 Combined Pairing Strategies

As the most successful pairing strategy identified in Sect. 5.2 (AN-NA) only seeks to impose restrictions on the origins of the two parents (from the archive or from the current population) it is worth investigating whether this can be combined beneficially with any of the other successful (better than random) strategies found, all of which take into account some measure of the dissmilarity between solutions in making pairing decisions.

The four combined strategies tested were:

- *AN-NA+Gmax*: AN-NA pairing restriction combined with maximization of genotypic dissimilarity;

- *AN-NA+Pmax*: AN-NA pairing restriction combined with maximization of phenotypic dissimilarity;
- *AN-NA+Nmax*: AN-NA pairing restriction combined with maximization of ranking dissimilarity;
- *AN-NA+Nmin*: AN-NA pairing restriction combined with minimization of ranking dissimilarity.

Table 3. $\overline{\Delta F}$ comparisons of combined pairing strategies

	AN-NA (direct)	AN-NA (complete)	Gmax (direct)	Gmax (complete)
AN-NA+Gmax	-0.0456	-0.0601	-0.0315	-0.0414
	AN-NA (direct)	AN-NA (complete)	Pmax (direct)	Pmax (complete)
AN-NA+Pmax	-0.0494	-0.0523	-0.0111	-0.0111
	AN-NA (direct)	AN-NA (complete)	Nmax (direct)	Nmax (complete)
AN-NA+Nmax	-0.0039	-0.0047	0.0422	0.0396
	AN-NA (direct)	AN-NA (complete)	Nmin (direct)	Nmin (complete)
AN-NA+Nmin	0.0065	-0.0216	0.0284	0.0148

Table 3 gives the results for NSC comparisons between the various combined strategies and the strategies concerned when used individually. For the first two combined strategies (AN-NA+Gmax and AN-NA+Pmax) the algorithm performance obtained is worse than that obtained when either of the two strategies concerned are used individually. For AN-NA+Nmax pairing the performance is better than for ranking dissimilarity maximization alone, but slightly (though not significantly) worse than for AN-NA alone. For AN-NA+Nmin a very small overall performance improvement compared to AN-NA pairing alone is seen for the direct comparison, but a performance deterioration is seen for the complete comparison. AN-NA+Nmin pairing performance is somewhat better than for ranking dissimilarity minimization alone.

Thus, combining ranking dissimilarity maximization or minimization with AN-NA pairing seems to have a relatively small, but slightly negative, impact on algorithm performance. The benefit gained through using AN-NA pairing is not greatly affected by the relative rankings of the solutions paired.

However, although AN-NA pairing and maximizing genotypic and phenotypic dissimilarity all individually appear to be beneficial (compared to random pairing) strategies, combining them results in a deterioration in performance. Thus, it seems that a certain amount of diversity between parents is beneficial to the multiobjective search process, but too much becomes counterproductive. This conclusion indicates that there is further work to be done to identify the strategy that provides the optimal amount of dissimilarity between parents.

6 Conclusions

The effectiveness of a variety of pairing strategies in a highly elitist multiobjective GA applied to the problem of PWR reload core design has been investigated. Our studies show that pairing strategies that seek to pair parents that are dissimilar yield better performance than a random pairing strategy. This conclusion applies to all the measures of dissimilarity we have investigated (genotypic, phenotypic and nondominated ranking). The most successful strategy we have identified is one where parents drawn from the archive of nondominated solutions found are paired with parents selected from the current population.

Somewhat surprisingly, pairing strategies that combine the features of two individually successful (better than random pairing) strategies do not seem to offer any further performance improvements, indeed in most cases algorithm performance deteriorates. We conclude that trying to enforce too much dissimilarity between parents can be counterproductive.

Although this paper has considered a single (rather specialised) application of multiobjective GA optimization, our findings may well apply to other multiobjective GA applications and implementations, and these issues are the subjects of ongoing research.

Acknowledgements. The authors gratefully acknowledge the collaborative co-operation and support of the North Carolina State University (NCSU) Electric Power Research Center.

References

1. Deb, K.: Multi-Objective Optimization Using Evolutionary Algorithms, John Wiley & Sons Ltd., Chichester (2001)
2. Coello Coello, A.A., Van Veldiuzen, D.A., Lamont, G.B.: Evolutionary Algorithms for Solving Multi-Objective Problems, Kluwer Academic/Plenum Publishers, New York (2002)
3. Turinsky, P.J., Parks, G.T.: Advances in Nuclear Fuel Management for Light Water Reactors. Adv. Nucl. Sci. Tech. **26** (1999) 137–165
4. Parks, G.T.: Multiobjective PWR Reload Core Design by Nondominated Genetic Algorithm Search. Nucl. Sci. Eng. **124** (1996) 178–187
5. Parks, G.T., Miller, I.: Selective Breeding in a Multiobjective Genetic Algorithm. In: Eiben, A.E., Bäck, T., Schoenauer, M., Schwefel, H-P. (eds.): Parallel Problem Solving from Nature – PPSN V. Lecture Notes in Computer Science, Vol. 1498. Springer-Verlag, Berlin Heidelberg New York (1998) 250–259
6. Parks, G.T., Li, J., Balazs, M.E., Miller, I.: An Empirical Investigation of Elitism in Multiobjective Genetic Algorithms. Found. Comput. Decision Sci. **26** (2001) 51–74
7. Poon, P.W., Parks, G.T.: Application of Genetic Algorithms to In-core Nuclear Fuel Management Optimization. In: Küsters, H., Stein, E., Werner, W. (eds.): Proc. Joint Int. Conf. Mathematical Methods and Supercomputing in Nuclear Applications. Kernforschungszentrum, Karlsruhe, Vol. 1 (1993) 777–786

8. Kropaczek, D.J., Turinsky, P.J., Parks, G.T., Maldonado, G.I.: The Efficiency and Fidelity of the In-core Nuclear Fuel Management Code FORMOSA-P. In: Ronen, Y., Elias, E. (eds.): Reactor Physics and Reactor Computations. Ben Gurion University of the Negev Press (1994) 572–579
9. Kropaczek, D.J., Parks, G.T., Maldonado, G.I., Turinsky, P.J.: Application of Simulated Annealing to In-core Nuclear Fuel Management Optimization. In: Proc. 1991 Int. Top. Mtg. Advances in Mathematics, Computations and Reactor Physics. American Nuclear Society, LaGrange Park, Vol. 5 (1991) 22.1 1.1–1.12
10. Goldberg, D.E.: Genetic Algorithms in Search, Optimization, and Machine Learning. Addison Wesley, Reading MA (1989)
11. Srinivas, N., Deb, K.: Multiobjective Optimization Using Nondominated Sorting in Genetic Algorithms. Evol. Comp. **2** (1994) 221–248
12. Baker, J.E.: Adaptive Selection Methods for Genetic Algorithms. In: Grefenstette, J.J. (ed.): Proc. Int. Conf. Genetic Algorithms and Their Applications. Lawrence Erlbaum Associates, Hillsdale NJ (1985) 101–111

An Empirical Study on the Effect of Mating Restriction on the Search Ability of EMO Algorithms

Hisao Ishibuchi and Youhei Shibata

Department of Industrial Engineering, Osaka Prefecture University,
1-1 Gakuen-cho, Sakai, Osaka 599-8531, Japan
{hisaoi, shibata}@ie.osakafu-u.ac.jp

Abstract. This paper examines the effect of mating restriction on the search ability of EMO algorithms. First we propose a simple but flexible mating restriction scheme where a pair of similar (or dissimilar) individuals is selected as parents. In the proposed scheme, one parent is selected from the current population by the standard binary tournament selection. Candidates for a mate of the selected parent are winners of multiple standard binary tournaments. The selection of the mate among multiple candidates is based on the similarity (or dissimilarity) to the first parent. The strength of mating restriction is controlled by the number of candidates (i.e., the number of tournaments used for choosing candidates from the current population). Next we examine the effect of mating restriction on the search ability of EMO algorithms to find all Pareto-optimal solutions through computational experiments on small test problems using the SPEA and the NSGA-II. It is shown that the choice of dissimilar parents improves the search ability of the NSGA-II on small test problems. Then we further examine the effect of mating restriction using large test problems. It is shown that the choice of similar parents improves the search ability of the SPEA and the NSGA-II to efficiently find near Pareto-optimal solutions of large test problems. Empirical results reported in this paper suggest that the proposed mating restriction scheme can improve the performance of EMO algorithms for many test problems while its effect is problem-dependent and algorithm-dependent.

1 Introduction

Since Schaffer's study [13], evolutionary algorithms have been applied to various multiobjective optimization problems for finding their Pareto-optimal solutions (e.g., see Coello et al. [1] and Deb [3]). Those algorithms are often referred to as EMO (evolutionary multiobjective optimization) algorithms. Recent EMO algorithms usually share some common ideas such as elitism, fitness sharing and Pareto ranking. While mating restriction has been often discussed in the literature, it has not been used in many EMO algorithms as pointed out in some reviews on EMO algorithms [6, 14, 18]. The aim of this paper is to examine the effect of mating restriction on the search ability of EMO algorithms. More specifically, we demonstrate how the search ability of EMO algorithms to find Pareto-optimal or near Pareto-optimal solutions can be improved by mating restriction.

Mating restriction was suggested by Goldberg [7] for single-objective genetic algorithms. Hajela & Lin [8] and Fonseca & Fleming [5] used it in their EMO algorithms. The basic idea of mating restriction is to ban the crossover of dissimilar parents from which good offspring are not likely to be generated. In the implementation of mating restriction, a user-definable parameter σ_{mating} called the mating radius is usually used for banning the crossover of two parents whose distance is larger than σ_{mating}. The distance between two parents is measured in the decision space or the objective space. The necessity of mating restriction in EMO algorithms was also stressed by Jaszkiewicz [11] and Watanabe et al. [15]. On the other hand, Zitzler & Thiele [17] reported that no improvement was achieved by mating restriction in their computational experiments. Van Veldhuizen & Lamont [14] mentioned that the empirical evidence presented in the literature could be interpreted as an argument either for or against the use of mating restriction. Moreover, there was also an argument for the selection of dissimilar parents. Horn et al. [9] argued that information from very different types of tradeoffs could be combined to yield other kinds of good tradeoffs. Schaffer [13] examined the selection of dissimilar parents but observed no improvement.

In this paper, we examine the effect of mating restriction on the search ability of EMO algorithms through computational experiments on multiobjective knapsack and permutation flowshop scheduling problems. As EMO algorithms, we use the SPEA [18] and the NSGA-II [4] because their high search ability was empirically demonstrated in the literature [3, 4, 16, 18]. We first propose a simple but flexible mating restriction scheme for implementing the selection of dissimilar parents as well as similar parents in a unified framework. Next we examine the effect of mating restriction on the search ability of the SPEA and the NSGA-II to find all Pareto-optimal solutions of small test problems. Then we examine their search ability to efficiently find near Pareto-optimal solutions of large test problems. Experimental results clearly show that the search ability of those EMO algorithms on some test problems can be improved by mating restriction. It is also shown that the effect of mating restriction is problem-dependent and algorithm-dependent.

2 Mating Restriction Scheme without Mating Radius

In general, an n-objective optimization problem can be written as

$$\text{Optimize } \mathbf{f}(\mathbf{x}) = (f_1(\mathbf{x}), f_2(\mathbf{x}), ..., f_n(\mathbf{x})), \quad (1)$$
$$\text{subject to } \mathbf{x} \in \mathbf{X}, \quad (2)$$

where $\mathbf{f}(\mathbf{x})$ is the objective vector, $f_i(\mathbf{x})$ is the i-th objective to be minimized or maximized, $\mathbf{x}$ is the decision vector, and $\mathbf{X}$ is the feasible region in the decision space.

Let us denote the distance between two solutions **x** and **y** as $|\mathbf{x}-\mathbf{y}|$ in the decision space and $|\mathbf{f}(\mathbf{x})-\mathbf{f}(\mathbf{y})|$ in the objective space. In this paper, the distance $|\mathbf{f}(\mathbf{x})-\mathbf{f}(\mathbf{y})|$ in the objective space is measured by the Euclidean distance as

$$|\mathbf{f}(\mathbf{x})-\mathbf{f}(\mathbf{y})| = \sqrt{|f_1(\mathbf{x})-f_1(\mathbf{y})|^2 + \cdots + |f_n(\mathbf{x})-f_n(\mathbf{y})|^2}. \quad (3)$$

On the other hand, the definition of the distance $|\mathbf{x}-\mathbf{y}|$ in the decision space totally depends on the representation of solutions in a particular problem. For example, we use the Hamming distance for *m*-item knapsack problems as

$$|\mathbf{x}-\mathbf{y}| = |x_1 - y_1| + \cdots + |x_m - y_m|, \quad (4)$$

where **x** and **y** are binary strings of the length m: $\mathbf{x} = x_1 x_2 \cdots x_m$ and $\mathbf{y} = y_1 y_2 \cdots y_m$. On the other hand, solutions of permutation flowshop scheduling problems are permutations of given jobs. In this case, we use the sum of the distance between the positions of each job as the distance of two solutions. The calculation of the distance is illustrated in Fig. 1. The distance between the positions of Job 1 (denoted by J1 in Fig. 1) is 4 since it is placed in the first position of String 1 and the fifth position of String 2. The distance between the positions of the other jobs is calculated in the same manner (i.e., 1 for Job 2, 0 for Job 3 and Job 4, and 3 for Job 5). Thus the distance between the two strings in Fig. 1 is calculated as 8 (i.e., $4+1+0+0+3$).

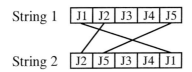

Fig. 1. Distance between two strings for five-job flowshop scheduling problems

In this paper, we propose a mating restriction scheme for examining the effect of mating restriction on the search ability of EMO algorithms. The proposed mating restriction scheme is illustrated in Fig. 2 where open circles at the bottom denote individuals randomly drawn from the current population with replacement. One parent (i.e., Parent A in Fig. 2) is chosen by the standard binary tournament selection. In the selection of a mate for Parent A (i.e., Parent B in Fig. 2), first the standard binary tournament selection is iterated β times for finding β candidates. Each candidate is the winner of a tournament. Then a mate is chosen among the β candidates by measuring the distance from each candidate to Parent A. The distance is measured in the decision or objective space. The most similar (or dissimilar) candidate with the minimum (or maximum) distance to Parent A is selected as its mate. In this manner, a mate for Parent A is selected through two-stage tournament selection. In the first stage, the fitness-based binary tournament selection is iterated for finding β candidates. In the second stage, the distance-based tournament selection of the tournament size β is performed for choosing a single individual as a mate for Parent A from the β winners in the first stage. Our mating restriction scheme has the following flexibility in its implementation:

(a) The choice between the decision space and the objective space in which the distance is measured.
(b) The choice between the similarity (i.e., minimum distance) and the dissimilarity (i.e., maximum distance) as the mate selection criterion in the distance-based tournament selection in the second stage.
(c) The value of β, i.e., the number of candidates from which a mate is chosen based on the mate selection criterion.

The user-definable parameter β can be viewed as the strength of mating restriction. That is, the strength of mating restriction is controllable through the value of β. When $\beta = 1$, our mating restriction scheme is the same as the standard binary tournament selection with no mating restriction. As the value of β increases, more similar (or dissimilar) parents are selected and recombined.

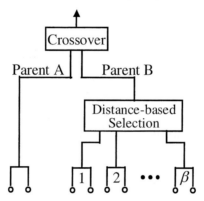

Fig. 2. Our mating restriction scheme

3 Examination of the Effect of Mating Restriction

In this section, we examine the effect of the proposed mating restriction scheme on the performance of EMO algorithms through computational experiments on multiobjective knapsack and permutation flowshop scheduling problems. For this purpose, we combined our mating restriction scheme with recently developed popular EMO algorithms: the SPEA [18] and the NSGA-II [4]. As mentioned in the previous section, our mating restriction scheme is the same as the standard binary tournament selection when $\beta = 1$. Thus the modified SPEA and the modified NSGA-II are the same as their original versions when $\beta = 1$.

3.1 Test Problems

In our computational experiments, we used four knapsack problems in Zitzler & Thiele [18]: two-objective 250-item, three-objective 250-item, two-objective 500-item, and three-objective 500-item test problems. We also generated 10 small test problems with two objectives and 30 items in the same manner as [18]. The small test problems were used for examining the search ability of the EMO algorithms to find all Pareto-optimal solutions while the large test problems were used for examining their search ability to efficiently find near Pareto-optimal solutions. We also generated 10 two-objective permutation flowshop scheduling problems with 10 machines and 12 jobs in the same manner as Ishibuchi & Murata [10]. The two objectives are the minimization of the makespan and the maximum tardiness. As large permutation flowshop scheduling problems, we generated four test problems with 20 machines: two-objective 40-job, three-objective 40-job, two-objective 80-job, and three-objective 80-job problems. In the three-objective problems, the minimization of the total flow time was used in addition to the minimization of the makespan and the maximum tardiness.

3.2 Parameter Specifications

We examined all the four combinations related to the distance definition and the mate selection criterion in our mating restriction scheme: {decision space, objective space} × {minimum distance, maximum distance}. We also examined ten different values of β: $\beta = 1,2,3,...,10$. The SPEA and the NSGA-II combined with our mating restriction scheme were applied to knapsack problems with m items under the following parameter specifications:

      Crossover probability: 0.8,
      Mutation probability: $1/m$,
      Population size in NSGA-II: 200,
      Population size in SPEA: 100,
      Population size of the secondary population in SPEA: 100,
      Stopping condition: 2000 generations.

The above specifications of the population size seem to somewhat favor the NSGA-II because 200 solutions were examined in each generation of the NSGA-II while 100 solutions were examined in the SPEA. This is, however, not a serious problem because our aim in this paper is to examine the effect of mating restriction on the search ability of each algorithm (not to compare them with each other).

We also used the same parameter specifications for flowshop scheduling except for the mutation probability. The mutation probability was defined for each string as 0.5 (for details of genetic operations for flowshop scheduling, see [10]). It should be noted that the mutation was applied to each string in flowshop scheduling while it was applied to each bit in knapsack problems.

3.3 Performance Measures

Various performance measures have been proposed in the literature for evaluating a set of non-dominated solutions. As explained in Knowles & Corne [12], no single performance measure can simultaneously evaluate various aspects of a solution set. Moreover, some performance measures are not designed for simultaneously comparing many solution sets but for comparing two solution sets with each other.

For the small test problems (i.e., 30-item knapsack problems and 12-job flowshop scheduling problems), we used the ratio of undiscovered Pareto-optimal solutions as a performance measure. This ratio is referred to as the undiscovered solution ratio in this paper. For calculating this ratio, all Pareto-optimal solutions of the small test problems were found by an enumeration method. The average number of the Pareto-optimal solutions was 17.4 in the knapsack problems and 13.6 in the flowshop scheduling problems. For the large test problems, we used the average distance from each Pareto-optimal solution to its nearest solution in a solution set as a performance measure. This measure was used in Czyzak & Jaszkiewicz [2] and referred to as $D1_R$ in Knowles & Corne [12]. For any multiobjective optimization problem, it is reasonable for the decision maker (DM) to choose a final solution $\mathbf{x}^*$ from the Pareto-optimal solution set. The final solution $\mathbf{x}^*$ is the best solution with respect to the DM's preference. When the true Pareto-optimal solution set is not given, the DM will choose a final solution $\mathbf{x}$ from an available solution set. When the available solution set is a good approximation of the true Pareto-optimal solution set, the chosen solution $\mathbf{x}$ may be close to $\mathbf{x}^*$. In this case, the loss due to choosing $\mathbf{x}$ instead of $\mathbf{x}^*$ can be approximately measured by the distance between $\mathbf{x}$ and $\mathbf{x}^*$ in the objective space. Since $\mathbf{x}$ and $\mathbf{x}^*$ are unknown, we cannot directly measure the distance. The expected value of the distance, however, can be roughly estimated by the average value of the distance from each Pareto-optimal solution to its nearest available solution. The $D1_R$ measure corresponds to this approximation.

The $D1_R$ measure needs all Pareto-optimal solutions of each test problem. Since all Pareto-optimal solutions of the two-objective 250-item and 500-item knapsack problems were available from the homepage of the first author of [18], we used them. For the three-objective 250-item and 500-item knapsack problems, we found near Pareto-optimal solutions using the SPEA and the NSGA-II. These algorithms were applied to each test problem using longer CPU time and larger memory storage (i.e., 30000 generations with the population size 200 and the secondary population of the same size in the SPEA, and 30000 generations with the population size 400 in the NSGA-II) than the other computational experiments (see Subsection 3.2). We also used a single-objective genetic algorithm with a secondary population where all the non-dominated solutions were stored with no size limitation. Each of the three objectives was used in the single-objective genetic algorithm. This algorithm was applied to each three-objective test problem 30 times (10 times for each objective using the same stopping condition as the NSGA-II: 30000 generations with the population size 400). The SPEA and the NSGA-II were also applied to each test problem 10 times. Thus we obtained 50 solution sets for each test problem. Then we chose non-dominated solutions from the obtained 50 solution sets as near Pareto-optimal solutions. For 40-job and 80-job flowshop scheduling problems, near Pareto-optimal solutions were found in the same manner (the stopping condition was specified as 50000 generations). The number of Pareto-optimal or near Pareto-optimal solutions for each test problem is summarized in Table 1. While each objective of the knapsack problems has the same order of magnitude, they are not the same in the flowshop scheduling problems. Thus the objective space of each flowshop scheduling problem was normalized using the obtained near Pareto-optimal solutions when the $D1_R$ measure was calculated. More specifically, the objective space was normalized so that the minimum and maximum values of each objective among the near Pareto-optimal solutions became 0 and 100, respectively.

As a performance measure of a solution set (say S_j), we also used the ratio of non-dominated solutions $|S_j^*|/|S_j|$ where S_j^* is a set of solutions in S_j that are not dominated by any other solutions in other solution sets when multiple solution sets are compared with each other.

Table 1. The number of Pareto-optimal or near Pareto-optimal solutions

Knapsack problems				Flowshop scheduling problems			
Two-objective		Three-objective		Two-objective		Three-objective	
250-item	500-item	250-item	500-item	40-job	80-job	40-job	80-job
567	1427	2158	2142	98	87	973	974

3.4 Results on Small Test Problems

In Fig. 3 (a), we show the average ratio of undiscovered Pareto-optimal solutions by the SPEA with our mating restriction scheme. The average ratio was calculated over 50 runs of the SPEA with each specification of β on each of the 10 small knapsack problems (i.e., over 500 solution sets in total for each value of β). The horizontal axis of this figure is the value of β. As shown in this figure, $\beta = 1$ corresponds to the original SPEA. The maximum distance was used in the left half of this figure as the mate selection criterion while the minimum distance was used in the right half. That is, the most dissimilar solution was selected among β candidates in the left half while the most similar solution was selected in the right half. Open circles and closed circles show the results when the distance was measured in the decision space and the objective space, respectively. In this figure, we cannot observe any significant improvement by mating restriction (by specifying β as $\beta > 1$).

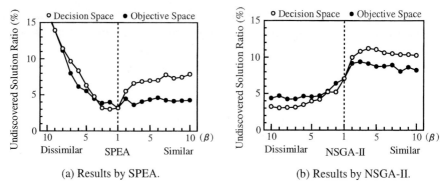

Fig. 3. Results on the two-objective 30-item knapsack problems

Fig. 3 (b) shows the average ratio of undiscovered Pareto-optimal solutions by the NSGA-II with our mating restriction scheme. While the search ability of the SPEA was not improved by our mating restriction scheme in Fig. 3 (a), we can observe clear improvement in the search ability of the NSGA-II in Fig. 3 (b). Large improvement was achieved in a wide range of β in the left half of Fig. 3 (b) independent of the choice between the decision space and the objective space (while somewhat better results were obtained from the distance in the decision space than the objective space). When the distance was calculated in the decision space, the average ratio of undiscovered solutions was improved from 7.05% in the case of $\beta = 1$ (i.e., the original NSGA-II) to about 3% by our mating restriction scheme (see open circles in the left half of Fig. 3 (b)). The improvement is statistically significant with the 99% confidence level for the results by $\beta \geq 2$ in the decision space and $\beta \geq 3$ in the objective space (the Mann-Whitney U test).

The average ratio of undiscovered Pareto-optimal solutions was also calculated over 50 runs of the SPEA and the NSGA-II for each of the ten small flowshop scheduling problems. Results by the SPEA and the NSGA-II were shown in Fig. 4 (a) and Fig. 4 (b), respectively. In Fig. 4 (a), the search ability of the SPEA was improved by our mating restriction scheme when the distance was measured in the decision space and dissimilar parents were chosen (i.e., open circles in the left half). The improvement is statistically significant with the 95% confidence level for the results by $\beta = 5, 6, 7, 8, 9$. On the other hand, the search ability of the NSGA-II was improved by our mating restriction scheme when the distance was measured in the objective space as well as in the decision space in the left half of Fig. 4 (b). The improvement is statistically significant with the 99% confidence level for the results by $\beta \geq 4$ in the decision space and $\beta = 4, 6, 7, 8, 9, 10$ in the objective space.

The experimental results in this subsection show that the search ability of the NSGA-II to find all Pareto-optimal solutions of the small test problems was improved by choosing dissimilar parents using our mating restriction scheme. On the other hand, the search ability of the SPEA was not improved by our mating restriction scheme except for the case of the 12-job flowshop scheduling problems where the distance was measured in the decision space (i.e., open circles in the left half of Fig. 4 (a)).

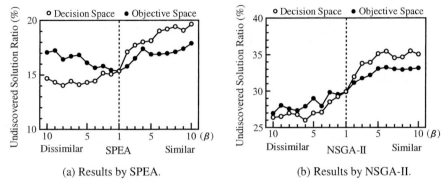

Fig. 4. Results on the two-objective 12-job permutation flowshop scheduling problems

3.5 Results on Large Test Problems

We examined the search ability of the SPEA and the NSGA-II to efficiently find near Pareto-optimal solutions through computational experiments on larger knapsack problems. Each algorithm was applied to each test problem 50 times using each specification of β. Due to the page limitation, we only report experimental results for the case where the distance was measured in the objective space. In Fig. 5 and Fig. 6, we show experimental results for the two-objective problems and the three-objective problems, respectively. Similar results to those figures were also obtained when the distance was measured in the decision space.

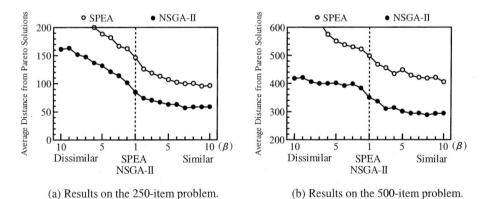

Fig. 5. Results on the two-objective knapsack problems

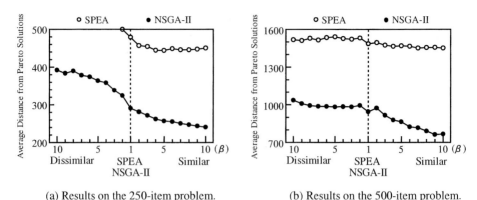

(a) Results on the 250-item problem.   (b) Results on the 500-item problem.

Fig. 6. Results on the three-objective knapsack problems

In Fig. 5 and Fig. 6, we can observe the improvement of the search ability of the SPEA and the NSGA-II by choosing similar parents. We examined the statistical significance of the improvement from the case of $\beta = 1$ (i.e., the original SPEA and NSGA-II) to the case of $\beta = 10$ (i.e., the right-most open and closed circles in each figure) for three confidence levels: 95%, 97.5% and 99%. Confidence levels of the improvement are summarized in Table 2 where each knapsack problem is denoted by the number of objectives and the number of items. From this table, we can conclude that the performance of the SPEA and the NSGA-II for the 250-item and 500-item knapsack problems was significantly improved by choosing similar parents.

Table 2. Confidence level of the improvement of each algorithm for each knapsack problem by choosing similar parents using our mating restriction scheme with $\beta = 10$

Problem	SPEA	NSGA-II
2/250	99	99
2/500	99	99
3/250	99	99
3/500	97.5	99

We show experimental results on the flowshop scheduling problems in Fig. 7 and Fig. 8 where the search ability of the SPEA was clearly improved by choosing similar parents. The improvement for the NSGA-II, however, was not clear. In the same manner as Table 2, we examined the statistical significance of the improvement from the case of $\beta = 1$ to the case of $\beta = 10$. Confidence levels of the improvement are summarized in Table 3 where each flowshop scheduling problem is denoted by the number of objectives and the number of jobs. From this table, we can conclude that the performance of the SPEA for the 40-job and 80-job flowshop scheduling problems was significantly improved by choosing similar parents. On the other hand, the performance of the NSGA-II was not significantly improved except for the case of the three-objective 80-job test problem.

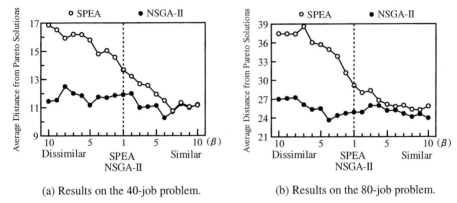

(a) Results on the 40-job problem.   (b) Results on the 80-job problem.

Fig. 7. Results on the two-objective permutation flowshop scheduling problems

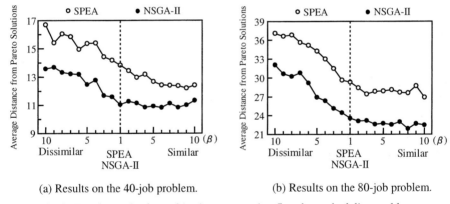

(a) Results on the 40-job problem.   (b) Results on the 80-job problem.

Fig. 8. Results on the three-objective permutation flowshop scheduling problems

Table 3. Confidence level of the improvement of each algorithm for each flowshop scheduling problem by choosing similar parents using our mating restriction scheme with $\beta = 10$

Problem	SPEA	NSGA-II
2/40	99	-
2/80	99	-
3/40	99	-
3/80	99	99

3.6 Discussions on Experimental Results

The experimental results on the small test problems in Subsection 3.4 suggest that the choice of dissimilar parents has a positive effect on the search ability of EMO algorithms to find a variety of Pareto-optimal solutions (i.e., a positive effect on the diversity of solutions). On the other hand, the experimental results on the large test problems in Subsection 3.5 suggest that the choice of similar parents has a positive effect on the search ability of EMO algorithms to efficiently find near Pareto-optimal solutions (i.e., a positive effect on the convergence speed to the Pareto-front).

These positive effects can be more clearly shown by the application to a larger permutation flowshop scheduling problem. We applied the SPEA and the NSGA-II to a two-objective 100-job permutation flowshop scheduling problem in the same manner as Subsection 3.5. We used the three variants of each algorithm: the choice of dissimilar parents with $\beta = 10$, the original algorithm with $\beta = 1$, and the choice of similar parents with $\beta = 10$. Experimental results by the SPEA and the NSGA-II are shown in Fig. 9 and Fig. 10, respectively. Each variant was applied to the 100-job permutation flowshop scheduling problem just once. We show a single solution set obtained by a single run of each variant in those figures. It should be noted that the axes of each figure are not the same because they are adjusted to the range of solution sets depicted in each figure.

In Fig. 9 (a) and Fig. 10 (a), we can observe both positive and negative effects of the choice of dissimilar parents: the increase in the diversity of solutions and the deterioration in the convergence speed to the Pareto-front. On the other hand, we can observe the positive effect of the choice of similar parents in Fig. 9 (b): the increase in the convergence speed to the Pareto-front. Actually, many solutions obtained by the original SPEA (i.e., closed circles) are clearly dominated by solutions obtained by the modified SPEA with the choice of similar parents (i.e., open circles) in Fig. 9 (b). Such improvement is not so clear for the NSGA-II in Fig. 10 (b).

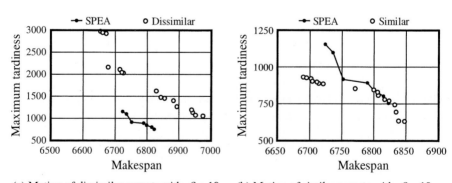

(a) Mating of dissimilar parents with $\beta = 10$.   (b) Mating of similar parents with $\beta = 10$.

Fig. 9. Solution sets obtained by the three variants of the SPEA for a 100-job problem

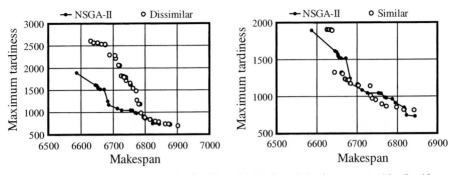

(a) Mating of dissimilar parents with $\beta = 10$. (b) Mating of similar parents with $\beta = 10$.

Fig. 10. Solution sets obtained by the three variants of the NSGA-II for a 100-job problem

The negative effect of the choice of dissimilar parents is the deterioration in the convergence speed to the Pareto-front as shown in Subsection 3.5, Fig. 9 (a) and Fig. 10 (a). On the other hand, the negative effect of the choice of similar parents is the decrease in the diversity of solutions as shown in Subsection 3.4 for the small test problems. This negative effect, however, was not clear for the large test problems.

For further examining the effect of our mating restriction scheme on the convergence speed to the Pareto-front, we calculated the average ratio of non-dominated solutions over 50 runs where five variants of each EMO algorithm were compared with each other. Experimental results are summarized in Table 4 for the SPEA and Table 5 for the NSGA-II. In those tables, K and FS mean knapsack and flowshop scheduling, respectively. Five variants in each table were compared with one another for calculating the average ratio of non-dominated solutions. From this table, we can see that the choice of similar parents increased the convergence speed while the choice of dissimilar parents decreased it.

Table 4. Average ratio of non-dominated solutions for each variant of the SPEA (%)

Problem	Dissimilar parents		SPEA	Similar parents	
	$\beta = 10$	$\beta = 5$	$\beta = 1$	$\beta = 5$	$\beta = 10$
K-2/30	89.5	96.7	99.1	97.7	97.8
K-2/250	0.2	0.0	17.5	54.7	48.2
K-2/500	0.0	0.1	25.4	61.2	31.9
K-3/250	2.4	10.2	59.4	62.4	51.8
K-3/500	1.1	4.2	67.9	67.0	36.7
FS-2/12	91.0	92.5	93.4	92.0	92.1
FS-2/40	5.4	10.9	21.1	43.4	49.6
FS-2/80	1.7	4.4	33.4	54.5	48.3
FS-3/40	4.4	9.8	36.7	52.8	54.2
FS-3/80	0.4	5.0	36.3	49.0	49.7

Table 5. Average ratio of non-dominated solutions for each variant of the NSGA-II (%)

Problem	Dissimilar parents		NSGA-II	Similar parents	
	$\beta = 10$	$\beta = 5$	$\beta = 1$	$\beta = 5$	$\beta = 10$
K-2/30	98.6	98.4	96.4	94.7	94.7
K-2/250	0.0	0.0	20.9	53.0	51.0
K-2/500	0.5	0.4	22.8	59.8	36.8
K-3/250	1.7	3.7	54.0	61.1	61.6
K-3/500	0.8	0.7	51.1	68.2	57.2
FS-2/12	87.8	86.8	87.0	84.5	84.8
FS-2/40	18.2	24.9	27.5	34.1	30.2
FS-2/80	10.9	17.3	31.7	33.1	45.2
FS-3/40	11.4	17.1	45.6	49.5	44.0
FS-3/80	2.4	6.7	41.5	47.6	50.1

4 Concluding Remarks

We examined the effect of mating restriction on the performance of the SPEA and the NSGA-II through computational experiments on multiobjective knapsack and permutation flowshop scheduling problems. Experimental results showed that the performance of these EMO algorithms on many test problems was significantly improved by mating restriction. The effect of mating restriction, however, was problem-dependent and algorithm-dependent. For example, the search ability of the NSGA-II to find all Pareto-optimal solutions of small knapsack problems was improved by choosing dissimilar parents while its search ability to efficiently find near Pareto-optimal solutions of large knapsack problems was improved by choosing similar parents. Experimental results suggest that the positive and negative effects of choosing dissimilar parents are the increase in the diversity of solutions and the deterioration in the convergence speed to the Pareto-front, respectively. Experimental results also suggest that the positive and negative effects of choosing similar parents are the increase in the convergence speed and the decrease in the diversity of solutions. If we want to improve the performance of an EMO algorithm with respect to the convergence speed to the Pareto-front, it may be worth examining the use of the proposed mating restriction scheme with the similarity as the mate selection criterion in the EMO algorithm. One of future research topics is to devise a mating restriction scheme that can improve the convergence speed to the Pareto-front without decreasing the diversity of solutions.

Acknowledgments. The authors would like to thank the financial support from Japan Society for the Promotion of Science (JSPS) through Grand-in-Aid for Scientific Research (B): KAKENHI (14380194).

References

1. Coello Coello, C. A., Van Veldhuizen, D. A., and Lamont, G. B.: *Evolutionary Algorithms for Solving Multi-Objective Problems*, Kluwer Academic Publishers, Boston (2002).
2. Czyzak, P., and Jaszkiewicz, A.: Pareto-Simulated Annealing – A Metaheuristic Technique for Multi-Objective Combinatorial Optimization, *Journal of Multi-Criteria Decision Analysis* 7 (1998) 34–47.
3. Deb, K.: *Multi-Objective Optimization Using Evolutionary Algorithms*, John Wiley & Sons, Chichester (2001).
4. Deb, K., Pratap, A., Agarwal, S., and Meyarivan, T.: A Fast and Elitist Multiobjective Genetic Algorithm: NSGA-II, *IEEE Trans. on Evolutionary Computation* 6 (2002) 182–197.
5. Fonseca, C. M., and Fleming, P. J.: Genetic Algorithms for Multiobjective Optimization: Formulation, Discussion and Generalization, *Proc. of 5th International Conference on Genetic Algorithms* (1993) 416–423.
6. Fonseca, C. M., and Fleming, P. J.: An Overview of Evolutionary Algorithms in Multiobjective Optimization, *Evolutionary Computation* 3 (1995) 1–16.
7. Goldberg, D. E.: *Genetic Algorithms in Search, Optimization, and Machine Learning*, Addison-Wesley, Reading (1989).
8. Hajela, P., and Lin, C. Y.: Genetic Search Strategies in Multicriterion Optimal Design, *Structural Optimization* 4 (1992) 99–107.
9. Horn, J., Nafpliotis, N., and Goldberg, D. E.: A Niched Pareto Genetic Algorithm for Multi-Objective Optimization, *Proc. of 1st IEEE International Conference on Evolutionary Computation* (1994) 82–87.
10. Ishibuchi, H., and Murata, T.: A Multi-Objective Genetic Local Search Algorithm and Its Application to Flowshop Scheduling, *IEEE Trans. on Systems, Man, and Cybernetics – Part C: Applications and Reviews* 28 (1998) 392–403.
11. Jaszkiewicz, A.: Genetic Local Search for Multi-Objective Combinatorial Optimization, *European Journal of Operational Research* 137 (2002) 50–71.
12. Knowles, J. D., and Corne, D. W.: On Metrics for Comparing Non-Dominated Sets, *Proc. of 2002 Congress on Evolutionary Computation* (2002) 711–716.
13. Schaffer, J. D.: Multiple Objective Optimization with Vector Evaluated Genetic Algorithms, *Proc. of 1st International Conference on Genetic Algorithms and Their Applications* (1985) 93–100.
14. Van Veldhuizen, D. A., and Lamont, G. B.: Multiobjective Evolutionary Algorithms: Analyzing the State-of-the-Art, *Evolutionary Computation* 8 (2000) 125–147.
15. Watanabe, S., Hiroyasu, T., and Miki, M.: LCGA: Local Cultivation Genetic Algorithm for Multi-Objective Optimization Problem, *Proc. of 2002 Genetic and Evolutionary Computation Conference* (2002) 702.
16. Zitzler, E., Deb, K., Thiele, L.: Comparison of Multiobjective Evolutionary Algorithms: Empirical Results, *Evolutionary Computation* 8 (2000) 173–195.
17. Zitzler, E., Thiele, L.: Multiobjective Optimization using Evolutionary Algorithms – A Comparative Case Study, *Proc. of 5th International Conference on Parallel Problem Solving from Nature* (1998) 292–301.
18. Zitzler, E., Thiele, L.: Multiobjective Evolutionary Algorithms: A Comparative Case Study and the Strength Pareto Approach, *IEEE Transactions on Evolutionary Computation* 3 (1999) 257–271.

Using Simulated Annealing and Spatial Goal Programming for Solving a Multi Site Land Use Allocation Problem

Jeroen C.J.H. Aerts[1], Marjan van Herwijnen[1], and Theodor J. Stewart[2]

[1] Institute for Environmental Studies (IVM), Vrije Universiteit Amsterdam
De Boelelaan 1087, 1081 HV Amsterdam, The Netherlands
Tel: +31-20-4449 555, Fax: +31-20-4449 553
{Jeroen.Aerts Marjan.van.Herwijnen}@ivm.falw.vu.nl

[2] Department of Statistical Sciences, University of Cape Town
Rondebosch 7701, South Africa
tjstew@maths.uct.ac.za

Abstract. Many resource allocation issues, such as land use- or irrigation planning, require input from extensive spatial databases and involve complex decision-making problems. Recent developments in this field focus on the design of allocation plans that utilize mathematical optimization techniques. These techniques, often referred to as multi criteria decision-making (MCDM) techniques, run into numerical problems when faced with the high dimensionality encountered in spatial applications. In this paper, it is demonstrated how both Simulated annealing, a heuristic algorithm, and Goal Programming techniques can be used to solve high-dimensional optimization problems for multi-site land use allocation (MLUA) problems. The optimization models both minimize development costs and maximize spatial compactness of the allocated land use. The method is applied to a case study in The Netherlands.

1 Introduction

The area of 'resource allocation' deals with the spatial distribution of (natural-) resources such as water or land. Land use allocation problems are complex as they often involve multiple stakeholders with conflicting goals and objectives. Therefore, much attention has been paid to solving land use allocation problems with multi criteria decision-making techniques (MCDM). Recent research focused on combining MCDM with a geographic information system (GIS). This appears to be a powerful combination, since land use allocation problems both involve multiple objectives and criteria as well as geographically dependent spatial attributes [2], [3], [8], [9] and [10].

Both GIS and MCDM techniques are derived from relatively technical areas of geography and operations research. Practical use of such techniques often requires a thorough understanding, and non-technical decision makers may find using these techniques difficult. However, combined GIS-MCDM techniques can be operationalized for non-technical users by integrating these techniques in a (spatial-) decision support system (SDSS) that is dedicated to a user. A series of SDSSs were built only using GIS based mapping and visualization tools to provide background information on the planning area. Some SDSSs successfully combined multi criteria analysis (MCA) techniques (a subset of MCDM) with GIS enabling the interactive evaluation of spatial land use plans against a pre-defined set of criteria, e.g. [13], [22]. The combination MCA-GIS can be useful in an SDSS environment when both the alternatives have been clearly defined and the total set of alternative solutions is limited to say six and rarely ten [23].

A challenge lies in further developing combined MCDM and GIS techniques suitable for implementation in an SDSS for land use allocation problems. For instance, in case alternative solutions are *not* defined, optimization techniques (another subset of MCDM techniques) can be combined with GIS to support the spatial design of land use allocation plans. Optimization techniques with GIS may therefore be referred to as spatial *design* techniques [2], [3], [26]. In this paper we examine the use of spatial - GIS based- optimization techniques for multi site land use allocation problems (MLUA). 'Multi site' refers to the problem of allocating more than one land use type in an area. A crucial element in the model is to introduce a spatial compactness objective. Spatial compactness objectives are used to address the problem of allocating the same land use not only at lowest cost but also at maximum compactness, e.g. [2], [10], [32], [33].

From the above, we arrive at the following objectives for this research:
(1) To develop a goal-programming model based on a reference point approach that can solve an MLUA problem. The model will be solved using a simulated annealing algorithm.
(2) To develop different spatial compactness objectives in order to provide a decision maker with different options
(3) To test the model on the basis of its efficacy to encourage spatial compactness.

A case study in The Netherlands illustrates how the optimization model can be used in a decision environment.

2 MLUA Model

2.1 The Basic Allocation Problem

As an example, consider a rectangular area to be allocated with land use. First, the area is divided to a grid with N rows and M columns. Let there be K different land use types. We now introduce a binary variable x_{ijk} which equals 1 when land use k is assigned to cell (i,j) and equals 0 otherwise. Furthermore, development costs (C_{ijpk}) are involved with each land use type k in cell (i,j). These costs vary with location because they may depend on specific cost attributes p (for p=1,...,P) of the area, such as soil type, elevation and management costs.

The objective is to minimize costs associated with allocating land uses k (for $k=1,\ldots,K$) to a map $\mathbf{u}$. Accordingly, the problem may be written as follows:

Minimize

$$f_p(u) = \sum_{k=1}^{K}\sum_{i=1}^{N}\sum_{j=1}^{M} C_{ijpk} x_{ijk} \qquad \forall\ p = 1,\ldots,P \qquad (1)$$

Subject to

$$\sum_{k=1}^{K} x_{ijk} = 1 \qquad \forall\ i = 1,\ldots,N,\ j = 1,\ldots,M \qquad (2)$$

$$x_{ijk} \in \{0,1\}$$

$$L_k \leq A_k \leq U_k \qquad (3)$$

where:

$$\sum_{i=1}^{N}\sum_{j=1}^{M} x_{ijk} = A_k \qquad \forall\ k = 1,\ldots,K \qquad (4)$$

and

$$\sum_{k=1}^{K} A_k = N \cdot M \qquad (5)$$

Equation 2 specifies that one and only one land use must be assigned to each cell. Because decision variable x_{ijk} must be either 0 or 1, the model is defined as an integer program (IP). Equations 3, 4 and 5 bound the number of cells A_k allocated to a certain land use type k between an upper and lower bound, expressed as L_k and U_k, respectively. This as compared to a stricter formulation:

$$\sum_{i=1}^{N}\sum_{j=1}^{M} x_{ijk} = N \cdot M \cdot O_k \qquad \forall\ k = 1,\ldots,K \qquad (6)$$

where a required fixed required area proportion for each land use type is represented by parameter O_k [3]

2.2 Spatial Objectives

The additive objective function described in Section 2.1 can be expanded with a second, spatial objective. This objective refers to spatial attributes as compactness or contiguity of land use of equal type. Note the difference between contiguity and compactness in this respect (Fig. 1) [11]. Contiguity requires all cells of equal land use to be connected (Fig. 1, *middle*). Compactness merely encourages cells of equal land use to be allocated next to one another, but this may result in divided patches (Fig. 1, *right*). In this paper, we will restrict ourselves to the compactness objective.

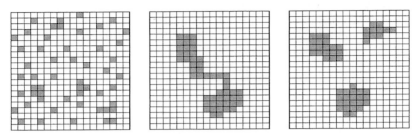

Fig. 1. Area with single land use (*light gray*) covering 52 cells. These cells are randomly placed before optimization (*left*). The cells are allocated by optimizing contiguity (*middle*) and compactness (*right*)

Spatial compactness objectives are, for instance, found in forestry research harvest schedules, which deal with strict adjacency constraints e.g. [14], 19], [21]. Some research in geographic information science have approached spatial compactness in optimization modeling by rewarding cases where neighboring cells have equal land use [2], [3], [4].

In order to accommodate flexibility for the user, it is proposed to develop a mix of three spatial compactness measures, from which a user can choose or can use combinations. The spatial compactness objectives merely address commonly used compactness characteristics as size, perimeter and area of a cluster [8], [11]. We here restrict ourselves to the following spatial compactness objectives:
- Spatial objective 1: minimizing the number of clusters per land use type. Less clusters of a certain land use type points to higher compactness and less fragmentation. Hence, the ideal compactness value would be 1.
- Spatial objective 2: maximizing the largest cluster relatively to the other clusters identified under spatial objective 1. It is preferred having at least one large compact cluster, rather than all clusters being compact but small. The ideal compactness value would be again 1.
- Spatial objective 3: minimizing the perimeter of a cluster. In order to transform this measure size and scale independent, the perimeter is divided by the square root of the cluster area. The ideal compactness value would be 4.

A simple algorithm, that counts the number of clusters, the area and perimeter of each cluster (per land use type) has been developed by [29]. The calculation of the spatial objectives is illustrated in Fig. 2. Here, the value for spatial objective 1 is 4, because 4 clusters of the same land use can be identified. The value for spatial objective 2 is 0.25, which can be calculated by dividing 1 (for identifying 1 largest cluster) by 4 (total number of clusters). The value for spatial objective 3 is calculated using the following equation:

$$\sum_{k=1}^{K}\sum_{r=1}^{R}\frac{H_{kr}}{\sqrt{L_{kr}}} \quad (7)$$

where H_{kr} stands for the perimeter of an identified cluster r for land use k and L_{kr} represents the area for each identified cluster r per land use k. The values for the perimeters for cluster a, b, c and d in Fig. 2 are 20, 10, 12, and 22 respectively. The values for the area of all clusters are 19, 6, 5 and 25. Hence by applying Equation 7, the value for spatial objective function 3 becomes 4.59 + 4.08 + 5.37 + 4.4 = 18.44.

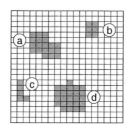

Fig. 2. Illustration of spatial compactness objectives for four clusters labeled a, b, c and d, in an area. The values for the perimeters for clusters a, b, c and d are 20, 10, 12, and 22 respectively. The values for the area are 19, 6, 5 and 25 for clusters a, b, c and d, respectively

In order to exclude single cells as a cluster in an optimal result, we may introduce an extra constraint for a minimum cluster area value S_k for any cluster L_{rk} by:

$$L_{rk} \geq S_k \qquad (F_k \in A_k) \; \forall \; k = 1,\ldots, K \quad (8)$$

The total spatial compactness objective can be calculated using a weighted sum of the above-described three spatial compactness measures. The difference of the spatial objectives as opposed to the cost objective is the non-linearity in the spatial formulations.

3 Goal Programming

3.1 Constraints and Criteria

The MLUA problem formulated above is clearly a multi objective problem, where costs and compactness objectives have to be traded off against each other. In a situation where decision makers know their goals but have difficulties with valuing or weighting the relevant attributes involved in the multi criteria analysis, goal programming is a commonly known technique to aid decision makers with their task. We have chosen to use a generalized goal programming approach (reference point) approach based on Wierzbicki [31]. For each objective, we define some goal or reference point, say γ_p for all cost attribute related goals and λ_q for all spatial objectives. The model should find a land use map **u** for which:

$$f_p(u) \leq \gamma_p \tag{9}$$

$$s_{kq}(u) \leq \lambda_{kq} \tag{10}$$

Where $f(u)_p$ is the total value for all cost attributes p ($p = 1, ..., P$) and $s_{kq}(x)$ the total of spatial measures q ($q=1,...,Q$), which in this case is set to 3 (see section 2.2).

Wierzbicki [31] uses a 'scalarizing' function, which measures under-achievement relative to the goals, but placing the greatest weight on the least well satisfied goal. Another commonly used scalarizing function can be found in the Tschebycheff approach [27] where the goal is to minimize the sum of deviations relative to the goals defined. We here use another scalarizing approach [28], which also minimizes the sum of deviations but then relative to an ideal value. This approach can be defined as follows:

Minimize:

$$\sum_{p=1}^{P} \left[\frac{f_p(u) - I_p}{\gamma_p - I_p} \right]^\rho + \sum_{k=1}^{K} \sum_{q=1}^{Q} \left[\frac{s_{kq}(u) - I_{kq}}{\lambda_{kq} - I_{kq}} \right]^\rho \tag{11}$$

Subject to: Equations (2), (3), (4) and (5)

In Equation 11, I_p is the best possible ideal value for each objective p if optimized on its own, and ρ is a suitably large power. A value of $\rho = 4$ has been found to yield good results [28]. Advantages of this approach is (1) to avoid the use of preferences or weights, which are often difficult to interpret by users, and (2) the function is scale free, which rules out the need for finding the worst performance levels to provide a normalized scaling.

Constraint in Equation 3 is fuzzy. We may treat this constraint in the same manner as goals, which is a useful contribution as this research aims at developing a goal programming approach for MLUA problems. Treating fuzzy constraints as goals may be implemented as follows. One should view the nominal constraint as an ideal, and the goal then should minimize the deviation from this ideal:

Lower bound objective:
$$\left[\frac{\max\{0; L_k - A_k\}}{\beta_k^0} \right]^\rho \tag{12}$$

Upper bound objective:
$$\left[\frac{\max\{0; A_k - U_k\}}{\beta_k^0} \right]^\rho \tag{13}$$

The scaling factor β_k is chosen such that a deviation of this magnitude corresponds to the same level of satisfying as achieving the assumed goals for the costs and spatial compactness objectives. It is proposed to handle these goals as extra penalty terms in Equation 11.

4 Optimization Algorithms

4.1 Heuristic Algorithms

Since both an MLUA problem can be classified as a combinatorial optimization problem, and our MLUA model is non-linear in its spatial objective formulations, we need to find a heuristic optimization algorithm for solving the model. Note that various researchers have developed linear MLUA models solved with an LP solver, but ran against a limitation in the size of the spatial area that could be optimized [2], [3], [4], [8], [9], [10]. Heuristic approaches, however, are robust, fast and capable of solving large combinatorial problems, but they do not guarantee the optimal solution. Applications of such algorithms for MLUA problems are simulated annealing, greedy growing algorithms, genetic algorithms and tabu search [2], [3], [5], [6], [19], [21]. We here focus on using simulated annealing to solve the above described optimization model.

4.2 Simulated Annealing

Kirkpatrick [15] introduced the concept of annealing in combinatorial optimization. This concept is based on a strong analogy between combinatorial optimization and the physical process of crystallization. This process has inspired Metropolis [20] to propose a numerical optimization procedure known as the Metropolis algorithm (Fig. 3). This works as follows for our problem. The initial situation is the current land use map **u**. The associated total development cost is denoted by $f(0)$. Following the flow diagram of Fig. 3, we now swap the land use of a randomly chosen cell into another randomly chosen land use. The new land use must be both another land use as compared by the land use in the current situation and one that is permitted by a transition matrix (Table 2). This yields a new situation, with new development costs $f(1)$. Whether we accept the change from state 0 to state 1 depends on the difference in costs $f(1)-f(0)$. Once this is decided we repeat the swapping procedure, decide whether the change is accepted, generate a new swap, and so on. Whenever the costs $f(1)$ are smaller than the costs $f(0)$, the cell change is accepted. When $f(1)>f(0)$, costs are accepted with a certain probability following the Metropolis criterion Equation (14) (see e.g. [1], [25]). This is achieved by comparing the value of the Metropolis criterion with a random number drawn from a uniform [0,1) distribution (Fig. 3).

$$P(accept\ change) = \exp\left(\frac{f(0)-f(1)}{s_0}\right) \qquad (14)$$

where s_0 is a *control* or *freezing* parameter.

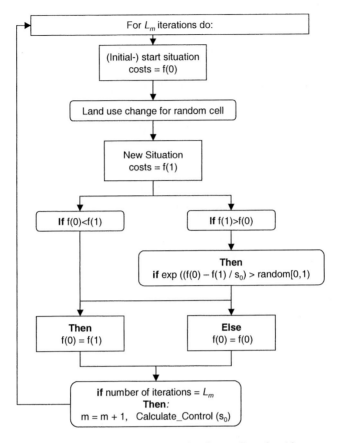

Fig. 3. Flow diagram of the simulated annealing algorithm

A crucial element of the procedure is the gradual decrease of the freezing parameter s_i [17]. Usually, this is done using a constant multiplication factor:

$$s_{i+1} = r \cdot s_i \qquad (15)$$

where $0<r<1$. Examples of simulated annealing studies for spatial optimization are [5], [7], [19], [30].

5 Case Study Jisperveld

5.1 Introduction

Jisperveld is the largest connected brackish fen-meadow area of Western Europe. It is situated in the Northwest of the Netherlands and measures about 2000 ha. The area is well known by its high natural values represented by a selection of rare bird species and wetland vegetation types. The Jisperveld area is subject to a debate on how to both plan and manage the area in the future. It appears that the governmental planning policy for land use is changing from predominantly agriculture to a combined agriculture, nature area. Therefore, planners need to develop an optimal location for two new land use types 'extensive agriculture' and 'water (limited access)', both of which are not yet present. In order to apply our goal programming approach, we've simplified the current land use map of the area.

Both the number of land use types present in the area and the resolution of the grid cells was reduced (Fig. 4)

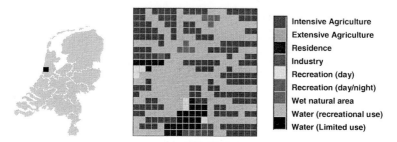

Fig. 4. Location of the Jisperveld area in The Netherlands indicated with the black dot (*left*) and the simplified current land use map of the Jisperveld, measuring 20 x 20 cells (*right*)

Experts made clear that closed patches of land use, like large closed areas of extensive agriculture and water (Limited use), represent a higher natural value than fragmented areas. Also, less fragmented areas have an increased potential for recreational activities. A simplification was to ignore the spatial requirement for creating corridors. This requirement would involve an adjustment of the compactness objective.

5.2 Objectives

Following our model formulations, the six objectives distinguished in the Jisperveld can be categorized in additive cost attributes and objectives that relate to maximizing compactness of land use of the same type. The objectives are to maximize natural and recreational values, minimize cost for changing land use and to maximize compactness following the three describes spatial objectives. Note that both natural and recreational values can be seen as costs that contribute positively to the overall objective function. The values related to objectives 1 and 2 are defined as C_{ijpk}, summed over all cells in the area, for each land use type k (for k=1,..., K). Values are defined by value maps, which either have a uniform value (a number in Table 1) or a variable value ('map' in Table 1). All values are scaled between 1 and 10.

Furthermore, transition costs for changing current land use k_c into future land use k_f are presented in Table 2. Management costs for maintaining certain land use types are not considered in this model, but can be easily integrated following the same approach as with objectives 1 and 2, but then by minimizing management costs.

Table 1. Nature values and recreation values for each land use type

Land use type (k)		Nature value Range: [1, 10]	Recreation value Range: [1, 10]
1.	Intensive agriculture	5	6
2.	Extensive agriculture	Map	Map
3.	Residence	2	4
4.	Industry	1	1
5.	Recreation (day trippers)	6	Map
6.	Recreation (overnight)	6	Map
7.	Wet natural area	Map	7
8.	Water (recreational use)	6	Map
9.	Water (limited access)	Map	3

Table 2. Transition matrix, showing costs [euro/cell] to change current land use k_c into future land use k_f

	Current land use k_c								
Future Land use type (k_f)	Int. agri.	Ext. agri.	Resi-dence	In-dus-try	Recr. (day tripp.)	Recr. (over night)	Wet natu-re	Wa-ter (recr)	Wa-ter (limit)
1. Intensive agriculture	0	1000	10000	500	–	7000	–	–	–
2. Extensive agriculture	–	0	–	–	–	–	–	–	–
3. Residence	–	–	0	–	–	–	–	–	–
4. Industry	–	–	–	0	–	–	–	–	–
5. Recreation (day trippers)	–	–	9000	–	0	5000	–	–	–
6. Recreation (overnight)	–	–	–	–	–	0	–	–	–
7. Wet natural area	–	–	–	–	–	–	0	–	–
8. Water (recreational use)	–	–	–	–	–	–	–	0	1000
9. Water (limited access)	–	–	–	–	–	–	–	–	0

The three spatial objectives are related to the extent to which the different land uses are connected or fragmented across the region. For this, the *number of clusters* for each land use type, as well as the *area* and *perimeter* of each cluster is measured with a simple counting algorithm (see for details [29]).

5.3 Constraints

The values for the constraints are listed in Table 3. They refer to Equations 8, 12 and 13. Land use type 'Intensive agriculture' covers 157 grid cells but is limited to 130 grid cells in the new design. Extensive agriculture, conversely, does not occur in the original map but has to cover at least 27 grid cells in the new design. Setting a minimum cluster size assumes that it is not realistic to have a connected area that is smaller than the minimum cluster size. Note, for example, that the minimum cluster size for 'extensive agriculture' is smaller than for 'intensive agriculture'.

Areas can be fixed to a certain land use type, by indicating areas that are not allowed to change. These constraints can be set through a map holding the fixed land use types. Fig. 5 shows the map with fixed land use types for this case study.

Table 3. Various spatial constraints for each land use type k

Land use type k	Lower bound	Upper bound	Current	Minimum cluster size
Intensive agriculture	100	130	157	4
Extensive agriculture	27	57	0	3
Residence	28	35	28	3
Industry	2	4	7	2
Recreation (day trippers)	3	10	6	3
Recreation (overnight)	1	5	1	1
Wet natural area	4	20	8	3
Water (recreational use)	150	193	193	4
Water (limited access)	0	43	0	4

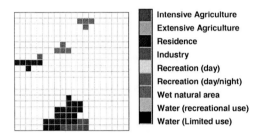

Fig. 5. Land use map showing the areas with fixed land use types

6 Results

6.1 Initial Conditions

Three model runs are selected, each having different parameter settings, in order to evaluate the model on its efficacy for generating compact patches of land-use. The initial simulated annealing values are the same for all runs. The start value of the freezing parameter was determined with a trial run, following [30] and [3]. The freezing parameter was chosen such that within 500 trial iterations, 80% of all calculated costs were greater than the original situation. The decrease parameter r was set to 0.85 and the iteration length L per temperature stage was set to 1,000. Both r and L are kept constant across all four runs. About ten runs using exactly the same parameter settings were executed to test the consistence of the model. Three of those runs are presented in Fig. 6. Although some runs start with considerably higher costs as compared to other runs, all runs eventually arrive within approximately 38,000 iterations at a scalarizing value of about 2000.

6.2 Spatial Designs

Run 1 evaluates compactness according to a standard parameter setting where cost objectives and spatial compactness objectives are equally preferred. Fig. 7 shows different stages in the iteration process using the standard parameter set. The map at the far *left* shows the initial – random – situation.

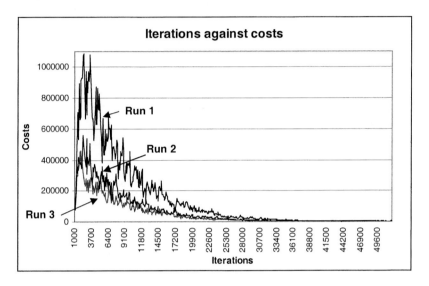

Fig. 6. Costs calculated by using the scalarizing function in Equation 11, against the total number of iterations by the simulated annealing algorithm

The map to the far *right* shows the final situation, achieved after a total of 80,000 iterations. At this stage, the objective function could no longer be improved and the iteration was terminated

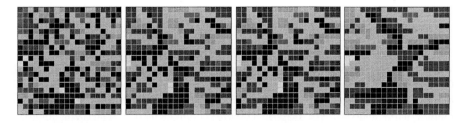

Fig. 7. Different stages in the optimization process. *Far left*: initial situation; *center left*: after 5,000 iteration stages; *center right*: after 20,000 iteration stages; *far right*: final situation after 80,000 iteration stages

Within the Run 2, compactness is stimulated by increasing the weights for all spatial objectives. The weights on the spatial objectives are set to twice the value for the weights on the cost objectives. The result is presented in Fig. 8, *left*. Table 4 shows indeed that the spatial objective values of Run 2 improve as compared to Run 1, but at the cost of lower natural and recreational values and higher land use change costs.

For Run 3, an increased weight is set to allocate one specific land use, in this case 'Water limited use'. It is shown that indeed a large compact cluster is formed in the middle of the area. This is probably the least expensive area to allocate a compact cluster of this land use type (Fig. 8, *right*).

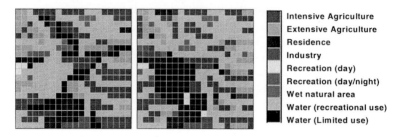

Fig. 8. Results for runs 2 (*left*) and 3 (*right*)

Table 4. Objectives values for the three test runs

	Nat Value	Rec. Val.	Change land use	Min #clusters	Largest cluster	Min perimeter
Run 1	1917	2389	175500	42	4.55	51.13
Run 2	1878	2258	179000	40	4.87	50.66
Run 3	1919	2412	198500	43	4.92	53.78

7 Conclusions and SDSS Implementation

The main goal of this paper was to investigate whether goal programming (GP) combined with simulated annealing, is an attractive alternative for designing spatial resource allocation alternatives.

We have developed a general GP approach to solve a MLUA problem. It is thought GP is a well-known approach in a situation where decision makers know their goals but have difficulties with valuing or weighting the relevant attributes involved in the multi criteria analysis. We here successfully implemented a scalarizing approach developed by [28], which minimizes the sum of deviations relative to an ideal value, this as opposed to the Tschebycheff goal programming approach where the goal is to minimize the sum of deviations relative to defined goals.

Three spatial compactness objectives have been developed, based on commonly used compactness characteristics that address size, perimeter and area of a cluster of the same land use. These objectives have been made operational within the above-mentioned scalarizing function. Then model has been solved using the simulated annealing algorithm that has been used in similar studies by e.g. [3]. The efficacy of the model for generating compact land use designs was evaluated in four model runs, all having different parameter settings. It appeared that the model indeed generated compact patches of land uses, which increased in size when the weights on the compactness objectives was increased.

The case study in The Netherlands, to which the model was applied, clearly shows the potential of the approach in a decision support setting. More attention is needed, however, to further develop the GP approach in an SDSS, accessible for non-technical users. Therefore, an important feature of such an SDSS will be to interactively specify goal levels of the development cost and spatial attributes. With respect to the spatial attributes, however, this is more difficult, and goal setting in the SDSS should be done by:

- pre define the worst and ideal levels of the spatial objectives
- allowing the user to interactively select his preference for one of the three spatial objectives
- allowing the user to select spatial goals as a proportion of the distance between worst and ideal values. E.g. by using a slide bar, which allows selecting values between worst and ideal.

Other applications of the model approach can be found in forestry services and water allocation issues. Further research on larger, more realistic maps is recommended as well as to assess the possibilities to solve the model with other heuristic methods.

References

1. Aarts, E., Korst, J.: Simulated Annealing and Boltzman Machines. A Stochastic Approach to Combinatorial Optimisation and Neural Computing, J. Wiley, New York (1989)
2. Aerts, J.C.J.H.: Spatial Decision Support for Resource Allocation. Integration of Optimization, Uncertainty analysis and Visualization techniques. PhD thesis, University of Amsterdam, Thela Thesis publishers, Amsterdam (2002)
3. Aerts, J.C.J.H. and Heuvelink, G.B.M.; Using Simulating Annealing for Resource Allocation. International Journal of Geographic Information Science **16** (2002) 571–587
4. Aerts, J.C.J.H., Eisinger, E, Heuvelink G.B.M, Stewart T.J.: Using Linear Integer Programming for Multi Site Land Use Allocation. Geographical Analysis. (Accepted Aug. 2002)
5. Boston, K. and Bettinger, P.: An Analysis of Monte Carlo Integer Programming, Simulated Annealing, and Tabu Search Heuristics for Solving Spatial Harvest Scheduling Problems. Forest Science **45** (1999) 292–301
6. Brookes, C. J.: A parameterized region-growing program for site allocation on raster suitability maps. International Journal of Geographical Information Science **11** (1997) 375–396
7. Church, R.L., Stoms D. M., Davis, F.W.: Reserve Selection as a Maximal Covering Allocation Problem. Biological conservation **76** (1996) 105–112
8. Cova, T. J.: A General Framework For Optimal Site Search. PhD thesis, University of California, Santa Barbara (1999)

9. Cova, T.J. and Church, R.L.: Contiguity Constraints for Single-Region Site Search Problems. Geographical Analysis **32** (2000) 306–329
10. Cova, T.J. and Church R.L. Exploratory spatial optimization in site search: a neighborhood operator approach. Computer, Environment and Urban Systems, 24 (2000) 401–419
11. Diamond, J. T. and Wright, J.R.: Efficient Land Allocation. Journal of Urban Planning and Development, **115** (1989) 81–96
12. Greenberg, H.J.: Mathematical Programming Glossary. http://www.cudenver.edu/~hgreenbe/glossary/glossary.html. Last checked, Jan. 2002.
13. Janssen, R.: Multi objective decision support for environmental problems. PhD thesis, Free University, Amsterdam (1991)
14. Jones, G.J., Meneghin, B.J., Kirby, M.W.: Formulating Adjacency Constraints in Linear Optimisation Models for Scheduling Projects in Tactical Planning. Forest Science, **37** (1991) 1283–1297.
15. Kirkpatrick, S., Gelatt, C.D., Vecchi, M.P.: Optimisation by Simulated Annealing, Science **220** (1983) 671–680.
16. Kurttila, M.: The spatial structure of forests in the optimization calculations of forest planning – a landscape ecological perspective. Forest Ecology and Management **142** (2002) 129–142
17. Laarhoven van P.J.M.: Theoretical and Computational Aspects of Simulated Annealing. PHD Thesis, Erasmus University Rotterdam (1987).
18. Levine, B.: Presentation for EE652, April15th, 1999. http://microsys6.engr.utk.edu/~levine/EE652/Slides/tsld001.htm. Last checked, May. 2002.
19. Lockwood, C. and Moore, T.: Harvest scheduling with spatial constraints: a Simulated Annealing approach. Can. J. of For. Res. **23** (1993) 468 – 478
20. Metropolis, N., Rosenbluth, A., Rosenbluth, M., Teller, A., and Teller, E.: Equation of state calculations by fast computing machines. Journal of Chemical Physics **21** (1953) 1087–1092.
21. Murray, A.T. and Church, R.L.: Measuring the efficacy of adjacency constraint structure in forest planning models. Canadian Journal of Forest Resources, **25** (1995)1416–1424.
22. Pereira, J.M.C., Duckstein, L.: A Multiple criteria decision making approach to GIS based land suitability evaluation, International Journal of Geographical Information Systems, vol. **7** (1993) 407–424.
23. Ridgley, M.R., Heil, G.W.: Multi Criterion planning of protected-area buffer zones. In: E. Beinat and P. Nijkamp (eds), Multi-criteria Evaluation in Land-Use Management, Kluwer (1998).
24. Ridgley, M. Penn, D.C., Tran, L.: Multicriterion Decision Support for a conflict over Stream Diversion and Land-Water Reallocation in Hawaii. Applied Mathematics and Computation **83** (1997) 153–172
25. Rogowski, A.S., Engman, E.T.: Using a SAR Image and a Decision Support System to Model Spatial Distribution of Soil Water in a GIS Framework. Integrating GIS and Environmental Modeling. National Center for Geographic Information and Analysis (NCGIA), Santa Fe (1996.)
26. Seppelt, R., Voinov A.: Optimization methodology for land use patterns using spatially explicit landscape models. Ecological modeling **151** (2002) 125–142.
27. Steuer, R.: The Tchebycheff procedure of interactive multiple objective programming. In: Karpak, B., Zionts, S (eds.) Multiple Objective Decision Making and Risk Analysis Using Micro Computers. Springe-Verlag, Berlin (1986).
28. Stewart, T.J.: A multi criteria decision support system for R&D project selection. Journal of the operational Research Society **42** (1991) 1369–1389.

29. Stewart, T.J., Herwijnen, M. van, Janssen, J. A Genetic Algorithm Approach to Multi Objective Land use Planning. In prep. (2002)
30. Sundermann, E.: PET Image Reconstruction Using Simulated Annealing, Proceedings of the SPIE Medical Imaging 1995 Conference, Image Processing (1995) 378–386
31. Wierzbicki, A.P.: Reference point approaches. In: T. Gal, T.J. Stewart and T. Hanne (eds.): Multi Criteria Decision Making: Advances in MCDM models, Algorithms, Theory and Aplications. Kluwer Academic Publishers, Boston (1999)
32. Williams, J.C. and Revelle, C.S.; Reserve assemblage of critical areas: A zero-one programming approach. European Journal of Operational Research **104** (1998) 497–509.
33. Wright, J., Revelle, C.S., Cohon, J.: A Multi objective Integer Programming model for the land acquisition problem. Regional Science and Urban Economics **12** (1983) 31–53.

Solving Multi-criteria Optimization Problems with Population-Based ACO

Michael Guntsch[1] and Martin Middendorf[2]

[1] Institute for Applied Computer Science and Formal Description Methods
University of Karlsruhe, D-76128 Karlsruhe, Germany
guntsch@aifb.uni-karlsruhe.de
[2] Department of Computer Science
University of Leipzig, Augustusplatz 10-11, D-04109 Leipzig, Germany
middendorf@informatik.uni-leipzig.de

Abstract. In this paper a Population-based Ant Colony Optimization approach is proposed to solve multi-criteria optimization problems where the population of solutions is chosen from the set of all non-dominated solutions found so far. We investigate different maximum sizes for this population. The algorithm employs one pheromone matrix for each type of optimization criterion. The matrices are derived from the chosen population of solutions, and can cope with an arbitrary number of criteria. As a test problem, Single Machine Total Tardiness with changeover costs is used.

1 Introduction

The Ant Colony Optimization (ACO) metaheuristic (see Dorigo and Di Caro [1]) has recently been applied to solve multi-criterion optimization problems (see [2, 3] for an overview over metaheuristics for multi-criteria optimization). In most of the earlier works it is assumed that the optimization criteria can be weighted by importance. Mariano and Morales [4] proposed a multi colony ACO approach where for each objective there exists one colony of ants. They studied a problem where every objective is influenced only by parts of a solution, so that an ant from colony i receives a (partial) solution from ant of colony $i-1$ and then tries to improve or extend this solution with respect to criterion i. A final solution that has passed through all colonies is allowed to update the pheromone information when is is part of the non-dominated front. Gambardella et al. [5] developed an ant algorithm for a bi-criterion vehicle routing problem where they also used one ant colony for each criterion. Criterion 1 — the number of vehicles — is considered to be more important than criterion 2 — the total travel time of the tours. The two colonies share a common global best solution which is used for pheromone update in both colonies. Colony 1 tries to find a solution with one vehicle less than the global best solution, while colony 2 tries to improve the global best solution with respect to criterion 2 under the restriction that the solution is not worse than the global best solution with respect to the first criterion. Whenever colony 1 finds a new global best solution both colonies start anew (with the new global best solution).

Gagné et al. [6] tested a multi-criterion approach for solving a single machine total tardiness problem with changeover costs and two additional criteria. In their approach the changeover costs were considered to be most important. Heuristic values for the decisions of the ants were used that take all criteria into account. However, the amount of pheromone that an ant adds to the pheromone matrix depends solely on the changeover costs of the solution. A similar approach was used in [7] for a four criterion industrial scheduling problem.

In [8,9] Doerner et al. proposed to solve a transportation problem were the aim was to minimize the total costs by searching for solutions that minimize two different criteria. The general approach was two use two colonies of ants where each colony concentrates on a different criterion by using different heuristics. In [8] one criterion was considered the main criterion. Every k iterations, the master population which minimizes the main criterion updates its pheromone information according to the good solutions found in the slave population, which minimizes the minor criterion. However, no information flow occurs from the slave to the master colony. In [9], both criteria were considered to be of equal importance. The size of both populations was adapted so that the colony that found the better solution with respect to costs became larger. Information exchange between the colonies is done by so called spy ants that base their decisions on the pheromone matrices in both colonies.

The only ACO approaches so far that aim to cover the pareto-front of a multiobjective optimization problem have been proposed by Doerner at al. [10, 11] and Iredi at al. [12].

Doerner at al. [10,11] studied a portfolio optimization problem with more than two criteria. For each criterion, a separate pheromone matrix is used. Instead of a population of ants for each criterion each ant assigns weights to the pheromone information for all criteria according to a random weight vector when constructing a solution. Pheromone update is done by ants that found the best or the second best solution with respect to one criterion. A problem with this approach is that solutions in the non-dominated front that are not among the best with respect to a single criterion do not update the pheromone information.

Iredi et al. [12] studied an approach to solve bi-criterion optimization problems with a multiple colony ant algorithm were the colonies specialize to find good solutions in different parts of the front of non-dominated solutions. Cooperation between the colonies is established by allowing only ants with solutions in the global front of non-dominated solutions to update the pheromone information (i.e. in contrast to [10,11], all solutions in the non-dominated front influence the future search process). Two methods for pheromone update in the colonies were proposed. In the update by origin method an ant updates only in its own colony. For the other method the sequence of solutions along the non-dominated front is split into p parts of equal size. Ants that have found solutions in the ith part update in colony i, $i \in [1,p]$. This update method is called update by region in the non-dominated front. It was shown that cooperation between the colonies allows to find good solutions along the whole Pareto front. Heterogeneous colonies were used where the ants have different preferences between the criteria when constructing a solution. For the SMTTP with changeover costs test problem, two pheromone matrices were used: $M = (\tau_{ij})$ for the total tardiness criterion, where τ_{ij} is the desirability that job j is on place i of the schedule, and $M' = (\tau'_{ij})$ for the changeover cost criterion, where τ'_{ij} is the desirability to schedule job j immediately after job i.

In this paper, a Population-based Ant Colony Optimization (PACO) approach to solve multi-criteria optimization problems is proposed where the population of solutions is chosen from the set of all non-dominated solutions found so far (see [13] for the concept of Population-based ACO). The aim is to find a set of different solutions which covers the Pareto-optimal front. One advantage of the proposed algorithm is that it can be applied to problems with more than two criteria and is not biased to solutions that are the best for one criterion.

The PACO approach for single-criteria problems is described in Section 2. In Section 3, we introduce the new methods for applying PACO to multi-criterial problems. The test instances and parameters are described in Section 4. The Results are discussed in Section 5 and conclusions are given in Section 6.

2 Monocriterial Optimization and Population-Based ACO

In this section, we describe the general principle employed by ACO to build solutions for single-criteria optimization problems and the modifications to the standard approach by PACO (see [13,14] for more details). As example problems, we use two Single-Machine Scheduling problems that are also used for later for evaluating the proposed methods. We also describe the Summation Evaluation method for pheromone evaluation as introduced by Merkle and Middendorf in [15], which is included in our algorithm.

2.1 Solution Construction

When constructing solutions to an optimization problem with ACO, (artificial) ants proceed in an iterative fashion, making a number of decisions until the global solution is completed ([16]). Ants that found a good solution mark their paths through the decision space by putting some amount of pheromone along the path. The following ants of the next generation are attracted by the pheromone so that they will search in the solution space near good solutions. For a single machine scheduling problem, an ant will choose an initial job and proceed by deciding which job to place next until all jobs have been scheduled. The decisions an ant makes are probabilistic in nature and influenced by two factors: the pheromone information, which is gained from the choices made by previous good ants, and heuristic information, which indicates the immediate benefit of making the corresponding choice. Depending on the type of problem being processed, the pheromone and heuristic information have different interpretations. Consider the Single Machine Total Tardiness Problem (SMTTP) which is defined as follows:

- Given: n jobs, where job $j \in [1, n]$ has a processing time p_j and a due date d_j.
- Find: A non-preemptive one machine schedule that minimizes the value of $\mathcal{T} = \sum_{j=1}^{n} \max\{0, C_j - d_j\}$, where C_j is the completion time of job j.

$\mathcal{T}$ is called the total tardiness of the schedule. For this problem, the pheromone information τ_{ij} and the heuristic information η_{ij} usually give information about the expected benefit of assigning job j to place i in the schedule, with $i, j \in [1, n]$. In this context we speak of the pheromone being stored in a job×place pheromone matrix. The heuristic information is defined via the modified due date rule ([15]):

$$\eta_{ij} = \frac{1}{\max\{T + p_j, d_j\} - T} \quad (1)$$

where T is the total processing time of all jobs already scheduled. The Single Machine problem with changeover costs is defined as follows:

- Given: n jobs, where for every pair of jobs i, j, $i \neq j$ there are changeover costs $c(i, j)$ that have to be paid when j is the direct successor of i in a schedule.
- Find: A non-preemptive one machine schedule that minimizes the sum of the changeover costs $C = \sum_{i=1}^{n-1} c_{\pi(i)\pi(i+1)}$, where the permutation π is the sequence of jobs in the schedule.

C is called the cost of the schedule. For this problem, the actual place in the schedule is no longer important for a given job. Rather, its predecessor determines the cost incurred. Hence, the pheromone information τ_{ij} and heuristic information η_{ij} refer to placing job j after job i in the schedule, again with $i, j \in [1, n]$. For this case we say that the pheromone is located in a job×job pheromone matrix. The heuristic information is defined by

$$\eta_{ij} = \frac{1}{c_{ij} + 1} \quad (2)$$

for non-negative changeover costs. Note that this problem is closely related to the Travelling Salesman Problem. Also, in contrast to the dynamic heuristic information for tardiness, η_{ij} is constant. Although only the predecessor of a job $j \in [1, n]$ is important for determining the resulting cost, it can still make sense to gain information about which job is placed first in the schedule, since this job has no predecessor and thereby no incurred changeover cost. To realize this, a dummy-job 0 is included, with $\forall j \in [1, n] : c_{0j} = 0$. This job is always scheduled first and given a row in the pheromone matrix, so that τ_{0j} will contain the information how beneficial it has previously been to schedule job $j \in [1, n]$ as the first "real" job.

For any given place, the set of jobs that can still be assigned is denoted by S. With probability q_0, where $0 \leq q_0 < 1$ is a parameter of the algorithm, the ant chooses the job $j \in S$ which maximizes $\tau_{ij}^\alpha \cdot \eta_{ij}^\beta$, where $\alpha > 0$ and

$\beta > 0$ are constants that determine the relative influence of the heuristic and the pheromone values on the decision of the ant. With the probability of $1 - q_0$, an ant chooses according to the selection probability distribution over S defined by ([16]):

$$\forall j \in S: \ p_{ij} = \frac{\tau_{ij}^\alpha \cdot \eta_{ij}^\beta}{\sum_{h \in S} \tau_{ih}^\alpha \cdot \eta_{ih}^\beta} \tag{3}$$

2.2 Summation Evaluation

Merkle and Middendorf [15] have proposed an alternative method for evaluating the pheromone information stored in the matrix $[\tau_{ij}]$ when dealing with tardiness minimization scheduling problems. Instead of using only the pheromone value τ_{ij}, the sum over all pheromone values up to and including i, that is $\tau'_{ij} = \sum_{l=1}^{i} \tau_{lj}$ is used. A study of combining a weighted version of this summation evaluation and regular evaluation was performed for the problem of Single Machine Total Weighted Tardiness in [17] and shown to be superior to regular evaluation. In this combination, instead of τ_{ij}, the value

$$\tau_{ij}^* = c \cdot x_i \cdot \tau_{ij} + (1 - c) \cdot y_i \cdot \sum_{l=1}^{i} \gamma^{i-l} \tau_{lj} \tag{4}$$

is used in Formula 3. The parameters of τ_{ij}^* are c, which determines the relative influence of weighted summation evaluation, γ, which indicates the weight of previous pheromone values, and x_i and y_i, which are used for scaling, which is necessary since the value provided by the weighted summation evaluation can be a lot larger than the standard pheromone value. Specifically, the scaling values are $x_i = \sum_{h \in S} \sum_{l=1}^{i} \gamma^{i-l} \tau_{lh}$ and $y_i = \sum_{h \in S} \tau_{ih}$.

2.3 Pheromone Update

After m ants have constructed solutions, the pheromone information is updated. This is the point where PACO differs from the standard ACO heuristic. The standard ACO employs evaporation to reduce all pheromone values by a relative amount ρ, $\tau_{ij} \mapsto (1 - \rho)\tau_{ij}$ and afterwards performs a positive update with the ant(s) that found the best solutions. For each of these solutions all pheromone values τ_{ij} corresponding to the choices ij of the solution an update is done according to:

$$\tau_{ij} \mapsto \tau_{ij} + \Delta \tag{5}$$

PACO employs a population $P = \{\pi_1, \ldots, \pi_k\}$ of k good solutions, from which the pheromone information τ_{ij} is derived as follows. Each element of the pheromone matrix has an initial value τ_{init}. Whenever a solution enters the population P, a positive update is performed as in Formula 5. If a solution is removed from the population, its influence is explicitly removed from the pheromone matrix by performing a *negative* update, i.e. using $-\Delta$ in Formula 5. As a result, if $\pi(i) = j$ signifies that job j was positioned at place i, then a job×place interpretation of the population P would yield the pheromone matrix $[\tau_{ij}]$ with

$$\tau_{ij} = \tau_{init} + \Delta \cdot |\{\pi \in P | \pi(i) = j\}|. \tag{6}$$

We denote the maximum possible value an element of the pheromone matrix can achieve by Equation 6 as $\tau_{max} := \tau_{init} + k \cdot \Delta$. Reciprocally, if τ_{max} is used as a parameter of the algorithm instead of Δ, we can derive $\Delta := (\tau_{max} - \tau_{init})/k$ so that with Equation 6, τ_{max} is indeed the maximum attainable value for any τ_{ij}. Note that the actual value for τ_{init} is arbitrary, as τ_{max} could simply be scaled in accordance to achieve an identical probability distribution in Equation 3. For reasons of clarity, we wish the row/column-sum of initial pheromone values to be 1, which means that $\tau_{init} = 1/(n-1)$ for the job×job pheromone matrix where the diagonal is 0, and $\tau_{init} = 1/n$ for the job×place matrix.

3 Multi-criteria Optimization

In this section we introduce a PACO approach for finding solutions in multi-criteria environments. We also propose a new method for ants to make decisions based on pheromone and heuristic information originating from different criteria.

3.1 Population of Solutions

Some methods for updating the population of PACO for single-criteria optimization problems have been studied by Guntsch and Middendorf in [13,14]. In this subsection, we describe a novel way to employ the population for multi-criteria optimization problems.

Let Q denote the set of non-dominated solutions that have been found so far. This set will act as the super-population for PACO, from which the population $P \subseteq Q$ is derived to construct the pheromone matrices for the ants to work with. First, the algorithm chooses one starting solution π from Q at random. Then, the $k-1$ solutions in Q which are closest to π with respect to some distance measure are determined (if $|Q| \geq k-1$). Here distance is defined simply by the sum of absolute differences in solution quality over all criteria. Together, these k solutions form the population $P = \{\pi_1, \pi_2, \ldots, \pi_k\}$, with $\pi_1 = \pi$, from which the two pheromone matrices are determined according to Formula 6. After a solution has been constructed by an ant the set Q is updated. After m ants have constructed a solution a new population P is chosen.

3.2 Average-Rank-Weight Method

For multi-criteria problems the ants make their decisions based on pheromone and heuristic information originating from different criteria. The method proposed here differs from the one employed by Merkle and Middendorf [15], where each ant is assigned a weight $\lambda \in [0,1]$ which defines the relative influence of the two criteria on the decisions of an ant. Instead, we calculate a probability distribution p_{ij}^{ζ} for each criterion ζ (according to Formula 3 and using summation evaluation as described in subsection 2.2 when ξ is a tardiness criterion) and from these construct the final selection probability distribution

$$p_{ij}^{\Sigma} = \sum_{\zeta} w_{\zeta} \cdot p_{ij}^{\zeta}, \qquad (7)$$

with each individual weight w_{ζ} determining the influence of criterion ζ on the decision process, and $\sum_{\zeta} w_{\zeta} = 1$. This method has the advantage of remaining feasible for an arbitrary number of criteria and not requiring any corrective scaling of pheromone or heuristic values.

Population P is used to determine the weights $w_{\zeta} = w_{\zeta}(P)$ for each criterion ζ needed for Formula 7. The general idea is to give a criterion a higher weight the better the solutions in P are with respect to this criterion compared to all solutions in Q. Formally, to compute these weights we assign each solution $\pi \in P$ a reverse rank $r_{\zeta}(\pi) \in [0, |Q|-1]$ for each criterion ζ. By reverse rank we mean that $r_{\zeta}(\pi) = 0$ is worst and $r_{\zeta}(\pi) = |Q|-1$ is best. Let $q_{\zeta}(\pi)$ denote the quality of solution π with respect to criterion ζ, where, since we are minimizing, lesser values of $q_{\zeta}(\pi)$ indicate a better solution. Then

$$r_{\zeta}(\pi) = |Q| - |\{\sigma \in Q | q_{\zeta}(\sigma) < q_{\zeta}(\pi)\}| - 1 \qquad (8)$$

and using this reverse rank, we define the solution weights via

$$w_{\zeta}(\pi) = \frac{r_{\zeta}(\pi)}{\sum_{\xi} r_{\xi}(\pi)} \qquad (9)$$

Finally, from the individual solution weights, we calculate the combined weight for the population $P = \{\pi_1, \ldots, \pi_k\}$ by aggregating the weights of all solutions in P with respect to the criterion ξ

$$w_{\zeta}(P) = \frac{1}{|P|} \sum_{i=1}^{k} w_{\zeta}(\pi_i) \qquad (10)$$

4 Test Setup

We now describe the problem, instances, and parameter settings used to evaluate the methods proposed in Section 2. As mentioned previously, we let the algorithm run on a Single Machine Total Tardiness problem with changeover costs. This problem is a combination of the two scheduling problems defined in Subsection 2.1, and thereby a bi-criterial optimization problem, with one matrix $[c_{ij}]$ for changeover costs when switching from job i to job j, and for each job i a processing time and a due date $[p_i, d_i]$. However, it is possible to scale this problem to more criteria by utilizing several matrices for changeover costs, each representing one criterion, as well as having more than one processing time and due date for each job, which again leads to multiple criteria. Hence, for n jobs the quality $q_\zeta(\pi)$ of a solution π with respect to criterion ζ is defined as

$$q_\zeta(\pi) = \begin{cases} \sum_{i=1}^{n-1} c^\zeta_{\pi(i)\pi(i+1)} & \text{if } \zeta \text{ is a changeover criterion,} \\ \sum_{i=1}^{n} \max(C^\zeta_{\pi(i)} - d^\zeta_{\pi(i)}, 0) & \text{if } \zeta \text{ is a tardiness criterion} \end{cases} \quad (11)$$

where $C^\zeta_{\pi(i)}$ ($c^\zeta_{\pi(i)\pi(i+1)}, d^\zeta_{\pi(i)}$) is the completion time (respectively, changeover cost, deadline) of job i according to the processing times of criterion ζ, that is $C^\zeta_{\pi(i)} = \sum_{j=1}^{i} p^\zeta_{\pi(j)}$. In both cases a lower value of $q_\zeta(\pi)$ is better.

We used two bi-criterial test instances, with one changeover and one due date criterion, from Iredi et al. in [12]. From these two instances a four-criterial instance with two changeover and two due date criteria was constructed. For the one bi-criterial instance, called instance A from here on, the changeover costs between the jobs were chosen randomly from the interval $[1, 100]$, while for the other one, dubbed instance B, interval $[50, 100]$ was used. The processing times and due dates were chosen according to an often employed scheme from [18]: for each job $j \in [1, 100]$, an integer processing time p_j is drawn randomly from the interval $[1, 100]$, and after all jobs have been assigned a processing time, the due dates for each job j are drawn randomly from the interval $\left[\sum_{j=1}^{100} p_j \cdot (1 - TF - \frac{RDD}{2}), \sum_{j=1}^{100} p_j \cdot (1 - TF + \frac{RDD}{2}) \right]$. RDD is the relative Range of Due Dates and determines the size of the interval from which the due dates are drawn. TF is the Tardiness Factor and indicates where the center of the above interval is located. For both instances, $RDD = 0.6$ was used; in instance A, we set $TF = 0.4$ and in instance B, $TF = 0.6$. The four-criterial instance that is a combination of instance A and B and is called instance AB.

For the algorithm, we used several parameter configurations. All combinations of population sizes $k \in \{1, 3, 5\}$, $q_0 \in \{0.0, 0.5, 0.9\}$ (see Subsection 2.1), and $\tau_{max} \in \{1, 5, 25, 125, 500, 2500\}$ (see Subsection 2.3) were tested for each of the two bi-criterial instances. The ants used $\alpha = 1, \beta = 5$ for the changeover based probability distributions and $\alpha = 1, \beta = 1$ for the tardiness based ones respectively. The reason for choosing different values of β is that these values are often used in the literature for the corresponding single-criterion versions of the two problems (i.e., TSP and SMTTP). Unless otherwise stated, only one ant constructed a solution in each iteration before a new population was constructed, i.e. the number of ants m per generation is one. For some configurations, using more than one ant was also tested. The four-criterial problem composed of the two bi-criterial ones was only studied for $q_0 = 0.9$, $\tau_{max} = 1$ and $k \in \{1, 3\}$ since these values performed well for the bi-criterial problems. Each run of the algorithm was stopped after 50000 solutions have been constructed.

In the following section we present and compare the median attainment surfaces of the fronts of non-dominated solutions found for 15 runs of the ant algorithm with different randoms seeds (the median attainment surface is the median line of all the attainment surfaces connecting the fronts of non-dominated solutions in every of the 15 runs).

5 Results

We start the evaluation of the performance of the ant algorithm with the results for problem instance A. Figure 1 shows the influence of different values of q_0 on the behaviour of the algorithm for $k = 1$ and $\tau_{max} = 25$. It can be seen that a higher value of q_0 leads to a set of solutions which completely dominates those attained by a smaller q_0 (This behaviour can also be observed for the other parameter settings of τ_{max} and k). This is a good indication of the ants being able to find new good solutions by making only minor adjustments to a solution already located in Q.

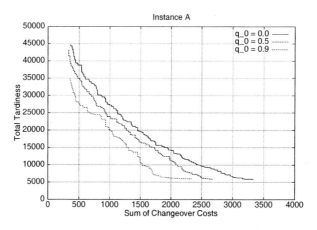

Fig. 1. Median attainment surface for instance A for $q_0 \in \{0.0, 0.5, 0.9\}$ after 50000 ants have built a solution. The other parameters are $\tau_{max} = 25$ and $k = 1$

The influence of population size k is shown in Figure 2. For instance A, the small population size 1, where exactly one pheromone value in each row equals τ_{max} and all other values are τ_{init}, performed best. Especially when combined with a good heuristic value and a high value of $q_0 = 0.9$, population size 1 keeps the ants very close to the solution from which the pheromone matrices were derived. Note, that the population sizes $k \in \{3,5\}$ also performed worse than $k = 1$ for $q_0 \in \{0.0, 0.5\}$.

Finally for instance A, we look at the impact of changing the maximum pheromone value τ_{max}, as shown for two different cases in Figures 3 and 4. In Figure 3, where we have a population size of $k = 1$, increasing the maximum pheromone value seems to have a beneficial effect on optimizing the tardiness criterion at the expense of some solution quality in the changeover criterion. A significant effect on the front of non-dominated solutions by different values of τ_{max} is only evident for $k = 1$ however (compare the results for a larger population size $k = 5$ in Figure 4).

It seems that the tardiness criterion and the changeover criterion require different values of τ_{max}. We therefore explored a combination of setting $\tau_{max} = 5$

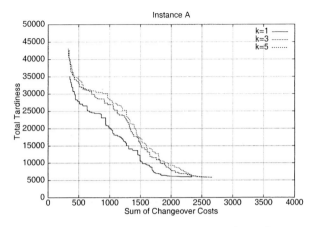

Fig. 2. Median attainment surface for instance A for $k \in \{1, 3, 5\}$ after 50000 ants have built a solution. The other parameters are $\tau_{max} = 25$ and $q_0 = 0.9$

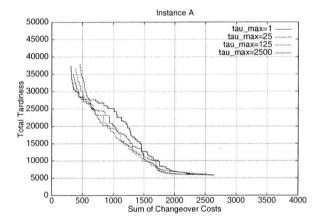

Fig. 3. Median attainment surface for instance A for $\tau_{max} \in \{1, 25, 125, 2500\}$ after 50000 ants have built a solution. The other parameters are $k = 1$ and $q_0 = 0.9$

for the job×job matrix and $\tau_{max} = 25$ for the job×place matrix, the result of which is shown in Figure 5. The median attainment surface for this case lies between those with the same value of τ_{max} for both criteria.

We now show the results for instance B. The effect of different values of q_0 on the median attainment surface is shown in Figure 6 for $\tau_{max} = 1$ and $k = 3$. Similar as for instance A, a higher value of q_0 outperforms a lower one (this holds also for other values of τ_{max} and k).

Considering the influence of the population size k, the results for instance B differ from those for A (see Figure 7). Whereas for instance A a population size of $k = 1$ performed best, here it performs worst everywhere except in the region of the median attainment surface with the low tardiness values. This

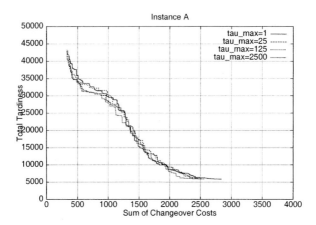

Fig. 4. Median attainment surface for instance A for $\tau_{max} \in \{1, 25, 125, 2500\}$ after 50000 ants have built a solution. The other parameters are $k = 5$ and $q_0 = 0.9$

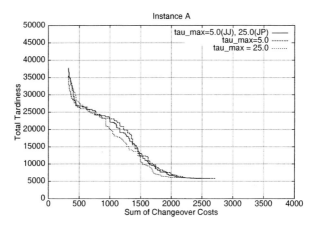

Fig. 5. Median attainment surface for instance A for $\tau_{max} \in \{5, 25\}$ and the combination $\tau_{max}^{JJ} = 5$ and $\tau_{max}^{JP} = 25$) for the job×job and job×place matrix respectively. Results are shown after 50000 ants have built a solution, with $k = 1$ and $q_0 = 0.9$

behaviour suggests that in comparison to instance A, a more diverse supply of pheromone is necessary to enhance different options for finding schedules with small changeover costs. The reason might be that the changeover costs for instance B are more similar and therefore possibly a relatively large set of different good solutions exist.

The influence of the maximum pheromone value τ_{max} on solution quality for instance B is shown in Figure 8 for $q_0 = 0$ (left) and $q_0 = 0.9$ (right). For $q_0 = 0.0$, a higher maximum pheromone value leads to a better median attainment surface, with $\tau_{max} = 1$ performing comparatively poor (this was observed also for $q_0 = 0.5$). This changes when setting $q_0 = 0.9$, where $\tau_{max} = 1$

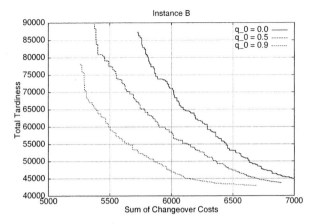

Fig. 6. Median attainment surface for instance B for $q_0 \in \{0.0, 0.5, 0.9\}$ after 50000 ants have built a solution. The other parameters are $\tau_{max} = 1$ and $k = 3$

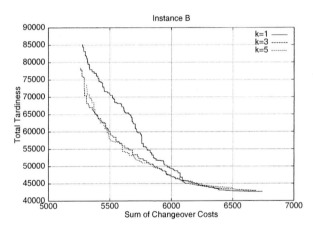

Fig. 7. Median attainment surface for instance B for $k \in \{1, 3, 5\}$ after 50000 ants have built a solution. The other parameters are $\tau_{max} = 1$ and $q_0 = 0.9$

outperforms the higher maximum pheromone values significantly. This differs from the results for instance A. A reason could be that the possibly different good solutions with respect to changeover costs for instance B might be difficult to find for the ants when using a combination of high τ_{max} and high q_0.

The progression of PACO over time for instance B is shown in Figure 9. It can be seen that in relation to the tardiness criterion, the relative improvement of the changeover costs criterion is by far larger, continuing to explore this outside edge of the set of non-dominated solutions for the entire runtime.

We now move our attention to the four-criterial instance AB. In order to evaluate the performance of the algorithm on this instance, we projected the resulting four-criterial median attainment surface to a 2-dimensional one for

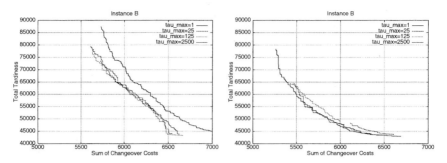

Fig. 8. Median attainment surface for instance B for $\tau_{max} \in \{1, 25, 125, 2500\}$ after 50000 ants have built a solution. The other parameters are $k = 3$ and $q_0 = 0.0$ (*left*) and $q_0 = 0.9$ (*right*)

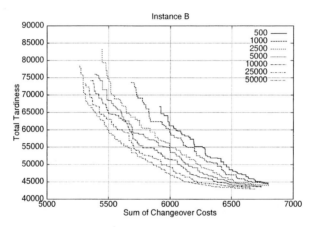

Fig. 9. Median attainment surface for instance B after an indicated number of ants have built solutions. The parameters are $\tau_{max} = 1$, $k = 3$ and $q_0 = 0.9$

each of the original instances A and B respectively. Thus we can compare the 2-dimensional fronts of the algorithm that ran on instance AB with the algorithm that, with the same parameter settings, was used to solve only A or B exclusively. The results are shown in Figure 10.

As can be seen, the 2-dimensional projection of the 4-dimensional front is worse for both original instances A and B. Note that this comparison is, of course, not completely fair for the algorithm working on instance AB, in the sense that it has to manage a much larger set Q then the algorithm working only on A or B; the size of Q was between 31 and 58 solutions for the bi-criterial instances and ranged from 2739 to 5102 for instance AB, after 50000 iterations. Therefore, it can be argued that each solution of set Q could not be exploited as exhaustively as in the case of a 2-dimensional instance. Despite performing worse than on the bi-criterial instances, the performance of PACO on the four-criterial

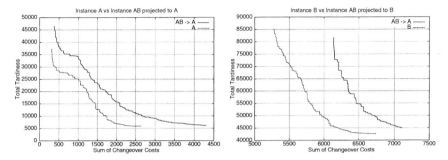

Fig. 10. Median attainment surface for instance A (*left*) and B (*right*) and instance AB projected to the corresponding two dimensions after 50000 ants have built a solution. The parameters are $\tau_{max} = 1$, $k = 1$ and $q_0 = 0.9$

instance is not actually bad, and signifies to us that the general approach is indeed feasible for more than bi-criterial instances.

6 Conclusion

We have successfully modified the PACO algorithm to deal with multi-criteria optimization problems, including the introduction of the Average-Weight-Rank method for constructing the selection probability distribution for the ants and the new derivation of the active population to determine the pheromone matrices. Specifically, the algorithm was tested when used in conjunction with the Single Machine Total Tardiness with changeover costs problem, and the influence of the different parameter settings on the behaviour of the algorithm was investigated.

For future work, we will focus more on many-dimensional problems and the specific handling they require when managing the candidate set for the population, like not biasing the algorithm towards dense parts of the non-dominated set of solutions and limiting the size of the set of solutions from which the population is derived. Also, different possibilities for combining solutions in the pareto-front exist and will be researched further.

References

[1] M. Dorigo and G. Di Caro. The ant colony optimization meta-heuristic. In D. Corne, M. Dorigo, and F. Glover, editors, *New Ideas in Optimization*, pages 11–32, London, 1999. McGraw-Hill.
[2] Kalyanmoy Deb. *Multi-Objective Optimization using Evolutionary Algorithms*. John Wiley & Sons, Chichester, 2001.
[3] Carlos A. Coello Coello, David A. Van Veldhuizen, and Gary B. Lamont, editors. *Evolutionary Algorithms for Solving Multi-Objective Problems*. Kluwer Academic Publishers, New York, 2002.

[4] Carlos E. Mariano and Eduardo Morales. MOAQ an ant-Q algorithm for multiple objective optimization problems. In W. Banzhaf and et al., editors, *Proceedings of the Genetic and Evolutionary Computation Conference*, volume 1, pages 894–901, Orlando, Florida, USA, 13-17 July 1999. Morgan Kaufmann.
[5] Luca Maria Gambardella, Éric Taillard, and Giovanni Agazzi. MACS-VRPTW: A multiple ant colony system for vehicle routing problems with time windows. In David Corne, Marco Dorigo, and Fred Glover, editors, *New Ideas in Optimization*, pages 63–76. McGraw-Hill, London, 1999.
[6] C. Gagné, M. Gravel, and W.L. Price. Scheduling a single machine where setup times are sequence dependent using an ant-colony heuristic. In *Abstract Proceedings of ANTS'2000, 7.-9. September Brussels, Belgium*, pages 157–160, 2000.
[7] M. Gravel, W.L. Price, and C. Gagné. Scheduling continous casting of aluminium using a multiple objective ant colony optimization heuristic. *European Journal of Operational Research*, 143:218–229, 2002.
[8] K. Doerner, R.F. Hartl, and M. Reimann. Cooperative ant colonies for optimizing resource allocation in transportation. In W. Boers and et al., editors, *Proceedings of the EvoWorkshops 2001*, pages 70 – 79, Berlin, Heidelberg, 2001. Springer.
[9] K. Doerner, R.F. Hartl, and M. Reimann. Are COMPETants more competent for problem solving? - the case of a multiple objective transportation problem. In L. Spector et al., editor, *Proceedings of the GECCO'01*, page 802, Berlin, Heidelberg, 2001. Morgan Kaufmann.
[10] K. Doerner, J.W. Gutjahr, R.F. Hartl, C. Strauss, and C. Stummer. Investitionsentscheidungen bei mehrfachen Zielsetzungen und künstliche Ameisen. In Chamoni and et al., editors, *Operations Research Proceedings*, pages 355–362, Berlin, Heidelberg. Springer.
[11] K. Doerner, J.W. Gutjahr, R.F. Hartl, C. Strauss, and C. Stummer. Pareto ant colony optimization: A metaheuristic approach to multiobjective portfolio selection. *Annals of Operations Research*. to appear.
[12] S. Iredi, D. Merkle, and M. Middendorf. Bi-criterion optimization with multi colony ant algorithms. In E. Zitzler and et al., editors, *Evolutionary Multi-Criterion Optimization, First International Conference (EMO'01)*, number 1993 in LNCS, pages 359–372, Berlin, Heidelberg. Springer.
[13] M. Guntsch and M. Middendorf. A population based approach for ACO. In S. Cagnoni and et al., editors, *Applications of Evolutionary Computing - EvoWorkshops 2002: EvoCOP, EvoIASP, EvoSTIM/EvoPLAN*, number 2279 in LNCS, pages 72–81. Springer, 2002.
[14] M. Guntsch and M. Middendorf. Applying population based aco to dynamic optimization problems. In *Ant Algorithms, Proceedings of Third International Workshop ANTS 2002*, number 2463 in LNCS, pages 111–122. Springer, 2002.
[15] D. Merkle and M. Middendorf. An ant algorithm with a new pheromone evaluation rule for total tardiness problems. In *Proceeding of the EvoWorkshops 2000*, number 1803 in LNCS, pages 287–296. Springer Verlag, 2000.
[16] M. Dorigo, V. Maniezzo, and A. Colorni. The ant system: Optimization by a colony of cooperating agents. *IEEE Trans. Systems, Man, and Cybernetics – Part B*, 26:29–41, 1996.
[17] D. Merkle, M. Middendorf, and H. Schmeck. Pheromone evaluation in ant colony optimization. In *Proceedings of the Third Asia-Pacific Conference on Simulated Evolution and Learning (SEAL2000)*, pages 2726–2731. IEEE Press, 2000.
[18] H.A.J. Crauwels, C.N. Potts, and L.N. Van Wassenhove. Local search heuristics for the single machine total weighted tardiness scheduling problem. *Informs Journal on Computing*, 10:341–350, 1998.

A Two-Phase Local Search for the Biobjective Traveling Salesman Problem

Luis Paquete and Thomas Stützle

Darmstadt University of Technology, Computer Science Department,
Intellectics Group, Alexanderstr. 10, D-64283 Darmstadt, Germany
{lpaquete,tom}@intellektik.informatik.tu-darmstadt.de

Abstract. This article proposes the Two-Phase Local Search for finding a good approximate set of non-dominated solutions. The two phases of this procedure are to (i) generate an initial solution by optimizing only one single objective, and then (ii) to start from this solution a search for non-dominated solutions exploiting a sequence of different formulations of the problem based on aggregations of the objectives. This second phase is a single chain, using the local optimum obtained in the previous formulation as a starting solution to solve the next formulation. Based on this basic idea, we propose some further improvements and report computational results on several instances of the biobjective TSP that show competitive results with state-of-the-art algorithms for this problem.

1 Introduction

Most successful, stochastic approaches to multiobjective combinatorial optimization problems apply a local search procedure embedded in some higher level search strategy to derive a good approximate set of efficient solutions. In order to guarantee that a dispersed set of efficient solutions is obtained, several techniques have been proposed; these can be classified into one of two main groups: Methods that aggregate the objective functions and dynamically modify the search direction during the search process or within several runs [3,6,7,9,10,19,23], and methods that are based on a Pareto-like dominance relation as an acceptance criterion in the search to distinguish between candidate solutions [24, 13]; this second type of methods avoid the aggregation of objectives. Recently, empirical evidence was gathered suggesting that methods belonging to the first group perform particularly well [10], the main reason being that local search algorithms can deal more easily with aggregated objective functions. However, most of the successful algorithms are rather complex search strategies.

In this article, the Two-Phase Local Search procedure (TPLS) is proposed to tackle biobjective combinatorial optimization problems. The first phase consists of finding a good solution to one single objective, using an effective single objective algorithm. This phase provides the starting solution for the second phase, in which a local search algorithm is applied to a sequence of different aggregations of the objectives, where each aggregation converts the biobjective problem into a single objective one. Important for the second phase is that successive aggregations *and* local searches are treated as a chain: an aggregation a_{i+1} modifies slightly the emphasis given to the different objectives when compared to aggregation a_i; the local search for aggregation a_{i+1} is started from the local optimum solution s_i^* that was returned for aggregation a_i. The main motivation for such an approach is to exploit the effectiveness of local search algorithms for single objective problems.

Algorithm 1 Two-Phase Local Search

Require: F: vector objective function
$\quad\quad\quad\;\; W$: aggregations of F
/*------------- phase one ----------------- */
$s_0^* \leftarrow$ GenerateInitialSolution()
$A \leftarrow$ UpdateArchive(s)
/*------------- phase two ----------------- */
for $i \leftarrow 1$ **to** $|W|$ **do**
$\quad f_i \leftarrow$ ModifyObjFunc($W_i, F, history$)
$\quad s_i^* \leftarrow$ LocalSearch(s_{i-1}^*, f_i)
$\quad A \leftarrow$ UpdateArchive(s_i^*)
end for
$A^* \leftarrow$ FilterArchive(A)

To gain insight into the behaviour and performance of TPLS, we applied it to the biobjective Traveling Salesman Problem (TSP). The experimental with a very basic version of TPLS show a very promising performance and with a rather straightforward improvement it was even able to achieve state-of-the-art performance for the biobjective TSP. This is a very promising result, because the procedure is rather simple and appears to be quite robust.

The article is organized as follows. Section 2 introduces the TPLS procedure. Section 3 describes the multiobjective TSP and in Section 4 we present how we adapted TPLS to the biobjective TSP. Section 5 introduces some further improvements over the basic TPLS procedure and analyses their performance. Additional analyses on this procedure are described in Section 6. We conclude and indicate directions for future work in Section 7.

2 The Two-Phase Local Search Procedure

The underlying ideas of the TPLS procedure are (i) to exploit the very good performance of local search algorithms for single-objective problems, (ii) to solve a biobjective problem by chains of related aggregations into single-objective ones *and* chains of good solutions for aggregations which provide starting solutions for a search regarding a next aggregation and (iii) to have an easily understandable but at the same time robust, flexible, and efficient procedure.

Aggregating multiple objective functions into several single-objective problems by utility functions has shown to be an effective approach for multiobjective combinatorial optimization problems as reported in [6,10,23]. Several such aggregations were proposed including weighted Tchebycheff utility functions as well as weighted linear utility functions. For the following, we assume that the aggregation is based on a weighted linear utility function with normalized weight vectors. In this case we have that the utility of a solution is rated by

$$f(s) = \sum_{n=1}^{N} w_n f_n(s) \tag{1}$$

where N is the number of objectives, w_n is the weight assigned to the n-th objective, and s is a feasible solution. Additionally, we have $\sum_{n=1}^{N} w_n = 1$ due to the normalization.

From a high level perspective, the TPLS procedure can be described as follows (see Algorithm 1):

First phase: Generate an initial solution s considering only one of the objectives (procedure GenerateInitialSolution);

Second phase: Apply a local search algorithm (LocalSearch), using an aggregation of the objective functions (the aggregated objective function is represented as f_i) generated by ModifyObjFunc, until a solution s_i^* is found, which is added to the archive. The argument s_{i-1}^* of LocalSearch indicates that s_{i-1}^*, the local optimum found in iteration $i-1$ is used as a starting solution for the local search in iteration i. The process is repeated until all aggregations of the objective functions are explored.

The GenerateInitialSolution plays an important role in the procedure, since it generates a first approximation to the Pareto global optima set, and it is the starting solution for the second phase. LocalSearch is the local search algorithm embedded in the procedure for solving the single-objective formulations. In fact, LocalSearch and GenerateInitialSolution can be the same procedure, since an efficient algorithm for the first phase could also be well suited for tackling the second phase; in that case, GenerateInitialSolution could be a local search starting from a random initial solution. In the second phase, the starting solution of LocalSearch is always taken to be the solution returned in the previous iteration of the algorithm.

Let us explicitely remark that there is no need to restrict ourselves to true "local search" procedures. In fact, some single-objective versions of multiobjective problems are solvable in polynomial time by exact methods; this is the case, for example, for the multiobjective linear assignment problem and the multiobjective shortest path problem [18]. When faced with such a problem, it obviously may be preferable to use the polynomial exact algorithm instead of a true local search.

The modification of the aggregated objective function ModifyObjFunc can be carried out by changing the weights assigned to each objective.[1] The change could be random or gradual and the decision can be taken from a previous search space analysis of the problem: if good solutions are clustered in the search space, gradual changes should be preferable, since it has the advantage that the local optimum for aggregation a_{i+1} is close to the one for aggregation a_i and that the local search for a_{i+1} needs only few improvement steps to identify a very good solution to a_{i+1}. If good solutions are spread all over the search space, a randomized change of weights may be more useful.

---

[1] In case of different ranges of the objectives, a normalization of the objective function values, for example, by range equalization factors [21] is needed.

Each time a local optimum is found by LocalSearch, it is stored in the archive A of the set approximating the Pareto global optima set. Since the solution to the multiobjective problem is a set of all non-dominated objective vectors in A, the FilterArchive procedure is applied in a post-processing step; it deletes all dominated solutions and returns A^*. This last procedure can be excluded if it is applied every time the UpdateArchive procedure is called.

One obvious advantage of TPLS is its modularity, which enables us to focus on the solution methods embedded in GenerateInitialSolution and LocalSearch. Once the choice for these operators is taken, the only numeric parameter needed is the number of different aggregations of the objective functions. In a weighted sum approach, a high number of weight combinations should return a better approximation to the Pareto global optima set. However, some care must be taken, since increasing the number of weight combinations may not be enough to escape from local optima, resulting in a waste of computation time. A study of the trade-off between computation time and solution quality should be carried out according to the real application and together with the decision maker.

3 The Multiobjective TSP

Given a complete, weighted graph $G = (N, E, c)$ with N being the set of nodes, E being the set of edges fully connecting the nodes, and c being a function that assigns to each edge $(i, j) \in E$ a vector $(c_{ij}^1, \ldots, c_{ij}^K)$, where each element c_{ij}^k corresponds to a certain measure like distance, cost, etc. between nodes i and j. For the following we assume that $c_{ij}^k = c_{ji}^k$ for all pairs of nodes i, j and objectives k, that is, we consider only symmetric problems. The multiobjective TSP is the problem of finding "minimal" Hamiltonian circuits of the graph, that is, a set of closed tours visiting each of the $n = |N|$ nodes of G exactly once; here "minimal" refers to the notion of Pareto optimality.

Usually there is not only one, but many Pareto global optimum solutions, which form the *Pareto global optima set*. This set contains all solutions that are not dominated by any other solution. The problem of finding the Pareto global optima set is $\mathcal{NP}$-hard [5] and, since for many problems determining exact solutions quickly becomes infeasible with increasing instance size, the goal typically shifts from identifying Pareto global optima solutions to obtaining a good approximation to this set. For this latter task, algorithms based on local search seem to be a suitable approach and already have shown to yield good performance [7,10].

In this article, we apply TPLS to the biobjective case, i.e., $K = 2$. As benchmark instances we use combinations of single-objective TSP instances that are available at TSPLIB via http://www.iwr.uni-heidelberg.de/groups/comopt/software/TSPLIB95/ with 100 cities (kroA100 and kroB100), 150 cities (kroA150 and kroB150) and 200 cities (kroA200 and kroB200) as defined in [7]. For convenience, were refer to them as instances kroAB100, kroAB150 and kroAB200, respectively. The first instance was also attacked in [2,7,10] and at least for the approach by Jaszkiewicz the solutions are publically available.

4 The Two-Phase Local Search for the Biobjective TSP

This section describes the details of how TPLS was adapted to the biobjective TSP.

4.1 First Phase

As argued before, it is likely to be best to generate the initial solution using a high performing algorithm for the single objective TSP. Research on the single-objective TSP has shown that Iterated Local Search (ILS) algorithms [14] are currently among the best available metaheuristics [11]. ILS is based on the observation that local search algorithms are easily trapped in a local optimum. Instead of restarting the local search from a new, randomly generated solution, it perturbs the current local optimum, moving it to a point beyond the neighborhood searched by the local search algorithm and, thus, allowing it to escape from local optima. An acceptance criterion decides from which local optimum solution the next perturbation is applied. We used an ILS algorithm that was extensively analysed in [22]; at a high level, it can be described as follows:

- It uses a *first improvement* local search based on a *3-exchange* neighborhood, where all tours are neighbors of some tour t that differ in at most three edges;
- The perturbation is a *double bridge* move [15] that cuts the current tour at four appropriately chosen edges into four sub-tours and reconnects these in a different order to yield a new starting tour for the local search;
- The acceptance criterion accepts only a tour if it is better than the best one found so far.

These steps are repeated for a given number of iterations. The local search applies two additional speed-up techniques, which are a fixed radius nearest neighbor within candidate lists of 40 nearest neighbors for each city [12] and *don't look bits* [1]. Some preliminary experiments showed that 50, 100 and 150 iterations for the number of perturbation steps was enough to obtain global optima solutions in short computation time (lower than 0.1 seconds) for kroAB100,kroAB150 and kroAB200, respectively. We refer to this procedure as 3.first.ils.

4.2 Second Phase

For the second step, various choices were tested for the LocalSearch. The reason was to study the trade-off between computation time and solution quality for using a powerful but time expensive local search or a less efficient but faster one. Specifically, for the biobjective TSP, the following algorithms were tested:

- 2.best, a *best improvement* local search in the *2-exchange* neighborhood (it tries to find the best solution in the neighborhood obtained by exchanging two edges in the current tour);
- 3.best, similar to 2.best but using the *3-exchange* neighborhood;
- 2.first, *first improvement* local search with *2-exchange* neighborhood;
- 3.first, similar to 2.first but using a *3-exchange* neighborhood ;
- 2.first.ils, similar to the 3.ils.first of phase one but using 2.first instead of the 3.first local search algorithm;

Table 1. Average and standard deviation of the R measure and CPU time on instances kroAB100, kroAB150, and kroAB200 for TPLS

kroAB100		2.first	2.best	3.first	3.best	2.first.ils	2.best.ils	3.first.ils	3.best.ils
R Measure	Avg.	0.5186	0.9288	0.9316	0.9319	0.9290	0.9288	**0.9339**	0.9318
	Std.	0.0029	0.0009	0.0004	0.0002	0.0007	0.0006	0.0001	0.0002
CPU	Avg.	0.41	0.41	0.53	0.72	0.69	0.72	6.95	6.01
time (sec)	Std.	0.02	0.02	0.01	0.03	0.01	0.01	0.10	0.09

kroAB150		2.first	2.best	3.first	3.best	2.first.ils	2.best.ils	3.first.ils	3.best.ils
R Measure	Avg.	0.5511	0.9364	0.9394	0.9395	0.9365	0.9365	**0.9412**	0.9395
	Std.	0.0029	0.0005	0.0003	0.0002	0.0004	0.0005	0.0001	0.0002
CPU	Avg.	1.47	1.44	1.75	2.19	2.58	2.86	29.91	26.71
time (sec)	Std.	0.09	0.14	0.03	0.19	0.03	0.02	0.32	0.31

kroAB200		2.first	2.best	3.first	3.best	2.first.ils	2.best.ils	3.first.ils	3.best.ils
R Measure	Avg.	0.5710	0.9404	0.9426	0.9428	0.9403	0.9403	**0.9444**	0.9428
	Std.	0.0020	0.0004	0.0002	0.0001	0.0004	0.0004	0.0001	0.0001
CPU	Avg.	4.04	4.13	4.79	6.30	7.13	8.55	120.30	110.37
time (sec)	Std.	0.24	0.07	0.13	0.11	0.10	0.13	1.57	1.44

- 3.first.ils, the same as in phase one;
- 2.best.ils similar to 2.first.ils but using the 2.best local search;
- 3.best.ils similar to 2.best.ils but using the 3.best local search.

All variants use nearest neighbor lists of length 40.

For the second phase a weighted sum approach was considered. The experiments reported in [2,16] showed a strong clustering of good solutions in these problems, which indicates that changes of the weights in ModifyObjFunc should be smooth. In particular, the weight of the first objective is decremented (*i.e.*, the single objective used in GenerateInitialSolutions) by $1/|W|$ and the weight of the second objective is incremented by the same amount, where $|W|$ is the number of aggregations tested. The latter was set to 100, 150 and 200 for kroAB100, kroAB150 and kroAB200, respectively. The nearest neighbor list was updated every time ModifyObjFunc was called. The code was written in C and 50 runs were performed on each instance, using a Dual Athlon with 1200 MHz and 512 MB of RAM running Suse Linux 7.3.

The expected value of the weighted Tchebycheff utility function measure (R measure) as proposed in [8] was computed for each of the runs.[2] This measure evaluates a non-dominated set by the expected value of the weighted Tchebycheff utility function over a set of normalized weight vectors. The parameters for analysing the result of the algorithm on instance kroAB100 were set following [10] and for instances kroAB150 and

---

[2] We used the code available at
http://www-idss.cs.put.poznan.pl/~jaszkiewicz/mokp and adapted it to compile it with gcc 2.95.3 under Linux.

Table 2. Average of the C measure for **3.first.ils** compared to the remaining variants of the TPLS for instances **kroAB100, kroAB150, kroAB200** (see text for details)

		2.first	2.best	3.first	3.best	2.first.ils	2.best.ils	3.best.ils
kroAB100								
3.first.ils	Covers	50%	87%	65%	59%	87 %	88%	59%
	Covered by	1%	1%	1%	2 %	1 %	1%	1%
kroAB150								
3.first.ils	Covers	50%	94%	79 %	75%	94%	94 %	71%
	Covered by	0%	1%	1%	1 %	1%	1%	1 %
kroAB200								
3.first.ils	Covers	62%	95%	82%	79 %	95%	96 %	80%
	Covered by	0%	0%	1%	1 %	0%	0 %	0%

kroAB200 the global optima values available on the TSPLIB were used. As worst tour lenghts for the two last instances, the pairs (280000, 280000) and (370000, 370000), respectively, were used. Table 1 presents the results regarding the R measure obtained by the various variants tested. The results show that **3.first.ils** gives the best performance in terms of this measure, though it takes large computation times. **3.best.ils** and **3.best** are almost equivalent, but the latter is much faster; both are closely followed regarding solution quality by **3.first**. The results also suggest that the variants based on the 2-exchange neighborhood are inferior to the variants based on the 3-exchange neighborhood.

The results according to the Coverage measure (C measure) [24] are presented in Table 2. This measure compares pairs of non-dominated sets by calculating the fraction of each set that is covered by the other set. In this case, the outcomes of **3.first.ils** were compared to each of the ones obtained by the remaining variants for **LocalSearch**. Therefore, the row *Covered by* indicates how much of the outcome of **3.first.ils** is dominated by another variant and the row *Covers* says how much of the outcome of the variant is covered by **3.first.ils**. The results are presented averaged over the pairwise comparison between all the runs of two variants. They clearly indicate a best performance of **3.first.ils** according to this measure.

One possible drawback of TPLS could be that non-supported solutions, *i.e.*, solutions which are not optimal for any weighted aggregation of the objectives, are not obtained due to its *aggressive* search. However, the **3.first.ils** obtained approximately an average of 12% non-supported points per run, which means that it still gets some points which are inside the convex hull of the non-dominated set. To verify how much of the remaining supported solutions were Pareto global optimaof solutions, we count how many were present in the set of supported Pareto global optimal solutions obtained by Borges and Hansen in [2] [3]. Approximately 55% of the supported solutions attained by the TPLS belong to the supported Pareto global optimal solutions. However, since the data from Borges and Hansen is an incomplete set of Pareto gobal optimal solutions, one does not know if the remaining supported solutions of the TPLS are also Pareto global optimal solutions not found by Borges and Hansen.

---

[3] The set of supported Pareto global optima solutions is extracted from the supported Pareto global optima solution to the three-objective instance combination of **kroA100, kroB100** and **kroC100**

Table 3. Average and standard deviation of the R measure, the C measure, and the CPU time on instances kroAB100, kroAB150, kroAB200 for the D-TPLS

kroAB100			3.first	3.best	2.first.ils	3.first.ils
R Measure	Avg.		0.9327	0.9329	0.9303	**0.9345**
	Std.		0.0002	0.0001	0.0004	0.0001
CPU	Avg.		1.06	1.58	1.40	13.95
time (sec)	Std.		0.02	0.04	0.04	0.19
C Measure						
TPLS	Covered by		42%	58%	42%	56%
	Covers		17%	23%	21%	34%

kroAB150			3.first	3.best	2.first.ils	3.first.ils
R Measure	Avg.		0.9399	0.9399	0.9375	**0.9416**
	Std.		0.0002	0.0002	0.0003	0.0000
CPU	Avg.		3.47	4.80	5.02	59.92
time (sec)	Std.		0.06	0.10	0.14	0.35
C Measure						
TPLS	Covered by		38%	45%	47%	42%
	Covers		20%	27%	18%	22%

kroAB200			3.first	3.best	2.first.ils	3.first.ils
R Measure	Avg.		0.9432	0.9432	0.9412	**0.9445**
	Std.		0.0001	0.0001	0.0002	0.0000
CPU	Avg.		9.19	13.44	13.99	245.28
time (sec)	Std.		0.10	0.28	0.16	4.74
C Measure						
TPLS	Covered by		38%	44%	46%	33%
	Covers		18%	25%	17%	17%

5 Improvements on the Two-Phase Local Search

In this section, two further improvements on the TPLS are introduced: the Double Two-Phase Local Search, and the Pareto Double Two-Phase Local Search.

5.1 The Double Two-Phase Local Search

TPLS starts from a very good solution for only one of the objectives. Hence, one may expect a bias of the non-dominated points generated towards this first objective in the second phase, resulting in a skewed set towards that objective in detriment of the other. This skewedness is exemplified by the fact that only 3.first.ils (but not the other local searches) was able, when starting the second phase from an optimal solution for the first objective, to obtain also the optimal solution of the second objective in most of the runs. One way of overcoming this skewedness effect is to apply the TPLS starting once from

Table 4. Average and standard deviation of the R measure, the C measure, and the CPU time, on instances kroAB100, kroAB150, kroAB200 for the PD-TPLS

kroAB100			3.first	3.best	2.first.ils	3.first.ils
R Measure	Avg.		0.9335	0.9337	0.9314	**0.9351**
	Std.		0.0002	0.0001	0.0003	0.0000
C Measure						
D-TPLS	Covered by		50%	60%	54%	62%
	Covers		22%	21%	20%	13%
CPU	Avg.		1.26	1.72	1.59	14.14
time (sec)	Std.		0.02	0.01	0.03	0.24

kroAB150			3.first	3.best	2.first.ils	3.first.ils
R Measure	Avg.		0.9404	0.9405	0.9382	**0.9420**
	Std.		0.0002	0.0001	0.0003	0.0000
C Measure						
D-TPLS	Covered by		51%	59%	52%	51%
	Covers		25%	22%	25%	13%
CPU	Avg.		4.32	5.62	5.99	60.64
time (sec)	Std.		0.11	0.18	0.14	0.79

kroAB200			3.first	3.best	2.first.ils	3.first.ils
R Measure	Avg.		0.9436	0.9436	0.9416	**0.9450**
	Std.		0.0001	0.0001	0.0003	0.0000
C Measure						
D-TPLS	Covered by		49%	56%	46%	47%
	Covers		23%	19%	29%	12%
CPU	Avg.		11.87	16.08	17.11	248.45
time (sec)	Std.		0.44	0.44	0.57	6.90

a solution for each single objective and then to filter the non-dominated solutions from the union of both sets of solutions. We call this procedure the *Double Two-Phase Local Search* (D-TPLS).

We run experiments with 2.first.ils, 3.first, 3.best and 3.first.ils for the LocalSearch using, as in Section 4, 50 trials per instance and the same number of aggregations. Table 3 presents the average and standard deviation of the R measure and computation time on the instances. In addition, the C measure [24] is presented given as the average of all pairwise comparisons between the TPLS and D-TPLS for each local search used in the second phase. Hence, the row *Covered by* indicates the relative frequency by which the outcome of TPLS is covered by D-TPLS and the row *Covers* gives the frequency by which the outcome of D-TPLS is covered by TPLS. The computational results indicate a general better performance of D-TPLS when compared to the same TPLS version, although at the cost of doubling the computation time. By analysing further the results of D-TPLS, approximately 25% of the solutions were considered as non-supported, and about 71% of the supported solutions were the same as in [2].

Table 5. Comparison of the PD-TPLS (3.first.ils) with GSL in terms of average and standard deviation of R measure and computation time, and C measure values on the set of instances tackled in [10]

		kroAB100	kroAC100	kroAD100	kroAE100	kroBC100
R Measure						
PD-TPLS	Avg.	0.9351	0.9323	0.9344	0.9380	0.9361
	Std.	0.0000	0.0000	0.0000	0.0001	0.0000
GLS		0.9351	0.9321	0.9342	0.9379	0.9359
C Measure						
GLS	Covered by	48%	59%	61%	55%	53%
	Covers	29%	21%	20%	25%	20%
CPU	Avg.	14.14	13.72	13.69	13.70	14.65
time (sec)	Std.	0.24	0.10	0.10	0.10	0.30

		kroBD100	kroBE100	kroCD100	kroCE100	kroDE100
R Measure						
PD-TPLS	Avg.	0.9347	0.9335	0.9390	0.9352	0.9339
	Std.	0.0000	0.0001	0.0000	0.0000	0.0000
GLS		0.9344	0.9334	0.9389	0.9350	0.9338
C Measure						
GLS	Covered by	56%	54%	55%	58%	54%
	Covers	23%	20%	23%	23%	20%
CPU	Avg.	14.86	14.50	14.02	13.23	14.14
time (sec)	Std.	0.28	0.32	0.15	0.08	0.06

5.2 The Pareto Double Two-Phase Local Search

A disadvantage of the D-TPLS and TPLS is that in the local neighborhood of each solution they return (there is a maximum of $2|W|$ tours returned by D-PTLS and $|W|$ in the case of TPLS), there may be additional, non-dominated solutions, which are missed by the aggregation used. Therefore, intuitively D-TPLS can be further enhanced by searching for non-dominated solutions in the neighborhood of the solutions returned by LocalSearch. For this aim a *2-opt* local search that accepts solutions which are not dominated by the current local optimum in both objectives was applied.[4] This local search is applied after a local optimum is found on each aggregation. We call the so enhanced D-TPLS the *Pareto Double Two-Phase Local Search* (PD-TPLS). A total of 50 runs of the PD-TPLS were performed using the same four variants of LocalSearch as in Section 5.1 and again analysed in terms of C and R measure.

The results are given in Table 4 and one can observe that, in fact, the PD-TPLS improves on D-TPLS and, hence, also on TPLS. Interestingly, the increase in computation time is rather small. Approximately 90% of the solutions were non-supported, and the same number of supported solutions as in D-TPLS were also found in [2].

---
[4] This algorithm was studied in [16]. An alternative to this approach would be to increase the number of aggregations; this is studied in Section 6.

5.3 Comparisons to State-of-the-Art Algorithms

One important aspect of a new algorithmic proposal is to know how competitive it is compared to *state-of-the-art* algorithms. Apparently, the Genetic Local Search by Jaszkiewicz (GLS) was the best performing algorithm for the biobjective TSP [10]. GLS combines ideas from evolutionary algorithms (recombination, population of solutions), local search (definition of neighborhood) with modifications of the aggregation of the objective functions. The algorithm proceeds by first generating an initial population of solutions where each solution is optimized by a local search method using randomly generated weights to aggregate the objective functions as indicated in Equation 1. Then, at each iteration, a sub-set of the best solutions from the full population, according to a random weighted aggregation of the objective function, is extracted. From this sub-set, two solutions are randomly chosen and recombined. A local search is then applied to the new solution and added to the population if it is better then the worst solution in the full population. This procedure was able to outperform several multiobjective metaheuristics on a set of 50 and 100 city instances [10].

The performance of PD-TPLS and GLS were compared in all possible pairwise combinations of the instances kroA100, kroB100, kroC100, kroD100 and kroE100. The outcomes of GLS were taken from http://www-idss.cs.put.poznan.pl/~jaszkiewicz/motsp. The PD-TPLS ran using the 3.first.ils variants for the LocalSearch 50 times per instance. The results, which are given in Table 5, show the average data and standard deviation of the R measure on the outcomes of PD-TPLS and GLS for each instance. The average and standard deviation of computation times for PD-TPLS are also reported. In addition the C measure was also reported for each pair. In this case, the row *Covered by* presents the frequency by which the outcome of GLS is covered by PD-TPLS and the row *Covers* gives the frequency by which the outcome of PD-TPLS is covered by GLS. The results indicate that, according to the R measure, PD-TPLS performs slightly better than GLS; in addition, when considering the C measure, the advantage of PD-TPLS over GLS appears to be even stronger. Hence, we can conclude that our PD-TPLS approach definitely appears to be competitive, if not even slightly better, than current state-of-the-art algorithms. For further comparisons, the outcomes of PD-TPLS are available at http://www.intellektik.informatik.tu-darmstadt.de/~lpaquete/TSP.

6 Further Analysis

This section presents results of a further analysis on the influence of the number of aggregations and of the number of iterations in 3.first.ils on the solution quality (as judged by the R and C measures) and on the computation time.

6.1 Study on the Number of Aggregations

An interesting question is if one can get significant further improvements by increasing the number of aggregations, which should result in a potentially larger number of non-dominated solutions. To investigate this effect on algorithm performance, 50 runs were performed for each of 3.first, 3.best, 2.first.ils and 3.first.ils as LocalSearch variants on instance kroAB100 using $|W| \in \{500, 1000, 1500, 2000\}$.

Table 6 shows the average and standard deviation of the R measure for each variant, number of aggregations and instance. One can observe that, when compared to results obtained by 100 aggregations, given in Table 4, only minor or no improvement at all was found, but that increasing $|W|$ leads to significantly larger computation times.

Table 6. Average and standard deviation of R measure and computation time on instance kroAB100 considering 500, 1000, 1500 and 2000 aggregations for PD-TPLS

3.first	500	1000	1500	2000	3.best	500	1000	1500	2000
R Measure					R Measure				
Avg.	0.9337	0.9338	0.9338	0.9338	Avg.	0.9338	0.9338	0.9338	0.9338
Std.	0.0001	0.0002	0.0002	0.0002	Std.	0.0001	0.0001	0.0001	0.0001
CPU time					CPU time				
Avg.	5.42	10.85	16.52	22.71	Avg.	5.96	11.45	17.39	23.55
Std.	0.19	0.27	0.32	0.98	Std.	0.11	0.27	0.66	0.60

2.first.ils	500	1000	1500	2000	3.first.ils	500	1000	1500	2000
R Measure					R Measure				
Avg.	0.9316	0.9315	0.9316	0.9316	Avg.	0.9352	0.9352	0.9352	0.9352
Std.	0.0004	0.0004	0.0004	0.0004	Std.	0.0000	0.0000	0.0000	0.0000
CPU time					CPU time				
Avg.	7.17	14.42	21.77	26.10	Avg.	69.53	139.30	209.01	279.88
Std.	0.20	0.54	0.91	1.29	Std.	0.46	0.64	0.58	0.96

6.2 Study of the Number of Search Steps of 3.first.ils on the LocalSearch

3.first.ils is used both in GenerateInitialSolution and LocalSearch, but so far no exploration of alternative parameter settings for the number of iterations of the ILS was carried out. Hence, it is interesting to observe how much the performance is affected by an increase of the number of iterations of 3.first.ils. A total of 50 runs were performed for each of the following parameters $\{st/5, st/2, st, 2st\}$, where st is the number of iterations used in the previous experiments on instances kroAB100, kroAB150 and kroAB200 (50, 100, and 150, respectively).

Table 7 shows the average and standard deviation of the R measure for each number of search steps and each instance. The C measure was also applied to compare between the algorithm using the original step value and the modified one. Therefore, the row *Covered by* presents how much of the outcome of the algorithm using the original number of steps is covered by the modified one, and the row *Covers* presents how much of the outcome of the modified version is covered by the original. Also average and standard deviation of the computation time is reported. The results with the original step value are presented for reference.

The results indicate that, according to the measures used, almost no performance degradation is observable when using half of the iterations for the ILS, but that, as expected, the computation time could be roughly halved. Allowing still less iterations for ILS degrades somewhat more the solution quality, but computation times drop further. However, the performance is still much better than with other choices for LocalSearch

Table 7. Average and standard deviation of R measure, C measure and computation time on the instances kroAB100, kroAB150, kroAB200 considering fifth ($^{st}/_5$), half ($^{st}/_2$) and double ($2 \cdot st$) of the steps used in GenerateInitialSolution (st) of the PD-TPLS

kroAB100	$^{st}/_5$	$^{st}/_2$	st	$2 \cdot st$	kroAB150	$^{st}/_5$	$^{st}/_2$	st	$2 \cdot st$
	10	25	50	100		20	50	100	200
R Measure					R Measure				
Avg.	0.9350	0.9351	0.9351	0.9351	Avg.	0.9419	0.9420	0.9420	0.9420
Std.	0.0001	0.0000	0.0000	0.0000	Std.	0.0000	0.0000	0.0000	0.0000
C Measure					C Measure				
Covered by	40%	57%	-	69%	Covered by	29%	47%	-	64%
Covers	63%	64%	-	69%	Covers	67%	60%	-	57%
CPU time					CPU time				
Avg.	3.68	7.58	14.14	26.97	Avg.	15.05	32.12	60.64	116.87
Std.	0.02	0.04	0.24	0.13	Std.	0.07	0.29	0.79	0.50

kroAB200	$^{st}/_5$	$^{st}/_2$	st	$2 \cdot st$
	30	75	150	300
R Measure				
Avg.	0.9449	0.9450	0.9450	0.9450
Std.	0.0000	0.0000	0.0000	0.0000
C Measure				
Covered by	21%	40%	-	61%
Covers	70%	58%	-	48%
CPU time				
Avg.	57.37	127.68	248.45	477.89
Std.	0.90	2.42	6.90	10.22

(compare with Table 4). A further increase of the number of ILS iterations seems not to significantly improve performance, probably because the algorithm is already obtaining near-optimal solutions.

7 Conclusions and Further Work

A new two-phase local search algorithm was proposed for tackling biobjective combinatorial optimization problems. There are three main ideas behind this approach. Firstly, one exploits to the maximum possible the excellent performance of local search algorithms for single-objective problems. Secondly, one exploits a chain of aggregations of biobjective problems into single-objective ones *and* previously obtained good solutions to generate new, non-dominated solutions. Thirdly, we want to have a procedure that is at the same time easy to understand and flexible enough to allow for enhancements that can convert it by minor modifications into higher performance algorithms.

As a first step, the impact of the local search algorithm in the second phase was performed. We found that more powerful local search algorithms results indeed also in better solutions, but at the cost of more computation time. We also proposed two enhancements of the TPLS, which finally lead to the PD-TPLS, which was shown to perform on par or even slightly better than the so far best known algorithm for the biobjective TSP.

The good results and low computation time obtained by this approach can be explained by the high level of clustering of good solutions found in these problems, as already observed in [2,16]. We intend to extend this approach to other biobjective combinatorial optimization problems in which this clustering does not hold or study problems with different correlations between the cost matrices.

We also intend to extend this approach to more than two objectives. One can, indeed, see that a generalization of the current approach to n objectives can be easily formalized by performing $\frac{n(n-1)}{2}$ runs of PD-TPLS. Although the evident quadratic increase on the number of runs, one should also not expect to find practical applications with a large number of objectives. If this is the case, the Pareto global optima set could be of an infeasible size, and it should be preferable to perform a pre-defined number of runs of the ILS using several aggregations of the objectives with equally dispersed weights.

In addition, we would like to use other measures which allows for a sound statistical analysis of our results. To this aim we will adopt the attainment functions methodology [4] for experimental analysis, as already done in [17,20].

Acknowledgements. We would like to thank Michael Hansen and Pedro Borges for the providing the supported Pareto global optimal solutions, and Andrzej Jaszkiewicz for discussions about the R measure code. Finally, we also thank the suggestions given by the referees of this paper. This work was supported by the Metaheuristics Network, a Research Training Network funded by the Improving Human Potential programme of the CEC, grant HPRN-CT-1999-00106. The information provided is the sole responsibility of the authors and does not reflect the Community's opinion. The Community is not responsible for any use that might be made of data appearing in this publication.

References

1. J.L. Bentley. Fast algorithms for geometric traveling salesman problems. *ORSA Journal on Computing*, 4(4):387–411, 1992.
2. P. C. Borges and P. H. Hansen. A study of global convexity for a multiple objective travelling salesman problem. In C.C. Ribeiro and P. Hansen, editors, *Essays and Surveys in Metaheuristics*, pages 129–150. Kluwer, 2000.
3. P. Czyzak and A. Jaszkiewicz. Pareto simulated annealing - a metaheuristic technique for multiple objective combinatorial optimization. *Journal of Multi-Criteria Decision Analysis*, 7:34–47, 1998.
4. V. G. da Fonseca, C. Fonseca, and A. Hall. Inferential performance assessment of stochastic optimisers and the attainment function. In E. Zitzler and et al., editors, *Evolutionary Multi-Criterion Optimization (EMO'2001)*, LNCS 1993, pages 213–225. Springer Verlag, 2001.
5. M. Ehrgott. Approximation algorithms for combinatorial multicriteria problems. *International Transactions in Operations Research*, 7:5–31, 2000.
6. X. Gandibleux, N. Mezdaoui, and A. Freville. A tabu search procedure to solve multiobjective combinatorial optimization problems. In R. Caballero et al., editor, *Advances in Multiple Objective and Goal Programming*, LNEMS, pages 291–300. Springer Verlag, 1997.
7. M.P. Hansen. Use of subsitute scalarizing functions to guide a local search base heuristics: the case of moTSP. *Journal of Heuristics*, 6:419–431, 2000.
8. M.P. Hansen and A. Jaszkiewicz. Evaluating the quality of approximations to the non-dominated set. Technical Report IMM-REP-1998-7, Institute of Mathematical Modelling, Technical University of Denmark, Lyngby, Denmark, 1998.

9. H. Ishibuchi and T. Murata. Multi-objective genetic local search algoritm. In T. Fukuda and T. Furuhashi, editors, *Proceedings of the 1996 International Conference on Evolutionary Optimization*, pages 119–124, Nagoya , Japan, 1996. IEEE.
10. A. Jaszkiewicz. Genetic local search for multiple objective combinatorial optimization. *European Journal of Operational Research*, 1(137):50–71, 2002.
11. D. S. Johnson and L. A. McGeoch. Experimental analysis of heuristics for the STSP. In G. Gutin and A. Punnen, editors, *The Traveling Salesman Problem and its Variations*, pages 369–443. Kluwer Academic Publishers, 2002.
12. D.S. Johnson and L.A. McGeoch. The travelling salesman problem: A case study in local optimization. In E.H.L. Aarts and J.K. Lenstra, editors, *Local Search in Combinatorial Optimization*, pages 215–310. John Wiley & Sons, Chichester, UK, 1997.
13. J. Knowles and D. Corne. The pareto archived evolution strategy: A new baseline algorithm for pareto multiobjective optimisation. In *Proceedings of CEC'99*, pages 98–105, 1999.
14. H. R. Lourenço, O. Martin, and T. Stützle. Iterated local search. In F. Glover and G. Kochenberger, editors, *Handbook of Metaheuristics*, volume 57 of *International Series in Operations Research & Management Science*, pages 321–353. Kluwer Academic Publishers, Norwell, MA, 2002.
15. O. Martin, S.W. Otto, and E.W. Felten. Large-step markov chains for the traveling salesman problem. *INFORMS Journal on Computing*, 8(1):1–15, 1996.
16. L. Paquete, M. Chiarandini, and T. Stützle. A study of local optima in the biojective travelling salesman problem. Technical Report AIDA-02-07, FG Intellektik, FB Informatik, TU Darmstadt, Germany, 2002.
17. L. Paquete and C. Fonseca. A study of examination timetabling with multiobjective evolutionary algorithms. In *4th Metaheuristics International Conference (MIC 2001)*, pages 149–154, Porto, 2001.
18. P. Serafini. Some considerations about computational complexity for multiobjective combinatorial problems. In *Recent Advances and Historical Development of Vector Optimization*, LNEMS, pages 222–231. Springer-Verlag, 1986.
19. P. Serafini. Simulated annealing for multiple objective optimization problems. In *Multiple Criteria Decision Making*, LNEMS, pages 283–292. Springer-Verlag, 1994.
20. K. Shaw, C. Fonseca, A. Nortcliffe, M. Thompson, J. Love, and P. Fleming. Assessing the performance of multiobjetive genetic algorithms for optimization of a batch process scheduling problem. In *Proceeding of CEC'99*, pages 37–45, 1999.
21. R.E. Steuer. *Multiple Criteria Optimization: Theory, Computation and Application*. Wiley Series in Probability and Mathematical Statistics. John Wiley & Sons, New York, 1986.
22. T. Stützle and H. Hoos. Analyzing the run-time behaviour of iterated local search for the TSP. In *III Metaheuristic International Conference (MIC'99)*, pages 1–6, 1999.
23. D. Tuyttens, J. Teghem, P. Fortemps, and K. Van Nieuwenhuyze. Performance of the MOSA method for the bicriteria assignment problem. *Jornal of Heuristics*, 6:295–310, 2000.
24. E. Zitzler and L. Thiele. Multiobjective evolutionary algorithms: a comparative case study and the strength pareto approach. *IEEE Trans. on Evol. Comput.*, 4(3):257–271, 1999.

PISA — A Platform and Programming Language Independent Interface for Search Algorithms

Stefan Bleuler, Marco Laumanns, Lothar Thiele, and Eckart Zitzler

ETH Zürich, Computer Engineering and Networks Laboratory (TIK),
CH–8092 Zürich, Switzerland,
{bleuler,laumanns,thiele,zitzler}@tik.ee.ethz.ch,
http://www.tik.ee.ethz.ch/pisa/

Abstract. This paper introduces an interface specification (PISA) that allows to separate the problem-specific part of an optimizer from the problem-independent part. We propose a view of the general optimization scenario, where the problem representation together with the variation operators is seen as an integral part of the optimization problem and can hence be easily separated from the selection operators. Both parts are implemented as independent programs, that can be provided as ready-to-use packages and arbitrarily combined. This makes it possible to specify and implement representation-independent selection modules, which form the essence of modern multiobjective optimization algorithms. The variation operators, on the other hand, have to be defined in one module together with the optimization problem, facilitating a customized problem description. Besides the specification, the paper contains a correctness proof for the protocol and measured efficiency results.

1 Introduction

The interest in the field of multiobjective optimization mainly comes from two research directions:

- The application side, where engineers face difficult real-world problems and therefore would like to apply powerful optimization methods.
- The algorithmic side, where researchers aim to develop optimization algorithms that can be successfully applied to a wide range of problems.

Since modern optimization methods have become increasingly complex the work in both areas requires a considerable programming effort. Often code for optimization methods and test problems is available, but the usage of these implementations is restricted to one programming language. A thorough understanding of the code is needed in order to integrate it with own work. This raises the question whether it is possible to divide the implementation into an application part and an optimizer part reflecting the interests of the two groups mentioned. An ideal separation would provide ready-to-use modules on both sides and these modules would be freely combinable.

The application engineer would like to couple his implementation of the optimization problem to an existing optimizer. However, it is obviously not possible to provide a general optimizer which works well for all problems since many parts of an optimization method are highly problem specific. In many realistic application studies, the representation of candidate solutions is problem specific. Consequently, also the variation operators are very often problem dependent. Consider for example discrete optimization problems which involve network and graph representations combined with continuous variables and specific repair mechanisms. In addition, the neighborhood structure induced by the variation operator (either explicitly like in simulated annealing or tabu search or implicitly like in evolutionary algorithms) strongly influences the success of the optimization. As a consequence, the representation of candidate solutions and the definition of the variation operator comprise the major locations where problem specific and a priori knowledge can be inserted into the search process. Clearly, there are cases where standard representations such as binary strings and associated variation operators are adequate. For these situations, standard libraries are available to ease programming, e.g. [1,3]. In summary, it is the task of the application engineer to define appropriate representations and neighborhood structures.

In contrary, most optimizers in the multiobjective field work with a selection operator which is only based on objective values of the candidate solutions and are thus problem independent.[1] This allows to separate the selection mechanism from the problem-specific parts of the optimization method. As most of the research in the area of multiobjective optimization has focused on selection mechanisms these algorithms have become more and more complex. Freely combinable modules for selection algorithms on one side and applications with appropriate variation operators on the other side would thus be beneficial for application engineers as well as algorithm developers.

This paper proposes an approach to realize such a separation into an application-specific part and a problem-independent part as shown in Fig. 1. The latter contains the selection procedure, while the former encapsulates the representation of solutions, the generation of new solutions, and the calculation of objective function values. Since the two parts are realized by distinct programs that communicate via a text-based interface, this approach provides maximum independence of programming languages and computing platforms. It even allows to use precompiled, ready-to-use executable files, which, in turn, minimizes the implementation overhead and avoids the problem of implementation errors. As a result, an application engineer can easily exchange the selection method and try different variants, while an algorithm designer has the opportunity to test a selection method on various problems without additional programming effort (cf. Fig. 1). Certainly, this concept is not meant to replace programming libraries. It is a complementary approach that allows to build collections of selec-

---

[1] Nevertheless, PISA is extensible and allows for transmitting additional information needed to consider e.g. niching strategies in selection.

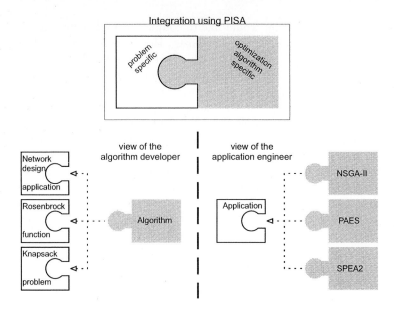

Fig. 1. Illustration of the concept underlying PISA. The applications on the left hand side and the multiobjective selection schemes on the right hand side are examples only and can be replaced arbitrarily

tion methods and applications including variation operators, all of them freely combinable across computing platforms.

2 Design Goals and Requirements

Our aim is to design a standardized, extendible and easy to use framework for the implementation of multiobjective optimization algorithms. For the development of such a framework, we follow several design goals:

Separation of concerns. The algorithm-specific and the problem-specific component should have a maximum independence from each other. It should be possible to implement only the part of interest, while the other part is treated as a ready-to-use black box.

Small overhead. The additional effort necessary to implement interfaces and communication mechanisms has to be as small as possible. The extra running time due to the data exchange between the components of the system should be minimized.

Simplicity and flexibility. The approach should have a simple and comprehensible way of handling input and output data and setting parameters, but should hide all implementation details from the user. The specification of the flow control and the data exchange format should state minimal requirements for all implementations, but still leave room for future extensions and optional elements.

Portability and platform independence. The framework itself, and hence the possibility to embed any existing algorithm into it, should not depend on machine types, operating systems or programming languages. The different components must interconnect seamlessly. It is obvious that running a module on a different operating system might require re-compilation, but porting an existing program to another operating system or machine type should not be complicated by the interface implementation. Furthermore, when porting is difficult, it must be possible to run the two processes on different machines with possibly different operating systems, letting them communicate over a network link.

Reliability and safety. A reliable and correct execution of the different components is very important for the broad acceptance of the system. For instance, unusual parameter settings must not cause a system failure.

Given these design goals, the development of a programming framework becomes a multiobjective problem itself, and it is impossible to reach a maximum satisfaction in all design aspects. Therefore, a compromise solution is sought. Here, we focus on simplicity and small overhead and are willing to accept less flexibility. The motivation behind this is that the system will only be employed by many people if it is easy to use and does not require excessive programming work. To compensate for the lack of flexibility, we will make the format extendible to a certain degree so that it will still be possible for interested users to adapt it to specific needs and features. How all these design goals are realized will be described in the next section.

3 Architecture

The proposed architecture is characterized by the following main features:

- Separation of selection on the one hand and variation and calculation of objective function values on the other.
- Communication via file system which allows for platform, programming language and operating system independence.
- Correct communication under weak assumptions about shared file system.
- Small communication overhead by avoiding to transmit decision space data.

We consider optimization methods that generate an initial set of candidate solutions and then proceed by an iterative cycle of evaluation, selection of promising candidates and variation. The selection is assumed to operate only in objective space. Nevertheless, the final specification of PISA is extensible and allows for transmitting additional information needed to consider e.g. niching strategies in selection.

Based on these assumptions, a formal model for our framework can be established. Our model will be based on Petri nets, because in contrast to state machines, Petri nets allow to describe the data flow and the control flow within a single computational model. The resulting architecture is depicted in Fig. 2,

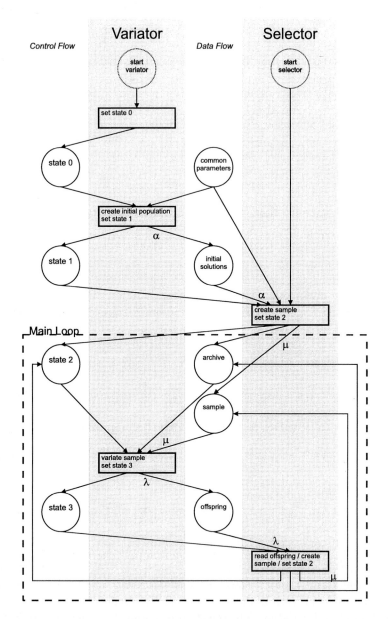

Fig. 2. The control flow and data flow specification of PISA using a Petri net. The transitions (rectangular boxes) represent the operations by the processes implementing the variator and the selector. The places in the middle represent the data flow and correspond to the data files which both processes read and write. The places at the left margin represent reading and writing of the state variable that is stored in a common state file and hence direct the control flow

Table 1. Stop and reset states

State	Action	Next State
State 4	Variator terminates.	State 5
State 6	Selector terminates.	State 7
State 8	Variator resets. (Getting ready to start in state 0)	State 9
State 10	Selector resets. (Getting ready to start in state 0)	State 11

where the term *variator* is used to denote the problem-dependent part and *selector* the problem-independent part. A transition (rectangular boxes) can fire if all inputs are available. On firing a transition performs the stated operations, consumes the input data and provides all outputs.

3.1 Control Flow

The model ensures that there is a consistent state for the whole optimization process and that only one module is active at any time. Whenever a module reads a state that requires some action on its part, the operations are performed and the next state is set. The implementation of the flow control is discussed in Section 4.1.

The core of the optimization process consists of state 2 and state 3: In each iteration the selector chooses a set of parent individuals and passes them to the variator. The variator generates new individuals on the basis of the parents, computes the objective function values of the new individuals, and passes them back to the selector.

In addition to the core states two more states are shown in Fig. 2. State 0 and state 1 trigger the initialization of the variator and the selector, respectively. In state 0 the variator reads the necessary parameters (the common parameters are shown in Fig. 2 and local parameters are not shown). For more information on parameters refer to Section 4.3. Then, the variator creates an initial population, calculates the objective values of the individuals and passes the initial population to the selector. In state 1, the selector also reads the required parameters, then selects a sample of parent individuals and passes them to the variator.

The above mentioned states provide the basic functionality of the optimization. To improve flexibility in the use of the modules states for resetting and stopping are added (see Table 1). The actions taken in states 5, 7, 9 and 11 are not defined. This allows a module to react flexibly, e.g., if the selector reads state 5, which signals that the variator has just terminated, it could choose to set the state to 6 in order to terminate as well. Another selector module could instead set the state to 10, thus, causing itself to reset.

3.2 Data Flow

The data transfer between the two modules introduces some overhead compared to a traditional monolithic implementation. Thus, the amount of data exchange for each individual should be minimized. Since all representation-specific operators are located in the variator, the selector does not have to know the representation of the individuals. Therefore, it is sufficient to convey only the following data to the selector for each individual: an index, which identifies the individual in both modules, and its objective vector. In return, the selector only needs to communicate the indices of the parent individuals to the variator. The proposed scheme allows to restrict the amount of data exchange between the two modules to a minimum. In the following we will refer to passing the essential information as passing a population or a sample of individuals.

As to objective vectors the following semantics is used: An individual is superior to another with regard to one objective, if the corresponding element of the objective vector is smaller, i.e., objective values are to be *minimized*. Furthermore, the two modules need to agree on the sizes of the three collections of individuals passed between each other: the initial population, the sample of parent individuals, and the offspring individuals. These sizes are denoted as α, μ and λ in Fig. 2. Instead of using some kind of automatic coordination, which would increase the overhead for implementing the interface we have decided to specify the sizes as parameter values. Setting μ and λ as parameters requires that they are constant during the optimization run. Most existing algorithms comply with this requirement. Nevertheless, dynamic population sizes could be implemented using the facility of transferring auxiliary data (cf. Section 4.2).

As described in Section 3.1, a collection of parent individuals is passed from the selector to the variator and a collection of offspring individuals is returned. The actual representation of the individuals is stored on the variator side. Since the selector might use some kind of archiving method, the variator would have to store all individuals ever created, because one of them might be selected as a parent again. This can lead to unnecessary memory exhaustion and can be prevented by the following mechanism: the selector provides the variator with a list of all individuals that could ever be selected again. This list is denoted as archive in Fig. 2. The variator can optionally read this list, delete the respective individuals and re-use their indices. Since most individuals in a usual optimization run are not archived, the benefit from this additional data exchange is much larger than its cost. Section 4.2 describes how the data exchange is implemented.

4 Implementation Aspects

After describing the architecture of the interface based on Petri nets in the previous section, this section discusses the most important issues of implementation.

4.1 Synchronization

In order to reach the necessary separation and compatibility, the selector and the variator are implemented as two separate processes. These two processes can be located on different machines with possibly different operating systems. This complicates the implementation of a synchronization method. Most common methods for interprocess communication are therefore not applicable.

Closely following the Petri net model (cf. Fig. 2), a common state variable which both modules can read and write is used for synchronization. The two processes regularly read this state variable and perform the corresponding actions. If no action is required in a certain state, the respective process sleeps for a specified amount of time and then rereads the state variable.

Coherent with our decision for simplicity and ease of implementation, the common state variable is implemented as an integer number written to a text file. In contrast to the alternative of using sockets, file access is completely portable and familiar to all programmers. The only requirement is access to the same file system. On a remote machine this can for example be achieved through simple `ftp put` and `get` operations. As another benefit of using a text file for synchronization it is possible for the user to manually edit the state file. The underlying assumptions about the file system and the correctness of this approach will be discussed in Section 4.4.

4.2 Data Exchange

Another important aspect of the implementation is the data transfer between the two processes. Following the same reasoning as for synchronization, all data exchange is established through text files. Using text files with human readable format allows the user to monitor data exchange easily, e.g., for debugging. For the same reason, a separate file is used for each collection of individuals shown in Fig. 2. The resulting set of files used for communication between the two modules and for parameters is shown in Fig. 3. Simple examples of possible contents are shown as well to illustrate to file format.

To achieve a reliable data exchange through text files, the receiving module should be able to detect corrupted files. For instance, a file could be corrupted because the receiving process tries to read the file before it is completely written. The detection of corrupted files is enabled by adding two control elements to the data elements: The first element specifies the number of data elements following. After the data elements an END tag ensures that the last element has been completely written. The receiving module can read the specified number of elements without looking for a END and then check if the END tag is at the expected place.

Additionally, the reading process is expected to replace a file's content with '0' after reading it and the writing process must check if no old data remains that would be overwritten. This represents the production and consumption of the data as shown in the Petri net in Fig. 2 and it prevents re-reading of old data which could happen if the writing of the new state can be seen earlier by the reading process then the writing of the data. For additional information on the requirements for the file system see Section 4.4.

Between the two control elements blocks of data are written, describing one individual each. In this example, such block consists of an index and two objec-

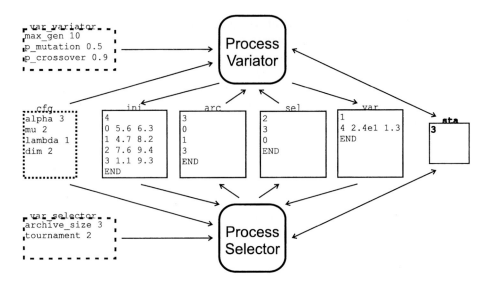

Fig. 3. Communication between modules through text files. Four files for the data flow: The initial population in ini, the archive of the selector in arc, the sample of parent individuals in sel and the offspring in var. The cfg file contains the common parameters and sta contains the state variable. Additionally two examples for local parameter files are shown

tive values for the files written by the variator and only one index for the files written by the selector.

The file format described so far provides the exchange of the data necessary for all optimization methods. This might not be sufficient for every module since some techniques, e.g. mating restrictions and constraint handling, require the exchange of additional data. Therefore, the specification allows for optional data blocks after the first END tag. A module which expects additional data can read on after the first END, whereas a simple module is not disturbed by data following after the first END. A block of optional data has to start with a name. Providing a name for blocks of optional data allows to have several blocks of optional data and therefore makes one module compatible with many other modules which require some specific data each. The exact specifications of the file formats are given in the appendix.

4.3 Parameters

Several parameters are necessary to specify the behavior of both modules. Following the principle of separation of concern, each module specifies its own parameter set (examples are shown in Fig. 3). As an exception, parameters that are common to both modules are given in a common parameter file. This prevents users from setting different values for the same parameter on the variation and the selection side. The set of common parameters consists of the number of objectives (dim) and the sizes of the three different collections of individuals that are passed between the two modules (see Fig. 2).

The author of a module must specify, which α, μ and λ combinations and which dim values the module can handle. A module can be flexible in accepting different settings of these parameters or it can require specific values. To ensure reliable execution, each module must verify the correct setting of the common parameters.

Two parameters, however, are needed in the part of each module which implements control flow shown in Fig. 2: i) the filename base specifying the location of the data exchange files as well as the state file and ii) the polling interval specifying the time for which a module in idle state waits before rereading the file. The values of these parameters need to be set before the variator and the selector can enter state 0 and state 1, respectively, for example as command line arguments.

4.4 Correctness

It is not obvious that the proposed method for synchronization and data transfer works correctly. One problem arises for example from the fact that on a ordinary file system it cannot be assumed that the two successive write operations by one process to different files can be read in the same order by another process, i.e., changes in files can overtake each other. It is therefore necessary to state the necessary assumptions made about the file system and to show that the proposed system works correctly under these assumptions.

We assume that the file system is used by two processes $P1$ and $P2$ and has the following properties :

A) Writing of characters to a file is serial.
B) Writing of a character to a file is atomic.
C) A read by a process $P1$ from a file F that follows a write by $P1$ to F with no writes of F by another process $P2$ occurring between the write and the read by $P1$, always returns the data written by $P1$.
D) A read by a process $P1$ from a file F that follows a write by another process $P2$ to F after some sufficiently long time, returns the data written by $P1$ if there are no writes of F in between.

A possible underlying scenario is presented in Fig. 4. Two Processes $P1$ and $P2$ execute blocks of operations and communicate through a file system which can be modeled by two caches $C1$ and $C2$ and a memory M (see Figure 4). The time x needed to copy a file from a local memory to M is arbitrary but bounded. It may be different for each file access and for each distinct file.

There are two major properties that can be proven and which guarantee the correctness of the protocol with respect to the algorithm in Fig. 2.

The data transfer using the data files (archive, sample and offspring) is atomic. This property is a consequence of the properties A and B and the particular protocol used. The writer of a file puts a special END tag at the end of

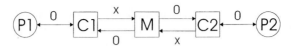

Fig. 4. Memory model for the file system. Two processes P1 and P2 communicate through the file system modeled by two local memories C1 and C2 and a global memory M. On some paths it takes zero time for the data to propagate from one element to the next. On others it takes a time x which is a positive bounded random variable

its data which can be recognized uniquely (see property B). As the writing of characters is serial, the reader of a file can determine when all data have been read. In addition, the reader starts reading if the file contains any data and it deletes all data after having read them. This way, there is a definite start and end time of the read process and all data are transmitted because of property A.

The sequence of operations according to Fig. 2 is guaranteed by the protocol. As shown in Fig. 2 two conditions are necessary for a process to resume operation: The presence of the respective state and the availability of the data. As a consequence of properties C and D the waiting process can only resume execution after the state has been changed by the other process. Property B then ensures that this state is unambiguously read. Access to the correct data is assured by the requirement that the reader deletes the data after reading it and the writer checks that the old data has been deleted before writing new data. Together with properties C and D this guarantees that no old data from the previous iteration can be read.

5 Experimental Results

The interface specification has been tested by implementing sample variators and selectors on various platforms.

In a first set of experiments, an interface has been written in the programming language C and extended with the simple multi-objective optimizer SEMO (selector) and the LOTZ problem (variator), see [2]. They have been tested on various platforms (Windows, Linux, Solaris) where the two processes have been residing as well on different machines as on the same machine.

In a second experiment, a large application written in Java was tested with the well known multiobjective optimizer SPEA2 [5] written in C++ using the library TEA [1]. The purpose of the optimization was the design space exploration of a network processor including architecture selection, binding of tasks and scheduling, see [4]. The interface worked reliably again, even if the application program and the optimizer ran on two different computing platforms, i.e., Windows and Solaris.

In a final set of experiments, the intention was to estimate the expected runtime overhead caused by the interface. Based on the cooperation between the two processes variator and selector, one can derive that the overhead caused by the interface for each generation can be estimated as

$$P + T_{comm} + (N + \lambda(1+D) + \mu)K_{comm}$$

where P denotes the polling interval chosen as well in the variator as in the selector, N, λ and μ denote the size of the archive, sample and offspring data sets, respectively, and D denotes the number of objectives. The rationale behind this estimation is that the time overhead consists of three parts, namely the average time to wait for a process to recognize a relevant state change, the overhead caused by opening and closing all relevant files including the state file, and a part that is proportional to the number of tokens in the data files. Note that besides the polling for a state change, the two processes do not compete for the processor, as the variation and selection are executed sequentially, see Fig. 2. It is not considered that in the variator as well as in the selector we need to store and process the population. On the other hand, the corresponding time overhead can be expected to be much smaller than the time to communicate via a file-based interface.

The parameters of this estimation formula have been determined for a specific platform and a specific interface implementation and good agreement over a large range of polling times and archive sizes has been found. In order to be on the pessimistic side, we have chosen to use the interface written in Java. The underlying platform for both processes was a Pentium Laptop (600 MHz) running Linux and we obtained the parameters

$$T_{comm} = 10\,\text{ms} \qquad K_{comm} = 0.05\,\text{ms/token}$$

For example, if we take an optimization problem with two objectives $D = 2$, a polling interval of $P = 100\,\text{ms}$, a population size of $N = 500$, and a sample and offspring size of $\lambda = \mu = 250$, then we obtain 185 ms time overhead for each generation. For any practically relevant optimization application, this time is much smaller than the computation time within the population-based optimizer and the application program. Note that for each generation, at least the 250 new individuals must be evaluated in the variator.

Clearly, these values are very much dependent on many factors such as the platform, the programming language and other processes running on the system. Nevertheless, we can summarize that the overhead caused by the interface is negligible for any practically relevant application.

6 Summary

In this paper, we have proposed a platform and programming language independent interface for search algorithms (PISA) that uses a well-defined text file format for data exchange. By separating the selection procedure of an optimizer from the representation-specific part, PISA allows to maintain collections of pre-compiled components which can be arbitrarily combined. That means on the one hand that application engineers with little knowledge in the optimization domain can easily try different optimization strategies for the problem at hand; on the other hand, algorithm developers have the opportunity to test optimization techniques on various applications without the need to program the problem-specific parts. This concept even works on distributed files systems across different operating systems and can also be used to implement application servers using the file transfer protocol over the Internet.

This flexibility certainly does not come for free. The data exchange via files increases the execution time, and the implementation of the interface requires some additional work. As to the first aspect, we have shown in Section 5 that the communication overhead can be neglected for practically relevant applications; this also holds for comparative studies, independent of the benchmark problems used, where we are mainly interested in relative run-times. Also concerning the implementation aspect, the overhead is small compared to the benefits of PISA. The interface is simple to realize, and most existing optimizers and applications can be adapted to the interface specification with only few modifications. Furthermore, the file format leaves room for extensions so that particular details such as diversity measures in decision space can be implemented on the basis of PISA.

Crucial, though, for the success of the proposed approach is the availability of optimization algorithms and applications compliant with the interface. To this end, the authors maintain a Web site at http://www.tik.ee.ethz.ch/pisa/ which contains example implementations for download.

Acknowledgment. This work has been supported by the Swiss National Science Foundation (SNF) under the ArOMA project 2100-057156.99/1 and the SEP program at ETH Zürich under the poly project TH-8/02-2.

References

1. M. Emmerich and R. Hosenberg. TEA - a C++ library for the design of evolutionary algorithms. Technical Report CI-106/01, SFB 531, Universität Dortmund, 2000.
2. M. Laumanns, L. Thiele, E. Zitzler, E. Welzl, and K. Deb. Running time analysis of multi-objective evolutionary algorithms on a simple discrete optimization problem. In *Parallel Problem Solving From Nature — PPSN VII*, 2002.
3. K. Tan, T. H. Lee, D. Khoo, and E. Khor. A Multiobjective Evolutionary Algorithm Toolbox for Computer-Aided Multiobjective Optimization. *IEEE Transactions on Systems, Man, and Cybernetics—Part B: Cybernetics*, 31(4):537–556, August 2001.
4. L. Thiele, S. Chakraborty, M. Gries, and S. Künzli. *Network Processor Design 2002: Design Principles and Practices*, chapter Design Space Exploration of Network Processor Architectures. Morgan Kaufmann, 2002.
5. E. Zitzler, M. Laumanns, and L. Thiele. SPEA2: Improving the strength pareto evolutionary algorithm for multiobjective optimization. In K. Giannakoglou, D. Tsahalis, J. Periaux, K. Papailiou, and T. Fogarty, editors, *Evolutionary Methods for Design, Optimisation, and Control*, pages 19–26, Barcelona, Spain, 2002. CIMNE.

Appendix

The formats of all files used in the interface are specified in the following. Note that the stated limits (e.g. largest integer) give minimal requirements for all modules. It is possible to state larger limits in the documentation of each module.

Common Parameter File (cfg)

All elements (parameter names and values) are separated by white space.

```
cfg := 'alpha' WS PosInt WS 'mu' WS PosInt WS 'lambda' WS PosInt
       WS 'dim' WS PosInt
```

State File (sta)

An integer i with $0 \leq i \leq 11$.

```
Statefile := Int
```

Selector Files (sel and arc)

The first element specifies the number of data elements following before the first END. The data contains only white space separated indices. Optional data blocks start with a name followed by the number of data elements before the next END.

```
SelectorFiles := PosInt WS SelData 'END' SelOptional*
SelOptional := Name WS PosInt WS SelData 'END'
SelData := (Int WS)*
```

Variator Files (ini and var)

The first element specifies the number of data elements m following before the first END. The data consists of one index and dim objective values (floats) per individual. If n denotes the number of individuals: $m = (dim + 1) \cdot n$. Optional data blocks start with a name followed by the number of data elements before the next END.

```
VariatorFiles := PosInt WS VarData 'END' VarOptional*
VarOptional := Name WS PosInt WS VarData 'END'
VarData := (Int WS (Float WS)*)*
```

Names for Optional Data

Names for optional data consist of maximally 127 characters, digits and underscores.

```
Name := Char (Digit | Char)*
Char : 'a-z' | 'A-Z' | '_'
```

White Space

```
WS := (Space | Newline | Tab)+
```

Integers

The largest integer allowed is equal to the largest positive value of a signed integer in a 32 bit system: $maxint = 32767$

```
Int := '0' | PosInt
PosInt: '1-9' Digits*
```

Floats

Floats are non-negative floating point numbers with optional exponents. The total number of digits before and after the decimal point can maximally be 10. The largest possible float is: $maxfloat = 1e37$. For the exponent value exp applies: $-37 \leq exp \leq 37$

```
Float := (Digit+ '.' Digit*) | ('.' Digit+ Exp?) | (Digit+ Exp)
Exp   := ('E'|'e') ('+'? | '-'?) Digit+
```

Digit

```
Digit := '0-9'
```

A New Data Structure for the Nondominance Problem in Multi-objective Optimization

Oliver Schütze

Department of Mathematics and Computer Science
http://math-www.uni-paderborn.de/~agdellnitz
University of Paderborn
Paderborn, Germany

Abstract. We propose a new data structure for the efficient computation of the nondominance problem which occurs in most multi-objective optimization algorithms. The strength of our data structure is illustrated by a comparison both to the linear list approach and the quad tree approach on a category of problems. The computational results indicate that our method is particularly advantageous in the case where the proportion of the nondominated vectors versus the total set of criterion vectors is not too large.

1 Introduction and Background

In most computational algorithms for the solution of a multi-criteria optimization problem[1]

$$\min F : Q \subset \mathbb{R}^n \to \mathbb{R}^k \tag{1.1}$$

the problem arises to sort out the *nondominated* vectors from a given finite (but large) set of criterion vectors $P \subset \mathbb{R}^k$. A vector $v \in \mathbb{R}^k$ is dominated by a vector $w \in \mathbb{R}^k$ if $w_i \leq v_i$ for all $i \in \{1, \dots, k\}$ and $v \neq w$ (i.e. there exists a $j \in \{1, .., k\}$ such that $w_j < v_j$). A vector v is called nondominated in P if there is no vector $p \in P$ which dominates v.

The *nondominance problem* can be divided into two main classes. First, there is the *static* nondominance problem. Here one has to find the subset N of nondominated vectors of a given set P at once. For details we refer e.g. to [5] and [8], where the problem is solved up to $k = 4$.

Second, there is the *dynamic* nondominance problem which occurs in most multi-objective optimization techniques and which we want to address in this paper. We are given a set of nondominated vectors P, and, in addition to this set, there is a sequence of candidates (which is generated by the optimization procedure). For every vector v of this sequence the *archive* P has to be updated (see Figure 1).

---
[1] In this paper we assume that all objectives have to be minimized.

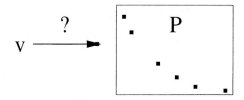

Fig. 1. Scheme of the dynamic nondominance problem: a given archive P of nondominated points has to be uptdated by arriving data

There are several alternative approaches for the solution of this problem. First, if all the candidates (and hence all elements of the archive P) lie in bounded (hyper-) rectangles, it is suitable to use *kd-trees* ([1,10]) or *range trees* ([2,10]). *Priority trees* ([11]) are suitable for the case where these rectangles are unbounded on a single side. If there are no restrictions to the range of the objectives the intuitive *linear list* approach (see e.g. [14]) can be used. Another way of attacking the problem is proposed in [9], where a clever usage of the data structure *quad tree* (see [7]) is utilized. These techniques were refined in [14] and [13]. We refer to [12] for the extensions of the quad tree approach to multi-objective evolutionary algorithms. Furthermore, there exists the *composite point* approach which is presented in [4] and finally discussed in [6]. The data structure we are proposing here is - like all the methods mentioned above except the linear list approach - tree-based. A slightly different use of this approach due to a different kind of problem is presented in [3].

2 Attacking the Nondominance Problem

Let us assume that we have a dynamic nondominance problem, i.e. a sequence of candidates $v^j \in \mathbb{R}^k$ for which a given archive $P \subset \mathbb{R}^k$ has to be updated. The basis for our approach is to store the nondominated vectors from P in the following tree:

> DEFINITION A k-ary tree T is called a *dominance decision tree*, if for every node
> $p = (p_1, \ldots, p_k) \in T$ and for each existing i-th son $s = (s_1, \ldots, s_k)$ from p the following holds:
>
> $$s_j \leq p_j \quad \forall j = 1, .., i-1 \qquad (2.1)$$
> $$s_i > p_i \qquad (2.2)$$

A simple example of a dominance decision tree for three objectives is shown in Figure 2.
For a given archive P and a new candidate $v \in \mathbb{R}^k$ the following steps have to be performed:

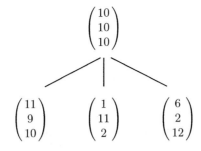

Fig. 2. Example of a dominance decision tree for $k = 3$

1.) If there exist one vector $D \in P$ which dominates v, then STOP, else go to step 2.
2.) Detect and delete all elements $d \in P$ which are dominated by v.
3.) Insert v into the archive P.

In the following we describe how to realize these steps and how to take advantage of the structure of the dominance decision tree.

ad 1.) Assume that a vector v and an archive P – stored in a dominance decision tree with root r – are given. First, v has to be compared to the root r. If r dominates v, we stop and v has to be discarded. If v and r are mutually non-dominating, then the algorithm has to make the comparisons recursively in some subtrees of r. Due to (2.2) this comparison has to be made only in the i-th subtrees of r where $r_i \leq v_i$. The algorithm DetectDomination reads as follows:

Algorithm DetectDomination
Input: root r, vector $v \in \mathbb{R}^k$.
Task: returns 1, if there exists a vector $p \in P$ (given by root r) which dominates v, else 0.

```
DetectDomination (root r,node v)
    if r dominates v
        return 1
    for i = 1,...,k
        if r_i ≤ v_i AND the i-th son of r exists (denote it by r → son_i)
            if DetectDomination(r → son_i, v) == 1
                return 1
    return 0
```

ad 2.) Assume again that a vector v and a dominance decision tree P are given. First we have to discuss which nodes have to be checked for domination, i.e. in which subtrees of P the algorithm has to look for dominated points. With given $p \in P$ it follows by conditions (2.1) and (2.2) that the search has to be continued in the first i subtrees of k where $i \in \{1, \ldots, k\}$ is the smallest index where $v_i > p_i$. In order to see this let s be a vector from the l-th subtree where $i < l <= k$. Then by construction of the dominance decision tree:

$$s_i \stackrel{(2.1)}{\leq} p_i < v_i,$$

and hence s cannot dominate v.

We illustrate this by an example: let p, v_1, v_2 and v_3 be given by

$$p = \begin{pmatrix} 10 \\ 10 \\ 10 \end{pmatrix}, \quad v_1 = \begin{pmatrix} 12 \\ 8 \\ 5 \end{pmatrix}, \quad v_2 = \begin{pmatrix} 2 \\ 12 \\ 1 \end{pmatrix}, \quad v_3 = \begin{pmatrix} 3 \\ 3 \\ 3 \end{pmatrix}.$$

In case of $v = v_1$ the search has only to be continued in the first subtree of v_1, whereas with the choice of $v = v_2$ the algorithm has to search in the first two subtrees of v_2. Eventually in case of $v = v_3$ the data structure has no advantage because all three subtrees have to be scanned.

A deletion of a node $p \in P$ can be done as follows: if one son s of p does exist, then it can be moved to the position of p. The other nodes of the subtree of p have to be reinserted – into the lifted subtree with root s. The deletion of the dominated nodes can be done via one postorder run through the tree:

Algorithm DeleteDominated
Input: root r, vector $v \in \mathbb{R}^k$.
Task: deletes every vector $p \in P$ which is dominated by v.

```
DeleteDominated (root r, vector v)
    for i = 1, ... , k
        if the i-th son of r exists (denote it by r → son_i)
            DeleteDominated (r → son_i, v)
        if v_i > r_i
            break
    if v dominates r
        if r is leave
            delete r and STOP
        j := arg min { the j-th son of r exists (denote it by r → son_j) }
        Move r → son_j to the position of r
        for l = j + 1, ... , k
            if the l-th son of r exists (denote it by r → son_l)
                TreeInsert (r → son_l)
        delete r
```

The algorithm `TreeInsert` used above reads as follows:

Algorithm TreeInsert
Input: root r, root s.
Task: inserts every vector of the tree given by root s into the tree given by root r.

```
TreeInsert (root r, root s)
    for i = 1, ..., k
        if the i-th son of s exists (denote it by s → son_i)
            TreeInsert (r, s → son_i)
    Insert (r, s)
    delete s
```

ad 3.) By its definition there is only one possible way for the insertion of a vector v into a given dominance decision tree P (given by root r):

Algorithm Insert
Input: root r, vector v.
Task: inserts v into the archive P (given by root r).

```
Insert(root r, vector v)

    i := arg min {v_i > r_i}
    if the i-th son of r exists (denote it by r → son_i)
        Insert(r → son_i, v)
    else
        r → son_i := v
```

Now the main algorithm for the update of an archive P (with root r) by a candidate v can be stated. Note that the root of the dominance decision tree can be changed in the algorithm DeleteDominated.

Algorithm Update
Input: root r, vector v.
Task: updates the archive P (given by root r) by the candidate v.

```
Update (root r, vector v)
    if( P is empty)
        P := {v}   (r := v)
        STOP
    if DetectDomination (r, v) == 1
        STOP
    DeleteDominated (r, v)
    Insert ((root of the archive P), v);
```

The algorithm presented above has the average case complexity of $O(n^2)$ for vector comparisons. However, since there is no algorithm with a provable complexity better than $O(n^2)$, we will discuss the particular advantages of the dominance decision tree approach in the following section.

3 Computational Results

Here we make a comparison of the approaches which need no restrictions to the range of the objective values. Regrettably, due to time limitations of the proceedings, it was not able for the author to involve the recently proposed composite point approach ([6]) to the comparison, though this would be very interesting and has to be done in the near future. Hence here we compare the linear list approach, the quad tree approach and the dominance decision tree approach.

For the comparison we proceed as in [14] and take test points generated by an annulus as criterion vectors because by this category of problems the particular advantages of the three approaches can be demonstrated.

We choose a sequence of vectors $v^j \in \mathbb{R}^k$ which have to be inserted to the archive P given by the nondominated vectors of the set $\{v^1, .., v^{j-1}\}$. The components of every vector $v^j = (v_1, .., v_k)$ are of the following form:

$$v_i := -\frac{\tilde{r}_i}{\|w\|} w_i, \qquad (3.1)$$

where $\tilde{r}_i \in [r, 1]$ and $w \in \mathbb{R}^k_+$ are chosen at random. This choice of criterion vectors allows to adjust not only the number k of "objectives" but also the proportion p_n of the nondominated vectors versus the total number of criterion vectors: it is easy to see that the larger the value of $r \in [0, 1)$ is the larger the value of p_n will typically be. Exactly this proportion is important for the comparison of the two tree based approaches: the computational results indicate that the dominance decision tree approach is advantageous in the case where the proportion p_n is "moderate". The larger the value of p_n the better is the performance of the quad tree approach and it gets eventually faster than the dominance decision tree approach. Of course it is barely possible to detect an exact "balance proportion" p_n^b where the two approaches have the same running time, but at least it seems to be possible to give some guidelines.

In Figure 3 and Table 1 we show that for $k = 3$ the proportion where the running time of the two approaches is basically the same is approximately $p_n^b = 1/3$ (with $r_b = 0.95$). That means that the dominance decision tree approach is faster when the optimization algorithm generates in average at most every third time a nondominated vector, otherwise the quad tree approach is faster.

For $k = 4$ the proportion p_n^b seems to be 0.5 (see Figure 4 and Table 2), but because of the higher dimension the limit radius $r_b \approx 0.85$ is lower than for k equals 3. A similar observation was made for $k = 5$ ($p_n^b \approx 0.5$ but $r_b \approx 0.7$). This means that the efficiency of the quad tree approach increases with growing k in comparison to the dominance decision tree approach.

In the case where k equals 2 we figured out that the linear list approach is faster than the dominance decision tree approach as well as the quad tree approach. This is possibly due to the overhead given by the tree based approaches.

Table 1. Annulus generated test problem results for $k = 3$

	N	2000	4000	6000	8000	10000
r = 0.3	T_{LL}	0.01	0.02	0.03	0.05	0.06
	T_{QT}	0.01	0.03	0.04	0.06	0.07
	T_{DDT}	0.01	0.01	0.02	0.03	0.03
	p_n	0.07	0.04	0.04	0.03	0.03
r = 0.5	T_{LL}	0.01	0.03	0.05	0.07	0.10
	T_{QT}	0.02	0.03	0.06	0.09	0.10
	T_{DDT}	0.01	0.02	0.03	0.04	0.05
	p_n	0.10	0.07	0.06	0.05	0.04
r = 0.7	T_{LL}	0.02	0.06	0.10	0.16	0.21
	T_{QT}	0.03	0.06	0.10	0.15	0.21
	T_{DDT}	0.02	0.03	0.05	0.06	0.08
	p_n	0.15	0.11	0.09	0.07	0.07
r = 0.9	T_{LL}	0.09	0.28	0.56	0.90	1.29
	T_{QT}	0.08	0.19	0.31	0.42	0.62
	T_{DDT}	0.05	0.11	0.19	0.26	0.33
	p_n	0.37	0.28	0.23	0.20	0.18
$r_b = 0.95$	T_{LL}	0.29	0.85	1.66	2.66	3.91
	T_{QT}	0.13	0.41	0.69	1.05	1.33
	T_{DDT}	0.15	0.40	0.65	0.95	1.31
	p_n	0.61	0.47	0.41	0.36	0.33

Table 2. Annulus generated test problem results for $k = 4$

	N	2000	4000	6000	8000	10000
r = 0.3	T_{LL}	0.05	0.11	0.21	0.30	0.39
	T_{QT}	0.06	0.12	0.19	0.28	0.38
	T_{DDT}	0.03	0.07	0.11	0.16	0.20
	p_n	0.18	0.14	0.12	0.10	0.10
r = 0.5	T_{LL}	0.08	0.21	0.38	0.55	0.77
	T_{QT}	0.09	0.21	0.34	0.40	0.52
	T_{DDT}	0.05	0.12	0.19	0.25	0.32
	p_n	0.26	0.20	0.17	0.15	0.14
r = 0.7	T_{LL}	0.17	0.47	0.88	1.38	1.98
	T_{QT}	0.14	0.34	0.53	0.72	1.06
	T_{DDT}	0.10	0.22	0.38	0.52	0.68
	p_n	0.42	0.32	0.28	0.25	0.22
$r_b = 0.85$	T_{LL}	0.20	0.71	1.80	3.36	5.25
	T_{QT}	0.13	0.29	0.56	0.76	1.20
	T_{DDT}	0.11	0.30	0.51	0.74	1.05
	p_n	0.68	0.56	0.50	0.46	0.43
r = 0.9	T_{LL}	0.45	1.68	3.82	6.98	11.16
	T_{QT}	0.23	0.64	1.15	1.65	2.54
	T_{DDT}	0.26	0.81	1.49	2.29	3.11
	p_n	0.85	0.77	0.71	0.66	0.63

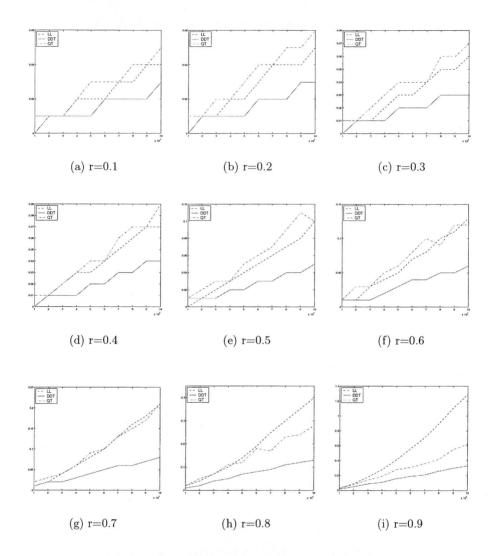

Fig. 3. Annulus generated test problem results for $k = 3$. In the Figures the number N of criterion points versus the running time of the three approaches is plotted for different values of the radius r. Here we have chosen $N = \{1000, 2000, \ldots, 10000\}$. For details see Table 1

4 Summary

In this paper we have presented a new data structure for the computation of the dynamic nondominance problem and have shown its efficiency by a particular category of problems. Compared to the linear list approach and the quad tree approach the new method is advantageous in the case where the proportion

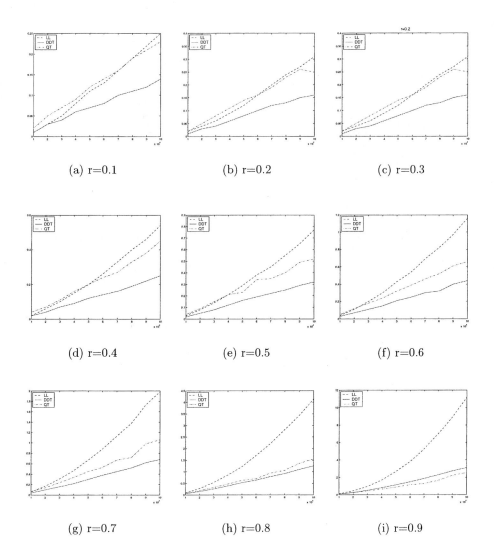

Fig. 4. Annulus generated test problem results for $k = 4$. For details see Figure 3 and Table 2

of the number of nondominated points to the number of all criterion vectors is "moderate". In future work the data structure has to be compared to the composite point approach ([6]).

Acknowledgements. The author thanks Robert Elsässer for helpfull discussions on the contents of this paper.

References

1. J. L. Bentley. Multidimensional binary search trees used for associative searching. *Communications of the ACM*, pages 509–517, 1975.
2. J. L. Bentley and J. H. Friedmann. Data structures for range searching. *Computing Surveys*, 4:398–409, 1979.
3. M. Dellnitz, R. Elsässer, T. Hestermeyer, O. Schütze, and S. Sertl. Covering Pareto sets with multilevel subdivision techniques. to appear, 2002.
4. R. M. Everson, J.E. Fieldsend, and S. Singh. Full elite sets for multi-objective optimization. In I. C. Parmee, editor, *Adaptive Computing in Design and Manufacture V*, Springer, 2002.
5. F. P. Preparata and M. I. Shamos. *Computational Geometry - An Introduction*. Springer Verlag, 1988.
6. J. Fieldsend, R.M. Everson, and S. Singh. Using unconstrained elite archives for multi-objective optimisation. to appear, 2003.
7. R. A. Finkel and J. L. Bentley. Quad trees, a datastructure for retrieval on composite keys. *Acta Informatica*, 4:1–9, 1974.
8. P. Gupta, R. Janardan, M. Smid, and B. Dasgupta. The rectangle enclosure and point-dominance problems revisited. *Int. J. Comput. Geom. Appl.*, 5:437–455, 1997.
9. W. Habenicht. Quad trees, a datatructure for discrete vector optimization problems. *Lecture Notes in Economics and Mathematical Systems*, 209:136–145, 1983.
10. M. de Berg and M. van Kreveld and M. Overmars and O. Scharzkopf. *Computational Geometry: algorithms and applications*. Springer Verlag, 1997.
11. E. M. McCreight. Priority search trees. *SIAM Journal on Computing*, 14:257–276, 1985.
12. S. Mostaghim, J. Teich, and A. Tyagi. Comparison of data structures for storing pareto-sets in moeas. *Int. J. Comput. Geom. Appl.*, 5:437–455, 2002.
13. M. Sun and R. E. Steuer. InterQuad: An interactive quad tree based procedure for solving the discrete alternative multiple criteria problem. *European Journal of Operational Research*, 89:462–472, 1996.
14. M. Sun and R. E. Steuer. Quad trees and linear lists for identifying nondominated criterion vectors. *INFORMS J. Comp.*, 8:367–375, 1996.

The Measure of Pareto Optima
Applications to Multi-objective Metaheuristics

M. Fleischer

Institute for Systems Research
University of Maryland, College Park
mfleisch@isr.umd.edu

Abstract. This article describes a set function that maps a set of Pareto optimal points to a scalar. A theorem[1] is presented that shows that the maximization of this scalar value constitutes the necessary and sufficient condition for the function's arguments to be maximally diverse Pareto optimal solutions of a discrete, multi-objective, optimization problem. This scalar quantity, a hypervolume based on a Lebesgue measure, is therefore the best metric to assess the quality of multiobjective optimization algorithms. Moreover, it can be used as the objective function in simulated annealing (SA) to induce convergence in probability to the Pareto optima. An efficient, polynomial-time algorithm for calculating this scalar and an analysis of its complexity is also presented.

1 Introduction

This article describes a measure theoretic approach for defining a set function that can be utilized for solving *multi*-objective optimization problems (MOPs). Zitzler *et al.* introduced the foundation for this set function in the following passage:

> In the two dimensional case each Pareto optimal solution **x** covers an area, a rectangle, defined by the points $(0,0)$ and $(f_1(\mathbf{x}), f_2(\mathbf{x}))$. The union of all rectangles covered by the Pareto optimal solutions constitutes the space totally covered, its size is used as measure. This concept may be canonically extended to multiple dimensions [2].

This article embellishes this notion of a set-cover measure by 1) extending it to an arbitrary number of dimensions, 2) rigorously proving that the maximization of the associated set function's scalar output is the necessary and sufficient condition for its arguments to be Pareto optimal solutions to a multi-objective optimization problem, and finally, 3) using insights from this proof to develop an efficient algorithm for computing the value of this set function. Analysis of the algorithm's complexity is also provided.

As the reader will no doubt discover, the intuition for this scalar is quite simple, yet its first appearance was surprisingly quite recent (much to the frustration

---

[1] See [1] for a formal proof.

of the author) [2,3,4]. Although in the two dimensional case (two objective functions) proving the validity of this measure seems almost trivial, for an arbitrary number of objective functions the proof seems less obvious, but is useful as it serves four purposes:

1. it establishes, with mathematical rigor, that the maximum value of the set function is a necessary and sufficient condition for the Pareto optimality of the function's arguments, hence, is *the best*[2] measure for evaluating heuristics that seek to find Pareto optima;
2. it therefore provides a sound mathematical basis for comparisons to other similar measures or approximations;
3. the proof points the way to a simple approach for calculating the measure;
4. it provides a mechanism for generalizing *any* optimization metaheuristic to handle multiple objectives.

With regard to Point 4, this scalar can be used, *e.g.*, as the objective function in simulated annealing (SA). Because it is well-known that SA converges in probability to the global optima ([7]), using this set function as the objective function in SA induces SA to *converge in probability to Pareto optima!*[3]

This article is organized as follows: Section 2 provides background on approaches for solving multi-objective optimization problems and recent results in the literature. This includes a philosophical discussion of the issues surrounding the relative merits of using genetic algorithms (GAs) versus SA. Formal definitions of Pareto optimality and other mathematical elements are described in Section 3. Section 3.1 describes the hypervolume in MOPs, its mathematical characteristics and presents the main results. Section 4 presents an efficient algorithm for computing this scalar and an analysis of its complexity. Finally, Section 5 discusses issues for future research and provides concluding remarks.

2 Background

2.1 Considerations of GAs vs. SA

Quite a few multi-objective algorithms have been described in recent years often motivated by design optimization problems. Most of these approaches have been based on GAs (see *e.g.*, [9,10]) although some have been based on SA [11]. While the level of research into multi-objective GAs seems to dominate similar research using an SA approach, SA may provide some untapped potential and have several advantages over GAs for solving MOPs.

One clear advantage of SA is its mathematical convergence properties described earlier. As will become clear later on, using the set function as the

---

[2] Regarding Points 1 and 2, practicalities may suggest other measures with more utility depending upon the nature of the problem and the algorithm used to solve it (see *e.g.*, [5,6]).
[3] This *mathematical* convergence is implied from the well-known converge in probability of SA coupled with the results presented here. cf. [8].

objective function in SA forces convergence to distinct Pareto optima.[4] This means that SA will converge to solutions with as good a 'spread' as possible. This *diversity* of solutions has been cited in a number of articles as an indicator of good performance of multi-objective optimization algorithms [3,5,10].

Notwithstanding the mathematical convergence of SA, GAs offer advantages in problems where some structure exists in the underlying domain. Indeed, the very elements of the GAs, in particular the crossover operator, lend themselves toward propagating those features in a chromosome that tend to be associated with high fitness values [12]. These particular advantages of GAs may however be diminished when compared to an SA-based approach *in the context of MOPs*.

Quite often in MOPs features associated with the underlying structure are masked and confounded by the interplay of several competing objective functions. Pareto optima may correspond to solutions that are not local optima of *any* of the objective functions—they may lie on the sides of hills rather than at their tops or bottoms. In other words, the pre-image of Pareto optima may be scattered throughout the domain space in an apparently haphazard or random manner rendering any structure within it to be of little or no consequence. As such, the *thermodynamic* approach inherent in SA may be more suitable for solving MOPs than GAs.

Before an elegant SA approach for solving MOPs is a realistic possibility, however, an efficient method for calculating this set function value is needed. To date, this has not been done even in the context of GAs although some GA approaches have indirectly utilized this notion of a scalar. Wu et al. [5] quantify a *hyperarea difference* metric closely related to the hypervolume based on the Lebesgue measure of the set of dominated points. Fonseca, et al. [6] describe the concept of the "attainment surface" which attempts to quantify how different chromosomes contribute to a performance metric linked to this scalar. Other methods involve archiving or updating solutions that are non-dominated [13], hence indirectly maximize this scalar. All of these methods, in various ways, attempt to produce solutions that ultimately maximize the value of this scalar, but avoid dealing with or computing it directly ostensibly for various reasons: either it is not computationally feasible, it is too difficult given some formulations, it can be well approximated, or other indirect methods are simpler (in some sense).

To illustrate why, perhaps, this measure has not been used directly consider the application of the *inclusion-exclusion* formula described in [14] (see also [5]). For sets $F_1, F_2, \ldots, F_n$ (using William's notation here) with μ the Lebesgue measure:

$$\mu\left(\bigcup_{i\leq n} F_i\right) = \sum_{i\leq n} \mu(F_i) - \sum\sum_{i<j\leq n} \mu(F_i \cap F_j) + \\ \sum\sum\sum_{i<j<k\leq n} \mu(F_i \cap F_j \cap F_k) \\ - \cdots + (-1)^{n-1} \mu(F_1 \cap F_2 \cap \ldots \cap F_n) \quad (1)$$

successive partial sums alternating between over- and under-estimates. [14, p.21].

---
[4] This depends on how many decision variables relative to Pareto optima are used in SA.

This rather ugly and unwieldy expression (in terms of computation) is due to the number of combinations of intersection sets the measures of which must be added and subtracted from the sum of the unions. Indeed, Wu et al. [15, p. 47] show a closed form solution for their *hyperarea difference* measure, something directly related to the hypervolume described here, that accounts for just three points that was quite "cumbersome" [5, see *e.g.*, Eq.(16) on p. 21]. Accounting for more points with more objective functions further complicates this approach. See also [3,4,5] for related works on quality metrics. Knowles [16, p.74] provides an exponential algorithm for calculating it. Figure 2 illustrates the potential difficulties in computing this hypervolume that can arise from the topological complications in higher dimensions. It also provides hints for the efficient,*i.e.,* polynomial-time computation of this scalar inspired by the proof in Section 3.1 and the area of *computational geometry*.

2.2 Computational Geometry

Computation of the hypervolume has a similar flavor to problems in *computational geometry*, a relatively new area of computer science. It turns out that Pareto optimal points constitute a *maximal set* of points (they have identical definitions). In the field of computer science, many algorithms pertaining to maximal sets have been studied. For example, articles have focused on dynamically maintaining a list of maximal points [17,18].

Notwithstanding the research on maximal sets, there does not seem to be any literature from the computer science community concerning the hypervolume of space covered by maximal sets even in the basic texts cited earlier (cf. [16]). This seems to be the state-of-affairs in the optimization community as well except for those references cited herein that refer to Zitzler, *et al.* [2].

3 Mathematical Preliminaries and Definitions

The main results require the domain space to be composed of a finite set of points. The reasons for this requirement will become clear later on. Let $\mathcal{X} \subset \mathrm{R}^d$ be a *finite* set of s feasible points in R^d for some MOP with objective functions f_i, $i = \{1, \ldots, n\}$ where for each i

$$f_i : \mathrm{R}^d \longrightarrow \mathrm{R}^1.$$

Also, let $\mathcal{X}_P \subseteq \mathcal{X}$ be the set of Pareto optima[5] in set $\mathcal{X}$ with $p \leq s$ the number of Pareto optima. Thus for MOPs with n objectives, each vector $\mathbf{x} \in \mathcal{X}$ produces n (real) objective function values $\mathbf{f}(\mathbf{x}) = \{f_1(\mathbf{x}), \ldots, f_n(\mathbf{x})\}$ corresponding to a single point $\mathbf{p_x}$ in objective function space. The image of set $\mathcal{X}$ is therefore a finite set of points $\mathbf{p_x} \in \mathrm{R}^n$ and denoted by S with $s' \leq s$ elements. Thus, the vector valued mapping $\mathbf{f}(\mathcal{X}) = S$ is onto and not necessarily one-to-one.

It is useful to define certain relationships between points $\mathbf{p}_x$ and a set S.

---

[5] See [1] for a formal definition of a Pareto optimal point.

Definition 1. *A point $p_x \in S$ corresponding to solution x is* **non-dominated with respect to set** *S if and only if for all other points $p_y \in S$ there exists an i such that $f_i(x) < f_i(y)$ (assume that if S has a single element it is non-dominated with respect to set S). A set S is a* **non-dominated set** *if all points $p \in S$ are non-dominated with respect to set S.*

3.1 The Hypervolume

The goal of this section is to define a set function that maps a set of points to a scalar and that has certain desirable properties. Let m be the number of arguments $x \in \mathcal{X}$ of a set function F. These m points in $\mathcal{X}$ map to m points $p \in S$ in objective function space which must then map to a *single* scalar, the hypervolume μ. Eqn. (2) makes this mapping clear:

$$\left. \begin{array}{ccc} x_1 \mapsto \{f_1(x_1), f_2(x_1), \ldots, f_n(x_1)\} & = & p_1 \\ \vdots & \vdots & \vdots \\ x_l \mapsto \{f_1(x_l), f_2(x_l), \ldots, f_n(x_l)\} & = & p_l \\ \vdots & \vdots & \vdots \\ x_m \mapsto \{f_1(x_m), f_2(x_m), \ldots, f_n(x_m)\} & = & p_m \end{array} \right\} \mapsto \mu. \qquad (2)$$

Such a mapping is made possible by generalizing the concept of optimality using a measure theoretic approach and extending the associated measures of performance to the multi-dimensional case. Zitzler [2] in effect[6] uses the interval length $M - f(x)$ as a measure of performance where M is some upper bound on $f(x)$, and $f(x)$ is some objective function value.[7] This interval captures the important feature of a performance measure, the ability to rank solutions according to their desirability. For given M, the larger the interval length, the smaller the objective function value, hence, the better the solution. Generalizing Zitzler's *et al.* [2] notion of interval as a set measure and establishing the mappings in (2) requires the following formal definitions. For n objective functions, a solution x defines the following dominance set and measure.

Definition 2. Lebesgue Measure of the Deleted Dominated Set: *Let $p = (f_1, f_2, \ldots, f_n)$[8] represent a point in objective function space where, without loss of generality, $i = 1, \ldots, j$ are indices of minimization functions and $i = j +$*

---

[6] The quote in Section 1 was obviously referring to a maximization problem. Here we extend this notion by defining upper bounds on minimizing objective functions.

[7] Stated this way is subtly different than stating that this measure is the size of the set cover of dominated solutions (which it of course is—see [4]). It is this subtle difference that indicates we can use this measure as the objective function in an optimization algorithm provided an efficient algorithm exists to calculate its value.

[8] Note that $p_x \equiv (f_1(x), f_2(x), \ldots, f_m(x))$ and will often be denoted using the simpler notation $(f_1, f_2, \ldots, f_m)$ and p where it is sufficiently clear that we mean $(f_1(x) \ldots)$ and p_x. Also, depending on the context, the dominance set associated with some point p_x or p_i will be denoted as $D_{p_x} \equiv D_x$ and $D_{p_i} \equiv D_i$, respectively.

$1, \ldots, n$ are the indices of maximization functions. Let $f_i(\mathbf{x})$ be a particular value of f_i produced by solution $\mathbf{x}$ where M_i and m_i are the upper and lower bounds for minimization and maximization objective functions, respectively. Then the deleted dominated set $D_\mathbf{x} = \{\mathbf{p} : \forall\ i,\ f_i \in ([m_i, f_i(\mathbf{x})] \cup [f_i(\mathbf{x}), M_i]) \wedge \mathbf{p} \neq \mathbf{p_x}\}$[9] and constitutes a set of points $\mathbf{p}$ **strictly inferior** to $\mathbf{p_x}$. The Lebesgue measure of this set is $\mu(D_\mathbf{x}) = \left(\prod_{i=1}^{j}[M_i - f_i(\mathbf{x})]\right)\left(\prod_{i=j+1}^{n}[f_i(\mathbf{x}) - m_i]\right)$.

The following lemmas[10] will be useful in proving the main result. Formal proofs of them may be obtained in [1].

Lemma 1. *Given a finite set of points S in objective function space, point $\mathbf{p_x} \in S$ is dominated if and only if there exists a $\mathbf{p_y} \in S$ such that $\mathbf{p_x} \in D_\mathbf{y}$.*

Corollary to Lemma 1: *Point $\mathbf{p_x} \in S$ is non-dominated with respect to S if and only if for all $\mathbf{p_y} \in S$, $\mathbf{p_x} \notin D_\mathbf{y}$.*

Definition 3. *Let $S_m = \{\mathbf{p}_1, \ldots, \mathbf{p}_m\} \subseteq S$, a set of m feasible points in objective function space. Then the dominance set D_{S_m} is the union of the dominance sets of each element of set S_m. That is, $D_{S_m} = \bigcup_{i=1}^{m} D_{\mathbf{p}_i}$ and the measure of this set is $\mu(D_{S_m}) = \mu\left(\bigcup_{i=1}^{m} D_{\mathbf{p}_i}\right)$.*

Lemma 2. Dominance Calculus: *From the previous definitions the following statements are true for any point sets A and B, $D_A \bigcup D_B \equiv D_{A \cup B}$. If $A \cap B \neq \emptyset$, then $D_A \bigcap D_B \equiv D_{A \cap B}$.*

The following definitions and lemmas show important relationships among points in a set S and bounds on objective function values and will be used to prove the main result in Theorem 1. For notational simplicity and without loss of generality, we shall assume all objective functions are to be minimized.

Definition 4.

$\mathbf{F}_i = \{m_i, f_i(\mathbf{x}_1), f_i(\mathbf{x}_2), \ldots, f_i(\mathbf{x}_m), M_i\}$, the set of the ith objective function values among points in set S and their lower and upper bounds.

$u_i(\mathbf{p_x})$, the least upper bound of $f_i(\mathbf{x})$ in set $\mathbf{F}_i$.

$l_i(\mathbf{p_x})$, the greatest lower bound of $f_i(\mathbf{x})$ in set $\mathbf{F}_i$.

$D'_\mathbf{x} = \{(f_1, f_2, \ldots, f_m) : \forall i, f_i(\mathbf{x}) < f_i < u_i(\mathbf{p_x})\}$, the set of points in a minimization problem exclusive to set $D_\mathbf{x}$. See Lemma 3 below.

---

[9] Note that since $m_i < f_i(\mathbf{x})$, then the interval $[f_i(\mathbf{x}), m_i] = \emptyset$ and similarly for the case where M_i is an upper bound for a minimizing function f_i. This notation effectively deals with both minimization and maximization objectives.

[10] The essence of Lemmas 1 and 4 can be found in other texts. Their form here facilitates the proofs of Lemmas 2 and 3 and Theorem 1 which are new.

For example, given $\mathbf{p}_1 = (1,3,2)$ $\mathbf{p}_2 = (4,1,6)$ $\mathbf{p}_3 = (4,5,1)$ with $m_i = 0$ and $M_i = 7$ for all i, then $\mathbf{F}_2 = \{0,3,1,5,7\}$, $u_2(\mathbf{p}_1) = 5$, $u_2(\mathbf{p}_2) = 3$ and

$$D'(\mathbf{p}_2) = \{(f_1, f_2, f_3) : 4 < f_1 < 7, \ 1 < f_2 < 3, \ 6 < f_3 < 7\}. \quad (3)$$

a hypercube of a set of points in objective function space exclusive to $D_{\mathbf{p}_2}$ with dimensions $3 \times 2 \times 1$. Figure 1 illustrates set $D'_{\mathbf{x}}$ for the two-dimensional case and the relationships of the definitions above. These will help to clarify elements of the proof. Note that the shaded area indicated by hash marks associated with $\mathbf{p}_{\mathbf{x}}$ shows a set of points exclusive to $D_{\mathbf{x}}$ that add to the measure of set S_m.

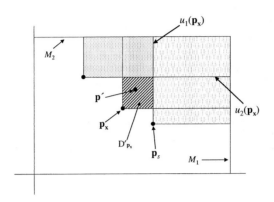

Fig. 1. Relationships of Points, $D'_{\mathbf{x}}$ and Upper bounds

Lemma 3 is a key element in proving the main result and provides the basic idea behind the algorithm described in Section 4. In fact, these results are based on the fact that there are a finite number of feasible points $\mathbf{x}$ in the domain. Without this requirement, Lemma 3 and its corollary would not be available to help prove Theorem 1. It is for this reason that the main results are restricted to a finite set of Pareto optima.

Lemma 3. *For every non-dominated point $\mathbf{p}_{\mathbf{x}} \in S$ there exists a set of points $D'_{\mathbf{x}} \subset D_{\mathbf{x}}$ such that for all $\mathbf{p}' \in D'_{\mathbf{x}}$, $\mathbf{p}'$ is non-dominated with respect to set $S \setminus \mathbf{p}_{\mathbf{x}}$ and such that for all $\mathbf{p}_{\mathbf{y}} \in S$, $D'_{\mathbf{x}} \cap D_{\mathbf{y}} = \emptyset$.*

Corollary to Lemma 3: For all sets $D'_{\mathbf{x}} \subset D_S$, $\mu(D'_{\mathbf{x}}) > 0$.

Lemma 4. *Given points $\mathbf{p}_{\mathbf{x}}, \mathbf{p}_{\mathbf{y}} \in S$, $\mathbf{p}_{\mathbf{x}}$ dominates $\mathbf{p}_{\mathbf{y}}$ if and only if $D_{\mathbf{y}} \subset D_{\mathbf{x}}$.*

Corollary to Lemma 4: If $\mathbf{p}_{\mathbf{x}}$ dominates $\mathbf{p}_{\mathbf{y}}$ then $\mu(D_{\mathbf{x}} \cup D_{\mathbf{y}}) = \mu(D_{\mathbf{x}}) > \mu(D_{\mathbf{y}})$.

Theorem 1 shows that the measure of set S_m achieves its maximum value if and only if points $\mathbf{p} \in S_m$ are Pareto optimal. The reader will note that the proof of the statement *if points $\mathbf{p} \in S_m$ are Pareto optima, then the measure of the set of dominated points is at its maximum value* is much easier to prove than the statement *if the measure of the set of dominated points is at its maximum value, then the points* $\mathbf{p} \in S_m$ *are Pareto optima*. This latter statement however seems important to justify in light of how random search algorithms such as SA operate. SAs search mechanism depends on the objective function value associated with the random guesses of decision variables. Thus, we want to know that if the objective function (this set measure) is as high as can be, does that suffice to claim the decision variables are Pareto optima? This proof says that it is indeed sufficient and, further, that it is necessary (depending on the number of variables—see the cases below).

Theorem 1. *Let S_m be a given set of m points in objective function space in an MOP with p Pareto optimal solutions. Let M_i be the given bounds for f_i (for the sake of clarity and without loss of generality, we assume each objective function is to be minimized and the M_i are therefore upper bounds on f_i). Let $F(\mathbf{x}_1, \mathbf{x}_2, \ldots, \mathbf{x}_m) \equiv \mu(D_{S_m})$ be a set function mapping a subset of points $S_m \subseteq S$ to the Lebesgue measure of the dominance set. Then the following are true:*

Case 1 ($m < p$): *If F is at its maximum value, then all m points in S_m are Pareto optimal and for all $\mathbf{p}_k, \mathbf{p}_l \in S_m$, $k \neq l \Rightarrow \mathbf{p}_k \neq \mathbf{p}_l$.*

Case 2 ($m \geq p$): *F is at its maximum value* **if and only if** *there is a subset $S' \subseteq S_m$ of size p such that all $\mathbf{p} \in S'$ are Pareto optimal and for all $\mathbf{p}_k, \mathbf{p}_l \in S'$, $k \neq l \Rightarrow \mathbf{p}_k \neq \mathbf{p}_l$.*

Proof. See [1].

4 Calculating the Hypervolume

As noted earlier, calculating the hypervolume based on the inclusion-exclusion formula can be quite messy. The basic idea behind the approach describe here stems directly from Lemma 3, its corollaries, and Theorem 1: the algorithm successively lops off hybercubes containing points in sets $D'_\mathbf{x}$ (points that do not intersect with any other sets $D_{\mathbf{p}_i}$) and adds its volume to a partial sum (*LebMeasure*). This 'lopping off' procedure continues until there is nothing left with positive measure.

The algorithm works by first storing the original set of points S in a list L. The lopped off hypercube is 'removed' from L by substituting the first point with a set of new points created by the 'removal' process. This is done using the SpawnData procedure. The 'spawned' points that are nondominated with respect to the remaining points in L are added to L. Thus, the size of L grows and shrinks as this process continues inevitably halting when the last point's (vector) volume is added to the partial sum. This procedure avoids the necessity of dealing with intersection sets and works for an arbitrary number of objective functions and finite set of points. In the following pseudocode, two data structures, *List* and *SpawnData*, hold the original and spawned vectors respectively, and *Size* equals the number of vectors in *List*.

The LebMeasure Algorithm

```
Initialize:   LebMeasure = 0.0;
              newSize = Size;
while(newSize > 1) do: {
        lopOffVol := 1.0;
        get first vector p_1 in List
        for(i = 0; i < n; i++) do {
              b_i := getBoundValue(f_i(x_1)) = {u_i(p_1), l_i(p_1)}
              spawnVector(p_1, i, b_i);  //Add spawned vectors to SpawnData
              lopOffVol *= |f_i(x_1) - b_i|;
        }
        LebMeasure += lopOffVol;
        delete p_1 from List
        newSize = ndFilter(List, SpawnData);
              //Check if and List vectors dominate SpawnData vectors
        clear SpawnData
} end of while loop.
lastVol = 1.0;
for(i = 0; i < n; i++){
        lastVol *= |f_i(x_1) - {m_i, M_i}|;
}
return(LebMeasure);
```

getBoundValue($f_i(\mathbf{x}_1)$): This routine compares $f_i(\mathbf{x}_1)$ to the corresponding elements in vectors 2 through *Size* and returns either $u_i(\mathbf{p}_1)$ or $l_i(\mathbf{p}_1)$ depending on whether f_i is to be minimized or maximized, respectively and assigns this value to b_i (note the notation in the pseudocode). Its time complexity is therefore $(L-1)n$ where L is the number of original vectors in *List* and n the number of objective functions. Thus, in the first while loop, the time complexity is $(m-1)n$.

spawnVector($\mathbf{p}_1, i, b_i$): This routine creates the following n vectors (is executed n times) based on the removal of $\mathbf{p}_1$ from *List*:[11,12]

$$\begin{aligned}
\mathbf{p}_{11} &= \{l_1(\mathbf{p}_1), f_2, \ldots, f_n\} = \{b_1, f_2, \ldots, f_n\} \\
\mathbf{p}_{12} &= \{f_1, l_2(\mathbf{p}_1), \ldots, f_n\} = \{f_1, b_2, \ldots, f_n\} \\
&\vdots \qquad\qquad \vdots \qquad\qquad \vdots \\
\mathbf{p}_{1n} &= \{f_1, f_2, \ldots, l_n(\mathbf{p}_1)\} = \{f_1, f_2, \ldots, b_n\}
\end{aligned} \quad (4)$$

The notation $\mathbf{p}_{1j}$ refers to the jth spawned vector by the removal of $\mathbf{p}_1$ from *List*. These n vectors comprise the *SpawnData* data structure. Note that all vectors in *SpawnData* are non-dominated with respect to *SpawnData*.

ndFilter: This routine compares the vectors in *List* to those in *SpawnData* and deletes any vectors in *SpawnData* that are either dominated by a vector in *List* or has an m_i or M_i as one of its elements. The remaining vectors in *SpawnData* are then inserted into *List* replacing $\mathbf{p}_1$.

---
[11] Note in this example that all objective functions are to be maximized.
[12] To simplify the expression, we will use f_i for objective function values $f_i(\mathbf{x}_1)$.

The following example illustrates this algorithm's operation on *List* where the *SpawnData* data structure is produced by the removal of vector $\{2, 2, 2\}$ after the first while loop is completed:

$$\text{List} = \begin{bmatrix} 2, 2, 2 \\ 1, 3, 1 \\ 1, 1, 3 \\ 3, 1, 1 \end{bmatrix} \quad \text{SpawnData} = \begin{bmatrix} 1, 2, 2 \\ 2, 1, 2 \\ 2, 2, 1 \end{bmatrix}.$$

The routine first initializes $Size = 4$. The first for loop selects the vector $\{2, 2, 2\}$, computes bounds for each vector element (1 in this case), creates *SpawnData* (as above), and calculates the incremental hypervolume based on the dimensions of the lopped off hypercube, *i.e.*, $lopOffVol = 1 \times 1 \times 1 = 1$ which is added to *LebMeasure*. Vector $\{2, 2, 2\}$ is then deleted from List.

The routine **ndFilter** deletes any vectors in *SpawnData* dominated by the *remaining* vectors in *List* or that have elements equal to m_i or M_i (in that case, the edge of a hypercube would be of zero length). The resulting *SpawnData* is then added to *List* and $newSize = 6$. *List* and its *SpawnData* are now

$$\text{List} = \begin{bmatrix} 1, 2, 2 \\ 2, 1, 2 \\ 2, 2, 1 \\ 1, 3, 1 \\ 1, 1, 3 \\ 3, 1, 1 \end{bmatrix} \quad \text{SpawnData} = \begin{bmatrix} 0, 2, 2 \\ 1, 1, 2 \\ 1, 2, 1 \end{bmatrix}.$$

Although *List* is larger, eventually all the vectors in *SpawnData* will either contain m_i or will be dominated by vectors in *List*. Now vector $\{1, 2, 2\}$ heads *List* and produces the *SpawnData* above all of which will be deleted since the first vector has an element at the lower bound (0), and the second and third vectors are dominated by the remaining vectors in *List*. *List* thus shrinks from 6 vectors to 5 and eventually to 1 breaking the while loop and ending the computation with the last vector's hypervolume being added to *LebMeasure*.

The following figures depict the operation of this algorithm. Initially, *List* corresponds to Figure 2a. After the first while loop, *List* contains 6 vectors corresponding to Figure 2b where a hypercube with dimensions $1 \times 1 \times 1$ has been "lopped off". Its volume of 1 is added to the partial sum of *LebMeasure*. Continuing with this procedure until it ends yields $LebMeasure = 11$. The reader can verify the result by starting the procedure with any of the four points (*i.e.*, the order of vectors in *List* makes no difference to the value of *LebMeasure*).

4.1 Complexity of Computing LebMeasure

Analyzing the long-run behavior of LebMeasure first requires an assessment of the size of a problem instance. Assuming all vectors in *List* are non-dominated, the size of the problem instance involves $m \times n$ values. Including the bounds for

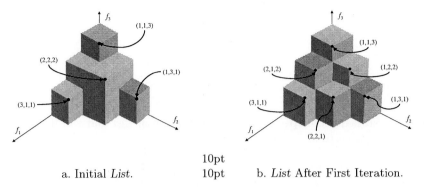

a. Initial *List*.   b. *List* After First Iteration.

Fig. 2. Evolution of the Hypervolume

each objective, the $(m+1)n$ numbers are sufficient to compute the hypervolume, hence constitutes the size of the problem instance.

The time complexity of the algorithm can be determined by first observing, as in the example, that the size of *List* may grow and shrink at various stages of the algorithm. The key to analyzing the complexity of LebMeasure therefore is in determining the manner and extent to which spawned vectors are added to and deleted from *List*. Thus, understanding the spawning procedure and how it works is critical towards understanding the complexity of LebMeasure. Again, perusal of the pseudocode and study of Figures 2a,b should help.

Before proceeding further, an important distinction is made among the vectors in *List*—those that are original members of *List*, and those that are spawned from these original members of *List*. Those vectors having only one subscript, e.g., $\mathbf{p}_1$ are the original members of *List* while those with more than one subscript, e.g., $\mathbf{p}_{11}$ are spawned descendants from the original members.

One important property of the spawning procedure is that the total number of spawned vectors in *List* can never exceed n, the number of objective functions. This fact is useful in performing a worst-case analysis. To gain insight into the developing patterns, the first few iterations of the while loop are closely examined.

Without loss of generality, assume each objective is to be maximized. In the first iteration of the while loop, and for a worst-case analysis, assume that all the spawned vectors of $\mathbf{p}_1$ in *SpawnData* are non-dominated with respect to the vectors in *List*. In this case, all of these vectors are added to *List* and $\mathbf{p}_1$ is removed. These spawned vectors are indicated in (4). The length of *List* therefore increases from m to at most $m + n - 1$ and evolves thusly:

$$\mathbf{p}_1, \mathbf{p}_2, \ldots, \mathbf{p}_m$$
$$\downarrow$$
$$\mathbf{p}_{11}, \ldots, \mathbf{p}_{1n}, \mathbf{p}_2, \ldots, \mathbf{p}_m$$

requiring the following computational effort:

Table 1. Complexity of first while loop

Subroutine	Complexity
BoundVal	$(m-1)n$
Spawn	n
ndFiler	$(m-1)n$
Total	$2(m-1)n + n$

Now the spawned vector $\mathbf{p}_{11}$ (see (4)) is at the top of *List* and will itself spawn the following n vectors:

$$\begin{aligned} \mathbf{p}_{111} &= \{\, l_1(\mathbf{p}_{11}),\ f_2,\ \ldots,\ f_n\,\} \\ \mathbf{p}_{112} &= \{\, l_1(\mathbf{p}_1),\ l_2(\mathbf{p}_1),\ \ldots,\ f_n\,\} \\ &\vdots \\ \mathbf{p}_{11n} &= \{\, l_1(\mathbf{p}_1),\ f_2,\ \ldots,\ l_n(\mathbf{p}_1)\,\} \end{aligned} \quad (5)$$

Note that the vector $\mathbf{p}_{111}$ is the same as vector $\mathbf{p}_{11}$ except for the first component ($l_1(\mathbf{p}_{11})$ versus $l_1(\mathbf{p}_1)$). Note also that by definition, $l_1(\mathbf{p}_{11}) < l_1(\mathbf{p}_1)$. The other vectors in (5) are the same as the corresponding vectors in (4) except for the first components. In (5), the first components of all the vectors (except the first) have the value $l_1(\mathbf{p}_1)$ instead of f_1. This means that those vectors in (4) dominate the corresponding ones in (5)—*i.e.*, for all i, $\mathbf{p}_{1i}$ dominates $\mathbf{p}_{11i}$. These vectors will therefore not be added to *List*. But because $\mathbf{p}_{11}$ has been removed from *List*, no point in *List necessarily* dominates $\mathbf{p}_{111}$. Consequently, the only possible point among the spawned points $\mathbf{p}_{11i}$ (*i.e.*, in (5)) that *may* be non-dominated is $\mathbf{p}_{111}$ which for purposes of a worst-case analysis is added to *List* in effect replacing $\mathbf{p}_{11}$. Thus, the size of *List* remains at $m + n - 1$. *List* evolves thusly

$$\mathbf{p}_{11}, \mathbf{p}_{12}, \ldots, \mathbf{p}_{1n}, \mathbf{p}_2, \ldots, \mathbf{p}_m$$
$$\downarrow$$
$$\mathbf{p}_{111}, \mathbf{p}_{12}, \ldots, \mathbf{p}_{1n}, \mathbf{p}_2, \ldots, \mathbf{p}_m$$

requiring the following computational effort:

Table 2. Complexity of while loop 2 to $(m-1)$

Subroutine	Complexity
BoundVal	$(m-1)$
Spawn	n
ndFiler	$(m-1)n$
Total	$(m-1) + (m-1)n + n$

Now, $\mathbf{p}_{111}$ is at the top of *List* and may spawn another set of vectors, but again, the same property as described above holds and may yield a maximum of only one non-dominated $\mathbf{p}_{1111}$ and so on. Thus, for all these iterations, the size of *List* remains at $m + n - 1$. Eventually, some descendant of $\mathbf{p}_{11}$ of the form $\mathbf{p}_{1\ldots1i}$ becomes dominated by vectors in *List* or the $l_1(\mathbf{p}_{111\ldots})$ becomes equal to the lower bound m_i at which point no vectors from *SpawnData* are added to *List* and the last of all the descendants of $\mathbf{p}_{11}$ along with $\mathbf{p}_{11}$ are removed from *List*. The question arises as to how many successive generations of $\mathbf{p}_{11}$ are possible in the worst-case. Obviously, it cannot be greater than $|List| - 1$. Consequently, an upper bound on the number of generations that the first *spawned* vector in *List* may spawn is $m - 1$. Thus, the complexity of the while loop 3 to $(m - 1)$ is the same as given in Table 2.

At this point in the algorithm, the next while loop starts with only $m + n - 2$ vectors $\mathbf{p}_{12}, \ldots, \mathbf{p}_{1n}, \mathbf{p}_2, \ldots, \mathbf{p}_m$. Now the spawned vector $\mathbf{p}_{12}$ is at the top of *List* and spawns its descendants:

$$\begin{aligned}
\mathbf{p}_{121} &= \{\, l_1(\mathbf{p}_1),\ l_2(\mathbf{p}_1),\ \ldots,\ f_n\ \} \\
\mathbf{p}_{122} &= \{\quad f_1,\ \ l_2(\mathbf{p}_{12}),\ \ldots,\ f_n\ \} \\
&\vdots \\
\mathbf{p}_{12n} &= \{\quad f_1,\ \ l_2(\mathbf{p}_1),\ \ldots,\ l_n(\mathbf{p}_1)\,\}
\end{aligned} \tag{6}$$

and $\mathbf{p}_{12}$ is itself removed leaving $m + n - 3$ vectors in *List* to which are added the non-dominated vectors in *SpawnData*, some subset of (6). Again, the same pattern is present. That is, for all k, $\mathbf{p}_{1k}$ dominates vectors $\mathbf{p}_{12k}$. But now there are only $k - 2$ vectors from the first spawned set, $\mathbf{p}_{13} \ldots \mathbf{p}_{1k}$ left in *List* to dominate the $\mathbf{p}_{12i}$ generation. Consequently, it is possible that *two* vectors, $\mathbf{p}_{121}$ and $\mathbf{p}_{122}$, may be non-dominated with respect to vectors in *List*. The size of *List* therefore increases from $m + n - 2$ back to at most $m + n - 3 + 2 = m + n - 1$. *List* becomes:

$$\underbrace{\mathbf{p}_{121}, \mathbf{p}_{122}, \mathbf{p}_{13}, \ldots, \mathbf{p}_{1n}}_{n \text{ vectors}},\ \underbrace{\mathbf{p}_2, \ldots, \mathbf{p}_m}_{m-1 \text{ vectors}}$$

and its length is again $m + n - 1$. The following general observation is made:

Each time one of the first spawned vectors (e.g., vectors with 2 subscripts such as $\mathbf{p}_{1i}$) is removed from List, the successive generations of the remaining vectors of the form $\mathbf{p}_{1i}$, add vectors to List in sufficient numbers so that the length of List remains the same.

The decrease by 1 observed earlier happens at various points, but for purposes of a worst-case analysis can be ignored (it also simplifies the analysis). Thus, each of the first n spawned vectors, have the following number of basic computations where the length of *List* is m:

$$[\underbrace{2(m-1)n + n}_{\text{first iteration}} + \underbrace{(m-1)((m-1) + (m-1)n + n)}_{\text{iterations 2 to } m-1}]n$$

which is of order $m^2 n^2$. After these calculations, the process begins with the original vector $\mathbf{p}_2$ at the top of *List* with the size decremented. Accounting for this decrease in the length of *List* we have

$$T(m, n) \approx \sum_{i=0}^{m-1} (m-i)^2 n^2 = n^2 \sum_{i=0}^{m-1} (m-i)^2$$
$$= n^2 \left(\frac{m(m+1)(2m+1)}{6} \right)$$

from the Sum of Squares Formula [19, p.199]. Consequently, the time complexity of LebMeasure is $T(m, n) \in O(m^3 n^2)$.

5 Future Research and Conclusion

This article described a set function $F(\mathbf{x}_1, \mathbf{x}_2, \ldots, \mathbf{x}_m) = \mu(D_{S_m})$, a hypervolume on a point set S_m, that maps the arguments to a scalar and achieves its maximum value only when these arguments are distinct Pareto optima. Mapping Pareto optima to a scalar unifies the concepts of single and multi-objective optimization. A polynomial algorithm for calculating this hypervolume was also described.

Because this scalar provides necessary and sufficient conditions for the arguments to be Pareto optima, it is also the best measure for evaluating different multi-objective evolutionary algorithms. The many different GAs for example can all be evaluated according to the magnitude of this scalar quantity. By using a scalar quantity to evaluate performance, the average rate of convergence can also be assessed, hence, the performance of evolutionary algorithms quantified using appropriate statistics to estimate the average hypervolume over many independent trials.

Finally, this result shows how *any* multi-objective optimization problem can be put into standard math programming form with a single scalar objective function. As such, many global optimization metaheuristics can be recast to solve multi-objective problems. Simulated annealing, for example, and its parallel variants can be fashioned to converge in probability to Pareto optima. Future research will therefore describe the relative merits of using parallel versions of SA [20], with the hypervolume as the objective function, and various GA implementations.

Acknowledgement and Disclaimer. The author was supported in part by the Center for Satellite and Hybrid Communications Networks in the Institute for Systems Research at the University of Maryland, College Park, and through collaborative participation in the Collaborative Technology Alliance for Communications & Networks sponsored by the U.S. Army Research Laboratory under Cooperative Agreement DAAD19-01-2-0011 and the National Aeronautics and Space Administration under award No. NCC8-235.

The views and conclusions contained in this document are those of the author and should not be interpreted as representing the official policies, either expressed or implied, of the Army Research Laboratory, the National Aeronautics and Space Administration, or the U.S. Government.

References

1. Fleischer, M.A.: The measure of pareto optima: Applications to multiobjective metaheuristics. Technical Report 2002-32, Institute for Systems Research, University of Maryland, College Park, MD. (2002) This is an unabridged version of the instant article.
2. Zitzler, E., Thiele, L.: Multiobjective optimization using evolutionary algorithms–a comparative case study. In A.E. Eiben, T. Bäck, M.S.H.S., ed.: Proceedings of the Fifth International Conference on Parallel Problem Solving from Nature—PPSN V, Berlin, Germany (1998)

3. M. Laumanns, G.R., Schwefel, H.: Approximating the pareto set: Diversity issues and performance assessment. (1999)
4. Zitzler, E.: Evolutionary Algorithms for Multiobjective Optimization: Methods and Applications. Ph.d. dissertation, Swiss Federal Institute fo Technology (ETH), Zurich, Switzerland (1999) The relevant part pertaining to our Lebesgue measure is discussed in Ch. 3.
5. Wu, J., Azarm, S.: Metrics for quality assessment of a multiobjective design optimization solution set. Transactions of the ASME **123** (2001) 18–25
6. Fonseca, C., Fleming, P.: On the performance assessment and comparison of stochastic multiobjective optimizers. In G. Goos, J.H., van Leeuwen, J., eds.: Proceedings of the Fourth International Conference on Parallel Problem Solving from Nature—PPSN IV, Berlin, Germany (1998) 584–593
7. Mitra, D., Romeo, F., Sangiovanni-Vincentelli, A.: Convergence and finite-time behavior of simulated annealing. Advances in Applied Probability **18** (1986) 747–771
8. M. Laumanns, L. Thiele, K.D., Zitzler, E.: Archiving with guaranteed convergence and diversity in multi-objective optimization. In: Proceedings of the Genetic and Evolutionary Computation Conference (GECCO). (2002) 439–447
9. Coello, C.: Bibliograph on evolutionary multiobjective optimization. http://www.lania.mx/~coello/EMOO/EMOObib.html (2001)
10. Veldhuizen, D.V., Lamont, G.B.: Multiobjective evolutionary algorithms: Analyzing the state-of-the-art. Evolutionary Computation **8** (2000) 125–147
11. Czyzak, P., Jaskiewicz, A.: Pareto simulated annealing—a metaheuristic technique for multiple-objective combinatorial optimization. Journal of Multi-criteria Decision Analysis **7** (1998) 34–47
12. Reeves, C.: Modern Heuristic Techniques for Combinatorial Problems. John Wiley & Sons, Inc., New York, NY (1993)
13. Knowles, J.D., Corne, D.W.: Approximating the nondominated front using the pareto archived evolution strategy. Evolutionary Computation **8** (2000) 149–172
14. Williams, D.: Probability with Martingales. Cambridge University Press, Cambridge, England (1991)
15. Wu, J.: Quality Assisted Multiobjective and Multi-disciplinary Genetic Algorithms. Department of mechanical engineering, University of Maryland, College Park, College Park, Maryland (2001) S. Azarm, Ph.D. advisor.
16. Knowles, J.: Local Search and Hybrid Evolutionary Algorithms for Pareto Optimization. PhD thesis, The University of Reading, Reading, UK (2002) Department of Computer Science.
17. Kapoor, S.: Dynamic maintenance of maxima of 2-d point sets. SIAM Journal on Computing **29** (2000) 1858–1877
18. Frederickson, G., Rodger, S.: A new approach to the dynamic maintenance of maximal points in a plane. Discrete and Computational Geometry **5** (1990) 365–374
19. Hildebrand, F.B.: Introduction to Numerical Analysis. 2nd edn. Dover Publications, Inc., Mineola, NY (1987)
20. Fleischer, M.A.: 28: Generalized Cybernetic Optimization: Solving Continuous Variable Problems. In: Metaheuristics: Advances and Trends in Local Search Paradigms for Optimization. Kluwer Academic Publishers (1999) 403–418

Distributed Computing of Pareto-Optimal Solutions with Evolutionary Algorithms

Kalyanmoy Deb, Pawan Zope, and Abhishek Jain

Kanpur Genetic Algorithms Laboratory (KanGAL)
Indian Institute of Technology Kanpur, PIN 208 016, India
deb@iitk.ac.in

Abstract. In this paper, we suggest a distributed computing approach for finding multiple Pareto-optimal solutions. When the number of objective functions is large, the resulting Pareto-optimal front is of large dimension, thereby requiring a single processor multi-objective EA (MOEA) to use a large population size and run for a large number of generations. However, the task of finding a well-distributed set of solutions on the Pareto-optimal front can be distributed among a number of processors, each pre-destined to find a particular portion of the Pareto-optimal set. Based on the guided domination approach [1], here we propose a modified domination criterion for handling problems with a convex Pareto-optimal front. The proof-of-principle results obtained with a parallel version of NSGA-II shows the efficacy of the proposed approach.

1 Introduction

The importance and efficacy of using multi-objective evolutionary algorithms (MOEAs) have been well established in the recent past [3,2]. Not only does there exist quite a few efficient MOEAs, there also exist a number of interesting applications [5]. However, like in the single-objective optimization studies, the computational time needed for solving multi-objective optimization is usually large in solving real-world optimization problems. In solving such computationally expensive problems, distributed computing with multiple processors are often used in the context of single-objective optimization. However, not many studies have been dedicated in using distributed computing for multi-objective optimization. A review of the existing studies can be found in [2].

In this paper, we suggest a parallel MOEA approach based on non-dominated sorting GA (or NSGA-II), which attempts to distribute the task of finding the entire Pareto-optimal front among participating processors. This way, each processor is destined to find a particular portion of a convex Pareto-optimal front. With the help of a number of test problems, the efficacy of the proposed procedure is demonstrated.

2   Need for Distributed Computing in EMO

The act of optimization demands comparison of a number of solutions before arriving at or near the optimal solution(s). Since the evaluation of a solution can be time-consuming for real-world problems, the use of distributed computing has always been a major thrust in the area of optimization. Thus, it is needless to reiterate the importance of distributed computing in the case of multi-objective optimization. In fact, the importance of distributed computing is even greater in multi-objective optimization for the following reason.

With the increase of the number of objectives, the dimension of the true Pareto-optimal front increases. For a fixed number N of randomly created solutions in an objective space $f_i \in [0, 1]$, the proportion of solutions in the non-dominated front is plotted in Figure 1. The figure shows how quickly the size

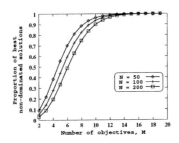

Fig. 1. The proportion of the best non-dominated solutions versus the number of objective functions (Taken from [3])

of the non-dominated front increases with the number of objectives. When the task is to find a well-distributed set of solutions on the entire Pareto-optimal front, the number of solutions needed to represent a multi-dimensional surface is also large. This requires an MOEA to use a large population size, thereby increasing the overall computational time. Thus, it is all the more necessary to use a distributed computing environment for solving multi-objective optimization problems.

A good description of different parallel MOEA implementations is presented in [2]. Most past studies have made a straightforward extension of parallel EA studies in solving multi-objective optimization problems. The studies mostly concentrate on three different approaches: (i) master-slave model, (ii) island model, and (iii) diffusion model. In the master-slave model, one master processor runs the MOEA and slave processors are used for evaluation purposes only. With such a model, the computational time can be expected to be reduced by at most P times with P processors. The solutions obtained by this procedure would be the same as that with a single processor, except that the computational time will be smaller. In the island model, different MOEAs are run on different processors and some solutions (called the migration rate) are migrated between processors after every few generations (called the migration frequency). The diffusion model consists of spatially distributed population in which there is one processor per individual and the evaluation of fitness is performed simultaneously for all the individuals. Thus, the diffusion model is a fine-grained approximation of the island model. It is the island model on which our proposed approach is based. The island model provides a number of flexibilities to be tried:

1. First, each processor can be specifically assigned to search a particular portion of the entire search space. Care should be taken to ensure that no feasible search space is left out in the assignment for all processors. Since each processor works on a smaller search region, a computationally faster search is possible. However, it is not trivial to make such an assignment in an arbitrary problem and this is a serious bottleneck of this approach.

2. The above suggestion makes the search in each processor independent to each other. However, with the help of a migration plan each processor can be assigned to work on the entire search space and made to communicate their best solutions with other processors once in a while. By doing this we are having an advantage that even if some processors get stuck at some suboptimal location, the migration of better solutions from other processors can break the statis and help improve the proceedings. In the context of multi-objective optimization, since our goal is to find a number of solutions, there is no guarantee that each processor will find solutions in different parts of the Pareto-optimal region. Because of the overlapping search effort among multiple processors, the overall computational time may be large with such a model.

3. The migration policy can be used with a scheme in which each processor is allocated to find a particular region of the Pareto-optimal front, although each processor searches the entire search space. Unlike the difficulties involved in systematically dividing the entire search space among multiple processors, the allocation of different portions of the Pareto-optimal front to different processors is not a difficult proposition. In such a scheme, care should be taken to ensure that no feasible Pareto-optimal solution is left unallotted to any processor. A number of such schemes are certainly possible, and in the next section we discuss one such procedure for problems having convex Pareto-optimal fronts. Since each processor searches the entire search space, a migration plan will help all processors, but since each processor is destined to find a particular portion of the search space, the agglomeration of solutions obtained by different processors will constitute a good diverse set of solutions.

Most past parallel MOEA studies based on the island model uses the second approach discussed above. Although some special action can be taken in each processor to ensure that there is a minimal overlap in the region searched by different processors, we do not pursue this approach further here. Instead, we suggest a procedure implementing the third concept discussed above.

3 Distributed Computing of Pareto-Optimal Solutions

Although not directly suggested for distributed computing of Pareto-optimal solutions, a number of techniques have been proposed for finding a part of the Pareto-optimal front, instead of the entire Pareto-optimal front with a single processor. In [3], a number of different such biasing techniques are discussed. Here, we present how one of them can be used for the distributed computing purpose.

3.1 Guided Domination Approach

In this approach [1], a weighted function of the objectives is defined as follows:

$$\Omega_i(\mathbf{f}(\mathbf{x})) = f_i(\mathbf{x}) + \sum_{j=1, j \neq i}^{M} a_{ij} f_j(\mathbf{x}), \quad i = 1, 2, \ldots, M. \tag{1}$$

where a_{ij} is the amount of gain in the j-th objective function for a loss of one unit in the i-th objective function. The above set of equations require fixing the matrix $\mathbf{a}$, which has a one in its diagonal elements. Now, we define a different domination concept for minimization problems as follows.

Definition 1 *A solution $\mathbf{x}^{(1)}$ dominates another solution $\mathbf{x}^{(2)}$, if $\Omega_i(\mathbf{f}(\mathbf{x}^{(1)})) \leq \Omega_i(\mathbf{f}(\mathbf{x}^{(2)}))$ for all $i = 1, 2, \ldots, M$ and the strict inequality is satisfied at least for one objective.*

Let us illustrate the concept for two ($M = 2$) objective functions, as we shall modify this approach for distributed computing of Pareto-optimal solutions. The two weighted functions are as follows:

$$\Omega_1(f_1, f_2) = f_1 + a_{12} f_2, \tag{2}$$
$$\Omega_2(f_1, f_2) = a_{21} f_1 + f_2. \tag{3}$$

The above equations can also be written as

$$\Omega = \begin{bmatrix} 1 & a_{12} \\ a_{21} & 1 \end{bmatrix} \mathbf{f}, \quad \text{or, } \Omega = \mathbf{a}\mathbf{f}. \tag{4}$$

Figure 2(b) shows the contour lines corresponding to the above two linear functions passing through a solution A in the objective space. All solutions in the hatched region are dominated by A according to the above definition of domination. It is interesting to note that when using the usual definition of domination (Figure 2(a)), the region marked by a horizontal and a vertical line will be dominated by A. Thus, it is clear from these figures that the modified definition of domination allows a larger region to become dominated by any solution than the usual definition. Since a larger region is now dominated, the complete Pareto-optimal front (as per the original domination definition) may not

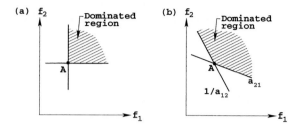

Fig. 2. The regions dominated by solution A: (a) the dominated region using the usual definition; (b) the dominated region using definition 1

be non-dominated according to the new definition. Thus, by choosing appropriate values for the elements of the matrix **a** given in equation 4, a part of the Pareto-optimal region can be emphasized. In the following subsection, we extend this guided domination approach to achieve a distributed computing of Pareto-optimal solutions.

3.2 Proposed Approach

The idea of distributed computing here is to allocate a processor a task of finding only a particular region of the Pareto-optimal front. However, while distributing the task, the following three aspects should be keep in mind:

1. No Pareto-optimal solution is left out by all processors,
2. The overlap of solutions allocated by any two processors should be as small as possible for an optimum allocation of computing resources, and
3. The procedure must be easy to extend for any number of objectives.

While using the guided domination approach for this purpose, the parameters a_{ij} must be chosen so as to fulfill the above aspects. The first two aspects can be ensured by choosing appropriate **a**-matrices for different processors. For example, for a two-objective problem, the following two **a**-matrices can be used

$$\begin{bmatrix} 1 & 0 \\ a & 1 \end{bmatrix} \text{ representing} \quad , \quad \begin{bmatrix} 1 & 1/a \\ 0 & 1 \end{bmatrix} \text{ representing}$$

for two processors. The slope a can be any positive real number. However, the third aspect mentioned above is difficult to achieve using the guided domination approach directly. For example, with three objectives, the **a**-matrix has 3×3 elements and six different off-diagonal entries in the matrix need to be fixed. Moreover, since in this case the modified Pareto-optimal front would be bounded by tangential planes, instead of lines, it becomes difficult to determine the slopes a_{ij} of the planes easily. In the following, we suggest a modified technique for this purpose.

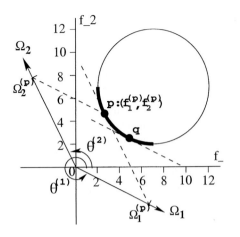 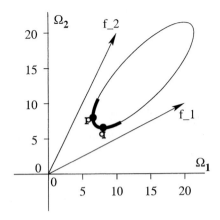

Fig. 3. The f_1-f_2 objective space **Fig. 4.** The Ω_1-Ω_2 objective space

Let us consider a feasible objective space bounded by a circle, as shown in Figure 3. Let us also consider two other axes Ω_1 and Ω_2 inclined to the objective axes. The direction vectors of these two inclined axes are given as follows:

$$\eta_1 = d_1^{(1)} \hat{e}_1 + d_2^{(1)} \hat{e}_2, \tag{5}$$

$$\eta_2 = d_1^{(2)} \hat{e}_1 + d_2^{(2)} \hat{e}_2, \tag{6}$$

where $d^{(i)}$ is the direction cosine vector of η_i (where $d_1^{(i)} = \cos\theta^{(i)}$ and $d_2^{(i)} = \sin\theta^{(i)}$) and $\hat{e}_i$ is the i-th coordinate direction. Now, any point $(f_1^{(p)}, f_2^{(p)})$ on the circle can also be represented by the Ω-coordinate system as $(\Omega_1^{(p)}, \Omega_2^{(p)})$ and we have the identity:

$$f_1^{(p)} \hat{e}_1 + f_2^{(p)} \hat{e}_2 = \Omega_1^{(p)} \eta_1^{(p)} + \Omega_2^{(p)} \eta_2^{(p)},$$
$$= \left(\Omega_1^{(p)} d_1^{(1)} + \Omega_2^{(p)} d_1^{(2)}\right) \hat{e}_1 + \left(\Omega_1^{(p)} d_2^{(1)} + \Omega_2^{(p)} d_2^{(2)}\right) \hat{e}_2.$$

Comparing terms for $\hat{e}_i$, we can write the following matrix equation:

$$\begin{bmatrix} f_1^{(p)} \\ f_2^{(p)} \end{bmatrix} = \begin{bmatrix} d_1^{(1)} & d_1^{(2)} \\ d_2^{(1)} & d_2^{(2)} \end{bmatrix} \begin{bmatrix} \Omega_1^{(p)} \\ \Omega_2^{(p)} \end{bmatrix}, \tag{7}$$

$$\text{or,} \quad \mathbf{f} = \mathbf{T}\boldsymbol{\Omega}. \tag{8}$$

From the above relationship, we can write

$$\boldsymbol{\Omega} = \mathbf{T}^{-1}\mathbf{f}. \tag{9}$$

Now relating the direction cosine vectors with respect to a_{ij}, we have

$$d^{(1)} = \frac{1}{\sqrt{1+a_{21}^2}}(1, -a_{21})^T, \quad d^{(2)} = \frac{1}{\sqrt{1+a_{12}^2}}(-a_{12}, 1)^T.$$

Forming the **T** matrix and taking the inverse, we obtain

$$\begin{bmatrix} \frac{1-a_{12}a_{21}}{\sqrt{1+a_{21}^2}} \Omega_1^{(p)} \\ \frac{1-a_{12}a_{21}}{\sqrt{1+a_{12}^2}} \Omega_2^{(p)} \end{bmatrix} = \begin{bmatrix} 1 & a_{12} \\ a_{21} & 1 \end{bmatrix} \begin{bmatrix} f_1^{(p)} \\ f_2^{(p)} \end{bmatrix}. \tag{10}$$

The Ω_1-Ω_2 space for the f_1-f_2 objective space is shown in Figure 4. The region pq on the original circle is bounded by tangents with Ω_1 and Ω_2 direction vectors. Thus, for the transformed objectives, the entire Pareto-optimal front is bounded within the arc pq. The transformation of the objective space and the location of the new Pareto-optimal solutions are clearly shown in the figure. Comparing this equation with equation 4, we observe that the right sides are equal and the Ω-axes are multiplied by different numbers. However, it is interesting to note that multiplying Ω_1 and Ω_2 values by different non-zero constants may translate or magnify the Pareto-optimal front in the Ω_1-Ω_2 space, but it does not change the Pareto-optimal solutions in terms of f_1 and f_2 values. Thus, the original a-matrix can be replaced with the $\mathbf{T}^{-1}$ matrix, constructed simply from the direction cosines of the two inclined lines. Here is the procedure:

1. Form the a-matrix from direction cosines of transformed axes,
2. Calculate the inverse of the **a** matrix and use the domination test with the transformed objectives (with Ω vector calculated using equation 9).

In order to ensure the first property of not leaving any feasible solution by the mapping, we need to ensure that two processors share a common direction cosine vector. In Figure 5, we show the allocation plan for choosing direction cosines for 1, 2, 3, and 4 processors in a two-objective problem. These direction cosines constitute an allocation plan for different processors. The intersecting point of the two adjacent axis-directions indicate the allocation plan of a particular processor. Note that the left-most plot indicates a single processor with $\Omega_1 = f_1$

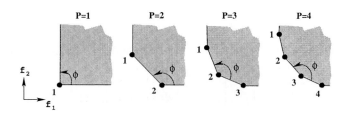

Fig. 5. Allocation plan for two-objective problems

and $\Omega_2 = f_2$. The next figure indicates that two processors share one common direction (the inclined direction), and so on. Although different schemes can be adopted for assigning common inclined directions, we suggest a methodology by which equal inscribed angles are assigned to each processor. For P processors solving a two-objective optimization problem, the included angle (in radians) between any two adjacent direction vectors is

$$\phi = \pi\left(1 - \frac{1}{2P}\right). \tag{11}$$

Extension to higher objectives. Interestingly, the above argument can be extended to more than two dimensions and can be applied easily to any number of objectives. For example, in a three-objective space, the entire Pareto-optimal front can be determined by tangential properties of three planes: f_1-plane, f_2-plane and f_3-plane. If three arbitrary (but non-identical) directions ($\Omega_1, \Omega_2, \Omega_3$) are chosen, the modified Pareto-optimal front will be bounded by the three new planes: Ω_1-Ω_2 plane, Ω_2-Ω_3 plane, and Ω_3-Ω_1 plane.

The procedure of generating an allocation plan for two objectives illustrated in Figure 5 can be considered as successive 'chopping-off' of the bottom-left corner (or origin) of a square by straight edges. Each corner at the bottom-left part of the square corresponds to the a processor for which the direction cosines are computed from the straight edges forming the corner. When an existing corner is chopped off, two new corners are added, thereby allowing one more processor to be used. For example, the $P = 2$ case is obtained from $P = 1$ case by chopping off the right-angle corner. Larger P cases can be considered as further chopping off the bottom-left corner of the square. The only restriction to the chopping process for P processors is that there must be exactly P corners generated by the chopping process. This chopping principle can be extended for higher objectives, except that the bounded square now becomes a larger-dimensional hyper-box, and hyper-planes are used to chop-off the origin. Figures 6 to 9 show a number of allocation plans for different number of processors in a three-objective spherical Pareto-optimal front. Although the systematic chopping of each corner will generate $P = (M-1)k + 1$ corners (where k is any non-negative integer), Figures 10 and 11 also show some interesting allocation plans for different numbers of processors.

Need for migration. In this study, we have started each processor with a population initialized from the same bounded region. In order to keep the total number of function evaluations the same, we have also kept a constant combined population size, although this may not be an optimal scheme. Thus, for a study with P processors, we have used N/P population members in each processor, where N is the population size used for the single-processor run. As the number of processors increase, it may become difficult for each processor to overcome identical hurdles to reach to the Pareto-optimal front with a smaller population size. In order to get assistance from other processors, we have introduced a migration policy among the processors. A certain number of solutions (migration rate) from one processor are sent to each other processor after every few generations (migration frequency). Each processor is chosen systematically for its turn to migrate solutions to other processors. The new solutions replace a randomly chosen set of members from the existing population.

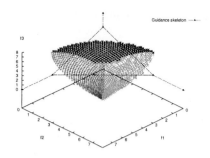
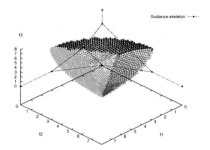

Fig. 6. 3 parts of Pareto-optimal front   **Fig. 7.** 5 parts of Pareto-optimal front

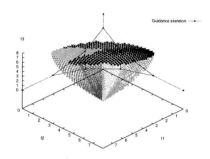
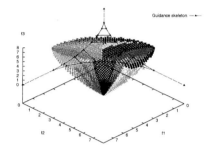

Fig. 8. 7 parts of Pareto-optimal front   **Fig. 9.** 21 parts Pareto-optimal front

4 Simulation Studies

Here, we present simulation results on two and three-objective test problems. We have used a Pentium-III cluster running MPI on the Linux operating system.

4.1 Two-Objective Test Problems

First, we study the 30-variable ZDT1 test problem [3]:

$$\begin{aligned}&\text{Minimize } f_1(\mathbf{x}) = x_1\\ &\text{Minimize } f_2(\mathbf{x}) = g(\mathbf{x})h(f_1,g),\\ &\text{where } g(\mathbf{x}) = 1 + \sum_{i=2}^{30}(x_i - 0.5)^2, \quad h(f_1,g) = 1 - \sqrt{f_1/g}.\end{aligned} \quad (12)$$

Here, each x_i lies $[0,1]$. Figure 12 shows the hyper-area [6] for a serial (with one processor) NSGA-II run with a population size of 200. The simulated binary crossover operator (with $\eta_c = 10$ and probability 0.9) and polynomial mutation operator (with $\eta_m = 50$ and probability 1/30) are used [3]. With a reference point taken as $(1.0646, 1.0646)$, the maximum hyper-area calculated for an infinite population lying on the true Pareto-optimal front is 0.8. The NSGA-II run is

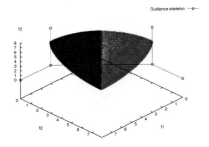 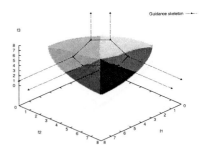

Fig. 10. 2 parts of Pareto-optimal front   **Fig. 11.** 6 parts of Pareto-optimal front

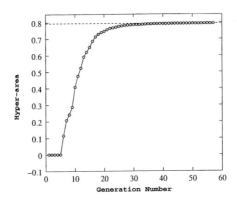

Fig. 12. Hyper-area calculated with (1.0646,1.0646) as the reference point

terminated when an hyper-area of 0.794 is achieved. In the particular run shown in Figure 12, 57 generations were needed to achieve this hyper-area.

Figure 13 shows the distribution of parallel real-parameter NSGA-II with two processors. The following two transformation matrices are used for the two processors:

$$\begin{bmatrix} 1 - \cos(0.75\pi) \\ 0 \quad \sin(0.75\pi) \end{bmatrix}, \quad \begin{bmatrix} \cos(0.75\pi) & 0 \\ -\sin(0.75\pi) & 1 \end{bmatrix}.$$

Each processor is initialized with a population of size 100. Other GA parameters are kept the same as in the serial NSGA-II run. In this case, 30 solutions from a processor are migrated after every 5 generations to the other processor. At the end of 52 generations, the hyper-area calculated for the combined population of two processors is 0.794. Thus, it is clear that a similar distribution can be achieved by the two-processor NSGA-II with more or less similar number of function evaluations.

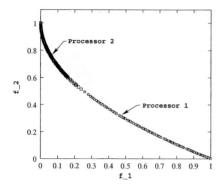 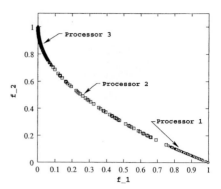

Fig. 13. Obtained solutions with two processors for ZDT1. Allocation plan is the same as that in Figure 5 with $P = 2$

Fig. 14. Obtained solutions with three processors for ZDT1. Allocation plan is the same as that in Figure 5 with $P = 3$

Figure 14 shows the distribution of solutions (with a hyper-area of 0.794) obtained with a three-processor NSGA-II after 57 generations. The following three transformation matrices are used:

$$\begin{bmatrix} 1 - \cos(\frac{5}{6}\pi) \\ 0 \quad \sin(\frac{5}{6}\pi) \end{bmatrix}, \quad \begin{bmatrix} \cos(\frac{5}{6}\pi) & -\cos(\frac{2}{3}\pi) \\ -\sin(\frac{5}{6}\pi) & \sin(\frac{2}{3}\pi) \end{bmatrix}, \quad \begin{bmatrix} \cos(\frac{2}{3}\pi) & 0 \\ -\sin(\frac{2}{3}\pi) & 1 \end{bmatrix}.$$

In this case, 30 solutions from a processor is sent to two other processors after every 5 generations.

In order to investigate the effect of the migration policy on the distribution of solutions, we have performed a parametric study on the ZDT1 test problem. The outcome of this study is presented in Table 1. In each case, the average and standard deviation of the number of generations needed to achieve a hyper-area of 0.794 in ten runs are calculated. The first row shows the single processor result and multi-processor results without any migration. It is observed that the parallel NSGA-II does not perform well without migration. Although the outcome of the parallel NSGA-II depends on the chosen migration rate and frequency, most of the two-processor runs achieved the required hyper-area (meaning that a similar combined convergence and diversity measure is achieved) in a smaller number of generations than that required with a single processor NSGA-II. The best performance occurs when 30 solutions are migrated after every 5 generations. In this case, on an average two-processor NSGA-II requires about 18% lesser function evaluations to achieve a similar performance.

With three processors, however, the required number of generations to achieve a similar performance is slightly more. Although the best performance is achieved when 10 solutions are migrated after every 5 generations, the runs with migration rate of 30 and frequency 5 (as that obtained to be the best in the two-processor case) is also very close to this optimum performance. In the three-processor case, each processor is allocated 68 population members. Since

Table 1. Parametric study of migration rate and frequency on ZDT1 to achieve a hyper-area of 0.794.

Migration Rate	Migration Frequency	Required Generations					
		Single Processor		Two Processors		Three Processors	
		Average	Std. Dev.	Average	Std. Dev.	Average	Std. Dev.
0	0	58.9	3.91	66.1	9.63	116.3	45.71
10	5			50.1	3.05	**59.6**	5.10
10	10			52.5	4.34	66.3	6.10
20	5			48.9	1.51	60.3	3.52
20	10			51.2	3.57	60.8	3.43
30	5			**48.3**	1.35	59.8	3.66
30	10			50.6	2.46	63.8	3.12
30	15			55.8	5.31	68.7	4.88
40	10			49.3	3.13	61.7	4.43
30	5			57.2	2.99	70.2	4.58

all processors tackle the same ZDT1 problem (but each using a different dominance relationship), the population size allocated to each processor may not be enough for it to reach near the Pareto-optimal solutions. For every problem, there exists an optimal population size (even with a migration policy) below which the NSGA-II may not be able to reach the desired goal. For ZDT1, the two-processor NSGA-II with 100 population members in each processor seems to have produced an optimal performance. It remains a future research task to repeat the above investigation by using optimal population sizes in each processor and by varying the migration parameters and connectivity among processors.

The last row in the table also shows the performance of an island two and three-processor NSGA-II without the transformation procedure. Here, all three processors search the entire Pareto-optimal front, but 30 solutions are migrated at every 5 generations. Since all processors independently find solutions on the Pareto-optimal front, the agglomerated set of obtained solutions is not guaranteed to have adequate diversity. It is clear from the table that this procedure requires more generations to achieve an identical hyper-area of 0.794 than the proposed transformation approach with both two and three processors.

The CTP7 test problem is a constrained test problem which makes the Pareto-optimal front a combination of a number of disconnected regions [3]:

$$\begin{aligned}
&\text{Minimize } f_1(\mathbf{x}) = x_1, \\
&\text{Minimize } f_2(\mathbf{x}) = g(\mathbf{x})\left(1 - \frac{f_1(\mathbf{X})}{g(\mathbf{X})}\right), \\
&\text{subject to } C(\mathbf{x}) \equiv \cos(\theta)[f_2(\mathbf{x}) - e] - \sin(\theta)f_1(\mathbf{x}) \geq \\
&\qquad a\left|\sin\left\{b\pi\left[\sin(\theta)(f_2(\mathbf{x}) - e) + \cos(\theta)f_1(\mathbf{x})\right]^c\right\}\right|^d.
\end{aligned} \tag{13}$$

The following parameter values are used:

$$\theta = -0.05\pi, \quad a = 40, \quad b = 5, \quad c = 1, \quad d = 6, \quad e = 0.$$

Each of the five variables x_i lies in $[0,1]$. Figure 15 shows the obtained front with a single processor. It is clear that not all regions are found by the NSGA-II with a population size of 200. This result agrees with the original result reported in [4]. However, when NSGA-II is run with three processors, each processor has a shorter Pareto-optimal region to discover. As a result, the parallel NSGA-II (with processor-wise population of 70) is able to find all the disconnected Pareto-optimal regions (Figure 16). Here, 30 solutions are migrated after every 5 generations. This example illustrates how the proposed approach can also be used in problems having a disconnected set of Pareto-optimal regions.

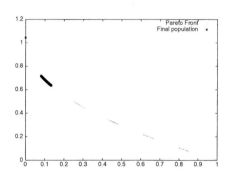
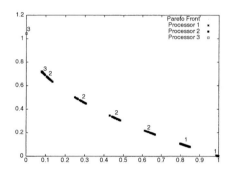

Fig. 15. Obtained non-dominated solution with single-processor NSGA-II on CTP7 problem

Fig. 16. Obtained non-dominated solution with 3-processor parallel NSGA-II on CTP7 problem. Allocation plan is given in Figure 5 with $P = 3$

4.2 Three-Objective Test Problem

Next, we consider the following three-objective modified DTLZ2 test problem having $n = 12$ variables:

$$\begin{aligned}
\text{Minimize } & f_1(\mathbf{x}) = 2 - (1 - g(\mathbf{x})) \cos(x_1\pi/2) \cos(x_2\pi/2), \\
\text{Minimize } & f_2(\mathbf{x}) = 2 - (1 - g(\mathbf{x})) \cos(x_1\pi/2) \sin(x_2\pi/2), \\
\text{Minimize } & f_2(\mathbf{x}) = 2 - (1 - g(\mathbf{x})) \sin(x_1\pi/2), \\
\text{where } & g(\mathbf{x}) = \frac{1}{n-2} \sum_{i=3}^{n} \left(\frac{x_i - 0.5}{0.5}\right)^2.
\end{aligned} \qquad (14)$$

Here, each variable x_i lies in $[0,1]$. The modification makes the Pareto-optimal front a convex spherical surface. Figure 17 shows the solutions obtained with a single processor NSGA-II using a population of size 300. Other NSGA-II parameters are the same as before, except that the crowding operator of NSGA-II is replaced with a *clustering* operator similar to that used in the SPEA [6]. Although this modified algorithm requires $O(N^3)$ (where N is the population size) computational time, the obtained diversity in solutions is much better than the original NSGA-II. Figure 18 shows a typical distribution of solutions obtained by a three-processor clustered NSGA-II with an allocation plan given in Figure 6. The following transformation matrices are used:

$$\begin{bmatrix} 1 & -\frac{1}{\sqrt{2}} & \frac{1}{\sqrt{2}} \\ 0 & \frac{1}{\sqrt{2}} & 0 \\ 0 & 0 & \frac{1}{\sqrt{2}} \end{bmatrix}, \begin{bmatrix} \frac{1}{\sqrt{2}} & 0 & 0 \\ -\frac{1}{\sqrt{2}} & 1 & -\frac{1}{\sqrt{2}} \\ 0 & 0 & \frac{1}{\sqrt{2}} \end{bmatrix}, \begin{bmatrix} \frac{1}{\sqrt{2}} & 0 & 0 \\ 0 & \frac{1}{\sqrt{2}} & 0 \\ -\frac{1}{\sqrt{2}} & -\frac{1}{\sqrt{2}} & 1 \end{bmatrix}.$$

In each processor, 100 population members are used and 30 solutions are migrated after every 10 generations. In both cases, NSGA-IIs are run for 200 generations so that solutions converge very close to the Pareto-optimal front. The figure shows that each processor finds an adequate number of well-distributed set of solutions in its own designated region in the Pareto-optimal front. The computational time needed with three processors is about 25 times lesser than that of the single processor, because of the use of one-third population members in each processor and the use of an $O(N^3)$ algorithm.

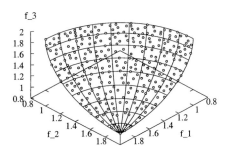

Fig. 17. Obtained solutions with a single processor for the modified DTLZ2

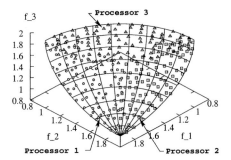

Fig. 18. Obtained solutions with three processors for the modified DTLZ2

To investigate further, we have applied the clustered NSGA-II with six processors (each using 50 population members) with an allocation plan shown in Figure 11. Figure 19 shows the agglomeration of solutions found in all six processors. Once again, a migration plan with 30 solutions migrating after every 10 generations is used. Each processor finds a well-distributed set of solutions in a particular region on the Pareto-optimal front.

5 Conclusions and Extensions

The study of multi-objective EAs with multiple processors has received a lukewarm interest so far. In this paper, we have proposed a technique in which

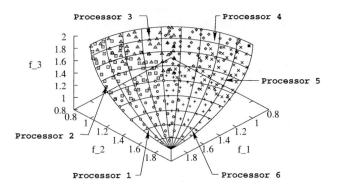

Fig. 19. Obtained solutions with six processors for the modified DTLZ2. Allocation plan is given in Figure 11

each processor has been assigned the task of finding only a particular portion of the Pareto-optimal region. A simple yet effective allocation plan has been suggested by using direction cosines of the transformed coordinate system. Since all processors have to handle the same complexity of the search space (starting from the same region in the search space and overcoming similar hurdles to reach near the Pareto-optimal front), a migration policy of sending a few good non-dominated solutions from one processor to other processors has also been suggested. On a number of two and three objective test problems, the efficacy of the proposed technique has been demonstrated using the NSGA-II. It has been observed that the use of multiple processors is beneficial in finding a widely distributed set of solutions with a good computational speed-up.

The transformation technique proposed in this paper can also be applied to some problems having a non-convex Pareto-optimal front with appropriate choice of coordinate transformations. However, a different transformation technique may also be tried with Tchebyshev metrics [3] and with other biasing techniques presented in Section 3. In an another island approach, an emphasis of the population members dissimilar to the migrated individuals may lead to formation of non-overlapping clusters of Pareto-optimal solutions. Nevertheless, since in most real-world applications convex Pareto-optimal fronts are encountered [3], the technique proposed in this paper should have a wide-scale applicability.

References

1. J. Branke, T. Kaußler, and H. Schmeck. Guidance in evolutionary multi-objective optimization. *Advances in Engineering Software*, 32:499–507, 2001.
2. C. A. C. Coello, D. A. VanVeldhuizen, and G. Lamont. *Evolutionary Algorithms for Solving Multi-Objective Problems.* Boston, MA: Kluwer Academic Publishers, 2002.
3. K. Deb. *Multi-objective optimization using evolutionary algorithms.* Chichester, UK: Wiley, 2001.

4. K. Deb, A. Pratap, and T. Meyarivan. Constrained test problems for multi-objective evolutionary optimization. In *Proceedings of the First International Conference on Evolutionary Multi-Criterion Optimization (EMO-01)*, pages 284–298, 2001.
5. E. Zitzler, K. Deb, L. Thiele, C. A. C. Coello, and D. Corne, editors. *Evolutionary Multi-Criterion Optimization (Lecture Notes in Computer Science 1993)*. Heidelberg: Springer, 2001.
6. E. Zitzler and L. Thiele. Multiobjective evolutionary algorithms: A comparative case study and the strength pareto approach. *IEEE Transactions on Evolutionary Computation*, 3(4):257–271, 1999.

Multiobjective Capacitated Arc Routing Problem

P. Lacomme[1], C. Prins[2], and M. Sevaux[3]

[1] Université Blaise Pascal
CNRS, UMR 6158 - LIMOS/Laboratoire d'Informatique
Campus Universitaire des Cézeaux - BP 10125 - F-63177 Aubière Cedex, France
lacomme@isima.fr
[2] Université de Technologie de Troyes
Laboratoire d'Optimisation des Systèmes Industriels
12 Rue Marie Curie - BP 2060 - F-10010 Troyes Cedex - France
prins@utt.fr
[3] Université de Valenciennes
CNRS, UMR 8530 - LAMIH/Systèmes de Production
Le Mont Houy - F-59313 Valenciennes Cedex 9 - France
msevaux@univ-valenciennes.fr

Abstract. The Capacitated Arc Routing Problem (CARP) is a very hard vehicle routing problem raised for instance by urban waste collection. In addition to the total route length (the only criterion minimized in the academic problem), waste management companies seek to minimize also the length of the longest trip. In this paper, a bi-objective genetic algorithm is presented for this more realistic CARP, never studied before in literature. Based on the NSGA-II template, it includes two-key features: use of good constructive heuristics to seed the initial population and hybridization with a powerful local search procedure. This genetic algorithm is appraised on 23 classical CARP instances, with excellent results. For instance, for a majority of instances, its efficient solutions include an optimal solution to the academic CARP (minimization of the total route length).

1 Introduction: The CARP and Its Applications

The Capacitated Arc Routing Problem (CARP) is defined in the literature on an undirected graph $G = (V, A)$ with a set V of n nodes and a set A of m edges. A fleet of identical vehicles of capacity W is based at a depot node s. A subset R of edges require service by a vehicle. All edges can be traversed any number of times. Each edge (i, j) has a traversal cost c_{ij} and a demand q_{ij}. The goal is to determine a set of vehicle trips (routes) of minimum total cost, such that each trip starts and ends at the depot, each required edge is serviced by one single trip, and the total demand handled by any vehicle does not exceed W.

Since the CARP is $\mathcal{NP}$-hard, large scale instances must be solved in practice with heuristics. Among fast constructive methods, one can cite for instance

Path-Scanning [10], Augment-Merge [11] and Ulusoy's splitting technique [19]. Metaheuristics available include tabu search methods [2,12] and the memetic algorithm of Lacomme, Prins and Ramdane-Chérif [14], which is currently the most efficient solution method. All these heuristic algorithms can be evaluated thanks to very good lower bounds [1,3].

CARP problems are raised by operations on street networks, e.g. urban waste collection, snow plowing, sweeping, gritting, etc. Economically speaking, the most important application certainly is municipal refuse collection. In that case, nodes in G correspond to crossroads, while edges in A model street segments linking crossroads. The demands are amounts of waste to be collected. The cost on each edge is generally a time.

The academic CARP deals with only one objective, minimizing the total duration of the trips. In fact, waste management companies are also interested in balancing the trips. For instance, in Troyes (France), all trucks leave the depot at 6 a.m. and waste collection must be finished as soon as possible to assign the crews to other tasks, e.g. sorting the waste at a recycling facility. So, the company wishes to solve in fact a bi-objective version of the CARP, in which both the total duration of the trips and the duration of the longest trip (*makespan*) are to be minimized.

In the last decade, there has been a growing interest in multi-objective vehicle routing problems, but all published papers deal with node routing problems, *i.e.*, in which the tasks to be performed are located on the nodes of a network [16, 5,13,18]. This paper is the first study devoted to an arc routing problem with multiple objectives.

The remainder of the paper is organized as follows. Section 2 recalls some definitions on multi-objective optimisation. Section 3 introduces the NSGA-II template and our specific components, while numerical experiments are conducted in Section 4. Conclusion is drawn in Section 5.

2 Multi-objective Optimisation

The growing interest in multi-objective optimisation results from the numerous industrial applications where solving a single objective optimisation problem is an abusive simplification. In such cases, computing an optimal solution for one objective results in poor values for other criteria. Hence, developing some algorithms giving a compromise between several criteria is better suited to industrial reality.

Multi-objective problems in combinatorial optimization are often very hard to solve: even their single-objective versions are generally $\mathcal{NP}$-hard. This is why lots of papers have been recently published on evolutionary techniques that are able to handle several objectives at a time and give efficient solutions by the mean of an evolving population.

For an extensive information on multi-objective optimisation see [4]. A recent survey and annotated bibliography [9] can help for finding the best suitable method to tackle a multi-objective optimisation problem (see also [6,7]).

A multi-objective optimisation problem (MOOP) can be formulated as follows :

$$\begin{aligned}\text{minimize} &\quad \{f_1(x), f_2(x), \ldots, f_{\text{nc}}(x)\} \\ \text{subject to} &\quad x \in S\end{aligned} \quad (1)$$

where nc is the number of criteria, $x = (x_1, x_2, \ldots, x_{\text{nv}})^T$ a vector of nv decision variables and S a set of feasible solutions.

Definition 1. *A solution x **dominates** a solution y, $x \succ y$ if x is at least as good as y for each objective: $\forall k \in [1, \ldots, nc], f_k(x) \leq f_k(y)$ and x is better than y for at least one citerion: $\exists k \in [1, \ldots, nc], f_k(x) < f_k(y)$.*

According to this definition, x is *non-dominated* if there is no solution y that dominates x. The set P of all non-dominated solution is called *non-dominated set* or *efficient set (E)* or *pareto-optimal set/front*. And for a set of solutions $U \supset P$, $\forall x \in P, \forall y \in P, x \not\succ y, y \not\succ x$, and $\forall y \in U \setminus P, \exists x \in P, x \succ y$. Hence a criterion of a solution in the non-dominated set cannot be improved without detariorating another criterion.

3 NSGA-II Implementation

Today, many multi-objective optimisation algorithms are available and selecting the best one for a given application is not obvious. For an up-to-date description and evaluation of these algorithms the reader can refer to [20]. Nevertheless, this survey evaluates the algorithms on a specific set of problems and it is difficult to generalize the conclusions to other problems.

Our choice then turned to one of the most widely used multi-objective genetic algorithm (MOGA), NSGA-II [8]. This MOGA has been adapted to match our requirements concerning the bi-objective CARP. Minor bugs from [6,7] were also corrected. The bug-free algorithms are provided in the next section.

3.1 NSGA-II for Two Objectives

The NSGA-II algorithm is based on the notion of non-dominated fronts. It improves an earlier algorithm called NSGA in terms of complexity for calculating the non-dominated fronts. Each iteration of NSGA-II starts by sorting the current set of solutions U by non-domination. This step calculates the non-dominated subset of U, remove its solutions from U and repeats this process until U becomes empty.

The set U is in practice a table or a list of solutions. $U(i)$ denotes the ith solution of U. To save space, the fronts store solution indexes, not the solutions themselves. Dominance between two solutions x and y is computed by the $\text{dom}(x, y)$ boolean function. `nb_better(i)` counts the solutions which dominates the solution $U(i)$, `set_worse(i)` is the set of solutions dominated by

$U(i)$. The non-dominated sorting partitions the solutions into nf fronts denoted front(1), front(2), ..., front(nf). rank(i) denotes the index of the front that stores solution $U(i)$.

The first step of the sorting process computes in $O(nc.ns^2)$ nb_better and set_worse values, the second phase in $O(ns^2)$ assigns solutions to their respective fronts and computes nf.

```
procedure non-dominated-sort()
  front(1) := emptyset
  for i := 1 to ns do
    nb_better(i) := 0
    set_worse(i) := emptyset
    for j := 1 to ns do
      if dom(U(i),U(j)) then add j to set_worse(i)
      elseif dom(U(j),U(i)) then nb_better(i) = nb_better(j)+1 endif
    endfor
    if nb_better(i) = 0 then add i to front(1) endif
  endfor
  nf := 1
  loop
    for each i in front(nf) do
      rank(i) := nf
      for each j in set_worse(i) do
        nb_better(j) = nb_better(j)-1
        if nb_better(j) = 0 then add j to front(nf+1) endif
      endfor
    endfor
    exit when front(nf+1) = emptyset
    nf = nf+1
  endloop
```

Comparing solutions in NSGA-II is done by two means. If two solutions belongs to two differents fronts, the smaller rank gives the better solution. If two solutions belong to the same front, the algorithm privileges the most isolated one by computing a crowding distance. The idea is to favor the best solutions while keeping the fronts well scattered to prevent clusters of solutions.

Let call *margin* the crowding distance function. The best solution among two candidates can then be computed thanks to a boolean function better(i,j), for which a solution $U(i)$ is better than a solution $U(j)$ if (rank(i) < rank(j)) or (rank(i) = rank(j) and margin(i) > margin(j)).

For a front R of nr solutions and nc = 2 criteria, margins are calculated as follows. R is sorted in increasing values of the first criteria. Let $R(k)$ the kth solution of the sorted front. If c is one of the criteria, let f_c^{kmin} and f_c^{kmax} the minimum and maximum values of f_c in R. For $k = 1$ or $k = $ nr, let margin(k) = ∞ to be sure that these two extreme points will be chosen if necessary. For $1 < k < $ nr,

$$\text{margin}(k) = \frac{f_1(R(k)+1) - f_1(R(k)-1)}{f_1^{max} - f_1^{min}} + \frac{f_2(R(k)-1) - f_2(R(k)+1)}{f_2^{max} - f_2^{min}} \quad (2)$$

The overall structure of NSGA-II is given below. The population, denoted by pop, is composed of ns chromosome records. Each record includes also the two criteria values, the rank and the margin of the individual. The procedure first_pop(pop,ns) builds the initial population.

The procedure non_dominated_sort(pop,ns,front,nf) sorts pop by non-domination with the algorithm previously defined. It returns the number of different fronts nf and the fronts themselves (front(i) is the ith front and front(i,j) is the index of the jth solution of front(i)). This procedure computes also the rank of each solution in pop, *i.e.* the index of the front the solution belongs to.

The procedure add_first_children(pop,ns) creates ns offsprings which are added at the end of pop, thus doubling its size. Parents are selected by binary tournament.

```
procedure nsga2()
  first_pop(pop,ns)
  non_dominated_sort(pop,ns,front,nf)
  add_first_children(pop,ns)
  repeat
    non_dominated_sort(pop,ns,front,nf)
    for i := 1 to nf do
      get_margins(front,i,pop)
      margin_sort(front,i,pop)
    endfor
    newpop := emptyset
    i := 1
    while |newpop|+|front(i)| <= ns do
      add front(i) to newpop
      i := i+1
    endwhile
    missing := ns-|newpop|
    for j := 1 to missing do
      add front(i,j) to newpop
    endfor
    pop := newpop
    add_children(pop,ns)
  until (stopping_criterion)
```

Each iteration of the *repeat* loop consists of keeping the ns best solutions and adding the ns offsprings of the next generation. The first *for* loop computes the margins for each front i with the get_margins(front,i,pop) procedure, then, in each front, solution are sorted in decreasing margin order by the margin_sort(front,i,pop) procedure.

The ns best solutions of pop are copied into newpop front by front until the new population is filled. Finally, newpop replaces pop.

3.2 Components for the GA

This section describes the components used in the NSGA-II algorithm to solve the multi-objective CARP. Some of them were initially developped for a GA that is currently the most effective solution method for the single-objective CARP [14]. Only a short description is recalled here.

Chromosome Representation and Evaluation. Following [14], a chromosome σ is a permutation of the required arcs, without trip delimiters. It can be seen as a giant tour ignoring vehicle capacity. Implicit shortest paths are assumed between the depot and the first task, between two consecutive tasks, and between the last task and the depot.

Trip delimiters are not necessary because a procedure described in [14] is able to derive a least total cost CARP solution by splitting the chromosome (giant tour) into capacity-feasible tours. The procedure is based on the computation of a shortest path in an auxiliary graph, in which each arc corresponds to one sub-sequence of tasks (one feasible tour for the CARP) that can be extracted from the giant tour.

Note that the use of a splitting procedure to define trip limits allows a simple encoding of chromosomes and the use of traditional crossover operators designed for permutation chromosomes, e.g. OX or LOX.

Initial Population. Like in other MOGAs, an initial population has to be generated. Generating permutations randomly is a classical technique, but adding high-quality solutions always accelerates the convergence of the algorithm.

Several good solutions are added to the initial population. They are computed by extending three classical heuristics for the CARP, namely Path-Scanning [10], Augment-merge [11] and Ulusoy's heuristic [19]. These solutions will be denoted by PS, AM and UL and will be used in the numerical experiments (see details in Section 4.1).

Crossover Operators. For the single objective CARP, the standard LOX and OX crossover operators give the best results and following the conclusions of [14], the OX crossover operator has been selected for the multi-objective problem. Details on its implementation can be found in [14].

Stopping Conditions. Today, a major problem is raised when developing a MOGA. Among the different methods, there is no standard technique to define reliable stopping conditions. In combinatorial optimisation and in the single-objective case, metaheuristics are often stopped after a fixed number of iteration without improving the current best solution. Since it is difficult to extend this method to multi-objective optimisation, authors of published MOGAs generally stop their algorithms after a fixed number of iterations. For this study, we will do the same, although we are currently investigating more sophisticated stopping criteria.

3.3 Local Search Procedures

In single-objective optimization, it is well known that a standard GA must be hybridized with local search (LS) operators to become competitive [15] and this property is also true in multi-objective optimization. Of course, several adaptations are required to convert a local search procedure for the single-objective case into an efficient bi-objective local search procedure.

Without going into details, we apply a local search with a fixed probability (called the *rate* of the local search) to each offspring resulting from a crossover. The child chromosome is first converted into a valid CARP solution (using the splitting procedure) and five types of moves are examined for each pair of required edges (tasks): 1) inverse the traversal direction of a task in its trip, 2) move one task after one other, 3) move two adjacent tasks after one other, 4) swap two tasks and 5) perform 2-opt moves. More information on the single-objective local search procedure is given in [14]. A descent algorithm is used with these neighborhoods.

A descent algorithm with the same neighborhoods is also used for the bi-objective CARP. The main change resides in the conditions for accepting a move. The variations of each objective function are computed. CVar denotes the total cost variation and MVar the makespan variation. Three acceptance criteria are defined:

Accept1 is True if CVar <= 0.
Accept2 is True if MVar <= 0.
Accept3 is True if (CVar<=0 and MVar<0) or (CVar<0 and MVar<=0).

The last version, Accept3, leads to a true bi-objective local search, in which a move is accepted if and only if its result dominates the current solution. Accept1 and Accept2 are two single-objective versions that improve either the total cost or the makespan, but not both.

4 Computational Experiments

All developements are done in Borland ©Delphi 6 and tested on a Pentium Celeron 650 Mhz with Windows 98. The computational evaluation uses the standard benchmark problems from Golden, DeArmon and Baker [10] which are used by all authors working on the single-objective CARP. Files can be downloaded from http://www.uv.es/ belengue/carp.html.

This set of instances is composed of 25 files from 7 to 27 nodes and 11 to 55 edges. In fact, two instances (originally gdb8 and 9) are inconsistent and never used. The remaining instances are re-numbered gdb1 to gdb23. All edges are required.

4.1 Test Protocol and Parameters

A first execution of the MOGA has been done and the results are used to provide a reference front for the other test protocols. This first run is done without the local search procedure but with the initial heuristics PS, AM and UL presented in Section 3.2. This reference experiment will be denoted by MO1. Table 1 lists the test protocols and parameters used for the other numerical experiments. The protocols are numbered from MO1 to MO6, the population size is always fixed to 60 individuals and the crossover operator is OX. The maximum number of iterations is 100, except for the test MO5 where no local search and no initial heuristics are used. The last two columns specify the type of local search procedure used (corresponding to the "Accept" function in the LS procedure) and the LS rate.

Table 1. Test protocol and parameters used

Test Protocol	Pop. size	XOver op.	Max # it.	Heuristics PS AM UL	Local Search LS	%LS
MO1	60	OX	100	Yes Yes Yes	No	–
MO2	60	OX	100	Yes Yes Yes	1	10
MO3	60	OX	100	Yes Yes Yes	2	10
MO4	60	OX	100	Yes Yes Yes	3	10
MO5	60	OX	300	No No No	No	–
MO6	60	OX	100	Yes Yes Yes	3	20

A MOGA with 60 individuals and 100 iterations performs a total of 6000 crossover operations. On a specific instance, the evolution of the different number of iterations will be shown.

4.2 Measuring the Deviation from a Reference Front

Comparing two different methods for solving the same multi-objective problem is very difficult, especially when no information is known concerning the final choices of the decider. Nevertheless, this step cannot be avoided in this study, at least to decide which test protocol is best suited for solving our problem.

A measure, presented in [17], has been enhanced and applied to compare each of the test protocol with the reference protocol MO1. The front obtained by MO1 will be considered as the reference front. Given a reference front, a kind of distance is measured between each solution to be compared and its projection onto the extrapolated reference front. The extrapolated reference front is a piecewise continuous line (for the bi-objective case) going through each solution of the front and extended beyond the two extreme solutions. Figure 1 gives an example of the extrapolated front and such a measure.

For the example of Figure 1, two measures can be given, the simplest is: Measure $= d1 + d2 + d3 + d4$, but in some cases, it could be interesting to normalize this measure by dividing it by the number of efficient solutions in front E to compare: Norma = Measure $/|E|$. If one of the solution of the front to compare is under the (extrapolated) reference front, the distance will be of course negative.

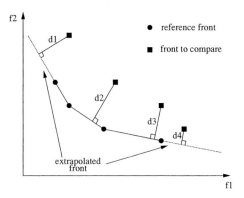

Fig. 1. Example of the distance to the extrapolated front

This measure is a bit arbitrary and of course subject to criticism. For example, what can be the measure if the reference front is reduced to only one solution? How can we measure the difference between two fronts where one is well spread while the other presents several "holes" between clusters of solutions? Nevertheless, its merit is to exist and to help us to rank the tested protocols.

Because of no a-priori knowledge of the solutions which lie on the efficient front, this measure should be completed by a visual inspection.

4.3 Lower Bounds and Heuristics

For the two criteria, lower bounds and heuristic results are presented in Table2. Column "File" gives the filename of the instance and columns "n" and "τ" give the number of nodes and the number of required edges for each instance. Columns "LBC" and "LBM" denote the lower bounds for the total cost and the makespan. Columns "PSC" and "PSM" report the total cost and the makespan given by the Path-Scanning heuristic (AM stand for Augment-Merge heuristics and UL for Ulusoy's heuristic, the last letter C - resp. M - is used for the total cost - resp. makespan). See Section 3.2 for more details.

The lower bound LBC for the total cost is given by Belenguer and Benavent [2] except for gdb14 given by Amberg and Voß [1]. The lower bound LBM for the makespan is defined by Lacomme et al. [14].

Best-known solutions for the single objective (total cost) optimization are not reported because they are equal to the LBC lower bound for 21 files out of 23. The memetic algorithm developped in [14] gives the 21 best solutions and all are equal to their lower bounds.

Note that the heuristics already find four least-cost solutions (files gdb4, gdb7, gdb15 and gdb17) and two solutions of minimal makespan (files gdb13 and gdb19). These solutions are used to provide the NSGA-II algorithm with very good starting points.

As mentioned in [14], the lower bound LBM computed for the makespan criteria is not tight and the deviation to this bound can be large. Results of the next section will emphasize this point.

Table 2. Lower bounds and heuristic results for the *gdb* files (*optimal solution for one criterion)

File	n τ	Lower Bounds				Heuristics			
		LBC	LBM	PSC	PSM	AMC	AMM	ULC	ULM
gdb1	12 22	316	63	350	91	349	91	330	107
gdb2	12 26	339	59	366	79	370	73	353	84
gdb3	12 22	275	59	293	79	319	71	297	84
gdb4	11 19	287	64	287*	91	302	104	320	91
gdb5	13 26	377	64	438	108	423	88	407	130
gdb6	12 22	298	64	324	87	340	84	318	102
gdb7	12 22	325	57	363	87	325*	74	330	91
gdb8	27 46	344	38	463	73	393	41	388	53
gdb9	27 51	303	37	354	57	352	44	358	53
gdb10	12 25	275	39	295	84	300	89	283	81
gdb11	22 45	395	43	447	114	449	97	413	121
gdb12	13 23	448	93	581	122	569	103	537	99
gdb13	10 28	536	128	563	203	560	128*	552	190
gdb14	7 21	100	15	114	26	102	28	104	37
gdb15	7 21	58	8	60	21	60	16	58*	27
gdb16	8 28	127	14	135	40	129	30	132	40
gdb17	8 28	91	9	93	31	91*	20	93	23
gdb18	9 36	164	19	177	44	174	42	172	41
gdb19	8 11	55	17	57	24	63	17*	63	25
gdb20	11 22	121	20	132	46	129	39	125	33
gdb21	11 33	156	15	176	44	163	34	162	48
gdb22	11 44	200	12	208	32	204	35	207	48
gdb23	11 55	233	13	251	50	237	31	239	52

4.4 Numerical Evaluation for the Test Protocols

To study the efficiency of the method through each test protocol and without any information on the optimal efficient front, we have based our analysis on a deviation from the known lower bounds and also on the measure defined in Section 4.2. Average results are reported for the 23 *gdb* files.

Table 3 presents the results for each test protocol. Deviations (in %) from the lower bounds are reported for the extreme points of the approximate efficient front (these points are named LEFT and RIGHT). Again, C denotes the total cost objective and M the makespan objective. For LEFTC (the minimum values obtained for the total cost objective) average deviations are reported but also the number of optimal solutions found (when the lower bound is reached). MO1 is the reference front used for the measure. Two columns report the measure in its standard and normalized forms defined in Section 4.2.

For instance, in the test protocol MO4, the average number of efficient solutions on the last front is 3.26 (average among the 23 instances files); the average deviation from the total cost lower bound is only 0.56% and 16 out of 23 optimal solutions are retrieved. The average deviation from the makespan lower bound of the best solution for the total cost criterion is 46.35%. The average deviation from the makespan lower bound is 22.07% for a deviation of the total cost criterion of 9.72%. 11 out of 23 optimal solutions were retrieved for the makespan criterion. Using our measure, the improvement of the reference front is indexed by -22.82 and -6.48 if this measure is normalized (divided by the number of solution in the efficient front).

Table 3. Results of each test protocol

Test Protocol	Aver. $|E|$	LEFTC dev.	LEFTC # opt.	LEFTM dev.	RIGHTC dev.	RIGHTM dev.	RIGHTM # opt.	Measure Std.	Measure Norm.	CPU time (s)
MO1	3.87	3.81	5	77.02	15.44	29.64	9	–	–	7.14
MO2	3.52	0.47	17	57.58	12.70	30.43	7	-18.15	-4.62	11.39
MO3	3.78	3.40	5	66.85	15.14	24.97	11	-5.56	-1.74	8.61
MO4	3.26	0.56	16	46.35	9.72	22.07	11	-22.82	-6.48	12.64
MO5	4.35	10.15	2	71.14	20.42	31.98	0	+35.41	+7.36	12.59
MO6	3.61	0.36	16	47.87	10.53	19.57	10	-26.17	-6.47	22.68

Looking at the measure columns, it appears that MO4 and MO6 test protocols better improve the reference front given by MO1. These two test protocols use the bi-objective LS procedure with the Accept3 function.

Less significant improvements are also obtained by test protocols MO2 and MO3. Local search is applied in one direction only (one criterion) and overall best solutions cannot be reached. As previously mentioned, the makespan lower bound is weak and gaps are rather large on average (from about 20% to 32%, column RIGHTM, dev.). However, some optimal solutions are found for this objective. The LBM bound corresponds in fact to the trivial solution in which each required edge is treated by a separate trip. A careful analysis of successive fronts show that this very particular solution is very unstable and quickly disappears. This is not very important because it is unlikely to be used in practice, due to its excessive number of vehicles.

The use of initial heuristics (as in the single criteria optimization) is very effective, and results of MO5 test protocol without these heuristics are far from the reference front, even with a larger number of iterations.

Increasing the local search rate from 10% to 20% while keeping the same number of iterations (test protocols MO4 and MO6) gives better solutions in terms of deviations to the lower bounds for the extreme points, at the expense of a very small increase for LEFTM and RIGHTC. The standard measure is also improved but, since the number of solutions in the efficient front is increased, the normalized measure is almost identical. Another price to pay for the higher LS rate is an increased average running time.

4.5 Influence of the Local Search Procedure

To highlight the importance of the use of the local search procedure, the second but largest instance *gdb11* is intensively tested. Two executions are compared: 1) No local search is applied, and 2) Systematic local search with rate 100%. Every 100 iterations the solutions of the approximate efficient front are scanned. The execution is stopped when two fronts (separated by 100 iterations) are identical.

Figure 2 presents the efficient solutions computed by the MOGA with and without local search procedure, after 100 iterations and at the end. 900 iterations are performed for the run without local search and 700 for the run with local search.

It clearly appears that the LS procedure speeds up the convergence of the algorithm. The final solutions without local search are quite equivalent to the ones obtained after only 100 iterations with local search. If we compare the front

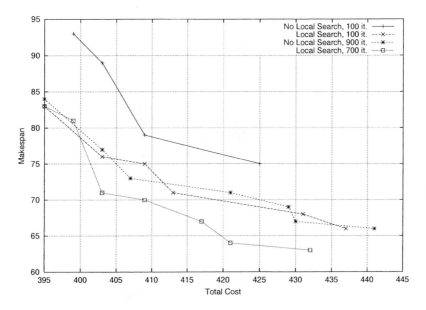

Fig. 2. Impact of local search on the efficient solutions for instance *gdb11*

obtained after 100 iterations when the local search procedure is activated and the final front whitout local search procedure, only two solutions of the former front are dominated by the solution of the latter front after 700 iterations.

Letting the algorithm with local search procedure running until the stopping conditions are met improves again the efficient front and provides high-quality results.

4.6 Evolution of the Efficient Fronts

Figure 3 shows the persistence of a solution in the efficient fronts during the iterations. Again, instance *gdb11* without local search procedure has been used here. As in Section 4.5, the algorithm is scanned every 100 iterations and stopped when no new solutions are found on the efficient front.

A grey box indicates that a solution has appeared in the efficient front. A hatched box indicates that the same solution 100 iterations before still belongs to the efficient front. For example, the fifth solution at the 200th iteration is in all efficient fronts until the end. The first solution at iteration 300 stays in the population for 200 iterations more and is replaced by a better solution at iteration 600. This recent solution remains in the population until iteration 900.

Figure 4 is identical to Figure 3 but the local search procedure is systematically applied to offspring (rate 100%). 700 iterations are performed before stopping the execution. More solutions are generated at iteration 100, 6 instead of 4 without local search.

Less iterations are necessary and moreover less solutions appear during the execution. From iteration 400, 11 solutions appear in Figure 3 whereas only 6

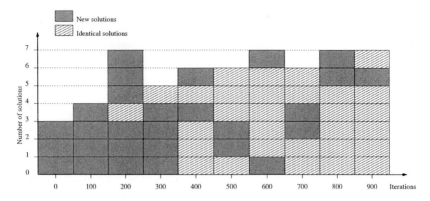

Fig. 3. Evolution of the number of efficient solutions for *gdb11* without local search procedure

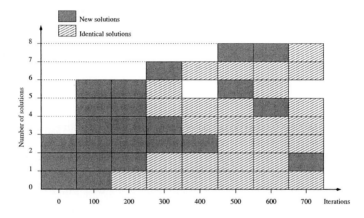

Fig. 4. Evolution of the number of efficient solutions for *gdb11* with local search

solutions appear in Figure 4. Again, this shows the benefit of incorporating a local search procedure into a MOGA. The convergence is accelerated by the local search.

5 Conclusion

Solving efficiently bi-objective real life problems is a challenging task. In addition to the total cost criterion, which is today well optimized by the best algorithms for the academic CARP, it is now possible to minimize the makespan simultaneously and this paper provides an efficient MOGA for this purpose.

In this approach, the implementation of the NSGA-II algorithm alone could not efficiently solve the bi-objective problem. Two key features are necessary: using good heuristics (Path-Scanning, Augment-Merge and Ulusoy's heuristics) in the initial population and adding a local search able to improve the solutions for both criteria. An intensive benchmark testing, with a strict comparison based on the best-known lower bounds and on a distance measure, proves the validity and the efficiency of the proposed approach.

It is foreseen to apply the MOGA to the town of Troyes. Although the algorithm is ready, this work cannot be done now, because the municipality must complete first its computerized representation of the street network: for instance, many forbidden turns are not yet entered.

References

1. A. Amberg and S. Voß. A hierarchical relaxations lower bound for the capacitated arc routing problem. In R.H. Sprague (Hrsg.), editor, *Proceedings of the 35th Annual Hawaï International Conference on Systems Sciences*, pages DTIST02:1–10, Piscataway, 2002. IEEE.
2. J.M. Belenguer and E. Benavent. A cutting plane algorithm for the capacitated arc routing problem. *Computers and Operations Research*, 2003. To appear.
3. J.M. Belenguer, E. Benavent, and F. Cognata. Un metaheuristico para el problema de rutas por arcos con capacidades. In *Proocedings of the 23th national SEIO meeting*, Valencia, Spain, 1997.
4. C.A. Coello Coello. An updated survey of ga-based multiobjective optimization techniques. *ACM Computing Surveys*, 32(2):109–143, 2000.
5. A. Corberan, E. Fernandez, M. Laguna, and R. Martí. Heuristic solutions to the problem of routing school buses with multiple objectives. *Journal of the Operational Research Society*, 53(4):427–435, 2002.
6. K. Deb. Multi-objective genetic algorithms: Problem difficulties and construction of test problems. *Evolutionary Computation*, 7(3):205–230, 1999.
7. K. Deb. *Multi objective optimization using evolutionary algorithms*. Wiley, 2001.
8. K. Deb, A. Pratap, S. Agarwal, and T. Meyarivan. A fast and elitist multi-objective genetic algorithm: NSGA-II. *IEEE Transactions on Evolutionary Computation*, 6(2):182–197, 2002.
9. M. Ehrgott and X. Gandibleux. *Multiobjective Combinatorial Optimization*, volume 52 of *International Series in Operations Research and Management Science*, pages 369–444. Kluwer, 2002.
10. B.L. Golden, J.S. DeArmon, and E.K. Baker. Computational experiments with algorithms for a class of routing problems. *Computers and Operations Research*, 10(1):47–59, 1983.
11. B.L. Golden and R.T. Wong. Capacitated arc routing problems. *Networks*, 11:305–315, 1981.
12. A. Hertz, G. Laporte, and M. Mittaz. A tabu search heuristic for the capacitated arc routing problem. *Operations Research*, 48(1):129–135, 2000.
13. S.C. Hong and Y.B. Park. A heuristic for bi-objective vehicle routing problem with time windows constraints. *International Journal of Production Economics*, 62:249–258, 1999.
14. P. Lacomme, C. Prins, and W. Ramdane-Chérif. Competitive memetic algorithms for arc routing problems. *Annals of Operations Research*, 2002. To appear. See also Research report LOSI-2001-01, Université de Technologie de Troyes, France.
15. P. Moscato. *New ideas in optimization*, chapter Memetic algorithms: a short introduction, pages 219–234. MacGraw-Hill, 1999.
16. Y.B. Park and C.P. Koelling. An interactive computerized algorithm for multicriteria vehicle routing problems. *Computers and Industrial Engineering*, 16:477–490, 1989.

17. A. Riise. Comparing genetic algorithms and tabu search for multi-objective optimization. In *Abstract conference proceedings*, page 29, Edinburgh, UK, July 2002. IFORS.
18. W. Sessomboon, K. Watanabe, T. Irohara, and K. Yoshimoto. A study on multi-objective vehicle routing problem considering customer satisfaction with due-time (the creation of pareto optimal solutions by hybrid genetic algorithm). *Transaction of the Japan Society of Mechanical Engineers*, 1998.
19. G. Ulusoy. The fleet size and mixed problem for capacitated arc routing. *European Journal of Operational Research*, 22:329–337, 1985.
20. E. Zitzler, K. Deb, and L. Thiele. Comparison of multiobjective evolutionary algorithms: Empirical results. *Evolutionary Computation*, 8(2):173–195, 2000.

Multi-objective Rectangular Packing Problem and Its Applications

Shinya Watanabe[1], Tomoyuki Hiroyasu[2], and Mitsunori Miki[2]

[1] Graduate Student, Department of Knowledge Engineering and Computer Sciences,
Doshisha University, 1-3 Tatara Miyakodani, Kyo-tanabe,
Kyoto, 610-0321, JAPAN
sin@mikilab.doshisha.ac.jp
http://mikilab.doshisha.ac.jp/~sin/

[2] Department of Knowledge Engineering and Computer Sciences,
Doshisha University, 1-3 Tatara Miyakodani, Kyo-tanabe,
Kyoto, 610-0321, JAPAN
tomo@is.doshisha.ac.jp,mmiki@mail.doshisha.ac.jp

Abstract. In this paper, Neighborhood Cultivation GA (NCGA) is applied to the rectangular packing problem. NCGA is one of the multi-objective Genetic Algorithms that includes not only the mechanisms of effective algorithms such as NSGA-II and SPEA2, but also the mechanism of the neighborhood crossover. This model can derive good non-dominated solutions in typical multi-objective optimization test problems. The rectangular packing problem (RP) is a well-known discrete combinatorial optimization problem in many applications such as LSI layout problems, setting of plant facility problems, and so on. The RP is a difficult and time-consuming problem since the number of possible placements of rectangles increase exponentially as the number of rectangles increases. In this paper, the sequent-pair is used for representing the solution of the rectangular packing and PPEX is used as the crossover. The results were compared to the other methods: SPEA2, NSGA-II and non-NCGA (NCGA without neighborhood crossover). Through numerical examples, the effectiveness of NCGA for the RP is demonstrated and it is found that the neighborhood crossover is very effective both when the number of modules is small and large.

1 Introduction

Recently, the study of the evolutionary computation of multi-objective optimization has been actively researched and has made great progress [1,9]. The genetic algorithm (GA) is standout among the many approaches that have been proposed [1]. Since GA is one of multi point search methods, it can approximate a set of Pareto-optimum solutions in a trial. This phenomenon is one of the reasons why GA is studied in the field of multi-objective optimization problems.

In recent years, several new algorithms that find good Pareto-optimum solutions with small calculation costs have been developed [1]. They are NSGA-II [1], SPEA2 [9], and so on. These new algorithms have the same search mechanisms:

the preservation scheme of excellent solutions found in the search and the sharing scheme without parameters.

On the other hand, we proposed the new model of multi-objective GA that called Neighborhood Cultivation GA (NCGA) [8]. NCGA includes not only the mechanisms of NSGA-II and SPEA2 that derive the good solutions but also the mechanism of neighborhood crossover.

This model derives good Pareto solutions in typical multi-objective optimization test problems. From the results of the test functions, it is theorized that NCGA can derive good solutions in complicated problems like large-scale or real-world problems.

In this paper, NCGA is applied to the rectangular packing problem (RP). Because RP is a NP-hard problem, good heuristics are generally solicited. The RP can be found in a setting problem of LSI floor plan problem [4,5,6], plant facilities [7], and so on. The sequent-pair is used for representing a solution of the rectangular packing and PPEX is used as a crossover.

In numerical experiments, the results of NCGA are compared with those of NSGA-II, SPEA2 and non-NCGA. Non-NCGA is the same algorithms as NCGA but without neighborhood crossover.

2 Multi-objective Optimization Problems by Genetic Algorithms and Neighborhood Cultivation GA

2.1 Multi-objective Optimization Problems and Genetic Algorithm

Several objectives are used in multi-objective optimization problems. These objectives usually cannot be minimized or maximized at the same time due to a trade-off relationship between them [1]. Therefore, one of the goals of the multi-objective optimization problem is to find a set of Pareto-optimum solutions.

The Genetic Algorithm is an algorithm that simulates the heredity and evolution of living things [1]. Because it is a multi point search method, an optimum solution can be determined even when the landscape of the objective function is multi modal. It can also find a Pareto-optimum set with one trial in multi-objective optimization. As a result, GA is a very effective tool for multi-objective optimization problems. Many researchers are researching multi-objective GA and are developing many algorithms of multi-objective GA [1,2,3,9].

The algorithms of multi-objective GA are roughly divided into two categories: algorithms that treat Pareto-optimum solutions implicitly and algorithms that treat Pareto-optimum solutions explicitly [1]. Many of the latest methods treat Pareto-optimum solutions explicitly.

The typical algorithms that treat Pareto-optimum solutions explicitly are NSGA-II [1] and SPEA2 [9]. These algorithms have the following similar schemes:
1) Reservation mechanism of the excellent solutions
2) Reflection to search solutions mechanism of the reserved excellent solutions
3) Cut down (sharing) method of the reserved excellent solutions
4) Unification mechanism of values of each objective

These mechanisms derive the good Pareto-optimum solutions. Consequently, the developed algorithms should have these mechanisms.

2.2 Neighborhood Cultivation Genetic Algorithm

In this paper, we extend GA and develop a new algorithm called Neighborhood Cultivation Genetic Algorithm (NCGA). NCGA has the neighborhood crossover mechanism in addition to the mechanisms of GAs that are explained in the former chapter. In GAs, exploration and exploitation are very important. By exploration, an optimum solution can be found around the elite solution. By exploitation, an optimum solution can be found in a global area. In NCGA, the exploitation factor of the crossover is reinforced. In the crossover operation of NCGA, a pair of the individuals for crossover is not chosen randomly, but individuals who are close to each other are chosen. Because of this operation, child individuals that are generated after the crossover may be close to the parent individuals. Therefore, the precise exploitation is expected.

The following steps demonstrate the overall flow of NCGA where

P_t : search population at generation
A_t : archive at generation .

Step 1: Initialization: Generate an initial population P_0. Population size is N. Set $t = 0$. Calculate fitness values of the initial individuals in P_0. Copy P_0 into A_0. Archive size is also N.

Step 2: Start new generation: set $t = t + 1$.

Step 3: Generate new search population: $P_t = A_{t-1}$.

Step 4: Sorting: Individuals of P_t are sorted according to the values of the focused objective. The focused objective is changed at every generation. For example, when there are three objectives, the first objective is focused in the first generation and the third objective is focused in the third generation. The first objective is focused again in the fourth generation.

Step 5: Grouping: P_t is divided into groups consisting of two individuals. These two individuals are chosen from the top to the bottom of the sorted individuals.

Step 6: Crossover and Mutation: In a group, crossover and mutation operations are performed. From two parent individuals, two child individuals are generated. Here, parent individuals are eliminated.

Step 7: Evaluation: All of the objectives of individuals are derived.

Step 8: Assembling: All the individuals are assembled into one group and this becomes new P_t.

Step 9: Renewing archives: Assemble P_t and A_{t-1} together. The N individuals are chosen from $2N$ individuals. To reduce the number of individuals, the same operation of SPEA2 (Environment Selection) is performed. In NCGA, this environment selection is applied as a selection operation.

Step 10: Termination: Check the terminal condition. If it is satisfied, the simulation is terminated. If not, the simulation returns to Step 2.

In NCGA, most of the genetic operations are performed in a group consisting of two individuals.

The following features of NCGA are the differences from SPEA2 and NSGA-II.
1) NCGA has the neighborhood crossover mechanism.
2) NCGA has only the environment selection. It does not have the mating selection.

3 Formulation of Layout Problems and Configuration of Genetic Algorithm

The rectangular packing problem (RP) is a well-known discrete combinatorial optimization problem in many applications such as LSI layout problems [4,5,6], plant facilities [7], and so on.

The rectangular packing problem (RP) is known to be a difficult and time-consuming problem since the number of possible placements of rectangles increase exponentially as the number of rectangles increases.

A module(block) $m_i \in M, (0 \leq i < n)$ is a rectangle with a given height and width in real numbers. A packing of a set of modules is a non-overlapping placement of given modules. The problem of RP is to find a packing M with the minimum area. This problem is NP-hard; therefore, good heuristics are generally solicited.

In this paper, we treat the RP as two objective optimization problems. This multi-objective RP aims to minimize not the packing area but the width and height of the packing area. In this formulation, a decision maker can select the aspect ratio of packing area.

3.1 Genetic Approach for RP

Representation. Many approaches have been proposed to solve RP in a practical computation time [4,5]. One important key in the struggle to solve the problem is the representation of an instance of RP. Recently, sequence-pair [4] and BSG [5] have been proposed as a solution of this problem. Sequence-pair and BSG are particularly suitable for stochastic algorithms such as GA and simulated annealing(SA). These coding schemes can represent not only slicing structure but also non-slicing structure. Currently, these coding schemes are different reverse polish notation (RPN).

In this paper, we used sequence-pair as the representation of a solution, since sequence-pair can perform more effective searches than BSG. The number of all combination of sequence-pair is smaller than that of BSG.

Sequence-Pair. The sequence-pair is used for representing the solution of the rectangular packing. Each module has the sequence-pair (Γ_-, Γ_+). By comparing the sequence-pair of the two modules, the relative position of these modules are defined. Let module A and B have the sequence pairs (x_{a-}, y_{a+}) and (x_{b-}, y_{b+}) respectively. In this case, there is a relationship between the positions of the modules and the sequence pairs as follows,

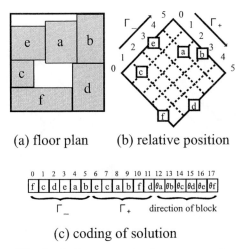

(a) floor plan (b) relative position

(c) coding of solution

Fig. 1. Coding example of sequence-pair

when $x_{a-} < x_{b-}$ and $y_{a+} < y_{b+}$, A is in the left side of B
when $x_{a-} > x_{b-}$ and $y_{a+} > y_{b+}$, A is in the right side of B
when $x_{a-} < x_{b-}$ and $y_{a+} > y_{b+}$, A is in the upper side of B
when $x_{a-} > x_{b-}$ and $y_{a-} < y_{b+}$, A is in the bottom side of B.

In addition to the sequence-pair, each module has the orientation information Θ. This information instructs the direction of the module arrangement.

Coding System. A gene of the GA consists of three parts; those are Γ_-, Γ_+, and Θ. Fig. 1 displays the coding for 6 modules.

From the coding information (Fig. 1(c)), the relative position (b) is derived. This position shows the floor plan (a). In this paper, each module is settled lengthwise or breadwith. Therefore, Θ takes 0 or 1.

Crossover Operator. In this paper, we use the Placement-based Partially Exchanging Crossover (PPEX) [6] that was proposed by Nakaya and et al. The PPEX makes a window-territory that locates in the neighborhood of modules chosen randomly. This window-territory means a continuous part of the oblique-grid that is defined by the sequence-pair. The PPEX performs as a crossover that exchanges modules within this window-territory. Therefore the PPEX can exchange modules within the neighborhood position. The procedure of the PPEX is illustrated as follows.

Step 1: Two modules are chosen randomly as parent modules.
Step 2: The window-territory is created in the neighborhood of the chosen modules. Let M_c be the set of modules within window-territory and M_{nc} be the set of the rest modules.

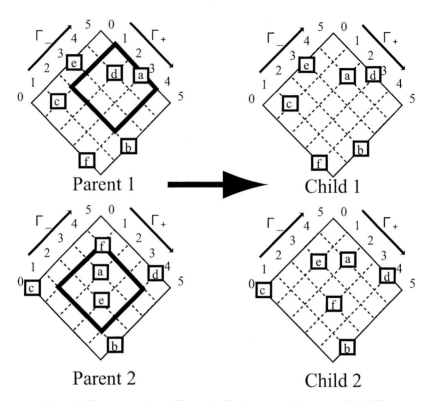

Fig. 2. Placement-based Partially Exchanging Crossover (PPEX)

Step 3: Each module of M_c is exchanged according to the sequence of its partner parent and is copied to the child.

Step 4: M_{nc} are directly copied to child.

Fig.2 displays PPEX when the window-territory size is 4.

In Parent 2, modules of a and e are chosen for M_c. Modules of M_c are exchanged. In this exchange, the relative position of the other parent is referenced. Then these modules are copied to the child. With the location information of Parent 1, a e and f are moved then copied to child 2.

Mutation Operator. In this paper, we use bit flip of orientation for module(θ). That is, if θ is 1, it let θ be 0, the opposite, if θ is 0, it let θ be 1.

3.2 Formulation of Layout Problems

In this paper, there are two objectives as follows,

$$\min f_1(x) = \text{width of packing area of modules}$$
$$\min f_2(x) = \text{length of packing area of modules}$$

Table 1. GA Parameters

population size	200
crossover rate	1.0
mutation rate	1/bit length
terminal generation	400

These two objects have trade-off relations with each other. A decision maker can select an aspect ratio of packing area.

4 Numerical Examples

In this paper, NCGA is applied to some numerical experiments. We used four instances of this problem: ami33, ami49, pcb146 and pcb500. The instances ami33 and ami49 whose data are in the MCNC benchmark consist of 33 and 49 modules (rectangles). The instances pcb146 and pcb500 were given by Kajitani [4]. These instances have 146 and 500 rectangles, respectively.

The results are compared with those of SPEA2 [9], NSGA-II [1] and non-NCGA. Non-NCGA is the same algorithm of NCGA without the neighborhood crossover.

4.1 Parameters of GAs

Table. 1 displays the used GA parameters. We used the above GA operator, PPEX and the bit flip of module orientation. The length of the chromosome is three times as long as the number of modules.

4.2 Evaluation Methods

To compare the results derived by each algorithm, the following evaluation methods are used.

Sampling of the Pareto Frontier Lines of Intersection(I_{LI}). This comparison method is presented by Knowles and Corne [3]. The concept of this method is shown in Fig. 3. This figure illustrates two solution sets of X and Y derived by the different methods.

At first, the attainment surfaces defined by the approximation sets are calculated. Secondly, the uniform sampling lines that cover the Pareto tradeoff area are decided. For each line, the intersections of the line and the attainment surfaces of the derived sets are obtained. These intersections are compared. Finally, the Indication of Lines of Intersection (I_{LI}) is derived. When the two approximation sets X and Y are considered, $I_{LI}(X,Y)$ indicates the average number of the points X are ranked higher than Y. Therefore the most significant outcome would be $I_{LI}(X,Y) = 1.0$ and $I_{LI}(Y,X) = 0.0$.

To focus only on the Pareto tradeoff area as defined by the approximation sets and to derive the intuitive evaluation value, the following terms are considered:

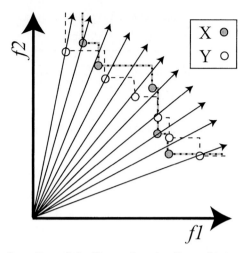

Fig. 3. Sampling of the Pareto frontier lines of intersection

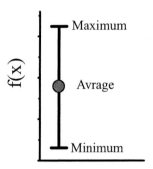

Fig. 4. An example of I_{MMA}

- The objective values of approximation sets are normalized.
- The sampling lines are located in the area where the approximation sets exist.
- Many sampling lines are prepared. In the following experiment, 1000 lines are used.

Maximum, Minimum and Average Values of Each Object of Derived Solutions (I_{MMA}). To evaluate the derived solutions, not only the accuracy but also the expanse of the solutions is important. To discuss the expanse of the solutions, the maximum, minimum and average values of each object are considered. Figure 4 is an example of this measurement. In this figure, the maximum and minimum values of objective function are illustrated. At the same time, the medium value is shown as a circle.

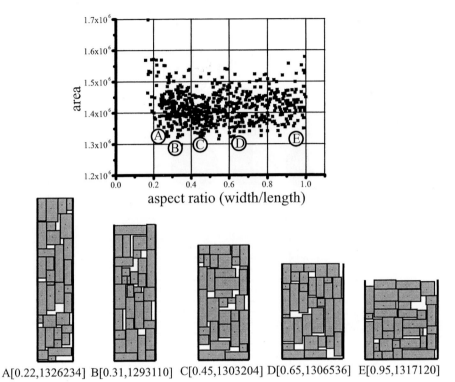

A[0.22,1326234] B[0.31,1293110] C[0.45,1303204] D[0.65,1306536] E[0.95,1317120]

Fig. 5. The placement of the modules(ami33)

4.3 Results

In this study, we tried four types of a problem: ami33, ami49, pcb146 and pcb500 modules. In this section, we discuss only the instances ami33 and pcb500.

Proposed NCGA, SPEA2, NSGA-II and non-NCGA (NCGA without neighborhood crossover) are applied to these problems. 30 trials have been performed and all the results are the average of the 30 trials.

Layout of the Solution. It should be verified whether solutions that are derived by the algorithm are opposite placement of modules. In this section, we focus on the ami33 which consist of 33 modules. The placement of ami33, which is presented by solutions of NCGA, is shown in Fig. 5.

In Fig. 5, some of the typical solutions are illustrated. Since this is the combination of $N! \times N! \times 2^N$ problem with N module, the real optimum solutions are not derived. In this paper, 80,000 function calls (200 population and 400 generations) are performed. These results may be reasonable, since there are very few blank spaces. We also use a sequence-pair and PPEX to derive good solutions since these techniques are very suitable for GA and RP.

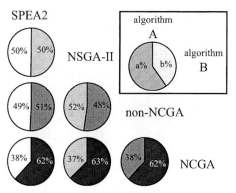

Fig. 6. Results of I_{LI}(ami33)

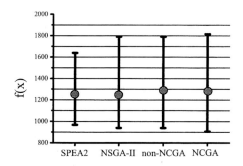

Fig. 7. I_{MMA} of ami33

ami33. In the results of ami33, I_{LI} are shown in Fig. 6, and I_{MMA} are shown in Fig. 7 . Fig. 8 shows the nondominated solutions of each algorithms. In this figure, all nondominated solutions derived from the 30 trials are plotted.

I_{LI} of Fig. 6 indicates that solutions of NCGA are closer to the real Pareto solutions than those of the other methods. This fact is also given from the plots of the nondominated solutions(Fig.8). It is also clear from I_{MMA} of Fig. 7 that NCGA and non-NCGA can find the wide spread nondominated solutions compared to the other methods.

Non-NCGA can get wide-spread nondominated solutions. However, compared to the real Pareto solutions, non-NCGA is not ideal. This result shows that the neighborhood crossover derives good solutions in RP.

pcb500. The results of pcb500 are shown in Fig. 9 and Fig. 10. Fig. 11 illustrates the nondominated solutions of the different algorithms.

The tendency of the results from this problem is similar to those of the previous problem. From Fig.9 and Fig.11, it is clear that NCGA obtained the better value of I_{LI}; namely, the solution of NCGA is much better than those of the other. Like the previous problem, the solutions of non-NCGA are so far from

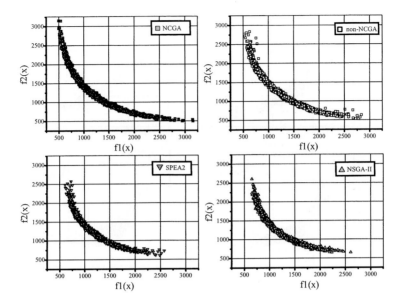

Fig. 8. Nondominated solutions(ami33)

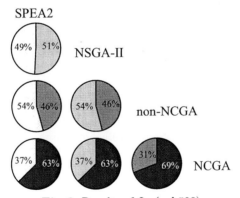

Fig. 9. Results of I_{LI}(pcb500)

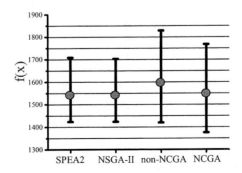

Fig. 10. I_{MMA} of pcb500

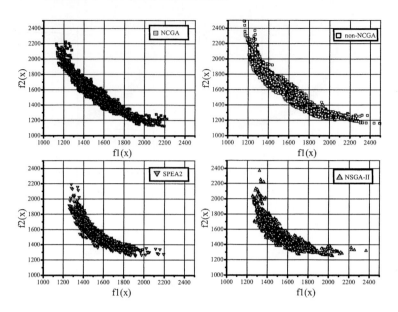

Fig. 11. Nondominated solutions(pcb500)

the real Pareto front. Therefore the neighborhood crossover is very effective to derive the good solutions in RP, irrespective of the number of modules.

On the other hand, in this problem, the solutions of SPEA2 and NSGA-II are gathered around the center of the Pareto front. These results emphasize that SPEA2 and NSGA-II tend to concentrate in one part of the Pareto front when the number of modules is very large. On the other hand, Fig10 and Fig11 indicate that NCGA and non-NCGA could keep the high diversity of the solution during the search even if the number of modules is very large.

5 Conclusion

In this paper, Neighborhood Cultivation GA (NCGA) is applied to the rectangular packing problem (RP). NCGA has not only the important mechanism of the other methods but also the mechanism of the neighborhood crossover selection. In this paper, the sequent-pair is used for representing a solution of the rectangular packing. PPEX is also used as a crossover.

To discuss the effectiveness of NCGA to RP, NCGA was applied to RP and its results were compared to the other methods: SPEA2, NSGA-II and non-NCGA (NCGA without neighborhood crossover). Through numerical examples, the following topics are clarified.

1) The RP that is used in this paper is a large scale problem. For this problem, a reasonable solution is derived with a small calculation cost. It is assumed that a sequence-pair and PPEX work well in this problem.

2) In almost all the test functions, the results of NCGA are superior to that of the others. From this result, it can be noted that NCGA is a good method for the RP.
3) Comparing NCGA and NCGA without the neighborhood crossover, the former is obviously superior to the latter in all the problems. The results emphasize that the neighborhood crossover acts to derive the good solutions in the RP.
4) When the number of modules is very large, the solutions of SPEA2 and NSGA-II tend to concentrate in the center of the Pareto front. However, NCGA and non-NCGA could keep the diversity of the solutions.

References

1. K. Deb. *Multi-Objective Optimization using Evolutionary Algorithms*. Chichester, UK : Wiley, 2001.
2. T. Hiroyasu, M. Miki, and S. Watanabe. The New Model of Parallel Genetic Algorithm in Multi-Objective Optimization Problems -Divided Range Multi-Objective Genetic Algorithms. In *IEEE Proceedings of the 2000 Congress on Evolutionary Computation*, pages 333–340, 2000.
3. J. D. Knowles and D. W. Corne. Approximating the Nondominated Front Using the Pareto Archived Evolution Strategy. volume 8, pages 149–172, 2000.
4. H. Murata, K. Fujiyoshi, S. Nakatake, and Y. Kajitani. VLSI Module Placement Based on Rectangle-Packing by the Sequence-Pair. In *IEEE Transactions on Computer Aided Design*, volume 15, pages 1518–1524, 1996.
5. S. Nakatake, H. Murata, K. Fujiyoshi, and Y. Kajitani. Module Placement on BSG-Structure and IC Layout Applications. In *Proc. of International Conference on Computer Aided Design '96*, pages 484–491, 1996.
6. S. Nakaya, S. Wakabayashi, and T. Koide. An adaptive genetic algorithm for vlsi floorplanning based on sequence-pair. In *2000 IEEE International Symposium on Circuits and Systems, (ISCAS2000)*, volume 3, pages 65–68, 2000.
7. Y. Shirai and N. Matsumoto. Performance Evaluation of ES Type Genetic Algorithms for Solving Block Layout Problems with Floor Constraints. In *PTransaction of JSME (C)65-634*, pages 296–304, 1999.
8. Shinya Watanabe, Tomoyuki Hiroyasu, and Mitsunori Miki. NCGA : Neighborhood Cultivation Genetic Algorithm for Multi-Objective Optimization Problems. In *Proceedings of the Genetic and Evolutionary Computation Conference (GECCO'2002) LATE-BREAKING PAPERS*, pages 458–465, 2002.
9. E. Zitzler, M. Laumanns, and L. Thiele. SPEA2: Improving the Performance of the Strength Pareto Evolutionary Algorithm. In *Technical Report 103, Computer Engineering and Communication Networks Lab (TIK), Swiss Federal Institute of Technology (ETH) Zurich*, 2001.

Experimental Genetic Operators Analysis for the Multi-objective Permutation Flowshop

Carlos A. Brizuela and Rodrigo Aceves

Computer Science Department, CICESE Research Center
Km 107 Carr. Tijuana-Ensenada, Ensenada, B.C., México
{cbrizuel, raceves}@cicese.mx, +52-646-175-0500

Abstract. The aim of this paper is to show the influence of genetic operators such as crossover and mutation on the performance of a genetic algorithm (GA). The GA is applied to the multi-objective permutation flowshop problem. To achieve our goal an experimental study of a set of crossover and mutation operators is presented. A measure related to the dominance relations of different non-dominated sets, generated by different algorithms, is proposed so as to decide which algorithm is the best. The main conclusion is that there is a crossover operator having the best average performance on a very specific set of instances, and under a very specific criterion. Explaining the reason why a given operator is better than others remains an open problem.

1 Introduction

Optimization problems coming from the real world are, in most cases, multi-objective (MO) in nature. The lack of efficient methodologies to tackle these problems makes them attractive both theoretically and practically.

Among these real-world MO problems, scheduling (a combinatorial one) seems to be one of the most challenging. In a real scheduling problem we are interested not only in minimizing the latest completion time (makespan) but also in minimizing the total time all jobs exceed their respective due dates as well as the total time the jobs remain in the shop (work-in-process). A few results on MO scheduling with a single-objective-like approaches reveal that there is much to do in this research area.

On the other hand, the available methodologies in genetic algorithms (GA's) have been focused more on function optimization rather than in combinatorial optimization problems (COP's). In order to fairly compare the performance of two given algorithms some metrics have been proposed. Recent results [17], [21] have shown that these metrics may mislead conclusions on the relative performance of algorithms. This situation forces us to find appropriate procedures to fairly compare two non-dominated fronts, at least in terms of dominance relations.

The remainder of the paper is organized as follows. Section 2 states the problem we are dealing with. Section 3 reviews some available results for this problem. Section 4 introduces the global algorithm used in the operators study. Section 5 explains the experimental setup and introduces the proposed performance measure. Section 6 shows the experimental results. Finally, section 7 presents the conclusions of this work.

2   Problem Statement

Let $\mathbf{K}_n$ be the set of the n first natural numbers, i.e. $\mathbf{K}_n = \{1, 2, \cdots, n\}$. The permutation flow-shop problem consists of a set $\mathbf{K}_n$ of jobs ($n > 1$) that must be processed in a set of machines $\mathbf{K}_m$ ($m > 1$). Each job $j \in \mathbf{K}_n$ has m operations. Each operation O_{kj}, representing the k-th operation of job j, has an associated processing time p_{kj}. Each machine must finish the operation once it is started to be processed (no preemption allowed). No machine can process more than one operation at the same time. No operation can be processed by more than one machine at a time. Each job j is assigned a readiness time r_j, and due date d_j. All jobs must have the same routing through all machines. The goal is to find a permutation of jobs that minimizes a given objective function (since the order of machines is fixed).

In order to understand the objective functions we want to optimize we need to set up some notation first. Let us denote the starting time of operation O_{kj} by s_{kj}. Define the permutations $\pi = \{\pi(1), \pi(2), \cdots, \pi(n)\}$ as a solution to the problem.

With this notation a feasible solution must hold the following conditions:

$$s_{kj} > r_j \quad \forall k \in \mathbf{K}_m, j \in \mathbf{K}_n \ , \tag{1}$$

$$s_{k\pi(j)} + p_{k\pi(j)} \leq s_{(k+1)\pi(j)} \forall k \in \mathbf{K}_{m-1}, j \in \mathbf{K}_n \ . \tag{2}$$

All pairs of operations, of consecutive jobs, processed by the same machine k must satisfy:

$$s_{k\pi(j)} + p_{k\pi(j)} \leq s_{k\pi(j+1)}$$

for every machine $k \in \mathbf{K}_m$, and every job $j \in \mathbf{K}_{n-1}$. $\tag{3}$

Now we are in position of defining the objective functions. First we consider the makespan, which is the completion time of the latest job, i.e.

$$f_1 = c_{max} = s_{m\pi(n)} + p_{m\pi(n)} \ . \tag{4}$$

The mean flow time, representing the average time the jobs remain in the shop, is the second objective.

$$f_2 = \overline{fl} = (1/n) \sum_{j=1}^{n} fl_j \ , \tag{5}$$

where $fl_j = \{s_{mj} + p_{mj}\} - r_j$, i.e. the time job j spends in the shop after its release. The third objective is the mean tardiness, i.e.

$$f_3 = \overline{T} = (1/n)\sum_{j=1}^{n} T_j \, , \tag{6}$$

where $T_j = \max\{0, L_j\}$, and $L_j = s_{mj} + t_{mj} - d_j$.
Thus, we have the following MO problem:

$$\text{Minimize } (f_1, f_2, f_3)$$
$$\text{subject to } (1) - (3) \, . \tag{7}$$

This problem, considering only the makespan (C_{max}) as the objective function, was proven to be NP-hard [11]. This implies that the problem given by (7) is at least as difficult as the makespan problem.

2.1 Tradeoff among Objectives

To show that "conflict" among objectives exists, it is necessary to define what "conflict" means. We give some definitions in order to understand why our problem should be treated as a multi-objective one.

Definition 1. The *Ideal Point or Ideal Vector* is defined as the point $\mathbf{z}^*$ composed of the best attainable objective values. This is,
$z_j^* = \max\{f_j(\mathbf{x})|\mathbf{x} \in A\}$, A is the set (without constraints) of all possible $\mathbf{x}$, $j \in \mathbf{K}_q$, where $q > 1$ is the number of objective functions.

Definition 2. *Two objectives are in conflict* if the Euclidean distance from the ideal point to the set of best values of feasible solutions is different from zero.

Definition 3. *Three or more objectives are in conflict* if they are in conflict pairwise.

To show that the objectives are in conflict we are going to construct a counter-example instance where we can enumerate all solutions and see that there is not a single solution with all its objective values being better than or equal to the respective objective values of the other solutions. This implies that the distance from the set of non-dominated solutions to the ideal point is positive.

Table 1 presents an instance of the PFSP with three jobs and two machines. All solutions for this problem are enumerated in Table 2, here we can see that solution 312 has the best makespan value, 231 the best tardiness, and 213 the best mean flow time. The ideal point $\mathbf{z}^*$ for this instance is given by $\mathbf{z}^* = (44, 16, 87)$. Therefore, the Euclidean distance from the set of values in Table 2 to the ideal point is positive.

It is not known whether or not, in randomly generated PFSP instances, it is enough to consider only two objectives, as the third one, in the Pareto front, will have the same behavior as one of the first two. We did not perform any experimental study regarding the answer for this question because of the required computational effort. However, in the experiments section we will see that for all generated instances there are two objective functions which are not in conflict. It will be very interesting to know which instances will have three conflicting objectives and which will not.

Table 1. Problem data for a 3-jobs 2-machines PFSP

Job (j)	t_{j1}	t_{j2}	Due Date (d_j)
1	17	8	30
2	5	4	15
3	15	11	30

Table 2. Counter-example for the PFSP

Π	f_1	f_2	f_3
321	45	30	101
312	44	39	110
231	45	16	105
213	48	18	87
132	47	45	115
123	48	32	102

3 Genetic Algorithm Approach

Surveys on the existing GA's methodologies for multi-objective optimization problems can be found in [7], [8], [10], and references therein. Almost any application uses the methodologies described in these documents.

The application of GA's to MO scheduling problems has been rather scarce. Two interesting ideas are those presented in [18] and [2].

In [18] the scheduling of identical parallel machines, considering as objective functions the maximum flow time among machines and a non-linear function of the tardiness and earliness of jobs, is presented. In [2] a natural extension of NSGA [19] is presented and applied to flow-shop and job-shop scheduling problems. Another, totally different approach is that presented by Isibuchi and Murata [15]. They use a local search strategy after the genetic operations without considering non-dominance properties of solutions. Their method is applied to the MO flow-shop problem. Basseur et al. [3] also presents a local search idea to deal with this problem. Research on how intensive the local search should be, for a very specific instances of the permutation flowshop problem, is presented by Ishibuchi et al. [16].

In the literature, the main idea when solving MO scheduling problems is to apply the existing GA's methodologies to the problem to solve. However, there are no traces of studies on how adequate these methodologies may be. Again, the lack of a fair methodology for comparing the results does not help to improve this situation.

Brizuela et al. [6] present a question related to the existence of a superior combination of genetic operators in terms of non-dominance relations, and whether or not this superiority will persist over some set of instances for a given flowshop problem. The work presented here try to answer this question.

The next section presents the algorithm we use to compare the performance of different operators.

4 The Algorithm

A standard GA for MO [19] is used with some modifications. The fitness share and dummy fitness assignment are standard. The main difference is in the replacement procedure, where an elitist replacement is used.

The specific NSGA we use here as a framework is stated as follows.
Algorithm 1. Multi-objective GA.

> **Step 1.** Set $r = 0$. Generate an initial population $POP[r]$ of g individuals.
> **Step 2.** Classify the individuals according to a non-dominance relation. Assign a dummy fitness to each individual.
> **Step 3.** Modify the dummy fitness by fitness sharing.
> **Step 4.** Set $i=1$.
> **Step 5.** Use RWS to select two individuals for crossover according to their dummy fitness. Perform crossover with probability p_c.
> **Step 6.** Perform mutation of individual i with probability p_m.
> **Step 7.** Set $i = i + 1$. If $i = g$ then go to Step 8 otherwise go to Step 5.
> **Step 8.** Set $r = r + 1$. Construct the new generation $POP[r]$ of g individuals. If $r = r_{max}$ then STOP; otherwise go to STEP 2.

The procedures involved at each step of this algorithm are explained in the following subsections.

4.1 Decoding

Each individual is represented by a string of integers representing job numbers to be scheduled (a permutation of job numbers). In this representation individual r looks like:

$$\mathbf{i}_r = (i_1^{(r)} \, i_2^{(r)} \cdots i_n^{(r)}), \quad r = 1, 2, \cdots, g \quad,$$

where $i_k^{(r)} \in \mathbf{K}_n$.

The schedule construction method for this individual is as follows:

> 1) Enumerate all machines in $\mathbf{K}_m$ from 1 to m.
> 2) Select the first job $(i_1^{(r)})$ of $\mathbf{i}_r$ and route it from the first machine (machine 1) to the last (machine m).
> 3) Select iteratively the second, third, $\cdots$, n-th job and route them through the machines in the same machine sequence adopted for the first job $i_1^{(r)}$ (machines 1 to m). This must be done without violating the restrictions imposed in (1) to (3).

4.2 Selection and Replacement

The selection operator we use here is standard to GA's, like those proposed elsewhere [14]. Two selection processes are distinguished here.

Selection for mating - *SELECTION* (Step 5). This is the way we choose two individuals to undergo reproduction (crossover and mutation). In our algorithm roulette wheel selection (RWS) is used. This selection procedure works based on the dummy fitness function assigned to each individual. The way to compute the dummy fitness (Step 2) and the way to do the fitness sharing (Step 3) are standard (see [10], pages 147-148).

Selection after reproduction - *REPLACEMENT* (Step 8). This is the way to choose individuals to form the new generation from a set given by all parents and all offsprings. In this paper, the best elements are selected from the pool of parents and offsprings.

To define "the best," all individuals (parents and offsprings) are sorted in fronts. Then the first g individuals are copied from the first front (the nondominated front) following the next fronts until g individuals are obtained. After this the g individuals are ordered according to the following rules: in each front, individuals with better makespans have higher priority followed by individuals with higher tardiness, and finally by the mean flow time. After sorting all individuals repeated ones are replaced by randomly selected individuals from the pool of parents and offsprings. Notice that this does not avoid having repeated individuals in the new generation. The average running time of the above procedure is $O(n \lg n)$, the average running time of Quicksort. The selection of makespan in the ordering is irrelevant to the overall algortihm. This ordering is only to have an efficient algorithm for replacing repeated individuals.

4.3 Crossover

The crossover operators described here are the most frequently used in the literature.

OBX. This is the well known order-based crossover (see [13], page 239) proposed by Syswerda for the TSP. The position of some genes corresponding to parent 1 are preserved in the offspring and the not yet copied elements in parent 2 are copied in the order they appear in this parent. For this crossover a random mask is generated as shown in Fig. 1 (left side).

PPX. Precedence Preservative Crossover [5]. A subset of precedence relations of the parents genes are preserved in the offspring. Fig. 1 (center) shows how this operator works. The 1's in the mask indicate that genes from parent 1 are to be copied and the 0's indicate that genes from parent 2 are to be copied, in the order they appear from left to right.

OSX. One Segment crossover. This is similar to the well known two point crossover ([13], page 408) where also two random points are selected. An example of how this operator work is illustrated in Fig. 1 (right), here genes from loci 1 to s_1 of parent 1 are copied to loci 1 to s_1 in the offspring, and loci $s_1 + 1$ to s_2 (in the offspring) are copied from parent 2 starting at locust 1 and copying

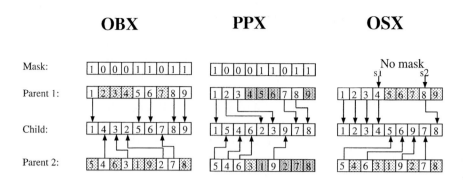

Fig. 1. Example of OBX, PPX, and OSX for nine jobs. In OSX the point s_1 indicates that all genes of P1 from loci 1 to s_1 are going to be copied in the offspring and from loci $s_1 + 1$ to s_2, in the offspring, genes come from P2. The rest will come from P1 copied from left to right considering only those that were not copied from P1 or P2

all those genes that have not been copied from parent 1. From loci $s_2 + 1$ to n, genes are copied from parent 1 considering only genes that were not copied into the offspring.

4.4 Mutation

The mutation operators that we use here are also standard (see [12]). The following mutation operators are used (Step 6).

INSERT. Two loci (s_1, s_2) are randomly selected if $s_1 < s_2$ then the gene corresponding to s_1 is placed on s_2 and all genes from $s_1 + 1$ to s_2 are shifted one position towards s_1. If $s_1 > s_2$ then the shift operation is performed towards s_2 as it is depicted in Fig. 2 (left side).

SWAP. Two loci are randomly selected and their genes interchanged. An example of this is shown in Fig. 2 (center).

SWITCH. A single interchange of two adjacent genes is performed. The locus to swap is randomly selected. Once this locus is selected the corresponding gene is interchanged with its immediate successor to the right, if the last gene (locus n) is selected then this is interchanged with the first one (locus 1). An example is shown in Fig. 2 (right side).

5 Performance Measures and Experimental Setup

Many performance measures for MO algorithms have been proposed (for a survey see [8]-[10]), these metrics try to identify three different characteristics [20] of the attained non-dominated front: *i*) how far the non-dominated front is from the true Pareto front, *ii*) how widely extended the set of solutions is, in the non-dominated front, and *iii*) how evenly distributed the solutions are, in the non-dominated front. Researchers have already noticed that, for many problems, the

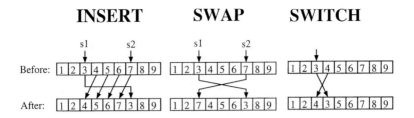

Fig. 2. INSERT, SWAP, and SWITCH mutation operators for a 9-jobs flowshop problem

proposed metrics are not "compatible" with the non-dominance relation among two different non-dominated fronts [17], this is especially true in combinatorial problems. For this kind of problems, properties such as expansion or evenly distributed may not be important. It will depend, of course, on the way the true Pareto front is extended and distributed. All the difficulties mentioned above motivate us to directly work with the nondominance relations between the fronts of the algorithms we want to compare.

In our approach, the non-dominated front to be used for comparisons, is obtained by applying the following procedure.

For each run i of algorithm j denote the set of solutions in the last generation as S_j^i. Calculate $Out(j) = \text{NDS}(\bigcup_{i=1}^{Numruns} S_j^i)$, where $\text{NDS}(X)$ is an operator obtaining the set of non-dominated solutions from the input set X. Let $ns = |Out(j)|$, na = number of algorithms to compare, and let $ss(r, j)$ denote element r of $Out(j)$. Take each solution of $Out(j)$ and compare it against all other solutions in the sets $Out(p)$ for $p = 1, 2, \cdots, na$ and $p \neq j$. For each pair of solutions belonging to different sets we can obtain the following results: i) no dominance relation can be established between $ss(r, j)$ and $ss(k, p)$ ($ss(r, j) \succ\prec ss(k, p)$), ii) solution $ss(r, j)$ is dominated by solution $ss(k, p)$ ($ss(r, j) \prec ss(k, p)$), iii) solution $ss(r, j)$ dominates solution $ss(k, p)$ ($ss(r, j) \succ ss(k, p)$), and iv) the solutions are the same ($ss(r, j) = ss(k, p)$).

We propose as measures counting the number of times each of these situations happen, i.e. m_j^1 counts the number of times a solution in $Out(j)$ can not establish a dominance relation with solutions in the other sets, m_j^2 counts the number of solutions in other sets that dominate solutions in $Out(j)$, m_j^3 counts the number of times solutions in $Out(j)$ dominates solutions in $Out(p)$, finally m_j^4 counts the number of times solution $ss(r, j)$ has copies of itself in other sets.

5.1 Statistical Considerations

Since the previous experiments are taken over the whole set of solutions generated in different runs, we need to know if this behavior statistically holds when considering the output of randomly generated runs. In order to verify this, we compare the set of non-dominated solutions of each run and compute the statistics for each of the four measures (m_j^1, m_j^2, m_j^3, and m_j^4). For computing these statistics we randomly select, without replacement, NDS(S_j^i) uniformly on $i \in \{1, \cdots, NumRuns\}$ for each $j \in \{1, \cdots, na\}$ until we have $NumRuns$ nondominated fronts to compare. The average for each measure m_j^i is computed over $NumRuns$ different samples.

The experiment is performed for each combination of crossover and mutation. Three different crossover operators and three different mutation operators are considered (see subsections 4.3 and 4.4).

5.2 Benchmark Generation

In order to generate a set of benchmark instances we take one of the problem instances in the library maintained by Beasly [4] and apply a procedure proposed by Armentano and Ronconi [1] for generating due-dates. The authors in [1] propose a way to systematically generate controlled due dates, following a certain structure. The procedure is based on an estimation of a lower bound for the makespan (f_1). This lower bound is defined as P, then due dates are randomly and uniformly generated for each job in the range of $P(1 - T - R/2)$ to $P(1 - T + R/2)$, where T and R are parameters for generating different scenarios of tardiness and due date ranges, respectively.

6 Results

The results on the experiments previously described are presented here. The main conclusion drawn from these results is that the crossover operator denominated OBX outperforms all other studied operators in terms of the dominance measures we propose in this work.

Table 3. Dominance relations, mean spacing (S), and ONVG for non-dominated fronts generated by nine algorithms. Instance 6 (75 jobs 20 machines)

Operator	(m_j^4)	(m_j^3)	(m_j^2)	(m_j^1)	(S)	(ONVG)
Algo 1	0.00	49.22	33.04	17.74	17.98	23.86
Algo 2	0.00	47.49	34.21	18.30	18.86	18.62
Algo 3	0.00	2.92	88.36	8.72	30.54	32.96
Algo 4	0.00	87.42	1.29	11.29	19.40	10.70
Algo 5	0.00	80.14	4.42	15.44	15.01	10.02
Algo 6	0.00	69.05	8.29	22.66	21.25	13.66
Algo 7	0.00	48.22	30.14	21.64	17.97	9.88
Algo 8	0.00	43.14	34.53	22.33	36.35	9.40
Algo 9	0.00	11.06	76.76	12.18	28.03	17.56

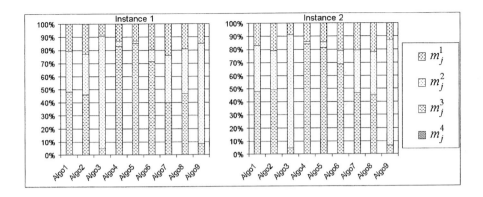

Fig. 3. Performance measures (m_j^i, $j \in \{1, \cdots, na\}, i \in \{1, \cdots, 4\}$) for the 75 jobs 20 machines flowshop problem. Instances 1 and 2

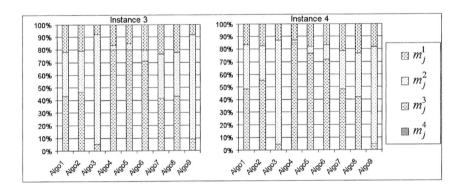

Fig. 4. Performance measures (m_j^i, $j \in \{1, \cdots, na\}, i \in \{1, \cdots, 4\}$) for the 75 jobs 20 machines flowshop problem. Instances 3 and 4

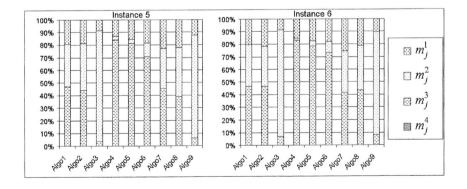

Fig. 5. Performance measures (m_j^i, $j \in \{1, \cdots, na\}, i \in \{1, \cdots, 4\}$) for the 75 jobs 20 machines flowshop problem. Instances 5 and 6

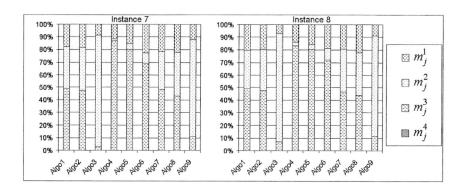

Fig. 6. Performance measures (m_j^i, $j \in \{1, \cdots, na\}, i \in \{1, \cdots, 4\}$) for the 75 jobs 20 machines flowshop problem. Instances 7 and 8

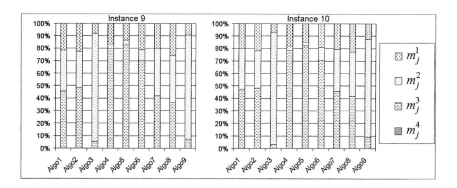

Fig. 7. Performance measures (m_j^i, $j \in \{1, \cdots, na\}, i \in \{1, \cdots, 4\}$) for the 75 jobs 20 machines flowshop problem. Instances 9 and 10

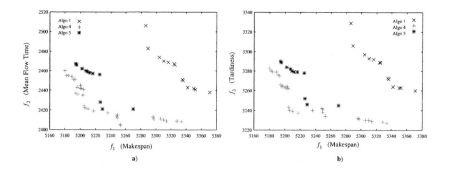

Fig. 8. Union of all nondominated solutions in the last generation ($Out(j)$) for Algo 1, 4, and 5 for Instance 6. a) Mean Flow Time and Makespan relations. b) Tardiness and Makespan relations

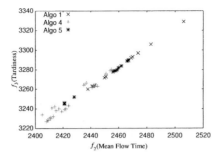

Fig. 9. Tardiness and Mean Flow Time relations. Union of all nondominated solutions in the last generation ($Out(j)$) for Algo 1, 4, and 5 for Instance 6

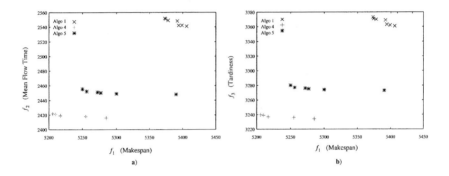

Fig. 10. Nondominated solutions in the last generation of a randomly selected run (NDS(S_j^i)) for Algo 1, 4, and 5 for Instance 6. a) Mean Flow Time and Makespan relations. b) Tardiness and Makespan relations

The maximum number of generations is 2000. The total number of runs for each algorithm is 50. The crossover rate was fixed to 1.0, the mutation rate to 0.1, and the population size to 100. Ten instances of 75 jobs and 20 machines are studied. The 75-jobs and 20-machines was taken from the benchmark library maintained by Beasly [4]. The due-dates were generated by setting T and R to 0.4 and 1.2, respectively. The operator combinations to study are denominated as follows: OSX-INSERT is Algo 1, OSX-SWAP (Algo 2), OSX-SWITCH (Algo 3), OBX-INSERT (Algo 4), OBX-SWAP (Algo 5), OBX-SWITCH (Algo 6), PPX-INSERT (Algo 7), PPX-SWAP (Algo 8), and PPX-SWITCH is Algo 9.

All generated instances have their processing times as in the benchmark instance in [4], the only difference is in the due-dates (the original problem in [4] considers no due dates).

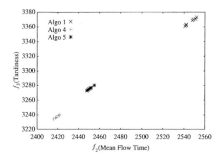

Fig. 11. Tardiness and Mean Flow Time relations. Nondominated solutions in the last generation of a randomly selected run (NDS(S_j^i)) for Algo 1, 4, and 5 for Instance 6

Table 4. Comparison of dominance relations between Algo 4 - Algo 5, and Algo 4 - Algo 1 over all instances. Dominance relations, mean and standard deviation

Algo 4 vs. Algo 5	(m_j^4)	(m_j^3)	(m_j^2)	(m_j^1)
Mean	0.00	27.26	20.12	52.62
Std. Dev.	1.00	6.87	6.55	4.63
Algo 4 vs. Algo 1				
Mean	0.00	84.61	1.42	13.97
Std. Dev.	0.00	4.84	1.80	3.30

Table 3 presents the results for a single instance and 50 runs, it is clear to see that the combinations OBX with INSERT (Algo 4) and OBX with SWAP (Algo 5) have a small number of solutions dominated by the other combinations (less than 4.5%) and the highest percentage of domination, over 80% each. Table 3 also presents results for the Spacing (S) and the Overall Nondominated Vector Generation (ONVG) as they are defined in [20]. In this table we can see that the ONVG measure is not a good indicator for the performance of algorithms. ONVG indicates that Algo 3 is the best one, but considering dominance relations it is one of the worst ($m_3^2 = 88.36\%$). Figures 3 to 7 show the performance measure for each combination of operators, and for each of the 10 different generated instances.

Figures 8 to 9 show the projections in two axes of the overall nondominated front for Algo 1, Algo 4, and Algo 5 (instance 6). It is easy to see that Algo 4 and 5 clearly outperforms Algo 1, in every projection. Figure 9 shows that the mean flow time (f_2) and the tardiness (f_3) are not in conflict. This is because of the range of generated due dates.

Figures 10 to 11 show the same two axes projection of three randomly selected non dominated fronts (one for each algorithm) of a single run, and for a particular problem instance (Instance 6). Again, Algo 4 and 5 outperform Algo 1.

By looking at Figs. 8 to 11 we can see that Algo 4 outperforms Algo 1 and Algo 5. To study the significance of the difference between these algorithms, dominance relations between Algo 4 - Algo 1, and Algo 4 - Algo 5 are computed for all instances. The results for Algo 4 and Algo 5 are presented in Table 4, these algorithms uses the same crossover operator (OBX) and different mutation operators, INSERT and SWAP, respectively. In this case we see that the domination rate produced by Algo 4 (27%) is not far away from the domination rate of Algo 5 (20%) and the nondominance is more than 50%. The lower part of Table 4 shows a different result, two different crossover operators are compared (OBX and OSX), Algo 4 - Algo 1. In this case we see that solutions in the nondominated front generated by Algo 4 dominates most of the times (84.61%) solutions in the front generated by Algo 1. This result shows the superiority of the OBX operator.

7 Conclusions

An experimental analysis of genetic operators (crossover and mutation) for a set of instances of the multi-objective permutation flowshop problem is presented. Outperformance measures, related to dominance relations of the nondominated set generated at each run, are proposed.

The main conclusion of the work is that the OBX crossover operator outperforms all others in terms of dominance relations. This outperformance is verified on a specific set of instances of the PFSP. The reason why this operator outperforms the others remains to be solved.

As future work we need to verify whether the properties for the best operator combinations still hold under different settings of the parameters T and R, different problems size, and different algorithms. The most important point will be to understand why some operators outperform others. This will help us to better understand this class of problems.

References

1. Armentano, V. A., and Ronconi, D. P.: Tabu search for total tardiness minimization in flowshop scheduling problems. Computers & Operations Research, Vol. 26 (1999) 219–235
2. Bagchi, T. P.: Multiobjective Scheduling by Genetic Algorithm. Kluwer Academic Publishers (1999)
3. Basseur, M., Seynhaeve, F., Talbi E.: Design of multi-objective evolutionary algorithms: Application to the flow-shop scheduling problem. Congress on Evolutionary Computation (2002) 459–465
4. http://mscmga.ms.ic.ac.uk/info.html
5. Bierwirth, C., Mattfeld, D. C., Kopfer, H.: On Permutation Representations for Scheduling Problems. In Proceedings of Parallel Problem Solving from Nature. Lecture Notes in Computer Science, Vol. 1141. Springer-Verlag, Berlin Heidelberg New York (1996) 310–318

6. C. Brizuela, Y. Zhao, N. Sannomiya.: Multi-Objective Flowshop: Preliminary Results. In Zitzler, E., Deb, K., Thiele, L., Coello Coello, C. A., Corne D., eds., Evolutionary Multi-Criterion Optimization, First International Conference, EMO 2001, vol. 1993 of LNCS, Berlin:Springer-Verlag (2001) 443–457
7. Coello Coello, C. A.: A Comprehensive Survey of Evolutionary-Based Multiobjective Optimization Techniques. Knowledge and Information Systems, Vol. 1, No. 3 (1999) 269–308
8. Coello Coello, C. A., Van Veldhuizen, D. A., and Lamont, G. B.: Evolutionary Algorithms for Solving Multi-Objective Problems. Kluwer Academic Publishers (2002)
9. Deb, K.: Multi-objective Genetic Algorithms: Problem Difficulties and Construction of Test Problems. Evolutionary Computation, 7(3) (1999) 205–230
10. Deb, K.: Multi-Objective Optimization using Evolutionary Algorithms. John Wiley & Sons (2001)
11. M. R. Garey, D. S. Johnson and Ravi Sethi.: The Complexity of Flowshop and Jobshop Scheduling. Mathematics of Operations Research, Vol. 1, No. 2 (1976) 117–129
12. Gen, M. and Cheng, R.: Genetic Algorithms & Engineering Design. John Wiley & Sons (1997)
13. Gen, M. and Cheng, R.: Genetic Algorithms & Engineering Optimization. John Wiley & Sons (1997)
14. Golberg, D.: Genetic Algorithms in Search, Optimization, and Machine Learning, Addison-Wesley (1989)
15. Isibuchi, H. and Murata, T.: Multi-objective Genetic Local Search Algorithm. Proceedings of the 1996 International Conference on Evolutionary Computation (1996) 119–124
16. Isibuchi, H. and Murata, T.: Multi-objective Genetic Local Search Algorithm and its application to flowshop scheduling. IEEE Transactions on Systems, Man, and Cybernetics – Part C: Applications and Reviews, 28(3) (1998) 392–403
17. Knowles J. and Corne D.: On Metrics for Comparing Nondominated Sets. Proceedings of the 2002 Congress on Evolutionary Computation (2002) 711–716
18. Tamaki, H., and Nishino, E.: A Genetic Algorithm approach to multi-objective scheduling problems with regular and non-regular objective functions. Proceedings of IFAC LSS'98 (1998) 289–294
19. Srinivas, N. and Deb, K.: Multi-Objective function optimization using nondominated sorting genetic algorithms. Evolutionary Computation, 2(3) (1995) 221–248
20. Van Veldhuizen, D. and Lamont, G.: On measuring multiobjective evolutionary algorithm performance. Proceedings of the 2000 Congress on Evolutionary Computation (2000) 204–211
21. Zitzler E., Laumanns M., Thiele L., Fonseca C. M., Grunert da Fonseca V.: Why Quality Assesment of Multiobjective Optimizers Is Difficult. Proceedings of the 2002 Genetic and Evolutionary Computation Conference (GECCO2002) (2002) 666–673

Modification of Local Search Directions for Non-dominated Solutions in Cellular Multiobjective Genetic Algorithms for Pattern Classification Problems

Tadahiko Murata[1], Hiroyuki Nozawa[2], Hisao Ishibuchi[3], and Mitsuo Gen [4]

[1] Department of Informatics, Kansai University,
2-1-1 Ryozenji-cho, Takatsuki, Osaka 569-1095, Japan
murata@res.kutc.kansai-u.ac.jp
http://www.res.kutc.kansai-u.ac.jp/~murata/

[2] Department of Industrial and Information Systems Engineering,
Ashikaga Institute of Technology,
268 Omae-cho, Ashikaga, Tochigi 326-8558, Japan

[3] Department of Industrial Engineering, Osaka Prefecture University,
1-1 Gakuen-cho, Sakai, Osaka 599-8531, Japan
hisaoi@ie.osakafu-u.ac.jp
http://www.ie.osakafu-u.ac.jp/~hisaoi/ci_lab_e/

[4] Graduate School of Information, Production and Systems, Waseda University
2-2 Hibikino, Wakamatsu-ku, Kitakyushu, Fukuoka 808-0135, Japan

Abstract. Hybridization of evolutionary algorithms with local search (LS) has already been investigated in many studies. Such a hybrid algorithm is often referred to as a memetic algorithm. Hart investigated the following four questions for designing efficient memetic algorithms for single-objective optimization: (1) How often should LS be applied? (2) On which solutions should LS be used? (3) How long should LS be run? (4) How efficient does LS need to be? When we apply LS to an evolutionary multiobjective optimization (EMO) algorithm, another question arises: (5) To which direction should LS drive? This paper mainly addresses the final issue together with the others. We apply LS to the set of non-dominated solutions that is stored separately from the population governed by genetic operations in a cellular multiobjective genetic algorithm (C-MOGA). The appropriate direction for the non-dominated solutions is attained in experiments on multiobjective classification problems.

1 Introduction

Since Schaffer's study [1], evolutionary algorithms have been applied to various multiobjective optimization problems for finding their Pareto-optimal solutions. Recently evolutionary algorithms for multiobjective optimization are often referred to as EMO (evolutionary multiobjective optimization) algorithms. The task of EMO algorithms is to find Pareto-optimal solutions as many as possible. In recent studies (e.g., [2-6]), emphasis was placed on the convergence speed to the Pareto-front as well as the diversity of solutions. In those studies, some form of elitism was used as an important ingredient of EMO algorithms. It was shown that use of elitism improved the convergence speed to the Pareto-front [5].

One promising approach for improving the convergence speed to the Pareto-front is the use of local search in EMO algorithms. Hybridization of evolutionary algorithms with local search has already been investigated for single-objective optimization problems in many studies (e.g., [7], [8]). Such a hybrid algorithm is often referred to as a memetic algorihm. See Moscato [9] for an introduction to this field and [10]-[12] for recent developments.

Hart [13] investigated the following four questions for designing efficient memetic algorithms for continuous optimization:

(1) How often should local search be applied?
(2) On which solutions should local search be used?
(3) How long should local search be run?
(4) How efficient does local search need to be?

The above four issues should be considered when a local search is hybridized with an evolutionary algorithm for single-objective optimization. When we try to apply a local search to an EMO algorithm, another question arises:

(5) To which direction should local search drive?

We briefly review EMO algorithms with local search from this aspect.

The hybridization with local search for multiobjective optimization was first implemented in [14], [15] as a multiobjective genetic local search (MOGLS) algorithm where a scalar fitness function with random weights was used for the selection of parents and the local search for their offspring. That is, the direction of the local search in their MOGLS algorithm was determined by *random weights*. Jaszkiewicz [16] improved the performance of the MOGLS by modifying its selection mechanism of parents. While his MOGLS still used the scalar fitness function with *random weights* in selection and local search, it did not use the roulette wheel selection over the entire population. A pair of parents was randomly selected from *a pre-specified number of the best solutions* with respect to the scalar fitness function with the current weights. This selection scheme can be viewed as a kind of mating restriction in EMO algorithms. The mating restriction mechanism is explicitly utilized in Cellular MOGLS (C-MOGLS) algorithms [17]. In the C-MOGLS, each individual solution resides in a cell uniformly distributed in a spatially structured space. The selection of parents for a cell is restricted within neighboring cells. The local search is applied to non-dominated solutions using *the weight vector* of the cell from which the solution is taken as a non-dominated solution. Knowles and Corne [18] combined their Pareto archived evolution strategy (PAES [2], [4]) with a crossover operation for designing a memetic PAES (M-PAES). In their M-PAES, *the Pareto-dominance relation* and *the grid-type partition of the objective space* were used for determining the acceptance (or rejection) of new solutions generated in genetic search and local search. The M-PAES had a special form of elitism inherent in the PAES. The performance of the M-PAES was examined in [19] for multiobjective knapsack problems and [20] for degree-constrained multiobjective MST (minimum-weight spanning tree) problems. In those studies, the M-PAES was compared with the PAES, the MOGLS of Jaszkiewicz [16], and an EMO algorithm. In the above-mentioned hybrid EMO algorithms (i.e., multiobjective memetic algorithms [14]-[20]), local search was applied to individuals in every generation. In some studies [21], [22], local search was applied to individuals only in the final generation. While Deb and Goel [21] used local search for decreasing the number of non-dominated solutions (i.e., for decreasing the diversity of final solutions), Talbi et al. [22] intended to increase the diversity of final solutions by the application of local search.

In this paper, we modify the local search directions for the C-MOGLS [17]. In the former C-MOGLS, we implemented the local search for each of non-dominated solutions with the weight vector of the cell from which the solution is taken. This specification is not always appropriate. Using Figs. 1 and 2, we illustrate the drawback of this specification method in the case of a two-objective maximization problem. Let us assume that seven cells are distributed in a two-dimensional space for the problem. As shown in Fig. 1, a weight vector is assigned to each cell according to its location in the space. Let us consider a weight vector of each cell satisfying the following conditions.

$$w_1 + w_2 = 1.0, \tag{1}$$

$$0 \le w_i \le 1. \tag{2}$$

Using this weight vector, the fitness function for selecting parent solutions is calculated as follows:

$$f(\mathbf{x}) = w_1 f_1(\mathbf{x}) + w_2 f_2(\mathbf{x}). \tag{3}$$

In [17], we applied a local search to each of non-dominated solutions, and replace the current solution with a candidate solution that improves the value of the scalar fitness function in (3). Suppose that we obtain three solutions *a*, *b*, and *c* from the cells A, B,

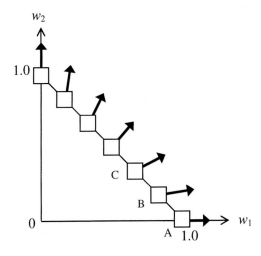

Fig. 1. Location of each cell in the two-dimensional space and its weight vector

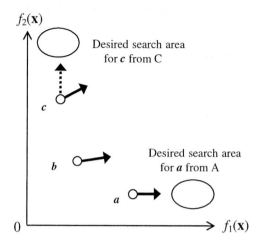

Fig. 2. Specification of a local search direction for a non-dominated solution

and C as non-dominated solutions, respectively. If we have only these three solutions in the set of non-dominated solutions, they are distributed as shown in Fig. 2. Because the local search direction for a solution is determined by the weight vector of the cell from which the solution is taken, local search directions for a, b, and c are the same as the weight vectors of A, B, and C (bold arrows in Fig. 2). When a non-dominated solution is located in a similar position among the set of non-dominated solutions to that of its cell in the structured space (e.g., solution a in Fig. 2), the weight vector of the cell is appropriate as the local search direction for the solution. On the contrary, when the non-dominated solution has the location which is dissimilar to its originated cell (e.g., solution c in Fig. 2), the weight vector is not appropriate as its local search direction. As we can see from Fig. 2, an appropriate local search direction for each non-dominated solution depends on its location in the set of non-dominated solutions. For example, $(w_1, w_2) = (0.0, 1.0)$ seems to be much more appropriate for the solution c than its original weight vector as its local search direction. These discussions suggest the importance of the specification of an appropriate local search direction for each solution according to its location.

In this paper, we examine the effectiveness of a weight specification method according to the location of the solution among the set of non-dominated solutions. In the proposed local search procedure, we replace the current solution with a better candidate solution with respect to the scalar fitness function using the adjusted weight vector. We modify the former C-MOGLS algorithm by adjusting weight vector in this manner. Of course, it is possible to use other replacement rules. One rule is to replace the current solution with candidates that are not dominated by the current solution. Another rule is to accept only better candidates that dominate the current solution. We show simulation results on three-objective pattern classification problems to compare these three replacement rules. We also examine the effect of replacement rules in C-MOGLS algorithms.

2 Multiobjective Optimization by Cellular MOGLS Algorithms

In this paper, we employ the following notations to describe a multiobjective optimization problem with n objectives:

$$\text{Maximize} \quad \mathbf{z} = (f_1(\mathbf{x}), f_2(\mathbf{x}), ..., f_n(\mathbf{x})), \quad (4)$$

$$\text{subject to} \quad \mathbf{x} \in \mathbf{X}, \quad (5)$$

where $\mathbf{z}$ is the objective vector, $\mathbf{x}$ is the decision vector, and $\mathbf{X}$ is the feasible region in the decision space. $f_1(\cdot)$, $f_2(\cdot)$, ..., $f_n(\cdot)$ are n objectives to be maximized. When the following inequalities hold between two solutions $\mathbf{x}$ and $\mathbf{y}$, the solution $\mathbf{y}$ is said to dominate the solution $\mathbf{x}$:

$$\forall i : f_i(\mathbf{x}) \leq f_i(\mathbf{y}) \text{ and } \exists j : f_j(\mathbf{x}) < f_j(\mathbf{y}). \quad (6)$$

If a solution is not dominated by any other solutions of the multiobjective optimization problem, that solution is said to be a Pareto-optimal solution. The task of EMO algorithms in this paper is not to select a single final solution but to find all Pareto-optimal solutions of the multi-objective optimization problem in (4). When we use heuristic search algorithms such as taboo search, simulated annealing, and genetic algorithms for finding Pareto-optimal solutions, we usually can not confirm the optimality of obtained solutions. We only know that each of the obtained solutions is not dominated by any other solutions examined during the execution of those algorithms. Therefore obtained solutions by heuristic algorithms are referred to as "non-ominated" solutions. For a large-scale multiobjective optimization problem, it is impossible to find all Pareto-optimal solutions. Thus our task is to find many near-optimal non-dominated solutions in a practically acceptable computational time. The performance of different EMO algorithms is compared based on several quality measures of obtained non-dominated solutions.

The concept of cellular genetic algorithms was proposed by Whitley [23]. In cellular genetic algorithms, each individual (i.e., a chromosome) resides in a cell of a spatially structured space. Genetic operations for generating new individuals are locally performed in the neighborhood of each cell. While the term "cellular genetic algorithm" was introduced by Whitley, such algorithms had already been proposed by Manderik and Spiessens [24]. Since each cell in our cellular algorithms is related to a uniformly generated weight vector, it is allocated in an n-dimensional weight space of an n-objective optimization problem. In this section, we first describe the former C-MOGLS [17]. Then we show a modified replacement rule in the local search procedure in detail. We also show the other two replacement rules based on the dominance relation.

2.1 Former C-MOGLS

We explain the former C-MOGLS [17] using the n-objective maximization problem in (4) and (5). One issue to be considered in the hybridization of EMO algorithms with local search is the direction of local search for each solution. In the former C-

MOGLS, the following scalar fitness function to be maximized was used in both the selection of parents and the local search.

$$f(\mathbf{x}) = w_1 f_1(\mathbf{x}) + w_2 f_2(\mathbf{x}) + \cdots + w_n f_n(\mathbf{x}). \tag{7}$$

The weight w_i ($w_i \geq 0$, $i = 1,2,...,n$ and $\sum_i w_i = 1$) was specified for each cell. In order to generate uniformly distributed weight vectors for multiobjective optimization problems with n-objectives, we proposed a weight specification method on an n-dimensional grid-world [25]. We generated weight vectors using integer factors w_i' $i = 1,2,...,n$ satisfying the following conditions.

$$w_1' + w_2' + \cdots + w_n' = d, \tag{8}$$
$$w_i' \in \{0,1,2,...,d\}, \tag{9}$$
$$w_i = w_i'/d. \tag{10}$$

These conditions show that weight vectors are generated using n non-negative integers where the sum of them is d. In our cellular algorithm, each cell is related to each weight vector satisfying the above conditions. Thus the number of cells (i.e., the population size) is equal to the total number of weight vectors satisfying the above conditions. This means that the population size is determined by d. For example, when we specify d as $d = 10$ in (8) for the case of two-objective problems, we will have eleven weight vectors (10, 0), (9, 1), ..., (0, 10) (see [25] for details).

Selecting parent solutions for a cell was governed by the weight vector related to that cell. In cellular algorithms, selecting candidate parent solutions is restricted within the neighboring cells of the current cell. That is, parent solutions were selected from only neighboring cells.

The former C-MOGLS used a simple form of elitism where all non-dominated solutions obtained during its execution were stored in a secondary population separately from the current population. When we stored a non-dominated solution to the secondary population, the weight vector of the cell from which the solution is taken was also stored with the solution. A local search procedure was applied to each non-dominated solution in the secondary population using the same scalar fitness function (i.e., the same weight vector) used in its originating cell. A few non-dominated solutions were randomly selected from the secondary population and their copies were added to the current population. The former C-MOGLS is written as follows:

Step 0) Initialization: Randomly generate an initial population of N_{pop} solutions (i.e., N_{pop} equals to the number of the cell).
Step 1) Evaluation: Calculate the n objectives for each solution in the current population. Then update the secondary population where non-dominated solutions are stored separately from the current population.

Step 2) Local search: Apply a local search procedure to each of the N_{pop} solutions in the secondary population using the scalar fitness function in (7). For each solution, the weight vector of its originating cell is used in local search. Local search is terminated when no better solution is found among k neighbors that are randomly selected from the neighborhood of the current solution.

Step 3) Selection: Repeat the following procedures to randomly selected ($N_{pop} - N_{elite}$) cells. Select a pair of parents based on the scalar fitness function in (7) using the weight vector related to the cell. The selection probability $p_S(\mathbf{x})$ of each solution $\mathbf{x}$ in the neighboring cells Ψ is specified by the following roulette wheel selection scheme with the linear scaling:

$$p_S(\mathbf{x}) = \frac{f(\mathbf{x}) - f_{min}(\Psi)}{\sum_{\mathbf{y} \in \Psi}(f(\mathbf{y}) - f_{min}(\Psi))}, \quad (11)$$

where $f_{min}(\Psi)$ is the minimum (i.e., worst) fitness value among the neighboring cells Ψ. Please note that the same weight vector of the current cell is used for calculating the fitness value of the candidate solutions in the neighboring cells.

Step 4) Crossover and mutation: Apply a crossover operation to each of the selected ($N_{pop} - N_{elite}$) cells with the crossover probability p_C. A new solution is generated from each pair. When the crossover operation is not applied, one parent is randomly chosen and handled as a new solution. Then apply a mutation operation to each new solution with the mutation probability p_M.

Step 5) Elitist strategy: Randomly select N_{elite} solutions from the secondary population. Then add their copies to the ($N_{pop} - N_{elite}$) solutions generated in Step 4 to construct a population of N_{pop} solutions.

Step 6) Return to Step 1.

This algorithm is terminated when a pre-specified number of solutions are examined during its execution. In the local search part (i.e., Step 2), a candidate solution is randomly generated by the local search operation from the current solution. If the generated candidate solution is better than the current solution, the current solution is replaced. That is, the first improvement strategy is used in the local search part instead of the best improvement strategy. When the current solution is updated, local search continues for the new current solution in the same manner. The same early termination strategy was used in the M-PAES [18]. On the other hand, all candidates were examined in the MOGLS of Jaszkiewicz [16]. In Knowles and Corne [19], the early termination strategy was used in Jaszkiewicz's MOGLS as well as the M-PAES in their computational experiments on multiobjective knapsack problems.

In this algorithm, all non-dominated solutions are stored in the secondary population with no restriction (i.e., no upper bound) on its size. In general, the restriction is necessary from the viewpoint of memory storage and computation time (e.g., see the SPEA [3]). We use, however, no restriction because we did not encounter any difficulties related to the maintenance of the secondary population in our computational experiments on pattern classification problems reported in this paper. Of course, there may be many application fields where the restriction on the size of the secondary population is necessary.

Randomly selected N_{elite} solutions from the secondary population in Step 5 work as elite solutions. It was shown in [15] that the performance of this algorithm was deteriorated by specifying the value of N_{elite} as $N_{elite} = 0$ (i.e., no elitism). It was also shown that the performance was not sensitive to the value of N_{elite} when $N_{elite} \geq 2$. In this paper, the value of N_{elite} is specified based on preliminary computational experiments.

2.2 Replacement Rule in LS

In the former C-MOGLS, we replace the current solution with the better candidate solution with respect to the scalar fitness function using the weight vector of the cell from which the solution was taken. As explained using Figs. 1 and 2, directions of the local search procedure should be adjusted according to the location of the solution among the set of non-dominated solutions. We can adjust the weight vectors using the pseudo-weight vector [26]. The pseudo-weight w_i for the i-th objective is defined for the current solution $\mathbf{x}$ as

$$w_i = \frac{f_i(\mathbf{x}) - f_i^{min}}{f_i^{max} - f_i^{min}} \bigg/ \sum_{j=1}^{n} \frac{f_j(\mathbf{x}) - f_j^{min}}{f_j^{max} - f_j^{min}}, \; i = 1,2,...,n, \quad (12)$$

where f_i^{max} and f_i^{min} are the maximum and minimum values of the i-th objective in the current set of non-dominated solutions, respectively. The scalar fitness function with the pseudo-weight vector $\mathbf{w} = (w_1, ..., w_n)$ determined by (12) is used in this paper. The determination of the weight vector by (12) is illustrated in Fig. 3 where open circles are the non-dominated solutions obtained at t-th generation and closed circles are those at $(t + 1)$-th generation. The arrow attached to each circles shows the weight vector for the corresponding solution. From this figure, we can see that an appropriate weight vector is assigned to each solution by (12). For example, the solution *a* has the appropriate direction among open circles. On the other hand, the solution *b* has different direction even if it is located close to *a*. This is because the direction of each arrow is determined according to the location of the solution among the set of non-dominated solutions. We modify the former C-MOGLS algorithm [17] using this adjusted weight vector.

It is possible to use other replacement rules for C-MOGLS algorithms. One rule is to replace the current solution with neighbors that are not dominated by the current solution. Let us consider Fig. 4 where the current solution and its candidates are denoted by a closed circle (i.e., *a*) and open circles (i.e., *b*, *c*, *d*, *e*, *f* and *g*), respectively. The current solution *a* can move to the five candidates except for *g* because only *g* is dominated by *a*. A drawback of this acceptance rule is that the current solution can be degraded by multiple moves. For example, the current solution *a* can move to the candidate *f*, from which the current solution can further move to *g*. Another replacement rule is to accept only better candidates that dominate the current solution. In this case, the current solution *a* can move only to the candidate *b* in Fig. 4. A drawback of this acceptance rule is that the movable area is very small especially when the number of objectives is large. We also examine these two replacement rules in C-MOGLS algorithms.

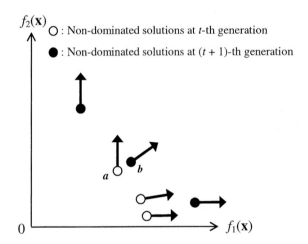

Fig. 3. Local search directions defined using the pseudo-weight vector method

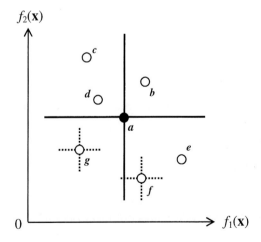

Fig. 4. Illustration of replacement rules

3 Pattern Classification Problems

In this paper, we apply the C-MOGLS algorithm with the modified local search to multiobjective classification problems. The aim of classification problems is to design classification systems which classify a pattern vector with several attributes into one of the already-known classes. Therefore, a classification system can be viewed as a mapping from a pattern vector in a multi-dimensional pattern space to a class label. Fuzzy theory [27], neural networks [28, 29], decision trees [30] are often employed to design classification systems.

We apply the modified C-MOGLS algorithm to design fuzzy classification systems. A fuzzy classification system consists of several fuzzy if-then rules in the following format:

If $input_1$ is A_{j1} and ... and $input_m$ is A_{jm}
then Class C_j with CF_j, $j = 1, 2, ..., N$, (13)

where **input** $= (input_1, ..., input_m)$ is a pattern vector with m attributes, A_{ji} is a fuzzy set related to a linguistic term (see Fig. 5) of the i-th attribute of the j-th rule, C_j is a consequent class, and CF_j is a certainty factor of the rule. We employ a heuristic method to calculate the consequent class C_j and the certainty factor CF_j for each rule [27].

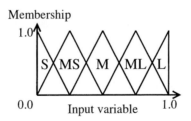

Fig. 5. Linguistic fuzzy sets (S: *small*, MS: *medium small*, M: *medium*, ML: *medium large*, and L: *large*)

We employ the following objective functions related to the learning ability and the interpretability.

Maximize $f_1(S) = NCP(S)$, (14)
Minimize $f_2(S) = |S|$, (15)
Minimize $f_3(S) = Length(S)$, (16)

where S is a classification system, $NCP(S)$ is the number of correctly classified training patterns by S, $|S|$ is the number of fuzzy rules in S, and $Length(S)$ is the total number of antecedent conditions of all rules in S. $NCP(S)$ is related to the learning ability. As for the interpretability we employed two criteria: $|S|$ and

$Length(S)$. $|S|$ and $Length(S)$ are related to the interpretability for a rule base, and for each rule, respectively. That is, a classification system with a small number of rules is easily understood by users or decision makers. Moreover we consider the interpretability of each rule since it is difficult for us to understand each rule with a lot of conditions. We try to find a better set of non-dominated solutions by the modified C-MOGLS algorithm.

4 Computer Simulations on Pattern Classification Problems

In this paper, we try to find a better set of non-dominated solutions with respect to the three objectives in (14)-(16). We apply the modified C-MOGLS algorithm to wine and glass data sets. The wine data set has 178 training patterns with 13 attributes and 3 classes. The glass data set has 214 patterns with 9 attributes and 6 classes.

4.1 Effect of the Modified C-MOGLS

In order to compare the performance of the modified C-MOGLS for finding better non-dominated solutions, we also apply a MOGA [31], a C-MOGA [32] and the former C-MOGLS on the two classification problems. The MOGA is a C-MOGLS algorithm that has no cellular structure nor local search procedure. The C-MOGA algorithm is a C-MOGLS algorithm without the local search procedure. In the C-MOGLS algorithms, we implement a local search operation by reversing bit values. Since each bit corresponds to a fuzzy rule where "0" and "1" show the exclusion and the inclusion of the corresponding rule in the classification system S, the bit reverse local search tries to find a system including another rule with the better classification performance, or to find a system excluding a rule with the same or the better classification ability. As we have already described in Subsection 2.1, we restricted the number of examined solutions in our local search procedure. We specified k as $k = 2$. That is, if there is no solution which improves the fitness function in two randomly selected candidates of the current solution, the local search for the solution is terminated and the local search for another solution is executed.

We specified the parameter d in the cellular algorithms as $d = 13$. Thus, the number of the cells in the cellular algorithms was 105. This parameter was most effective in simulations on three objective flowshop scheduling problems in [32]. In order to compare the MOGA with the cellular algorithms under the same conditions, we specified the population size as 105 in the MOGA. The stopping condition of each algorithm was the evaluation of 105,000 classification systems. Thus 1,000 generations are updated in the MOGA and the C-MOGA. We compared four sets of non-dominated solutions obtained by the four algorithms from the same initial population. We generated ten different initial populations to be applied by each algorithm, and then obtained average results.

In order to compare four algorithms, we show the average number of obtained non-dominated solutions for each algorithm, and the number of solutions which are not dominated by the non-dominated solutions obtained by other algorithms. Table 1 shows the simulation results for the wine data and the glass data. From Table 1, we can see that the average number of obtained solutions by the modified C-MOGA was the best among the four algorithms for both the data sets. As for the number of non-dominated solutions among four solution sets, the average values of the former C-MOGLS and the modified C-MOGLS are almost the same at the final generation. In order to see online performance of each algorithm, we compared the four sets of non-dominated solutions at the same number of evaluated solutions. Figs. 6 and 7 show the number of non-dominated solutions over the number of evaluated solutions. From Figs. 6 and 7, we can observe that the convergence speed of the modified C-MOGLS is faster than that of the former C-MOGLS in both the problems.

Table 1. MOGA, C-MOGA, Former C-MOGLS and Modified C-MOGLS

Data set	Wine			Glass		
	Obtained solutions (A)	Non-dominated (B)	B/A	Obtained solutions (A)	Non-dominated (B)	B/A
MOGA	11.6	6.8	0.63	18.9	14.9	0.79
C-MOGA	12.9	10.2	0.79	22.7	13.6	0.61
Former C-MOGLS	14.2	11.5	0.82	25.9	21.0	0.82
Modified C-MOGLS	14.4	11.4	0.80	27.1	23.7	0.88

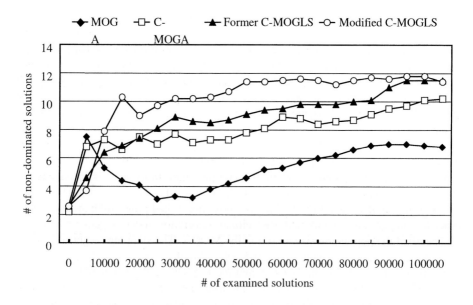

Fig. 6. The number of non-dominated solutions (Wine data set)

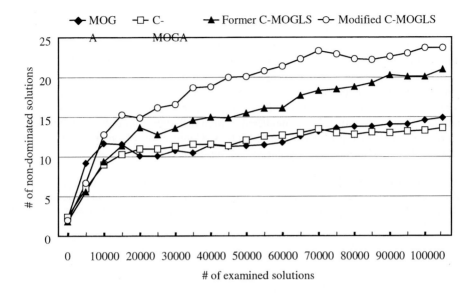

Fig. 7. The number of non-dominated solutions (Glass data set)

4.2 Effect of Replacement Rules in LS

We also examined the effect of choice of replacement rules in the local search procedure. We compared the former C-MOGLS, the modified C-MOGLS, C-MOGLS algorithms based on the dominance relation (two replacement rules: replacing with a non-dominated solution and/or a better solution). Table 2 shows that the replacement rules based on the scalar fitness function are better than those based on the dominance relation. We can see that the local search procedure with replacement rules based on the dominance relation was not effective in C-MOGLS algorithms.

Table 2. Former C-MOGLS, Modified C-MOGLS, C-MOGLS with non-dominated replacement, and C-MOGLS with better replacement

Data set	Wine			Glass		
	Obtained solutions (A)	Non-dominated (B)	B/A	Obtained solutions (A)	Non-dominated (B)	B/A
Former C-MOGLS	14.2	12.5	0.89	25.9	21.2	0.83
Modified C-MOGLS	14.4	12.3	0.86	27.1	23.9	0.89
Non-dominated	13.6	8.4	0.64	22.5	10.7	0.49
Better	12.6	8.2	0.67	20.0	15.0	0.76

5 Conclusion

In this paper, we modified the local search procedure in C-MOGLS algorithms. We adjusted the weight vector for the local search according to the location of a non-dominated solution among the set of non-dominated solutions. Simulation results on pattern classification problems clearly show that the modification of the local search direction is effective. The performance evaluation of our C-MOGLS in this paper is not complete. We compared the modified C-MOGLS algorithm only with the former algorithms such as the MOGA, the C-MOGA. Comparison of our C-MOGLS with other multiobjective memetic algorithms is a future research topic.

Acknowledgements. The authors would like to thank the financial support from Kansai University through the Kansai University Grand-in-Aid for the Faculty Joint Research Program, 2002. Special thanks are due to the anonymous reviewers for their valuable comments.

References

1. Schaffer, J.D.: Multi-objective optimization with vector evaluated genetic algorithms. *Proc. of 1st International Conference on Genetic Algorithms* (1985) 93–100.
2. Knowles, J.D. and Corne, D.W.: The Pareto archived evolution strategy: A new baseline algorithm for Pareto multiobjective optimization, *Proc. of 1999 Congress on Evolutionary Computation* (1999) 98–105.
3. Zitzler, E. and Thiele, L.: Multiobjective evolutionary algorithms: A comparative case study and the strength Pareto approach, *IEEE Trans. on Evolutionary Computation* **3**, 4 (1999) 257–271.
4. Knowles, J.D. and Corne, D.W.: Approximating the nondominated front using the Pareto archived evolution strategy, *Evolutionary Computation* **8**, 2 (2000) 149–172.
5. Zitzler, E., Deb, K. and Thiele, L.: Comparison of multiobjective evolutionary algorithms: Empirical results, *Evolutionary Computation* **8**, 2 (2000) 173–195.
6. Deb, K., Pratap, A., Agarwal, S. and Meyarivan, T.: A fast and elitist multiobjective genetic algorithm: NSGA-II, *IEEE Trans. on Evolutionary Computaiton* **6**, 2 (2002) 182–197.
7. Merz, P. and Freisleben B.: Genetic local search for the TSP: New results, *Proc. of 4th IEEE International Conference on Evolutionary Computation* (1997) 159–164.
8. Krasnogor, N. and Smith, J.: A memetic algorithm with self-adaptive local search: TSP as a case study, *Proc. of 2000 Genetic and Evolutionary Computation Conference* (2000) 987–994.
9. Moscato, P.: Memetic algorithms: A short instruction, in Corne, D., Glover F. and Doring M. (eds.) *New Ideas in Optimization.* McGraw-Hill, Maidenhead (1999) 219–234.
10. Hart, W. E., Krasnogor, N., and Smith, J. (eds.), *First Workshop on Memetic Algorithms (WOMA I)*, in *Proc. of 2000 Genetic and Evolutionary Computation Conference Workshop Program* (2000) 95–130.
11. Hart, W. E., Krasnogor, N., and Smith, J. (eds.), *Second Workshop on Memetic Algorithms (WOMA II)*, in *Proc. of 2001 Genetic and Evolutionary Computation Conference Workshop Program* (2001) 137–179.
12. Hart, W. E., Krasnogor, N., and Smith, J. (eds.), *Proc. of Third Workshop on Memetic Algorithms* (WOMA III) (2002) (in press).
13. Hart, W. E.: Adaptive global optimization with local search, *Ph. D. Thesis*, University of California (1994).

14. Ishibuchi, H. and Murata, T.: Multi-objective genetic local search algorithm, *Proc. of 3rd IEEE International Conference on Evolutionary Computation* (1996) 119–124.
15. Ishibuchi, H. and Murata, T.: A multi-objective genetic local search algorithm and its application to flowshop scheduling, *IEEE Trans. on Systems, Man, and Cybernetics – Part C: Applications and Reviews* **28**, 3 (1998) 392–403.
16. Jaszkiewicz, A.: Genetic local search for multi-objective combinatorial optimization, *European Journal of Operational Research* **137**, 1 (2002) 50–71.
17. Murata, T., Nozawa, H., Tsujimura, Y., Gen, M. and Ishibuchi, H.: Effect of local search on the performance of cellular multi-objective genetic algorithms for designing fuzzy rule-based classification systems, *Proc. of the 2002 Congress on Evolutionary Computation* (2002) 663–668.
18. Knowles, J.D. and Corne, D.W.: M-PAES: A memetic algorithm for multiobjective optimization, *Proc. of 2000 Congress on Evolutionary Computation* (2000) 325–332.
19. Knowles, J.D. and Corne, D.W.: A comparison of diverse approaches to memetic multiobjective combinatorial optimization, *Proc. of 2000 Genetic and Evolutionary Computation Conference Workshop Program* (2000) 103–108.
20. Knowles, J.D. and Corne, D.W.: A comparison of diverse approaches to memetic multiobjective combinatorial optimization, *Proc. of 2001 Genetic and Evolutionary Computation Conference Workshop Program* (2001) 162–167.
21. Deb, K. and Goel, T.: A hybrid multi-objective evolutionary approach to engineering shape design, *Proc. of 1st International Conference on Evolutionary Multi-Criterion Optimization* (2001) 385–399.
22. Talbi, E., Rahoual, M., Mabed, M. H. and Dhaenens, C.: A hybrid evolutionary approach for multicriteria optimization problems: Application to the flow shop, *Proc. of 1st International Conference on Evolutionary Multi-Criterion Optimization* (2001) 416–428.
23. Whitley, D.: Cellular genetic algorithms, *Proc. of 5th International Conference on Genetic Algorithms* (1993) 658.
24. Manderick, B. and Spiessens, P.: Fine-grained parallel genetic algorithms, *Proc. of 3rd International Conference on Genetic Algorithms* (1989) 428–433.
25. Murata, T., Ishibuchi, H. and Gen, M.: Specification of genetic search directions in cellular multi-objective genetic algorithms, *Proc. of First International Conference on Evolutionary multi-criterion optimization* (2001) 82–95.
26. Deb, K.: *Multi-Objective Optimization Using Evolutionary Algorithms*, John Wiley & Sons, Chichester (2001).
27. Ishibuchi, H, Nozaki, K. and Tanaka, H.: Distributed representation of fuzzy rules and its application to pattern classification, *Fuzzy Sets and Systems* **52** (1992) 21–32.
28. Hammerstrom, D.: Neural networks at work, *IEEE Spectrum* June (1993) 26–32.
29. Hammerstrom, D.: Working with neural networks, *IEEE Spectrum* July (1993) 46–53.
30. Quinlan, J.R.: Introduction of decision trees, *Machine Learning* **1** (1985) 71–99.
31. Murata, T. and Ishibuchi, H.: MOGA: Multi-objective genetic algorithms, *Proc. of The 2nd IEEE International Conference on Evolutionary Computing* (1995) 289–294.
32. Murata, T., Ishibuchi, H. and Gen, M.: Specification of genetic search directions in cellular multi-objective genetic algorithms, *Proc. of First International Conference on Evolutionary multi-criterion optimization* (2001) 82–95.

Effects of Three-Objective Genetic Rule Selection on the Generalization Ability of Fuzzy Rule-Based Systems

Hisao Ishibuchi and Takashi Yamamoto

Department of Industrial Engineering, Osaka Prefecture University,
1-1 Gakuen-cho, Sakai, Osaka 599-8531, Japan
{hisaoi, yama}@ie.osakafu-u.ac.jp

Abstract. One advantage of evolutionary multiobjective optimization (EMO) algorithms over classical approaches is that many non-dominated solutions can be simultaneously obtained by their single run. This paper shows how this advantage can be utilized in genetic rule selection for the design of fuzzy rule-based classification systems. Our genetic rule selection is a two-stage approach. In the first stage, a pre-specified number of candidate rules are extracted from numerical data using a data mining technique. In the second stage, an EMO algorithm is used for finding non-dominated rule sets with respect to three objectives: to maximize the number of correctly classified training patterns, to minimize the number of rules, and to minimize the total rule length. Since the first objective is measured on training patterns, the evolution of rule sets tends to overfit to training patterns. The question is whether the other two objectives work as a safeguard against the overfitting. In this paper, we examine the effect of the three-objective formulation on the generalization ability (i.e., classification rates on test patterns) of obtained rule sets through computer simulations where many non-dominated rule sets are generated using an EMO algorithm for a number of high-dimensional pattern classification problems.

1 Introduction

Fuzzy rule-based systems are universal approximators of nonlinear functions as multi-layer feedforward neural networks. These two models have been applied to various problems such as control, function approximation and pattern classification. The main advantage of fuzzy rule-based systems is their comprehensibility because each fuzzy rule is linguistically interpretable. In many studies on the design of fuzzy rule-based systems, however, emphasis has been mainly placed on their accuracy rather than their comprehensibility. Thus the performance maximization has been the primary objective. Recently the tradeoff between the accuracy and the comprehensibility was discussed in some studies [20], [21], [26]-[29]. While those studies took into account several criteria related to the accuracy and the comprehensibility, the design of fuzzy rule-based systems was handled in the framework of single-objective optimization. That is, those studies tried to find a single fuzzy rule-based system by considering both the accuracy and the comprehensibility. One of the first studies on fuzzy rule-based systems in the framework of multiobjective optimization was a two-objective rule selection [9] where genetic algorithms were used for finding non-dominated rule sets with respect to the classification accuracy and the number of fuzzy rules. The two-objective rule selection was extended to the case of three objectives [12] where the total rule length was considered as the third objective in addition to the above-mentioned two objectives in [9]. See [2] for further discussions on the tradeoff between the accuracy and the comprehensibility of fuzzy rule-based systems.

If compared with standard optimization problems, an additional difficulty in the design of classification systems is that the maximization of any accuracy measure does not always mean the maximization of their actual performance. This is because the accuracy of classification systems is measured on training patterns while their actual performance should be measured on unseen test patterns. That is, any accuracy measure is just an estimation of the actual performance. The maximization of any accuracy measure often leads to the overfitting to training patterns, which degrades the actual performance of classification systems on test patterns. Thus we need some sort of safeguard for preventing the overfitting. A weighted sum of accuracy and complexity measures is often used as a safeguard against the overfitting to training patterns. This paper examines the usefulness of multiobjective formulations as a safeguard. In the three-objective formulation in [12], the number of fuzzy rules and their total length were used as complexity measures together with an accuracy measure. While those complexity measures were originally introduced for obtaining comprehensible fuzzy rule-based systems, we examine their usefulness as a safeguard against the overfitting. That is, we examine the effect of those complexity measures in the three-objective formulation on the generalization ability (i.e., classification rates on test patterns) of obtained fuzzy rule-based classification systems.

In this paper, we first briefly describe fuzzy rules and fuzzy reasoning for fuzzy rule-based classification in Section 2. Then we explain our two-stage approach [16] to the design of fuzzy rule-based systems in Section 3. In the first stage, a pre-specified number of fuzzy rules are generated as candidate rules from training patterns using a data mining technique. In the second stage, non-dominated rule sets are found from the candidate rules by an EMO algorithm. Simulation results on several data sets are reported in Section 4 where the generalization ability of obtained rule sets on test patterns is examined. Simulation results clearly show that the two complexity measures improve not only the comprehensibility of obtained rule sets but also their generalization ability on test patterns. Finally Section 5 summarizes this paper.

2 Fuzzy Rule-Based Classification Systems

Let us assume that we have m training patterns $\mathbf{x}_p = (x_{p1}, ..., x_{pn})$, $p = 1,2,...,m$ from M classes where x_{pi} is the attribute value of the p-th training pattern for the i-th attribute ($i = 1,2,...,n$). For our n-dimensional M-class pattern classification problem, we use fuzzy rules of the following form:

$$\text{Rule } R_q : \text{If } x_1 \text{ is } A_{q1} \text{ and } ... \text{ and } x_n \text{ is } A_{qn} \text{ then Class } C_q \text{ with } CF_q, \quad (1)$$

where R_q is the label of the q-th rule, $\mathbf{x} = (x_1, ..., x_n)$ is an n-dimensional pattern vector, A_{qi} is an antecedent fuzzy set (i.e., linguistic value such as *small* and *large*), C_q is a class label, and CF_q is a rule weight. Fuzzy rules of this type were first used for classification problems in [13]. For other types of fuzzy rules, see [4], [11], [23].

We define the compatibility grade of each training pattern $\mathbf{x}_p$ with the antecedent part $\mathbf{A}_q = (A_{q1}, ..., A_{qn})$ using the product operator as

$$\mu_{\mathbf{A}_q}(\mathbf{x}_p) = \mu_{A_{q1}}(x_{p1}) \cdot \mu_{A_{q2}}(x_{p2}) \cdot ... \cdot \mu_{A_{qn}}(x_{pn}), \quad p = 1, 2, ..., m, \quad (2)$$

where $\mu_{A_{qi}}(\cdot)$ is the membership function of A_{qi}. For determining the consequent class C_q, we calculate the confidence of the fuzzy association rule "$\mathbf{A}_q \Rightarrow \text{Class } h$" for each class as an extension of its non-fuzzy version [1] as follows [8], [19]:

$$c(\mathbf{A}_q \Rightarrow \text{Class } h) = \sum_{\mathbf{x}_p \in \text{Class } h} \mu_{\mathbf{A}_q}(\mathbf{x}_p) \Big/ \sum_{p=1}^{m} \mu_{\mathbf{A}_q}(\mathbf{x}_p), \quad h = 1, 2, ..., M. \quad (3)$$

The confidence is the same as the fuzzy conditional probability [30]. The consequent class C_q is specified by identifying the class with the maximum confidence:

$$c(\mathbf{A}_q \Rightarrow \text{Class } C_q) = \max\{c(\mathbf{A}_q \Rightarrow \text{Class } h) \mid h = 1, 2, ..., M\}. \quad (4)$$

On the other hand, the rule weight CF_q is specified as follows:

$$CF_q = c(\mathbf{A}_q \Rightarrow \text{Class } C_q) - \sum_{\substack{h=1 \\ h \neq C_q}}^{M} c(\mathbf{A}_q \Rightarrow \text{Class } h). \quad (5)$$

The rule weight of each fuzzy rule has a large effect on the classification ability of fuzzy rule-based systems [10]. There are several alternative definitions of rule weights (see [17]). Better results were obtained in [17] from the above definition in (5) than the direct use of the confidence (i.e., $CF_q = c(\mathbf{A}_q \Rightarrow \text{Class } C_q)$) when we used a single winner-based method for classifying new patterns.

In this paper, we use a single winner-based fuzzy reasoning method [13]. For other fuzzy reasoning methods for pattern classification, see [4], [11], [23]. Let S be the set of fuzzy rules in our fuzzy rule-based system. A single winner rule R_w is chosen from the rule set S for an input pattern $\mathbf{x}_p$ as

$$\mu_{\mathbf{A}_w}(\mathbf{x}_p) \cdot CF_w = \max\{\mu_{\mathbf{A}_q}(\mathbf{x}_p) \cdot CF_q \mid R_q \in S\}. \quad (6)$$

Since the winner rule is chosen based on not only the compatibility grade but also the rule weight, high classification accuracy can be achieved by adjusting the rule weight of each fuzzy rule without modifying each antecedent fuzzy set [10].

3 Heuristic Rule Extraction and Genetic Rule Selection

Genetic rule selection was proposed for designing fuzzy rule-based classification systems with high accuracy and high comprehensibility in [14], [15]. A small number of fuzzy rules were selected from a large number of candidate rules based on a scalar fitness function defined as a weighted sum of the number of correctly classified training patterns and the number of fuzzy rules. A two-objective genetic algorithm was used in [9] for finding non-dominated rule sets. Genetic rule selection was further extended to the following three-objective optimization problem in [12]:

$$\text{Maximize } f_1(S), \text{ minimize } f_2(S), \text{ and minimize } f_3(S), \qquad (7)$$

where S is a subset of candidate rules, $f_1(S)$ is the number of correctly classified training patterns by the rule set S, $f_2(S)$ is the number of fuzzy rules in S, and $f_3(S)$ is the total rule length of fuzzy rules in S. The number of antecedent conditions of each fuzzy rule is referred to as the rule length in this paper. As clearly shown in [12], the use of the average rule length as the third objective $f_3(S)$ leads to counter-intuitive results. Thus we use the total rule length as $f_3(S)$ in (7).

When we use K linguistic values and *"don't care"* as antecedent fuzzy sets, the total number of possible combinations of antecedent fuzzy sets is $(K+1)^n$. In early studies [9], [14], [15], all combinations were examined for generating candidate rules. Thus genetic rule selection was applicable only to low-dimensional problems (e.g., iris data with four attributes). On the other hand, only short fuzzy rules were examined for generating candidate rules in [12] where genetic rule selection was applied to higher-dimensional problems (e.g., wine data with 13 attributes).

In our former study [16], we suggested the use of a data mining technique for extracting a pre-specified number of candidate rules in a heuristic manner. That is, genetic rule selection was extended to a two-stage approach with heuristic rule extraction and genetic rule selection. Our two-stage approach is applicable to high-dimensional problems (e.g., sonar data with 60 attributes).

3.1 Heuristic Rule Extraction

In the field of data mining, association rules are often evaluated by two rule evaluation criteria: support and confidence. In the same manner as the fuzzy version of confidence in (3), the definition of support [1] can be also extended to the case of fuzzy association rules as follows [8], [19]:

$$s(\mathbf{A}_q \Rightarrow \text{Class } h) = \frac{1}{m} \sum_{\mathbf{x}_p \in \text{Class } h} \mu_{\mathbf{A}_q}(\mathbf{x}_p). \qquad (8)$$

The product of the confidence and the support was used in our former study [16] on the two-stage approach. Seven heuristic criteria were compared with each other in [18] where good results were obtained from the following criterion:

$$f_{\text{SLAVE}}(R_q) = s(\mathbf{A}_q \Rightarrow \text{Class } C_q) - \sum_{\substack{h=1 \\ h \neq C_q}}^{M} s(\mathbf{A}_q \Rightarrow \text{Class } h). \tag{9}$$

This is a modified version of a rule evaluation criterion used in an iterative fuzzy GBML (genetics-based machine learning) algorithm called SLAVE [3], [7].

In our heuristic rule extraction, a pre-specified number of candidate rules with the largest values of the SLAVE criterion are found for each class. For designing fuzzy rule-based systems with high comprehensibility, only short rules are examined as candidate rules. The restriction on the rule length is consistent with the third objective (i.e., the total rule length) of our three-objective rule selection problem in (7).

3.2 Genetic Rule Selection

Let us assume that N fuzzy rules have been extracted as candidate rules using the SLAVE criterion. A subset S of the N candidate rules is handled as an individual in EMO algorithms, which is represented by a binary string of the length N as

$$S = s_1 s_2 \cdots s_N, \tag{10}$$

where $s_j = 1$ and $s_j = 0$ mean that the j-th candidate rule is included in S and excluded from S, respectively.

A simple multiobjective genetic algorithm [24] based on a scalar fitness function with random weights was used in our former studies [9], [12], [16]. Recently several EMO algorithms with much higher search ability have been proposed (for example, NSGA-II [5], PAES [22], and SPEA [32]). Since each rule set is represented by a binary string in our three-objective rule selection problem in (7), most EMO algorithms are applicable. In this paper, we use the NSGA-II because its high search ability has been demonstrated in [5] and its implementation is relatively easy.

We use two problem-specific heuristic tricks in the NSGA-II. One is biased mutation where a larger probability is assigned to the mutation from 1 to 0 than that from 0 to 1. This is for efficiently decreasing the number of fuzzy rules in each rule set. The other is the removal of unnecessary rules. Since we use the single winner-based method for classifying each pattern, some fuzzy rules in S may be chosen as winner rules for no patterns. We can remove those fuzzy rules without degrading the first objective (i.e., the number of correctly classified training patterns). At the same time, the second objective (i.e., the number of fuzzy rules) and the third objective (i.e., the total rule length) are improved by removing unnecessary rules. Thus we remove all fuzzy rules that are not selected as winner rules for any training patterns from the rule set S. The removal of unnecessary rules is performed after the first objective is calculated for each rule set and before the second and third objectives are calculated.

4 Computer Simulations

4.1 Data Sets

We used six data sets in Table 1 available from the UCI ML repository (http://www.ics.uci.edu~mlearn/). Data sets with missing values are marked by "*" in the third column of Table 1. Since we did not use incomplete patterns with missing values, the number of patterns in the third column does not include those patterns with missing values. As benchmark results, we cited simulation results by Elomaa and Rousu [6] in Table 1. They applied six variants of the C4.5 algorithm [25] to 30 data sets in the UCI ML repository. The performance of each variant was examined by ten iterations of the whole ten-fold cross-validation (10-CV) procedure [25], [31]. We show in the last two columns of Table 1 the best and worst error rates on test patterns among the six variants reported in [6] for each data set.

Table 1. Data sets used in our computer simulations

Data set	Number of attributes	Number of patterns	Number of classes	Error rate by C4.5 in [6]	
				Best	Worst
Breast W	9	683*	2	5.1	6.0
Diabetes	8	768	2	25.0	27.2
Glass	9	214	6	27.3	32.2
Heart C	13	97*	5	46.3	47.9
Sonar	60	208	2	24.6	35.8
Wine	13	178	3	5.6	8.8

* Incomplete patterns with missing values are not included.

4.2 Simulation Conditions

We applied our two-stage approach to six data sets in Table 1. All attribute values were normalized into real numbers in the unit interval [0, 1]. As antecedent fuzzy sets, we used 14 triangular fuzzy sets generated from four fuzzy partitions with different granularities in Fig. 1 because we did not know an appropriate granularity of the fuzzy partition for each attribute. In addition to the 14 triangular fuzzy sets, we also used *"don't care"* as an additional antecedent fuzzy set. We generated 300 fuzzy rules of the length two or less for each class of the sonar data set as candidate rules in a greedy manner using the SLAVE criterion. That is, the best 300 candidate rules with the largest values of the SLAVE criterion were found for each class. For the other data sets, we generated 300 fuzzy rules of the length three or less for each class. Thus the total number of candidate rules was $300M$ where M is the number of classes.

The NSGA-II was employed for finding non-dominated rule sets from $300M$ candidate rules. We used the following parameter values in the NSGA-II:

Population size: 200 strings,
Crossover probability: 0.8,
Biased mutation probabilities: $p_m(0 \to 1) = 1/300M$ and $p_m(1 \to 0) = 0.1$,
Stopping condition: 5000 generations.

We also examined the combination of 2000 strings and 500 generations. Almost the same results were obtained from this combination and the above parameter values.

For evaluating the generalization ability of obtained rule sets, we used the 10-CV technique as in [6]. First each data set was randomly divided into ten subsets of the same size. One subset was used as test patterns while the other nine subsets were used as training patterns. Our two-stage approach was applied to training patterns for finding non-dominated rule sets. The generalization ability of obtained rule sets was evaluated by classifying test patterns. This train-and-test procedure was iterated ten times so that all the ten subsets were used as test patterns. As in [6], we iterated the whole 10-CV procedure ten times using different data partitions. Thus our two-stage approach was executed 100 times in total for each data set.

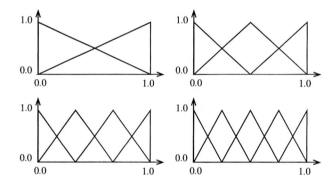

Fig. 1. Four fuzzy partitions used in our computer simulations

4.3 Simulation Results

Wisconsin Breast Cancer Data Set. The NSGA-II was applied to the Wisconsin breast cancer data set (Breast W in Table 1) 100 times. From each run of the NSGA-II, 11.5 non-dominated rule sets were obtained on the average. We calculated error rates of each non-dominated rule set on training patterns and test patterns. Simulation results are summarized in Table 2 where the last column shows the number of runs from which the corresponding rule sets (with respect to the number of fuzzy rules and the average rule length) were obtained. For example, rule sets including four rules of the average length 1.50 were obtained from 72 out of 100 runs. We omit from Table 2 some rare combinations of the number of fuzzy rules and the average rule length that were obtained from only 30 (out of 100) runs or less.

Table 2. Performance of obtained rule sets for the Wisconsin breast cancer data set

Number of rules	Average length	Average error rate Training	Average error rate Test	Number of runs
0	0.00	100.00	100.00	100
1	1.00	35.43	35.43	100
2	1.00	5.25	6.13	100
2	1.50	3.34	3.47	100
2	2.00	3.15	3.87	92
3	1.33	2.85	4.19	79
3	1.67	2.64	4.33	92
4	1.50	2.42	4.41	72
4	1.75	2.32	5.09	36
5	1.40	2.21	4.43	35
5	1.60	2.05	4.51	61
5	1.80	2.07	4.02	35
6	1.50	1.91	4.19	35
6	1.67	1.87	3.97	45

We can see from Table 1 and Table 2 that the generalization ability of many rule sets outperforms the best result of the C4.5 algorithm in Table 1 (i.e., 5.1% error rate). For visually demonstrating the tradeoff between the accuracy and the complexity, error rates on training patterns in Table 2 are shown in Fig. 2 (a) where the smallest error rate is denoted by a closed circle for each number of fuzzy rules. Thus closed circles in Fig. 2 (a) can be viewed as simulation results obtained from the two-objective formulation without the third objective (i.e., total rule length). From this figure, we can observe a clear tradeoff between the error rate on training patterns and the number of fuzzy rules. If we use a weighted sum of the accuracy on training patterns and the number of fuzzy rules as a scalar fitness function, one of the closed circles is obtained as a single optimal solution. For example, the right-most closed circle may be obtained when the weight for the accuracy is very large. On the other hand, error rates on test patterns are shown in Fig. 2 (b). Rule sets corresponding to closed circles in Fig. 2 (a) are also denoted by closed circles in Fig. 2 (b). From Fig. 2 (b), we can observe the overfitting due to the increase in the number of fuzzy rules. That is, error rates on test patterns in Fig. 2 (b) tend to increase with the number of fuzzy rules while error rates on training patterns in Fig. 2 (a) monotonically decrease. Moreover we can notice another kind of overfitting in Fig. 2 (b) from the difference between the closed circle and the smallest error rate on test patterns for each number of fuzzy rules. That is the overfitting due to the increase in the average rule length.

For demonstrating the overfitting due to the increase in the average rule length, a part of the simulation results in Table 2 are depicted in Fig. 3. It should be noted that the horizontal axis of Fig. 3 is the average rule length while it was the number of fuzzy rules in Fig. 2. Fig. 3 (a) and Fig. 3 (b) show error rates of obtained rule sets with two and four fuzzy rules, respectively. From Fig. 3, we can see that error rates on test patterns increased as the average rule length increased in some cases.

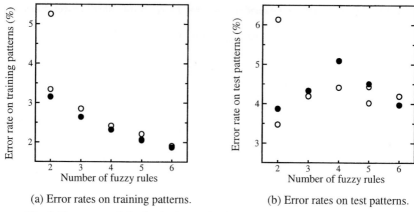

Fig. 2. Error rates of obtained rule sets for the Wisconsin breast cancer data set

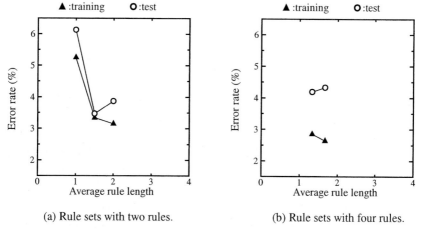

Fig. 3. Error rates of obtained rule sets with the same number of fuzzy rules and different average rule length for the Wisconsin breast cancer data set

Diabetes Data Set. In the same manner as Fig. 2 and Fig. 3, simulation results on the diabetes data set (Diabetes in Table 1) are summarized in Fig. 4 and Fig. 5. In Fig. 4 (a), we can observe a clear tradeoff between the accuracy on training patterns and the number of fuzzy rules. On the other hand, error rates on test patterns increase in some cases in Fig. 4 (b) as the number of fuzzy rules increases. That is, we can observe the overfitting due to the increase in the number of fuzzy rules in Fig. 4 (b). The overfitting due to the increase in the average rule length is clear in Fig. 5 (b) where we show error rates by obtained rule sets including four rules. We can see from the comparison between Fig. 4 (b) and Table 1 that the generalization ability of many rule sets is slightly inferior to the best result of the C4.5 algorithm (i.e., 25.0% error rate) and slightly superior to its worst result (i.e., 27.2% error rate).

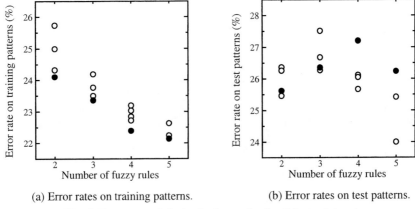

(a) Error rates on training patterns. (b) Error rates on test patterns.

Fig. 4. Error rates of obtained rule sets for the diabetes data set

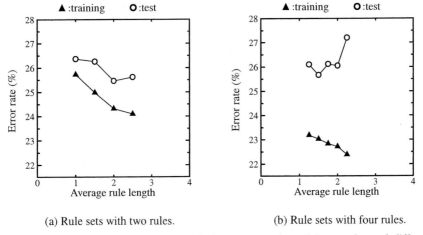

(a) Rule sets with two rules. (b) Rule sets with four rules.

Fig. 5. Error rates of obtained rule sets with the same number of fuzzy rules and different average rule length for the diabetes data set

Glass Identification Data Set. Simulation results on the glass identification data set (Glass in Table 1) are summarized in Fig. 6 and Fig. 7. In Fig. 6 (b), the overfitting due to the increase in the number of fuzzy rules is not clear. This result may suggest that the generalization ability of fuzzy rule-based systems can be further improved by using more fuzzy rules and/or adjusting each fuzzy rule (e.g., adjusting the rule weight). This is also suggested from the fact that the generalization ability on test patterns in Fig. 6 (b) is significantly inferior to the best result of the C4.5 algorithm in Table 1 (i.e., 27.3% error rate). On the other hand, we can observe the overfitting due to the increase in the average rule length in Fig. 7. For this data set, Sanchez et al.[27] reported a 42.1% error rate on test patterns by fuzzy rule-based systems with 8.5 rules on the average. Many rule sets in Fig. 6 (b) outperform the reported result in [27].

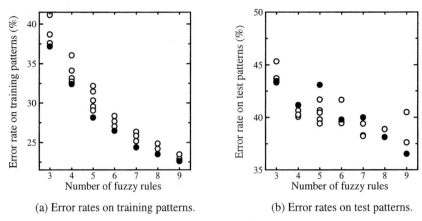

Fig. 6. Error rates of obtained rule sets for the glass identification data set

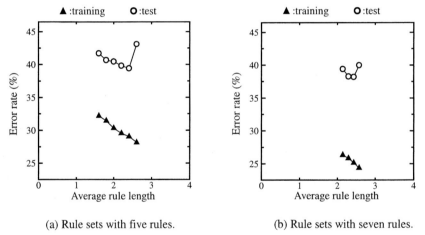

Fig. 7. Error rates of obtained rule sets with the same number of fuzzy rules and different average rule length for the glass identification data set

Cleveland Heart Disease Data Set. Simulation results on the Cleveland heart disease data set (Heart C in Table 1) are summarized in Fig. 8 and Fig. 9. In Fig. 8 (a), we can observe a clear tradeoff between the accuracy on training patterns and the number of fuzzy rules. On the other hand, the overfitting due to the increase in the number of fuzzy rules is clear in Fig. 8 (b). That is, error rates on test patterns tend to increase with the number of fuzzy rules in Fig. 8 (b) while error rates on training patterns in Fig. 8 (a) monotonically decrease. The worst result on test patterns in Fig. 8 (b) corresponds to the best result on training patterns in Fig. 8 (a). The overfitting due to the increase in the average rule length is also clear in Fig. 9. The generalization ability of some rule sets in Fig. 8 (b) outperforms the best result of the C4.5 algorithm in Table 1 (i.e., 46.3% error rate).

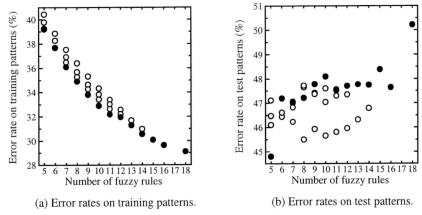

(a) Error rates on training patterns.   (b) Error rates on test patterns.

Fig. 8. Error rates of obtained rule sets for the Cleveland heart disease data set

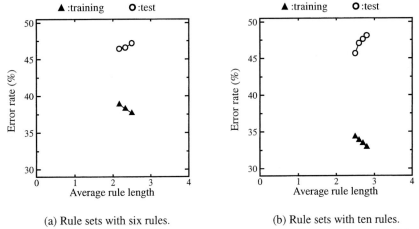

(a) Rule sets with six rules.   (b) Rule sets with ten rules.

Fig. 9. Error rates of obtained rule sets with the same number of fuzzy rules and different average rule length for the Cleveland heart disease data set

Sonar Data Set. Simulation results on the sonar data set (Sonar in Table 1) are summarized in Fig. 10. We can observe the tradeoff between the accuracy and the number of fuzzy rules in Fig. 10. The overfitting due to the increase in the number of fuzzy rules is not observed in Fig. 10 (b). The overfitting due to the increase in the average rule length is observed in the case of three fuzzy rules in Fig. 10 (b). The generalization ability of some rule sets in Fig. 10 (b) outperforms the best result of the C4.5 algorithm in Table 1 (i.e., 24.6% error rate).

Wine Recognition Data Set. Simulation results on the wine recognition data set (Wine in Table 1) are summarized in Fig. 11. The generalization ability of some rule sets in Fig. 11 (b) outperforms the best result of the C4.5 algorithm in Table 1 (i.e., 5.6% error rate).

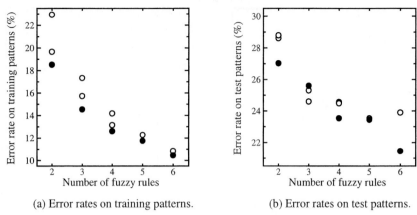

(a) Error rates on training patterns. (b) Error rates on test patterns.

Fig. 10. Error rates of obtained rule sets for the sonar data set

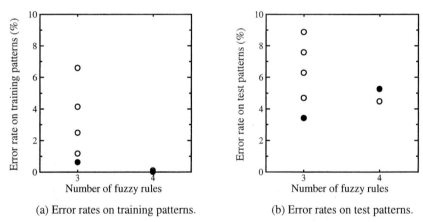

(a) Error rates on training patterns. (b) Error rates on test patterns.

Fig. 11. Error rates of obtained rule sets for the wine recognition data set

5 Concluding Remarks

We demonstrated the effect of a three-objective formulation of fuzzy rule selection on the generalization ability of obtained rule sets through computer simulations on six data sets. We observed clear overfitting to training patterns due to the increase in the number of fuzzy rules in computer simulations on three data sets: Wisconsin, diabetes and Cleveland. For those data sets, the second objective of our three-objective formulation (i.e., minimization of the number of fuzzy rules) can work as a safeguard against the overfitting. We also observed the overfitting due to the increase in the rule length in computer simulations on all the six data sets. The two-objective formulation is not enough for those data sets where the third objective (i.e., minimization of the total rule length) is necessary as a safeguard against the overfitting. Except for the glass identification data set and the sonar data set, the maximization of the accuracy on training patterns did not lead to the maximization of the accuracy on test patterns. Thanks to the three-objective formulation, we found fuzzy rule-based systems with high generalization ability for many data sets. Empirical analysis in this paper on the relation between the generalization ability of fuzzy rule-based systems and their complexity strongly relied on the ability of EMO algorithms to simultaneously find many non-dominated rule sets. Without this ability of EMO algorithms, we could not efficiently examine many non-dominated rule sets. Simulation results reported in this paper suggest the potential usefulness of EMO algorithms in the field of knowledge discovery and data mining.

Acknowledgments. The authors would like to thank the financial support from Japan Society for the Promotion of Science (JSPS) through Grand-in-Aid for Scientific Research (B): KAKENHI (14380194).

References

1. Agrawal, R., Mannila, H., Srikant, R., Toivonen, H., and Verkamo, A. I.: Fast Discovery of Association Rules, in Fayyad, U. M., Piatetsky-Shapiro, G., Smyth, P., and Uthurusamy, R. (eds.) *Advances in Knowledge Discovery and Data Mining*, AAAI Press, Metro Park (1996) 307–328.
2. Casillas, J., Cordon, O., Herrera, F., and Magdalena, L. (Eds.): *Accuracy Improvement in Linguistic Fuzzy Modelling*, Physica-Verlag (2003 in press).
3. Castillo, L., Gonzalez, A., and Perez, R.: Including a Simplicity Criterion in the Selection of the Best Rule in a Genetic Fuzzy Learning Algorithm, *Fuzzy Sets and Systems* 120 (2001) 309–321.
4. Cordon, O., Del Jesus, M. J., and Herrera, F.: A Proposal on Reasoning Methods in Fuzzy Rule-Based Classification Systems, *International Journal of Approximate Reasoning* 20 (1999) 21–45.
5. Deb, K., Pratap, A., Agarwal, S., and Meyarivan, T.: A Fast and Elitist Multiobjective Genetic Algorithm: NSGA-II, *IEEE Trans. on Evolutionary Computation* 6 (2002) 182–197.
6. Elomaa, T., and Rousu, J.: General and Efficient Multisplitting of Numerical Attributes, *Machine Learning* 36 (1999) 201–244.
7. Gonzalez, A., and Perez, R.: SLAVE: A Genetic Learning System Based on an Iterative Approach, *IEEE Trans. on Fuzzy Systems* 7 (1999) 176–191.
8. Hong, T. -P., Kuo, C. -S., and Chi, S. -C.: Trade-off between Computation Time and Number of Rules for Fuzzy Mining from Quantitative Data, *International Journal of Uncertainty, Fuzziness and Knowledge-Based Systems* 9 (2001) 587–604.
9. Ishibuchi, H., Murata, T., and Turksen, I. B.: Single-Objective and Two-Objective Genetic Algorithms for Selecting Linguistic Rules for Pattern Classification Problems, *Fuzzy Sets and Systems* 89 (1997) 135–149.
10. Ishibuchi, H., and Nakashima, T.: Effect of Rule Weights in Fuzzy Rule-Based Classification Systems, *IEEE Trans. on Fuzzy Systems* 9 (2001) 506–515.
11. Ishibuchi, H., Nakashima, T., and Morisawa, T.: Voting in Fuzzy Rule-Based Systems for Pattern Classification Problems, *Fuzzy Sets and Systems* 103 (1999) 223–238.
12. Ishibuchi, H., Nakashima, T., and Murata, T.: Three-Objective Genetics-Based Machine Learning for Linguistic Rule Extraction, *Information Sciences* 136 (2001) 109–133.

13. Ishibuchi, H., Nozaki, K., and Tanaka, H.: Distributed Representation of Fuzzy Rules and Its Application to Pattern Classification, *Fuzzy Sets and Systems* 52 (1992) 21–32.
14. Ishibuchi, H., Nozaki, K., Yamamoto, N., and Tanaka, H.: Construction of Fuzzy Classification Systems with Rectangular Fuzzy Rules Using Genetic Algorithms, *Fuzzy Sets and Systems* 65 (1994) 237–253.
15. Ishibuchi, H., Nozaki, K., Yamamoto, N., and Tanaka, H.: Selecting Fuzzy If-Then Rules for Classification Problems Using Genetic Algorithms, *IEEE Trans. on Fuzzy Systems* 3 (1995) 260–270.
16. Ishibuchi, H., and Yamamoto, T.: Fuzzy Rule Selection by Data Mining Criteria and Genetic Algorithms, *Proc. of Genetic and Evolutionary Computation Conference* (2002) 399–406.
17. Ishibuchi, H., and Yamamoto, T.: Comparison of Heuristic Rule Weight Specification Methods, *Proc. of 11th IEEE International Conference on Fuzzy Systems* (2002) 908–913.
18. Ishibuchi, H., and Yamamoto, T.: Comparison of Fuzzy Rule Selection Criteria for Classification Problems, in A. Abraham et al. (eds.): *Soft Computing Systems: Design, Management and Applications* (Frontiers in Artificial Intelligence and Applications, Volume 87), pp. 132–141, IOS Press, 2002.
19. Ishibuchi, H., Yamamoto, T., and Nakashima, T.: Fuzzy Data Mining: Effect of Fuzzy Discretization, *Proc. of 1st IEEE International Conference on Data Mining* (2001) 241–248.
20. Jin, Y.: Fuzzy Modeling of High-dimensional Systems: Complexity Reduction and Interpretability Improvement, *IEEE Trans. on Fuzzy Systems* 8 (2000) 212–221.
21. Jin, Y., Von Seelen, W., and Sendhoff, B.: On Generating FC^3 Fuzzy Rule Systems from Data Using Evolution Strategies, *IEEE Trans. on Systems, Man and Cybernetics – Part B: Cybernetics* 29 (1999) 829–845.
22. Knowles, J. D., and Corne, D. W.: Approximating the Nondominated Front Using the Pareto Archived Evolution Strategy, *Evolutionary Computation* 8 (2000) 149–172.
23. Kuncheva, L. I.: *Fuzzy Classifier Design*, Physica-Verlag, Heidelberg (2000).
24. Murata, T., and Ishibuchi, H.: MOGA: Multi-Objective Genetic Algorithms, *Proc. of 2nd IEEE International Conference on Evolutionary Computation* (1995) 289–294.
25. Quinlan, J. R.: *C4.5: Programs for Machine Learning*. Morgan Kaufmann, San Mateo (1993).
26. Roubos, H., and Setnes, M.: Compact and Transparent Fuzzy Models and Classifiers Through Iterative Complexity Reduction, *IEEE Trans. on Fuzzy Systems* 9 (2001) 516–524.
27. Sanchez, L., Couso, I., and Corrales, J. A.: Combining GA Operators with SA Search to Evolve Fuzzy Rule Base Classifiers, *Information Sciences* 136 (2001) 175–191.
28. Setnes, M., Babuska, R., Kaymak, U., and Van Nauta Lemke, H. R.: Similarity Measures in Fuzzy Rule Base Simplification, *IEEE Trans. on Systems, Man, and Cybernetics – Part B: Cybernetics* 28 (1998) 376–386.
29. Setnes, M., Babuska, R., and Verbruggen, B.: Rule-based Modeling: Precision and Transparency, *IEEE Trans. on Systems, Man, and Cybernetics – Part C: Applications and Reviews* 28 (1998) 165–169.
30. Van den Berg, J., Kaymak, U., and Van den Bergh, W. -M.: Fuzzy Classification Using Probability Based Rule Weighting, *Proc. of 11th IEEE International Conference on Fuzzy Systems* (2002) 991–996.
31. Weiss, S. M., and Kulikowski, C. A.: *Computer Systems That Learn*, Morgan Kaufmann Publishers, San Mateo (1991).
32. Zitzler, E., and Thiele, L.: Multiobjective Evolutionary Algorithms: A Comparative Case Study and the Strength Pareto Approach, *IEEE Trans. on Evolutionary Computation* 3 (1999) 257–271.

Identification of Multiple Gene Subsets Using Multi-objective Evolutionary Algorithms

A. Raji Reddy and Kalyanmoy Deb

Kanpur Genetic Algorithms Laboratory (KanGAL)
Indian Institute of Technology Kanpur
Kanpur, PIN 208 016, India
{arreddy,deb}@iitk.ac.in

Abstract. In the area of bioinformatics, the identification of gene subsets responsible for classifying available samples to two or more classes (for example, classes being 'malignant' or 'benign') is an important task. The main difficulties in solving the resulting optimization problem are the availability of only a few samples compared to the number of genes in the samples and the exorbitantly large search space of solutions. Although there exist a few applications of evolutionary algorithms (EAs) for this task, we treat the problem as a multi-objective optimization problem of minimizing the gene subset size and simultaneous minimizing the number of misclassified samples. Contrary to the past studies, we have discovered that a small gene subset size (such as four or five) is enough to correctly classify 100% or near 100% samples for three cancer samples (Leukemia, Lymphoma, and Colon). Besides a few variants of NSGA-II, in one implementation NSGA-II is modified to find multi-modal non-dominated solutions discovering as many as 630 different three-gene combinations providing a 100% correct classification to the Leukemia data. In order to perform the identification task with more confidence, we have also introduced a threshold in the prediction strength. All simulation results show consistent gene subset identifications on three disease samples and exhibit the flexibilities and efficacies in using a multi-objective EA for the gene identification task.

1 Introduction

The biological information contained in a genome is divided into discrete units called genes. Proteins read the information contained in a gene and initiate a series of biochemical reactions called as *gene expression*. Its level indicates the amount of mRNA or proteins to be made from a gene. Recent advances in microarray technologies based on cDNA hybridization or high density oligonucleotide probes enable to monitor the expression patterns of thousands of genes in parallel, and revolutionized the way in which researchers analyze gene expression in cells and tissues. DNA micro-array is an orchestrated arrangement of thousands of different single-stranded DNA probes in the form of cDNAs or oligonucleotides immobilized onto a glass or silicon substrate. The underlying principle of microarray technology is hybridization or base-paring (i.e., A-T and G-C). An array chip, hybridized to a labeled unknown cDNA extracted from a particular tissue of interest, makes it possible to measure simultaneously the expression level in a cell or tissue sample for each gene represented on the chip. DNA micro-arrays [7] enable to determine which genes are being expressed in a given cell type at a particular time and under particular conditions, to compare the gene expression in two different cell types or tissue samples, to examine changes in gene expression at different stages in the cell cycle and to assign probable functions to newly discovered genes with the expression patterns of known genes. Moreover, DNA array provides a global perspective of gene expression levels, which in turn find applications in gene clustering [1], tissue classification [9], identification of new targets for therapeutic drugs [3], and others. In this paper, we have concentrated on the gene subset identification problem.

The gene subset identification problem reduces to an optimization problem consisting of a number of objectives [11,12]. Although the optimization problem is a multi-objective one, all of these past studies have scalarized multiple objectives into one. In this paper, we have used a multi-objective evolutionary algorithm (MOEA) to find the optimum gene subset for three commonly-used cancer samples – Leukemia, Lymphoma and Colon. By using three objectives for minimization – gene subset size, number of misclassifications in training and number of misclassifications in test samples, several variants of a particular MOEA (NSGA-II) are applied to investigate if gene subsets with 100% correct classifications in both training and test samples exist. Since the gene subset identification problem may involve multiple gene subsets of the same size causing identical number of misclassifications [10], in this paper for the first time, we have proposed and developed a multi-modal NSGA-II for finding multiple gene subsets simultaneously in one single simulation run. One other important matter in the gene subset identification problem is the confidence level in which the samples are classified. We introduce the classification procedure based on the prediction strength consideration, suggested in [9], in the proposed multi-modal NSGA-II to find gene subsets with 20% and 30% prediction strength thresholds.

In the reminder of the paper, we briefly discuss the procedure of identifying gene subsets in a set of cancer samples and then discuss the underlying optimization problem. Thereafter, we discuss the procedure of using NSGA-II to this problem. Finally, we present simulation results on three disease samples using variants of NSGA-II and the proposed multi-modal NSGA-II, and discuss the merits of using an MOEA to the gene subset identification problem.

2 Identification of Gene Subsets

In this study, we concentrate on classifying samples for two classes only, although modifications can be made to generalize the procedure for any number of classes. In most problems involving identification of gene subsets in bioinformatics, the number of samples available compared to the gene pool size is very small. This aspect makes it difficult to identify which and how many genes are responsible for causing different classifications. It is a common practice in machine learning algorithms to divide the available data sets into two groups – one used for training purposes for generating a classifier and the other used for testing the developed classifier. Although most classification methods will perform well on samples used during training, it is necessary and important to test the developed classifier on unseen samples which were not used during training to get a realistic estimate of performance of the classifier and to avoid any training error. Most commonly employed method to estimate the accuracy in such situations is the cross-validation approach [2]. In cross-validation, the training data set (say T of them) is partitioned into k subsets, $C_1, C_2, \ldots, C_k$ (k is also known as the number of cross-validation trials). Each subset is kept roughly of the same size. Then a classifier is constructed using $T_i = T - C_i$ samples to test the accuracy on the samples in C_i. The construction procedure is described a little later. Once the classifier is constructed using T_i samples, each of the C_i samples is tested using the classifier for its class A or B. Since these T_i samples are used as training samples, we can compare the classification given by the above procedure with the actual class in which the sample belongs. If there is a mismatch, we increment the training sample mismatch counter τ_{train} by one. This procedure is repeated for all C_i samples in the i-th subset. Thereafter, this procedure is repeated for all k subsets and the overall training sample mismatch counter τ_{train} is noted. Cross-validation has several important properties such as the classifier is tested on each sample exactly once. One of the most-commonly used method in cross-validation is *leave-one-out-cross-validation* (LOOCV), in which only one sample in the training set is withheld and the classifier is constructed using the rest of the samples to predict the class of withheld sample. Thus, in the LOOCV there are $k = T$ subsets. In this study, we have used LOOCV to estimate the number of mismatches in the training set. The classifiers obtained from the training samples are used to predict the class of each test sample. The number of mismatches τ_{test} obtained by comparing the predicted class with the actual class of each sample is noted. Note that the LOOCV procedure is not used with the test samples, instead the classifier obtained using the training samples is directly used to find the number of mismatches in the test samples [9].

2.1 Class Prediction Procedure

For a given gene subset G, we can predict the class of any sample x (whether belonging to A or B) with respect to a known set of S samples in the following manner. Let us say that S samples are composed of two subsets S_A and S_B, belonging to class A and B, respectively. First, for each gene $g \in G$, we calculate the mean μ_A^g and standard deviation σ_A^g of the gene expression levels of all S_A samples. The same is repeated for the class B samples and μ_B^g and σ_B^g are found. Thereafter, we calculate the class of sample x as follows [9]:

$$\text{class}(x) = \text{sign}\left\{\sum_{g \in G}\left(\frac{\mu_A^g - \mu_B^g}{\sigma_A^g + \sigma_B^g}\right)\left(x_g - \frac{\mu_A^g + \mu_B^g}{2}\right)\right\}, \quad (1)$$

where x_g is the expression level of gene g in sample x. If the right term of the above equation is positive, the sample belongs to class A and if it is negative, it belongs to class B.

2.2 Resulting Optimization Problem

One of the objectives of the above task is to identify the smallest size of a gene subset for predicting the class of all samples correctly. Although not obvious, when a too small gene subset is used, the classification procedure becomes erroneous. Thus, minimizations of class prediction mismatches in the training and test samples are also important objectives. Here, we use these three objectives in a multi-objective optimization problem: The first objective f_1 is to minimize the size of gene subset in the classifier. The second objective f_2 is to minimize the number of mismatches in the training samples calculated using the LOOCV procedure and is equal to τ_{train} described above. The third objective f_3 is to minimize the number of mismatches τ_{test} in the test samples.

Like in [11], we use an ℓ-bit binary string (where ℓ is the number of genes in a data set) to represent a solution. For a particular string (individual), the positions marked with a 1 are included in the gene subset for that solution. We initialize each population member by randomly choosing at most 10% of string positions to have a 1. Since the gene subset size is to be minimized, this biasing against 1 in a string allows an EA to start with good population members. To handle three objectives, we have used NSGA-II [5] with a single-point crossover and a bit-wise mutation operator. The procedure of evaluating a string is as follows. We first collect all genes for which there is a 1 in the string in a gene subset G. Thereafter, we calculate f_1, f_2, and f_3 as described above as three objective values associated with the string.

3 Simulation Results

In this section, we show the application of NSGA-II and its variants on three different cancer samples: Leukemia, Lymphoma, and Colon. The flexibility in using different modifications to NSGA-II in the identification task is mostly demonstrated for the well-studied Leukemia samples.

The Leukemia data set is a collection of gene expression measurements from 72 leukemia (composed of 62 bone marrow and 10 peripheral blood) samples reported by Golub et al. [9]. It contains an initial training set composed of 27 samples of acute lymphoblastic leukemia (ALL) and 11 samples of acute myeloblastic leukemia (AML), and an independent test set composed of 20 ALL and 14 AML samples. The gene expression measurements were taken from high density oligonucleotide micro-arrays containing 7,129 probes for 6,817 human genes. This data is available at http://www.genome.wi.mit.edu/MPR.

The Lymphoma data set is a collection of expression measurements from 96 normal and malignant lymphocyte samples reported by Alizadeh et al. [6]. It contains 42 samples of diffused large B-cell lymphoma (DLBCL) and 54 samples of other types. The lymphoma data containing 4,026 genes is available at http://llmpp.nih.gov/lymphoma/data/figure1.cdt.

The Colon data set is a collection of 62 expression measurements from Colon biopsy samples reported by Alon et al. [1]. It contains 22 normal and 40 Colon cancer samples. The colon data having 2,000 genes is available at http://microaaray.princeton.edu/oncology.

3.1 Minimization of Gene Subset Size

First, we apply the standard NSGA-II on 50 genes which are used in another Leukemia study [9] to minimize two objectives: (i) the size of gene-subset (f_1), and (ii) the sum of mismatches ($f_2 + f_3$) in the training and test samples. We choose a population of size 500 and run NSGA-II for 500 generations. With a single-point crossover with a probability of 0.7 and a bit-wise mutation with a probability of $p_m = 1/50$, we obtain five non-dominated solutions, as shown in Figure 1. We have found a solution with zero mismatches in all training and test samples. This solution requires only four (out of 50) genes to correctly identify all 72 samples. The obtained solution has the following gene accession numbers: (M31211, M31523, M23197 and X85116). The non-dominated set also has other solutions with reduced number of genes but with non-zero mismatches.

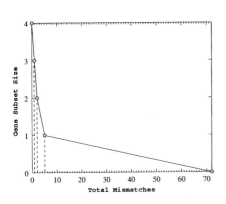
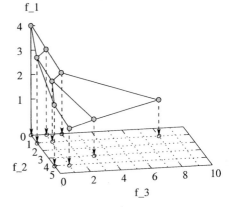

Fig. 1. Two-objective solutions obtained using NSGA-II for the Leukemia samples

Fig. 2. Three-objective solutions obtained using NSGA-II for the Leukemia samples

One difficulty with the above two-objective problem is that it is not clear from solutions with non-zero mismatches whether the mismatches occur in the training or in the test samples. To differentiate this matter, we now consider all three objectives – the gene subset size, the mismatches in the training samples, and the mismatches in the test samples. Figure 2 shows the corresponding non-dominated solutions with identical parameter settings. For clarity, we have not shown the solution having no genes (causing 38 and 34 mismatches in training and test samples, respectively) in this figure. It is clear from the figure that smaller gene subsets cause more mismatches. Interestingly, four different four-gene solutions are found to provide 100% classification and these solutions are different from that obtained in the two-objective case. This indicates that there exist multiple gene-combinations for a 100% perfect classification, a matter we discuss in more detail later. The three-objective case also clearly shows the break-up in the mismatches. For example, Figure 1 shows that the two-gene solution has two mismatches, whereas Figure 2 shows that there exist three two-gene solutions – four mismatches in training samples only, two mismatches in test samples only, and one mismatch each in training and test samples.

An ideal research in bioinformatics requires a collaboration between computer scientists and biologists. The above study shows that a four-gene classifier is enough to correctly classify all available Leukemia samples. But the obtained four-gene set is discovered purely from the algorithmic point of view. Due to the availability of only a few samples, this algorithmically best solution may not be biologically (or medically) attractive. In order to find if there exists any larger size gene subset resulting in a 100% correct classification, we first attempt to identify the largest size gene subset which can correctly classify all 72 samples. For this task, we apply NSGA-II with the above three objectives, but instead of minimizing the first objective we maximize it. With identical NSGA-II parameters, the obtained non-dominated set has a solution having 37 genes which can correctly classify all 72 samples (Figure 3). If more genes are added, the classification becomes less than perfect.

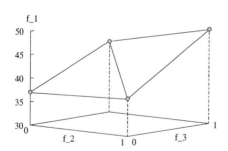 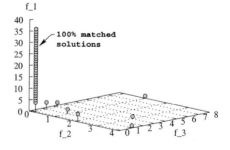

Fig. 3. Maximization of gene subset size for the Leukemia samples

Fig. 4. NSGA-II solutions with modified domination criterion for the Leukemia samples

3.2 Modified Domination Criterion for Multiple Gene Subset Sizes

The above subsection demonstrated how solutions with smallest and largest number of gene subsets can be found by simply using minimization and maximization of the first objective, respectively. However, for the reason given above, ideally it could be better to find the entire range of gene subsets for 100% classifications. For this purpose, we modify the domination criterion in NSGA-II.

Definition 1 (Bias-dominance $\prec_i$ criterion) *Solution* $\mathbf{x}^{(1)}$ *bias-dominates ($\prec_i$) solution* $\mathbf{x}^{(2)}$ *if* $f_j(\mathbf{x}^{(1)}) \leq f_j(\mathbf{x}^{(2)})$ *for all objectives* ($j = 1, 2, \ldots, M$) *and* $f_k(\mathbf{x}^{(1)}) < f_k(\mathbf{x}^{(2)})$ *for at least one objective other than the i-th objective* ($k = 1, 2, \ldots, M$ *and* $k \neq i$).

This bias-domination definition differs from the original dominance definition [4] in that any two solutions with identical f_j values (where $j \neq i$) will not dominate each other. This way, multiple solutions lying along f_i axis direction can be all non-dominated to each other. When we apply NSGA-II with identical parameter settings as in the previous subsection and with the above bias-dominance criterion for f_1 (or $\prec_1$), all solutions ranging from four-gene subset to 36-gene subset (in the interval of one) are found to produce a 100% perfect classification. Figure 4 shows the obtained non-dominated solutions. This shows that instead of using two NSGA-II applications as shown in the previous two subsections, a whole spectrum of solutions can be obtained in one single simulation run. This illustrates the flexibility and efficiency of using NSGA-II in the gene subset identification task.

3.3 Multi-modal MOEAs for Multiple Solutions

We have observed in subsection 3.1 that the gene subset identification task with multiple objectives may involve multi-modal solutions, meaning that for a point in the objective space (on Figures 1 to 4), there may exist more than one solutions in the decision variable space (on the Hamming string space). When this happens, it becomes important and useful for a biologist to know which different gene combinations may provide an identical classification. In this subsection, we suggest a modified NSGA-II procedure for identifying such multi-modal solutions. We define two multi-modal solutions in the context of multi-objective optimization as follows:

Definition 2 (Multi-modal solutions) *If for two different solutions* $\mathbf{x}^{(1)}$ *and* $\mathbf{x}^{(2)}$ *satisfying* $\mathbf{x}^{(1)} \neq \mathbf{x}^{(2)}$, *all objective values are the same, or* $f_i(\mathbf{x}^{(1)}) = f_i(\mathbf{x}^{(2)})$ *for all* $i = 1, 2 \ldots, M$, *then solutions* $\mathbf{x}^{(1)}$ *and* $\mathbf{x}^{(2)}$ *are multi-modal solutions.*

In such problems, more than one solutions in the decision variable space map to one point in the objective space. We are not aware of any MOEA which has been attempted for this task or for any other similar problems. Fortunately, the gene subset identification problem is an ideal problem in which multiple gene subset combinations produce an identical outcome in terms of their mismatches in the training and test data sets.

Recall that in an NSGA-II iteration [5], the parent P_t and offspring Q_t populations are combined to form an intermediate population R_t of size $2N$. Thereafter, R_t is sorted according to a decreasing order of non-domination level $(\mathcal{F}_1, \mathcal{F}_2, \ldots)$. In the original NSGA-II, solutions from each non-dominated level (starting from the best set) are accepted till a complete set cannot be included without increasing the designated population size. A crowding operator was then used to choose the remaining population members. We follow an identical procedure here till the non-dominated sorting is done. But before we describe the procedure, let us present two definitions:

Definition 3 (Duplicate solutions) *Two solutions $\mathbf{x}^{(1)}$ and $\mathbf{x}^{(2)}$ are duplicates to each other if $\mathbf{x}^{(1)} = \mathbf{x}^{(2)}$.*

It follows that duplicate solutions have identical objective values.

Definition 4 (Distinct objective solutions) *Two solutions $\mathbf{x}^{(1)}$ and $\mathbf{x}^{(2)}$ are distinct objective solutions if $f_i(\mathbf{x}^{(1)}) \neq f_i(\mathbf{x}^{(2)})$ for at least one i.*

First, we delete the duplicate solutions from each non-domination set in R_t. Thereafter, each set is accepted as usual till the last front $\mathcal{F}_l$ which can be accommodated. Let us say that solutions remaining to be filled before this last front is considered is N' and the number of non-duplicate solutions in the last front is N_l ($> N'$). We also compute the number of distinct objective solutions in the set $\mathcal{F}_l$ and let us say it is n_l (obviously, $n_l \leq N_l$). If $n_l \geq N'$, we use the usual crowding distance procedure to choose N' most dispersed and distinct solutions from n_l solutions. In this case, even if there exist multi-modal solutions to any n_l solutions, they are ignored due to lack of space in the population. The major modification to NSGA-II is made when $n_l < N'$. This means that although there are fewer distinct solutions than the population slots, the distinct solutions are multi-modal. However, the total number of multi-modal solutions of all distinct solutions (N_l) is more than the remaining population slots. Thus, we need to make a decision of choosing a few solutions. The purpose here is to have at least one copy of each distinct objective solution and as many multi-modal copies of them so as to fill up the population. Here, we choose a strategy in which every distinct objective solution is allowed to have a proportionate number of multi-modal solutions as they appear in $\mathcal{F}_l$. To avoid losing any distinct objective solutions, we first allocate one copy of each distinct objective solution, thereby allocating n_l copies. Thereafter, the proportionate rule is applied to the remaining solutions ($N_l - n_l$) to find the accepted number of solutions for the i-th distinct objective solution as follows: $\alpha_i = \frac{N'-n_l}{N_l-n_l}(m_i - 1)$, where m_i is the number of multi-modal solutions of the i-th distinct objective solution in $\mathcal{F}_l$, such that $\sum_{i=1}^{n_l} m_i = N_l$. It is true that $\sum_{i=1}^{n_l} \alpha_i = N' - n_l$. The final task is to choose $(\alpha_i + 1)$ multi-modal solutions from m_i copies for the i-th distinct objective solution. Although a sharing strategy [8] can be used to choose the maximally different multi-modal solutions, here we simply choose them randomly. Along with the duplicate-deletion strategy, the random acceptance of a specified number multi-modal solutions to each distinct objective solution ensures a good spread of solutions in both objective and decision variable space. In the rare occasions of having less than N non-duplicate solutions in R_t, new random solutions are used to fill up the population. For a problem having many multi-modal solutions, the latter case will occur often and the above systematic preservation of distinct objective solutions and then their multi-modal solutions will maintain a rich collection of multi-modal Pareto-optimal solutions.

We apply the multi-modal NSGA-II to the 50-gene Leukemia data set first. All three objectives are minimized here. With 500 population sizes running for 500 generations, we obtain the same non-dominated front as shown in Figure 2. However, each distinct objective solution has a number of multi-modal solutions. For the solution with four-gene subset causing 100% classification on both training and test samples, the multi-modal NSGA-II has found 26 different solutions. All these four-gene solutions are shown in Figure 5. The figure brings out an interesting aspect. Among different four-gene combinations, three genes (accession numbers M31211, M31303, and M63138) frequently appear in the obtained gene subsets. Of the 26 solutions, these three genes appear together in eight of them. Such information about frequently appearing genes in high-performing classifiers is certainly useful to biologists. Interestingly, two of these three genes also appear quite frequently in other trade-off non-dominated solutions with non-zero mismatches, as shown in Figure 5.

It is also interesting to note that when multi-modal NSGA-II was not used (in subsection 3.1), only four distinct solutions with 100% correct classification were obtained. These four solutions are also rediscovered in the set of 26 multi-modal solutions shown in Figure 5. This illustrates the efficiency of the proposed multi-modal NSGA-II approach in finding and maintaining multi-modal Pareto-optimal solutions.

3.4 Complete Leukemia Data Set

Now, we apply the multi-modal NSGA-II to the entire Leukemia data set having 7,129 samples. Because of the large string length requirement, we have chosen 1,000 population size and run NSGA-II for 2,000 generations. We have used a mutation probability of 1/7129, so that on an average one bit gets mutated in the complete string of size 7,129. The perfect classification is obtained with only three genes. However, the number of such three-gene combinations is 630, meaning that any one of these 630 three-gene combinations will produce a 100% classification of training as well as test data. Recall that the 50-gene study above has required four genes for this task. Interestingly, all 630 three-gene combinations required 557 independent genes, of which *only* 14 genes are common with the 50-gene study. It is also observed that in 606 three-gene combinations, the gene X95735 from the 50-gene study is found together with a gene (HG1612) outside the 50-gene set. Surprisingly, the gene (X95735) did not appear even once in the

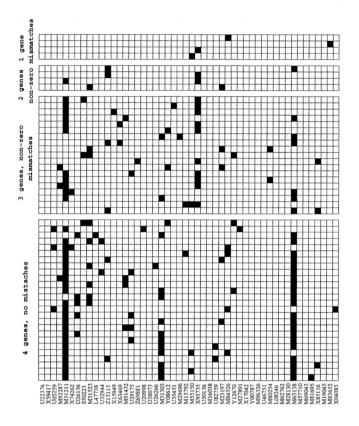

Fig. 5. Multi-modal solutions obtained using NSGA-II for the Leukemia samples

four-gene combinations found in the 50-gene study. Only with the presence of the gene (HG1612), this gene is discovered to be a good combination for a 100% classification. This clearly indicates that the 50-gene subset used in [9] did not include some crucial genes, the presence of which required one less gene (with only three genes) to make a 100% classification here. Table 1 shows other lower gene-subsets with non-zero mismatches. Another interesting aspect to note here is that the solutions found have much smaller gene subsets compared to those reported in the literature. A study [11] used a single-objective GA with $p_m = 0.001$ and 500 population size run for 500 generations achieved a 16-gene subset as the optimum solution with 97% matches in 72 samples. We rerun NSGA-II with $p_m = 0.001$ and 500 population size and 500 generations but using the three objectives as before and obtain a solution with 17-gene subset and having 100% matches in the training and test samples. Since these solutions are dominated by the solutions shown in Table 1, we argue that the mutation probability used in [11] was not adequate. With a large mutation probability (about 10 times larger than usual), more bits gets mutated in a string, thereby

Table 1. Multi-modal solutions for three disease samples. Parameters f_1, f_2, f_3, and α represent gene subset size, mismatches in training samples, mismatches in test samples, and the number of multi-modal solutions obtained with (f_1, f_2, f_3) values

Leukemia Samples				Lymphoma Samples				Colon Samples			
f_1	f_2	f_3	α	f_1	f_2	f_3	α	f_1	f_2	f_3	α
3	0	0	630	5	0	0	121	7	0	1	25
2	0	1	2	4	0	1	17	6	1	1	4
2	3	0	1	3	1	0	1	4	0	2	3
1	9	1	1	3	0	3	2	3	4	3	1
1	2	12	1	2	1	2	1	3	0	4	1
1	5	2	1	2	2	1	1	2	3	4	2
1	3	3	1	1	4	5	2	2	2	6	2
								1	4	5	1

not allowing smaller sized gene-subsets to exist in the population. The fact that our implementation with a similar large mutation probability has also found a solution similar to that reported in [11] and the fact that we could get better solutions (with 100% matches) with the same implementation but with a smaller mutation probability indicate the sub-optimal use of mutation probability in [11]. Interestingly, if we continue to run our NSGA-II beyond 500 generations, smaller gene subsets with 100% classification appeared. However, the former study [11] used the LOOCV procedure to all 72 samples for calculating the mismatches, whereas we have used the LOOCV procedure to the training samples only. The classifier developed from the training set is simply used to classify test samples.

3.5 Complete Lymphoma Data Set

Next, we apply the multi-modal NSGA-II to the Lymphoma data set having 4,026 genes. There are a total of 96 samples available. We have randomly divided 50% of them for training and 50% of them for testing purposes. With the same NSGA-II parameters as in the 7129-gene Leukemia case, we obtain a total of 144 solutions, which are tabulated in Table 1. As small as five (out of 4,026) genes are enough to correctly identify all 96 samples. Interestingly, there are 121 such five-gene combinations to classify the task perfectly. Some other solutions with smaller gene subsets are also shown in the table.

3.6 Complete Colon Data Set

For the colon data set having 62 samples each with 2,000 genes, we have applied the multi-modal NSGA-II. We have randomly chosen 50% samples for training and the rest for testing. Identical NSGA-II parameters are used here. A total of 39 solutions are obtained. Of them, 25 solutions contain seven genes and each can correctly identify all 31 training samples but misses to identify only one test sample. Table 1 shows these and other more mismatched solutions obtained using the multi-modal NSGA-II.

Interestingly, in this data set, NSGA-II has failed to identify a single solution with 100% correct classification. Although we have chosen different combinations of 31 training and 31 test samples and rerun NSGA-II, no improvement to this solution is observed. In order to investigate if there exists at all any solution (gene combination) which will correctly classify all training and test samples, we have used a binary-coded genetic algorithm for minimizing the sum of mismatches in the training and test samples and without any care of minimizing the gene subset size. The resulting minimum solution corresponds to one of the 25 solutions found using NSGA-II having seven genes and with overall mismatch of one. This application supports the minimum mismatched solution obtained using the multi-modal NSGA-II.

4 Classification with Confidence

One of the difficulties with the above classification procedure is that the sign of the right term in equation 1 is checked to identify if a sample belongs to one class or another. For each (g) of the 50 genes in a particular Leukemia sample x, we have calculated the term (say the statistic $S(x, g)$) inside the summation in equation 1. For this sample, a correct prediction has been made. The statistic $S(x, g)$ values are plotted in Figure 6. It can be seen from the figure that for 27

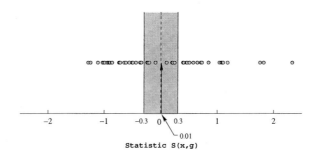

Fig. 6. The statistic $S(x, g)$ of a sample for the 50-gene Leukemia data

genes, negative values of $S(x, g)$ emerged, thereby classifying individually that the sample belongs to AML (class B), whereas only 23 genes detects the sample to be ALL (class A). Equation 1 finds the right side value to be 0.01, thereby declaring the sample to be an ALL sample (which is correct). But it has been argued elsewhere [9] that a correct prediction with such a small strength does not allow to make the classification with a reasonable confidence.

For a more confident classification, we may fix a prediction strength threshold θ and modify the classification procedure slightly. Let us say that the sum of the positive $S(x,g)$ values is S^+ and the sum of the negative $S(x,g)$ values is S^-. Then, the prediction strength $|(S^+ - S^-)/(S^+ + S^-)|$ [9] is compared with θ. If it is more than θ, the classification is accepted, else the sample is considered to be undetermined for class A or B. For simplicity, we assume these undetermined samples to be identical to mismatched samples and include this sample to increment τ_1 or τ_2, as the case may be. This way, a 100% correctly classified gene subset will ensure that the prediction strength is outside $(-\theta, \theta)$. Figure 6 illustrates this concept with $\theta = 0.3$ (30% threshold) and demands that a match will be scored only when the prediction strength falls outside the shaded area. In the sample illustrated in the figure, there are eight genes which will case a undetermined classification with less than 30% threshold, thereby causing this sample to be a 'mismatched' sample.

Here, we show the effect of classification with a threshold in prediction strength on the 50-gene Leukemia samples. All 72 samples are used in the study. With identical parameter settings as in subsection 3.3, the multi-modal NSGA-II with $\theta = 30\%$ threshold prediction strength finds a solution with four genes and one mismatch in the test samples. The smallest prediction strength observed in any sample is 32.1%. The sample on which the mismatch occurs, the prediction strength is 41.7%. Thus, the obtained four-gene subset does its best keeping the total mismatch down to one (in the test sample). Taking any more genes in the subset increases the number of mismatches. The multi-modal NSGA-II has discovered two solutions having four genes and one mismatch in the test samples. These gene subsets are (X95735, M23197, X85116, M31211) and (X95735, M23197, X85116, M31523). Since these solutions have one mismatch in the test sample, they were not found to be non-dominated solutions in subsection 3.3. However, some of the above genes are also found to be common to the frequently-found classifier genes in Figure 5. However, if the threshold is reduced to $\theta = 20\%$, we obtain a solution with 100% correct classification with a gene subset of size five. In this case, there are eight samples in which the prediction strength is between 20% and 30%. Thus, while optimizing for the 30% threshold prediction strength, this five-gene solution was dominated by the four-gene solution and hence did not appear as one of the non-dominated solutions. Recall that the study presented in subsection 3.3 used a prediction strength threshold of $\theta = 0\%$ and we obtained a 100% correct classification with only four genes. There are two five-gene solutions found with a 100% correct classification. They are (U05259, M31211, M23197, N96326, M83652) and (U05259, Y08612, M23197, X85116, M83652), which have three genes in common and share some common genes with the four-gene solutions found in the 30% threshold study. This study shows that the outcome of the classification depends on the chosen prediction strength threshold. Keeping a higher threshold makes a more confident classification, but at the expense of some mismatches, while keeping a low threshold may make a 100% classification, but the classification may be performed with a poor confidence level.

Since no gene combination with 50 selected genes could identify all 72 samples with 30% prediction strength, we apply the multi-modal NSGA-II to the complete Leukemia data set having 7,129 genes and apply 30% prediction strength threshold. The multi-modal NSGA-II with 1,000 population size is run for 2,000 generations. 10 multi-modal solutions with only four-gene subsets are found to correctly classify all 72 samples with 30% threshold. Recall that in subsection 3.4 only three-gene subsets were enough for 100% classification, but with 0% prediction strength threshold. There were eight samples in which the prediction strengths were less than 30% and the minimum prediction strength was as low as 3.5%. This study shows that in order to have more confidence in classification (meaning using a larger prediction strength), one additional gene is necessary.

5 Conclusions

The identification of gene subsets responsible for classifying disease samples to fall in one category or another has been dealt in this paper. By treating the resulting optimization problem with two or more objectives, we have applied NSGA-II and a number of its variants to find optimal gene subsets for different classification levels to three available disease samples – Leukemia, Lymphoma, and Colon. One remarkable finding is that compared to past studies our study has discovered only three- or four-gene combinations are enough to perfectly classify all samples involving Leukemia and Lymphoma two-class samples. For the Colon samples, we have found a gene subset which classify all samples except one. Different variants of NSGA-II have exhibited the flexibility with which such identifications can be made.

In this paper, we have also suggested a multi-modal NSGA-II by which multiple multi-modal non-dominated solutions can be found simultaneously in one single simulation run. These solutions have identical objective values but they differ in their phenotypes. This study shows that in the gene subset identification problem there exist a large number of such multi-modal solutions, even corresponding to the 100% correctly classified gene subset. In the Leukemia data set, the multi-modal NSGA-II has discovered as many as 630 different three-gene combinations which correctly classified all samples. Investigating these multiple solutions may provide intriguing information about crucial gene combinations responsible for classifying samples into one class or another. In the Leukemia data set, we have observed that among all 50 genes there are three genes which appear frequently on 26 different four-gene solutions capable of perfectly classifying all 72 samples. Investigating these frequently appearing genes further from a biological point of view should provide crucial information about the causes of different classes of a disease. The proposed multi-modal NSGA-II is generic and can also be applied to similar other multi-objective optimization problems in which multiple multi-modal solutions are desired.

Finally, a more confident gene subset identification task is performed by using a minimum threshold in the prediction strength. With a 30% threshold, it has been found that one extra gene is needed to make a 100% correct classification compared to that needed with the 0% threshold. Interestingly, although there exist some common genes in the subsets obtained in the two cases, all genes found in any subset in the 0% threshold case are *not* found in any gene subset found in the 30% threshold case.

All these results are interesting and exhibit the flexibility with which NSGA-II (or other MOEA techniques) can be applied to discover important gene subsets reliably for an adequate level of classification. These results open up a number of avenues of further research in this area, which we are currently pursuing.

References

1. U. Alon, N. Barkai, D. A. Notterman, K. Gish, S. Ybarra, D. Mack, and A. J. Levine. Broad patterns of gene expression revealed by clustering analysis of tumor and normal colon tissues probed by oligonucleotide arrays. In *Proceedings of National Academy of Science, Cell Biology*, volume 96, pages 6745–6750, 1999.
2. A. Ben-Dor, L. Bruhn, N. Friedman, I. Nachman, M. Schummer, and Z. Yakhini. Tissue classification with gene expression profiles. *Journal of Computational Biology*, 7:559–583, 2000.
3. P. A. Clarke, M. George, D. Cunningham, I. Swift, and P. Workman. Analysis of tumor gene expression following chemotherapeutic treatment of patients with bowl cancer. In *Proceedings of Nature Genetics Microarray Meeting – 99*, page 39, 1999.
4. K. Deb. *Multi-objective optimization using evolutionary algorithms*. Chichester, UK: Wiley, 2001.
5. K. Deb, S. Agrawal, A. Pratap, and T. Meyarivan. A fast and elitist multi-objective genetic algorithm: NSGA-II. *IEEE Transactions on Evolutionary Computation*, 6(2):182–197, 2002.
6. A. A. Alizadeh et al. Distinct types of diffuse large B-cell lymphoma identified by gene expression profiling. *Nature*, 403:503–511, 2000.
7. D. Gershon. Microarray technology an array of opportunities. *Nature*, 416:885–891, 2002.
8. D. E. Goldberg and J. Richardson. Genetic algorithms with sharing for multimodal function optimization. In *Proceedings of the First International Conference on Genetic Algorithms and Their Applications*, pages 41–49, 1987.
9. T. R. Golub, D. K. Slonim, P. Tamayo, C. Huard, M. Gaasenbeek, J. P. Mesirov, H. Coller, M. L. Loh, J. R. Downing, M. A. Caligiuri, C. D. Bloomfield, and E. S. Lander. Molecular classification of cancer: Class discovery and class prediction by gene expression monitoring. *Science*, 286:531–537, 1999.
10. R. Kohavi and G. H. John. Wrappers for feature subset selection. *Artificial Intelligence Journal, Special Issue on Relevance*, 97:234–271, 1997.
11. J. Liu and H. Iba. Selecting informative genes using a multiobjective evolutionary algorithm. In *Proceedings of the World Congress on Computational Intelligence (WCCI-2002)*, pages 297–302, 2002.
12. J. Liu, H. Iba, and M. Ishizuka. Selecting informative genes with parallel genetic algorithms in tissue classification. *Genome Informatics*, 12:14–23, 2001.

Non-invasive Atrial Disease Diagnosis Using Decision Rules: A Multi-objective Optimization Approach

Francisco de Toro[1], Eduardo Ros[2], Sonia Mota[2], and Julio Ortega[2]

[1] Departamento de Ingeniería Electrónica, Sistemas Informáticos y Automática,
Universidad de Huelva, Spain
ftoro@uhu.es

[2] Departamento de Arquitectura y Tecnología de Computadores,
Universidad de Granada, Spain
{eduardo,sonia,julio}@atc.ugr.es

Abstract. This paper deals with the application of multi-objective optimization to the diagnosis of Paroxysmal Atrial Fibrillation (PAF). The automatic diagnosis of patients that suffer PAF is done by analysing Electrocardiogram (ECG) traces with no explicit fibrillation episode. This task presents difficult problems to solve, and, although it has been addressed by several authors, none of them has obtained definitive results. A recent international initiative to study the viability of such an automatic diagnosis application has concluded that it can be achieved, with a reasonable efficiency. Furthermore, such an application is clinically important because it is based on a non-invasive examination and can be used to decide whether more specific and complex diagnosis testing is required. In this paper we have formulated the problem in order to be approached by a multi-objective optimisation algorithm, providing good results through this alternative.

1 Introduction

Most real-world optimisation problems are multi-objective in nature, since they normally have several (usually conflicting) objectives that must be satisfied at the same time. These problems are known as MOP (*Multi-objective Optimisation Problems*) [1]. The notion of *optimum* has to be re-defined in this context, as we no longer aim to find a single solution; a procedure for solving MOP should determine a set of good compromises or *trade-off* solutions, generally known as *Pareto optimal solutions* from which the decision maker will select one. These solutions are optimal in the wider sense that no other solution in the search space is superior when all objectives are considered.

Evolutionary Algorithms (EAs) have the potential to find multiple Pareto optimal solutions in a single run and have been widely used in this area [2,3]. After the first studies on evolutionary multi-objective optimisation (EMO) in the mid-1980s, a number of Pareto-based techniques were proposed, e.g. MOGA [4], NSGA [5] and NPGA [6]. These approaches did not explicitly incorporate elitism: recently, however, the importance of this concept in multi-objective searching has been recognized and supported experimentally [2,7,8]. In this sense, the present work continues exploring the benefits of elitism by applying the *Single Front Genetic Algorithms* [15,16], and the *Strengh Pareto Evolutionary Algorithm* [7] to a real world multiobjective optimization problem.

The Atrial Fibrillation is the heart arrhythmia that causes most frequently embolic events that may generate cerebrovascular accidents. In this paper, the multiobjective Evolutionary Algorithms aforementioned are applied to Paroxysmal Atrial Fibrillation (PAF) diagnosis. The automatic diagnosis of patients that suffer Paroxysmal Atrial Fibrillation is done by analysing Electrocardiogram (ECG) traces with no explicit fibrillation episode. This task presents difficult problems to solve, and, although it has been addressed by several authors[9], none of them has obtained definitive results. Thus the automatic PAF diagnosis remains an open problem. A recent international initiative to study the viability of such an automatic diagnosis application has concluded that it can be achieved, with a reasonable efficiency [9,10]. Furthermore, such an application is clinically important because it is based on a non-invasive examination and can be used to decide whether more specific and complex diagnosis testing is required. In this paper we have formulated the problem in order to be approached by a multi-objective optimisation algorithm, providing good results through this alternative.

In this paper, Section 2 introduces the MOPs and provides a brief survey about the evolutionary multi-objective techniques proposed in the literature. Section 3 reviews the Single Front Genetic Algorithms for multi-objective optimisation proposed previously by some of the authors, while Section 4 describes the way PAF diagnosis has been solved by using a multi-objective optimisation approach. Finally, experimental results including comparisons of the different multi-objective optimization algorithms are given in Section 5, and the concluding remarks are summarized in Section 6.

2 Evolutionary Algorithms for Multi-objective Optimization

A multi-objective optimisation problem (MOP) can be defined [1] as one of finding a vector of *decision variables* that satisfies a set of constraints and optimises a vector function whose elements represent the objectives. These functions form a mathematical description of performance criteria, and are usually in conflict with each other.

The problem can be formally stated as finding the vector $\mathbf{x}^* = [x_1^*, x_2^*, \ldots, x_n^*]$ which satisfies the *m* inequality constraints

$$g_i(\mathbf{x}) \geq 0 \qquad i=1,2,\ldots,m \tag{1}$$

the *p* equality constraints

$$h_i(\mathbf{x}) = 0 \qquad i=1,2,\ldots,p \tag{2}$$

and optimizes the vector function

$$\mathbf{f}(\mathbf{x}) = [f_1(\mathbf{x}), f_2(\mathbf{x}), \ldots, f_k(\mathbf{x})]^T \tag{3}$$

where $\mathbf{x} = [x_1, x_2, \ldots, x_n]^T$ is a vector of *decision variables*.

The constraints given by (1) and (2) define the *feasible region* Φ: any point **x** in Φ is a *feasible solution*.

The vector function f (**x**) is one that maps the set Φ in the set χ of all possible values of the objective functions. The k components of the vector **f** (**x**) represent the *non-commensurable criteria* to be considered. The constraints g_i (**x**) and h_i (**x**) represent the restriction imposed on the decision variables. The vector **x*** is reserved to denote the optimal solutions (normally there will be more than one).

The meaning of *optimum* is not well defined in this context, since in these problems it is difficult to have an **x*** where all the components of f_i (**x**) have a minimum in Φ. Thus, a point **x***∈ Φ is defined as *Pareto Optimal*

$$\text{If } \forall \mathbf{x} \in \Phi, \forall i := 1..k, \ (f_i(\mathbf{x}) = f_i(\mathbf{x}^*)) \ \ or \ \ f_i(\mathbf{x}^*) < f_i(\mathbf{x}) \tag{4}$$

The inequality equation in (4) must be fulfilled by at least one component i.

This means that **x*** is Pareto optimal if there exists no feasible vector **x** which would decrease one criterion without causing a simultaneous increase in at least one of the others. The notion of Pareto optimum almost always gives, not a single solution, but rather a set of solutions called non-inferior or *non-dominated* solutions (Figure1). This set of non-dominated solutions is known as the *Pareto front*. As, in general, it is not easy to find an analytical expression for the *Pareto front*, the usual procedure is to determine a set of *Pareto optimal* points that provide a good approximate description of the *Pareto front*.

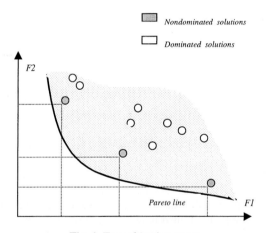

Fig. 1. Two-objective space

Evolutionary algorithms seem to be especially suited to multi-objective optimisation because they are able to capture multiple Pareto-optimal solutions in a single run, and may exploit similarities of solutions by recombination. Indeed, some research suggests that multi-objective optimisation might be an area where EAs perform better than other search strategies. The considerable amount of research related to MOEAs currently reported in the literature is evidence of present interest in this subject.

Pareto-based fitness assignment in a genetic algorithm (GA) was first proposed by Goldberg [11]. The basic idea is to find a set of Pareto non-dominated individuals in the population. These individuals are then assigned the highest rank and eliminated from further competition. Then, another set of Pareto non-dominated individuals are determined from the remaining population and are assigned the next highest rank. This process continues until the whole population is suitably ranked. Goldberg also suggested the use of a *niching technique* [12] to preserve diversity in the population, in order to converge to different solutions.

The *Non-dominated Sorting Genetic Algorithm (NSGA)* [5] uses several layers of ranked individuals. Before selection is performed, the population is ranked on the basis of non-domination: all non-dominated individuals are classified into one category (with a dummy fitness value, which is proportional to the population size, to provide an equal reproductive potential for these individuals). To maintain the diversity of the population, those so classified are shared with their dummy fitness values. Then this group of classified individuals is ignored and another layer of non-dominated individuals is considered. The process continues until all individuals in the population have been classified. Then a stochastic remainder proportionate selection is used, followed by the usual cross and mutation operators. An improved version of NSGA (called NSGA-II) that uses less parameter has been proposed more recently [8].

Fonseca and Fleming [4] have proposed an algorithm called *Multiple Objective Genetic Algorithm (MOGA)* where the rank of each individual is obtained from the number of individuals in the current population that dominate it. Thus, if at generation t, an individual x_i is dominated by $p_i(t)$ individuals, its current rank can be given by:

$$Rank(x_i, t) = 1 + p_i(t)$$

All non-dominated individuals are assigned rank 1, while dominated ones are penalized according to the population density of the corresponding region of the trade-off surface. In this way, the fitness assignment is performed by the following steps:

1. - Sort population according to the rank of the individuals.
2. - Assign fitness to individuals by interpolating from the best (rank 1) to the worst (rank n<=N), according to a function (not necessarily linear)
3. - Average the fitness of individuals with the same rank, so that all of them will be sampled at the same rate.

In their *Niched Pareto Genetic Algorithm (NPGA)*, Horn and Nafpliotis [6] proposed a tournament selection scheme based on Pareto dominance. Instead of limiting the comparison to two individuals, a number of other individuals (usually about 10) in the population are used to help determine dominance. Whether the competitors are dominated or non-dominated, the result is decided through fitness sharing.

Zitzler and Thiele [7] suggested an elitist multi-criterion EA with the concept of non-domination in their *Strength Pareto Evolutionary Algorithm (SPEA)*. They suggested maintaining an external population at every generation storing all non-dominated solutions discovered so far beginning from the initial population. This external population participates in genetic operations. At each generation, a combined population with the external and the current population is first constructed. All non-dominated solutions in the combined population are assigned a fitness based on the number of solutions they dominate, and dominated solutions are assigned fitness worse than the worst fitness of any non-dominated solution. This fitness assignment assures that the search is directed towards the non-dominated solutions. A deterministic clustering technique is also used to maintain diversity among non-dominated solutions. Although the implementation suggested is $O(mN^3)$, with appropriate bookkeeping the complexity of SPEA can be reduced to $O(mN^2)$. An improved version of SPEA known as SPEA2 [13] has recently been proposed. This incorporates additionally fine-grained fitness assignment strategy, a density estimation technique, and an enhanced archive truncation method.

Knowles and Corne [14] suggested a simple MOEA using an evolutionary strategy (ES). In their *Pareto-Archived ES (PAES)*, a parent and a child are compared. If the child dominates the parent, the child is accepted as the next parent and the iteration continues. On the other hand, if the parent dominates the child, the child is discarded and a new mutated solution (a new child) is found. However, if the child and the parent do not dominate each other, the choice between the child and the parent considers the second objective of maintaining diversity among obtained solutions. To achieve this diversity, an archive of non-dominated solutions is created. The child is compared with the archive to determine whether it dominates any member of the archive. If so, the child is accepted as the new parent and the dominated solution is eliminated from the archive. If the child does not dominate any member of the archive, both parent and child are examined for their proximity to the solutions of the archive. If the child resides in an uncrowded region in the parameter space among the members of the archive, it is accepted as a parent and a copy is added to the archive. Knowles and Corne later, suggested a *multiparent PAES* with similar principles to the above. The authors have calculated the worst case complexity of PAES for N evaluations as $O(amN)$ where a is the archive length. Since the archive size is usually chosen proportional to the population size N, the overall complexity of the algorithm is $O(mN^2)$.

3 Single Front Genetic Algorithms

The *Single Front Genetic Algorithm* (SFGA) [15] (figure 2.a) , implements a super elitist procedure in which only the non-dominated (and well-diversified) individuals in the current population are copied to the mating pool for recombination purposes and all non-dominated individuals in the current population are copied to the next population (see Figure 2.a). The rest of the individuals required to complete the population are obtained by recombination and mutation of these non-dominated individuals. The preservation of diversity in the population is ensured by means of a filtering function, which prevents the crowding of individuals by removing individuals according to a given *grid* in the *objective* space. The filtering function uses the distance evaluated in the objective space. This approach has proved very effective when applied to Zitzler test functions in comparison to other similar algorithms that use a more complex selection scheme to produce the mating pool

In the *New Single Front Genetic Algorithm* (NSFGA) [16] (Figure 2.b), some features have been added to the original SFGA. Firstly an external archive keeps track of the best current solutions found during the running of the algorithm. A selection procedure produces a mating pool of size S by randomly choosing individuals from the external set and the filtered current population. The variation operators produce the offspring set that is copied to the next population. The updating procedure adds the first front of non-dominated individuals to the current population and deletes the dominated individuals from the archive.

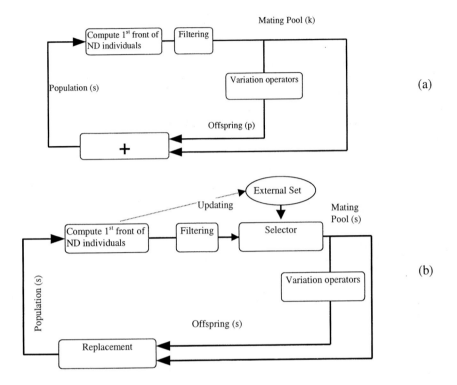

Fig. 2. Description of SFGA (*a*) and NSFGA (*b*)

4 The Problem of PAF Diagnosis

A public database for PAF diagnosis applications is available [17], comprising registers obtained from 25 healthy individuals and 25 patients diagnosed with PAF. In this paper, multi-objective optimisation is applied to improve the ability to automatically discriminate registers of these two groups with a certain degree of accuracy. For this purpose, 48 parameters were extracted from each ECG register [18] to obtain a 48 component vector that characterizes each subject ($p_1 \ldots p_{48}$). We have implemented a diagnosis scheme based on weighted threshold dependent decision rules that the multi-objective optimisation procedure will determine. For each parameter we can apply 4 different decision rules:

$$\text{If } p_i < U_{i(Low\_1)} \text{ then } C_{PAF} = C_{PAF} + W_{i1}$$
$$\text{If } p_i < U_{i(Low\_2)} \text{ then } C_{PAF} = C_{PAF} - W_{i2}$$

$$\text{If } p_i > U_{i(High\_1)} \text{ then } C_{PAF} = C_{PAF} + W_{i3}$$
$$\text{If } p_i > U_{i(High\_2)} \text{ then } C_{PAF} = C_{PAF} - W_{i4}$$

Where U represents different thresholds, C_{PAF} is a level that will determine the final diagnosis and the weights (W_{ij}) are constrained in the interval [0,1]. Thus, the first and third rules increment C_{PAF} (leading to a positive PAF diagnosis) while the second and fourth rules decrement C_{PAF} (leading to a negative PAF diagnosis).

In the diagnosis procedure, the C_{PAF} level is finally compared with a security interval [-F, F]. If C_{PAF} remains within this interval we consider there is not enough certainty about the diagnosis and we leave this case undiagnosed. If $C_{PAF} > F$ the subject is diagnosed positive (as PAF patient), while if $C_{PAF} < F$ it is diagnosed negative.

The multi-objective optimisation procedure uses two objectives to be applied to this diagnosis application: the Classification Rate (CR) and the Coverage Level (CL), defined in (5) and (6), respectively.

$$CR = \frac{Correct\ Diagnosed\ Cases}{Number\ of\ Diagnosed\ Cases} \quad (5)$$

$$CL = \frac{Number\ of\ Diagnosed\ Cases}{Total\ Number\ of\ Cases} \quad (6)$$

High levels of F increment CR because the certainty of the diagnosis rises but it leads to a low CL because it leaves more undiagnosed cases.

This approach could take into consideration 48 parameters with 4 decision rules associated with each of them. In each decision rule the variables to be fixed are the threshold (U) and the weight (W). Therefore we can configure a chromosome with 192 weights and 192 thresholds to be optimised.

In order to reduce the complexity of the optimisation problem, and after a statistical study, an expert selected a subset of 32 rules and their associated 32 thresholds that apparently maximize the discrimination power. In this way, the chromosome length is reduced to 32 (weights) if the expert thresholds are adopted or to 64 if weights and thresholds are optimised.

5 Experimental Results

For performance comparison, the hypervolume metric [7] for *maximization* problems has been used (see Table1). For the sake of simplicity just S metric is shown. S(A) is the volume of the space that is *dominated* by the pareto optimal solution set A. All of the algorithms were executed with the same initial population. The filter parameter, ft, is set to 0.01, the mutation probability per gene is 0.01 and the crossover probability is 0.6. Each Algorithm is executed for 1000 iterations and the population size is set to 200. In Fig 3-5 are shown the solutions in the objective space: CR (Vertical Axis) *versus* CL (horizontal axis) for all SFGA, NSFGA and SPEA. We have considered two cases:

(1) PAF diagnosis: optimisation of weights of decision rules given by an expert.
In this case the chromosome length is 32, and only the weights associated with the decision rules given by an expert are optimised (see Figures 3a-5a)

(2) *PAF diagnosis: optimisation of threshold and weights of the decision rules given by an expert.* In this case the chromosome length is 64. The parameters to be optimised are the 32 weights and 32 thresholds for the rules selected by an expert (see Fig. 3b-5b). Solutions found by an expert are available for F=1 and included in Fig.4b.

6 Concluding Remarks

In this paper the NSFGA, SFGA and SPEA are applied to the PAF diagnosis problem. All algorithms show a similar performance. The experiments show that NSFGA and SPEA provide very similar results, and that better results are obtained when both threshold and weights are optimised (SPEA$_{64}$, NSFGA$_{64}$ in Table 1). Moreover, our procedure has improved SPEA results in some cases (threshold given by an expert with F=1.5 and threshold not given by an expert and F=1). The application of multi-objective optimisation to PAF diagnosis is a new approach to tackle this problem. This procedure achieves classification results that are similar to classic schemes but searches optimised values for weights and thresholds that would otherwise have to be given by an expert. In this way, the integration of new rules is made automatic by applying EAs. Furthermore, these MO optimisation schemes lead to multiple solutions that can be of interest for certain patients who suffer from other disorders. In these cases some decision rules may become unreliable and certain solutions are more suitable than others.

Acknowledgements. This paper has been supported by the Spanish *Ministerio de Ciencia y Tecnología* under grant TIC2000-1348.

Table 1. Performance of the algorithms for the 32-gene and 64-gene problems

F	SFGA$_{32}$	NSFGA$_{32}$	SPEA$_{32}$	SFGA$_{64}$	NSFGA$_{64}$	SPEA$_{64}$
0.5	0.82016	0.81849	**0.83010**	0.92555	0.93026	**0.96605**
1	0.80538	0.79240	**0.81015**	0.94943	**0.96698**	0.95733
1.5	0.77278	**0.78852**	0.78331	0.86934	0.89811	**0.92828**

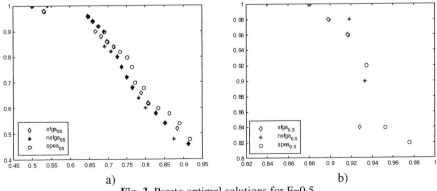

Fig. 3. Pareto optimal solutions for F=0.5

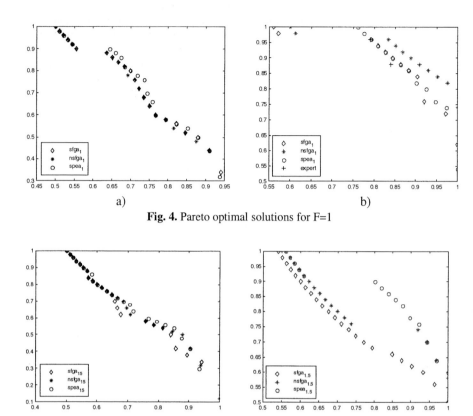

Fig. 4. Pareto optimal solutions for F=1

Fig. 5. Pareto optimal solutions for F=1.5

References

1. Carlos A. Coello Coello. An Updated Survey of GA-Based Multiobjective Optimization Techniques, Technical Report Lania-RD-98-08, Laboratorio Nacional de Informática Avanzada (LANIA), 1998.
2. Parks, G.T. and I. Miller. *"Selective breeding in a multiobjective genetic algorithm"*. In A.E. Eiben, T. Bäck, M. Schoenauer, and H.-P. Schwefel (Editos).5th International Conference on Parallel Problem Solving from Nature (PPSN-V), Berlin, Germany, pp.250–259. Springer.
3. Fonseca, C.M. and P.J. Fleming. Multiobjective optimization and multiple constraint handling with evolutionary algorithms-part i: A unified formulation. IEEE Transactions on Systems, Man, and Cybernetics 28(1), 38–47.
4. Carlos M. Fonseca and Peter J. Fleming. *Genetic Algorithms for Multiobjective Optimization: Formulation, Discussion and Generalization*, In Stephanie Forrest, editor, Proceedings of the Fifth International Conference on Genetic Algorithms, pages 416–423, San Mateo, California, 1993. University of Illinois at Urbana-Champaign, Morgan Kauffman Publishers.

5. N.Srinivas and Kalyanmoy Deb, *"Multiobjective Optimization Using Nondominated Sorting in Genetic Algorithms"*. Evolutionary Computation, Vol. 2, No. 3, pages 221–248
6. Jeffrey Horn and Nicholas Nafpliotis.*Multiobjetive Optimization Using the Niched Pareto Genetic Algorithm*. Proceedings of the first IEEE Conference on Evolutionary Computation. Vol 1, 1994.
7. Eckart Zitzler, Kalyanmoy Deb, and Lothar Thiele. *Comparison of Multiobjective Evolutionary Algorithms: Empirical Results*, Technical Report 70, Computer Engineering and Networks Laboratory (TIK), Swiss Federal Institute of Technology (ETH) Zurich, Gloriastrasse 35, CH-8092 Zurich, Switzerland, December 1999.
8. Kalyanmoy Deb, Samir Agrawal, Amrit Pratab, and T. Meyarivan. A Fast Elitist Non-Dominated Sorting Genetic Algorithm for Multi-Objective Optimization: NSGA-II, KanGAL report 200001, Indian Institute of Technology, Kanpur, India, 2000.
9. http://www.cinc.org/LocalHost/CIC2001_1.htm
10. Goldberger AL, Amaral LAN, Glass L, Hausdorff JM, Ivanov PCh, Mark RG, Mietus JE, Moody GB, Peng CK, Stanley HE. PhysioBank, PhysioToolkit, and Physionet: Components of a New Research Resource for Complex Physiologic Signals. *Circulation* 101(23):e215-e220 [Circulation Electronic Pages; http://circ.ahajournals.org/cgi/content/full/101/ 23/ e215]; 2000 (June 13).
11. D.E. Goldberg, *"Genetic Algorithms in Search, Optimization and Machine Learning"*, New York: Addison Wesley, 1989.
12. B.Sareni and L. Krähenbühl, *"Fitness Sharing and Niching Methods Revisited"*. IEEE Transaction on Evolutionary Computation, Vol 2, No. 3, 1998.
13. Eckart Zitzler, M. Laumannas; L.Thiele. SPEA2: Improving the Strengh Pareto Evolutionary Algorithm. TIK – Report 103, May 2001.
14. Corne, D.W., Knowles, J.D., and Oates, M.J. (2000). The Pareto envelope-based selection algorithm for multiobjective optimization. Proceedings of the Parallel Problem Solving from Nature VI Conference, pp. 839–848.
15. F. de Toro; J.Ortega.; J.Fernández; A.F.Díaz. PSFGA: A parallel Genetic Algorithm for Multiobjective Optimization. 10th Euromicro Workshop on Parallel and Distributed Processing. Gran Canaria, January 2002
16. F. de Toro; E.Ros, S.Mota, J. Ortega: Multiobjective Optimization Evolutionary Algorithms applied to Paroxismal Atrial Fibrillation diagnosis based on the k-nearest neighbours classifier. Lecture Notes in Artificial Intelligence, Vol 2527, pp. 313–318, November 2002.
17. http://physionet.cps.unizar.es/physiobank/database/afpdb/
18. Mota S., Ros E., Fernández F.J., Díaz A.F., Prieto, A.: ECG Parameter Characterization of Paroxysmal Atrial Fibrillation. 4th International Workshop on Biosignal Interpretation (BSI2002), 24th -26th June , 2002, Como, Italy.

Intensity Modulated Beam Radiation Therapy Dose Optimization with Multiobjective Evolutionary Algorithms

Michael Lahanas[1], Eduard Schreibmann[1,2], Natasa Milickovic[1], and Dimos Baltas[1,3]

[1] Department of Medical Physics and Engineering, Strahlenklinik, Klinikum Offenbach, 63069 Offenbach, Germany.
[2] Medical Physics Department, Medical School, Patras University,
[3] Institute of Communication and Computer Systems, National Technical University of Athens, 15773 Zografou, Athens, Greece.

Abstract. We apply the NSGA-II algorithm and its controlled elitist version NSGA-IIc for the intensity modulated beam radiotherapy dose optimization problem. We compare the performance of the algorithms with objectives for which deterministic optimization methods provide global optimal solutions. The number of parameters to be optimized can be up to a few thousands and the number of objectives varies from 3 to 6. We compare the results with and without supporting solutions. Optimization with constraints for the target dose variance value provides clinical acceptable solutions.

1 Introduction

Every year more than one million patients only in the United States will be diagnosed with cancer. Half of these will be treated with radiation therapy [1]. In teletherapy or external radiotherapy beams of penetrating radiation are directed at the tumor. Along their path through the patient body the beams deposit energy. Cancer cells have a smaller probability than healthy normal cells to survive the radiation damage. The dose is the amount of energy deposited per unit of mass. The physical and biological characteristics of the patient anatomy and of the source, such as intensity and geometry are used for the calculation of the dose function, i.e. the absorbed dose as a function of the location in the body. A physician depending on the patient, the size etc. prescribes the so called desired dose function. The objectives of dose optimization are to deliver a sufficient high dose in the cancerous tissue and to protect the surrounding normal tissue (NT) and sensitive structures from excessive radiation. The problem is to determine a intensity distribution for the radiation sources so that the resulting dose function is equal to the desired dose function. The calculation of the dose function for a given intensity distribution is possible with a high accuracy, whereas the inverse problem, i.e. the determination of the intensity distribution for a given dose function is with some exceptions not possible as the inverse dose operator produces non-physical solutions with negative intensities. Optimization algorithms are used to minimize the difference between the desired and the obtained dose function.

In the past the multiobjective (MO) dose optimization problem has been transformed into a single objective problem using as a score function the weighted sum of the individual objective functions. The weights, called also importance factors, have been determined by trial and error methods. The treatment planner was required to often repeat the optimization with other importance factors until some satisfactory solution was obtained. We have recognized that a better method is to produce a representative set of solutions and to select out of these the best possible. Even if methods have been proposed to select automatically optimal weights [2], these require additional importance factors which are *a priori* not known. The MO optimization method is important as it provides the treatment planner information about the trade-off between the objectives and the limitations of the available solutions. Gradient based optimization algorithms can be used only for variance based objectives [3]. The most used optimization method is simulated annealing (SA)[4] which can be applied for all types of objectives but requires a very large number of iterations and it is practical not possible in clinical relevant time to produce a sufficient large number of solutions. We used in the past MO evolutionary algorithms for dose optimization in brachytherapy [5], which is another radiation based cancer treatment method. The resulting dose distribution must satisfy similar objectives. In radiotherapy the sources are outside the patients body and the problem is to find the beam directions and intensity of the beams so that the resulting dose distribution satisfies various criteria. The number of parameters is much larger than in brachytherapy and it can be as large as 2000 - 10000. Previous methods used in radiotherapy include SA, iterative approaches and filtered back-projection [6]. We applied successfully gradient based optimization algorithms which are fast enough to produce a large number of solutions [3]. Problems such as the selection of beams in two dimensions have been considered with MO evolutionary algorithms by Haas *et al* [7],[8]. Various single objective genetic algorithms have also been used. Knowles *et al* [9],[10]used EA for training neural networks which bypass the optimization problem by learning from previous optimization results.

From our experience with deterministic MO algorithms we know that high quality solutions can only be obtained by analyzing the trade-off information provided by a representative non-dominated set. We use realistic three-dimensional cases with a large number of optimization parameters. We study the possibility of the use of MO evolutionary algorithms for the intensity modulated beam radiotherapy (IMRT) dose optimization problem.

2 Methods

2.1 Intensity Modulated Beam Radiotherapy

In IMRT each beam is divided in a number of small beamlets (bixels), see Fig. 1. The intensity of each beamlet can individually be adjusted. The geometry of the planning target volume (PTV) which includes the tumor and a margin and organs at risk (OAR) is specified by contours. Tools can be used to select the number of beams and their directions based on geometric criteria. The dose at each sampling point in the PTV and other structures is calculated from the contributions of each beamlet of each beam and its corresponding intensity. A sparse dose matrix is precalculated and contains the dose value at each sampling point from each bixel with a unit radiation intensity. The intensity (weights) of the beamlets have to be determined such that the produced dose distribution is "optimal".

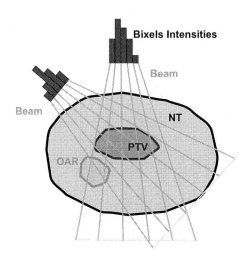

Fig. 1. Principle of IMRT dose optimization. The contours of the body, the PTV and one OAR are shown. The problem is to determine the intensities of the tiny subdivisions (bixels) of each beam, so that the resulting dose distribution is optimal

Radiation oncologist use for the evaluation of the dose distribution quality a cumulative dose volume histogram (DVH) for each structure (PTV, NT or OARs), which displays the fraction of the structure that receives at least a specified dose level. If the objectives are expressed in terms of DVHs related values, then the objectives are called DVH based objectives.

2.2 Variance Based Objectives

The optimization goals can be expressed also with variance based objectives which are only indirect related to the DVHs values but they are used because deterministic gradient based optimization algorithms can be applied. The objective functions are: for the PTV the dose variance f_{PTV} around the prescription dose D_{ref}, for NT the sum of the squared dose values f_{NT} and for each OAR the variance f_{OAR} for dose values above a specific critical dose value D_{cr}^{OAR}.

$$f_{PTV} = \frac{1}{N_{PTV}} \sum_{j=1}^{N_{PTV}} (d_j^{PTV} - D_{ref}^{PTV})^2 \qquad (1)$$

$$f_{NT} = \frac{1}{N_{NT}} \sum_{j=1}^{N_{NT}} (d_j^{NT})^2 \qquad (2)$$

$$f_{OAR} = \frac{1}{N_{OAR}} \sum_{j=1}^{N_{OAR}} \Theta(d_j^{OAR} - D_{cr}^{OAR})(d_j^{OAR} - D_{cr}^{OAR})^2 \qquad (3)$$

$\Theta(x)$ is the Heaviside step function. d_j^{PTV}, d_j^{NT} and d_j^{OAR} are the calculated dose values at the j-th sampling point for the PTV, the NT and each OAR respectively. N_{PTV}, N_{NT} and N_{OAR} are the corresponding number of sampling points.

2.3 L-BFGS

We generate a representative set of non-dominated global optimal solutions with the L-BFGS algorithm [11]. This deterministic gradient based optimization algorithm requires derivatives of the objective function. We produce a representative set of solutions using a weighted sum $f(x)$ of the M single-objective functions $f_j(x)$, $j = 1, \ldots, M$. Normalized and uniformly distributed weights w_j are taken from the set of importance vectors W.

$$f(x) = \sum_{j=1}^{M} w_j f_j(x) \qquad (4)$$

$$W = \left\{ [w_1, \ldots, w_M]; \sum_{j=1}^{M} w_j = 1; w_i \in [\frac{0}{k}, \frac{1}{k}, \ldots, \frac{k-1}{k}, 1] \right\} \qquad (5)$$

We call k the sampling parameter. L-BFGS is especially suited for high-dimensional problems as for N parameters it uses indirectly an approximation of the Hessian matrix using only an order of $5N$ instead N^2 operations. A representative set of solutions is generated by repeating the optimization with L-BFGS each time with a different vector of importance factors from the set W. The stopping criteria are 300 iterations or a tolerance value of 10^{-6}.

2.4 NSGA-II and NSGA-IIc

For the MO evolutionary optimization we use the non-dominated sorting genetic algorithm NSGA-II [12] and the controlled elitist version NSGA-IIc [13]. The population of these algorithms can be initialized by a specific number of solutions generated with L-BFGS. We call these solutions *supporting solutions* [14],[15]. This is necessary as the number of objectives depending on the number of OARs to be considered is in the range 3-6. The calculation of the objective values requires a significant fraction of the optimization time due to the large number of sampling points where the dose has to be calculated. It is practical not possible to allow the algorithm to evolve for thousands of generations. In clinical practice time is important and a representative set of the Pareto front has to be found in a few minutes. In order to obtain a sufficient large set of solutions we archive all non-dominated solutions found during the optimization. Dominated solutions are removed from this archive. This external archived population is not used in the optimization directly like in the PAES [16] or SPEA [17] algorithm. We include a comparison of NSGA-II with its controlled elitist version NSGA-IIc where a lateral diversity is kept by allowing a sufficient number of individuals to survive in various non-dominated fronts. The distribution is specified by the geometric parameter G. For the crossover operator we use simulated binary crossover SBX [18], [19]. This operator produces near parent solutions with increasing probability as the population evolves. The probability of generating near parent solutions increases with a parameter, the distribution index η_c. For mutation we use a similar operator specified by the corresponding index η_m. The mutation and crossover probability used was 0.01 and 0.9 respectively.

3 Results

3.1 Comparison of Unconstrained Optimization with NSGA-II and NSGA-IIc

We compare the results from L-BFGS with NSGA-II and NSGA-IIc for two clinical cases. The first is a brain tumor case, see Fig. 2. We consider only a two-dimensional dose optimization, i.e. only the dose distribution in a slice is optimized. Four beams are used and each beam is divided in 22 bixels. The number of parameters to be optimized is 88. We have five objectives by considering the PTV, the left and right lobus temporalis, the left eye and the NT, i.e. the brain as OARs. We use this clinical case to study the dependence of the evolutionary algorithms results on the population size, number of generations and other genetic parameters. The second case is a prostate tumor case, see Fig. 3. Again four beams are used. Each beam is divided in 22·22 bixels. The number of parameters for a full three-dimensional optimization is 1936. The dose is calculated at 15000 sampling points. For this case we have with the PTV the NT and two OARs four objectives. We use this case in order to look at the performance of NSGA-II, if the number of parameters is very large. The calculations were performed using a 933 MHz Intel III Windows NT computer with 512 MB RAM.

The population size was 200. For the crossover and mutation we set $\eta_c = \eta_m = 10$. The number of accumulated non-dominated solutions in the archive obtained by NSGA-IIc as a function of the geometric factor G after 200 generations is shown in Fig. 4. A maximum at 0.6 is observed which is close to G=0.65 found in [20] for a completely different problem. We use therefore G=0.65 for

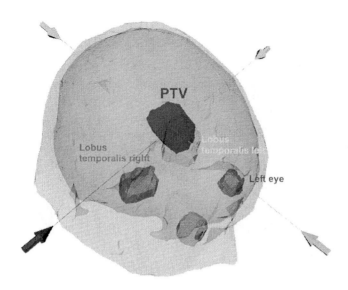

Fig. 2. 3D anatomy of the head tumor case. The PTV and the OARs are shown. The orientation of the four beams is shown

NSGA-IIc. The archived population from NSGA-II and NSGA-IIc without supporting solutions after 1000 generations is shown in Fig. 5. NSGA-IIc covers a larger part of the Pareto front found by L-BFGS, whereas for NSGA-II stagnation or premature convergence is observed.

The distribution of the archived non-dominated solutions of NSGA-II and NSGA-IIc with and without support is shown after 200 generations in Fig. 6 for the two out of ten two-dimensional projections of the Pareto front. The number of supported solutions were 35 (k=3) generated by L-BFGS. Without the supporting solutions parts of the Pareto front are not accessible even after a few thousand generations. Even if the supporting solutions improve the performance of NSGA-II its controlled elitist version is superior in the coverage and number of accumulated non-dominated solutions.

The six two-dimensional projections for the prostate tumor case are shown in Fig. 7. The results of NSGA-II and NSGA-IIc with supporting solutions are compared with the corresponding 2380 generated L-BFGS solutions. The population size was increased to 1000 for this high dimensional problem. The sampling parameter $k = 4$ was used with corresponds to 35 supported solutions.

3.2 Constraint Optimization with NSGA-II

In IMRT only a part of the Pareto front is of practical interest. The dose variance in the PTV for clinical acceptable solutions is very small whereas for the OARs and the NT it can be much larger. It is important to apply constraints for the value of f_{PTV}.

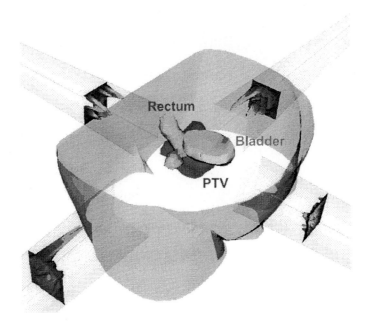

Fig. 3. 3D anatomy of the prostate tumor case. The PTV and the two OARs are shown. The orientation of the four beams is shown together with the intensity profile of the beams for one selected solution

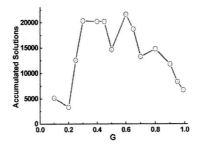

Fig. 4. Number of archived non-dominated solutions of NSGA-IIc as a function of the geometric factor G

3.3 Comparison of Deterministic and Evolutionary Optimization Results

We compare the spectrum of solutions (non-dominated set) obtained by a sequential application of L-BFGS and constrained optimization with the supported NSGA-II algorithm. Our current implementation of NSGA-IIc does not support constraint optimization. The optimization with NSGA-II is performed with a

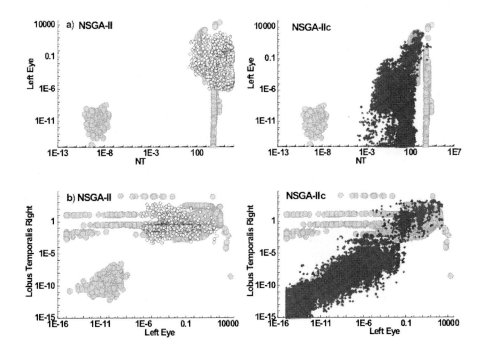

Fig. 5. Two examples of two-dimensional projections of the Pareto front of the archived non-dominated solutions for the five-dimensional dose optimization problem for the brain case. The solutions of NSGA-II and NSGA-IIc after 1000 generations are shown. The result of L-BFGS is included. The objectives values shown are the dose variances in organs at risk

population size of 500 and 100 generations. The resulting 2D projections of the Pareto front using 30 supported L-BFGS solutions is shown in Fig. 8. The result is compared with the Pareto front obtained by L-BFGS with 815 solutions. The calculation time for L-BFGS was 3860 s and for NSGA-II 271 s. We perform a 3D-dimensional dose optimization for the brain tumor with 856 bixels and nine beams. A display table with a list of objectives and dosimetric values of all solutions is used for the selection of the best solution. Constraints on these values can be applied. Solutions that satisfy these constraints are marked and their corresponding DVHs highlighted.

The spectrum of DVHs of solutions from NSGA-II is shown in Fig. 9 for the PTV, the left and right eye. The best solution has been selected as the solution with the smallest product of dose variances for the NT and the eyes. This solution obtained from the L-BFGS algorithm and NSGA-II is marked in Fig. 9.

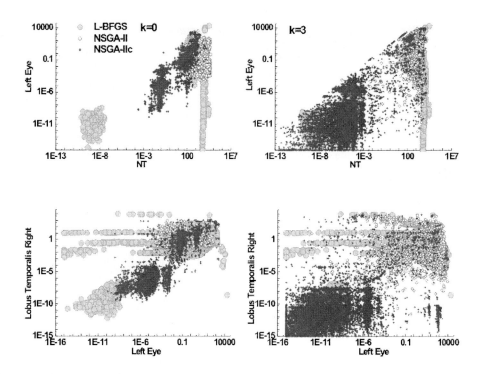

Fig. 6. The result as in Fig. 5 but only after 200 generations. The result from NSGA-II and NSGA-IIc with and without supporting solutions is shown

4 Discussion and Conclusions

In the past many single-objective optimization methods have been proposed for the dose optimization problem in radiotherapy. These methods were single-objective whereas the dose optimization problem is a MO problem. Gradient based optimization algorithms can be used with variance based objectives but if other objectives such as radiobiological or DVH based are used then stochastic algorithms such as SA are not efficient to produce a representative set of solutions. With a support of a small fraction of solutions by deterministic algorithms or SA the MO algorithm NSGA-IIc is able to produce efficiently a representative set of non-dominated solutions. The supporting solutions are necessary to guide a fraction of the population in parts of the objective space which are accessible only after a very large number of generations with Pareto ranking algorithms. A similar idea uses the genetic local search algorithm MOGLS [21] which using a scalarization of the individual objective functions performs a repeated local search in randomly specified directions. Also the local search hybrid method used with NSGA-II in [22] increased its performance. The supported solution approach is applied only at the start of the algorithm and requires a small number of solutions to be initialized, whereas the local search approach would require

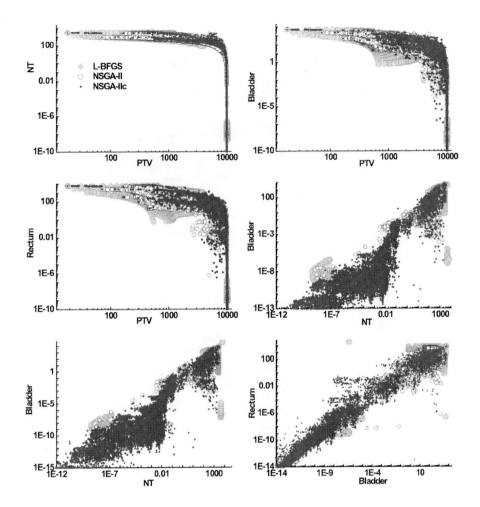

Fig. 7. Example of a the six two-dimensional projections of the four-dimensional Pareto front for the prostate tumor case. The result from L-BFGS, NSGA-II and NSGA-IIc with supported solutions

for the high dimensional and MO problem a very large number of function evaluations. The number of supported solutions depends on the number of objectives. A sampling parameter at least $k = 3$ is required.

The archiving of non-dominated solutions found is necessary in order to have a sufficient large set of representative solutions for the large number of objectives. It additionally allows the population size to be reasonable small and reduces the optimization time in order to obtain an approximate similar number of solutions as the archive contains. The number of accumulated solutions in 200 generations was 15000-20000 for both cases studied so that finally a filter has to be applied

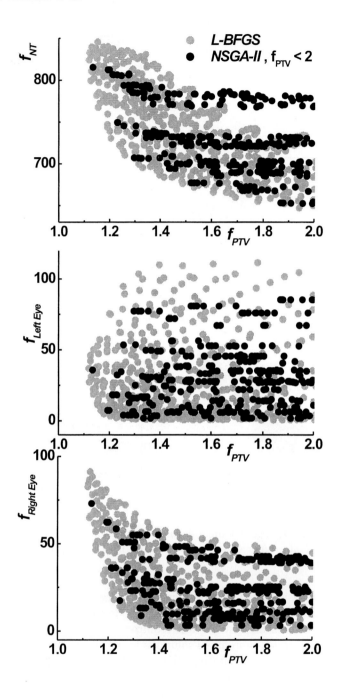

Fig. 8. Two-dimensional projections of the four-dimensional Pareto front for a brain tumor case. The result using L-BFGS and NSGA-II with supported solutions and the constraint $f_{PTV} < 2$ is shown

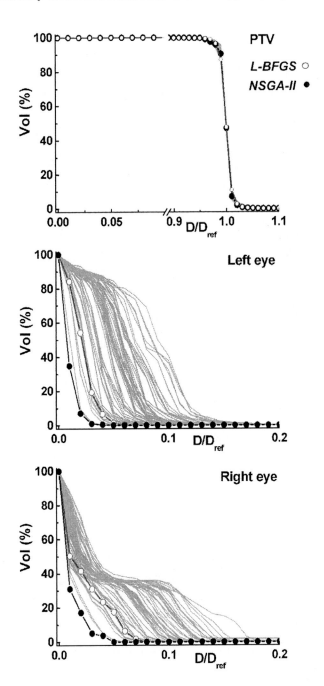

Fig. 9. Spectrum of DVHs of solutions obtained by NSGA-II for the PTV, the left and right eye for a brain tumor case. The best selected solution obtained by L-BFGS and NSGA-II is shown

to reduce the number to an acceptable level. The results show that a lateral diversity is important to avoid a premature convergence and the controlled elitist algorithm benefits more from the support. Even without support NSGA-IIc is able to approach the global Pareto front better than NSGA-II which prematurely converges. For the problem with 1936 parameters a sufficiently large population with at least 500-1000 members is required. IMRT requires the use of constraint optimization with a PTV dose variance value less than 10. The spectrum of the obtained DVHs shows that we have a true MO optimization problem. NSGA-II with constraints for the f_{PTV} value produces clinical acceptable solutions even if the best result from L-BFGS is better. MO dose optimization for IMRT requires only a few minutes, whereas L-BFGS needs more than 10 times more time to produce a comparable number of solutions. The Pareto front obtained from NSGA-II is close to the global optimal Pareto front and this distance can possibly be reduced if constrained optimization is applied using NSGA-IIc. We hope that MO optimization with evolutionary algorithms will be applied in IMRT where currently only single-objective optimization algorithms are used. MO optimization provides not only a satisfactory solution but the best possible. This reduces the dose in the OARs and in the NT to a minimum possible level.

We want to use MO evolutionary optimization algorithms for inverse planning in IMRT where the optimal number of beams and their orientation additional must be found. For readers interested in experimenting with the IMRT dose optimization problem an extension of the MOMHLIB library [23] used in this study, together with dosimetric data sets and additional information, is available at the website: www.mlahanas.de/IMRTOpt.html.

Acknowledgments. The Research was Supported through a European Community Marie Curie Training Fellowship (HPTM -2000-00011).

References

1. Peres, C. A., and Brady, L. W.: Principles and practice of radiotherapy. Lippincott-Raven, Philadelphia, 3rd edition, 1998.
2. Xing, L., Li, J.G., Donaldson, S., Le, Q.T. and Boyer, A. L.: Optimization of importance factors in inverse planning. Phys Med. Biol., **44** 2525–2536, 1999.
3. Cotrutz, C., Lahanas, M., Kappas, C. and Baltas, D.: A multiobjective gradient based dose optimization algorithm for conformal radiotherapy. Phys. Med. Biol. **46** 2161–2175, 2001.
4. Webb, S.: Optimization of conformal radiotherapy dose distributions by simulated annealing. Phys. Med. Biol., **34** 1349–1370, 1990.
5. Lahanas, M., Baltas, D. and Zamboglou, N.: A hybrid evolutionary algorithm for multiobjective anatomy based dose optimization in HDR brachytherapy, to be published in Phys. Med. Biol. 2003.
6. Bortfeld, T., Ürkelbach, J., Boesecke, R. and Schlegel, W.: Methods of image reconstruction from projections applied to conformation therapy, Phys. Med. Biol. **35** 1423–1434, 1990.

7. Haas, O. C. L., Burnham K. J. and Mills J. A.: On Improving the selectivity in the treatment of cancer: a systems modelling and optimisation approach . J. Control Engineering Practice, **5** 1739–45, 1997
8. Haas, O. C. L.: Radiotherapy Treatment Planning: New System Approaches. Springer Verlag London, Advances in Industrial Control Monograph, ISBN 1-85233-063-5, 1999.
9. Knowles, J. D., Corne, D. and Bishop J. M.: Evolutionary Training of Artificial Neural Networks for Radiotherapy Treatment of Cancers in Proceedings of the 1998 IEEE International Conference on Evolutionary Computation IEEE Neural Networks Council, 0-7803-4871-0, pp 398–403, pp. 398–403
10. Knowles, J. D. and Corne, D.: Evolving Neural Networks for Cancer Radiotherapy, in Chambers, L.(ed.), Practical Handbook of Genetic Algorithms: Application 2nd Edition, Chapman Hall/CRC Press, pp. 443–448. ISBN L-58488-240-9, 2000.
11. Liu, D.C., and Nocedal, J.: On the limited memory BFGS method for large scale optimization. Mathematical Programming **45** 503–528, 1989.
12. Deb, K., Agrawal, S., Pratap, A., and Meyarivan, T.: A fast and elitist multiobjective genetic algorithm: NSGA-II. Technical Report 20001, Indian Institute of Technology, Kanpur, Kanpur Genetic Algorithms Laboratory (KanGAL), 2000.
13. Deb, K. and Goel, T.: Controlled elitist non-dominated sorting genetic algorithms for better convergence, in Proceedings of the first international conference. EMO 2001, Zurich, Switzerland, edited by E. Zitzler, K. Deb, L. Thiele, C. A. Coello Coello, D. Corne, Lecture Notes in Computer Science Vol. 1993, Springer 67–81, 2001
14. Gandibleaux, X., Morita, H. and Katoh, N.: The Supported Solutions Used as a Genetic Information in Population Heuristic, in Proceedings of the first international conference. EMO 2001, Zurich, Switzerland, edited by E. Zitzler, K. Deb, L. Thiele, C. A. Coello Coello, D. Corne, Lecture Notes in Computer Science Vol. 1993, Springer 429–42. 2001.
15. Milickovic, N., Lahanas, M., Baltas, D. and Zamboglou, N.: Comparison of Evolutionary and Deterministic Multiobjective Algorithms for Dose Optimization in Brachytherapy, in Proceedings of the first international conference. EMO 2001, Zurich, Switzerland, edited by E. Zitzler, K. Deb, L. Thiele, C. A. Coello Coello, D. Corne, Lecture Notes in Computer Science Vol. 1993, Springer 167–180. 2001.
16. Knowles, J.D., Corne, D.W.: Approximating the nondominated front using the Pareto Archived Evolution Strategy. Evolutionary Computation **8** 149–172, 2000.
17. Zitzler, E., Thiele, L.: Multiobjective Evolutionary Algorithms: A Comparative Case Study and the Strength Pareto Approach. IEEE Transactions on Evolutionary Computation **37** 257–271, 1999.
18. Deb, K. and Agrawal, R.B.: Simulated binary crossover for continuous search space. Complex Systems **9** 115–148, 1995.
19. Deb, K. and Beyer, H.G.: Self-Adaptive genetic Algorithms with Simulated Binary Crossover. Evolutionary Computation **9** 197–221, 2001.
20. Deb, K., Multi-Objective Optimization using Evolutionary Algorithms, Chichester, Wiley, UK, 2001.
21. Jaszkiewicz, A. Genetic local search for multiple objective combinatorial optimization, Technical Report RA-014/98, Institute of Computing Science, Poznan University of Technology, 1998.
22. Goel, T. and Deb, K.: Hybrid Methods for Multi-Objective Evolutionary Algorithms. KanGAL Report Number 2001004, 2001.
23. http://www-idss.cs.put.poznan.pl/~jaskiewicz/MOMHLIB/

Multiobjective Evolutionary Algorithms Applied to the Rehabilitation of a Water Distribution System: A Comparative Study

Peter B. Cheung[1], Luisa F.R. Reis[1], Klebber T.M. Formiga[1], Fazal H. Chaudhry[1], and Waldo G.C. Ticona[2]

[1] São Carlos School of Engineering, University of São Paulo, Brazil
Av. do Trabalhador São Carlense 400, C. P. 359, São Carlos, SP, Brazil, 13560-250
[2] Institute of Mathematics and Computer Science, University of São Paulo, Brazil
{pcheung@bol.com.br ; fernanda@sc.usp.br;
klebberformiga@uol.com.br; fazal@sc.usp.br;
wcancino@icmc.sc.usp.br}

Abstract. Recognising the multiobjective nature of the decision process for rehabilitation of water supply distribution systems, this paper presents a comparative study of two multiobjective evolutionary methods, namely, multiobjective genetic algorithm (MOGA) and strength Pareto evolutionary algorithm (SPEA). The analyses were conducted on a simple hypothetical network for cost minimisation and minimum pressure requirement, treated as a two-objective problem. For the application example studied, SPEA outperforms MOGA in terms of the Pareto fronts produced and processing time required.

1 Introduction

Most of the existing water supply distribution systems were developed to operate over a determined planning period. Along time, however, failures caused by the deterioration of pipes and hydraulic components become frequent in such systems. Besides, the increasing levels of urbanisation and demand for water lead to problems such as insufficient discharges to meet demand and low-pressure levels in the network. Thus, the decision process for rehabilitation and replacement of existing components to meet current and future demands constitutes a subject of great interest.

Improvements in a distribution system performance can be achieved through rehabilitation of some pipes or other components and/or adding new components to the existing network. Generally limited funding is available to modify the systems in order to guarantee a satisfactory level of the water supply service. Several researchers [1]-[5] have applied optimisation techniques in rehabilitation of water distribution systems, focusing on the economic considerations. Techniques such as linear, integer, non-linear and dynamic programming have been exhaustively used in water distribution system optimisation.

Many researchers [6]-[7] have pointed out the disadvantages of the conventional optimisation methods. The rehabilitation of water distribution systems is a complex and discontinuous problem with many local optima. Many conventional optimisation methods do not guarantee that the global optimum shall be found. Further, they are based on single objectives, whereas many real situations require simultaneous optimisation of multiple objectives.

Optimisation has evolved over the last years by the introduction of a number of non-conventional algorithms as the Genetic Algorithms (GAs), which mimic the evolutionary principles of nature to drive the search towards optimal solutions. One of the most striking differences between classical search methods and GAs is the use of populations of solutions instead of only one solution [8].

In single objective optimisation, this algorithm attempts to obtain the best design or operational strategy, which is usually the global minimum or maximum, depending on the nature of the problem to be solved. Based on a very different concept, a typical multiobjective method seeks for a set of solutions that are superior to the remainder solutions in the search space. This set is denominated Pareto optimal front [8]. Because GAs work with populations of points, a number of optimal solutions can be captured during their iterative search process. Thus they are naturally well-suited to treat the multiobjective problems.

Evolutionary techniques for multiobjective optimisation can be classified into several classes [9]: objective reduction approaches, classified population approaches, weight-randomising approaches, preference relationship approaches and Pareto-based approaches. In Pareto-based approaches the objectives are dealt with simultaneously however in a different manner as compared to other cited classes which need some simplification process.

Three types of implementations based on non-dominance concept of Pareto [10] were first proposed as: multiobjective GA (MOGA) [11], niched Pareto GA (NPGA) [12] and non-dominated sorting GA (NSGA) [13]. Later, many others methods are being proposed, which can be classified as non-elitist and elitist [14]-[15] and still others based on tournament approaches [16]. In spite of such variety of methods, there are few comparisons available in literature for water resources engineering problems.

Many authors [17]-[20] have used GAs applied to the water engineering problem, focusing on sizing and layout of water distribution networks. However, few studies have applied multiobjective optimisation techniques in water distribution system problems. Halhal et al. [19] and Walters et al. [18] developed a structured messy genetic algorithm for a multiobjective approach. These studies incorporate the multiple objectives into the single objective formulation using weighting factors for each objective or constraint. More recently, MOGA was applied [20] to the problem of rehabilitation of a hypothetical network, considering economic and reliability criteria.

In order to compare the performance of the non-elitist (MOGA) and elitist (SPEA) methods, this paper uses a hypothetical network from literature. EPANET 2 [21] is used for the hydraulic evaluation in terms of nodal heads and pipe discharges.

2 The Problem of Water Distribution System Rehabilitation

The performance of water networks can be improved in terms of their hydraulic capacity by cleaning, relining, duplicating or replacing existing pipes; increasing their physical integrity by replacing structurally weak pipes; increasing system flexibility by additional pipe links; improving water quality by removing or relining old pipes [19].

Generally, high costs are involved in remedial works and available economic resources are limited to implementation of such task. Thus, there is need for implementation and development of optimal rehabilitation plans since the funding must be optimally invested over the planning period.

3 Multiobjective Optimisation Model

Five objectives can be pointed out as regards the operation of water distribution networks, namely, hydraulic capacity, physical integrity, flexibility, water quality and economy, each of which can be expressed by means of several attributes, constituting a complex multiobjective problem.

In the absence of any preference information among the objectives, the goal of a multiobjective optimisation method is to arrive at a set of Pareto optimal designs. In addition to a number of Pareto optimal designs, a widely varying set of solutions is usually required to allow the decision-maker to choose from the set [22].

Classical methods are not efficient for multiobjective problems as they often lead to a single solution instead of a set of Pareto optimal solutions. Multiple runs cannot guarantee generation of different points on the Pareto front each time and some methods cannot even handle problems with multiple optimal solutions [8],[22].

Evolutionary methods, on the other hand, maintain a set of solutions as a population during the course of search and thus result in a set of Pareto optimal solutions in a single run. Widely differing Pareto optimal solutions can also be generated by using a diversification strategy within the evolutionary algorithms [8],[11].

3.1 Constraints

In GAs, one of the most important issues is the manner in which the constraints are incorporated into the fitness function to guide the search properly. In the last years, researchers [10],[23],[24] have proposed penalty functions to incorporate constraints into the fitness function. In many engineering problems this approach has been used and the results have demonstrated consistency. However, penalty functions have some known limitations of which the most significant is the difficulty of defining good penalty factors. These penalty factors are normally found by trial and error as their definition may affect the results produced by the GAs.

3.2 Problem Formulation

Water distribution systems frequently require rehabilitation (cleaning, lining, reinforcement among others) to maintain the satisfactory services for the society. However, the rehabilitation of an existing system is a complex task if it is to be implemented in the most effective and economic manner. It necessitates a systematic and thorough approach, backed up by skilful engineering judgement, and significant capital resources. The examination and evaluation for design alternatives is a field in which optimisation models can play an important role, particularly when finances are limited and the problems are of a large size [18].

Many objectives can be incorporated in rehabilitation decision models. We prefer to formulate a two-objective network rehabilitation problem in order to compare our results with those of a similar one-objective problem [17] which dealt with the remaining objective as a constraint include in objective function. Thus the present paper formulates the rehabilitation problem as that of minimisation of cost (1) as well as pressure deficit (2) considering various combinations of rehabilitation choices. The individual objectives are:

$$\text{Minimise Cost, } F_1 = \sum_{\ell \in \mathfrak{I}} c_\ell L_\ell + \sum_{k \in \pi} c_k L_k \qquad (1)$$

where ℓ is the index of the pipes to be rehabilitated (cleaned or left unaltered); k is the index of the new pipes (replaced or duplicated); $\mathfrak{I}$ is the set of alternatives related to the pipes requiring rehabilitation; π is the set of alternatives for new pipes; L is length of the pipe; c_j are rehabilitation unit costs and c_k are unit costs of new pipes. The decision problem corresponds to the identification of pipes to be added in parallel or as a new pipe.

$$\text{Minimise Pressure Deficit, } F_2 = \sum_{i=1}^{LC} \max(H_j - H_{j\min})_i \quad j=1,2,...,nn \qquad (2)$$

where pressure deficit is the sum of maximum nodal deficits on the network for each demand pattern; j is the index that represents the nodes; nn is the total number of nodes in the system; H_j is the energy and $H_{j\min}$ is the required minimum energy at node j. LC denotes the number of demand patterns considered. In this study three demand patterns shall be investigated: peak, average and minimum demands. One can observe that F_2 replaces a constraint of the rehabilitation problem.

3.3 Multiobjective Evolutionary Algorithms

Fonseca and Fleming [11] introduced multiobjective genetic algorithms where the ranking of a solution was based on non-dominated classification. The multiobjective genetic algorithm (MOGA) classifies each solution assigning a rank value that means the number of solutions that dominates it plus one. After the rank is assigned, the algorithm tries to distribute the points evenly over the Pareto optimal region using a sharing mechanism in the objective function domain. The solutions are proportionally selected and submitted to the crossover and mutation operators.

Ziztler and Thiele [14] introduced elitism by explicitly maintaining an external population. This population stores a fixed number of the non-dominated solutions that are found until the beginning of a simulation. At every generation, newly found non-dominated solutions are compared with existing external population and the resulting non-dominated solutions are preserved. This algorithm is called strength Pareto evolutionary (SPEA). It does more than just preserve the best solutions, it also uses these elite solutions to participate in the genetic operations along with the current population in the hope of influencing the population to steer towards good regions in the search space.

In this paper, these two multiobjective evolutionary algorithm techniques (MOGA and SPEA) are applied to compare non-elitist and elitist methods in the rehabilitation water distribution system problem (2), whose details can be found in [8].

3.3.1 Evolutionary Algorithms Libraries
We implemented the MOGA algorithm through a C++ code supported by the GAlib library [25] basic operators written by Matthew Wall at Massachusetts Institute of Technology.

The MOMHLib++ library [26] was used to obtain the results for the SPEA algorithm developed by Andrzej Jaszkiewicz at Poznan University of Technology.

3.4 Hydraulic Simulator Model

A steady-state hydraulic analysis was used to evaluate the consequences of a rehabilitation plan in terms of the objectives, F_1 and F_2 using the EPANET 2 [21] which was linked to our C++ code. It should be noted that EPANET 2 represents an efficient code for hydraulic calculations related to water distribution networks.

3.5 Comparative Study

The set coverage metric [8] was adopted as performance index to compare the applied methods (MOGA and SPEA). The metric is used to get an idea of the relative spread of solutions between two sets of solution vectors A and B. The set coverage metric C(A,B) calculates the proportion of solutions in B, which are weakly dominated by solutions of A (3):

$$C(A,B) = \frac{|\{b \in B | \exists a \in A : a \preceq b\}|}{|B|} \quad (3)$$

The expression $a \preceq b$ denotes that a dominates b. The metric value C(A,B) = 1 means that all the members of B are weakly dominated by A. On the other hand, C(A,B) = 0 expresses that no member of B is weakly dominated by A. Since the domination operator is not a symmetric operator, C(A,B) is not necessarily equal to 1 − C(B,A). Thus, one must calculate both C(A,B) and C(B,A) to understand how many solutions of A are covered by B and vice versa [8].

4 Application Example

The rehabilitation study of the hypothetical network in Fig.1 was initially proposed in [27]. The network in Fig.1 has 14 pipes, 2 constant level reservoirs (nodes 1 and 5) and 9 demand nodes (2, 3, 6, 7, 8, 9, 10, 11 and 12), where the solid lines represent the existing system and dashed lines depict new pipes. The pipe and node data are presented in Tables 1 and 2.

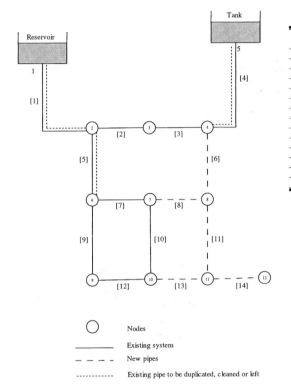

Table 1. Pipe characteristics

Pipe	Diameter (mm)	Length (m)	C_{hw}
1	356	4828	75
2	254	1609	80
3	254	1609	80
4	254	6437	80
5	254	1609	80
6	New	1609	120
7	203	1609	100
8	New	1609	120
9	254	1609	80
10	102	1609	100
11	New	1609	120
12	203	1609	100
13	New	1609	120
14	New	1609	120

Fig. 1. Hypothetical Network [27]

The problem as posed in [27] has some interesting features that include: selection of diameters for five new pipes; three existing pipes may be cleaned, duplicated, or may remain unaltered; three demand patterns are considered; and two supply sources are available, whose options are described in Tables 3 and 4. The respective costs are presented in Tables 5 and 6.

Table 2. Demand patterns, associated minimum pressures and node elevation for example network

Node	Elevation (m)	Demand Pattern 1		Demand Pattern 2		Demand Pattern 3	
		Demand (L/s)	Minimum Pressure (m)	Demand (L/s)	Minimum Pressure (m)	Demand (L/s)	Minimum Pressure (m)
1	365.76	Reservoir	-	Reservoir	-	Reservoir	-
2	320.40	12.62	28.18	12.62	14.09	12.62	14.09
3	326.14	12.62	17.61	12.62	14.09	12.62	14.09
4	323.23	0	17.61	0	14.09	0	14.09
5	371.86	Tank	-	Tank	-	Tank	-
6	298.70	18.93	35.22	18.93	14.09	18.93	14.09
7	295.66	18.93	35.22	82.03	10.57	18.93	14.09
8	292.61	18.93	35.22	18.93	14.09	18.93	14.09
9	289.56	12.62	35.22	12.62	14.09	12.62	14.09
10	289.56	18.93	35.22	18.93	14.09	18.93	14.09
11	292.61	18.93	35.22	18.93	14.09	18.93	14.09
12	289.56	12.62	35.22	12.62	14.09	50.48	10.57

Table 3. Decision options - rehabilitation

Rehabilitation Option	Real Code
Leave as existing	0
Clean existing pipe	1
Duplicate with 152 mm	2
Duplicate with 203 mm	3
Duplicate with 254 mm	4
Duplicate with 305 mm	5
Duplicate with 356 mm	6
Duplicate with 407 mm	7

Table 4. Decision options - new pipes

New Pipe Diameter (mm)	Real Code
152	0
203	1
254	2
305	3
356	4
407	5
458	6
509	7

Table 5. Rehabilitation Costs [12]

Diameter (mm)	Cleaning of Pipe ($/m)
152	47.57
203	51.51
254	55.12
305	58.07
356	60.70
407	63.00
458	-
509	-

Table 6. Costs of new pipes [12]

Diameter (mm)	New Pipe Cost ($/m)
152	49.54
203	63.32
254	94.82
305	132.87
356	170.93
407	194.88
458	232.94
509	264.10

4.1 Genetic Algorithm Implementation

It is important in GA applications to find the appropriate representation of decision variables by strings of fixed length. While many coding schemes are possible, it is convenient to avoid the decoding phase in order to reduce processing time. Several authors [23], [28] have suggested the use of real code instead of the binary one in order to keep one gene-one variable correspondence. Hence, this study has preferred the use of real code for decision variables representing the rehabilitation options to be implemented in the network to improve its hydraulic performance. The first three variables in the string refers to the decision in pipes 1, 4 and 5, for which values in the range from 0 to 7 have to be determined, according to the options defined in Table 3. The next five variables in the string refer to the decision for pipes 6, 8, 11, 13 and 14, for which values in the range from 0 to 7 have to be determined according to the options defined in Table 4.

4.1.1 Genetic Algorithm Parameters

For this application example, three population sizes were considered: 100, 300 and 500; crossover probability of 1.0 and mutation probability of 0.1, following [17], although a check of other values is desirable. In MOGA algorithm, the sharing parameter value of was assumed to be 0.5. The algorithms were permitted to run for 50,100 and 200 generations, starting from thirty different initial populations [14] of solutions (random seeds) for each population size.

5 Results and Discussions

The results obtained from the application of MOGA and SPEA methods to the example problem (Fig.1) are presented in this section.

Firstly, we performed a sensitivity analysis on the number of generations, in order to identify the consequences of this parameter in the final Pareto fronts. Several simulations were developed for both methods starting from 30 distinct initial populations, producing similar final Pareto fronts. To illustrate this behavior, Fig. 2 presents some results found for population size of 300 solutions. The analysis suggests that no premature convergence was detected. Thus, only the results obtained for 200 generations are shown in this paper for reasons of limited space.

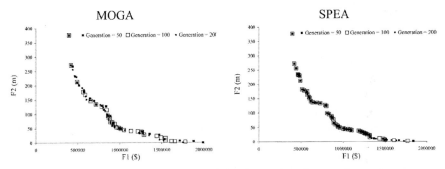

Fig. 2. Sensitivity analysis for the number of generations

Fig. 3, 4 and 5 show Pareto fronts obtained using the population sizes of 100, 300 and 500, where one can observe that SPEA produces solutions (represented by points) better defining the fronts than those met by MOGA. Considering the points in Pareto front corresponding to values for F_2 near zero as a region of special interest for our purposes, SPEA algorithm appears more efficient in defining such points.

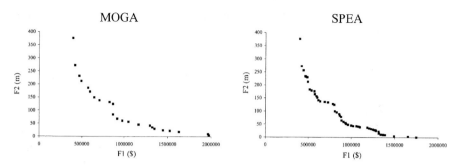

Fig. 3. Pareto fronts of population size 100 and generation number 200 of both methods

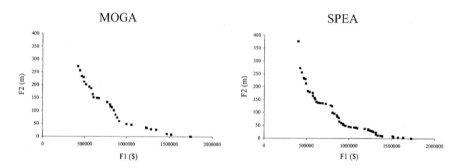

Fig. 4. Pareto fronts of population size 300 and generation number 200 of both methods

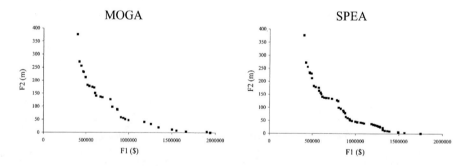

Fig. 5. Pareto fronts of population size 500 and generation number 200 of both methods

Fig. 6 presents the individual MOGA and SPEA solutions randomly chosen among thirty (30) final solutions produced for population sizes of 500 and 200 generations. These fronts were chosen for visual comparison effect, as they were obtained starting from different initial populations for both methods.

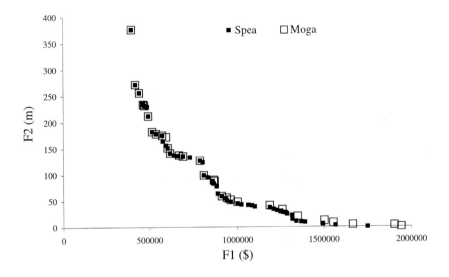

Fig. 6. Comparison of the methods for a run case

Even in this simple example of two objectives functions, the visual comparison is difficult, requiring a comprehensive metric to evaluate the relative merit of both algorithms. Once 30 runs were made for each algorithm starting from distinct initial populations, the comparisons were performed considering the combinations between all possible pairs of solutions obtained from both algorithms. The metric in (3) was used and the results presented in matrix box plot form in Fig.7, where each rectangle contain a box plot representing the distribution of the C values for all combinations (900) of pairs of two algorithms. The scale is 0 at the bottom and 1 at the top in each box plot. Each graph presents distribution of C calculated from results (A) obtained from the algorithm indicated in the row in combination with those (B) from the algorithm indicated in the column through definition in (3) for C(A,B).

Box plots are used to visualise the distribution of these samples. The upper and lower ends of the box are the upper and lower quartiles, while a thick line within the box encodes the median. Dashed appendages summarise the spread and shape of the distribution.

Fig.7 shows the direct comparison based on the measure C (3) for MOGA and SPEA methods. The Pareto fronts achieved by MOGA (population size 100, 300 and 500) are entirely dominated by fronts identified by SPEA for all the population sizes. The elitist method (SPEA) seems to perform better than the non-elitist multiobjective evolutionary algorithm (MOGA). However, there is no clear evidence of superior performance when the results obtained for several population sizes are compared among themselves for a given algorithm.

Table 7 presents the average processing times for MOGA and SPEA on Athlon XP 1.8 GHz with 512 MB RAM computer. Note that SPEA is faster than MOGA for all cases investigated.

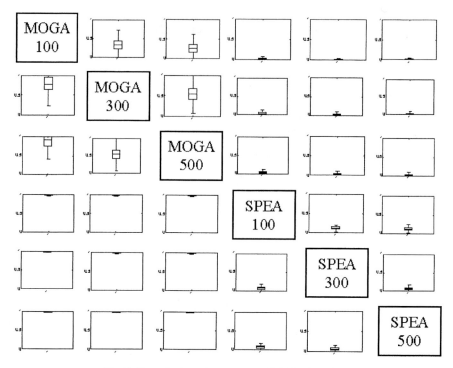

Fig. 7. Box plot based on measure C defined in (3)

Table 7. Average processing time

Population Size	Average Processing Time (Seconds)		
	Generation	MOGA	SPEA
100	50	12.4	7.6
	100	20.6	17.4
	200	47.3	30.1
300	50	36.5	26.3
	100	74.2	43.4
	200	249.6	86.5
500	50	65.2	44.2
	100	127.6	87.3
	200	249.6	165.1

Table 8 presents ten combinations of least cost alternatives with zero pressure deficit found by the SPEA algorithm. In this table, we can see the lowest value of $1,666,760 whereas [27], using selective enumeration to optimise this problem found a least cost network of $1,833,700. Simpson et al. [17] accomplished complete enumeration of all alternative solutions for each of the three demand patterns finding two least cost networks at a cost of $1,750,000. We attribute the difference in results obtained to the hydraulic simulator used by [17] based on Newton-Raphson solver. In this paper EPANET 2 was used as hydraulic simulator which is based on gradient method solver.

Table 8. Some solutions of lower costs found by SPEA algorithm

Solution	DECISION VECTOR Pipes								Objective Function	
	1	4	5	6	8	11	13	14	F1 ($)	F2 (m)
1	1	4	4	2	1	1	1	2	1,666,760	0
2	1	4	4	2	1	2	0	2	1,695,270	0
3	1	4	3	2	2	2	0	2	1,695,270	0
4	1	4	4	2	1	3	0	1	1,705,810	0
5	1	4	4	3	1	2	0	1	1,705,810	0
6	1	4	4	2	1	1	2	2	1,717,440	0
7	7	0	5	2	0	1	0	2	1,721,100	0
8	1	4	3	2	2	1	1	3	1,727,980	0
9	0	6	0	3	1	1	0	2	1,750,100	0
10	0	6	0	3	1	2	0	1	1,750,100	0

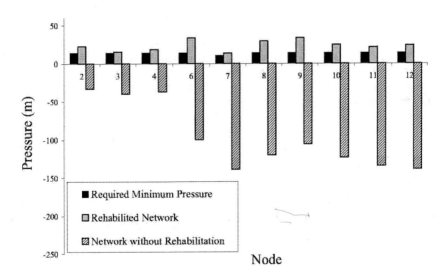

Fig. 8. Pressures in the example network considering demand pattern 2 (average)

As mentioned above, the points on Pareto front region which are of special interest, were analysed in detail from the hydraulic point of view in terms of nodal pressures. Fig. 8 reproduces the solution vector in terms of pressures corresponding to the alternative solution 1 in Table 8. The minimum pressures in Fig. 8 are the values stipulated in Table 2. The pressures in the rehabilitated network are the values obtained from the hydraulic simulations applying the decision vector of solution 1

(Table 8). The pressures in the network without rehabilitation refer to the values resulting from the hydraulic simulations assuming minimum diameters for pipes (6, 8, 11, 13 and 14) besides the existing pipes according to Table 1. One can observe that the network rehabilitation improved the hydraulic performance, making the nodal pressures acceptable for the demand patterns considered (peak, average and minimum).

6 Conclusions

This paper represents an effort to compare the performance of two approaches known as multiobjective genetic algorithm (MOGA) and strength Pareto evolutionary algorithm (SPEA) applied to the water supply network rehabilitation problem. The same problem was studied before [17] through the conventional approach that includes a penalty function in the single objective formulation.

A sensitivity analysis of the number of generations with respect to the final solutions defined in terms of Pareto fronts was conducted. Several simulations were made for both methods starting from distinct initial populations, producing similar final Pareto fronts. Direct comparison based on set coverage metric [8] shows that the Pareto fronts achieved by MOGA are entirely dominated by indicated by SPEA for various population sizes. Further, SPEA is faster than MOGA, requiring smaller processing time.

A least cost solution of $1,666,760 could be identified, whereas [17], using complete enumeration for each of the three demands, found two networks with a least cost of $1,750,000. It is presumed that this difference results from the use of a hydraulic simulator inferior to EPANET 2.

Finally, once the potentialities of the elitist multiobjective evolutionary algorithm SPEA are demonstrated, several future possibilities open for solution of engineering problems. Such possibilities include the treatment of more realistic and complex objectives than those dealt with here and comparative studies between more recent elitist methods and SPEA.

Acknowledgements. The present paper has resulted from a current research grant "Tools for the Rational and Optimised Use of Urban Water (CT-HIDRO 01/2001 – 550070/2002-8)" funded by Brazilian National Research Council (CNPq). The authors wish to express their gratitude also to CAPES and FAPESP for concession of scholarships to the first and third authors of this work, respectively. We are grateful to the anonymous EMO 2003 reviewers whose comments greatly improved the manuscript.

References

1. Shamir, U., Howard, C.D.D.: An Analytic Approach to Scheduling Pipe Replacement. Journal American Water Works Association, Vol. 71, No. 5 (1979) 248–258
2. Walski, T.M.: Economic Analysis of Rehabilitation of Water Mains. Journal of Water Resources Planning and Management, ASCE, Vol. 108, No. 3 (1982) 296–308

3. Clark, R.M., Stafford, C.L., Goodrich, J.A.: Water Distribution Systems: A Spatial Cost Evaluation. Journal of Water Resources Planning and Management, ASCE, Vol. 108 WR 3 (1982) 243–255
4. Kleiner, Y., Adams B. J., Rogers, J. S.: Long-Term Planning Metodology for Water Distribution System Rehabilitation. Water Resources Research, Vol. 34, No. 8 (1998) 2039–2051
5. Kim, J.H., Mays, L.W.: Optimal Rehabilitation Model for Water Distribution Systems. Journal of Water Resources Planning and Management, ASCE, Vol. 120, No. 5 (1994) 674–692
6. Walski, T.M.: The Wrong Paradigm, Why Water Distribution Optimization Doesn't Work. Journal of Water Resources Planning and Management (Editorial), ASCE, Vol. 123, No. 3 (2001) 203–205
7. Engelhardt, M.O., Skipworth, P.J., Savic, D.A., Saul, A.J., Walters, G.A.: Rehabilitation Strategies for Water Distribution Networks: a Literature Review with a UK Perspective. Urban Water, No. 2 (2000) 153–170
8. Deb, K.: Multi-Objective Using Evolutionary Algorithms. John Wiley & Sons, Ltd (2001)
9. Fonseca, C.M., Fleming, P. J.: An Overview of Evolutionary in Multiobjective Optimization. Evolutionary Computation, Vol. 3, No. 1 (1995) 1–16
10. Goldberg, D.E.: Genetic Algorithms in Search, Optimization, and Machine Learning. Addison-Wesley, Reading, Massachusetts (1989)
11. Fonseca, C.M., Fleming, P. J.: Genetic Algorithms for Multiobjective Optimization: Formulation, Discussion and Generalization. Proceedings of the Fifth International Conference, San Mateo (1993)
12. Horn, J., Nafpliotis, N.: Multiobjective Optimization using the Niched Pareto Genetic Algorithm. Proceedings of the First IEEE Conference on Evolutionary Computation, IEEE World Congress on Computational Intelligence, Vol. 1 (1994) 1–32
13. Srinivas, N., Deb, K.: Multiobjective Optimization using Nondominated Sorting in Genetic Algorithms. Evolutionary Computation, Massachusetts Institute of Technology, Vol .2, No 3 (1995) 221–248
14. Zitzler, E., Thiele, L.: Multiobjective Evolutionary Algorithms: A Comparative Case Study and the Strength Pareto Approach. IEEE Transactions on Evolutionary Computation, Vol. 3, No 4 (1999) 257–271
15. Deb, K., Pratap, A., Agarwal, S., Meyarivan, T.: A Fast and Elitist Multi-Objective Genetic Algorithm – NSGAII. KanGAL Report Number 2000001 (2000)
16. Veldhuizen, D. V.: Multiobjective Evolutionary Algorithms: Classifications, Analyses, and New Innovations. Ph. D. Thesis, Dayton, OH: Air Force Institute of Technology (1999)
17. Simpson, A. R., Dandy, G.C., Murphy, L. J.: Genetic Algorithms Compared to Other Techniques for Pipe Optimization. Journal of Water Resources Planning and Management, ASCE, Vol. 120, No. 4 (1994) 423–443
18. Walters, G.A., Halhal, D., Savic, D.A., Ouazar, D.: Improved Design of Anytown Distribution Network using Structured Messy Genetic Algorithms. Urban Water, Vol. 1 (1999) 23–38
19. Halhal, D., Walters, G.A., Savic, D.A., Ouazar, D.: Scheduling of Water Distribution System Rehabilitation using Structured Messy Genetic Algorithms. Evolutionary Computation, Vol. 7, No. 3 (1999) 311–329
20. Dandy, G.C., Engelhardt, M.O.: Optimum Rehabilitation of Water Distribution System considering Cost and Reliability. Proceedings of the World Water and Environmental Resources Congress, Orlando, Florida, (2001)
21. Rossman, L. A.: EPANET 2, USERS MANUAL. U. S. Environmental Protection Agency, Cincinnati, Ohio, (2000)
22. Ray, T., Tai K., Seow, K.C.: Multiobjective Design Optimization by an Evolutionary Algorithm. Engineering Optimization, Vol. 33 (2001) 399–424

23. Michalewicz, Z.: Genetic Algorithms + Data Structures = Evolution Programs. Springer-Verlag, New York, N.Y., (1992)
24. Gen e Cheng: Genetic Algorithms and Engineering Design. John Wiley & Sons, INC, (1996)
25. Wall, M.: Galib: A C++ Library of Genetic Algorithm Components (version 2.4). Mechanical Engineering Department, Massachusetts Institute of Technology, (1996)
26. Jaskiewicz, A.: MOMHLIB++: Multiple Objective Metaheuristics Library in C++. http://www-idss.cs.put.poznan.pl/~jaszkiewicz/MOMHLIB.
27. Gessler, J.: Pipe Network Optimization by Enumeration. Proceedings Computer Applications for Water Resources, ASCE, New York, N.Y., (1985) 572–581
28. Goldberg, D.E.: Real-Coded Genetic Algorithm, Virtual Alphabets, and Blocking. Complex Systems, Vol. 5, No. 2 (1991) 139–168

Optimal Design of Water Distribution System by Multiobjective Evolutionary Methods

Klebber T.M. Formiga, Fazal H. Chaudhry, Peter B. Cheung, and Luisa F.R. Reis

São Carlos School of Engineering, University of São Paulo
Av. do Trabalhador São Carlense 400, C.P. 359, São Carlos - SP - Brazil 13560-250
klebberformiga@uol.com.br pcheung@bol.com.br
{fazal,fernanda}@sc.usp.br

Abstract. Determination of pipe diameters is the most important problem in design of water supply networks. Several authors have focused on the methods capable of sizing the network considering uncertainty and other important aspects. This study presents an application of multiobjective decision making techniques using evolutionary algorithms to generate a series of nondominated solutions. The three objective functions considered here include investment costs, entropy system and system demand supply ratio. The determination of Pareto frontier employed the public domain library MOMHLib++ and a hybrid hydraulic simulator based on the method of Nielsen. This technique is found to be quite promising, the nondominated region being identified in a reasonably small number of iterations.

1 Introduction

Water distribution networks are hydraulic systems consisting of components such as: pipes, pumps, valves and reservoirs among others, in order to supply water in desired quantity and quality within preestablished pressure limits. These systems can be represented by a graph in which the nodes may symbolize points of consumption or sources and the links are the pipes, pumps or valves.

The planning process of water supply networks, in general, consists of three steps: definition of the layout, design of the components and the systems operation. This paper deals with the design phase which has been the object of the study for more than 30 years by researchers from various disciplines. One considers a water distribution network composed of pipes, nodes and circuits where the topology and topography are known, water supply is guaranteed by one or more sources and where demands at various node points are assumed known. The classical design problem of such systems involves finding the least cost network that meets the required demands satisfying some constraints. Within this framework, various researchers developed design procedures making use of different optimization methods such as Linear Programming ([1], [2]), Nonlinear Programming ([3]-[4]), and Genetic Algorithms ([5]-[6]). The problem posed in these studies had as unknowns pipe diameters in various links and constraints were placed on pressures.

The above design problem dealt with by classical optimization methods faces two difficulties [7]: 1) the decision involves pipe diameters, pumps and valves specification which are discrete in nature and must me chosen among sizes which are commercially available; 2) any realistic formulation of the problem involves nonlinear and nonconvex equations. Further, it is observed that the solutions obtained by these methods may not be feasible in practice as they tend towards branching networks, instead of looped network, choosing null pipe diameters and discharges.

In an effort to circumvent these difficulties, some researches ([8]-[11]) introduced new constraints, such as reliability, into the design problems. The use of traditional reliability measures as constraints in the optimization of networks, particularly those of medium and large sizes, did not prove to be fruitful despite the added computational complexity of exponential nature. In the search for a simple and efficient reliability measure, heuristic concepts ([12]-[15]), such as the concept of entropy of system, have been experimented with good results [15].

The traditional optimization problems look for a single optimal solution. However, this search considering a single aspect of design may not be convenient for water supply systems. For example, when one optimizes such a system, one obtains a least cost configuration. However, there may be a layout with slightly higher cost but a higher reliability. Thus, a more convenient solution may be non-optimal with respect to cost but more advantageous with respect to other aspects.

An alternative is to consider this problem in the multiobjective framework wherein the evolutionary techniques have differentiated themselves by their capacity to deal with a variety of problems. This paper evaluates the use of multiobjective evolutionary methods in the design of water supply systems.

2 Hydraulic Model

The hydraulic analysis of a pipe network determines two kinds of unknowns: hydraulic heads at the nodes (h_i) and discharges in various pipe reaches (q_j). Let there be a pipe network composed of n nodes, n_r sources nodes (reservoirs and tanks) and m pipes connecting the nodes. The hydraulic head loss, between two nodes i and j, can be expressed as:

$$h_{ji,2} - h_{ji,1} = K_i q_i^\alpha \tag{1}$$

in which K_i is the hydraulic resistance coefficient of link i, and α is the exponent of the head loss formula.

Mass conservation in the system is given by:

$$f_j = \sum_{i=1}^{m} a_{ij} q_i - Q_j = 0 \quad \forall \quad j=1\ldots n \tag{2}$$

where Q_j = demand concentrated at the node j, $a_{ij} = 1$ if the discharge is towards node j, $a_{ij} = -1$ if discharge is away from node j, and $a_{ij} = 0$ if the nodes i and j are not connected.

In the case of solution in terms of node heads, equation (2), can be written as:

$$f_j = \sum_{i=1}^{m} a_{ij} \frac{(h_{ji,2} - h_{ji,1})^{1/\alpha}}{K_i^{1/\alpha}} - Q_j = 0 \quad \forall \; j = 1 \ldots n \tag{3}$$

The incidence matrix $\hat{A}$ (m x n+nr) formed by the elements a_{ij} can be divided in two submatrices A (m x n) and A_r (m x nr). The first matrix corresponds to the interior nodes and the second to the source nodes.

The system of equations in (3) is nonlinear of order n. Iterative numerical methods such as Newton-Raphson (N-R) or its variations can be employed to solve this system for nodal heads.

2.1 Newton-Raphson Method

The N-R method for solution of nonlinear system of equations can be described as,

$$X_{k+1} = X_k + \Delta X_k \tag{4}$$

in which X is the vector of variables x in iteration k and ΔX_k can be obtained from the expression:

$$J_k \Delta X_k = -F_k \tag{5}$$

where the Jacobian matrix function is given by,

$$J_k = \left\{ \frac{\partial f_i}{\partial x_i} \right\}_{|x=x_k} \tag{6}$$

The system for finding the hydraulic characteristics of the network can be expressed as:

$$H_{k+1} = H_k + \Delta H_k \tag{7}$$

The Jacobian of the function f in (3) is given by,

$$J_k = A^T C' A \tag{8}$$

where,

$$C' = \text{diag}\left[\frac{1}{\alpha} \frac{\left|h_{j1,2} - h_{j1,1}\right|^{\frac{1}{\alpha}-1}}{K_1^{1/\alpha}}, \frac{1}{\alpha} \frac{\left|h_{j2,2} - h_{j2,1}\right|^{\frac{1}{\alpha}-1}}{K_2^{1/\alpha}}, \ldots, \frac{1}{\alpha} \frac{\left|h_{jm,2} - h_{jm,1}\right|^{\frac{1}{\alpha}-1}}{K_m^{1/\alpha}} \right] \tag{9}$$

Putting

$$C = \alpha C' \tag{10}$$

one obtains,

$$J_k = \frac{1}{\alpha} A^T CA \qquad (11)$$

The vector F_k is expressed mathematically as [16]:

$$F_k = [Q + A^T C(AH_k + A_r H_r)] \qquad (12)$$

H_r being the vector of reservoir elevations.

Thus the equation (7) becomes:

$$h_{k+1} = h_k - \alpha [A^T CA]^{-1}[Q + A^T C(Ah_k + A_r h_r)] \qquad (13)$$

3 Objective Functions

This study considers three objectives, namely, the network cost, system entropy, and the water supply capacity of the network. Water supply capacity was incorporated into the problem to avoid any kind of constraint. In the order words, the constraints are transformed into objectives.

3.1 Network Cost

The cost function includes expenses for the main hydraulic components of the water distribution system. These are divided into: investment costs and operation costs. The investment costs cover the acquisition costs and the costs of installation of the pipes, pumps and reservoirs. The operation costs are expressed in terms of energy consumption by motor-pumps. These costs are distributed along the useful life of the network, thus requiring the use of present value factor.

This paper considers the minimization of investment costs expressed in terms of diameters. Thus, the first objective function is:

$$F_1 = Min[Cost] = Min\left[\sum_{i=1}^{m} CT(D)_i L_i\right] \qquad (14)$$

where *Cost* refers to the investment cost, $CT(D)_i$ express the unit cost associated with the pipe reach i with diameter D_i and L_i is the length of reach i.

3.2 Entropy

The basic requirement of a looped water distribution network is that water demand at a given node is met not only by a preferential path, but also through a number of alternative paths directly connected to this node. Such paths would transport water in different amounts to meet the nodal demands. However, in order to increase the reliability of the network, it is desirable that this distribution of flow be as even as possible. If the distribution were very uneven, the failure of a reach that supplies the largest discharge to the node would cause a major impact on the water supply in the network.

Consider a network of N nodes. Let the entropy (or redundancy) of node j be S_j. The axioms proposed by [17] that make possible the calculation of reliability of water distribution system on the basis of entropy are:
1. S_j will be a function of the discharge fractions $X_{1j}, X_{2j}, X_{3j}, ... X_{n(j)j}$, given that:

$$\sum_{i=1}^{n(j)} X_{ij} = 1 \quad (15)$$

where $n(j)$ is the number of pipes that are incident on node j;
2. S_j will be zero if $n(j)$ is 1;
3. for given fixed number of nodes j, the redundancy will be maximum when the discharges in the incident pipes are equal.
4. the value of S_j is incremented by an increase in $n(j)$ which implies that, for equal discharges, redundancy can be increased by increasing the number of incident pipes.

There are many functions that help express entropy. However, we prefer the definition in [17], given that it produced consistent results for different types of networks (communication networks [17], electrical networks [18], hydraulic networks [12]). That expression for entropy is:

$$\frac{S}{K} = -\sum_{i=1}^{I} P_i \ln(P_i) \quad (16)$$

where K is an arbitrary positive constant generally taken to be 1, P_i is a system parameter and I is the number of subsystems.

This entropy function is the basis of all other functions used in the water supply systems ([12]-[15]). It can be seen that this equation satisfies all the axioms.

To begin with, it is necessary to define the parameter P for the network. Let this network be composed of N nodes which are its subsystems. For a given demand pattern, let the ith pipe incident on node j transport discharge q_{ij} then,

$$X_{ij} = \frac{q_{ij}}{Q_j} \quad (17)$$

where Q_j is the sum of $n(j)$ discharges incident on node j.

The parameter X_{ij} represents the relative contribution of pipe i to the total discharge that passes through node j and serves as a measure of its relative capacity incorporated in the entropy function. This measure is an indicator of the potential contribution of a pipe to the system's possibility of meeting the extra demand imposed by the failure of another pipe in this system. Thus, the entropy for the node j can be written as:

$$S_j = -\sum_{i=1}^{n(j)} \left(\frac{q_{ij}}{Q_j}\right) \ln\left(\frac{q_{ij}}{Q_j}\right) \quad (18)$$

Maximizing the function S_j in equation (18) is equivalent to maximizing redundancy at the node j because the highest value of S_j will happen when all q_{ij} values at j are equal, which is desirable discharge distribution. The expression for S_j in equation (18) gives entropy of one node only. The development of entropy equation for whole network follows similar procedure.

One might think that the entropy of a system is the sum of entropies at all the nodes in the network. However, this description does not take into account the discharge carried by given pipe in the general context because, in the event of a failure, it is not just the pipes adjacent to the node that will be exacted. Thus, the importance of a pipe with respect to the total discharge, and not just relative to the discharges incident on the node, is relevant to the determination of network entropy.

Let Q_o be the sum all the discharges incident on all the nodes of the network for a given demand pattern. It is important to observe that the value of Q_o shall be equal to or greater than the total demand on the network (Q_T), in view of the fact that a certain portion of water circulates through more than one pipe. Thus, the parameter P used to compute network entropy should be q_{ij}/Q_o. This definition leads to the following expression for entropy of the system [12]:

$$\hat{S} = \sum_{j=1}^{N}\left(\frac{Q_j}{Q_o}\right) - \sum_{i=1}^{n(j)}\left(\frac{q_{ij}}{Q_o}\right)\ln\left(\frac{q_{ij}}{Q_o}\right) \tag{19}$$

This expression can also be expressed as:

$$\hat{S} = \sum_{j=1}^{N}\left(\frac{Q_j}{Q_o}\right)S_j - \sum_{j=1}^{N}\left(\frac{Q_j}{Q_o}\right)\ln\left(\frac{Q_j}{Q_o}\right) \tag{20}$$

Due to the presence of the logarithm, the entropy of the system presents very close values. We opted for working with the exponential of the value of the entropy of the network. Then, the second objective function will be:

$$F_2 = \text{Max}\left[e^{\hat{S}}\right] \tag{21}$$

3.3 Demand Supply Ratio

The amount of water supplied by a given node depends upon the pressure at that node. In order to establish a correspondence between pressure and demand, it is necessary to determine a functional relationship between these two variables. The description used for this purpose is that proposed by [19], and modified by [11], is as follows.

$$\begin{aligned} Q_j^{dem} &= 0 & \text{se} \quad h_j < h_j^{min} \\ Q_j^{dem} &= Q_j^{req}\left(\frac{h_j - h_j^{min}}{h_j^{des} - h_j^{min}}\right)^{\frac{1}{\beta_j}} & \text{se} \quad h_j^{min} \leq h_j \leq h_j^{des} \\ Q_j^{dem} &= Q_j^{req} & \text{se} \quad h_j > h_j^{des} \end{aligned} \tag{22}$$

Here, Q_j^{dem} is the demand met at node j, Q_j^{req} is the demand required at the node j, h_j^{des} is the hydraulic head required at the node j to satisfy the demand completely, h_j^{min} corresponds to the zero demand satisfaction and β_j is the exponent associated with the nodal head set equal to 2.

On the basis of discharge supplied at the node, one can establish a function that determines the failure in the system. Thus F_3, the third objective function, is the maximization of average demand supply ratio which was proposed by [10], as,

$$F_3 = \text{Max}\left[\frac{1}{n}\sum_{i=1}^{n}\frac{Q_j^{req}}{Q_j^{dem}}\right] \qquad (23)$$

4 Multiobjective Evolutionary Methods

The multiobjective evolutionary methods appeared in the middle of the 1980's decade starting with the work of Schaffer [20]. Since then dozens of new such methods have been proposed in the order to obtain the Pareto frontier for diverse problems.

This study employs the method NSGAII - Elitist Non-Dominated Sorting Genetic Algorithm - [21] based on elitism implementing the code developed by Andrzej Jaskiewicz [22] in the MOMHLib++ - Multiobjective Methods Metaheuristic Library for C++.

The crossover method BLX-α [23], was used with crossover probability 1. The Random Mutation [24] was adopted with rate of 5%.

5 The Problem

The NSGAII multiobjective method was applied to two test networks used in literature. The first is a two-loop fictitious network [1], and the second is Hanoi-Vietnam water distribution system [25]. These networks were chosen with the purpose of evaluating the adaptation characteristics of networks of different sizes.

5.1 Example Problem 1 – Two-Loop Network

This network was proposed by [1] who applied Linear Programming to find its least cost solution. Later various authors ([2]-[7]) used this problem as a basis for comparison of their formulations. This network is shown in Fig 1 and its hydraulic data can be found in the original paper [1].

5.2 Example Problem 2 – Hanoi Network

The Hanoi Network presented in Fig.2 was studied by [25]. This is a medium size network composed of 35 pipes and 32 nodes, including one reservoir. Its hydraulic data are available in the above paper.

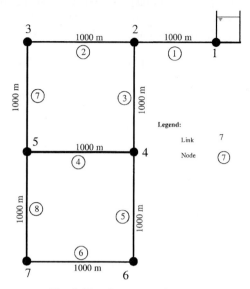

Fig. 1. Two-loop network [1]

The original data limited the diameters to 1000 mm (40 in) size. This diameter is considered insufficient to transport the design discharge of about 5.5 m³/s, as velocity in this pipe would be greater than 7 m/s, which is much higher than the usual maximum tolerance of 3 m/s. Thus, the diameter upper bound of 2000 mm (80 in) was used in this study.

5.3 Performance Measures

According to [26], at least two types of performance measures should be used in the multiobjective analysis. The first measure considers the proximity of the calculated solution to the Pareto region. The second evaluates the manner of distribution of solutions along the Pareto frontier.

Measure of Proximity of Solutions to the Pareto Front. The method used for this measure was proposed by [27]. This measure is well-suited to the comparison between 2 sets of solutions. Expressed as $C(A,B)$, it determines the proportion of solutions of B that are dominated by the set A. It is given by,

$$C(A,B) = \frac{|\{b \in B \mid \exists a \in A : a \preceq b\}|}{|B|}. \tag{24}$$

The value of $C(A,B)$ shall be 1 if all the members of B are dominated by A and 0, if no member of A dominates B. As the measure is not symmetrical, it follows that, in majority of the cases, C(B,A)≠1- C(A,B), then, the calculation of both C(A,B) and C(B,A) becomes necessary.

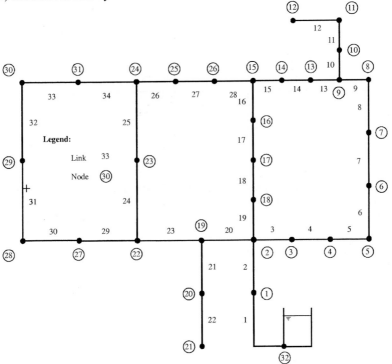

Fig. 2. Hanoi Network [24]

Measure of Diversity among Non-dominated Solutions. The measure of the diversity of solutions, proposed by [28], is calculated on the basis of relative distance between the solutions expressed as,

$$S_B = \sqrt{\frac{1}{|B|}\sum_{i=1}^{|B|}(d_i - \bar{d})^2} \qquad (25)$$

where d_i is the lowest Euclidean distance from the nearest solution, $\bar{d}$ is the average of d_i values and $|B|$ is the number of solutions in the set B.

6 Results

To study the influence of the population size and the number of generations on the nondominated solutions, a 30-population set was generated by NSGAII method for different populations sizes (50, 100, 250 and 500) and generations (50 and 100).

Table 1. Average run time (s) for various sizes of initial solutions and generations

Configuration	Number of Initial Solutions	Number of Generations	Number of Solutions Generated	Network 1	Network 2
1	50	50	2500	23	63
2	50	100	5000	46	112
3	100	50	5000	47	113
4	100	100	10000	89	202
5	250	50	12500	121	288
6	250	100	25000	233	521
7	500	50	25000	251	610
8	500	100	50000	488	1253

The average processing time for each type of generated set of solutions using an AMD Athlon Dual 1800+ is presented in Table 1. The average run time per evaluated individual was 0.009s for Two-loop network and 0.025s for Hanoi network showing a nearly linear relation between number of solutions sought and processing time in both networks.

In case of network in Example Problem 1, some solution results are similar to the optimal values considering single objective F_1 only. Savic and Walters [7] found the optimal value of \$419.000, considering one diameter in a pipe reach, for $F_3=1$. This result corresponds to a low value of entropy (F_2) of 5.13. For slightly higher solution cost (F_1), as compared to the above solution, a much higher value for F_2 is indicated, as for example the solution: F_1 = \$421.000 (only 0,5% higher) and F_2 = 5.84 (14% higher).

Fig. 3 presents a set of nondominated solutions found for configuration #8. It is seen that the relation between the functions F_1 and F_3 has a much concentrated distribution of the solutions along the Pareto frontier as compared to the relations that contain F_2. This happens because F_1 and F_3 are directly proportional. In others works, more expensive the network, greater is its tendency towards meeting demands. However, entropy does present a similar relation to cost especially for its intermediate values. One can also observe that for large capacity of demand satisfaction (F_3), equal to 1, entropy does not present large values which are of the order of 5.9. There is an inverse relation between F_2 and F_3 for large values of F_2.

In the Example Problem 2 (Fig. 4), one observes, in general, the same behavior as above for Example Problem 1. However some differences can be noted as the objective functions F_1 and F_3, now show a much neater relationship among themselves. Also, one can observe that as entropy, and consequently reliability increases, the network becomes more expensive. Another important point is that the maximum values of entropy now occur corresponding to the maximum demand supply ratio.

The difference in the behavior of entropy in the two problems is due to different degrees of redundancy in the two networks, given that the network in Example Problem 2 has higher number of loops and, at the same time, has greater capacity to face pipe failures.

The results of comparison of generated solutions in terms of the measure of proximity to the nondominated frontier are presented in Fig. 5 as Boxplots. The middle line represents the medium value and the upper and lower edges of the rectangles indicate the upper and lower quartiles, respectively, of the distribution of solutions. The upperworst and lowerworst lines indicate the maximum and the minimum values in the sample.

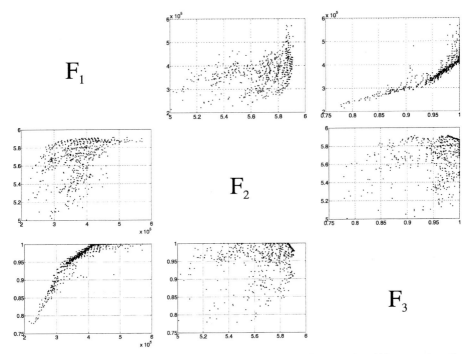

Fig. 3. Scatter-matrix of a generated set of nondominated solutions after 100 generations in Example Problem 1 starting with an initial set of 500 individuals

The results in Fig. 5 show that, in Example Problem 1, the initial population size has greater importance than the number of generation. This is obvious when one compares C values of the solutions for the same number of evaluations (2x3 and 6x7). This happens due to the fact that, in generation 50, the population presents higher degree of maturation because when one doubles the number of generations, there is no significant evolution of the solutions. This becomes evident when one compares the solutions obtained with the same population size.

In Example Problem 2, one observes (Fig. 6) that the number of generations is as important a factor as the population size, as shown clearly in the comparison of configurations 6 and 7. Nearly half of the final solutions of the sets 7 are dominated at least by one solution of the sets 6. Once again, it is observed that the degree of evolution of the population is relevant. The 50th generations in Hanoi network problem, are not sufficiently mature as can be seen from the comparison of the solutions corresponding to initial population sizes and different number of generations.

The measure of performance for the problems under study points to a more predictable behavior. It is seen that the initial population size has greater importance for the diversity of the final population.

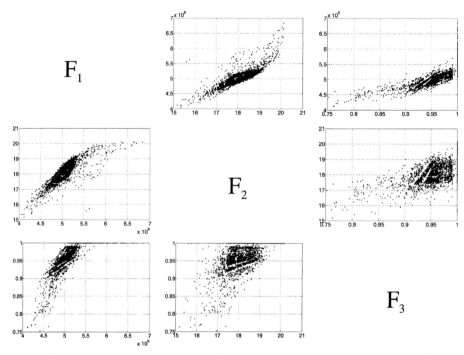

Fig. 4. Scatter-matrix of a generated set of nondominated solutions after 100 generations in Example Problem 2 starting with an initial set of 500 individuals

7 Conclusions

This paper employs an elitist multiobjective evolutionary method (NSGAII) for the determination of the nondominated region, considering the objectives of minimizations of hydraulic network costs and maximization of entropy and capacity of network to meet water demand. The MOMHLib++ Library of Andrzej Jaskiewicz and a hydraulic simulator based on the method of Nielsen [16] were used to model the multiobjective problem for two classic literature networks: a two-loop network and the Hanoi network.

It was found that the two-loop network problem shows greater sensitivity to the evolutionary parameter of initial population size, whereas the Hanoi network problem is more sensitive to the number of generations. Thus it is concluded that the number of generations is important to a certain point where there is no significant change in the Pareto frontier. When the number of generations reaches this point, the initial population size becomes more relevant.

In order to obtain satisfactory results using multiobjective evolutionary methods for water distribution networks, it is important to choose appropriate size of initial populations and the number of generations necessary for arriving at a stable Pareto frontier.

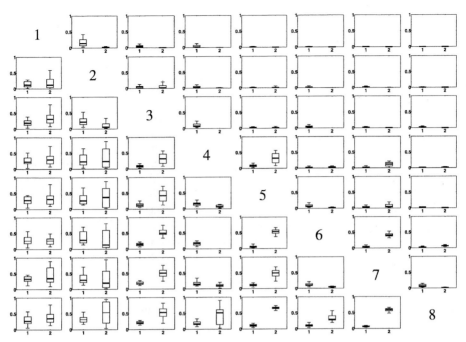

Fig. 5. Values of the metric C(A,B) for different initial population and generation sizes. The solution A and B refer to the solutions along the rows and columns respectively

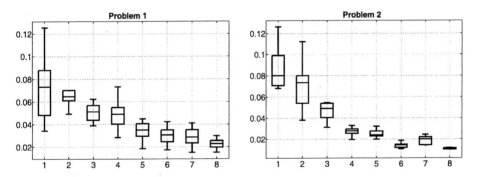

Fig. 6. Values of the measure of diversity of the nondominated solutions for different initial population and generation sizes.

Acknowledgements. The present paper has resulted from current research grant "Tools for the Rational and Optimized Use of Urban Water (CT-HIDRO 01/2001 - 550070/200-8)" funded by Brazilian National Research Council (CNPq). The authors wish to express their gratitude also to FAPESP and CAPES for concession of scholarships to the first and third authors of this work, respectively. We are grateful to anonymous reviewers from EMO 2003 whose comments greatly improved the manuscript.

References

1. Alperovits, E., Shamir, U.: Design of Optimal Water Distribution Systems. Water Resources Research, Vol. 13, No 6 (1977) 885–900
2. Quindry, G., Brill, E.D., Liebman, J.C.: Optimization of Looped Distribution Systems. Journal of Environmental Engineering, ASCE, Vol. 107, No 4 (1981) 665–679
3. Morgan, D.R., Goulter, I.C.: Optimal urban water distribution design. Water Resources Research, Vol. 21, No 5 (1985) 642–652
4. Formiga, K.T.M.: Looped Network Optimization by Nonlinear Programming. Master's Thesis. Federal University of Paraiba. Campina Grande, Brazil (1999)
5. Simpson, A. : Genetic Algorithms Compared to Other Techniques for Pipe Optimization. Journal of Water Resources Planning and Managament, Vol. 20, No. 4 (1994) 423–443
6. Savic, D.A., Walters, G.A.: Genetic Algorithms for Least-Cost of Water Distribution Networks. Journal of Water Resources Planning and Management, ASCE, Vol. 123, No 2 (1997) 67–77
7. Eiger, G., Shamir, U., Ben-Tal, A.: Optimal Design of Water Distribution Networks. Water Resources Research, Vol. 30, No 9 (1994) 2637–2646
8. Bhave, P.R.: Node Flow Analysis of Water Distribution Systems. Journal of Transport Engineering, ASCE, Vol. 107, No 4 (1981) 457–467
9. Jowitt, P.W., Xu, C.: Predicting Pipe Failure Effects in Water Distribution Networks. Journal of Water Resources Planning and Management, ASCE, Vol. 119, No 1 (1993) 18–31
10. Fujiwara, O., Li, J.: Reliability Analysis of Water Distribution Networks in Consideration of Equity, Redistribution, and Pressure Dependent Demand. Water Resources Research, Vol. 34, No 7 (1998) 1843–1850
11. Tanyimboh, T.T., Tabesh, M., Burrows, R.: Appraisal of Source Head Methods for Calculating Reliability of Water Distribution Networks. Journal of Water Resources Planning and Management, ASCE, Vol. 127, No 4. (2001) 206–213
12. Awmah, K., Goulter, I., Bhatt, S.K.: Entropy-Based Redundancy Measures in Water Distribution Design. Journal of Hydraulic Engineering, ASCE, Vol. 117, No 5 (1991) 595–614
13. Tanyimboh, T.T., Templeman, A..B.: Maximum Entropy Flows for Single-Source Network. Engineering Optimization, Vol. 22 (1993) 49–63
14. Walters, G.A: Discussion on: Maximum Entropy Flows for Single-Source Network. Engineering Optimization, Vol. 25 (1995) 155–163
15. Tanyimboh, T.T., Templeman, A.B.: A Quantified Assessment of Relationship Between the Reliability and Entropy of Water Distribution Systems. Engineering Optimization, Vol. 33 (2000) 179–199
16. Nielsen, H.B.: Methods for Analyzing Pipe Networks. Journal of Hydraulic Engineering, ASCE, Vol. 125, No 2 (1989) 139–157
17. Shannon, C.: A Mathematical Theory of Communication. Bell System Technical Journal, Vol. 27, No 3 (1948) 379–423, 623–659
18. Kapur, J.N.: Twenty-five years of maximum entropy. Journal of Mathematical Physics Science. Vol. 17, No 2 (1983) 103–156.
19. Chandapillai, J.: Realistic Simulation of Water Distribution System. Journal of Transportation Engineering, ASCE, Vol. 117, No 2 (1991) 258–263
20. Schafer, J.D.: Some Experiments in Machine Learning Using Vector Evaluated Genetic Algorithms. Ph. D. Thesis. Vanderbuilt University. Nashville, TN (1984)
21. Deb, K., Agrawal, S., Pratap, A., Meyarivan, T.: A Fast Elitist Multiobjective genetic Algorithm for Multi-Objective Optimization: NSGA-II. In: Proceedings of the Parallel Problem Solving from Nature VI, (2000) 849–858
22. Jaszkiewicz, A. Multiobjective Methods Metaheuristic Library for C++. http://www-idss.cs.put.poznan.pl/~jaszkiewicz/MOMHLIB/

23. Eshelman, L.J., Schaffer J.D.: Real-Coded Genetic Algorithm and Interval-Schemata. In: Foundations of Genetic Algorithms 2, (1991) 187–202
24. Michalewicz, Z.: Genetic Algorithms + Data Structures = Evolution Programs. Berlin Springer-Verlag. Berlin (1992)
25. Fujiwara O., Tung, D.B.: A Two-Phase Decomposition Method for Optimal Design of Looped Water Distribution Networks. Water Resources Research, Vol. 26, No 4 (1990) 539–549
26. Deb, K. Multi-Objective Optimization Using Evolutionary Algorithms. John Willey & Sons. Chichester, England (2001)
27. Zitzler, E., Deb, K., Thiele, L.: Comparison of Multiobjective Evolutionary Algorithms: Empirical Results. Evolutionary Computation Journal, Vol. 8, No 2 (2000) 125–128
28. Schott, J.R.: Fault Tolerant Design Using Multi-Criteria Genetic Algorithms. Master's Thesis. Department of Aeronautics and Astronautics, Massachusetts Institute of Technology, Boston, MA (1995)

Evolutionary Multiobjective Optimization in Watershed Water Quality Management

Jason L. Dorn[1] and S. Ranji Ranjithan[2]

[1] Research Assistant, Department of Civil Engineering, Campus Box 7908, North Carolina State University, Raleigh, NC 27695, USA
jldorn@eos.ncsu.edu

[2] Associate Professor, Department of Civil Engineering, Campus Box 7908, North Carolina State University, Raleigh, NC 27695, USA
ranji@eos.ncsu.edu

Abstract. The watershed water quality management problem considered in this study involves the identification of pollution control choices that help meet water quality targets while sustaining necessary growth. The primary challenge is to identify nondominated management choices that represent the noninferior tradeoff between the two competing management objectives, namely allowable urban growth and water quality. Given the complex simulation models and the decision space associated with this problem, a genetic algorithm-based multiobjective optimization (MO) approach is needed to solve and analyze it. This paper describes the application of the Nondominated Sorting Algorithm II (NSGA-II) to this realistic problem. The effects of different population sizes and sensitivity to random seed are explored. As the water quality simulation run times can become prohibitive, appropriate stopping criteria to minimize the number of fitness evaluations are being investigated. To compare with the NSGA-II results, the MO watershed management problem was also analyzed via an iterative application of a hybrid GA/local-search method that solved a series of single objective ε-constraint formulations of the multiobjective problem. In this approach, the GA solutions were used as the starting points for the Nelder-Mead local search algorithm. The results indicate that NSGA-II offers a promising approach to solving this complex, real-world MO watershed management problem.

1 Introduction

Public sector planning and management problems require, in general, consideration of multiple competing objectives that represent the interests of a diverse set of stakeholder groups. In addition to the consideration of cost during the evaluation of alternative solutions, the government agencies or elected officials that are responsible for making the final decision are compelled to incorporate the stakeholders' interests during the decision making process. As most public sector problems are complex to analyze and have numerous feasible solutions, systematic evaluation of the decision space in consideration of the multiple objectives becomes important. Thus the decision making process can potentially be improved by presenting the pertinent information associated with the nondominated solutions and the noninferior tradeoff among the competing objectives. In this paper, such a public sector problem associated with watershed management is analyzed using an evolutionary multiobjective optimization approach.

Watershed management is a broad term that generally refers to how a unit of land (i.e., the watershed) is managed with respect to its associated water resources primarily for economic, water supply, and recreational benefit to society. Generally, watershed management is only concerned with surface water, but in some instances (especially is coastal areas) consideration is also concurrently given to groundwater supplies. The primary problem watershed managers face is the impact of urban growth or land use changes on water quality. Urban growth impacts can be managed through proper land use planning, as well as through implementation of structural best management practices (BMPs), such as buffer strips or wet detention ponds, to treat urban runoff before it is discharged to the receiving water bodies.

While the land use allocation choices and BMP requirements affect the cost as well as property tax revenues, they contribute to water quality improvements, necessitating the examination of the multiobjective tradeoff. The physical-chemical relationships that characterize the impact of management choices on water quality are often complex, and are typically represented within water quality simulation programs. A systematic search for good solutions (i.e., efficient management choices) thus requires coupling of the search algorithms with these simulation programs. Given the complexity of the simulation programs and the decision space, such search is conducted typically via nongradient-based heuristic search procedures such as genetic algorithms (GAs).

1.1 Evolutionary Multiobjective Algorithms

Numerous evolutionary algorithm-based multiobjective optimization procedures are available [1]. Single-objective evolutionary algorithms may be used to solve multiobjective optimization (MO) problems by transforming, for example, via the ϵ-constraint method, the MO problem to a single-objective problem, and then solving it iteratively using an evolutionary algorithm. This approach is automated and made computationally more efficient in the Constraint Method-based Evolutionary Algorithm[2]. Alternatively, a number of multiobjective evolutionary algorithms (MOEAs) are designed to solve the problem in a single pass, where the population represents the set of nondominated solutions (e.g., Pareto-Archived Evolution Strategy (PAES) [3], Strength-Pareto EA (SPEA) [4], and the Nondominated Sorting Genetic Algorithm-II (NSGA-II) [5]).

One problem with applying MOEAs to real-world engineering problems is the immense computational burden associated with evaluating thousands of potential solutions using complex simulation models. This problem is exacerbated when the algorithm contains numerous "tuning" parameters (e.g., a sharing parameter in a niching-type algorithm) that affect the quality of the results. These parameters cannot be estimated with confidence *a priori*, requiring the execution of a number of test runs to explore good settings before analysis may begin in earnest.

Among the commonly used MOEAs, NSGA-II contains fewer tuning parameters, making it more attractive for real-world applications. Comparisons of performance of

PAES and SPEA with NSGA-II for a wide variety of test problems are presented in [5]. NSGA-II was generally found to provide a better spread of solutions and to converge better to the non-dominated set for the test problems. This paper investigates the use of NSGA-II in solving a realistic multiobjective optimization problem associated with watershed management.

1.2 Scope of the Paper

The purpose of this paper is to study the applicability of NSGA-II to solve a realistic watershed management problem in which urban growth and the resultant water quality are treated as conflicting objectives. In this two-objective problem, water quality is expressed in terms of the total mass of the nutrient phosphorus being discharged to the receiving river, and urban growth is expressed as the total area allocated to urban land use types. NSGA-II is coupled with a complex continuous water quality simulation model. The Pareto-front is also estimated through an iterative application, via the ε-constraint method, of a hybrid search technique utilizing a single-objective GA and a local search technique. This nondominated front is compared with that obtained using NSGA-II. The effect of population size and random seed on the performance of NSGA-II for this problem is also explored. Finally, the authors present conclusions and recommend future related activities.

2 Nondominated Sorting Genetic Algorithm II

In NSGA-II, fitness is based on the degree of dominance within the population (i.e., if a given solution is not dominated by any other solution, that solution has the highest possible fitness). In addition, the algorithm seeks to preserve diversity along the first non-dominated front so that the entire Pareto-optimal region is found. NSGA-II is designed to overcome criticisms of MOEAs that rely on nondominated sorting and sharing, including their computational complexity, reliance on a non-elitist approach, and the need for specifying a sharing parameter [5].

As stated previously, a significant practical limitation of all MOEAs is their potentially prohibitive run times when applied to real-world engineering design problems. Run times are proportional to the algorithmic computational complexity. Some nondominated sorting algorithms have computational complexity of $O(MN^3)$, where M is the number of objectives and N is the population size [5]. One of the ways by which computational complexity is managed in NSGA-II is through the rapid non-dominated sort [5] as described below.

At the beginning of every generation, the domination count, n_p, is calculated for every solution p, where n_p is the number of solutions that are dominated by p. All of the solutions in the first non-dominated front, therefore, will have n_p equal to 0. In addition, every solution has associated with it a set, S_p, of solutions that it dominates. After identifying the solutions that comprise the first non-dominated front, the second non-dominated front is found where, for every solution in the first non-dominated front, the domination count of each member in the associated domination set, S_p, is reduced by 1. After processing all members of the first domination front, the solutions with a domination count of zero form the second domination front. This process is repeated until all solutions are assigned to a domination front. The domination front to which a solution belongs is referred to as its rank. This procedure results in a computational complexity of $O(MN^2)$, a slight improvement over other MO algorithms.

In MOEAs that use a sharing operator to induce spread, the sharing parameter sets the extent of sharing desired along a given front. The performance of the MOEA in maintaining a good spread of solutions along the noninferior front, therefore, is a function of this parameter. It is difficult to know what constitutes a good value for this parameter *a priori*. When applying a sharing operator, each solution must be compared to every other solution in a given population, adding further to the computational burden.

Instead of a sharing parameter, NSGA-II uses an alternative quantitative measure that represents the degree of crowding. For a given solution, its density is estimated as the average distance between the adjacent solutions relative to each of the n objectives. This distance is referred to as the crowding distance, and may be thought of as the average length of the sides of the cuboid formed by adjacent solutions on the same domination front. The boundary solutions are assigned a crowding distance of infinity. A solution with a smaller value of the crowding distance is considered to be residing in a more crowded region of objective space than one with a larger crowding distance. Figure 1 illustrates the notion of a crowding distance for a two objective problem.

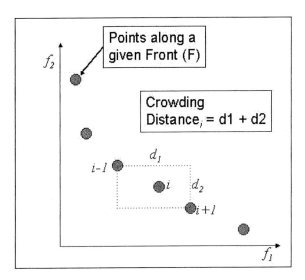

Fig. 1. Illustration of the calculation of the crowding distance for a given solution, i, along a given non-domination front, F

Elitism offers multiple benefits to a GA including improved efficiency, and preventing the loss of good solutions once found. Elitism is incorporated in NSGA-II through the combination of parent and offspring populations prior to selection. The combined population is sorted according to nondomination and the best solutions are allowed to survive to the next generation. The selection operator is applied as follows:

- If the two solutions selected are of different rank, the solution with the lower rank is selected to undergo recombination.
- If the two solutions are of the same rank, the solution with the higher crowding distance is selected to undergo recombination.

The overall NSGA-II algorithm is presented in Figure 2. In the implementation of NSGA-II used in the study presented in this paper, no explicit stopping criterion was implemented. The algorithm was terminated after a pre-specified number of generations were evaluated. However, as in other MOEAs, it is possible to track the generation number at which a new solution entered the population. Therefore, it is possible to implement a stopping criterion that would halt the algorithm when no new solutions have entered the first non-dominated front after a pre-specified number of generations.

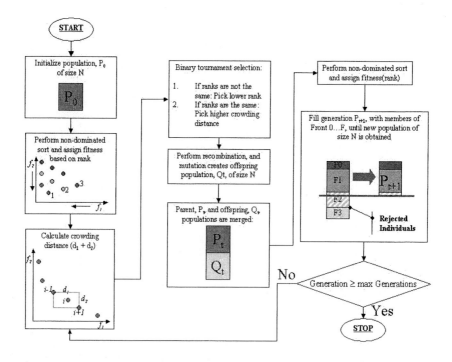

Fig. 2. Flow chart illustrating the NSGA-II algorithm

3 Watershed Water Quality Management Problem

Gwinnett County, located near Atlanta, Georgia, already the third largest urbanized area in the USA, is also one of the most rapidly growing areas in the country [6] (Figure 3). As a result, the water quality of streams and rivers has been detrimentally impacted by increased pollutant loadings due to urban growth, which may be anticipated to get worse as growth continues. County government agencies initiated a watershed management process to help gain insight into prudent land use planning and pollution control practices[7]. As part of this plan, a complex water quality simulation model, the Hydrologic Simulation Program – FORTRAN (HSPF), was developed that linked land use development and pollution control with the resultant in-stream water quality impacts. HSPF is a comprehensive model for simulation of watershed hydrology and water quality for both conventional and toxic organic pollutants that includes fate and transport in one-dimensional stream channels [8].

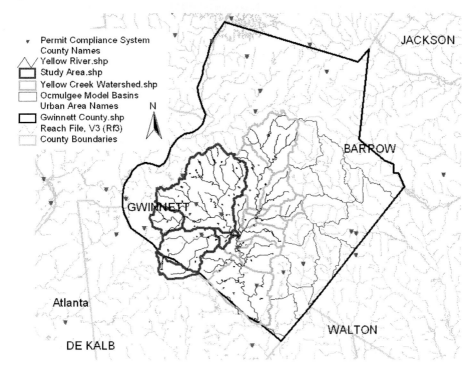

Fig. 3. The case study model watershed is located in Gwinnett County, northeast of Atlanta.

Only the five upstream subwatersheds in Southwestern Gwinnett County (highlighted in Figure 3) were considered in this case study [9]. In addition, eight different land use types were considered. Figure 4 shows the relative areas currently allocated to each of the eight different land use types in the study area. The combined areas of high-density residential, low-density residential, office parks, and roads are representative of urban development, resulting in a net urbanization of 68 percent currently in the watershed. As more development in this watershed is being considered, the goal of the study is to investigate how much urban development may be accommodated while protecting the water quality, resulting in two conflicting management objectives.

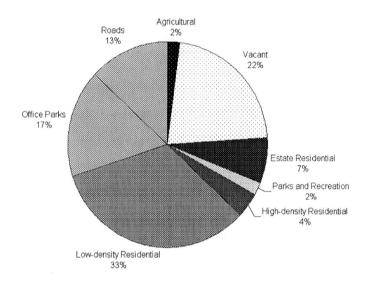

Fig. 4. Existing land use allocation in the study area in Gwinnett County, Georgia, USA

The problem discussed in this paper explores the trade off between urban land development and water quality. To enable more development while minimizing water quality impacts, future urban development that incorporates pollution mitigation options are also considered. Thus two new land uses, namely office parks with buffer and high-density residential with buffer, which simulate the use of stream buffers for pollution mitigation are included. Other land use types in the watershed, as shown in Figure 4, include vacant land, estate residential, agricultural, and parks. These land use types do not contribute to the overall level of urban development in the watershed, but existing areas of these land use may be redeveloped into urban land uses.

3.1 Multiobjective Optimization Problem Formulation

The two objective optimization problem can be mathematically represented as follows:

$$\text{Max } Z_1 = \sum_{i=1,N} \sum_{k=1,K} LU_{ik} \tag{1}$$

$$\text{Min } Z_2 = W(LU_{ik}) \tag{2}$$

Subject to:

$$\sum_{i=1,M} LU_{ik} = A_k, \forall\ k = 1, ..., K \qquad (3)$$

$$P^0_{ik} + \sum_{j=1,M} (p^k_{ij} - p^k_{ji}) \leq P^{max}_{ik},\ \forall\ k = 1, ..., K, \forall\ i = 1, ..., M, i \neq j \qquad (4)$$

$$P^0_{ik} + \sum_{j=1,M} (p^k_{ij} - p^k_{ji}) \geq P^{min}_{ik},\ \forall\ k = 1, ..., K, \forall\ i = 1, ..., M, i \neq j \qquad (5)$$

$$0 \leq p^k_{ij} \leq 1,\ \forall\ k = 1, ..., K, \forall\ i = 1, ..., M, j = 1, ..., M, i \neq j \qquad (6)$$

Where:
N = Total number of urban land use types
M = Total number of all land use types
K = Total number of subwatersheds
LU_{ik} = Dependent variable representing the area of land use, i, in the future land use plan in subwatershed k, where, for $1 \leq i \leq N$, corresponds to urban land uses and for $N < i \leq M$, corresponds to non-urban land uses.

$$LU_{ik} = [P^0_{ik} + \sum_{j=1,M} (p^k_{ij} - p^k_{ji})] * A_k,\ \forall\ k = 1, ..., K, \forall\ i = 1, ..., M \qquad (7)$$

$W(LU_{ik})$ = The function (implemented within a simulation program (i.e., HSPF)) that estimates the water quality in response to future land use allocations and pollution control options
A_k = Total area of subwatershed k
P^0_{ik} = Original percentage of land area dedicated to land use i in watershed k
P^{max}_{ik} = Maximum allowable percentage of land use i in watershed k
P^{min}_{ik} = Minimum allowable percentage of land use i in watershed k
p^k_{ij} = Decision variable that represents the percentage of area of subwatershed k to be converted from land use i to land use j. Note that p^k_{ij} is forced to zero if the associated land use conversion is infeasible. Furthermore, $i \neq j$, so that a land use cannot be converted into itself.

The constraint given in Equation 3 ensures conservation of land area. The constraint represented by Equations 4 and 5 limit the amount of a land use that could be allowed in any single subwatershed and ensures that the future land use plan provides a wide-variety of land uses to support an economically vibrant community such as Gwinnett County. The variable, p^k_{ij}, is required to maintain feasible land use conversions, thus preventing land use conversions that do not make sense or are impractical. For example, it is unlikely that an existing road would be replaced with another land use type.

The decision variable, p^k_{ij}, which represent fractions between 0 and 1, was encoded as real-valued variables in the GA. In this case study, where there are five subwatersheds and ten different land use types, a total of 500 of these real variables are used to represent the problem. The selection was carried out using a binary tournament procedure. Crossover was implemented using a uniform crossover operator. A linear recombination operator was employed to combine the values of the real-valued variables that underwent crossover. Mutation was implemented via random assignment of the real value between the allowable ranges for a given variable. All the land use conversion constraints (i.e., Equations 3-6) were handled implicitly through repair operators.

The goal of the MO analysis is to generate the noninferior tradeoff between urban development and water quality for different levels of point source pollution control. Typical tradeoff curves are shown in Figure 5. To generate each tradeoff curve, the level of point source control (i.e., percent reduction of the existing load) is assigned a fixed value, and then the noninferior tradeoff between urban development and water quality is identified by determining the efficient assignment of nonpoint source pollution via land use allocation. Solution of this problem was attempted previously [9] based on a single-objective GA that was used iteratively to identify noninferior solutions through a series of ε-constraint formulations. The following section describes the solution of this MO problem using NSGA-II. In this paper, the results and discussion are focused on only one of the tradeoff curves (corresponding to one fixed level of point source controls).

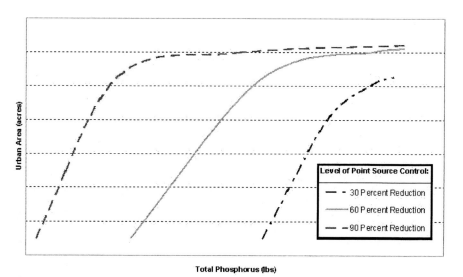

Fig. 5. Typical tradeoff between land development and water quality at different levels of point source control

4 Evolutionary Algorithm-Based Solution

The MO problem described by Equations 1-6 was solved using NSGA-II. To enable a comparison of the NSGA-II results, this MO problem was also solved using a single-objective GA via an ε-constraint approach. The single-objective GA solutions were then refined using a local-search method. The hybrid-GA approach is described below. The solutions obtained using the hybrid-GA approach are used as the baseline noninferior solutions when comparing the NSGA-II solutions.

4.1 Hybrid GA/Nelder-Mead Simplex Algorithm

The MO problem (Equations 1-6) was first converted to a series of single-objective problems, such that their solutions represent a noninferior point on the Pareto tradeoff. Using the ε-constraint approach, the water quality objective (Equation 2) was converted to a constraint with varying water quality targets. The transformed single objective problem (that maximizes the urban land use acreage) was solved iteratively for different values of the water quality target. A single-objective GA was used to solve a series of such transformed problems. The parameter values used for this single-objective GA are provided in Table 1. Based on these parameter settings, over 6,000 model evaluations were required to complete the 100 generations for each noninferior solution. Model run-times are approximately 30 seconds each, resulting in an overall run time of about two days for a single-objective GA solution. To generate the tradeoff curve, several separate single-objective GA runs were carried out, requiring a total of approximately 100,000 model evaluations.

Table 1. GA parameters and settings for the single-objective GA

GA Parameter	Value
Maximum number of generations	100
Max number of generations with no improvement	15
Population size	100
Crossover rate	0.6
Mutation rate	0.001

Each single-objective GA solution was then refined using a local search procedure based on the Nelder-Mead Simplex algorithm [10]. The starting solutions for the local search algorithm were seeded using the GA solutions. The Nelder-Mead Simplex algorithm is a numerical local search technique or "hill-climbing" strategy. The approach may be linked with a simulation model as discussed herein since it does not require any gradient information, which would be difficult to obtain for the HSPF model. This algorithm searches the decision space by forming a simplex (or convex hull) of n+1 points, where n is the number of decision variables. Each point is a potential solution to the problem and is evaluated relative to its fitness. The algorithm proceeds by finding the worst of the points and reflecting opposite of it in decision space to find a new point, which is then evaluated. If this point is an improvement, then an expansion step is carried out, extending along this vector, until a point of no improvement is found. A new simplex is formed with this point, discarding the previous worst, and the algorithm proceeds. Alternatively, if the expansion step does not lead to any improvement, a contraction step is carried out. The procedure continues until all points in the simplex are close to each other as defined by some tolerance, δ. The tolerance in this case was set at 10^{-4} acres.

The result of the application of the single-objective GA followed by application of the Nelder-Mead Simplex algorithm is an estimate of the nondominated front. As the run-time for each noninferior solution was long, only a handful of noninferior solutions were generated using the iterative approach described above. The noninferior front identified by this iterative approach using the Hybrid GA is shown in Figure 6.

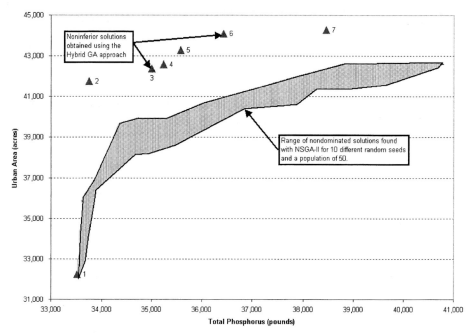

Fig. 6. Noninferior points (triangles) found with the Hybrid GA. Figure also shows the tight range of nondominated fronts found with the NSGA-II algorithm across 10 different random seeds, for a population size of 50

4.2 NSGA-II Results

The NSGA-II algorithm was applied to the same MO watershed management problem discussed above. The two objectives, maximizing urban land use and minimizing total phosphorus loadings to the stream, were treated as the two conflicting objectives in NSGA-II. The NSGA-II algorithm was implemented by building upon the single-objective GA described previously. This allowed the use of the same initial random populations, crossover, mutation, tournament selection and other operators and parameter settings common to both the single-objective GA and NSGA-II. While NSGA-II doesn't require setting of any new parameters, the underlying GA parameters were set as shown in Table 1, except for the crossover rate, which was set to 100 percent of the selected mating population.

NSGA-II was first run with a population of 50. The random seed was varied and the results were found not to vary greatly. A summary of the results corresponding to 10 different random seeds is shown graphically in Figure 6. Given the small variation in the results with different random seeds, the rest of the results are presented in terms of a representative set of noninferior solutions.

Figure 7 illustrates the progression of the non-dominated front in NSGA-II over the course of 100 generations for a population of 50. Initially only a few (8 of the 50 in this case) solutions are nondominated. As the search proceeds, the noninferior set gets increasingly larger with greater coverage. The noninferior set generated by the iterative approach using the Hybrid GA is also shown in this figure as a basis for comparing the NSGA-II solutions.

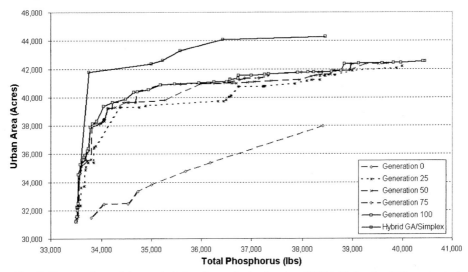

Fig. 7. Progression of the noninferior front obtained using NSGA-II for the MO watershed management problem

The NSGA-II algorithm was also run with different population sizes (30 and 100) for a number of random seeds. Representative results are shown in Figure 8, which illustrates the effect of varying the population size on the quality of the noninferior set. As expected, increasing population size leads to better approximations of the Pareto front over most of the range. However, increased populations also require additional model evaluations. For a run of 100 generations, the average number of model evaluations for population sizes of 30 and 100 is 3,000 and 10,000 respectively. These number of fitness evaluations are still an order of magnitude smaller than that required by the iterative approach using the hybrid GA. Furthermore, while only seven noninferior solutions were found by the hybrid GA approach, NSGA-II was able to find a larger number of noninferior solutions (i.e., the majority of the population) in the final nondominated set.

While all the runs conducted in this study are based on 100 generations, it is important to consider a stopping criterion that could potentially reduce the number of fitness evaluations, which are typically expensive for realistic MO problems. The age of the solutions in the final nondominated set was tracked to determine when the population converges. Table 2 summarizes the typical solution ages for the three different population sizes considered in this study. The maximum age indicates that the final nondominated set is identified significantly early, indicating the potential for an alternative-stopping criterion to terminate the search. One possible stopping criterion that is being considered is a specified number of generations within which little or no change in the nondominated set is observed. This investigation is underway, and the results will be reported in a later publication.

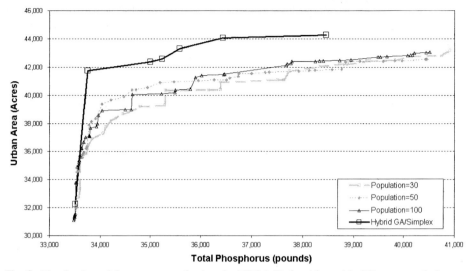

Fig. 8. Nondominated fronts generated using the NSGA-II algorithm with different population sizes

Table 2. Summary of ages (in generations) of the nondominated solutions obtained using NSGA-II that was run for a maximum of 100 generations

Population Size	Minimum Age	Average Age	Maximum Age
30	1	14	40
50	1	14	36
100	1	15	69

4.3 Comparison of Noninferior Solutions in Decision Space

The results of the MO analysis provide the decision maker with important insights into the nature of the problem and its solutions, which may potentially assist during the decision-making process. To highlight this fact, representative solutions from the noninferior set are examined in more detail. Noninferior solutions 1, 4, and 7 from the noninferior set shown in Figure 6 are considered. Table 3 summarizes the land use areas for all five subwatersheds for these three solutions.

The pollutant loading is more limited for noninferior solution 4 than noninferior solution 7. Examination of Table 3 shows that to achieve similar (within 3 percent) levels of overall urban development, more area of the buffered land uses is required for noninferior solution 4 (17,241 acres) than that for noninferior solution 7 (14,740 acres). This is because a buffered area discharges less pollution to the receiving stream than a non-buffered area. Further, a significant improvement in water quality (more than an 8 percent reduction) is achieved via noninferior solution 4 when compared to solution 7 with only a slight reduction in urban area, a trend that continues through noninferior solution 2. This observation would be critical to decision makers in developing a watershed management plan. The solution space also indicates that subwatershed 4 shows the greatest change in urban development. This could be attributed to the relative location and size of that subwatershed and thus its implications on water quality. Consequently, the noninferior tradeoff is greatly affected by the changes in that subwatershed and by those in subwatershed 2.

Table 3. Land use areas for solutions 1, 4, and 7 along the noninferior tradeoff (see Fig.6)

Sub-Watershed	Land Use	Area (acres)		
		Non-inferior Solution 1	Non-inferior Solution 4	Non-inferior Solution 7
1	Non-Urban	648	234	58
	Urban without buffers	2,970	2,993	3,261
	Urban with buffers	81	472	379
2	Non-Urban	3,149	1,168	18
	Urban without buffers	4,946	7,830	9,379
	Urban with buffers	4,597	3,694	3,295
3	Non-Urban	35	21	15
	Urban without buffers	111	143	171
	Urban with buffers	43	27	4
4	Non-Urban	8,321	303	12
	Urban without buffers	11,359	13,609	16,077
	Urban with buffers	7,151	12,918	10,741
5	Non-Urban	67	141	76
	Urban without buffers	731	779	654
	Urban with buffers	253	131	320
Totals	Non-Urban	12,220	1,867	179
	Urban without buffers	20,117	25,354	29,543
	Urban with buffers	12,124	17,241	14,740

5 Conclusions

NSGA-II was applied successfully to a real-world engineering multiobjective optimization problem dealing with watershed management. The results were confirmed using the approximate noninferior front obtained via the ε-constraint method, which was solved by a Hybrid GA/local-search procedure. The noninferior solutions obtained using the multiobjective evolutionary algorithms indicate a strong trade off between urban development and water quality, especially at tighter water quality targets. The NSGA-II approach was an order of magnitude more efficient at generating a good approximation of the Pareto front than the iterative application of a single-objective Hybrid GA/local-search algorithm. The NSGA-II results show little sensitivity to the random seed, and indicate that relatively smaller population sizes can yield the noninferior front with relatively minor sacrifice in accuracy. As no other algorithmic parameter needs to be set, NSGA-II promises to be a stable approach to solving this real world multiobjective optimization problem. Preliminary investigations suggest that different stopping criteria can be employed to avoid unnecessary fitness evaluations towards the end of the search. Future work will explore extending the problem discussed above through the consideration of additional objectives (e.g., other pollutants or levels of point source control) and constraints, as well as larger problem sizes that would constitute more subwatersheds and land use types. In addition, other MOEA approaches will be evaluated and their results will be compared with those reported in this paper.

Acknowledgments. This research was supported by NSF award BES-9733788, and by the Department of Civil Engineering at North Carolina State University.

References

1. Deb, K., *Multi-Objective Optimization Using Evolutionary Algorithms*. 1 ed. Wiley-Interscience series in systems and optimization. 2001, New York: John Wiley and Sons. 497.
2. Ranjithan, S., K. Chetan, and H. Dakshina. *Constraint Method-Based Evolutionary Algorithm (CMEA) for Multiobjective Optimization*. In Zitzler et al. (eds.), Evolutionary Multi-Criteria Optimization, Lecture Notes in Computer Science (LNCS) Vol. 1993, pp. 299–313, Springer-Verlag, Berlin, 2001
3. Knowles, J.D. and D.W. Corne, *Approximating the nondominated front using the Pareto Archived Evolution Strategy*. Evolutionary Computation, 1999. **7**(3): p. 1–26.
4. Zitzler, E., *Evolutionary Algorithms for Multiobjective Optimization: Methods and Applications PhD thesis*. 1999, Swiss Federal Institute of Technology (ETH): Zurich, Switzerland.
5. Deb, K., et al., *A Fast and Elitist Multiobjective Genetic Algorithm: NSGA-II*. IEEE Transactions on Evolutionary Computation, 2002. **6**(2): p. 182–197.
6. Anckorn, P.D. and M. Landers. *Lessons Learned from Turbidity Field Monitoring of 12 Metropolitan Atlanta Streams*. in *Turbidity and Other Sediment Surrogates Workshop*. 2002. Reno, NV: USGS.
7. CH2MHill, et al., *Gwinnett County Watershed Protection Plan*. 2000, Prepared for Gwinnett County, Georgia: Atlanta, GA.
8. Bicknell, B.R., *Hydrological Simulation Program – Fortran, User's manual for version 11*. 1997, U.S. Environmental Protection Agency, National Exposure Research Laboratory: Athens, GA.
9. Kuterdem, C.A., *Integrated Watershed Management Using a Genetic Algorithm-Based Approach*, in *Dept. of Civil and Environmental Engineering*. 2001, NCSU: Raleigh, NC.
10. Nelder, J.A. and R. Mead, *A Simplex Method for Function Minimization*. Computer Journal, 1965. **7**: p. 308–313.

Different Multi-objective Evolutionary Programming Approaches for Detecting Computer Network Attacks

Kevin P. Anchor*, Jesse B. Zydallis, Gregg H. Gunsch, and Gary B. Lamont

Dept of Electrical and Computer Engineering,
Graduate School of Engineering and Management,
Air Force Institute of Technology,
Dayton, OH 45433, USA
{first.last}@afit.edu
http://www.afit.edu

Abstract. Attacks against computer networks are becoming more sophisticated, with adversaries using new attacks or modifying existing attacks. This research uses three different types of multiobjective approaches, one lexicographic and two Pareto-based, in a multiobjective evolutionary programming algorithm to develop a new method for detecting such attacks. The approach evolves finite state transducers to detect attacks; this approach may allow the system to detect attacks with features similar to known attacks. Also, the approach examines the solution quality of each detector. Initial testing shows the algorithm performs satisfactorily in generating finite state transducers capable of detecting attacks.

1 Introduction

Attacks, or intrusions, against computer systems and networks have become commonplace events. Many intrusion detection systems and other tools are available to help counter the threat of these attacks; however, none of these tools is perfect, and attackers are continually trying to evade detection. This paper presents research into detecting attacks using an evolutionary algorithm, specifically evolutionary programming. The algorithm uses multiple objectives to learn to recognize computer attacks, thereby treating the problem of intrusion detection as a multiobjective search problem.

The paper is organized as follows. Section 2 briefly discusses intrusion detection systems, evolutionary programming (EP), and multiobjective evolutionary algorithms (MOEA). The problem addressed in this paper is defined and discussed in Section 3, and Section 4 discusses the EP algorithms chosen to solve the problem. The associated tests, results, and analysis are described in Section 5, followed by the conclusions and future work in Section 6.

---

* The views expressed in this article are those of the authors and do not reflect the official policy or position of the United States Air Force, Department of Defense, or the U.S. Government. This work is sponsored in part by the Air Force Office of Scientific Research.

2 Background

2.1 Intrusion Detection Systems

An intrusion detection system (IDS) helps detect and identify an attack, which is defined as any inappropriate, incorrect, or anomalous activity, on a computer system or network [9]. This definition is used only in the context of security, so violations of policy such as game-playing or viewing inappropriate web sites are not considered an attack. The IDS detects attacks by collecting and analyzing information; this information may consist of network traffic, data from a particular host, or both. This research focuses only on the network traffic, so it is a network-based IDS.

An IDS can also be categorized based on its approach for detecting an attack. The two broad categories are knowledge-based and behavior-based [9]. A knowledge-based system uses knowledge about an attack, such as the pattern or signature of the attack, to determine whether an attack is occurring. If the system does not recognize an attack pattern, then it assumes the data is acceptable. A main disadvantage to this type of system is that an attack that is not in the knowledge base is not detected [9]. The second technique is known as behavior-based intrusion detection. This technique uses a model of known good behavior and then detects deviations from this model. Any behavior that does not match the model is assumed to be an intrusion [9]. Thus, the system can detect new attacks because it does not rely on a knowledge base of known attack patterns; instead, it relies on a "knowledge base" of known good behaviors. For this type of system, the model of known good behavior must be accurate or the system generates many false detection warnings. The research system discussed in this paper is a hybrid of the two forms, as it uses knowledge about an attack and information based on a partial model of known good network traffic, or "self," to evolve finite state transducers (FSTs) that can detect the attack and other similar or related attacks.

2.2 Evolutionary Programming

The motivation for use of Evolutionary Algorithms (EA) originated from attempts to find more efficient and effective methods of solving optimization problems. A further motivation is to solve problems that cannot currently be solved due to a lack of computing power necessary to find the optimal solution in a reasonable amount of time. EAs offer the ability to find a "good" solution, and sometimes the optimal solution, to these problems in an acceptable amount of time. Even though the optimal solution is not guaranteed to be found by any EA, EAs provide a level of confidence in the discovery of a "good" solution that may be better than the solution produced by other methods. These difficult problems involve the maximization or minimization of an *objective* or *fitness function* over some feasible region. The overall goal of this optimization is to find the global optimum over the entire search space.

At a high-level, the standard EAs are all very similar in that they use a model of the biological process of evolution as a framework for the algorithm. However, each class of algorithm has its own representation, reproductive operators, and selection procedure. These differences explain why some EAs are better suited to certain problems; the differences also show that the evolutionary process can be modeled at many levels and in many ways.

One type of EA is EP, which is similar in concept to other EAs but differs in the generation of offspring from the parent population members. Typical EP algorithms utilize a mutation operator and generate one offspring for each parent population member without the use of recombination. A standard EP algorithm begins by initializing a population of individuals randomly. This process is meant to generate a wide spread of solutions within the search space. Once the starting population is generated, all of the members are evaluated based on the defined fitness function. The fitness value assigned to each of the population members is necessary for the selection operators that are utilized in a later step. A mutation operator is applied to each of the population members to "move" them throughout the search space. This is the "searching" process that the EP conducts to find the "best" solution. This offspring population of solutions are evaluated and a selection operator is applied over the combined population to determine which members are fit to become the parent population for the next generation. The algorithm terminates after some specified stopping criteria.

2.3 Multiobjective Evolutionary Algorithms

A relatively new and increased focus of much research is in Multiobjective Evolutionary Algorithms (MOEA) [13]. This area of the EA field is currently of interest to many researchers due to its applicability to real-world problems. In order to understand the concepts applied in the multiobjective version of our algorithm, some terminology must be defined. The process of finding the global maximum or minimum of a set of functions is referred to as Global Optimization. In general, this formulation must reflect the nature of multiobjective problems (MOP) where there may not be one unique solution but a set of solutions found through the analysis of associated Pareto Optimality Theory. MOPs typically consist of competing objective functions, which may be independent or dependent on each other. Many times MOPs force the decision maker to make a choice, which is essentially a tradeoff, of one solution over another in objective space. MOPs are those problems where the goal is to optimize n objective functions simultaneously. This may involve the maximization of all n functions, the minimization of all n functions or a combination of maximization and minimization of these n functions. The formal definition of an MOP is found in [11].

MOPs typically consist of competing objective functions, which may be independent or dependent on each other. An example of this is a company's quest to purchase a backbone for its computer network that provides the greatest throughput at the least monetary cost. These objectives are highly dependent on each other as increased cost brings increased throughput and vice-versa. The term *objective* is used to refer to the goal of the MOP to be achieved and *objective space* is used to refer to the coordinate space within which vectors resulting from the MOP evaluation are plotted [11].

There are three main evolutionary approaches taken to solve MOPs; aggregation approaches, population based non-Pareto approaches, and Pareto-based approaches [6]. In this paper, we use the latter two approaches. The non-Pareto based approach implemented in this paper is a lexicographic approach [6]. This approach involves the rank ordering of objectives based on the priority associated with each. Essentially, each of the fitness functions is applied sequentially to a given population member.

The other multiobjective approaches used here are Pareto-based approaches that utilize the concepts of Pareto Dominance in determining the set of solutions [6]. The concept of Pareto Optimality is integral to determining which members dominate each other. A way to determine if one solution is "better," or dominates another, is a necessity here as well as in all problems. Pareto concepts allow for the determination of a set of optimal solutions in MOPs. Although single-objective optimization problems may have a unique optimal solution, MOPs have a possibly uncountable set of solutions, which when evaluated produce vectors whose components represent trade-offs in decision space.

Pareto optimal solutions are those solutions within the search space whose corresponding objective vector components cannot be all simultaneously improved. These solutions are also termed *non-inferior*, *admissible*, or *efficient* solutions, and their corresponding vectors are termed *nondominated*; selecting a vector(s) from this vector set (the Pareto Front set) implicitly indicates acceptable Pareto optimal solutions (**genotypes**). These solutions may have no clearly apparent relationship besides their membership in the Pareto optimal set. It is simply the set of all solutions whose associated vectors are nondominated; it is stressed here that these solutions are classified as such based on their *phenotypical* expression. Their expression (the nondominated vectors), when plotted in criterion (**phenotype**) space, is known as the *Pareto Front* [11,16].

Solutions on the Pareto Front represent optimal solutions in the sense that improving the value in one dimension of the objective function vector leads to a degradation in at least one other dimension of the objective function vector. This forces the decision-maker to make a tradeoff when presented with the optimal solutions on the Pareto Front. The decision-maker will typically choose only one of the associated Pareto Optimal solutions as being the acceptable compromise solution, even though all of the Pareto Optimal solutions are optimal.

3 Problem Description

This section discusses the intrusion detection problem. The research goal and details of the problem domain are discussed.

3.1 Intrusion Detection Problem Statement

The goal of our research is to develop an innovative method for detecting new or stealthy attacks on the network. One type of stealthy attack, called a "low and slow" attack, is a probe or intrusion attempt that is stealthy in that it takes place over a long period of time, covers a large number of targets, or originates from a number of different, coordinated sources. Because these attacks are designed to be stealthy, they are hard to detect using current intrusion detection systems [15]. Current IDSs can be tuned to detect some stealthy attacks, but the resulting false alarm, or false detection, rate usually increases to an unacceptable level. Thus, new methods for detecting these types of attacks are needed.

New attacks may be modifications of existing attacks [15], so an approach for an ID system is to use knowledge of existing attacks to develop generalized detectors. These generalized detectors might have the ability to detect unknown attacks that are based on existing attacks or that are similar to existing attacks. Developing such generalized detectors is one aspect of the Intrusion Detection (ID) problem. This approach appears to map to the Time Series Prediction problem, in which a sequence of symbols is input and the correct output symbol must be predicted based on the input symbols. In this mapping, the input symbols are a sequence of network packets, and the output symbols represent whether the sequence is assumed to be an attack or not.

3.2 Approach

A network Internet Protocol (IP) packet is made up of a number of fields of information, including routing information, packet function, status flags, and content. Table 1 summarizes some of the main IP and Transmission Control Protocol (TCP) fields[1] that were found to be useful in earlier ID work [14]. Although the packet content or payload is an important part of each packet, it is not used in this research for two reasons. The main reason is that the size of the search space increases immensely if this field is used; the second reason is that existing signature-based detectors can be used to examine the content field in an efficient manner.

Network traffic consists of a sequence of packets, and an attack is also a sequence of packets. The packet features and relationships between features of multiple packets can be used to determine if a particular sequence of packets is an attack or not. Thus, the ID problem for this research focuses on the features shown in Table 1 along with the packet relationships shown in Table 2 to decide whether a particular sequence of packets is an attack or not.

The previous discussion motivates a new method for detecting attacks. This method is to use a finite state transducer[2] (FST) to examine the relationships between packets coming across the network to determine if any particular sequence contains an attack. This type of detection provides the ability to define patterns of known attacks and variations or modifications of known attacks. This method might also detect new attacks that have similar packet relationships as do existing attacks. In addition, this method allows for distinguishing between attack sequences and non-attack sequences because the FST can be built to accept an attack sequence while rejecting a non-attack sequence.

---

[1] Only TCP packets are examined in this effort; however, other IP sub-protocols could be examined in a similar manner since the nature of the algorithm does not specifically exclude any protocol.

[2] The term "finite state transducer" as used here is the same as a Mealy-type finite state machine.

The genotype, or internal representation, of a detector in this scheme is an FST, which represents some regular language or pattern. The phenotype, or outward expression, of the detector is a "Detect" or "Not Detect" signal, which corresponds to the FST rejecting or accepting the word, which represents the network packets that may constitute an attack. The fitness value of a particular FST is dependent on two factors: whether it detects an attack correctly and whether it does not detect a non-attack string as an attack. Because there are multiple factors involved, a multiobjective approach to solving this problem seems a natural fit; the particular multiobjective approach used is discussed in the next section.

Table 1. Packet Features [14]

Field Name	Possible Values
IP Fields	**(All packets)**
IP Identification Number	0-65535
IP Time to live (TTL)	0-255
IP Flags	0-65535
IP Overall Packet Length	0-65535
IP Source Address	Valid IP address
IP Destination Address	Valid IP address
TCP-Only Fields	**(TCP packets only)**
TCP Source Port	0-65535
TCP Destination Port	0-65535
TCP Sequence Number	0-4294967295
TCP Next Sequence Number	0-4294967295
TCP Acknowledgment Number	0-4294967295
TCP Packet Size	0-65535
Individual TCP Flags (*PCWR, Echo, Urgent, Ack, Push, Reset, Syn, Fin*)	Boolean

Figure 1 shows a pedagogical example FST as generated by this approach. The symbols inside the brackets on the transitions are the input symbols that cause a transition; these symbols represent which of the relationships from Table 2 are present between two sequential packets. The symbol after the colon represents the output symbol: "Detect" or "Not Detect."

4 Algorithm

Standard EP approaches have been modified to use representations such as finite state machines (FSMs) and FSTs, so appropriate reproductive operators have already been defined. EP has been repeatedly used to solve problems somewhat similar to the ID problem using only one objective. L. Fogel [5] used EP to

Table 2. Packet Relationships

IP Relationships (for all Packets)	TCP-Only Relationships
Same-IP-Class-A-Source	Same-TCP-Port-Source
Same-IP-Class-B-Source	Same-TCP-Port-Destination
Same-IP-Class-C-Source	Same-TCP-Flags
Same-IP-Class-D-Source	Same-TCP-Seq-Num
Same-IP-Class-A-Destination	Same-TCP-Next-Sequence-Number
Same-IP-Class-B-Destination	Same-TCP-Acknowledgement-Number
Same-IP-Class-C-Destination	Same-TCP-Size
Same-IP-Class-D-Destination	
Same-IP-Flags	
Same-Packet-Length	
Same-IP-Identification-Number	
Same-IP-TimeToLive	

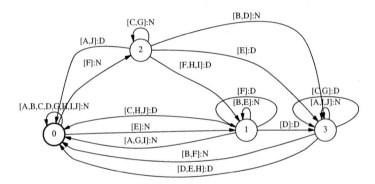

Fig. 1. Example FST

evolve FSMs that were capable of predicting a number in a series, while D. Fogel and Chellapilla [4] later modernized this work and added additional features. Spears [10] used EP to evolve FSMs that were capable of playing a game in which they defended network resources from an opponent; this work implies that EP can be applied to network resource problems.

Given an input attack string, each FST produces some output string. A "detect" signal is generated when an FST outputs the sequence (N,N,...,N,D), where the length of the sequence is the same as the input string length. Thus, any output string that is not in the form of the "detect" signal is considered a non-attack.

Detectors are evolved through the EP process by selecting and reproducing the FSTs that best match the input attack string to the detect sequence. In addition, the algorithm uses some number of self-strings, or known non-attack strings, which the FSTs must not detect as attacks. The fitness functions discussed in subsection 4.1 are the embodiment of this process.

The original EP algorithm for this problem used only the single objective of detecting a given attack string, while the second version added the ability to use a single self-string along with the attack string in a lexicographic multiobjective fashion [1]. The current implementation provides the ability to use an arbitrary number of self-strings along with the attack string in either a lexicographic or one of two pareto-based multiobjective optimization methods. The selection mechanisms used for the current version are described in subsection 4.1.

4.1 Evolutionary Operators for the Algorithm

Since the algorithm for the current project was inspired by the work of Fogel and Chellapilla [4], it is similar to their EP algorithm, with differences such as the selection mechanism and the addition of the ability to use multiple objectives. The particular evolutionary operators used are discussed in the following subsections.

Reproduction. Mutation is the only reproductive operator used in this algorithm. It can take one of the following five forms: change the output symbol, change a state transition, add a state, delete a state, or change the start state [4]. Some internal bookkeeping is required so that a state is not deleted from an FST with only one state or a state is not added to an FST that already has the maximum number of allowed states [4].

One of these five mutations is performed on each population element during every generation; the specific mutation is chosen uniformly. A second or third mutation for each population element is also allowed; the user sets the probability of occurrence for each of these other mutations.

Recombination is not used in the algorithm, since mutation is historically the primary reproductive operator used for EP. However, future versions might incorporate recombination of some sort to further explore the search space. Since the search landscape has not been characterized for this problem, experimentation is required to determine what reproduction operators work well.

Fitness Functions. The first fitness function measures the percentage of FST output symbols that match the sequence (N,N,...,N,D), given the input attack string. This is the only fitness function used in the single-objective version of the algorithm. This fitness function is similar to that of the Boolean satisfiability problem [8]. The input string causes a sequence of output symbols to be generated by each FST; these symbols are then compared to the expected or desired output string of (N,N,...,N,D) to give a fitness value of absolute error in the generated output versus the expected output.

The other fitness functions, used in the multiobjective versions of the algorithm, measure the percentage of FST output symbols that do not match the sequence (0,0,...,0,1), given the particular "self" or non-attack string. Note that the higher the fitness, the fewer symbols match the "detect" sequence, but any fitness value above zero indicates the FST did not detect the "self" input string as an attack. Each of the fitness functions is completely independent of the other fitness functions because each of the "self" strings are different.

Selection. Standard tournament selection with replacement is employed in all versions of the algorithm. Any number of competitors is allowed in the tournament, so the selection pressure can be adjusted as desired.

In the single-objective version of the algorithm, only the first fitness function is calculated and used for the tournament process. Similarly, the lexicographic multiobjective version of the algorithm only uses the first fitness value for the tournament selection procedure. However, when a FST has a fitness of 1.0, which means it detects the attack correctly, the other fitness values are calculated to determine if the FST detects any of the "self" strings as an attack. If any of the strings are detected as an attack, then the FST receives a user-specified penalty on its first fitness score and then continues into the evolutionary programming process with the penalty. On the other hand, the pareto-based multiobjective versions of the algorithm use all fitness values for every tournament.

The two pareto-based methods differ in that the second uses elitism. The next generation's parent population is filled by using any non-dominated solutions in the current parent or child populations, and then using the standard tournament selection procedure to fill any remaining positions.

4.2 Evolutionary Process

Solution Quality. Solution quality refers to the false positive rate and the false negative rate of a detector or set of detectors. The false positive rate, or false alarm rate, measures how often the solution incorrectly detects some set of non-attacks as attacks, while the false negative rate measures how often an actual attack is not detected. These rates are a key factor in determining the usefulness of an intrusion detection system, so they are important to test.

Process Overview. Figure 2 shows an overview of the process of evolving FSTs that detect some attack. The EP approach described in Section 4 is shown as the first step, while further testing and analysis of the solution quality is shown as the second step of the process. The EP approach, using the Lexicographic or one of the Pareto-based approaches, finds one or more solution FSTs that are on the Pareto Front; i.e., $F_1 = 1.0$. Then, the solution quality of each is tested.

This two-step approach is a novel way of using the Pareto-based idea. Any solution on the Pareto front is a valid solution which can be fed into the second part of the process; the resulting measures of solution quality can then be used along with external factors to make decisions about which solutions to put in the IDS. For example, based on the different solution quality measures, some solutions may be preferred when certain threats are expected, or some solutions may be preferred if the system administrator wants to have few false alarms (in a tradeoff with the number of false negatives).

5 Experimental Design, Results, and Analysis

Earlier testing showed that the single-objective, lexicographic, and non-elitist pareto-based versions of the multiobjective EP algorithms are capable of finding

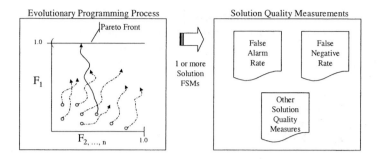

Fig. 2. Process Overview for Detecting Network Attacks

solutions, or good detectors, over a range of input attack strings [1,3]. The current tests are designed to examine how the elitist pareto-based version performs compared to the lexicographic and non-elitist pareto-based versions. In addition, the solution quality is examined for the different versions of the algorithm. The tests use an actual attack from a scan tool called *Queso*. Each test run also uses five self strings that are generated by randomly changing two positions of the input attack string. The number of positions to change was chosen to keep the self strings "close" to the original attack string in terms of a distance metric. Thus, each test uses one attack string and five self strings, for a total of six fitness functions to be maximized.

The time and number of generations needed to evolve "good" detectors are tracked; a "good" detector is one which detects the input string developed from the sequence of attack packets and generates a "Detect" signal, while not generating a "Detect" signal for any of the self-strings. The solution quality, as described in Section 4.2, is also tracked for these tests.

Each false alarm test uses 100,000 non-attack strings created by randomly changing a specified number of the symbols in the input attack string. Thus, each of the 100,000 test strings is "close" to the attack string but does not represent an actual attack. The false alarm tests are conducted ten times, and the number of changed symbols varies between one and ten. Also, each of the tests is conducted with ten replications.

The false negative tests also use 100,000 test strings. For these tests, between one and ten symbols are randomly inserted into the attack string to represent an attacker trying to perform a stealthy attack. Each new attack string still contains the attack string, so each should be detected as an attack. As with the false alarm tests, these tests are performed with ten replications.

5.1 Test Setup and Parameters

This testing is performed on a dedicated Athlon XP 1800+ computer with 512 MB RAM and the Windows XP Professional Edition operating system. The algorithm is implemented in Java 1.3.1_01 and is compiled to a native Windows executable using the JOVE 2.0 compiler. The random number generator is Sean

Luke's Java implementation of the Mersenne Twister algorithm, which has a longer period and is faster than the standard Java random number generator [7].

Since the three different multiobjective approaches are being compared, each needs to use the same parameters. The parameters used allow a maximum of 20 states for any FSM and a population size of 100. These parameters were chosen based on previous internal testing to show the feasibility of using the multiobjective approaches. Each experiment was performed with ten replications, where each replication is allowed to execute until a solution is found. Different random seeds were used for each of the test runs.

5.2 Test Results and Analysis

Table 3 summarizes the average time and number of generations needed to find a solution FST for the tests. Figure 3 shows the boxplots which present a graphical view of the test results for the number of generations and time required to find a solution and provide some appreciation for the distribution of the results. The boxplot provides a method for graphically depicting the distribution of a dataset and comparing multiple datasets. The box contains the middle 50% or the interquartile range (IQR) of the distribution. The unmarked line inside the box represents the median, while the line with square symbols represents the average. The lines extending from the top and bottom of the box contain any points within 1.5*IQR; any points outside of this range are shown as outliers [12].

Table 3. Summary of Test Results

Test Name	Avg Time (s)	Time StdDev	Avg Gen	Gen StdDev
Lexicographic	296.12	224.09	338.40	285.61
Pareto-based	5441.33	6962.71	6750.00	8692.65
Pareto-based with elitism	25.45	15.55	51.70	58.32

As can be seen in the figures, there is a large variance in the time to execute and the number of generations required to generate a solution for the tests. This is expected as the process is based on random events, but the variance and averages in the non-elitist Pareto-based testing are significantly larger than those noticed in the other tests. The outliers in these Pareto-based tests skews the average higher.

The box plots also illustrate the fact that the time and generations required for each set of tests seem to be related. This observation is expected, since the time to execute one generation is fairly constant during the course of a test run.

Additionally, we note that the Lexicographic and elitist Pareto-based tests converge to a solution in less time than the non-elitist Pareto-based tests, in all of the test runs. At an initial glance this would lead us to state a preference for using either the Lexicographic or elitist Pareto-based approach, but there is a

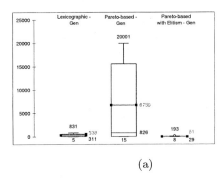 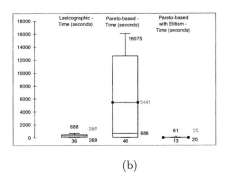

(a)   (b)

Fig. 3. Comparison of Tests Results for (a) Number of Generations, and (b) Run Time

need to further analyze the overall quality of the solutions found to determine which method performs statistically better or if they are equivalent in terms of solution quality.

Figure 4 (a) shows the average of the results for the false alarm testing. The results for all three methods are consistent; the false alarm rate is fairly high when the non-attack strings are "close" to the actual attack but drops off to zero as the non-attack strings become very different from the actual attack. The Pareto-based method with elitism consistently has the lowest average false alarm rate, with the Lexicographic method very close behind.

Figure 4 (b) shows the average of the results for the false negative testing. Again, the results for all three methods are similar; the false negative rate is low when only a few symbols are inserted but rises rapidly when several symbols are inserted. For these tests, the Lexicographic tests always produce the lowest false negative rate, while the Pareto-based method with elitism always produces the highest false negative rate.

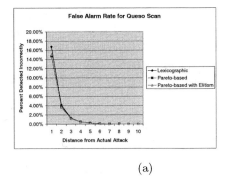 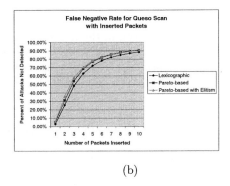

(a)   (b)

Fig. 4. Comparison of Tests Results for (a) False Alarm Rate, and (b) False Negative Rate

Figure 5 presents the false alarm and false negative rates when a process called costimulation is used. The idea of costimulation is derived from our work using Artificial Immune Systems [2]; it involves treating a collection of detectors in a system as cooperative detectors. In order for the system to "detect" a string as an attack, multiple individual detectors have to detect that string as an attack. If only a single FST detects the string as an attack, then the system ignores the string. The goal of costimulation is to reduce the amount of false alarms generated by the IDS; however, the tradeoff is that the false negative rate is likely to go up when the process is used. It can be considered one of the "Other Solution Quality Measures" shown in the Second Step of the Process in Figure 2.

The false alarm rates shown in Figure 5(a) and the false negative rates shown in Figure 5(b) follow the expected pattern. The false alarm rates drop dramatically when costimulation is used; for example, the false alarm rate for the same set of FSTs generated using the Lexicographic approach drops from about 17% (at distance metric 1) without costimulation to just over 5% when costimulation of two detectors is used. It drops even further to about 3% if costimulation of three detectors is used. However, the false negative rate rises when costimulation is used. Using the same Lexicographic example, the false negative rate rises from about 3% without costimulation to about 15% when costimulation of two detectors is used and to over 35% when costimulation of three detectors is used.

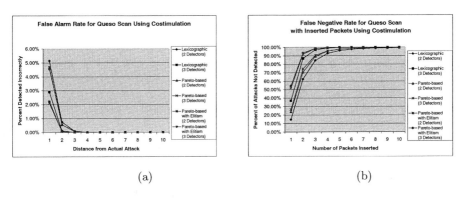

Fig. 5. Results using Costimulation for (a) False Alarm Rate, and (b) False Negative Rate

6 Conclusions and Future Work

This research presents our work in detecting computer network intrusions through the use of three different multiobjective evolutionary programming algorithms. The testing shows that the evolutionary programming technique generates finite state transducers, or Mealy-type finite state machines, capable of matching or detecting an input attack string, for the attack tested. The lexicographic approach allows the use of the attack and "self" strings while performing significantly faster than the naive Pareto-based approach; however, an improved Pareto-based algorithm that uses elitism in the selection mechanism generates solutions faster than the Lexicographic method. All of the methods produce solutions that perform acceptably in terms of solution qulaity, or false alarm and false negative rate. The Pareto-based method with elitism produces solutions with the lowest false alarm rate but the highest false negative rate; it is also the fastest method at producing solutions. The Lexicographic approach also performs quickly and produces solutions with the lowest false negative rate but a slightly higher false alarm rate than the Pareto-based method with elitism. Since both of these algorithms perform fairly well, a trade-off between false alarm and false negative rate could be used to determine which method to use for a given application or threat level, or both methods could be used to generate some detectors in the system if so desired.

Future research will test the algorithm on more real-world attacks, and an investigation into using more complex fitness functions is planned

References

1. Kevin Anchor, Gary Lamont, and Gregg Gunsch. An Evolutionary Programming Approach for Detecting Novel Computer Network Attacks. In *Proceedings of the 2002 Congress on Evolutionary Computation*, pages 1618–1623, Honolulu, 2002. IEEE Press.
2. Kevin Anchor, Paul Williams, Gregg Gunsch, and Gary Lamont. The Computer Defense Immune System: Current and Future Research in Intrusion Detection. In *Proceedings of the 2002 Congress on Evolutionary Computation*, pages 1027–1032, Honolulu, 2002. IEEE Press.
3. Kevin Anchor, Jesse Zydallis, Gary Lamont, and Gregg Gunsch. A Multiobjective Evolutionary Approach for Detecting Computer Network Attacks. In *Second Workshop on Multiobjective Problem Solving from Nature, in association with PPSN VII: The Seventh International Conference on Parallel Problem Solving from Nature*, Granada, Spain, 2002.
4. David B. Fogel and Kumar Chellapilla. Revisiting Evolutionary Programming. In S.K. Rogers, D.B. Fogel, J.C. Bezdek, and B. Bosacchi, editors, *SPIE Aerosense98, Applications and Science of Computational Intelligence*, pages 2–11, Orlando, FL, 1998.
5. L. J. Fogel, A. J. Owens, and M. J. Walsh. *Artificial Intelligence through Simulated Intelligence*. John Wiley, NY, 1966.
6. C M Fonseca and P J Fleming. Multiobjective Optimization. In Thomas Bäck, David B. Fogel, and Zbigniew Michalewicz, editors, *Evolutionary Computation 2 Advanced Algorithms and Operators*, volume 2, pages 25–37. Institute of Physics Publishing, Bristol (UK), 2000.
7. M. Matsumoto and T. Nishimura. Mersenne twister: A 623-dimensionally equidistributed uniform pseudorandom number generator. *ACM Transactions on Modeling and Computer Simulation*, 8(1):3–30, 1998.
8. Z. Michalewicz and D. Fogel. *How to Solve It: Modern Heuristics*. Springer-Verlag, Berlin, 2000.

9. SANS. *Intrusion Detection FAQ*. SANS Institute, 2001. World Wide Web Page. URL http://www.sans.org/newlook/resources/IDFAQ/ID_FAQ.htm.
10. William Spears and Diana Gordon. Evolving Finite-State Machine Strategies for Protecting Resources. In *Proceedings of the International Symposium on Methodologies for Intelligent Systems 2000*. ACM Special Interest Group on Artificial Intelligence, 2000.
11. David A. Van Veldhuizen. *Multiobjective Evolutionary Algorithms: Classifications, Analyses, and New Innovations*. PhD thesis, Department of Electrical and Computer Engineering. Graduate School of Engineering. Air Force Institute of Technology, Wright-Patterson AFB, Ohio, May 1999.
12. Stephen Vardeman and Marcus Jobe. *Statistical Qualtiy Assurance Methods for Engineers*. John Wiley and Sons, Inc., New York, 1999.
13. David A. Van Veldhuizen and Gary B. Lamont. Multiobjective Evolutionary Algorithms: Analyzing the State-of-the-Art. *Evolutionary Computation*, 8(2):125–147, 2000.
14. Paul Williams, Kevin Anchor, John Bebo, Gregg Gunsch, and Gary Lamont. CDIS: Towards a Computer Immune System for Detecting Network Intrusions. In *Proceedings of the 4th International Symposium, Recent Advances in Intrusion Detection 2001*, pages 117–133, Berlin, 2001. Springer-Verlag.
15. Paul D. Williams. Warthog: Towards a Computer Immune System for Detecting "Low and Slow" Information System Attacks. Master's thesis, AFIT/GCS/ENG/01M-15, Graduate School of Engineering and Management, Air Force Institute of Technology (AU), Wright-Patterson AFB, OH, March 2001.
16. Jesse B. Zydallis, David A. Van Veldhuizen, and Gary B. Lamont. A Statistical Comparison of Multiobjective Evolutionary Algorithms Including the MOMGA–II. In Eckart Zitzler, Kalyanmoy Deb, Lothar Thiele, Carlos A. Coello Coello, and David Corne, editors, *First International Conference on Evolutionary Multi-Criterion Optimization*, pages 226–240. Springer-Verlag. Lecture Notes in Computer Science No. 1993, 2001.

Safety Systems Optimum Design by Multicriteria Evolutionary Algorithms

David Greiner, Blas Galván, and Gabriel Winter

Evolutionary Computation and Applications Division-CEANI
Intelligent Systems and Numerical Applications in Engineering Institute-IUSIANI
35017, Campus de Tafira Baja, Las Palmas de Gran Canaria University, Spain
dgreiner@iusiani.ulpgc.es, {bgalvan, gabw}@step.es

Abstract. In this work new safety systems multiobjective optimum design methodologies are introduced and compared. Various multicriteria evolutionary algorithms are analysed (SPEA2, NSGAII and controlled elitist-NSGAII) and applied to a Containment Spray Injection System of a nuclear power plant. Influence of various mutation rates is considered. A double minimization is handled: unavailability and cost of the system. The comparative statistical results of the test case show a convergence study during evolution by means of certain metrics that measure front coverage and distance to the optimal front. Results succeed in solving the problem.

1 Introduction

The concepts considered in the Engineering discipline known as Systems Design have experienced considerable evolution in the last decade, not only because the growing complexity of the modern systems, but also by the change of criterion which implies the necessity to obtain the optimum designs instead of merely adequate ones. Design requirements are specially strict for systems whose failure implies damage to people, environment or facilities with social-technical importance. For such systems, modern safety requirements force to consider complex scenarios with many variables, normally developed under Probabilistic Risk Assessment frameworks. In such systems, special attention must be paid to the Safety Systems Design, considering design alternatives (at the design stage) and/or maintenance strategies (during the system operation), in order to perform a global optimization. In both cases different objectives under conflict will be considered at the same time by the decision-makers, so multiobjective methods are specially suited to be used.

Key points in Safety Systems Design Optimization (SSDO) are the use of advanced mathematical models of the systems, and powerful multiobjective optimization methods. The Fault Tree Analysis (FTA) is the most widely used mathematical model, while Evolutionary Algorithms have been successfully used for optimization. Although big advances have been introduced in last years, important problems require more development in order to facilitate the application to a broad variety of complex technical systems, specially the use of more efficient multiobjective optimization methods.

Systems Optimization considering Reliability has been studied from more than one decade [1], but SSDO using evolutionary optimizers have been introduced recently [2], and continue in last years [3] towards more complex systems. The mentioned contributions include both the system design and the maintenance strategy during the system operation. In certain problems, only one of both is required, therefore they have been also studied independently (maintenance strategy of existing systems [4][5]; and system design [6]). Unavailability and Cost have been considered as the main functions to optimize in both single objective (to minimize Unavailability subject to Cost constraints) and multiobjective (to minimize Unavailability and Cost)[7]. Evolutionary Algorithms [8] [9] have been the preferred optimization method by authors [10], mainly because of their ability to search for solutions in complex and big search spaces.

System design options have been considered using Fault Trees with appropriate classic logic gates to permit component and redundancy-diversity levels selection [11]. In last years new logic gates have been developed, permitting to model complex, Dynamic [12] or Fault Tolerant [13], systems. The main task concerned to Fault Tree models is the quantitative evaluation, which have been performed using the classical Minimal Cut Sets (MCS) approach [14], or the Binary Decision Diagrams (BDD) [15][16]. Both approaches have well known drawbacks: The former have exponential growth with the number of basic events and/or logic gates of the tree, while the later depends on a basic events ordering without mathematical foundations, supported only in a few existing heuristics [17][18]. Experiences with real systems demonstrate poor results when using MCS, and prohibitive computational cost of fault tree assessment when using BDD. Consequently, new approaches are now under development like the use of Boolean Expression Diagrams (BED) [19], Direct Methods based on an Intrinsic Order Criterion (IOC) of basic events [20][21], or Monte Carlo methods based on powerful Variance Reduction Techniques [22].

In this work a comparative study of the performance of the last and more advanced multiobjective methodologies is presented when apply to SSDO problems. The main characteristics of the approach are:

- Binary Decision Diagrams construction is bypassed using a direct and efficient method (the 'weight method' [21][22]) to evaluate fault trees.
- A proper codification of the problem is used integrating design constraints and allowing to reduce the search space and the non-feasible solutions appearance.
- An analysis of exploitation and exploration is done searching for an efficient multicriteria evolutionary optimization method, and taking into account the SPEA2 (with an improved truncation operator[23]), NSGAII, and two controlled elitist NSGAII algorithms.

In the following sections the main characteristics of the used methodology are described. In Section 2 the safety systems design is introduced. Section 3 presents the used multicriteria evolutionary algorithms. Section 4 describes two test cases handled for a Safety System of a Nuclear Power Plant, and particular considerations. In Section 5 the results are described and analysed, and the paper finalizes with the Conclusions section.

2 Safety Systems Optimum Design

The Safety Systems mission is to operate when certain events or conditions occur, preventing hazardous situations or mitigating its consequences. Where possible, design of such systems must ensure the mission success even in the case of single failure of its components, or when external events can prevent entire safety subsystems from functioning. The designer(s) must optimize the overall system performance deciding the best compromise among: Component Selection, Component/Subsystems allocation and, in some cases, Time Intervals between Preventive Maintenance.

In SSDO problems the system performance measure is normally the *System Unavailability*, being a second important variable the total *System Cost*. The whole optimization task is constrained, because practical considerations place limits on available physical and economic resources.

System components are chosen from possible alternatives, each of them with different nature, characteristics or manufacturers. Component/Subsystems allocation involves deciding the overall system arrangement as well as the levels of redundancy, diversity and physical separation. Redundancy duplicates certain system components, diversity is incorporated achieving certain function by means of two, or more, different components / subsystems. Physical separation among subsystems guarantees the mission success, even in the case of critical common causes of malfunctioning.

In automated SSDO problems, a computer program will be in charge to obtain the optimum design considering all abovementioned alternatives and restrictions. There are two main components in such software: The system model and the optimization method.

2.1 The System Model

The system model must reflect all possible design alternatives, normally using as starting point the knowledge of the design team and the physical constraints on resources. A set of binary variables termed "indicator variables" permit to select (switch "on") or reject (switch "of") the different configuration alternatives. For each configuration selected, the system Unavailability and the Cost must be computed. The Fault Tree Analysis (FTA) is the most widely accepted methodology to model safety systems for design optimization purposes. A Fault Tree is a Boolean logic diagram used to identify the causal relationships leading to a specific system failure mode [14]. The system failure mode to be considered is termed the "top event" and the fault tree is developed using logic gates below this event until component failure events, termed "basic events", are encountered. H.A. Watson of the Bell Telephone Labs developed the Fault Tree Analysis in 1961-62, and the first published papers were presented in 1965 [24]. Since these dates, the use of the methodology has become very widespread. Important theoretical [14][25], methodological [26] and practical [27] advances have been developed. Figure 1 shows a simplified Fault Tree of the system depicted in figure 3.

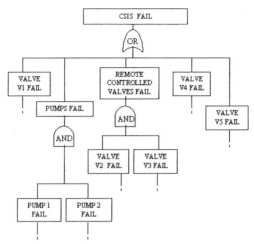

Fig. 1. Simplified Fault Tree of the Containment Spray Injection System (CSIS) of a Nuclear Power Plant (figure 3)

For SSDO problems one important contribution was the possibility to introduce design alternatives into the fault tree [11], by means of associate the "house events" used in Fault Tree logic to "indicator variables" of the system design. Figure 2 shows a Fault Tree section, which permits to choose between two models of pumps and his redundancy level (one or two pumps). Among other system measures [14], the FTA permits to calculate the system unavailability for every configuration of the "house events" (indicator variables). Along the time many methods have been developed for quantitative evaluation of fault trees.

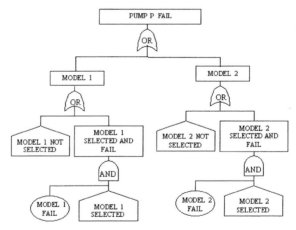

Fig. 2. Example of Fault Tree with design alternatives

During the 90´s was presented and developed, probably, the most important contribution to this task: the use of the FT conversion to Binary Decision Diagrams (BDD)[15][16] that are suitable for easy computation of the top event measure (in our case, the system unavailability). Despite the widespread development and use of the methodology, one important inconveniency still remains unresolved: the determination of the basic event ordering necessary to obtain the simplest BDD suitable for easy computation, among the set of BDDs associated to one Fault Tree. Many contributions have been presented in the field [17][18], but to date only a few heuristics support the selection of the adequate basic event ordering. That is the reason to evaluate fault trees. We use in this work an efficient direct algorithm, called the 'weight method', which provides exact lower and upper bounds on the system unavailability for a pre-specified maximum error. This algorithm is based on a Theorem that establishes an intrinsic order criterion (*IOC*) for elementary state probabilities. *IOC* "a priori" allows choosing the most relevant elementary states to compute the top event probability, with independence on the system basic probabilities. *IOC* is exclusively based on the positions of 0s and 1s in the binary n-tuples of the elementary states. This method was exhaustively described [21][22] in the past years.

2.2 The Optimization Method

Evolutionary Algorithms [8] [9] [10] have been the most used optimization method. An important matter previous to use evolutionary algorithms is to decide the coding scheme of candidate solutions. Binary coding or real coding can be used, being the operators to use with each scheme totally different. In SSDO problems, considering design alternatives and maintenance strategies, the most used coding scheme was binary [2]. Although there are many applications of advanced mixed-integer EAs and MOEAs in the literature, recently authors reported convergence problems in the SSDO field, [3] because the different nature of the variables involved, on the one hand strict binary variables (design alternatives), on the other hand real variables (maintenance strategies). In the following section the optimization method handled in the present article will be described deeply.

3 The Evolutionary Approach

If the improvement in one criteria implies the worsening in another objective, as happens with the unavailability and the cost of the system, which are our two minimising objective functions, a multiobjective optimization is required. Searching for solving efficiently the problem handled, an analysis of different multicriteria algorithms and mutation rates has been taken into account, focusing in the exploitation-exploration balance. Among the most recent algorithms [28], those which include elitism and parameter independence are outstanding [29]. A first comparison of multicriteria evolutionary methods can be seen in [30]. We have selected for our comparison here, the SPEA2 [31], NSGA-II [32], and NSGA-II with Controlled Elitism (CE) [33]:

The SPEA2 algorithm was developed by the authors of SPEA to withdraw their weaknesses. Compared with SPEA, NSGAII and PESA in five test functions, SPEA2 and NSGAII were the most competitive algorithms [34]. The SPEA2 uses a truncation operator to maintain the diversity along the non-dominated front. Here is implemented an improved adaptation of the truncation operator specially suited for two dimensional multicriteria problems, proposed in [23]. It takes advantage of the linearity of the distribution of the non-dominated solutions in the bicriteria case, allowing a lower time consuming. When ordering the non-dominated set by one of the criteria, we obtain a sequence of consecutive non-dominated solutions that are adjacent, permitting to reduce the calculations of the distance among non-dominated solutions. For a size of L individuals, we just need to calculate L-1 distances, and can deduce the rest by their relative positions. Therefore, the total computational cost of the truncation operator is reduced in $O(L/2)$.

Controlled Elitism offers a quantitative control over the algorithm selection pressure. Its activity is focused in two factors: a) It smoothes the elitism of the NSGA-II by substituting the tournament selection with a random procedure to create the offspring population from the parental one; b) It also limits the maximum number of individuals belonging to each front, by a geometric decreasing function governed by the reduction rate r. Being K the number of desired fronts, N the total number of individuals in the population, and r the selected reduction factor, $(r < 1)$, then the number of individuals n_i in each i front, is [33]:

$$n_i = N \frac{1-r}{1-r^K} r^{i-1} \qquad (1)$$

Two metrics are considered, defined on objective space, concerning about accuracy and coverage of the front. They are averaged from thirty independent runs of each algorithm and represented during the convergence process in figures 6 to 12.

The first metric (Metric 1) is the M1* metric of Zitzler [35], representative of the approximation to the optimal Pareto front. To evaluate this metric, belonging to the scalable metrics type, the best Pareto front should be known. Its expression is [35]:

$$M1^*(U) = \frac{1}{|U|} \sum_{u \in U} \min\{\|u-y\| \mid y \in Y_p\} \qquad (2)$$

where $U = f(A) \subseteq Y$ (being A a set of decision vectors, Y the objective space and Yp referred to the Pareto set).

The second metric (Metric 2) handles with the coverage of the front. The adopted criteria is a scaled value of the M3* metric of Zitzler [35]:

$$M3^*(U) = \sum_{i=1}^{popsize} \max\{\|u_i - v_i\| \mid u, v \in U\} \qquad (3)$$

The Metric 2 is computed as:

$$Metric\ 2 = \frac{M3^*}{Max\ M3^*} \qquad (4)$$

so it means that achieving a value of one, the maximum coverage of the front is obtained. In the test case 2, it is observed that the most difficult solutions to achieve are the lower right, those corresponding to the values of cost higher than 912, as can be seen in table 2 and at right part of figure 4. That means, that the lack of whole coverage of the front seen in the figure results, is due to the difficulty to reach to them.

4 Test Cases

4.1 Description

A simplified model diagram of the containment spray injection system (CSIS) of a nuclear power plant can be viewed in fig 3, where V means valve, and P means pump. The CSIS mission is the injection of the sufficient amount of borated water into the containment in order to wipe radioactive contamination released after a loss of coolant accident. The values of cost and unavailability of the system components are described in table 1, being C the value of the component cost and P, the unavailability probability of the component. For valves one and four, three different models are possible; for the rest of the valves and the pumps, two distinct models are considered. The whole solutions set and the Pareto set can be viewed at left of fig 4, where the x-axis represents the total cost of the system and the y-axis represents its unavailability.

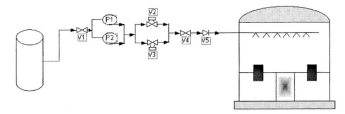

Fig. 3. First Test Case Diagram

Table 1. First Test Case Parameter Values

	V1, V4	V2, V3	V5	P1,P2
Model 1	P=2.9E-3 C=50	P=3.0E-3 C=65	P=5.0E-4 C=37	P=3.5E-3 C=90
Model 2	P=8.7E-3 C=35	P=1.0E-3 C=70	P=6.0E-4 C=35	P=3.8E-3 C=85
Model 3	P=4.0E-4 C=60			

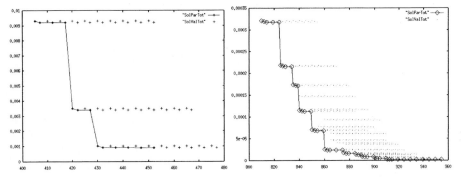

Fig. 4. Solution and Pareto Sets for Test Cases 1 and 2, respectively

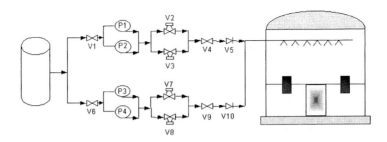

Fig. 5. Second Test Case Diagram

If we duplicate the line, due to safety reasons, we obtain the diagram represented in figure 5, where the second branch components have the same parameters as the first. The search space of the problem grows exponentially, as can be seen at right side of figure 4. The value of the Pareto solutions are represented in table 2. It can be observed, that the cost has increased respect to the first test case, but the unavailability of the system has decreased significantly.

Table 2. Pareto Set Values of Test Case 2

Cost	Unavailability	Cost	Unavailability
810	0.0003217533	877	0.0000161601
812	0.0003199908	879	0.0000160294
814	0.0003182379	884	0.0000159571
819	0.0003181330	885	0.0000138547
824	0.0003180280	887	0.0000128842
825	0.0002186917	889	0.0000127531
827	0.0002169189	890	0.0000091560
829	0.0002157307	892	0.0000085154
834	0.0002156252	894	0.0000083839
835	0.0001742693	899	0.0000083462
837	0.0001724920	900	0.0000056120
839	0.0001715472	902	0.0000052204
840	0.0001150281	904	0.0000050886
842	0.0001132449	909	0.0000050657
844	0.0001126247	910	0.0000020593
849	0.0001125186	912	0.0000019171
850	0.0000703462	914	0.0000017850
852	0.0000685586	919	0.0000017771
854	0.0000681832	924	0.0000017692
859	0.0000680768	929	0.0000017666
860	0.0000255526	934	0.0000017640
862	0.0000237605	939	0.0000017625
864	0.0000236305	944	0.0000017610
869	0.0000235239	949	0.0000017596
874	0.0000234884	954	0.0000017583
875	0.0000173781		

4.2 Evolutionary Considerations

In order to study the effect of both factors of CE independently, we use two values of the r parameter, zero and 0.4, imposing four fronts (K=4) and, therefore, considering four different algorithms in our analysis (SPEA2, NSGAII and two controlled elitist NSGAII).

Thirty independent executions per tested algorithm, and three different uniform mutation rates (0%, 0.4% and 3%) have been considered. So, it summarizes a total of twelve procedures, balancing exploitation and exploration.

The population size is one hundred individuals (in case of NSGA-II with CE and r=0.4, the imposed population size of the fronts are: 62, 25, 10 and 3). A uniform crossover with crossover rate of 1.0, and binary codification, are considered.

The constraint handling in this problem is of significant importance. It is related to respect the configuration of the described problem, taking into account the alternatives of design. The standard binary codification represents in each bit every single alternative of components design [3], considering 1 to selection and 0 to rejection. We propose to integrate the constraints into the coding of the chromosome, allowing to reduce the search space and the possible infeasible solutions, whose fault tree for the unavailability would not be evaluable [30]. Its significance in the test cases follows:

First Test Case: Feasible solutions: 288; Pareto Optimal: 17. Search space size for the naïve binary codification with 16 bits = 65536 solutions. Search space size for the constraint integrated binary codification with 9 bits = 512 solutions

Second Test Case: Feasible solutions: 82944; Pareto Optimal: 178. Search space size for the naïve binary codification with 32 bits = 4294967296 solutions. Search space size for the constraint integrated binary codification with 18 bits = 262144 solutions

Because of the bigger size of the search space, we concentrate the analysis in the second test case, being the first one very small with the constrained codification to be analyzed.

5 Analysis of Results

The average results of the thirty executions for each metric are presented. At the left side is represented Metric 1 (accuracy), and at the right side is represented Metric 2 (coverage). The x-axis corresponds to the generation number, and the y-axis to the value of the referred metric. Figures are organized in two groups. In the first section (figs. 6 to 9) are shown the comparative results grouped by algorithm and comparing the different studied mutation rates. In this figures, the symbol diamond, cross and square correspond to mutation rates of 3%, 0.4% and 0% respectively. A second section (figs. 10 to 14) shows the comparative results grouped by mutation rate and comparing the different algorithms. Figures 10 to 12 represent the metrics average, and figures 13 and 14 represent the metrics variance. In this figures, the symbol diamond, cross, square and blade correspond to algorithms NSGA-II with controlled elitism and r=0.4; NSGA-II with controlled elitism and r=0.0, NSGA-II and SPEA2, respectively.

5.1 Analysing Each Algorithm with Different Mutation Rates

From the observation of Metric 1 (left part of figures) from figures 6 to 9, it can be seen a common behaviour in the four tested algorithms. In the initial generations, the higher mutation rate (diamond) is the one which exhibits a slower convergence to the Pareto optimal front, and the one which lacks of mutation is the fast one (square). The algorithm with 0.4% mutation rate is between them. The lower exploration allows a higher exploitation that increases the selection pressure around the best individuals found. We observe however, that the difference is not significant, and about generation twenty all the cases reach similar approach to the optimum non-dominated front.

Metric 2 is indicative of the total coverage of the front. Here the differences between mutation rates are more noticeable. The common behaviour is presented in figures 6, 7 and 9, (right part of figures) where the required diversity introduced into the population by the higher mutation rate, is essential to localize the most right outer points of the front. The diamonds line is the one more proximal to the best value of one in all the algorithms studied. We observe that a mutation rate of 3% follows to a convergence speed towards the front not much away of the others, and allows a better coverage of the front.

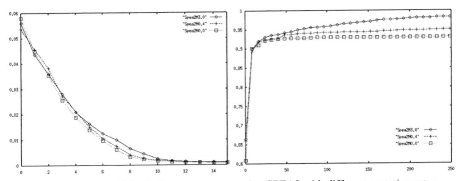

Fig. 6. Metric 1 and Metric 2 averages, comparing SPEA2 with different mutation rates

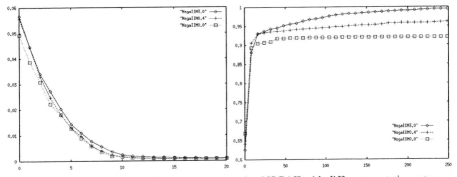

Fig. 7. Metric 1 and Metric 2 averages, comparing NSGAII with different mutation rates

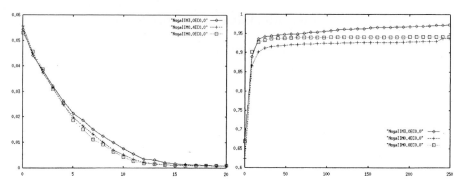

Fig. 8. Metric 1 and Metric 2 averages, comparing NSGAII with CE (r=0.0) with different mutation rates

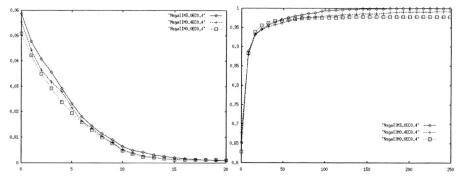

Fig. 9 Metric 1 and Metric 2 averages, comparing NSGAII with CE (r=0.4) with different mutation rates

5.2 Analysing the Algorithms for a Fixed Mutation Rate

From the observation of Metric1 (left part of figures) from figures 10 to 12, it can be seen a common behaviour in the four tested algorithms. In the initial generations, the two algorithms with controlled elitism (CE) (diamond and cross) are the one which experiments a slower convergence to the Pareto optimal front, and the NSGAII, which has the highest selection pressure (elitism and tournament selection), is the fast one (square). As has been deduced also in section 5.1, here again the figures show, that, increasing the exploitation balance favours a faster search in the initial stages. However, then the diversity is decremented, as observed in Metric 2. The algorithm which introduces more diversity into the population, NSGAII with CE and r=0.4, reaches the highest values of the coverage measure. It is remarkable the positive collaboration between higher mutation rates and the controlled elitism, as can be observed at the right part of figure 12, showing that controlled elitism and mutation can be complementary, interacting simultaneously in a positive way. Other studies, such as Laumanns et al. [36] applied to the bi-objective knapsack-problem, find that strong elitism and high mutation rates has the best overall performance, and there is a certain optimum combination of the two.

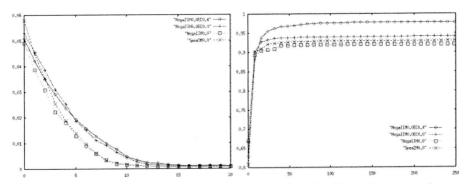

Fig. 10. Metric 1 and Metric 2 averages, comparing the four algorithms without mutation

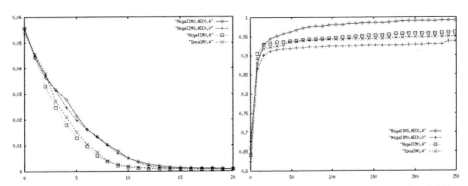

Fig. 11. Metric 1 and Metric 2 averages, comparing the four algorithms with mutation 0.4%

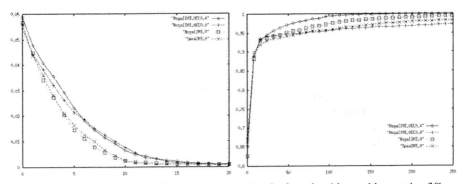

Fig. 12. Metric 1 and Metric 2 averages, comparing the four algorithms with mutation 3%

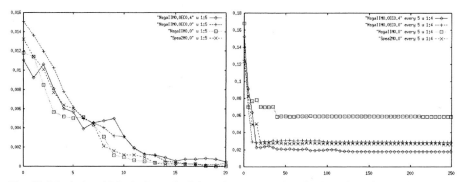

Fig. 13. Metric 1 and Metric 2 standard deviations, comparing the four algorithms without mutation

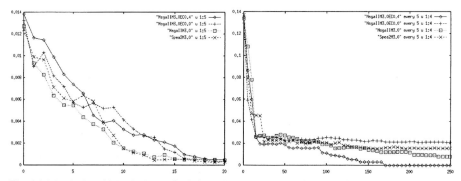

Fig. 14. Metric 1 and Metric 2 standard deviations, comparing the four algorithms with mutation 3%

6 Conclusions

A new, successful and efficient methodology has been introduced for safety systems design multiobjective optimization. This methodology implies: the bypassing of BDD construction using the 'weight method' [21][22]; a binary codification that integrates constraints of design avoiding many unfeasible solutions; and an efficient multiobjective algorithm.

A comparative study of some of the most recent multicriteria evolutionary algorithms is being handled, showing that the NSGAII with controlled elitism and lateral diversity has obtained the best overall average results in the nuclear power plant safety system test case.

An analysis of the exploration-exploitation equilibrium in a multiobjective safety system optimization design problem is discussed, taking into account both mutation and 'controlled elitism' operators in the NSGA-II. The obtained results show that the 'controlled elitism' operator offers advantages in the convergence of the algorithm, being capable to cooperate with the mutation operator with benefits to the convergence behaviour. A similar conclusion has also been extracted recently, applied to multiobjective frame optimization [37].

Acknowledgements. This investigation was funded by the Grant AP2001-3797 from the Ministry of Education and Culture of Spanish Government (Secretary of Education and Universities). The authors wish to acknowledge the anonymous reviewers for their valuable comments on the original manuscript.

References

1. Kuo W & Rajendra Prasad V, An Annotated Overview of System-Reliability Optimization. *IEEE Transactions on Reliability*, Vol. 49, No. 2, pp. 176–187, June 2000.
2. Andrews JD, Pattisson RL. Optimal Safety-system performance. In *Proceedings of the 1997 Reliability and Maintainability Symposium*, Philadelphia, Pennsylvania, January 1997, pp. 76–83.
3. Pattisson RL, Andrews JD. Genetic algorithms in optimal safety system design. In *Proc. Inst. Mech. Engrs.* Vol. 213, Part E, pp. 187–197. 1999.
4. Martorell, S., Muñoz, A. y Serradell. V., Age-dependent models for evaluating risks and costs of surveillance and maintenance of components. *IEEE Transactions on Reliability*, 1996, Vol. 45, No.3, pp. 433–442.
5. Martorell S, Sánchez C, Serradell V. Constrained optimization of test intervals using steady-state genetic algorithm. *Rel. Engng. & System Safety*, Vol. 67, 2000, pp. 215–232.
6. Galván B, Marin D, Benitez E, Alonso S, Juvier J, "Safety System Design Optimization using Genetic Algorithms with Incomplete Information", *Evolutionary Methods for Design, Optimization and Control with Applications to Industrial Problems*, CIMNE, Barcelona 2002.
7. Giuggioli Busacca B, Marseguerra M, Zio E, Multiobjective optimization by genetic algorithms: application to safety systems, *Reliability Engineering & System Safety*, Vol. 72, issue 1, 2001 pp. 59–74
8. Holland JH. *Adaptation in natural and artificial systems*. Ann Arbor: University of Michigan Press., 1975.
9. Goldberg DE. Genetic Algorithms in Search, Optimization, and Machine Learning. Reading, MA: Addison-Wesley, 1989.
10. Gen M, Cheng R. *Genetic Algorithms & Engineering Design*. Wiley Interscience, John Wiley & Sons, USA, 1997.
11. Andrews JD, Optimal safety system design using fault tree analysis. In Proc. *Instn. Mech. Engrs.* Vol. 208, pp. 123–131. 1994.
12. Dugan JB, Venkataraman B, Gulati R. [1997]. "A software package for the analysis of dynamics fault tree models", *IEEE Proceedings Annual Reliability and Maintainability Symposium*, pp. 64–70.
13. Dugan JB, Bavuso SJ, Boyd MA. [1992]. "Dynamic Fault-Tree models for fault-tolerant computer systems", *IEEE Transactions on Reliability* 41(3): 363–377.
14. Vesely WE, Goldberg FF, Roberts NH, Haals DF. Fault Tree Handbook. Systems and Reliability research office of Nuclear Regulatory Research, U.S. Nuclear Regulatory Commission, Washington D.C., January, 1981.
15. Bryant RE [1992]. "Symbolic boolean manipulation with ordered binary-decision diagrams", *ACM Computing Surveys* 24(3): 293–318
16. Coudert O, Madre JC [1993]. "Fault Tree analysis: 10^{20} Prime Implicants and beyond", *IEEE Proceedings Annual Reliability and Maintainability Symposium*, pp. 240–245.
17. Bouissou M. An ordering Heuristic for building Binary Decision Diagrams from Fault Trees. *Proceedings RAMS 96*, Las Vegas, Jan. 96.

18. Bartlett LM, Andrews JD. An ordering heuristic to develop the binary decision diagram based on structural importance. *Reliability Engineering & System Safety*, Vol. 72, issue 1, 2001, pp. 31–38.
19. Williams PF, Nikolskaia M, Rauzy A. "Bypassing BDD construction for reliability analysis". *Information Processing Letters*. Elsevier, 75 (2000) 85–89.
20. González L, García D, Galván BJ. "Sobre el análisis computacional de funciones Booleanas estocásticas de muchas variables". EACA95-*Actas del primer encuentro de Álgebra computacional y Aplicaciones* – Santander (Spain), 1995 Sep, pp. 45–55. (in Spanish)
21. González L, García D, Galván BJ. "An Intrinsic Order Criterion to Evaluate Large, Complex Fault Trees" IEEE Transactions on Reliability (accepted, to appear).
22. Galván BJ. "Contribuciones a la Evaluación Cuantitativa de Árboles de Fallos". *PhD Thesis*, Departamento de Física, Las Palmas de Gran Canaria University (Canary Islands-Spain), 1999. (in Spanish)
23. Greiner D, Winter G, Emperador JM, Galván B. An efficient adaptation of the truncation operator in SPEA2. In: *Proceedings of the First Spanish Congress on Evolutionary and Bioinspired Algorithms*. Eds: Herrera et al. Mérida, Spain, February 2002. (in Spanish)
24. Haasl DF. "Advanced concepts on Fault Tree Analysis". The Boeing company System Safety Symposium, 1965, USA.
25. Schneeweiss WG. Boolean Functions with Engineering Applications and Computer Programs. Springer-Verlag, 1989.
26. Lee WS, Grosh DL, Tillman FA, Lie CH. "Fault Tree Analysis, Methods and Applications – A review". *IEEE Transactions on Reliability*, R-34(3), 1985, pp. 194–203.
27. NUREG-75/014: WASH-1400. "Reactor Safety Study". U.S. Nuclear Regulatory Commission, 1975.
28. Coello Coello C, "A Short Tutorial on Evolutionary Multiobjective Optimization", pp. 21–40, in *Evolutionary Multi-Criterion Optimization*, Springer, 2001.
29. Purshouse R, Fleming P, "Why use Elitism and Sharing in a Multiobjective Genetic Algorithm?", *Proceedings of the Genetic and Evolutionary Computation Conference GECCO-2002*, pp. 520-527, New York, Morgan Kaufmann Publishers.
30. Greiner D, Winter G, Galván B. Multiobjective Optimization in Safety Systems: A Comparative between NSGA-II and SPEA2. In: *Proceedings of the IV Congress on Reliability: Dependability*. Eds: Galván, Winter, Cuesta and Aguasca. Spain, Sept. 2002.
31. Zitzler E, Laumanns M, Thiele L, "SPEA2: Improving the Strength Pareto Evolutionary Algorithm for Multiobjective Optimization", *Evolutionary Methods for Design, Optimization and Control with Applications to Industrial Problems*, CIMNE, Barcelona 2002.
32. Deb K, Pratap A, Agrawal S, Meyarivan T, "A fast and elitist multiobjective genetic algorithm: NSGA-II", *IEEE Transactions on Evolutionary Computation* 6 (2), 182–197, 2002.
33. Deb K, Goel T, "Controlled Elitist Non-dominated Sorting Genetic Algorithms for Better Convergence", pp. 67–81, in *Evolutionary Multi-Criterion Optimization*, Springer, 2001.
34. Zitzler E, Laumanns M, Thiele L, "*SPEA2: Improving the Strength Pareto Evolutionary Algorithm*", TIK-Report 103, (May 2001) Swiss Federal Institute of Technology, Zurich.
35. Zitzler E. Evolutionary Algorithms for Multiobjective Optimization : Methods and Applications. *PhD Thesis*. Swiss Federal Institute of Technology (ETH), Zurich 1999.
36. Laumanns M, Zitzler E, Thiele L. On the Effects of Archiving, Elitism, and Density Based Selection in Evolutionary Multi-objective Optimization. pp. 181–196, in *Evolutionary Multi-Criterion Optimization*, Springer, 2001.
37. Greiner D, Winter G, Emperador JM, Galván B. A Comparative Analysis of Controlled Elitism in the NSGA-II applied to Frame Optimization. In: *Proceedings of the IUTAM Symposium on Evolutionary Methods in Mechanics*. (September 2002) Krakow, Poland. Kluwer Academic Publishers (in print).

Applications of a Multi-objective Genetic Algorithm to Engineering Design Problems

Johan Andersson

Department of Mechanical Engineering
Linköping University
581 83 Linköping, Sweden
johan@ikp.liu.se

Abstract. This paper presents the usage of a multi-objective genetic algorithm to a set of engineering design problems. The studied problems span from detailed design of a hydraulic pump to more comprehensive system design. Furthermore, the problems are modeled using dynamic simulation models, response surfaces based on FE–models as well as static equations. The proposed method is simple and straight forward and it does not require any problem specific parameter tuning. The studied problems have all been successfully solved with the same algorithm without any problem specific parameter tuning. The resulting Pareto frontiers have proven very illustrative and supportive for the decision-maker.

1 Introduction

Many real-world engineering design problems involve simultaneous optimization of several conflicting objectives. In many cases, multiple objective problems are aggregated into one single overall objective function. However, design engineers are often interested in identifying a Pareto optimal set of alternatives when exploring a design space. Pareto optimality is defined as a set where every element is a problem solution for which no other solutions can be better in all design attributes. For the two-dimensional case, the Pareto front is a curve that clearly illustrates the trade-off between the objectives.

The objective of this paper is to present a multi-objective genetic algorithm and describe how it has been applied to a variety of real world applications, without any problem specific parameter tuning. The paper begins by discussing engineering design and its similarities to an ordinary optimization process. We go on to discuss genetic algorithms in general and a multi-objective genetic algorithm (MOGA) in particular. Thereafter, a set of different design problems are studied with the help of the proposed optimization strategy. These problems include design of a hydraulic actuation systems, detail design of a hydraulic pump, a crashworthiness design problem as well as the problem of determining which functionality to include in a mechatronic system. Finally the results are summarized in the conclusions.

1.1 Optimization and Engineering Design

Engineering design is a special form of problem solving where a set of frequently unclear objectives has to be balanced without violating any given constraints. Furthermore, the design process is an iterative process as have been stated by several authors, e.g. Ulrich and Eppinger [18] and Roosenburg and Eekels [15]. According to Roosenburg and Eekels, the iterative part consists of *analysis, synthesis, simulation, evaluation* and *decision*. For each provisional design the expected properties are compared to the criteria. If the design does not meet the criteria it is modified and evaluated again in the search for the best possible design. From this it could be seen that design is essentially an optimization process, as stated already in 1967 by Simon [17]. Therefore, it seems natural to look upon a design problem as an optimization problem. By employing modern modeling, simulation and optimization techniques, vast improvements could be achieved in design. However, there will always be parts of the design process that require human or inquantifiable judgment that is not suited for automation with any optimization strategy.

Figure 1 below depicts a system design process from [1] where modeling, simulation and optimization are introduced to support and speed up the design process. In the proposed system design process, the iterative part of the design process is formalized and partly automated with the help of an optimization algorithm.

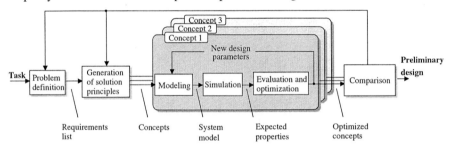

Fig. 1. System design process

The 'problem definition' in Figure 1 results in a requirements list which is used in order to generate different solution principles/concepts. Once the concepts have reached a sufficient degree of refinement, modeling and simulation are employed in order to predict the properties of particular system solutions. Each solution is evaluated with the help of an objective function, which acts as a figure of merit. Optimization is then employed in order to automate the evaluation of system solutions and to generate new system proposals. The process continues until the optimization is converged and a set of optimal systems are found. One part of the optimization is the evaluation of design proposals. The second part is the generation of new, and hopefully better designs. Thus, optimization consists of both analysis (evaluation) and synthesis (generation of new solutions).

Often the first optimization run does not result in the final design. If the optimization does not converge to a desired system, the concept has to be modified or the problem reformulated, which results in new objectives. In Figure 1 this is visualized by the two outer loops back to 'generation of solution principles' and 'problem definition' respectively.

Naturally the activity 'generation of solution principles' produces a number of conceivable concepts, which each one is optimized. Thus each concept is brought to maximum performance; optimization thereby provides a solid basis for concept selection. This will be illustrated in a study of hydraulic actuation systems.

One essential aspect of using modeling and simulation is to understand the system we are designing. The other aspect is to understand our expectations on the system, and our priorities among the objectives. Both aspects are equally important. It is essential to engineering design to manage the dialog between specification and prototype. Often simulations confirm that what we wish for is unrealistic or ill conceived. Conversely, they can also reveal that our whishes are not imaginative enough.

2 Genetic Algorithms

Genetic algorithms are modeled after mechanisms of natural selection. Each optimization parameter (x_n) is encoded by a gene using an appropriate representation, such as a real number or a string of bits. The corresponding genes for all parameters $x_1...x_n$ form a chromosome capable of describing an individual design solution. A set of chromosomes representing several individual design solutions comprise a population where the most fit are selected to reproduce. Mating is performed using crossover to combine genes from different parents to produce children. The children are inserted into the population and the procedure starts over again, thus creating an artificial Darwinian environment. For a general introduction to genetic algorithms, see work by Goldberg [8].

Additionally, there are many different types of multi-objective genetic algorithms. For a review of genetic algorithms applied to multi-objective optimization readers are referred to work by Deb [6].

2.1 The Proposed Method

In this paper the multi-objective struggle genetic algorithm (MOSGA) [4] is used for the Pareto optimization. MOSGA combines the struggle crowding genetic algorithm presented by Grueninger and Wallace [10] with Pareto-based ranking as devised by Fonseca and Fleming [7]. As there is no single objective function to determine the fitness of the different individuals in a Pareto optimization, the ranking scheme presented by Fonseca and Fleming is employed, and the "degree of dominance" in attribute space is used to rank the population. Each individual is given a rank based on the number of individuals in the population that are preferred to it, i.e. for each individual the algorithm loops through the whole population counting the number of preferred individuals. "Preferred to" is implemented in a strict Pareto sense, but one could also combine Pareto optimality with the satisfaction of objective goal levels, as discussed in [7]. The principle of the MOSGA algorithm is outlined below.

Step 1: Initialize the population.
Step 2: Select individuals uniformly from population.
Step 3: Perform crossover and mutation to create a child.
Step 4: Calculate the rank of the new child.
Step 5: Find the individual in the entire population that is most similar to the child. Replace that individual with the new child if the child's ranking is better, or if the child dominates it.
Step 6: Update the ranking of the population if the child has been inserted.
Step 7: Perform steps 2-6 according to the population size.
Step 8: If the stop criterion is not met go to step 2 and start a new generation.

Step 5 implies that that the new child is only inserted into the population if it dominates the most similar individual, or if it has a lower ranking, i.e. a lower "degree of dominance". Since the ranking of the population does not consider the presence of the new child it is possible for the child to dominate an individual and still have the same ranking. This restricted replacement scheme counteracts genetic drifts and is the only mechanism needed in order to preserve population diversity. Furthermore, it does not need any specific parameter tuning. The restricted replacement strategy also constitutes an extreme form of elitism, as the only way of replacing a non-dominated individual is to create a child that dominates it.

The similarity of two individuals is measured using a distance function. The method has been tested with distance functions based upon the Euclidean distance in both attribute and parameter space. A mixed distance function combining both the attribute and parameter distance has been evaluated as well. The result presented in this paper was obtained using an attribute based distance function. As can be seen from the description of the method there are no algorithm parameters that have to be set by the user. The inputs are only: population size, number of generations, genome representation and crossover and mutation methods, as in every genetic algorithm.

3 Design Examples

This section describes a set of engineering design problems which have been studied with the proposed method. It shall be pointed out that each problem has been solved without tuning any algorithm parameters. Real parameters are always real encoded whereas for the combinatorial problem we use binary encoding. Furthermore blend crossover has been used for both crossover and mutation of real encoding variables, whereas one point crossover and flip mutation have been used for binary genomes. All problems have been solved with a population size of 40 individuals which has been run for 400 generations.

3.1 Hydraulic Actuation Systems

The objects of study for this design problem are two different concepts of hydraulic actuation systems. Both systems consist of a hydraulic cylinder that is connected to a mass of 1000 kilograms. The objective is to follow a pulse in the position command with a small control error and simultaneously obtain low energy consumption. Naturally, these two objectives are in conflict with each other. The problem is thus to minimize both the control error and the energy consumption from a Pareto optimal perspective.

In the first more conventional system, the cylinder is controlled by a servo valve, which is powered from a constant pressure system. In the second concept, the cylinder is controlled by a servo pump. Thus, the systems have different properties. The valve concept has all that is required for a low control error, as the valve has a very high bandwidth. On the other hand, the valve system is associated with higher losses, as the valve constantly throttles fluid to the tank.

The different concepts have been modeled in the simulation package Hopsan [11]. The models of each component consist of a set of algebraic and differential equations taking aspects such as friction, leakage and non-linearities into account. The system models are depicted in Figure 2.

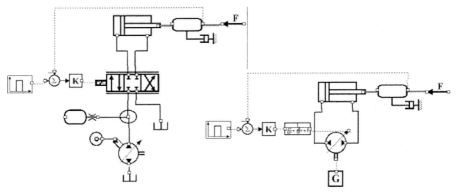

Fig. 2. Two different concepts of hydraulic actuation systems, left is the servo valve system and right is the servo pump system

The servo valve system consists of the mass and the hydraulic cylinder, the servo valve and a p-controller that is controlling the motion. The servo valve is powered by a constant pressure pump and an accumulator, which keeps the system pressure at a constant level. The optimization parameters are the sizes of the cylinder, valve and the pump, the pressure lever, the feedback gain and a leakage parameter that is necessary to dampen the system. Thus, this problem consists of six optimization parameters and two objectives.

The servo pump concept contains fewer components, the cylinder and the mass, the controller and the pump. A second order low-pass filter is added in order to model the dynamics of the pump. The servo pump system consists of only four optimization parameters.

The optimization is based on component size selection rather then component design, i.e. it is assumed that each component is a predefined entity. As a consequence of this assumption most component parameters are expressed as a function of the component size. Both systems where optimized in order to simultaneously minimize the control error f_1 and the energy consumption f_2. The control error is obtained by integrating the absolute value of the difference between reference and actual cylinder position, whereas the energy consumption is calculated by integrating the product of flow and the pressure difference over the pump.

As the Pareto optimization searches for all non-dominated individuals, the final population will contain individuals with a very high control error, as they have low energy consumption. It is possible to obtain an energy consumption close to zero if the cylinder does not move at all. However, these solutions are not of interest, as we want the system to follow the pulse. Therefore, a goal level for the control error is introduced. The ranking scheme is modified so that solutions, which are bellow the goal level for the control error are always preferred to solutions that are above it regardless of their energy consumption, as described by Fonseca and Fleming in [7]. In this manner, the population is focused on the relevant part of the Pareto front.

In order to achieve fast systems, and thereby low control errors, large pumps and valves are chosen by the optimization strategy. A large pump delivers more fluid, which enables higher speed of the cylinder. However, bigger components consume more energy, which explains the shape of the Pareto fronts in Figure 3. This problem has been analyzed in more detail in [2]. Furthermore, in [3] the problem was extended to include a mixture of real parameters and selection of valves and cylinders from catalogues.

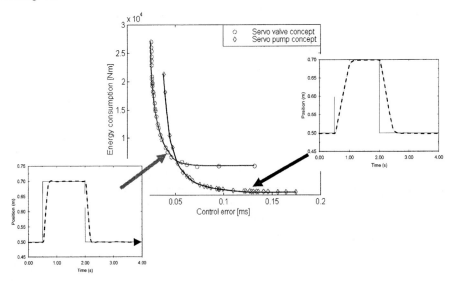

Fig. 3. Pareto front showing the trade-off between energy consumption and control error for the two concepts. Furthermore, the graph on the right shows the simulation result for a slow pulse response, whereas the graph on the left shows a fast pulse response

In figure 3 the obtained Pareto fronts for the two concepts are plotted in the same graph. In this manner, Pareto optimization constitutes an excellent support for concept selection, as it brings forth the properties of the different concepts. It is evident that the final design should preferably be on the overall Pareto front, which elucidates when to change between concepts. This is a very illustrative way of presenting the results from a multi-objective optimization. In this particular design example, the servo pump system consumes less energy, and is preferred if a control error larger then 0.05ms is acceptable. The servo valve system on the other hand, is fast but consumes more energy. If lower control error then 0.05ms is desired, the final design should preferably be a servo valve system.

3.2 Detail Design of a Hydraulic Piston Pump

One of the most important origins of noise and vibrations in hydraulic systems is the system pressure ripple. The system pressure ripple is a hydraulic response to introduced flow pulsations, of which the hydrostatic pump is a major source. Consequently, in order to lower the noise from hydraulic systems, the flow transients created by the pump must be reduced. The pump flow ripple constitutes two principal parts; the kinematic flow ripple, due to the limited number of pumping elements, and the compressible flow ripple, due to the compressibility of the fluid. At high pressure levels, the compressible flow ripple is normally the clearly dominating kind. However, with very small design modifications, the compressible flow ripple can be changed considerably, both with regard to amplitude and frequency content. Perhaps the most obvious measure to reduce the compressible flow transient is to equalize the cylinder pressure to the supply port pressure before the cylinder is connected to the supply port.

This is called pre-compression, and produces a rather satisfying flow ripple reduction if designed correctly. There are more refined ways of achieving this pressure equalization, for example the pressure relief groove explained in [13] and the pre-compression filter volume, see [14]. The exact design of these features can rather easily be tuned to minimize flow ripple at a specific operational condition, i.e. a certain displacement angle, rotational speed and discharge and inlet pressure levels. However, as the conditions are changed, the optimum drifts away, implying that a pump optimised for a certain condition may give severe flow ripple at other conditions. In this section a design feature called cross-angle, which reduces the pump's sensitivity to variations in displacement angles is analysed.

The cross-angle is a small (1-4°) fixed displacement angle around the axis perpendicular to the trunnion axis, see [12]. In practice, the cross-angle results in that the additional pre-compression required for achieving optimal pressure equalization is obtained for a wide range of displacement angles.

The pump studied is a seven piston, 40 cm^3, in-line axial piston pump at 1500 rpm with discharge and inlet pressure levels at 250 and 2 bar respectively. The pump has been studied using a very detailed simulation model developed in the HOPSAN simulation environment [11]. The accuracy of the model has been experimentally verified in for example work by Petterson [14]. In the model a large number of different states are available, for example cylinder pressure, flow from each cylinder and piston forces. In the previous design example the time resolution was in milliseconds, in this example it is microseconds.

There are several objectives that have to be considered regarding the minimization of pump related noise. Traditionally, the flow ripple peak-to-peak value has been the obvious objective, which is justified since it correlates well to the system pressure ripple amplitude. However, when running an optimisation, focusing solely on the flow ripple, unreasonable cylinder pressure-peaks may occur. Since the cylinder pressure is directly proportional to the piston forces, this will have a direct impact on the excitation of the pump casting vibrations and thus noise, i.e. the maximal cylinder pressure p_{max} should be below a certain limit p_{lim}. In addition, it is important to avoid cavitation in order to obtain low noise level and a long life of the pump, i.e the minimal system pressure p_{min} should be above a certain value p_{cav}. Thus, the optimisation problem is defined as:

$$
\begin{aligned}
&\text{Minimize} \quad f_1(\mathbf{x}) && \text{discharge flow ripple} \\
&\text{Minimize} \quad f_2(\mathbf{x}) && \text{inlet flow ripple} \\
&\text{Subjected to} \quad g_1(\mathbf{x}) = p_{min}(\mathbf{x}) \geq p_{cav} && \text{cavitation constraint} \\
&\quad g_2(\mathbf{x}) = p_{max}(\mathbf{x}) \leq p_{lim} && \text{pressure-peak constraint} \\
&\quad x_i^l \leq x_i \leq x_i^u && \text{design parameter limits}
\end{aligned} \tag{1}
$$

For this application the cavitation and pressure-peak constraints are added to the objective functions with the help of penalty functions. For each constraint, there is a penalty function that equals to zero if the constraint is not violated. The penalty function then increases exponentially with the degree of constraint violation. The sum of the penalty functions are finally added to both objectives. With this problem formulation a set of non-dominated individuals that do not violate any constraints are identified as shown in Figure 4.

In order to assure a low noise level for a wide range of displacement angles, the simulation model is executed at different displacement angles between zero and maximum displacement for each optimisation iteration. The flow peak-to-peak values for each displacement angle are summed to a total flow pulsation measure.

Furthermore, this problem has also been solved as a single objective problem, with $f_1(\mathbf{x})$ as well as $f_2(\mathbf{x})$ as the single objective using the same constraint formulation. It was then concluded that optimal discharge and inlet performance require different cross angles, as can also be seen in the Pareto front in Figure 4. The single objective problem was solved using the Complex method, see Box [5]. The Complex method is a non-gradient method that has been successfully used in other studies too; see [1] and [12]. Solving the single objective problem yielded the same results as the extreme values on the Pareto front. This is very encouraging since two different optimization methods have given the same optimal results, which increases the confidence that the true optima has really been found. However, as always when using this type of optimization methods, optimality of the final solutions can not be proven.

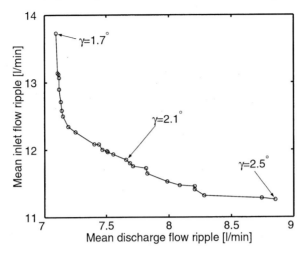

Fig. 4. Pareto front showing the trade-off between optimal discharge and inlet performance

The multi-objective genetic algorithm identifies the Pareto optimal front that visualizes the trade-off between the two objectives, see Figure 4. Along the x-axis, the mean discharge flow ripple amplitude from simulations at three different displacements ($Dp=100\%$, $Dp=50\%$ and $Dp=0\%$) is displayed, whereas the inlet mean flow ripple is displayed along the y-axis.

With a cross-angle of 1.7°, optimal discharge performance is obtained with mean discharge flow ripple amplitude of approximately 7.1 l/min, while the inlet performance is rather poor with mean flow ripple amplitude of 13.7 l/min. A cross-angle of 2.5°, on the other hand, implies optimal inlet performance with mean flow ripple amplitude of 11.2 l/min, while the discharge mean flow ripple is 8.8 l/min. By selecting a point in the middle of the curve, a fair compromise point where $\gamma = 2.1°$ is obtained. By choosing this point instead of $\gamma = 1.7°$, the mean inlet flow ripple is improved with 1.9 l/min (approximately 14% better), while the discharge flow ripple becomes only 0.5 l/min worse (7.0% worse). If the compromise point is chosen instead of $\gamma = 2.5°$, the gain in discharge flow ripple is 1.2 l/min (approximately 14% better) while the inlet flow ripple is deteriorated with only 0.5 l/min (4.4% worse). Altogether, it can be seen that the total gain from choosing the compromise point instead of the end-points is consequently higher than the loss.

Figures 5(a) and 5(b) show the simulated discharge and inlet flow peak-to-peak value for different displacement angles, for ordinary pre-compression ($\gamma = 0°$,), optimal discharge performance, optimal inlet performance and the compromise point. The diagrams constitute an illustrative way of describing the practical implication of the trade-off between Pareto optimal solutions. As can be seen the compromise point is just slightly worse then the individual optima in both objectives. However, the individual optima are much worse in the other objective. This is evident in figure 5(b) where the compromise solution (solid line) is close to the optimum for inlet performance, whereas the individual optimum for discharge performance has a very poor inlet performance.

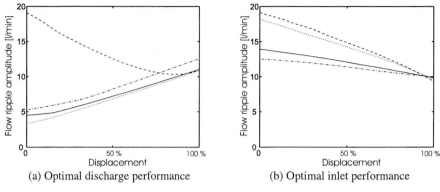

Fig. 5. Flow ripple sensitivity to changes in displacement angle. Ordinary pre-compression without any cross-angle ($\gamma=0°$) is shown as the dashed line. Optimal discharge performance ($\gamma=1.7°$) is obtained as the dotted line, while optimal inlet performance ($\gamma=2.5°$) is shown by the dash-dotted line. The solid line represents the compromise solution ($\gamma=2.1°$)

3.3 Crashworthiness Design Example

This section presents an example where the multi objective optimization technique is used together with response surface methods in order to support crashworthiness design. Here the conflicting objectives are exemplified by the desire to minimize the intrusion into the passenger compartment area and simultaneously obtain low maximum acceleration during vehicle impact. These two objectives are naturally conflicting, since low acceleration implies large intrusion.

The problem is solved by first creating quadratic response surfaces that captures the global performance of the objectives using a D-optimal experimental design setup. The crash behaviour of the vehicle is studied with a comprehensive FE model, see [15]. The FE model of the vehicle was a sub-model of a complete FE vehicle model consisting of 56.000 shell elements. The vehicle model impacts into a rigid wall with an initial velocity of 30 miles per hour (56 km/h) and the impact is simulated during 100 ms. The impact event is solved using the FE-code LS-DYNA in a LINUX cluster. Sheet thicknesses of parts such as the crash-box, midrail–closingplate, midrail-C-profile, rail-extension and the upper-rail, were used as design parameters. The FE model is evaluated in a set of experimental design points in order to establish quadratic response surfaces for both the intrusion and the acceleration.

Based on the two quadratic response surfaces a Pareto optimization was performed using the multi objective genetic algorithm. The outcome of this optimization is the Pareto front shown in Figure 6. By studying the trade-off among the Pareto optimal solutions, the final design is chosen at a point where a fair compromise between intrusion and maximum acceleration is obtained.

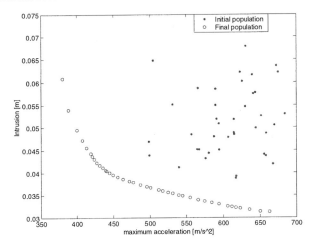

Fig. 6. Pareto front showing the trade-off between intrusion and maximum acceleration in the crashworthiness design problem

3.4 Optimization of the Product Specification for Mechatronic Systems

In this section a method is presented that supports the designer in determining which functionality that should be implemented in a product and which should not. The proposed method identifies the set of customer functions and technical implementations that maximize the possible product profit. The customer functions represent the functionality of the product, and the technical implementations are the hardware and software components needed to realize these functions. For industrial applications, the numbers of possible combinations of customer functions and technical implementations are extremely large.

The purpose of the method is to maximize the possible profit by selecting the optimal functionality to be designed into a mechatronic product with an open architecture. The profit consists of two parameters, value and cost, where the cost is subtracted from the value. The value is the amount the customers are prepared to pay for the set of customer functions in the product and the cost is the sum of the costs of implementing the technical implementations needed to realize the customer functions in the product. The assumption is that each customer function can be implemented separately and adds customer value. This is not to say that the different customer functions are independent since they can share technical implementations.

The customer functions are represented in the customer function vector (**CF**) and the technical implementations in the technical implementation vector (**TI**). A coupling vector cv_i expresses which technical implementations are necessary in order to realize the customer function CF_i. Each element of the coupling vector could be either one, representing that the corresponding technical implementation are needed, or zero otherwise. The coupling vectors for all customer functions, $\mathbf{cv}_i, i = 1...m$, make up the coupling matrix **CM**. Thus the problem could be described according to equation (2)

$$\mathbf{CF} = \mathbf{CM} \cdot \mathbf{TI} \quad \text{i.e.} \quad \begin{bmatrix} CF_1 \\ ... \\ ... \\ CF_m \end{bmatrix} = \begin{bmatrix} - & \mathbf{cv}_1 & - \\ & ... & \\ & ... & \\ - & \mathbf{cv}_m & - \end{bmatrix} \begin{bmatrix} TI_1 \\ ... \\ ... \\ TI_n \end{bmatrix} \quad (2)$$

where m is the number of potential customer functions and n the number of technical implementations. Each different combination of technical implementations yields one possible solution to the problem, or one concept. For a problem with n technical implementations there exist 2^n different concepts. A particular concept **X**, is expressed by a vector $\mathbf{X} = [x_1, x_2,....x_n]$, where x_i can be either one if the technical implementation $\mathbf{TI}_i$ is in the concept or zero otherwise.

In order to calculate the value for a specific concept, the customer functions that are possible to realize with the concept's technical implementations have to be found. The function realization vector, **W**, represents this. **W** is calculated according to equation (3).

$$\mathbf{W}(\mathbf{X}) = \begin{bmatrix} w_1 \\ ... \\ ... \\ w_m \end{bmatrix}, \text{ where } w_i(\mathbf{X}) = \left\lfloor \frac{\mathbf{cv}_i \bullet \mathbf{X}^T}{\mathbf{1}^T \bullet \mathbf{cv}_i^T} \right\rfloor \quad (3)$$

The notation $\lfloor a \rfloor$ denotes the largest integer less than or equal to a. $\mathbf{cv}_i \bullet \mathbf{X}^T$ represents the number of the necessary technical implementations for customer function i that are included in concept **X**. This number is divided by the sum of all functions needed in order to implement customer function i, i.e. $\mathbf{1}^T \bullet \mathbf{cv}_i^T$, where $\mathbf{1}^T$ is a vector of ones. If **X** contains all functions needed by CF_i this quotient equals 1, otherwise it is lees then one. Thus equation (3) returns 1 only for the customer functions that are implemented by **X**. The total value, of a concept is calculated by summing up the customer value for each customer function realized by concept **X**.

The cost of implementing each technical implementation is represented by the implementation cost vector **IC**, $\mathbf{IC}^T = [ic_1, ic_2,....,ic_n]$, where each implantation cost, ic_i, is made up of the development, material and production cost. The total cost, c, for the concept is thus simply obtained by multiplying **X** and **IC**.

In this simplified model the profit, p, for the company is expressed as the value the customer is prepared to pay for a particular concept minus the cost of developing and producing it. The problem could thus be described as to find the concept **X** that maximizes the profit $p(\mathbf{X})$ without exceeding the development budget. The objectives are then to maximize the profit and simultaneously minimize the cost, see equation (4).

$$\max_{X} p(\mathbf{X})$$
$$\min_{X} c(\mathbf{X}) \tag{4}$$
$$\text{s.t.} \quad \mathbf{X} = [x_1, x_2, ..., x_n]$$
$$x_i = 0 \vee 1, \forall i \{1, 2, ..., n\}$$

The method has been evaluated on an industrial case study of active safety system performed at Volvo Cars. The size of the problem is rather impressive, with 51 customer functions and 46 technical implementations, there are $7 \cdot 10^{13}$ different possible solutions to the problem. The resulting Pareto front is shown in figure 7. As can be seen it is possible to increase the profit by increasing the development budget. In reality there is however always a restriction on the development budget. Thus this type of optimization could be used as an argument when negotiating the development budget.

Based on this case study the proposed method shows a substantial profit potential compared to the methods presently used, see Figure 7. The method used today is experts choosing the concepts they "believe in", using the same data but disregarding the sharing of technical implementations by customer functions. This is necessary because it is impossible to grasp all the feasible combinations. The experts might however take other issues into account which are not accounted for in this simple model.

Figure 8 shows the optimal set of customer functions that should be implemented in the product depending on the development budget. As can be seen the number of functions increase as the budget is enlarged. For a more detailed description of this application readers are referred to [9].

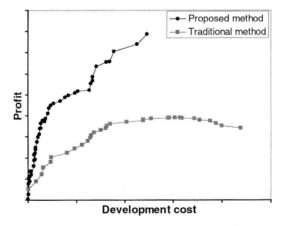

Fig. 7. Pareto fronts obtained by the proposed method, above, and the traditional method, below. As can be seen the profit could be increased by increasing the development budget. An interesting result is that the experts keep selecting functions even when the profit decreases

750   J. Andersson

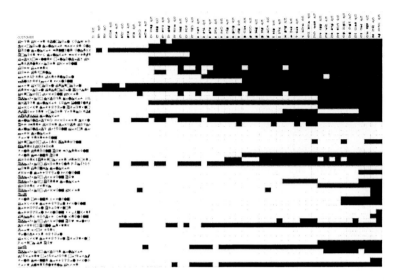

Fig. 8. Visualization of the combinations of customer functions which yield the highest profit for different development budgets. As can be seen the number of functions increases as the budget increases. Due to confidentiality, the names of the functions can not be shown

4   Discussion and Conclusion

This paper has shown how a simple MOGA could be used to support engineering design. The MOGA used in this paper seems very robust and it requires few algorithm specific parameters. The method has been applied to a wide range of engineering optimization problems spanning from detailed design via more comprehensive system design to overall determination of the product specification. The intention of the paper has been to exemplify how a simple MOGA could solve a set of very different real world problems without any problem specific parameter tuning. For a more detailed discussion of the problems studied, the reader is referred to the references listed. Numerical properties of the proposed MOGA could be found in reference [4].

It has also been shown how Pareto optimization could be a useful tool for concept selection. From an engineering perspective the goal of the optimization is not only finding "the optimal" solution, but gaining insight about the properties of the system being designed and the behavior of the system model. Another important lesson is defining the objectives, which forces the designer to make clear what is desired of the system and then to challenge the preferences of the decision-maker by visualizing the trade-off between the objectives. Therefore, what is crucial for a method that should be applied to engineering design optimization is that it is simple and yet robust, and that it could search vast design spaces and produce near optimal solutions that covers the whole Pareto front. The proposed multi-objective genetic algorithm is a good example of such a method.

Acknowledgements. The author whishes to thank the fellow researchers; Professor Petter Krus, Christian Grante, Andreas Johansson and Marcus Redhe for sharing their specific problems and their expertise. Part of this work has been financially supported by the Swedish Strategic Research Foundation through the ENDREA program. This support is gratefully acknowledged.

References

1. Andersson J., Multiobjective Optimization in Engineering Design – Application to Fluid Power Systems, Dissertation, Thesis No. 675, Linköping University, Sweden, (2001).
2. Andersson J., Krus P. and Wallace D., "Multi-objective optimization of hydraulic actuation systems", ASME Design Automation Conference, Baltimore, September 11–13, (2000).
3. Andersson J. and Krus P., Multiobjective Optimization of Mixed Variable Design Problems, in Proceedings of 1st International Conference on Evolutionary Multi Criteria Optimization, Zurich, Switzerland, March 7–9, (2001).
4. Andersson J. and Wallace D., Pareto optimization using the Struggle Genetic Crowding Algorithm, Engineering Optimization, vol. 34, No. 6, pp. 623–643, (2002).
5. Box M. J., "A new method of constraint optimization and a comparison with other methods", Computer Journal 8:42–52, (1965).
6. Deb K., Multi-objective Objective Optimization using Evolutionary algorithms, Wiley and Sons Ltd, (2001).
7. Fonseca C. M. and Fleming P. J., "Multiobjective optimization and multiple constraint handling with evolutionary algorithms – Part I: a unified formulation," IEEE Transactions on Systems, Man, & Cybernetics Part A: Systems & Humans, vol. 28, pp. 26–37, (1998).
8. Goldberg D., "Genetic Algorithms in Search and Machine Learning. Reading, Addison Wesley, (1989).
9. Grante C. and Andersson J., Optimisation of Design Specifications for Mechatronic Systems, submitted for publication, (2002).
10. Grüniger T. and Wallace D., Multi-modal optimization using genetic algorithms", Technical Report 96.02, CADlab, Massachusetts Institute of Technology, Cambridge, (1996).
11. Hopsan, , "Hopsan, a simulation package – User's guide", Technical report LiTH-IKP-R-704, Department of Mechanical Engineering, Linköping University, Linköping, Sweden.
12. Johansson A., Andersson J. and Palmberg J.-O., Optimal Design of the Cross-Angle for Pulsation Reduction in Variable Displacement Pumps, In Bath Workshop of Power Transmission and Motion Control, University of Bath, UK, September, (2002).
13. Palmberg J-O., Modelling of flow ripple from fluid power piston pumps. In 2nd Bath International Power Workshop, University of Bath, UK, September, (1989).
14. Pettersson M., Design of fluid power piston pumps, with special reference to noise reduction, PhD thesis, Dissertations No. 394, Department of Mechanical Engineering, Linköping University, Linköping, Sweden, (1995).
15. Redhe M., Simulation based design – structural optimization in early design stages, Licentiate Thesis, Dept. of Mech. Eng., Linköping University, Linköping, Sweden, (2001).
16. Roozenburg N. and Eekels J., Product Design: Fundamentals and Methods, John Wiley & Sons Inc, (1995).
17. Simon H., The Sciences of the Artificial, MIT Press, (1969).
18. Ulrich K. and Eppinger D., Product Design and Development, second edition, McGraw-Hill Companies, (1995).

A Real-World Test Problem for EMO Algorithms

A. Gaspar-Cunha and J.A. Covas

Dept. of Polymer Engineering, University of Minho, Guimarães, Portugal
{gaspar,jcovas}@dep.uminho.pt

Abstract. In this paper, a real-world test problem is presented and made available for use by the EMO community. The problem deals with the optimization of polymer extrusion, in terms of setting the operating conditions and/or the screw geometry. The binary code of a computer program that predicts the thermomechanical experience of a polymer inside the machine, as a function of geometry, polymer properties and operating conditions, is developed. The program can be used through input and output data files, so that the parameters to optimize and the criteria evaluated data is communicated in both directions. Two distinct EMO algorithms are used to illustrate and test the optimization of this problem.

1 Introduction

Most, if not all, real-world optimization problems are multiobjective and their solution is not unique. Instead, a set of optimal solutions, denoted as Pareto optimal set, explicits the trade-off between the criteria considered. The aim of multiobjective optimization is not only to improve the fitness of the solutions as the search proceeds, but also to distribute them uniformly along the Pareto frontier. Since Evolutionary Algorithms, EAs, use a population of points, this can be used with great advantage to find the Pareto frontier. After the initial work of Schaffer [1], a considerable number of different Multi-Objective Evolutionary Algorithms, MOEA, have been proposed, good reviews being available in the literature [2,3].

A significant number of problems have been proposed to test the performance of those algorithms [2,3]. Generally, they use relatively simple mathematical functions, whose solutions are known *a priori* and require little computational effort. The aim is to test the robustness of any optimization algorithm solving problems that have Pareto fronts with a wide range of characteristics [4]. The topography of the search space of each problem under study can cause difficulty in the convergence to the Pareto optimal front and/or in preserving the diversity of the Pareto optimal solutions. The former can be due to the existence of properties such as multi-modality, deception, isolated optimum or collateral noise, whereas the latter is associated with convexity or non-convexity, discontinuity and non-uniform distribution of solutions. Specific test problems dealing with the existence of these characteristics have been suggested [2,4]. Test problems having demanding constraints causing additional difficulties to the MOEAs have also been developed [4,5]. Deb *et al.* [6] suggested a method to build scalable test problems with any number of decision variables and criteria. Three different schemes were recommended, namely a multiple single-objective function approach, a bottom-up approach and a constraint surface approach. In the first case, different single-objective functions are used to create a multiobjective test problem, which corresponds to the usual practice by MOEAs researchers. The bottom-up approach uses a mathematical function describing the Pareto-optimal front in the objective space, which is used to define an overall search space and, therefore, to define the test problem. Finally, the constraint surface approach starts with the definition of the overall search space, for example a rectangular hyper-box. Then, constraints are added systematically to cut off regions from the box in order to create arbitrary Pareto-optimal hypersurfaces.

Since the ultimate goal of developing MOEAs is to solve real-world problems, it seems relevant to question whether the above test problems illustrate adequately the complex behavior of real-world applications. These, not only involve a considerable number of decision variables and criteria, as well as complicated relationships between them, but are also generally ill-posed (i.e., the inverse solution is often not unique) and combinatorial. Although EMO algorithms have been applied to practical problems [7,8], the current absence of accepted real-world test problems that can be used to test the algorithms in a real environment is evident.

Therefore, the aim of the present work is to propose a real-world test problem with attractive features in terms of EMOAs, more specifically, the optimization of the polymer extrusion process, as a test problem to be adopted by the scientific community. A modeling routine of the process in binary code and two data files are available through the Internet (*www.dep.uminho.pt/pp/index.php3?gaspar@dep.uminho.pt*). The first is used to transfer from the EMO algorithm to the modeling routine the values of the solutions proposed. The second file passes on from the modeling routine to the EMO algorithm the values of the criteria selected. The test problem proposed will be optimized using two different EMO algorithms.

2 Polymer Extrusion as a Test Problem

2.1 The Process

Extrusion is one of the most industrially relevant processing techniques used by companies producing plastics parts. It is used to manufacture mass-consumption products such as tubing, pipes and profiles for the building, automotive and medical industries, film and sheet for packaging and agricultural applications, filaments and fibers for textiles, electrical wires and cables. Plastics compounding, i.e., the preparation of raw materials with advanced properties, through the incorporation of additives to a polymer matrix, polymer modification, or polymer blending, are also carried out in extruders.

In a typical single screw extruder (see Figure 1) an Archimedes-type screw rotates at a given frequency inside a heated barrel. The aim of the process is to receive the solid pellets at the inlet (hopper), melt and mix the material, and pump it at a constant rate through the die, which is coupled to the other end of the barrel, in order to produce an extrudate with a prescribed cross-section. The material will be subsequently cooled downstream, printed/decorated, coiled or cut in regular lengths, using other machines that are not represented in Fig. 1.

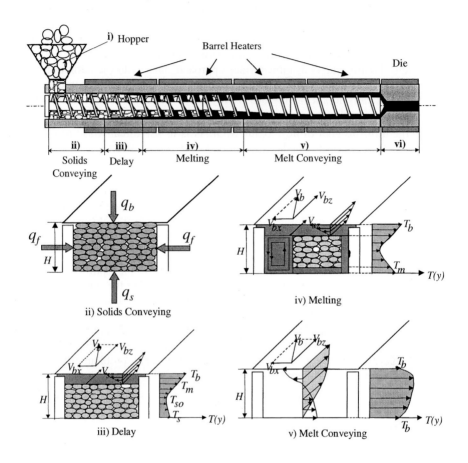

Fig. 1. Plasticating sequence inside a polymer extruder

From a thermomechanical point of view, as the material progresses along the screw it is subjected to various mechanisms, which are also illustrated in Figure 1. The polymer pellets in the hopper (i) flow into the barrel by action of gravity. Friction between the polymer and the adjacent metallic walls will drag the material downstream, along the helical screw channel (ii). Heat conduction and dissipation will melt the material closer to the inner barrel wall (this is often known as *delay* in melting) (iii). When the rate of melting becomes sufficiently significant, a specific melting mechanism develops, the melt and solid components segregating from each other (iv). The resulting melt is mixed and conveyed (v) towards the die (adopting a helical velocity profile along the screw channel), where it is shaped into the desired cross-section (vi). Detailed phenomenological and mathematical descriptions of this sequence of events can be found elsewhere [9-11]. In addition to the complexity of the process, polymer melts are viscoelastic, i.e., they exhibit a combination of solid-like and liquid-like behavior (which is time, temperature, deformation rate and stress dependent). For the sake of simplicity, a power law viscous constitutive equation is often adopted (where the power index, n quantifies the non- melt character).

For a given equipment geometry and polymer system the main operating variables are the screw rotation frequency and the barrel set axial temperature profile. Upon starting the manufacture of a new product, it is often necessary to define the best screw and the operating conditions that will process most efficiently the polymer selected. Process performance is usually evaluated in terms of mass output, mixing quality, length of screw required for melting, power consumption, average residence time inside the extruder and level of viscous dissipation.

2.2 The Modeling Routine

The solution of the direct problem involves predicting the value of the performance variables identified above for a given input set of material/geometry/operating conditions. Since each of the individual process stages has to be described mathematically, use of proven models was made, whenever possible [9-12].

Gravity flow in the hopper (Fig.1-i) was modeled considering the existence of a sequence of vertical and/or convergent columns subjected to static loading conditions [13]. The vertical pressure profile can be determined from a force balance on an elemental slice of the bulk [14]. We assumed the transport of a non-isothermal elastic solid plug with heat dissipation at all surfaces in the initial screw turns (Fig.1-ii). The computation of the solids temperature takes into account the contribution of conduction from the hot barrel, of friction near the polymer/metal interfaces, and of heat convection along the channel due to polymer motion [15,16]:

$$V_{sz} \frac{\partial T(y)}{\partial z} = \alpha_s \frac{\partial^2 T(y)}{\partial y^2} \tag{1}$$

where V_{sz} is the solid bed velocity, $T(y)$ is the cross temperature profile (direction y) and α_s is the thermal diffusivity of the solid plug. The pressure generated can be determined from force and torque balances made on differential down-channel elements [15].

$$P_2 = P_1 \exp\left[\int_{z_1}^{z_2}\left(\frac{B_1 - A_1 K}{B_2 + A_2 K}\right) dz\right] \tag{2}$$

where P_1 and P_2 are the pressures at down-channel distances z_1 and z_2, respectively, and A_1, A_2, B_1, B_2 and K are constants that are dependent on the friction coefficients and geometry.

The higher temperatures near to the inner barrel wall will cause the formation of a melt film (Fig. 1-iii) [12,17,18]. The film thickness and temperature can be computed from the momentum and energy equations, assuming heat convection in the down-channel and radial directions and heat conduction in the radial direction [19].

$$\frac{\partial P}{\partial x} = \frac{\partial}{\partial y}\left(\eta \frac{\partial V_x}{\partial y}\right) \tag{3}$$

$$\frac{\partial P}{\partial y} = 0 \tag{4}$$

$$\frac{\partial P}{\partial z} = \frac{\partial}{\partial y}\left(\eta \frac{\partial V_z}{\partial y}\right) \tag{5}$$

$$\rho_m \, c_p \, V_z(y) \frac{\partial T}{\partial z} = k_m \frac{\partial^2 T}{\partial y^2} + \eta \, \dot{\gamma}^2 \tag{6}$$

where ρ_m, c_p and k_m denote melt density, specific heat, and thermal conductivity, respectively, and η is the melt viscosity, which follows a temperature dependent power law:

$$\eta = k \, \exp[-a(T - T_0)] \, \dot{\gamma}^{n-1} \tag{7}$$

where k and n are the consistency and power law indices, respectively, a is the temperature shift factor, T_0 is the reference temperature, and $\dot{\gamma}$ is given by:

$$\dot{\gamma} = \left[\left(\frac{\partial V_x}{\partial y}\right)^2 + \left(\frac{\partial V_z}{\partial y}\right)^2\right]^{1/2} \tag{8}$$

$$\int_0^{\delta_C} V_x(y) \, dy = 0 \tag{9}$$

The melting step was described mathematically adopting the 5-regions model proposed Lindt et al [20,21], shown in Fig. 1-iv (development of a melt pool near the left screw flight, of melt films near the barrel and screw walls and a solid bed that is vanishing slowly). Treatment of each region requires different forms of the momentum and energy equations, coupled to the relevant boundary conditions. The flow of the melt films can be described by equations (3) to (8), together with specific boundary conditions and a condition for cross-channel flow (equation 9). Flow in the melt pool is taken as two-dimensional, hence equations (5), (6) and (8) are replaced, respectively, by:

$$\frac{\partial P}{\partial z} = \frac{\partial}{\partial y}\left(\eta \frac{\partial V_z}{\partial x}\right) + \frac{\partial}{\partial y}\left(\eta \frac{\partial V_z}{\partial y}\right) \tag{10}$$

$$\rho_m c_p V_z(y) \frac{\partial T}{\partial z} = k_m \left(\frac{\partial^2 T}{\partial x^2} + \frac{\partial^2 T}{\partial y^2}\right) + \eta \dot{\gamma}^2 \tag{11}$$

$$\dot{\gamma} = \left[\left(\frac{\partial V_x}{\partial y}\right)^2 \left(\frac{\partial V_z}{\partial y}\right)^2 + \left(\frac{\partial V_x}{\partial y}\right)^2\right]^{1/2} \tag{12}$$

Melt conveying (Fig.1-v) is identical to flow in the melt pool of the melting stage upstream, i.e. a non-isothermal two-dimensional flow of a power-law fluid develops. Hence, equations (3), (4), (10) and (11) remain valid, if coupled to updated boundary conditions [19]. Distributive mixing was taken into consideration. Its intensity is related to the growth of the interfacial area between any two adjacent fluid elements, which depends on the local melt shear strain and average residence time. We adopted a weighted-average total strain function, WATS [19,22,23], as a simple criterion to estimate the degree of mixing.

Finally, pressure flow in the die (Fig.1-vi) was computed assuming that the latter is made of successive channels with uniform cross-section in the downstream direction, the non-isothermal two-dimensional flow of a power law fluid taking place (equations (1), (9) and (10) are solved by finite differences).

The global plasticating extrusion model links sequentially the above stages, using appropriate boundary conditions, and ensuring coherence of the physical phenomena between any two adjacent zones. The calculations are performed for small down-channel increments along the screw and die, using a tentative output. Convergence is attained when the pressure drop predicted at the die exit becomes sufficiently small. Two algorithms are used in sequence. First, it is necessary to establish an interval (two values for output) that contains the solution, using a bracketing algorithm [24]. Then, Brent's algorithm is used to find the solution for the output inside this interval [25]. Using a Personal Computer with an Intel Celeron at 800 MHz, is takes typically one minute to compute one solution.

2.3 Typical Process Response

The Output – Pressure relationship for a typical extruder and die is shown in Figure 2 (Leistritz LSM 34 extruder and die described in Fig. 5, processing a High Density Polyethylene). The effect of increasing the screw speed and the die set temperature on the response of each of these components is also represented. The effective operating point corresponds to the intersection between the curves valid for each component, i.e., when the extruder (operating at specific conditions) is coupled to the die (kept at a certain temperature), one output is obtained as a result of a pressure generated by the extruder, which obviously equals the pressure drop in the die.

Table 1 presents the typical extruder response when the screw speed is gradually increased from 20 to 50 rpm, for constant system and polymer properties. The values of all the criteria considered increase with increasing screw speed, but in many cases in a nonlinear fashion. Temperature related criteria depend on material temperature development inside the extruder, which in turn results from the contribution of heat transfer and viscous dissipation. The former depends on the set temperatures and residence time (proportional to output), while the latter depends on average velocity (again, output) and material viscosity. Velocity related criteria (output, power consumption, WATS) depend on pressure generation and on the shape of the velocity fields.

Table 1. Influence of screw speed on the extruder performance

Criteria	Screw Speed (rpm)						
	20	30	50.0%	40	33.3%	50	25.0%
Output (kg/hr)	3.15	4.92	56.0%	6.67	35.8%	8.46	26.8%
Length for Melting (m)	0.46	0.61	32.9%	0.66	8.5%	0.69	5.1%
Melt Temperature (°C)	174.1	177.1	1.7%	180.2	1.8%	183.3	1.7%
Power Consumption (W)	794.5	1140.8	43.6%	1558.4	36.6%	2075.7	33.2%
WATS	195.2	219.4	12.4%	254.8	16.1%	240.4	-5.6%
T_{avg}/T_b	1.03	1.05	1.6%	1.06	1.5%	1.08	1.6%
T_{max}/T_b	1.08	1.12	4.1%	1.16	3.5%	1.20	3.3%

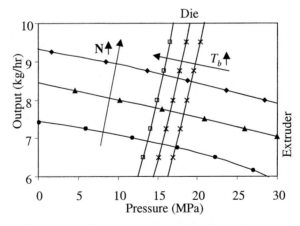

Fig. 2. Operating point of an extruder-die combination

Table 2 is similar to Table 1, but now the operating conditions have been kept fixed, the channel depth of the last section of the screw being increased. An increase in the cross-section of the screw channel favors output, but the corresponding decrease in residence time compromises the melting and mixing efficiencies.

Finally, Fig. 3 illustrates the evolution of pressure, melting (in terms of reduced solid bed width, X/W), average melt temperature and mechanical power consumption along the machine. The results produced by this modeling routine were assessed experimentally, using different polymers and operating conditions [19]. Generally, the predicted values are within ± 10% of the experimental data.

Table 2. Influence of channel depth on the extruder performance.

Criteria	Internal Screw Diameter/Channel Height (mm)							
	32	30	-6.3%	28	-6.7%	26	-7.1%	
	2	3	50.0%	4	33.3%	5	25.0%	
Output (kg/hr)	8.46	11.82	39.7%	15.45	30.7%	18.48	19.6%	
Length for Melting (m)	0.69	0.83	19.4%	0.89	7.0%	0.89	0.0%	
Melt Temperature (°C)	183.3	182.9	-0.2%	176.7	-3.4%	178.7	1.1%	
Power Consumption (W)	2075.7	1920.8	-7.5%	1763.6	-8.2%	1906.8	8.1%	
WATS	240.4	54.4	-77.4%	12.4	-77.2%	11.0	-11.3%	
T_{avg}/T_b	1.08	1.08	-0.2%	1.04	-3.4%	1.05	1.1%	
T_{max}/T_b	1.20	1.19	-1.3%	1.18	-0.2%	1.18	-0.3%	

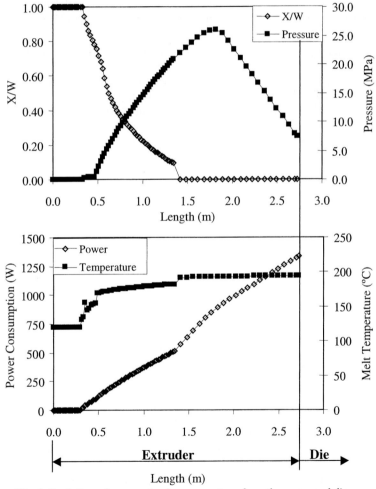

Fig. 3. Evolution of some extrusion parameters along the screw and die

2.4 Using the Test Problem

The polymer extrusion test problem is accessible at www.dep.uminho.pt/pp /index.php3?gaspar@dep.uminho.pt, where detailed instructions are also given. Figure 4 illustrates how it can be inserted in a MOEA sequence. After defining which variables are to be optimized and which process criteria are to be used, any MOEA needs to evaluate a proposed solution. In order to accomplish this, it transfers to the extrusion program, via the text file "extr.var", the input values of the variables corresponding to that solution and receives, via the text file "extr.cri", the corresponding values of the process criteria. The following criteria are available for consideration: mass output (Q), degree of mixing ($WATS$), length of screw required to melt the polymer (Z_T), mechanical power consumption (*Power*), melt temperature at die exit (T_{exit}) and viscous dissipation (which can be estimated from the average melt temperature/barrel temperature and/or maximum temperature/barrel temperature). Usually, the first two are to be maximized, while the remaining should be kept as low as possible.

The extrusion program "extr.exe" requires as input data operating conditions, namely screw speed (N) and barrel temperatures in three zones (T_{b1}, T_{b2}, T_{b3}), and geometrical parameters, such as the internal diameter of sections 1 and 3 (D_1, D_3), the length of sections 1 and 2 (L_1, L_2), the screw pitch (P) and the flight thickness (e) (see Figure 5). These are stored in the data file "extreq". The file "extrmat" contains the material properties, which include solid and melt densities, solid and melt thermal conductivities, heat capacity, friction coefficients and melt viscosity.

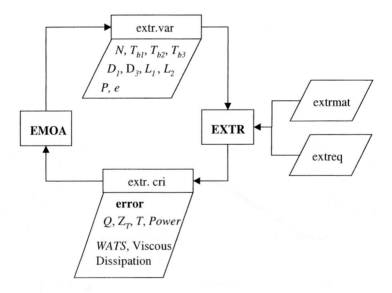

Fig. 4. Scheme used to link an MOEA with the extrusion test problem

In short, six files can be downloaded from the Internet: the modeling program "extr.exe", the polymer properties "extrmat" and the extruder geometry and operating conditions "extreq" databases, the parameters to optimize "extr.var" and the criteria "extr.cri" files, and the text file "readme" with instructions on the working of the test problem.

3 Optimization

3.1 Reduced Pareto Set Genetic Algorithm with Elitism (RPSGAe)

The test problem will be used by two optimisation algorithms, the NSGA-II, an elitist non-dominated sorting GA, developed by Deb *et al* [27], and the RPSGAe, proposed by the authors [28]. The first method uses simultaneously an elite preservation strategy and an explicit diversity preserving mechanism. First, an offspring population is created using the parent population, both of size N. These populations are combined together to form a single population of size $2N$. Then, a classification of the population using a non-dominated sorting is performed. Finally, the new population is filled with the individuals of the best fronts, until the size of the population becomes equal to N. If the population becomes bigger than N, a niching strategy is used to select the individuals of the last front. The RPSGAe algorithm incorporates a solution clustering technique, in order to reduce the number of solutions on the efficient frontier while maintaining their characteristics intact. In each generation, the N population individuals are reduced to a pre-defined number of ranks ($r=1,2,...,N_{Ranks}$), and the value of the objective function is calculated using a ranking function. Simultaneously, an external elitist population is maintained, with the aim of preserving the best solutions already found and of incorporating periodically some of them in the main population. The number of ranks influences the performance of the algorithm and must be set carefully [28].

3.2 Case Studies

Two case studies will be studied (see also Fig. 5). The first deals with the optimization of the operating conditions (for fixed polymer properties and extruder geometry). The aim is to define the screw speed (N) and the barrel temperature (in zones T_{b1}, T_{b2} and T_{b3}) that maximize the mass output (Q) and the mixing quality (*WATS*), and minimize the length of screw required to melt the polymer (Z_t), the mechanical power consumption (*Power*) and the melt temperature at die exit (T_{exit}). Fig. 5.A) indicates the range of variation of the parameters to optimize between square brackets. Three runs will be performed. The first uses the five criteria, while the remaining will be performed taking into account two criteria at a time, namely output and power consumption, and output and WATS, respectively. The NSGA-II algorithm will use a population size of 100, 50 generations, a roulette wheel and tournament selection strategies, a crossover probability of 0.8 and a mutation probability of 0.05 a random seed of 0.5. The RPSGAe algorithm introduces the number of ranks (N_{ranks}, 30) and the limits of indifference of the clustering technique, 0.01. These values have been defined in previous studies of the algorithm [28]. In both cases, a real codification of the variables is used, together with SBX crossover and polynomial mutation operators.

The second case study deals with the optimization of the screw geometrical parameters, *i.e.*, with screw design, which is an emerging research area in polymer processing [30]. The parameters to optimize are the screw length of zones 1 and 2 (L_1 and L_2), the internal screw diameter of sections 1 and 3 (D_1 and D_3), the screw pitch (P) and the flight thickness (e), whose range of variation is indicated between square brackets in Fig. 5.B). The operating conditions are fixed at N=50rpm and T_i=170°C. Output maximization and power consumption minimization will be used as criteria. The same algorithms parameters were adopted here.

In both case studies it was also necessary to consider some constraints for the criteria. For example, the mechanical power that is available by the extruder for screw rotation is limited (9200 W). The thermal stability of polymers is also limited, henceforth a maximum temperature of 270°C was allowed. The length of screw required for melting must also be constrained (0.9 m), since a sufficiently long melt conveying zone, for pressure generation and distributive mixing, must be guaranteed. A small value for the fitness is attributed to any solution violating these constraints.

3.3 Results and Discussion

The results obtained for the optimization of the operating conditions of the single screw extruder of Fig. 5.A) considering five criteria are presented in Fig. 6, for the two algorithms. All criteria are plotted against output, which was taken as the most important. As expected, output, length of screw required for melting and power consumption are conflicting. All Pareto frontiers are well defined, with the exception of that concerning melt temperature, since its interaction with other variables is very strong. There are some points that appear to be dominated in one Pareto frontier, but they are, certainly, non-dominated in another. Fig. 7.A) refers to the maximization of output and the minimization of power consumption. Both algorithms define clearly an optimal Pareto frontier evidencing the contradicting nature of these criteria. When mixing degree (WATS) and output are selected as the two criteria – Fig. 7.B) – the optimal Pareto frontier becomes discontinuous.

The total number of function evaluations was fixed at 4000 for both algorithms. The performance of the two algorithms was compared using the S-metric proposed by Zitzler [31]. Table 3 presents the results obtained after running each algorithm 5 times using different initial populations. The performance of NSGAII appears to be slightly better than that of RPSGAe, but the latter is more stable, as shown by the considerably lower standard deviation.

Table 3. Comparison results (S-metric)

	NSGAII with Roulette-wheel	NSGAII with Tournament selection	RPSGAe
S-metric average	0.076945	0.076497	0.07536
Standard deviation	0.026167	0.0268	0.000806

The results concerning screw design are presented in Fig. 8. The Pareto frontiers are quite distinct from the previous ones, because the geometry influences significantly the output, the power consumption and the mixing degree. However, both optimization algorithms produce coherent and similar results.

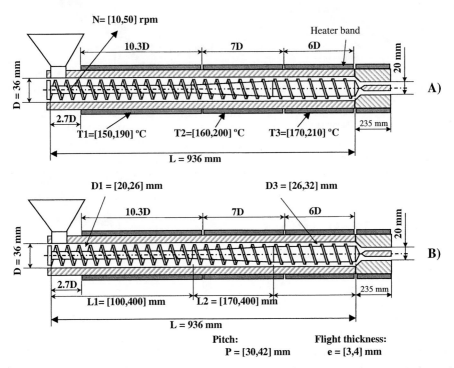

Fig. 5. Case studies A) Optimization of the operating conditions B) Screw design (L is the total screw length and D is the internal barrel diameter)

4 Conclusions

We have suggested and made available through the Internet a real-world test problem for multiobjective optimization, polymer extrusion. The problem can focus the optimization of the operating conditions, and/or the design of a screw for a particular application. As a test problem for EMO algorithms, it contains a variety of interesting characteristics, such as the possibility of using a variable number of criteria and parameters, of considering criteria with and without constraints, and of producing continuous, discontinuous and concave Pareto frontiers. Therefore, polymer extrusion could be a valid benchmark problem for multiobjective optimization algorithms, together with the available test functions.

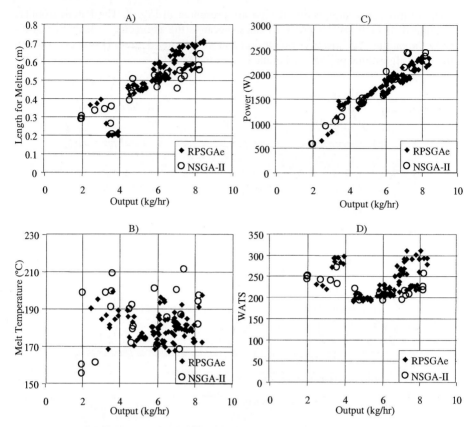

Fig. 6. Optimization of the operating conditions using five criteria

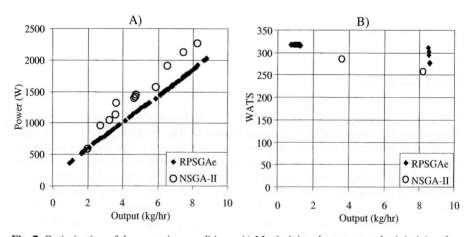

Fig. 7. Optimization of the operating conditions. A) Maximizing the output and minimizing the power consumption; B) Maximizing the output and the WATS

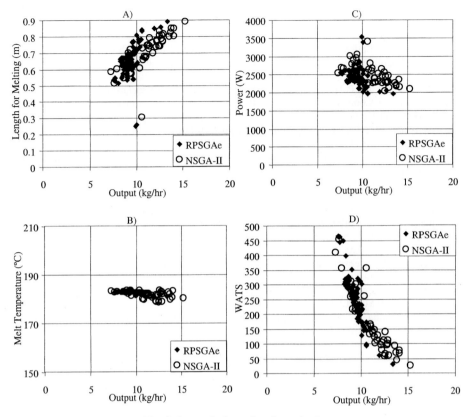

Fig. 8. Screw design using five criteria

References

1. Schafer, J.D.: Some Experiments in Machine Learning Using Vector Evaluated Genetic Algorithms, Ph. D. Thesis, Nashville, TN, Vanderbilt University (1984)
2. Deb, K.: Multi-Objective Optimisation using Evolutionary Algorithms, Wiley (2001)
3. Coello Coello, C.A., Van Veldhuizen, D.A., Lamont, G.B.: Evolutionary Algorithms for Solving Multi-Objective Problems, Kluwer (2002)
4. Deb, K.: Multi-objective Genetic Algorithms: Problems Difficulties and Construction of Test Problems, Evolutionary Computation Journal, 7 (1999) 205–230
5. Deb, K., Pratap, Meyarivan, T: Constrained Test Problems for Multiobjective Evolutionary Optimization, Proceedings of the First Int. Conf. On Evolutionary Multiobjective Optimization (EMO-2001), Zurich, Switzerland (2001) 284–298
6. Deb, K., Thiele, L., Laumanns, M., Zitzler, E.: Scalable Multi-Objective Optimization Test Problems, Proceedings of the 2002 IEEE Congress on Evolutionary Computation (CEC 2002) (2002)
7. Eckart Zitzler, Kalyanmoy Deb, Lothar Thiele, Carlos A. Coello Coello, David Corne (Eds.): Evolutionary Multi-Criterion Optimization, First International Conference, EMO 2001, Zurich, Switzerland, March 2001, Proceedings, Lecture Notes in Computer Science (LNCS) Vol. 1993, Springer-Verlag, Berlin (2001)

8. Deb, K., Thiele, L., Yen, G., Zitzler, E. (eds.): Special Track on Evolutionary Multi-Objective Optimization (EMO), Congress on Evolutionary Computation (CEC), Honolulu, Hawaii (2002)
9. Amellal, K., Lafleur, P.G., Arpin, B.: Computer Aided Design of Single-Screw Extruders, in A.A. Collyer, L.A. Utracki (eds): Polymer Rheology and Processing, Elsevier (1989) 277–317
10. Rauwendaal, C.: Polymer Extrusion, Hanser Publishers, Munich (1986)
11. O´Brian, K.: Computer Modelling for Extrusion and Other Continuous Polymer Processes, Carl Hanser Verlag, Munich (1992)
12. Agassant, J.F., Avenas, P., Sergent, J.: La Mise en Forme des Matiéres Plastiques, 3rd edn, Lavoisier, Paris (1996)
13. Stevens, M.J., Covas, J.A.: Extruder Principles and Operation, 2nd ed., Chapman & Hall, London (1995)
14. Walker, D.M.: An Approximate Theory for Pressures and Arching in Hoppers, Chem. Eng. Sci., 21 (1966) 975–997
15. E. Broyer, Z. Tadmor, Solids Conveying in Screw Extruders – Part I: A modified Isothermal Model, Polym. Eng. Sci., 12, pp. 12–24 (1972).
16. Tadmor, Z., Broyer, E.: Solids Conveying in Screw Extruders – Part II: Non Isothermal Model, Polym. Eng. Sci., 12 (1972) 378–386
17. Tadmor, Z., Klein, I.: Engineering Principles of Plasticating Extrusion, Van Nostrand Reinhold, New York (1970)
18. Kacir, L., Tadmor, Z.: Solids Conveying in Screw Extruders – Part III: The Delay Zone, Polym. Eng. Sci., 12 (1972) 387–395
19. Gaspar-Cunha, A.: Modelling and Optimisation of Single Screw Extrusion, Ph. D. Thesis, University of Minho, Guimarães, Portugal (2000)
20. Lindt, J.T., Elbirli, B.: Effect of the Cross-Channel Flow on the Melting Performance of a Single-Screw Extruder, Polym. Eng. Sci, 25 (1985) 412–418
21. Elbirli, B., Lindt, J.T., Gottgetreu, S.R., Baba, S.M.: Mathematical Modelling of Melting of Polymers in a Single-Screw Extruder, Polym. Eng. Sci., 24 (1984) 988–999
22. Pinto, G., Tadmor, Z.: Mixing and Residence Time Distribution in Melt Screw Extruders, Polym. Eng. Sci., 10 (1970) 279–288
23. Bigg, D.M.: Mixing in a Single Screw Extruder, Ph. D. Thesis, University of Massachusetts (1973)
24. Press, W.H., Teukolsky, A.A., Vetterling, W.T., Flannery, B.P.: Numerical Recipes in C: The Art of Scientific Computation, 2nd edition, Chapter 9, Cambridge University Press, Cambridge (1992)
25. Brent, R.P.: Algorithms for Minimization without Derivatives, Englewood Cliffs, Prentice-Hall, New Jersey (1973)
26. Web Page: www.dep.uminho.pt/pp/index.php3?gaspar@dep.uminho.pt (2002)
27. Deb, K., Pratap, A., Agrawal, S., Meyarivan, T.: A Fast and Elitist Multi-Objective Genetic Algorithm: NSGAII, IEEE Transactions on Evolutionary Computation, 6 (2002) 182–197
28. Gaspar-Cunha, A.: Reduced Pareto Set Genetic Algorithm (RPSGAe): A Comparative Study, The Second Workshop on Multiobjective Problem Solving from Nature (MPSN-II), Granada, Spain (2002)
29. Kanpur Genetic Algorithm Laboratory (KANGAL) web page: http://www.iitk.ac.in/kangal/soft.htm
30. Gaspar-Cunha, A.; Covas,J.A.: The Design of Extrusion Screws: An Optimisation Approach, Intern. Polym. Process., 16, pp. 229–240 (2001)
31. Zitzler, E.: Evolutionary Algorithms for Multiobjective Optimization: Methods and Applications, PhD Thesis, Swiss Federal Institute of Technology (ETH), Zurich, Switzerland (1999)

Genetic Methods in Multi-objective Optimization of Structures with an Equality Constraint on Volume

J.F. Aguilar Madeira[1,2], H. Rodrigues[1], and Heitor Pina[1]

[1] IDMEC/IST-Mechanical Engineering Department,
Instituto Superior Técnico, Av. Rovisco Pais,
1049-001 Lisbon, Portugal.
{jaguilar, hcr, hpina}@dem.ist.utl.pt
[2] ISEL-Instituto Superior de Engenharia de Lisboa.

Abstract. The method is developed for multi-objective optimization problems. Its purpose is to evolve an evenly distributed group of solutions to determine the optimum Pareto set for a given problem. The algorithm determines a set of solutions (a population), this population being sorted by its domination properties and a filter is defined in order to retain the Pareto solutions.

In most topology design problem volume is in general a constraint of the problem. Due to this constraint, all chromosomes used in the genetic algorithm must generate individuals with the same volume value; in the coding adopted this means that they must preserve the same number of ones and, implicitly, the same number of zeros, along the evolutionary process. It is thus necessary to define these chromosomes and to create corresponding operators of crossover and mutation which preserve volume.

To reduce computational effort, optimal solutions of each of the single-objective problems are introduced in the initial population.

Results obtained by the evolutionary and classical methods are compared.

1 Introduction

The binary chromosome design and global search capabilities of the Genetic Algorithm (GA) make is a powerful tool for solving topology design problem (see e. g. Hajela et all [9], Chapman et al [1]).

In contrast to single-objective optimization, where objective and fitness functions are often identical, both fitness assignment and selection must allow for several objectives when multi-objective optimization problems are considered. Hence, instead of a single optima, multi-objective optimization problems solutions consist often of a family of points, known as the Pareto optimal set, where each objective component of any point along the Pareto-front can only be improved by degrading at least one of its other objective components. In the total absence of information regarding the preferences of objectives, a ranking scheme based upon the Pareto optimality is regarded as an appropriate approach to represent the strength of each individual in an evolutionary algorithm for multi-objective optimization (Fonseca and Fleming [7], Srinivas and Deb [16]).

In this work a GA for multi-objective topology optimization of linear elastic structures, with an equality constraint on volume, is developed. Due to this constraint all the chromosomes used in the GA are constructed to ensure the same volume for all individuals (designs), hence all of them have the same number of ones (implicitly, the same number of zeros) along the evolutionary process.

For a typical chromosome defined by L genes there are a total of 2^L possible solutions in the representation scheme. Considering the volume equality constraint, a significant decrease of this value can be achieved. If a chromosome with a total of L genes is considered and K genes being one ($1 < K < L$, $(L-K)$ genes being zero), only $C_K^L = \frac{L!}{K!(L-K)!}$ possible configurations are possible. The present analysis explores this fact to reduce the computational effort required. It is thus necessary to define these chromosomes and to create accordingly new operators of crossover and mutation that preserve this constraint.

Since a great diversity of solutions exist, additional information obtained a priori from classical methods for single-objective optimization is incorporated in the algorithm to obtain a reduction in the computational effort. This information is introduced in the initial population through special individuals (single criterion solutions). This synergy between genetic and classical methods leads to a very powerful tool for solving multi-objective optimization problems.

2 Background

2.1 A General Multi-objective Optimization Problem (MOOP)

A general multi-objective optimization problem consisting of k competing objectives and $(m+p+q)$ constraints, defined as functions of decision variable set $\mathbf{x}$, can be represented as follows:

$$\begin{aligned}
&\text{Minimize/Maximize} && \mathbf{f}(\mathbf{x}) = (f_1(\mathbf{x}), f_2(\mathbf{x}), \ldots, f_k(\mathbf{x})) \in \mathbf{Y}; \\
&\text{subject to} && g_i(\mathbf{x}) \leq 0, \quad \forall i = 1, 2, \ldots, m; \\
& && h_j(\mathbf{x}) = 0, \quad \forall j = 1, 2, \ldots, p; \\
& && x_l^L \leq x_l \leq x_l^U, \quad \forall l = 1, 2, \ldots, q; \\
&\text{where} && \mathbf{x} = (x_1, x_2, \ldots, x_n) \in \mathbf{X}.
\end{aligned} \quad (1)$$

where $\mathbf{x}$ is the decision vector, $\mathbf{f}(\mathbf{x})$ the multi-objective vector, $f_j(\mathbf{x})$ the jth objective function, $\mathbf{X}$ denotes the decision space and $\mathbf{Y}$ the objective space.

The constraints $g_i(\mathbf{x})$ and $h_j(\mathbf{x})$ determine the set of feasible solutions. The remaining set of constraints are called variable bounds, restricting each decision variable x_l to values between a lower x_l^L and an upper x_l^U bond.

2.2 Concept of Domination

Most multi-objective optimization algorithms use the concept of domination referred as Pareto optimality [13]. A non-dominated set of solutions is formally defined as follows (Cohon [2]): "a feasible solution to a multi-objective problem is non-dominated if there exists no other feasible solution that will yield an improvement in one objective without causing a degradation in at least one other objective".

The set of solutions of a multi-objective optimization problem consist then of all decision vectors which cannot be improved in any objective without degradation in the other objectives, the Pareto-optima. Mathematically, the concept of Pareto-optima is as follows. Assuming, without loss of generality, a minimization problem and considering two decision vectors $\mathbf{a}, \mathbf{b} \in \mathbf{X}$, we say that $\mathbf{a}$ dominates $\mathbf{b}$ (also written as $\mathbf{a} \prec \mathbf{b}$) if and only if

$$\forall i \in \{1, 2, \ldots, k\} : f_i(\mathbf{a}) \leq f_i(\mathbf{b}) \wedge \exists j \in \{1, 2, \ldots, k\} : f_i(\mathbf{a}) < f_i(\mathbf{b}) \quad (2)$$

All decision vectors which are not dominated by any other decision vector are called non-dominated or Pareto-optimal. The family of all non-dominated solutions is denoted as Pareto-optimal set (Pareto set) or Pareto-optimal front.

2.3 Multi-objective Evolutionary Algorithms (MOEAs)

A number of multi-objective optimization techniques using evolutionary algorithms have been suggested since the pioneering work by Schaffer [14] [15](1984). Other important developments are described in the works by Fonseca and Fleming [7](1993), Horn et al [11](1994), Srinivas and Deb [16](1995), Knowles et al [12](1999), Zitzler et al [17](1999) and Deb et al [4](2000).

Evolutionary optimization algorithms work with a population of solutions instead of a single solution. Since multi-objective optimization problems give rise to a set of Pareto-optimal solutions, evolutionary optimization algorithms are ideal for handling multi-objective optimization problems.

In order to obtain effective results with an evolutionary algorithm for multi-objective optimization problems we need to guide the search toward the Pareto-optimal set and maintain a diverse population to prevent premature convergence and to achieve a well distributed population in the Pareto front.

3 Formulation of Multi-objective Optimization Problem (MOOP)

Consider a structural component, occupying the structural domain Ω, subjected to applied body forces $\mathbf{b}$, boundary traction $\mathbf{t}$ on Γ_t and essential boundary conditions on Γ_u (fig. 1.). To introduce the material based formulation, consider the structural component made of material with a variable volume fraction μ.

The k competing objectives for a multi-objective optimization problem are:

$$f_\alpha(\mathbf{x}) = \int_\Omega b_i u_i^\alpha d\Omega + \int_{\Gamma_t} t_i^\alpha u_i^\alpha d\Gamma \quad \text{with} \quad \alpha = 1, 2, \ldots, k \quad (3)$$

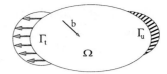

Fig. 1. General problem of elasticity

where $f_\alpha(\mathbf{x})$ with $\alpha = 1, 2, \ldots, k$ represent the compliance (equivalent to the energy norm of the total displacement) for the load case α (boundary traction $\mathbf{t}^\alpha$ and body forces $\mathbf{b}$) and where $\mathbf{u}^\alpha$ is the respective displacement field.

The multi-objective optimization problem is:

$$\text{Minimize} \quad \mathbf{f}(\mathbf{x}) = (f_1(\mathbf{x}), f_2(\mathbf{x}), \ldots, f_k(\mathbf{x})) \in \mathbf{Y}; \tag{4}$$

where $\mathbf{x} = (x_1, x_2, \ldots, x_n) \in \mathbf{X}$ (decision space) and $\mathbf{Y}$ is the objective space, $\mathbf{f}(\mathbf{x})$ the multi-objective vector, and $f_j(\mathbf{x})$ the jth objective function. Subjected to the volume constraint,

$$\int_\Omega \mu \, d\Omega = vol \tag{5}$$

and where the displacement $\mathbf{u}^\alpha$ is the solution of the integral equilibrium equation, in virtual displacement form,

$$\int_\Omega E_{ijkm}(\mu, q) e_{ij}(\mathbf{u}^\alpha) e_{km}(\mathbf{w}) - b_i w_i d\Omega - \int_{\Gamma_t} t_i^\alpha w_i d\Gamma = 0 \, , \forall \mathbf{w} \text{ admissible.} \tag{6}$$

4 Methodology

As previously stated, the equality constraint on volume requires that all chromosomes used in the GA lead to the same volume value, hence all of them should have the same number of ones along the evolutionary process. It is thus necessary to properly define these chromosomes and to create crossover and mutation operators that enforce this volume constraint.

4.1 The Chromosome Storage

The topology model for the structure is a grid that defines the design domain. A value of 1 assigned to an element or cell corresponds to the presence of material and a value 0 to a void in the domain. The fig. 2. illustrates the structural topology represented in the grid model (B) by the chromosome array (C), the binary chromosome string (D) (containing the numbering of the elements (A) and the volume value 0 or 1 for each element), and the chromosome viewed as a loop (E) formed by joining the ends together. A chromosome considered as a loop can represent more building blocks for the crossover operator, since they are able to "wrap around" at the end of the string. These two last representations depend on the elements numbering.

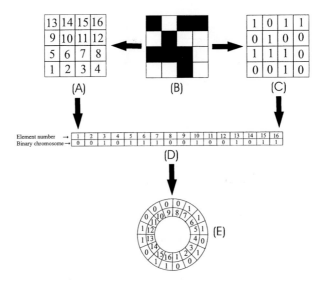

Fig. 2. Element numbering (A), structural topology in the grid model (B), binary chromosome array (C), binary chromosome string (D), and chromosome viewed as a loop (E)

4.2 Crossover Techniques

The selected progenitors exchange between them their genetic material producing individuals with new characteristics. To preserve volume the progenitors and the descendants must have the same number of ones in their chromosomes a requirement calling for new types of crossover to be described below.

Algorithm. To explain these crossover techniques the chromosome's representation is considered as a loop (fig. 3.). In the following Ngc, α, p_1, p_2, $n1s_1$, e $n1s_2$ are natural numbers representing:

Ngc - Number of genes in the chromosome (in fig. 3., $Ngc = 16$),
α - Number of genes to exchange in the crossover operator,
p_1 - Number of the gene of the first progenitor where the cut begins,
p_2 - Number of the gene of the second progenitor, where the cut begins,
$n1s_1$ - Number of ones of the first progenitor existing in the interval $[p_1, p_1 + (\alpha - 1)]$,
$n1s_2$ - Number of ones of the second progenitor existing in the interval $[p_2, p_2 + (\alpha - 1)]$.

The following algorithm describes a step-by-step procedure for finding the progenitors gene sequence to use in the crossover operator:

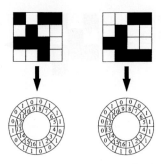

Fig. 3. Chromosome representation considered as a loop

Step 1 Fix α, the number of genes to exchange.
Step 2 Determine the initial p_1.
Step 3 Determine the initial p_2 (fig. 4.).

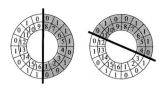

Fig. 4. Example with: $\alpha = 8$, $p_1 = 1$ and $p_2 = 4$

Step 4 The number of ones of the first progenitor existing in the interval $[p_1, p_1 + (\alpha - 1)]$ is counted. Let us say that there are $n1s_1$ ones.
Step 5 The number of ones of the second progenitor existing in the interval $[p_2, p_2 + (\alpha - 1)]$ is counted. Let us say that there are $n1s_2$ ones.
Step 6 If $\{(n1s_1 \neq n1s_2) \wedge (p_2 < Ngc)\}$ increment p_2 by one, and go to **Step 5**.
Step 7 If $\{(n1s_1 \neq n1s_2) \wedge (p_2 = Ngc)\}$ increment p_1 by one, and go to **Step 4**.
Step 8 If $n1s_1 = n1s_2$, make the exchange.

Types of Crossover with this Technique. In the **Step 2** and **Step 3**, the determination of p_1 and p_2 leads to three possible types of crossover:

Type 1 Start with $p_1 = p_2 = $ *fixed value*, for all the evolution.
Type 2 Start with $p_1 = p_2 = $ *random value*, this value to be determined whenever the operator is used.
Type 3 p_1 and p_2 are independently determined by random choices.

In the first type, whenever two individuals cross their descendants will always be the same. In the second type, the probability to get the same descendants is smaller, so we have a bigger diversity of descendants than in the previous one. The third type gives a higher diversity in the descendant population compared with the previous ones and is by this reason the one used in section 6.

Example. We illustrate the method with the example of fig. 3. Let us consider $\alpha = 8$, $p_1 = 1$ e $p_2 = 1$. The number of ones in the marked region is for the first progenitor, $n1s_1 = 4$ and, for the second, $n1s_2 = 5$ (fig. 5.). We are in **Step 5**. As it happens that $\{(n1s_1 \neq n1s_2) \wedge (p_2 < Ngc)\}$, then increment p_2 by one (fig. 6.) and return to **Step 5**. In summary,

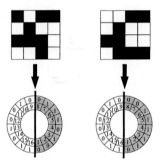

Fig. 5. $\alpha = 8$, $p_1 = 1$, $p_2 = 1$, $n1s_1 = 4$ and $n1s_2 = 5$

Fig. 6. $\alpha = 8$, $p_1 = 1$, $p_2 = 2$, $n1s_1 = 4$ and $n1s_2 = 5$

with
$\quad p_2 = 1 \quad \Longrightarrow \quad n1s_2 = 5 \neq n1s_1 = 4 \quad$ and $\quad p_2 < Ngc$;
$\quad p_2 = 2 \quad \Longrightarrow \quad n1s_2 = 5 \neq n1s_1 = 4 \quad$ and $\quad p_2 < Ngc$;
$\quad p_2 = 3 \quad \Longrightarrow \quad n1s_2 = 5 \neq n1s_1 = 4 \quad$ and $\quad p_2 < Ngc$;

with $p_2 = 4$ we finally get that, $n1s_1 = n1s_2 = 4$, then we make the exchange of genetic material of the two progenitors, taking into account the point of the selected cut, as suggested in fig. 7.

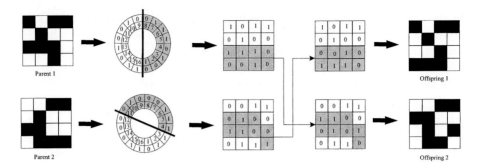

Fig. 7. Crossover example

4.3 Mutation

In spite its secondary importance as compared with the crossover operator, the mutation plays an important role in the GA procedure as it is responsible for the maintenance of the population genetic diversity (Holland [10]) and to prevent the algorithm to get locked in a local extrema by making it to explore other regions. The method is as follows:

Step 1 Determine randomly the positions of two genes in the chromosome.
Step 2 If they have equal values, they will remain in the same position. Return to **Step 1**.
Step 3 If they have different values, they will exchange positions.

5 Computational Model

A flow chart of the general procedure for a optimization problem with two objectives is outlined in fig. 8.

5.1 Populations

Assuming that the initial population has N individuals and MOOP has k competing objectives, the initial population is determined by the following procedure:

i) Solve k single-objective problems to determine the k optimal solutions for each objective, such that: necessary conditions for the optimum are derived analytically, approximated numerically through a suitable finite element discretization and solved by a first order method based on the optimization problem Lagrangian.
ii) The initial population is formed by these k optimal solutions of the single-objective problems together with $(N - k)$ different individuals randomly generated. Hence, the initial population has N different individuals.

To achieve low memory demand and computational effort reduction, the algorithm does not allow two equal individuals to be present in the population.

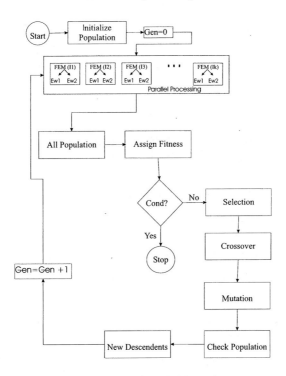

Fig. 8. Numerical model flow chart

5.2 Finite Element Method (FEM) and Parallel Processing

The work of the external loads is calculated for each objective using the finite element method. As the work of the external loads can be calculated independently for each individual parallel processing was used in its evaluation.

5.3 Fitness Assignment, New Population for Mating and Selection

The concept of calculating an individual's fitness on the basis of Pareto dominance was first suggested by Goldberg [8]. First, a ranking strategy based in non-domination properties is used (Deb [3]), that is, the non-dominated solution receives the rank 1 (front 1), then a new non-domination test is made in the population without these solutions and the new non-dominated solution receives the rank 2 (front 2). This procedure is repeated until all solutions are ranked. Second, the rank of an individual determines its fitness value. It is worth remarking that individual fitness is related to the whole population.

In order to avoid a possible premature convergence to a population where only non-dominated individuals (of front 1) exist, it is imposed that the new population for matting will be constituted by individuals of several fronts. A geometric function, proposed by Deb [5], the NSGA-II method, is employed for this purpose. This function is governed by a reduction factor $r < 1$ which leads to the introduction of a greater diversity in the population. Let the number of fronts to use be given by k, the total number of individuals in the population be N, then the number of individuals n_i in each front i is given by

$$n_i = N \frac{1-r}{1-r^k} r^{i-1}. \tag{7}$$

The mating individuals, in each front, are selected by the Clustering Method [6] and a tournament selection based in ranking position is used.

5.4 Population Check and New Descendants

After crossover and mutation, it is verified if the descendants already exist in the population and, if they do, they are eliminated to prevent unnecessary calculations. The new descendants are finally obtained.

6 Genetic Examples

The method was tested with two cases. The optimization problem for these two cases, is:

$$\text{Minimize} \quad \mathbf{f}(\mathbf{x}) = (f_1(\mathbf{x}), f_2(\mathbf{x})); \tag{8}$$

where $\mathbf{x} = (x_1, x_2)$, $f_1(\mathbf{x})$ and $f_2(\mathbf{x})$ are the objective functions, representing the compliance (or external work (EW)) for different boundary tractions, subjected to the volume constraint (5).

In the subsequent examples, analysis is performed by considering a domain discretization with 768 finite elements. The number of ones used in each chromosome is 236, corresponding to approximately 30% of the total design volume. The number of possible solutions is then given by C_{236}^{768}.

6.1 Example 1

For the present example, domain geometry and boundary conditions are shown in fig. 9. Loading conditions for both of the two objectives are also presented in

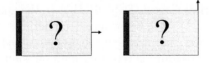

Fig. 9. Design domain for both objectives

the same figure. Two load cases are considered in the present application, for each one of the objectives:

1. A point load in the x direction with a magnitude of 200 N, for the first objective.
2. A point load in the y direction with a magnitude of 200 N, for the second objective.

Initially, the proposed single-objective problems were solved and the resulting solutions, for each one of the objectives are, depicted in fig. 10.:

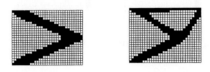

Fig. 10. Topology optimized for both objectives

These single-objective optimal solutions were introduced in the initial population. The initial population was composed of 200 individuals (in each generation only 100 individuals were selected for matting). The solutions obtained after 600 generations are presented in fig. 11.

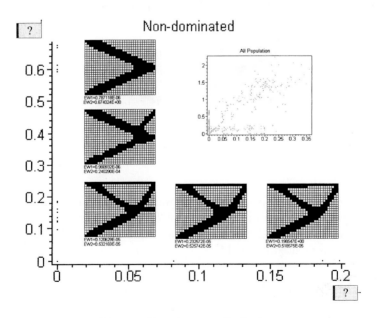

Fig. 11. Non-dominated solutions

6.2 Example 2

For the present example, the domain geometry and boundary conditions are shown in fig. 12. Loading conditions for both of the two objectives are also

Fig. 12. Design domain for both objectives

presented in the same figure. Two load cases are considered in the present application, for each one of the objectives:

1. A point load in the y direction with a magnitude of -200 N, for the first objective.
2. A point load in the y direction with a magnitude of 200 N, for the second objective.

Initially, the proposed single-objective problems were solved and the resulting solutions, for each one of the objectives are, depicted in fig. 13:

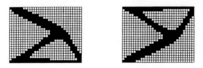

Fig. 13. Topology optimized for both objectives

These single-objective optimal solutions were introduced in the initial population. The initial population was composed of 200 individuals (in each generation only 100 individuals were selected for matting). The solutions obtained after 600 generations are presented in fig. 14.

7 Classical Examples

For comparison purposes the problems formulated in examples 1 and 2 have been solved through a weighting objective (classical) method (see e. g. Osyczka [18]). The problem is

$$\text{Minimize} \quad \alpha_1 \cdot f_1(\mathbf{x}) + \alpha_2 \cdot f_2(\mathbf{x}) \tag{9}$$

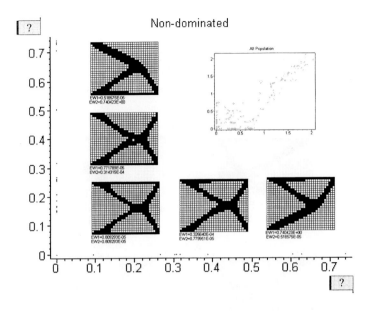

Fig. 14. Non-dominated solutions

where $\mathbf{x} = (x_1, x_2)$, $f_1(\mathbf{x})$ and $f_2(\mathbf{x})$ are the objective functions, representing the compliance (or external work (EW)) for different boundary tractions. The problem is subjected to the volume constraint (5), the displacement $\mathbf{u}^\alpha$ is the solution of the integral equilibrium equation (6) and the limit condition for the design variables are

$$0 \leq \mu \leq 1. \tag{10}$$

The necessary conditions for optimum are derived analytically, approximated numerically through a suitable finite element discretization and solved by a fixed point method based on optimization problem Lagrangian. The solutions obtained for examples 1 and 2 are, presented in fig. 15.

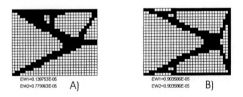

Fig. 15. A) Example 1, B) Example 2. With $\alpha_1 = \alpha_2 = 0.5$

8 Comparison of Results

When analyzing the results shown in fig. 16., it is observed that the genetic algorithm leads to better results, minimizing the objective functions more effectively and thus allowing stiffer structures. These results are consequence of the higher

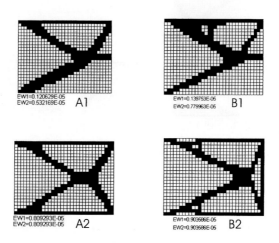

Fig. 16. Genetic (A1) versus classical (B1) results of example 1. Genetic (A2) versus classical (B2) results of example 2

capability of the genetic algorithms to explore the space of admissible solutions, while the classic methods can lead to a local minimum.

Acknowledgements. This work was supported by Portuguese Foundation for Science and Technology through Project POCTI/EME/43531/99. Partial support from project RTA P129 is also gratefully acknowledged.

References

1. C.D. Chapman, K. Saitou and M. J. Jakiela: Genetic Algorithms as an Approach to Configuration and Topology Design. ASME Journal of Mechanical Design, (1994).
2. J.L. Cohon: Multiobjective programming and planning. Matematics in Science and Engineering, Vol.140, Academic Press,Inc., (1978).
3. K.Deb: Evolutionary algorithms for multi-criterion optimization in engineering design. Proceedings of Evolutionary Algorithms inEngineering and computer Science,(EUROGEN'99), (1999).
4. K.Deb, A. Pratap, S. Agrawal and T. Meyarivan: A fast elitist Non-dominated sorting genetic algorithm for multi-objective: NSGA-II. Proceedings of the Parallel Problem Solving from Nature VI conference,pp 846–858, (2000).

5. K.Deb and T. Goel: Controlled Elitist Non-dominated Sorting Genetic Algorithms for Better Convergence. Evolutionary Multi-Criterion Optimization, EMO 2001, Springer,pp 67–81, (2001).
6. K.Deb: Multi-Objective Optimization using Evolutionary Algorithms. Wiley, (2002).
7. C.M.Fonseca and P.J.Fleming: Genetic algorithms for multi-objective optimization: Formulation, discussion, and generalization. Proceedings of the fifth international conferance on Genetic Algorithms, 416–423, (1993).
8. D.E. Goldberg: Genetic Algorithms in Search, Optimization, and Machine Learning. Reading, Massachusetts: Addison Wesley, (1989).
9. P. Hajela: Genetic Algorithms In Structural Topology Optimization. Topology Design of Structures, NATO Advanced Research Workshop, Sesimbra, Portugal, (1992).
10. Jonh Holland: Adaptation in Natural and Artificial Systems. Ann Arbor: University of Michigan Press, (1975).
11. J.Horn, N. Nafploitis and D.E. Goldberg: A niched Pareto genetic algorithm for multi-objective optimization. Proceedings of the first IEEE conference on evolutionary computation, 82–87.
12. J.Knowles and D.Corne: The Pareto archived evolution strategy: A new baseline algorithm for multi-objective optimization. Proceedings of the 1999 congress on evolutionary computation, Piscataway: New Jersey: IEEE Service Center, 98–105, (1999).
13. V.Pareto: Cours D'Economie Politique. Vol.I and II, F. Rouge, Lausanne, (1896).
14. J.D. Schaffer: Multiple objective optimization with vector evaluated genetic algorithms, Ph.D.Thesis, Vanderbilt University, (1984).
15. J.D. Schaffer: Multiple objective optimization with vector evaluated genetic algorithms and Their Applications, Proceedings of the first international conferance on Genetic Algorithms, pp.93–100, (1985).
16. N. Srinivas and K. Deb: Multi-objective funcion optimization using non-dominated sorting genetic algorithms, Evolutionary Computation(2), pp.221–248, (1995).
17. E. Zitzler and L. Thiele: Multiobjective evolutionary algorithm: a comparative case study and the strength Pareto approach. IEEE Transation on Evolutionary Computation, 3(4), pp.257–271, (1999).
18. A. Osyczka: Computer aided multicriterion optimization system (camos), software package in fortran. International Software Publisher (1992).

Multi-criteria Airfoil Design with Evolution Strategies

Lars Willmes[1] and Thomas Bäck[1,2]

[1] NuTech Solutions GmbH, Martin-Schmeißer-Weg 15,
44227 Dortmund, Germany
{willmes,baeck}@nutechsolutions.de
[2] Leiden Institute of Advanced Computer Science (LIACS),
Leiden University, Niels-Bohrweg 1,
2333 CA Leiden, The Netherlands

Abstract. In this paper we will describe the optimisation of a two-criteria wing-design problem where calculation of the objective function requires the solution of the two-dimensional Navier-Stokes equations. It will be shown that basic concepts of the Strength Pareto Evolutionary Algorithm 2 (SPEA2) and the Non dominated Sorting Genetic Algorithm II (NSGA-II) work well with Evolution Strategies. Results for the wing design problem are presented for the selection operators of SPEA2 and NSGA-II in combination with three different mutation operators. These results are compared with results found by a multi-objective Genetic Algorithm.

1 Introduction

Airfoil design provides a wealth of multi-criteria optimisation problems. The layout of a wing heavily influences its efficiency regarding e.g. fuel consumption, etc. Efficiency of an airfoil design can not be measured independently of the anticipated use of the wing, since even the most efficient design must still be able to produce enough lift at low speeds to allow a plane to take off. Different flight conditions like starting and landing or cruising at high altitudes, induce different conditions for optimality. This naturally leads to the formulation of a multi-criteria optimisation problem where each flight condition (flight point) states its own objective function. The airfoil design problem considered in this study and the resulting objective function are introduced in section 2 and 3. Section 4 introduces basic features of the *Strength Pareto Evolutionary Algorithm 2* (SPEA2) and the *Non-dominated Sorting Genetic Algorithm II* (NSGA-II) which for this study represent the state of the art in evolutionary multi-criteria optimisation. Section 5 gives a short overview of Evolution Strategies and the mutation operators used in this study. In section 6 the results generated by the multi-criteria Evolution Strategy are discussed.

2   Airfoil Design

One of the main characteristics of a wing design is its pressure profile, i.e. the distribution of pressure over the chord. The inverse design problem under study is to find the wing profile that produces a given pressure profile at given flow conditions. The test problem at hand was proposed for the AEROSHAPE[1] project and consists of two target wing designs, namely the standard NACA0012 wing at typical starting flow conditions and the standard NACA4412 wing at typical cruise flow conditions. The flow conditions for the two wings are given in table 1; the target wing designs and their respective pressure profiles are shown in figures 1 and 2.

Table 1. Flow conditions for high lift with the NACA4412 wing and for low drag with the NACA0012 wing.

	High Lift	Low Drag
Target wing	NACA0012	NACA4412
Mach number	0.2	0.77
Reynolds number	$5.1 \cdot 10^6$	10^7
Angle of Attack	10.8°	1.0°
c_w	$2.252 \cdot 10^{-2}$	$1.682 \cdot 10^{-2}$
c_a	1.252	0.5794

The optimisation goal is the identification of a set of wing designs whose pressure distributions provide a certain performance for the lift off situation at the expense of cruise condition efficiency. This set is supposed to contain designs very similar to the NACA0012 design on the one hand and the NACA4412 design on the other hand as extreme solutions. Ultimately, an engineer would select one design from this collection of wing profiles that fits best to a given aeroplane concept where neither the standard NACA0012 nor the standard NACA4412 would be an optimal choice.

The rational behind using the NACA0012 and NACA4412 designs is to construct a test case that contains all major difficulties of fluid dynamics and its simulation, but at the same time produces verifiable and comprehensive results. In a real world application, the target pressure distribution may be given independently from a standard airfoil.

3   Objective Function

The pressure distribution $p(s)$ at position s of the chord is computed by solving the two-dimensional Navier-Stokes Equations. Together with the pressure p_∞, the density ρ_∞ and the speed v_∞ of the surrounding stream the value of

---
[1] http://aeroshape.cira.it/

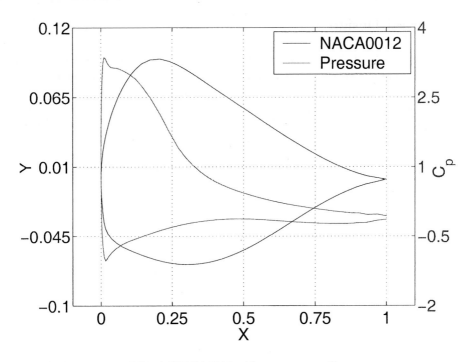

Fig. 1. NACA0012 with pressure profile

$$c_p(s) = \frac{p(s) - p_\infty}{\frac{\rho_\infty}{2} \cdot v_\infty^2} \quad (1)$$

is calculated, so that the two objective functions to be minimised are the cumulated squared differences between the actual pressure profile of the current configuration and the target pressure profiles:

$$f_1(x) = \int_0^1 \left[c_p^x(s) - c_p^{ld}(s)\right]^2 ds \quad (2)$$

$$f_2(x) = \int_0^1 \left[c_p^x(s) - c_p^{hl}(s)\right]^2 ds \quad (3)$$

where s is normalised by the chord length ($s \in [0,1]$), c_p^x is the pressure profile of the current design, c_p^{ld} is the low-drag pressure profile of the NACA4412 wing and c_p^{hl} is the high-lift pressure profile of the NACA0012 wing.

Since a single evaluation of the objective function is costly the total number of evaluations was restricted to 1000. Figures 3 and 4 give some indication of the quality of the problem: For both objective functions the function values are plotted against the first two design variable dimensions. It is quite clear from figures 3 and 4 that any optimisation algorithm may easily get trapped in local optima.

4 Multi-criteria Selection Operators

The main problem in evolutionary multi-criteria optimisation lies in the selection operator that chooses the parent individuals for the next reproduction cycle. The single-criteria selection operators of $(\mu + \lambda)$ and (μ, λ) Evolution Strategies rely on the total order of scalar fitness values. Since in general, there is no such total order given for vector valued objective functions, it must be derived by additional selection criteria. SPEA2 and NSGA-II both prefer non-dominated individuals to dominated ones and they both try to establish evenly spread parent populations by preferring parents in sparsely populated regions of the objective function space.

The SPEA2 operator accumulates information on dominance relationships by summing so called strength values that are computed for an individual by counting the number of individuals it dominates. To enable comparison of individuals with identical strength count, a density value $0 \leq \rho \leq 1$ based on a k-nearest neighbour method is added that is low for sparsely populated regions of the objective function space. Additionally, a special "exclude worst" method based on smallest pairwise distances is applied, if a Pareto front must be reduced to a given size. The details of the algorithm can be found in [6].

The NSGA-II operator uses the non-dominated sorting method to split a population in disjunct Pareto fronts, such that in each front individuals do not

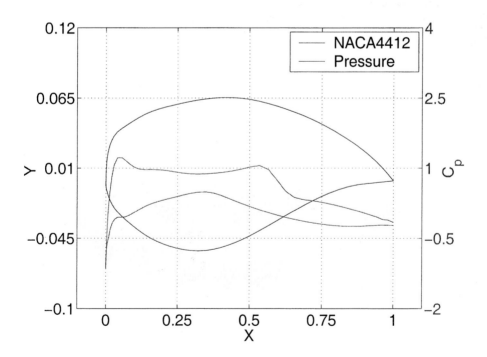

Fig. 2. NACA4412 with pressure profile

dominate each other. NSGA-II then adds new parent individuals front wise, starting with the global Pareto front. If the addition of the next Pareto front would yield more than the prescribed number of new parents, each individual of the current front is assigned a density value based on the city block distance of its closest neighbours. As with SPEA2, individuals from sparsely populated regions of the objective function space are preferred. Details on the methods used in NSGA-II can be found in [1,2].

5 Evolution Strategies

Evolution Strategies are a class of evolutionary algorithms that are particularly useful for engineering design optimisation, c.f. [9]. In this work the $(\mu + \lambda)$ Evolution Strategy is used, which denotes a strategy with μ parents that produce λ offspring in one reproduction cycle. The "+" in $(\mu+\lambda)$ indicates that the parent individuals of the $g+1$st reproduction cycle are selected from the union of parents and offspring of the g-th reproduction cycle. Since elitism is considered a valuable approach in multi-criteria evolutionary optimisation, c.f. [5], the $(\mu + \lambda)$-ES is the method of choice and automatically implements an archive similar to the original implementations of NSGA-II and SPEA2.

One of the most interesting features of Evolution Strategies is their ability to self adapt strategy parameters of the mutation operator. But especially for

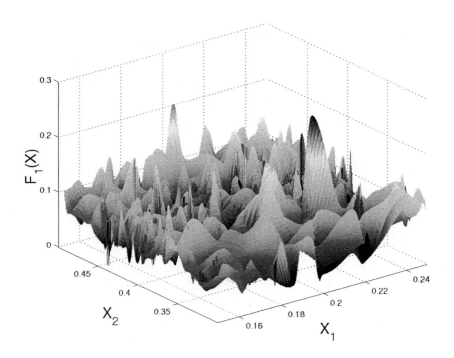

Fig. 3. Objective function landscape of f_1, projected on the first 2 dimensions

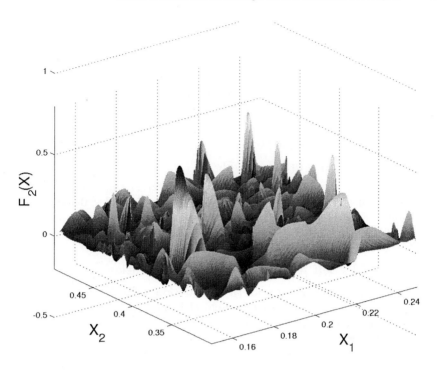

Fig. 4. Objective function landscape of f_2, projected on the first 2 dimensions

multi-criteria Evolution Strategies the process of self adaptation is not yet well understood and lacks major theoretical insight [4]. For this study three fundamentally different mutation operators were used. The derandomised step size adaptation (section 5.1) tries to accumulate knowledge on successful mutations and derives preferred mutation directions from that knowledge. The self adaptation method proposed by Schwefel (section 5.2) relies on the same mutation and selection mechanism as the design variables, and the pooling mutation (section 5.3) is a very simple concept that randomly chooses one of four different step sizes.

5.1 Derandomised Step Size Adaptation

The derandomised mutation operator was originally proposed in [3]. It performed very well in experiments with the single criteria pressure reconstruction test case analogous to the one under study in this work, c.f [7].

Let $\mathcal{N}(0,1)$ denote a normally distributed random variable with expected value 0 and variance 1. Then the derandomised step size adaptation can be expressed as

$$z'_i = (1-c) \cdot z_i + c \cdot m_i \qquad (4)$$

$$\sigma' = \sigma \cdot \left(\exp\left(\frac{|z'|}{\sqrt{n}\sqrt{\frac{c}{2-c}}} - 1 + \frac{1}{5n}\right)\right)^{\beta} \tag{5}$$

$$\sigma'_i = \sigma_i \cdot \left(\frac{|z'_i|}{\sqrt{\frac{c}{2-c}}} + 0.35\right)^{\beta'} \tag{6}$$

$$m'_i \sim \mathcal{N}(0,1) \tag{7}$$

$$x'_i = x_i + \sigma' \cdot \sigma'_i \cdot m'_i, \tag{8}$$

with $c = \beta = 1/\sqrt{n}$ and $\beta' = 1/n$. The vector $z = (z_1, \ldots, z_n)'$ can be interpreted as a memory of successful mutation directions. z is initialised to $\mathbf{0}$ and the update operation in equation 4 uses the most successful mutation vector m to adapt a new search direction.

5.2 Schwefel's Self Adaptive Step Size Adaptation

The classical mutation operator for (μ, λ) evolution strategies is the self adaptive mutation operator as introduced by Schwefel, c.f. [11]. For the single criteria test case the derandomised step size adaptation achieved better results than Schwefel's operator, c.f [7].

Schwefel's mutation operator can be expressed as follows

$$m \sim \mathcal{N}(0,1) \tag{9}$$
$$\sigma'_i = \sigma_i \cdot \exp(\tau' \cdot m + \tau \cdot \mathcal{N}(0,1)) \tag{10}$$
$$x'_i = x_i + \sigma'_i \cdot \mathcal{N}(0,1) \tag{11}$$

with $\tau = \tau' = 1/\sqrt{2n}$.

5.3 Pooling

The pooling mutation operator was inspired by a similar approach in [10]. The main idea is to only allow a small and limited number of fixed step sizes. Let $u \sim \mathcal{U}_\mathbb{R}(0,1)$ be a uniformelly distributed random variable with $0 \leq u \leq 1$, then the pooling mutation operator can be expressed as

$$u \sim \mathcal{U}_\mathbb{R}(0,1)$$
$$\sigma' = \begin{cases} \sigma & \text{if } 0 < u \leq 1/4 \\ c_1 & \text{if } 1/4 < u \leq 1/2 \\ c_2 & \text{if } 1/2 < u \leq 3/4 \\ c_3 & \text{else} \end{cases} \tag{12}$$

The constant step sizes c_1, c_2 and c_3 are derived as a fixed fraction of a variables design space by

$$c_1 = (ub_i - lb_i) \cdot 10^{-1}$$
$$c_2 = (ub_i - lb_i) \cdot 10^{-2}$$
$$c_3 = (ub_i - lb_i) \cdot 10^{-4}$$

where ub_i and lb_i denote the upper and lower bounds of the i-th design variable. A minimum of adaptation is introduced by keeping the step size unchanged with probability 1/2 so that successful parent solutions can maintain a successful step size.

6 Results

6.1 Pareto-Sets

Since the computational cost of the CFD-Simulation does not allow numerous experiments even for the two-dimensional models, we show representative results of single runs from a number of experiments. We do not try to average several runs in any way, because the small number of experiments that could be conducted does not allow meaningful statistics. Additionally, from visual inspection, the runs made did not significantly differ from each other.

The pareto-fronts displayed in figures 5, 6 and 7 contain a reference Pareto front (denoted "Reference") that is the common Pareto front of all optimisation experiments that were conducted for this study and results computed with a multi-criteria genetic algorithm (MOGA) as described in [8]. The reference set was included to show the quality of a single run optimisation compared to an aggregated pareto-set that needs much more fitness function evaluations and to supply a benchmark line the individual algorithms have to approach.

Figure 5 demonstrates that SPEA2 and NSGA-II are closer to the reference set than the MOGA. SPEA2 is slightly better in reconstructing the NACA4412 profile, while NSGA-II is slightly more successful for the NACA0012 profile. In the compromise region there is hardly a difference between SPEA2 and NSGA-II.

Figure 6 displays a rather surprising result: The SPEA2 selection operator in combination with the pooling mutation converges close to the reference pareto-set in the compromise region, but fails to place good solutions at the tails of the reference set. The NSGA-II selection operator, contrarily, has a better spread of solutions when combined with pooling mutation, while failing to closely approach the reference pareto-set. But still, both Evolution Strategies outperform the MOGA.

Figure 7 finally shows that both selection methods produce nicely spread solutions when combined with Schwefel's mutation which are not as close to the reference set as with the derandomised mutation. Once again, the pareto-set produced by the MOGA is worse than the Evolution Strategies' solutions.

It is difficult to make serious judgements on the basis of the few data available. Comparing the results available from the experiments, the derandomised mutation operator seems to work very well, but neither Schwefel's mutation nor the pooling mutation are clearly inferior. Clarification of the question which mutation operator to choose will involve further tests with extended computational effort that was beyond the scope of this study.

The same statement holds for the choice of selection method. The results with the pooling mutation as shown in figure 6, where the NSGA-II selection provides better spread of solutions than the SPEA2 selection and the SPEA2 selection converges closer to the reference set than the NSGA-II selection, should

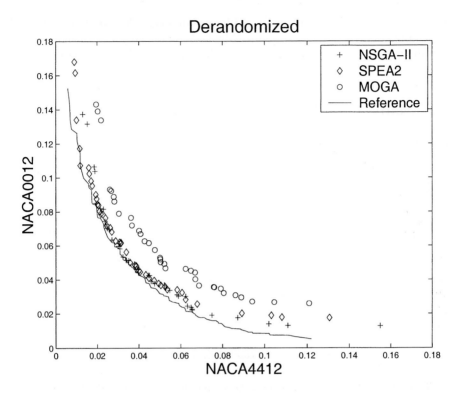

Fig. 5. Pareto front with derandomised mutation and NSGA-II- and SPEA2-Selection

be considered with some suspicion, as there seems to be no obvious reason for this behaviour that could be attributed to the selection method. In fact the effect was less obvious in the other experiments with pooling mutation. For the airfoil design problem, as stated in this study, SPEA2 and NSGA-II perform comparably well.

It can clearly be seen from the results, though, that using NSGA-II or SPEA2 in combination with an Evolution Strategy yields better performance than using the MOGA as described in [8]. This result is supported by all experiments that were made for the airfoil design problem.

6.2 Airfoils

Figures 8, 9, and 10 show three different wing designs and their respective pressure profiles. These profiles are taken from an optimisation with NSGA-II selection and derandomised step size adaptation. The P1 design is close to the NACA4412 pressure profile, i.e. it is from the top left corner of figure 5. This means, that the actual wing profile is close to the NACA4412 profile which can be seen by comparing the wing profile displayed in figure 2 and the P1-profile in figure 8. Therefore P1's low drag pressure distribution in figure 9 is also close to

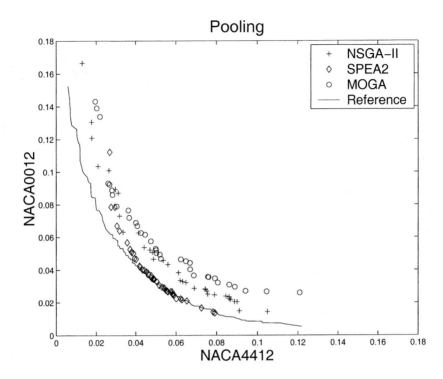

Fig. 6. Pareto front with pooling mutation and NSGA-II- and SPEA2-Selection

the accompanying target pressure distribution for the NACA4412 at low drag conditions which is also displayed in figure 2. The draw back of the NACA4412-like design can be seen from P1's high lift pressure distribution in figure 10. In contrast to the high lift target distribution (that looks similar to the P3 curve), P1's high lift pressure profile drops off at the leading edge of the wing very fast which translates to bad high lift performance.

P3, on the other hand, is located in the bottom right corner of figure 5, i.e. it is close to the NACA0012 pressure profile (figure 1). Figure 10 indicates that the P3 profile leads to a pressure profile that is close to the NACA0012's target distribution. The most obvious difference to P1 is the much slower decrease of the leading edge pressure. For the low drag conditions, figure 9 shows that for $X \in [0.05, 0.5]$ the pressure distribution of P3 has one single maximum close to the center whereas the P1 distribution has its minimum almost at the same location and two maxima at the borders of the interval.

P2 then is a compromise solution between P1 and P3. From figures 9 and 10 it can be seen that P2 mimics the low drag target pressure by redesigning the pressure at the trailing edge of the wing while at the same time being an obvious "in-between" solution at the leading edge for the high lift target.

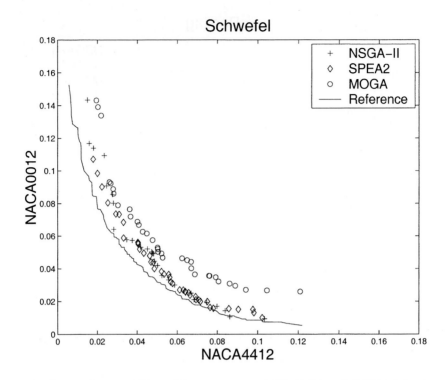

Fig. 7. Pareto front with Schwefel's mutation and NSGA-II- and SPEA2-Selection

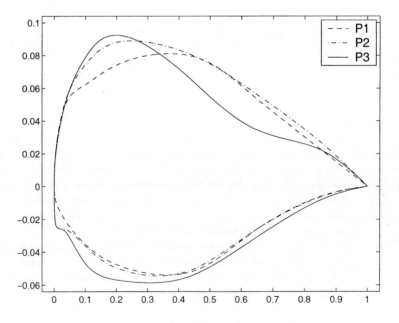

Fig. 8. Wing profiles of 3 different design configurations

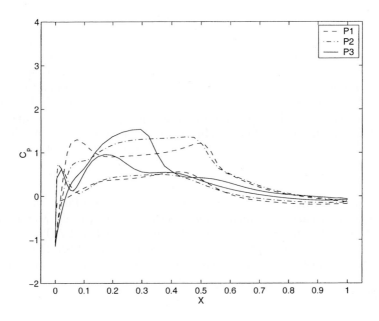

Fig. 9. Pressure profiles of 3 different design configurations for the low drag flow conditions

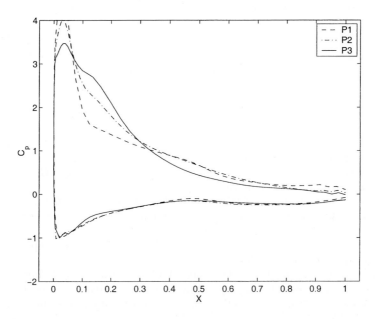

Fig. 10. Pressure profiles of 3 different design configurations for the high lift flow conditions

7 Conclusion

The results stated in this work show that multi-criteria Evolution Strategies can successfully be used for complex engineering design problems. The selection operators used perform comparably well. The quality of the mutation operators, though, cannot reasonably be judged on the basis of the experiments conducted. The derandomised step size adaptation appears to be beneficial, but the differences compared to the other two operators are quite small. The question of self adaptation in multi-criteria evolution strategies surly needs much further research and should ultimately give room for further improvement for real world applications.

Acknowledgements. The AEROSHAPE project (Multi-Point Aerodynamic Shape Optimisation) is a collaboration between Aerospatiale Matra Airbus, Alenia Aeronautica (Coordinator), DaimlerChrysler Airbus, EADS-M, Dassault Aviation, SAAB, SENER, SYNAPS, CIRA, QinetiQ, DLR, FFA, INRIA, HCSA, NLR, ONERA, and NuTech Solutions. The project is funded by the European Commission, DG Research, under the GROWTH initiative (Project Ref: GRD1-1999-10752).

References

[1] Kalyanmoy Deb. *Multi-Objective Optimization using Evolutionary Algorithms*. Wiley-Interscience Series in Systems and Optimization. John Wiley & Sons, Chichester, New York, 2001.
[2] Kalyanmoy Deb and Tushar Goel. Controlled elitist non-dominated sorting genetic algorithms for better convergence. In *Proceedings of the First International Conference on Evolutionary Multi-Criterion Optimization (EMO-2001)*, pages 67–81, 7–9 March 2001.
[3] Nikolaus Hansen, Andreas Ostermeier, and Andreas Gawelcyk. A derandomized approach to self-adaptation of evolution strategies. *Evolutionary Computation*, 4(2):369–380, 1994.
[4] Marco Laumanns, Günther Rudolph, and Hans Paul Schwefel. Mutation control and convergence in evolutionary multi-objective optimization. In *Proceedings of the 7th International Mendel Conference on Soft Computing (MENDEL 2001)*, Brno, Czech Republic, June 2001. -.
[5] Marco Laumanns, Eckart Zitzler, and Lothar Thiele. On the effects of archiving, elitism, and density based selection in evolutionary multi-objective optimization. In E. Zitzler et al., editor, *Evolutionary Multi-criterion Optimization (EMO 2001), First International Conference*, Proceedings. Lecture Notes on Computer Science, pages 181–196. Springer, March 7–9 2001.
[6] Marco Laumanns, Eckart Zitzler, and Lothar Thiele. Spea2: Improving the strength pareto evolutionary algorithm. TIK-Report 103, Computer Engineering and Networks Laboratory (TIK), Swiss Federal Institute of Technology (ETH) Zurich, May 2001. -.

[7] Boris Naujoks, Lars Willmes, Werner Haase, Thomas Bäck, and Martin Schütz. Multi-point airfoil optimization using evolution strategies. In *Proceedings of the European Congress on Computational Methods in Applied Sciences and Engineering (ECCOMAS'00) (CD-Rom und Book of Abstracts)*, page 948 (Book of Abstracts), Barcelona, 11.-14. September 2000. Center for Numerical Methods in Engineering (CIMNE).

[8] Carlo Poloni. Multi objective aerodynamic optimisation by means of robust and efficient genetic algorithm. In Kozo Fujii and George S. Dulikravich, editors, *Recent development of aerodynamic design methodologies : inverse design and optimization*, volume 68 of *Notes on numerical fluid mechanics*, pages 1–24. Vieweg, Braunschweig/Wiesbaden, 1999.

[9] Domenico Quagliarella, Jaques Périaux, Carlo Polnoi, and Gabriel Winter, editors. *Genetic Algorithms and Evolution Strategies in Engineering and Computer Science – Recent Advances and Industrial Applications*. John Wiley & Sons, Chichester, New York, 1998.

[10] M. Schütz and J. Sprave. Application of parallel mixed-integer evolution strategies with mutation rate pooling. In L.J. Fogel, P.J. Angeline, and T. Bäck, editors, *Proceedings of the 5th Annual Conference on Evolutionary Programming (EP-96), San Diego, CA, 29. February - 2. March*, pages 345–354, 1996.

[11] Hans Paul Schwefel. *Evolution and Optimum Seeking*. Wiley, New York, 1995.

Visualization and Data Mining of Pareto Solutions Using Self-Organizing Map

Shigeru Obayashi and Daisuke Sasaki

Institute of Fluid Science, Tohoku University,
Sendai, 980-8577 JAPAN
obayashi@ieee.org, sasaki@reynolds.ifs.tohoku.ac.jp

Abstract. Self-Organizing Maps (SOMs) have been used to visualize tradeoffs of Pareto solutions in the objective function space for engineering design obtained by Evolutionary Computation. Furthermore, based on the codebook vectors of cluster-averaged values of respective design variables obtained from the SOM, the design variable space is mapped onto another SOM. The resulting SOM generates clusters of design variables, which indicate roles of the design variables for design improvements and tradeoffs. These processes can be considered as data mining of the engineering design. Data mining examples are given for supersonic wing design and supersonic wing-fuselage design.

1 Introduction

Multiobjective Evolutionary Algorithms (MOEAs) are getting popular in many fields because they will provide a unique opportunity to address global tradeoffs between multiple objectives by sampling a number of non-dominated solutions. To understand tradeoffs, visualization is essential. Although it is trivial to understand tradeoffs between two objectives, tradeoff analysis in more than three dimensions is not trivial as shown in Fig. 1. To visualize higher dimensions, Self-Organizing Map (SOM) by Kohonen [1,2] is employed in this paper.

SOM is one of neural network models. SOM algorithm is based on unsupervised, competitive learning. It provides a topology preserving mapping from the high dimensional space to map units. Map units, or neurons, usually form a two-dimensional lattice and thus SOM is a mapping from the high dimensions onto the two dimensions. The topology preserving mapping means that nearby points in the input space are mapped to nearby units in SOM. SOM can thus serve as a cluster analyzing tool for high-dimensional data. The cluster analysis of the objective function values will help to identify design tradeoffs.

Design is a process to find a point in the design variable space that matches with the given point in the objective function space. This is, however, very difficult. For example, the design variable spaces considered here have 72 and 131 dimensions, respectively. One way of overcoming high dimensionality is to group some of design variables together. To do so, the cluster analysis based on SOM can be applied again.

Based on the codebook vectors of cluster-averaged values of respective design variables obtained from the SOM, the design variable space can be mapped onto another SOM. The resulting SOM generates clusters of design variables. Design variables in such a cluster behave similar to each other and thus a typical design variable in the cluster indicates the behaviour/role of the cluster. A designer may extract design information from this cluster analysis. These processes can be considered as data mining for the engineering design.

At first, SOM is applied to map objective function values of non-dominated solutions in four dimensions. This will reveal global tradeoffs between four design objectives. The multipoint aerodynamic optimization of a wing shape for SST at both supersonic and transonic cruise conditions has been performed by using MOEAs previously [3]. Both aerodynamic drags were to be minimized under lift constraints, and the bending and pitching moments of the wing were also minimized instead of imposing constraints on structure and stability. A high fidelity Computational Fluid Dynamics (CFD) code, a Navier-Stokes code, was used to evaluate the wing performance at both conditions. In this design optimization, planform shapes, camber, thickness distributions and twist distributions were parameterized in total of 72 design variables. To alleviate the required computational time, parallel computing was performed for function evaluations. The resulting 766 non-dominated solutions are analyzed to reveal tradeoffs in this paper. The resulting SOM is also used to create a new SOM of the cluster-averaged design variables.

Second, SOM is applied to map entire solutions evaluated during the evolution of two-objective optimization. Based on the wing design system mentioned above, an aerodynamic optimization system for SST wing-body configuration was developed in [4]. To satisfy severe tradeoff between high aerodynamic performance and low sonic boom, the present objectives were to reduce C_D at a fixed C_L as well as to satisfy the equivalent area distribution for low boom design proposed by Darden [5]. Wing shape and fuselage configuration were defined in total of 131 design variables. The SOM of the objective function values indicates the non-dominated front as edges of the map and the SOM of the cluster-averaged design variables reveals the role of the design variables for design tradeoffs.

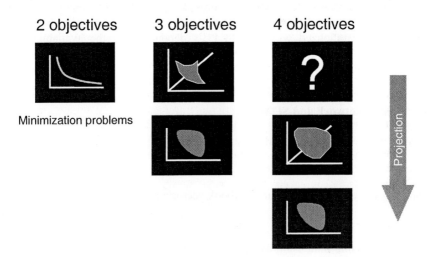

Fig. 1. Visualization of Pareto front

2 Evolutionary Multiobjective Optimization

2.1 MOGAs

The genetic operators used here are based on MOGAs [6,7]. Selection is based on the Pareto ranking method and fitness sharing. Each individual is assigned to its rank according to the number of individuals that dominate it. A fitness sharing function is used to maintain the diversity of the population. To find non-dominated solutions more effectively, the so-called best-N selection is employed.

For real function optimizations like the present research, however, it is more straightforward to use real numbers for encoding. Thus, the floating-point representation is used here. Accordingly, blended crossover (BLX-α) [8] is adopted at the crossover rate of 100%. This operator generates children on a segment defined by two parents and a user specified parameter α. The disturbance is added to new design variables within 10% of the given range of each design variable at a mutation rate of 20%. Crossover and mutation rates are kept high because the best-N selection gives a very strong elitism. Details for the present MOGA were given in Refs. 3, 4 and 7.

2.2 CFD Evaluation

To evaluate the design, a high fidelity Euler/Navier-Stokes code was used. Taking advantage of the characteristic of GAs, the present optimization is parallelized on SGI ORIGIN2000 at the Institute of Fluid Science, Tohoku University. The system has 640 Processing Elements (PE's) with peak performance of 384 GFLOPS and 640 GB of memory.

A simple master-slave strategy was employed: The master PE manages the optimization process, while the slave PE's compute the Navier-Stokes code. The parallelization became almost 100% because almost all the CPU time was dominated by CFD computations. The population size used in this study was set to 64 so that the process was parallelized with 32-128 PE's depending on the availability of job classes. The present optimization requires about six hours per generation for the supersonic wing case when parallelized on 128 PE's.

2.3 Neural Network and SOM

SOM [1,2] is a two-dimensional array of neurons:

$$\mathbf{M} = \{\mathbf{m}_1 \quad \cdots \quad \mathbf{m}_{p \times q}\} \tag{1}$$

One neuron is a vector called the codebook vector:

$$\mathbf{m}_i = [m_{i_1} \quad \cdots \quad m_{i_n}] \tag{2}$$

This has the same dimension as the input vectors (n-dimensional). The neurons are connected to adjacent neurons by a neighbourhood relation. This dictates the topology, or the structure, of the map. Usually, the neurons are connected to each other via rectangular or hexagonal topology. One can also define a distance between the map units according to their topology relations.

The training consists of drawing sample vectors from the input data set and "teaching" them to SOM. The teaching consists of choosing a winner unit by means of a similarity measure and updating the values of codebook vectors in the neighbourhood of the winner unit. This process is repeated a number of times.

In one training step, one sample vector is drawn randomly from the input data set. This vector is fed to all units in the network and a similarity measure is calculated between the input data sample and all the codebook vectors. The best-matching unit is chosen to be the codebook vector with greatest similarity with the input sample. The similarity is usually defined by means of a distance measure. For example in the case of Euclidean distance the best-matching unit is the closest neuron to the sample in the input space.

The best-matching unit, usually noted as $\mathbf{m}_c$, is the codebook vector that matches a given input vector $\mathbf{x}$ best. It is defined formally as the neuron for which

$$\|\mathbf{x}-\mathbf{m}_c\| = \min_i \|\mathbf{x}-\mathbf{m}_i\| \tag{3}$$

After finding the best-matching unit, units in SOM are updated. During the update procedure, the best-matching unit is updated to be a little closer to the sample vector in the input space. The topological neighbours of the best-matching unit are also similarly updated. This update procedure stretches the best-matching unit and its topological neighbours towards the sample vector. The neighbourhood function should be a decreasing function of time. In the following, SOMs were generated in the hexagonal topology by using Viscovery® SOMine 4.0 Plus [9].

2.4 Cluster Analysis

Once SOM projects input space on a low-dimensional regular grid, the map can be utilized to visualize and explore properties of the data. When the number of SOM units is large, to facilitate quantitative analysis of the map and the data, similar units need to be grouped, i.e., clustered. The two-stage procedure --- first using SOM to produce the prototypes which are then clustered in the second stage --- was reported to perform well when compared to direct clustering of the data [10].

Hierarchical agglomerative algorithm is used for clustering here. The algorithm starts with a clustering where each node by itself forms a cluster. In each step of the algorithm two clusters are merged: those with minimal distance according to a special distance measure, the SOM-Ward distance [9]. This measure takes into account whether two clusters are adjacent in the map. This means that the process of merging clusters is restricted to topologically neighbored clusters. The number of clusters will be different according to the hierarchical sequence of clustering. A relatively small number will be chosen for visualization (§3.2), while a large number will be used for generation of codebook vectors for respective design variables (§3.3).

3 Four-Objective Optimization for Supersonic Wing Design

3.1 Formulation of Optimization

Four objective functions used here are

1. Drag coefficient at transonic cruise, $C_{D,t}$
2. Drag coefficient at supersonic cruise, $C_{D,s}$
3. Bending moment at the wing root at supersonic cruise condition, M_B
4. Pitching moment at supersonic cruise condition, M_P

In the present optimization, these objective functions are to be minimized. The transonic drag minimization corresponds to the cruise over land; the supersonic drag minimization corresponds to the cruise over sea. Lower bending moments allow less structural weight to support the wing. Lower pitching moments mean less trim drag.

The present optimization is performed at two design points for the transonic and supersonic cruises. Corresponding flow conditions and the target lift coefficients are described as

1. Transonic cruising Mach number, $M_{\infty,t} = 0.9$
2. Supersonic cruising Mach number, $M_{\infty,s} = 2.0$
3. Target lift coefficient at transonic cruising condition, $C_{L,t} = 0.15$
4. Target lift coefficient at supersonic cruising condition, $C_{L,s} = 0.10$
5. Reynolds number based on the root chord length at both conditions, $Re = 1.0 \times 10^7$

Flight altitude is assumed at 10 km for the transonic cruise and at 15 km for the supersonic cruise. To maintain lift constraints, the angle of attack is computed for each configuration by using $C_{L\alpha}$ obtained from the finite difference. Thus, three Navier-Stokes computations per evaluation are required. During the aerodynamic optimization, wing area is frozen at a constant value.

Design variables are categorized to planform, airfoil shapes and the wing twist. Planform shape is defined by six design variables, allowing one kink in the spanwise direction. Airfoil shapes are composed of its thickness distribution and camber line. The thickness distribution is represented by a Bézier curve defined by nine polygons. The wing thickness is constrained for structural strength. The thickness distributions are defined at the wing root, kink and tip, and then linearly interpolated in the spanwise direction. The camber surfaces composed of the airfoil camber lines are defined at the inboard and outboard of the wing separately. Each surface is represented by the Bézier surface defined by four polygons in the chordwise direction and three in the spanwise direction. Finally, the wing twist is represented by a B-spline curve with six polygons. In total, 72 design variables are used to define a whole wing shape. A three-dimensional wing with computational structured grid and the corresponding CFD result are shown in Figs. 2 and 3. See Ref. 3 for more details for geometry definition and CFD information.

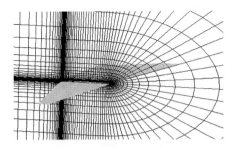 

Fig. 2. Wing grid in C-H topology

Fig. 3. Pressure contours on the upper surface of a wing computed by the CFD code

3.2 Visualization of Design Tradeoffs: SOM of Tradeoffs

The evolution was computed for 75 generations until all individuals become non-dominated. An archive of non-dominated solutions was also created along the evolution. After the computation, the 766 non-dominated solutions were obtained in the archive as a three-dimensional surface in the four-dimensional objective function space. By examining the extreme non-dominated solutions, the archive was found to represent the Pareto front qualitatively.

The present non-dominated solutions of supersonic wing designs have four design objectives. First, let's project the resulting non-dominated front onto the two-dimensional map. Figure 4 shows the resulting SOM with seven clusters. For better understanding, the typical planform shapes of wings are also plotted in the figure. Lower right corner of the map corresponds to highly swept, high aspect ratio wings good for supersonic aerodynamics. Lower left corner corresponds to moderate sweep angles good for reducing the pitching moment. Upper right corner corresponds to small aspect ratios good for reducing the bending moment. Upper left corner thus reduces both pitching and bending moments.

Figure 5 shows the same SOM contoured by four design objective values. All the objective function values are scaled between 0 and 1. Low supersonic drag region corresponds to high pitching moment region. This is primarily because of high sweep angles. Low supersonic drag region also corresponds to high bending moment region because of high aspect ratios. Combination of high sweep angle and high aspect ratio confirm that supersonic wing design is highly constrained.

3.3 Data Mining of Design Space: SOM of Design Variables

The previous SOM provides clusters based on the similarity in the objective function values. The next step is to find similarity in the design variables that corresponds to the previous clusters. To visualize this, the previous SOM is first revised by using larger number of clusters of 49 as shown in Fig. 6. Then, all the design variables are averaged in each cluster, respectively. Now each design variable has a codebook vector of 49 cluster-averaged values. This codebook vector may be regarded to represent focal areas in the design variable space. Finally, a new SOM is generated from these codebook vectors as shown in Fig. 7.

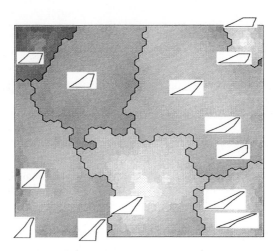

Fig. 4. SOM of the objective function values and typical wing planform shapes

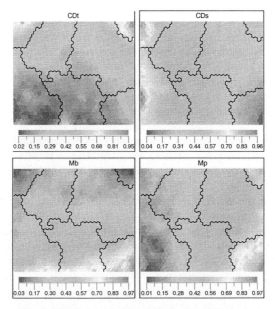

Fig. 5. SOM contoured by each design objective

This process can be done for encoded design variables (genotype) and decoded design variables (phenotype). In the earlier study, the genotype was used for SOM. However, the genotype and phenotype generated completely different SOMs. A possible reason is because the various scaling appears in phenotype, for example, one design variable is between 0 and 1 and another is between 35 to 70. The difference of order of magnitude in design variables may lead to different clusters. To avoid such confusion, the genotype is used for SOM here.

In Fig. 7, the labels indicate 72 design variables. DVs 00 to 05 correspond to the planform design variables. These variables have dominant influence on the wing performance. DVs 00 and 01 determine the span lengths of the inboard and outboard wing panels, respectively. DVs 02 and 03 correspond to leading-edge sweep angles. DVs 04 and 05 are root-side chord lengths. DVs 06 to 25 define wing camber. DVs 26 to 32 determine wing twist. Figure 7 contains seven clusters and thus seven design variables are chosen from each cluster as indicated. Figure 8 shows SOM's of Fig. 4 contoured by these design variables.

The sweep angles, DVs 02 and 03, make a cluster in the lower left corner of the map in Fig. 7 and the corresponding plots in Fig. 8 confirm that the wing sweep has a large impact on the aerodynamic performance. DVs 11 and 51 in Fig. 8 do not appear influential to any particular objective. By comparing Figs. 8 and 5, DV 01 has similar distribution with the bending moment Mb, indicating that the wing outboard span has an impact on the wing bending moment. On the other hand, DV 00, the wing inboard span, has an impact on the pitching moment. DV 28 is related to transonic drag. DV 04 and 05 are in the same cluster. Both of them have an impact on the transonic drag because their reduction means the increase of aspect ratio. Several features of the wing planform design variables and the corresponding clusters are found out in the SOMs and they are consistent with the existing aerodynamic knowledge.

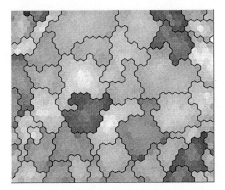

Fig. 6. SOM of objective function values with 49 clusters

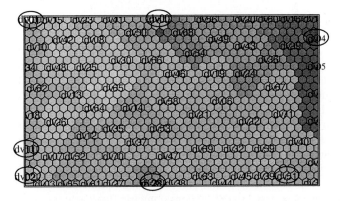

Fig. 7. SOM of cluster-averaged design variables

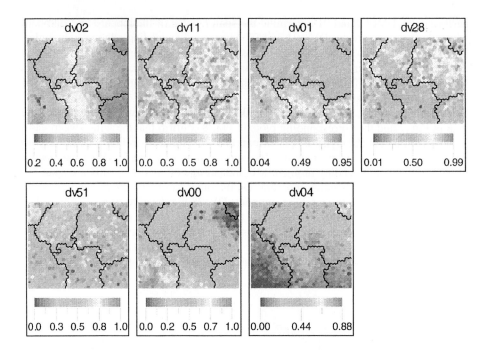

Fig. 8. SOM contoured by design variables selected from clusters in Fig. 7

4 Two-Objective Optimization for Supersonic Wing-Fuselage Design

4.1 Formulation of Optimization

In this study, SST wing-body configurations are designed to improve the aerodynamic performance and to lower the sonic boom strength. Therefore, design objectives are to reduce C_D at Mach number 2.0 at a fixed C_L (=0.10) and to match Darden's equivalent area distribution that can achieve low sonic boom. Multiblock Euler calculation was used to evaluate aerodynamic performance [11]. For the evaluation of sonic boom strength, an equivalent area distribution is matched to Darden's equivalent area distribution for 300 ft fuselage SST at Mach number 1.6 at $C_L = 0.125$.

To evaluate aerodynamic performances, aerodynamic evaluation has to be automatically performed for a given SST wing-body configuration. The wing definition was almost same as the previous wing optimization. Then, 55 additional design variables were used to define nonsymmetric fuselage configuration. Four more design variables represented the wing lofting. The total number of design variables is 131.

As body length and wing area is fixed to 300 ft and 9,000 ft^2, respectively, body volume, minimum diameter of body and wing volume must be greater than values given in Table 1. The other constraints are implemented to design variables as boundaries. As a result, the present SST wing-body design problem has two objective functions of minimization, three constraints and 131 design variables, and is optimized by real-coded MOGAs. Master-slave type parallelization was again performed to reduce the large computational time of each CFD evaluation in the optimization process. Figures 9 and 10 show typical computational grid and corresponding CFD result, respectively. See Ref. 4 for more details for geometry definition and CFD information.

Table 1. Constraints of SST wing-body configuration

Body volume $\geq 30,000\ ft^3$
Minimum diameter $\geq 11.8\ ft\ (0.23 \leq x/L \leq 0.70)$
Wing volume $\geq 16,800\ ft^3$

Fig. 9. Surface grid for SST wing-fuselage configuration (numbers indicate corresponding multiblock grids)

Fig. 10. Computed pressure distribution on the upper surface of SST wing-fuselage configuration

4.2 Visualization of Design Tradeoffs: SOM of Function Landscape

First, all the solutions obtained during the present evolutionary computation were mapped onto SOM according to the scaled objective function values. The resulting SOM is shown in Fig. 11. Several non-dominated solutions are indicated by * in the figure. The map consists of eight clusters. The lower left cluster contains the extreme non-dominated solution of the minimum drag. The upper right cluster contains the extreme non-dominated solution of the minimum boom. The corresponding objective functions values are then plotted in Fig. 12. Because only two objectives are used here, the map coordinates approximately matches to the objectives. The vertical direction corresponds to the drag and the horizontal axis corresponds to the sonic boom. The lower edge and the right edge of the map indicate the non-dominated front. Although the mapping is not essential to visualize tradeoffs here, the cluster analysis may be used to generate clusters of design variables.

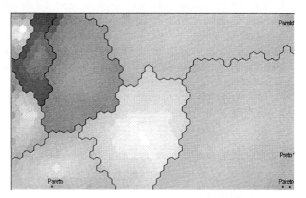

Fig. 11. SOM of the objective function values

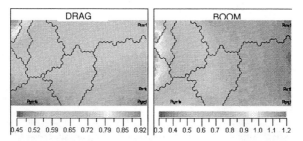

Fig. 12. SOM coloured by each design objective

4.3 Data Mining of Design Space: SOM of Design Variables

To generate SOM of the design variables, Fig. 11 was divided into 50 clusters as Fig. 13. Then, Fig. 14 was generated from codebook vectors of cluster-averaged design variables in Fig. 13. Figure 14 shows SOM of the design variables in five clusters. In Fig. 14, the labels indicate 131 design variables. Figure 14 can be interpreted from the behaviors of the design variables representing the corresponding clusters. Figure 15 shows the map of Fig. 11 contoured by the five design variables indicated in Fig. 14. A trend of the design variables in the left cluster of Fig. 14 is represented by DV 123 in Fig. 15. Its distribution appears the inverse of the sonic boom in Fig. 12. DV 123 determines the twist angle at the wing tip. It has an impact on the list distribution, leading to influences on the equivalent cross sectional distribution and thus on the sonic boom strength. The center cluster in Fig. 14 is represented by DV 2 and its distribution in Fig. 15 appears the inverse of the drag in Fig. 12. DV 2 is one of the design variables that define the sharpness of the nose of the fuselage. Blunt nose is known to increase drag for supersonic aircraft. The right cluster in Fig. 14 is represented by DV 28 and the corresponding distribution in Fig. 15 has a local minimum in the middle of the left, upper edge of the map. This is one of the design variables that determine the body radius distribution at the side of the fuselage, but it does not seem primarily related to either objective here. DV's 89 and 91 have opposite trends, but they are not influential to the non-dominated front, either.

Fig. 13. SOM of objective function values with 50 clusters

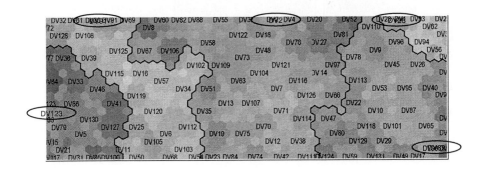

Fig. 14. SOM of cluster-averaged design variables

5 Concluding Remarks

Design tradeoffs have been investigated for two multiobjective aerodynamic design problems of supersonic transport by using visualization and cluster analysis of the non-dominated solutions based on SOMs. The first optimization is to design supersonic wings defined by 72 design variables with four objectives to be minimized. The second optimization is to design supersonic wing-body configurations represented by in total 131 design variables with drag and boom minimization. Design data were gathered by MOGAs.

SOM is first applied to visualize tradeoffs between design objectives. In the first design case, four objective functions were employed and 766 non-dominated solutions were obtained. Three-dimensional non-dominated front in the objective function space has been mapped onto the two-dimensional SOM where global tradeoffs are successfully visualized. In the second design case, entire solutions during the evolution have been mapped onto SOM to visualize function landscape, and the non-dominated front was found at the edges of the map. The resulting SOMs are further contoured by each objective, which provides better insights into design tradeoffs.

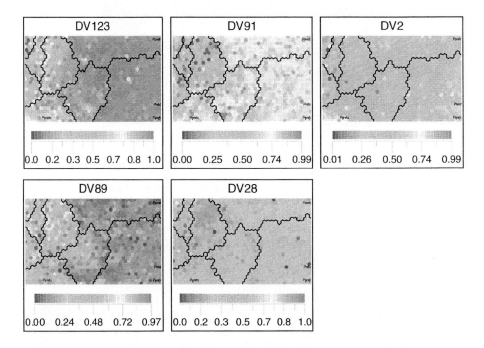

Fig. 15. SOM contoured by design variables selected from clusters in Fig. 14

Furthermore, based on the codebook vectors of cluster-averaged values for respective design variables obtained from the SOMs, the design variable space is mapped onto another SOM. Design variables in the same cluster are considered to have similar influences in design tradeoffs. Therefore, by selecting a member (design variable) from a cluster, the original SOM in the objective function space is contoured by the particular design variable. It reveals correlation of the cluster of design variables with objective functions and their relative importance. Because each cluster of design variables can be identified influential or not to a particular design objective, the optimization problem may be divided into subproblems where the optimization will be easier to lead to better solutions.

These processes may be considered as data mining of the engineering design. The present work demonstrates that MOGAs and SOMs are versatile design tools for engineering design.

References

1 Kohonen T.: *Self-Organizing Maps*. Springer, Berlin, Heidelberg (1995)
2 Hollmen J.: Self-Organizing Map, http://www.cis.hut.fi/~jhollmen/dippa/node7. html, last access on October 3, 2002

3. Sasaki D., Obayashi S. and Nakahashi K.: Navier-Stokes Optimization of Supersonic Wings with Four Objectives Using Evolutionary Algorithm. Journal of Aircraft Vol.39, No.4 (2002) 621–629
4. Sasaki D., Yang G. and Obayashi S.: Automated Aerodynamic Optimization System for SST Wing-Body Configuration. AIAA Paper 2002–5549 (2002)
5. Darden, C. M.: Sonic Boom Theory: Its Status in Prediction and Minimization. Journal of Aircraft, Vol.14, No.6 (1977) 569–576
6. Fonseca C. M. and Fleming P. J.: Genetic Algorithms for Multiobjective Optimization: Formulation, Discussion and Generalization. Proc. of the 5th ICGA (1993) 416–423
7. Obayashi S., Takahashi S. and Takeguchi Y.: Niching and Elitist Models for MOGAs. Parallel Problem Solving from Nature – PPSN V, Lecture Notes in Computer Science, Springer, Vol.1498, Berlin Heidelberg New York (1998) 260–269
8. Eshelman L. J. and Schaffer J. D.: Real-Coded Genetic Algorithms and Interval Schemata. Foundations of Genetic Algorithms 2, Morgan Kaufmann Publishers, Inc., San Mateo (1993) 187–202
9. Eudaptics software gmbh. http://www.eudaptics.com/technology/somine4.html, last access on October 3, 2002
10. Vesanto, J. and Alhoniemi, E.: Clustering of the Self-Organizing Map, IEEE Transactions on Neural Networks, Vol.11, No.3 (2000) 586–600
11. Yang, G., Kondo, M. and Obayashi, S.: Multiblock Navier-Stokes Solver for Wing/Fuselage Transport Aircraft. JSME International Journal, Series B, Vol.45, No.1 (2002) 85–90

Author Index

Abbass, Hussein A. 391
Aceves, Rodrigo 578
Aerts, Jeroen C.J.H. 448
Amato, P. 58, 311
Anchor, Kevin P. 707
Andersson, Johan 737
Azarm, S. 148, 405

Bäck, Thomas 782
Baets, Bernard De 31
Balling, Richard 1
Baltas, Dimos 648
Barichard, Vincent 88
Bleuler, Stefan 494
Botello Rionda, Salvador 73
Brizuela, Carlos A. 578
Büche, Dirk 267

Caswell, David J. 177
Chaudhry, Fazal H. 662, 677
Cheung, Peter B. 662, 677
Coello Coello, Carlos A. 73, 252
Corne, David W. 295, 327
Costa, Mario 282
Covas, J.A. 752

Deb, Kalyanmoy 222, 311, 376 391, 534, 623
Dellnitz, Michael 118
Dorn, Jason L. 692
Ducheyne, Els 31

Farhang-Mehr, A. 148, 405
Farina, M. 58, 311
Fleischer, M. 519
Fleming, Peter J. 16, 133
Formiga, Klebber T.M. 662, 677

Galván, Blas 722
Gandibleux, Xavier 43
Gaspar-Cunha, A. 752
Gen, Mitsuo 593
Greiner, David 722
Gunawan, S. 148
Gunsch, Gregg H. 707
Guntsch, Michael 464

Hao, Jin-Kao 88
Haubelt, Christian 162
Hernández Aguirre, Arturo 73
Herwijnen, Marjan van 448
Hiroyasu, Tomoyuki 565
Horn, Jeffrey 365
Hughes, Evan J. 102

Ishibuchi, Hisao 433, 593, 608

Jain, Abhishek 534
Jin, Yaochu 237

Kang, Lishan 342
Katoh, Naoki 43
Khare, V. 376
Knowles, Joshua D. 295, 327
Kort, Skander 192
Koumoutsakos, Petros 267

Lacomme, P. 550
Lahanas, Michael 648
Lamont, Gary B. 177, 707
Laumanns, Marco 494
Li, Xiaodong 207
Lin, Guangming 342
Lizárraga Lizárraga, Giovanni 73

Madeira, J.F. Aguilar 767
Middendorf, Martin 464
Miki, Mitsunori 565
Milickovic, Natasa 648
Minisci, Edmondo 282
Mishra, Shikhar 222
Mohan, Manikanth 222
Morita, Hiroyuki 43
Mostaghim, Sanaz 118, 162
Mota, Sonia 638
Müller, Sibylle 267
Murata, Tadahiko 593

Nozawa, Hiroyuki 593

Obayashi, Shigeru 796
Ortega, Julio 638

Paquete, Luis 479
Parks, Geoff 418
Pina, Heitor 767
Prins, C. 550
Purshouse, Robin C. 16, 133

Ranjithan, S. Ranji 692
Reddy, A. Raji 623
Reis, Luisa F.R. 662, 677
Rodrigues, H. 767
Ros, Eduardo 638

Sasaki, Daisuke 796
Schreibmann, Eduard 648
Schütze, Oliver 118, 509
Sendhoff, Bernhard 237
Sevaux, M. 550
Shibata, Youhei 433
Stewart, Theodor J. 448
Stützle, Thomas 479

Teich, Jürgen 118, 162

Thiele, Lothar 494
Thierens, Dirk 355
Ticona, Waldo G.C. 662
Toro, Francisco de 638
Toscano Pulido, Gregorio 252
Tyagi, Ambrish 162

Watanabe, Shinya 565
Wildman, Andrew 418
Willmes, Lars 782
Winter, Gabriel 722
Wulf, Robert De 31

Yamamoto, Takashi 608
Yan, Zhenyu 342
Yao, X. 376

Zhang, Linghai 342
Zitzler, Eckart 494
Zope, Pawan 534
Zydallis, Jesse B. 707

Lecture Notes in Computer Science

For information about Vols. 1–2547

please contact your bookseller or Springer-Verlag

Vol. 2548: J. Hernández, Ana Moreira (Eds.), Object-Oriented Technology. Proceedings, 2002. VIII, 223 pages. 2002.

Vol. 2549: J. Cortadella, A. Yakovlev, G. Rozenberg (Eds.), Concurrency and Hardware Design. XI, 345 pages. 2002.

Vol. 2550: A. Jean-Marie (Ed.), Advances in Computing Science – ASIAN 2002. Proceedings, 2002. X, 233 pages. 2002.

Vol. 2551: A. Menezes, P. Sarkar (Eds.), Progress in Cryptology – INDOCRYPT 2002. Proceedings, 2002. XI, 437 pages. 2002.

Vol. 2552: S. Sahni, V.K. Prasanna, U. Shukla (Eds.), High Performance Computing – HiPC 2002. Proceedings, 2002. XXI, 735 pages. 2002.

Vol. 2553: B. Andersson, M. Bergholtz, P. Johannesson (Eds.), Natural Language Processing and Information Systems. Proceedings, 2002. X, 241 pages. 2002.

Vol. 2554: M. Beetz, Plan-Based Control of Robotic Agents. XI, 191 pages. 2002. (Subseries LNAI).

Vol. 2555: E.-P. Lim, S. Foo, C. Khoo, H. Chen, E. Fox, S. Urs, T. Costantino (Eds.), Digital Libraries: People, Knowledge, and Technology. Proceedings, 2002. XVII, 535 pages. 2002.

Vol. 2556: M. Agrawal, A. Seth (Eds.), FST TCS 2002: Foundations of Software Technology and Theoretical Computer Science. Proceedings, 2002. XI, 361 pages. 2002.

Vol. 2557: B. McKay, J. Slaney (Eds.), AI 2002: Advances in Artificial Intelligence. Proceedings, 2002. XV, 730 pages. 2002. (Subseries LNAI).

Vol. 2558: P. Perner, Data Mining on Multimedia Data. X, 131 pages. 2002.

Vol. 2559: M. Oivo, S. Komi-Sirviö (Eds.), Product Focused Software Process Improvement. Proceedings, 2002. XV, 646 pages. 2002.

Vol. 2560: S. Goronzy, Robust Adaptation to Non-Native Accents in Automatic Speech Recognition. Proceedings, 2002. XI, 144 pages. 2002. (Subseries LNAI).

Vol. 2561: H.C.M. de Swart (Ed.), Relational Methods in Computer Science. Proceedings, 2001. X, 315 pages. 2002.

Vol. 2562: V. Dahl, P. Wadler (Eds.), Practical Aspects of Declarative Languages. Proceedings, 2003. X, 315 pages. 2002.

Vol. 2565: J.M.L.M. Palma, J. Dongarra, V. Hernández, A. Augusto Sousa (Eds.), High Performance Computing for Computational Science – VECPAR 2002. Proceedings, 2002. XVII, 732 pages. 2003.

Vol. 2566: T.Æ. Mogensen, D.A. Schmidt, I.H. Sudborough (Eds.), The Essence of Computation. XIV, 473 pages. 2002.

Vol. 2567: Y.G. Desmedt (Ed.), Public Key Cryptography – PKC 2003. Proceedings, 2003. XI, 365 pages. 2002.

Vol. 2568: M. Hagiya, A. Ohuchi (Eds.), DNA Computing. Proceedings, 2002. XI, 338 pages. 2003.

Vol. 2569: D. Gollmann, G. Karjoth, M. Waidner (Eds.), Computer Security – ESORICS 2002. Proceedings, 2002. XIII, 648 pages. 2002. (Subseries LNAI).

Vol. 2570: M. Jünger, G. Reinelt, G. Rinaldi (Eds.), Combinatorial Optimization – Eureka, You Shrink!. Proceedings, 2001. X, 209 pages. 2003.

Vol. 2571: S.K. Das, S. Bhattacharya (Eds.), Distributed Computing. Proceedings, 2002. XIV, 354 pages. 2002.

Vol. 2572: D. Calvanese, M. Lenzerini, R. Motwani (Eds.), Database Theory – ICDT 2003. Proceedings, 2003. XI, 455 pages. 2002.

Vol. 2574: M.-S. Chen, P.K. Chrysanthis, M. Sloman, A. Zaslavsky (Eds.), Mobile Data Management. Proceedings, 2003. XII, 414 pages. 2003.

Vol. 2575: L.D. Zuck, P.C. Attie, A. Cortesi, S. Mukhopadhyay (Eds.), Verification, Model Checking, and Abstract Interpretation. Proceedings, 2003. XI, 325 pages. 2003.

Vol. 2576: S. Cimato, C. Galdi, G. Persiano (Eds.), Security in Communication Networks. Proceedings, 2002. IX, 365 pages. 2003.

Vol. 2578: F.A.P. Petitcolas (Ed.), Information Hiding. Proceedings, 2002. IX, 427 pages. 2003.

Vol. 2580: H. Erdogmus, T. Weng (Eds.), COTS-Based Software Systems. Proceedings, 2003. XVIII, 261 pages. 2003.

Vol. 2581: J.S. Sichman, F. Bousquet, P. Davidsson (Eds.), Multi-Agent-Based Simulation II. Proceedings, 2002. X, 195 pages. 2003. (Subseries LNAI).

Vol. 2582: L. Bertossi, G.O.H. Katona, K.-D. Schewe, B. Thalheim (Eds.), Semantics in Databases. Proceedings, 2001. IX, 229 pages. 2003.

Vol. 2583: S. Matwin, C. Sammut (Eds.), Inductive Logic Programming. Proceedings, 2002. X, 351 pages. 2003. (Subseries LNAI).

Vol. 2584: A. Schiper, A.A. Shvartsman, H. Weatherspoon, B.Y. Zhao (Eds.), Future Directions in Distributed Computing. X, 219 pages. 2003.

Vol. 2585: F. Giunchiglia, J. Odell, G. Weiß (Eds.), Agent-Oriented Software Engineering III. Proceedings, 2002. X, 229 pages. 2003.

Vol. 2586: M. Klusch, S. Bergamaschi, P. Edwards, P. Petta (Eds.), Intelligent Information Agents. VI, 275 pages. 2003. (Subseries LNAI).

Vol. 2587: P.J. Lee, C.H. Lim (Eds.), Information Security and Cryptology – ICISC 2002. Proceedings, 2002. XI, 536 pages. 2003.

Vol. 2586: M. Klusch, S. Bergamaschi, P. Edwards, P. Petta (Eds.), Intelligent Information Agents. VI, 275 pages. 2003. (Subseries LNAI).

Vol. 2587: P.J. Lee, C.H. Lim (Eds.), Information Security and Cryptology – ICISC 2002. Proceedings, 2002. XI, 536 pages. 2003.

Vol. 2588: A. Gelbukh (Ed.), Computational Linguistics and Intelligent Text Processing. Proceedings, 2003. XV, 648 pages. 2003.

Vol. 2589: E. Börger, A. Gargantini, E. Riccobene (Eds.), Abstract State Machines 2003. Proceedings, 2003. XI, 427 pages. 2003.

Vol. 2590: S. Bressan, A.B. Chaudhri, M.L. Lee, J.X. Yu, Z. Lacroix (Eds.), Efficiency and Effectiveness of XML Tools and Techniques and Data Integration over the Web. Proceedings, 2002. X, 259 pages. 2003.

Vol. 2591: M. Aksit, M. Mezini, R. Unland (Eds.), Objects, Components, Architectures, Services, and Applications for a Networked World. Proceedings, 2002. XI, 431 pages. 2003.

Vol. 2592: R. Kowalczyk, J.P. Müller, H. Tianfield, R. Unland (Eds.), Agent Technologies, Infrastructures, Tools, and Applications for E-Services. Proceedings, 2002. XVII, 371 pages. 2003. (Subseries LNAI).

Vol. 2593: A.B. Chaudhri, M. Jeckle, E. Rahm, R. Unland (Eds.), Web, Web-Services, and Database Systems. Proceedings, 2002. XI, 311 pages. 2003.

Vol. 2594: A. Asperti, B. Buchberger, J.H. Davenport (Eds.), Mathematical Knowledge Management. Proceedings, 2003. X, 225 pages. 2003.

Vol. 2595: K. Nyberg, H. Heys (Eds.), Selected Areas in Cryptography. Proceedings, 2002. XI, 405 pages. 2003.

Vol. 2597: G. Păun, G. Rozenberg, A. Salomaa, C. Zandron (Eds.), Membrane Computing. Proceedings, 2002. VIII, 423 pages. 2003.

Vol. 2598: R. Klein, H.-W. Six, L. Wegner (Eds.), Computer Science in Perspective. X, 357 pages. 2003.

Vol. 2599: E. Sherratt (Ed.), Telecommunications and beyond: The Broader Applicability of SDL and MSC. Proceedings, 2002. X, 253 pages. 2003.

Vol. 2600: S. Mendelson, A.J. Smola, Advanced Lectures on Machine Learning. Proceedings, 2002. IX, 259 pages. 2003. (Subseries LNAI).

Vol. 2601: M. Ajmone Marsan, G. Corazza, M. Listanti, A. Roveri (Eds.) Quality of Service in Multiservice IP Networks. Proceedings, 2003. XV, 759 pages. 2003.

Vol. 2602: C. Priami (Ed.), Computational Methods in Systems Biology. Proceedings, 2003. IX, 214 pages. 2003.

Vol. 2604: N. Guelfi, E. Astesiano, G. Reggio (Eds.), Scientific Engineering for Distributed Java Applications. Proceedings, 2002. X, 205 pages. 2003.

Vol. 2606: A.M. Tyrrell, P.C. Haddow, J. Torresen (Eds.), Evolvable Systems: From Biology to Hardware. Proceedings, 2003. XIV, 468 pages. 2003.

Vol. 2607: H. Alt, M. Habib (Eds.), STACS 2003. Proceedings, 2003. XVII, 700 pages. 2003.

Vol. 2609: M. Okada, B. Pierce, A. Scedrov, H. Tokuda, A. Yonezawa (Eds.), Software Security – Theories and Systems. Proceedings, 2002. XI, 471 pages. 2003.

Vol. 2610: C. Ryan, T. Soule, M. Keijzer, E. Tsang, R. Poli, E. Costa (Eds.), Genetic Programming. Proceedings, 2003. XII, 486 pages. 2003.

Vol. 2611: S. Cagnoni, J.J. Romero Cardalda, D.W. Corne, J. Gottlieb, A. Guillot, E. Hart, C.G. Johnson, E. Marchiori, J.-A. Meyer, M. Middendorf, G.R. Raidl (Eds.), Applications of Evolutionary Computing. Proceedings, 2003. XXI, 708 pages. 2003.

Vol. 2612: M. Joye (Ed.), Topics in Cryptology – CT-RSA 2003. Proceedings, 2003. XI, 417 pages. 2003.

Vol. 2613: F.A.P. Petitcolas, H.J. Kim (Eds.), Digital Watermarking. Proceedings, 2002. XI, 265 pages. 2003.

Vol. 2614: R. Laddaga, P. Robertson, H. Shrobe (Eds.), Self-Adaptive Software: Applications. Proceedings, 2001. VIII, 291 pages. 2003.

Vol. 2615: N. Carbonell, C. Stephanidis (Eds.), Universal Access. Proceedings, 2002. XIV, 534 pages. 2003.

Vol. 2616: T. Asano, R. Klette, C. Ronse (Eds.), Geometry, Morphology, and Computational Imaging. Proceedings, 2002. X, 437 pages. 2003.

Vol. 2617: H.A. Reijers (Eds.), Design and Control of Workflow Processes. Proceedings, 2002. XV, 624 pages. 2003.

Vol. 2618: P. Degano (Ed.), Programming Languages and Systems. Proceedings, 2003. XV, 415 pages. 2003.

Vol. 2619: H. Garavel, J. Hatcliff (Eds.), Tools and Algorithms for the Construction and Analysis of Systems. Proceedings, 2003. XVI, 604 pages. 2003.

Vol. 2620: A.D. Gordon (Ed.), Foundations of Software Science and Computation Structures. Proceedings, 2003. XII, 441 pages. 2003.

Vol. 2621: M. Pezzè (Ed.), Fundamental Approaches to Software Engineering. Proceedings, 2003. XIV, 403 pages. 2003.

Vol. 2622: G. Hedin (Ed.), Compiler Construction. Proceedings, 2003. XII, 335 pages. 2003.

Vol. 2623: O. Maler, A. Pnueli (Eds.), Hybrid Systems: Computation and Control. Proceedings, 2003. XII, 558 pages. 2003.

Vol. 2625: U. Meyer, P. Sanders, J. Sibeyn (Eds.), Algorithms for Memory Hierarchies. Proceedings, 2003. XVIII, 428 pages. 2003.

Vol. 2626: J.L. Crowley, J.H. Piater, M. Vincze, L. Paletta (Eds.), Computer Vision Systems. Proceedings, 2003. XIII, 546 pages. 2003.

Vol. 2627: B. O'Sullivan (Ed.), Recent Advances in Constraints. Proceedings, 2002. X, 201 pages. 2003. (Subseries LNAI).

Vol. 2628: T. Fahringer, B. Scholz, Advanced Symbolic Analysis for Compilers. XII, 129 pages. 2003.

Vol. 2631: R. Falcone, S. Barber, L. Korba, M. Singh (Eds.), Trust, Reputation, and Security: Theories and Practice. Proceedings, 2002. X, 235 pages. 2003. (Subseries LNAI).

Vol. 2632: C.M. Fonseca, P.J. Fleming, E. Zitzler, K. Deb, L. Thiele (Eds.), Evolutionary Multi-Criterion Optimization. Proceedings, 2003. XV, 812 pages. 2003.

Vol. 2633: F. Sebastiani (Ed.), Advances in Information Retrieval. Proceedings, 2003. XIII, 546 pages. 2003.